VOLUME FORTY FIVE

METHODS IN GEOCHEMISTRY AND GEOPHYSICS

Principles of Electromagnetic Methods in Surface Geophysics

METHODS IN GEOCHEMISTRY AND GEOPHYSICS

(Volumes 1–28 are out of print)

29. V.P. Dimri – Deconvolution and Inverse Theory – Application to Geophysical Problems
30. K.-M Strack – Exploration with Deep Transient Electromagnetics
31. M.S. Zhdanov and G.V. Keller – The Geoelectrical Methods in Geophysical Exploration
32. A.A. Kaufman and A.L. Levshin – Acoustic and Elastic Wave Fields in Geophysics, I
33. A.A. Kaufman and P.A. Eaton – The Theory of Inductive Prospecting
34. A.A. Kaufman and P. Hoekstra – Electromagnetic Soundings
35. M.S. Zhdanov and P.E. Wannamaker – Three-Dimensional Electromagnetics
36. M.S. Zhdanov – Geophysical Inverse Theory and Regularization Problems
37. A.A. Kaufman, A.L. Levshin and K.L. Larner – Acoustic and Elastic Wave Fields in Geophysics, II
38. A.A. Kaufman and Yu. A. Dashevsky – Principles of Induction Logging
39. A.A. Kaufman and A.L. Levshin – Acoustic and Elastic Wave Fields in Geophysics, III
40. V.V. Spichak – Electromagnetic Sounding of the Earth's Interior
41. A.A. Kaufman and R.O. Hansen – Principles of the Gravitational Method
42. A.A. Kaufman, R.O. Hansen and Robert L.K. Kleinberg – Principles of the Magnetic Methods in Geophysics
43. Michael S. Zhdanov – Geophysical Electromagnetic Theory and Methods
44. A.A. Kaufman and B.I. Anderson – Principles of Electric Methods in Surface and Borehole Geophysics
45. A.A. Kaufman, D. Alekseev and M. Oristaglio - Principles of Electromagnetic Methods in Surface Geophysics

VOLUME FORTY FIVE

METHODS IN GEOCHEMISTRY AND GEOPHYSICS

Principles of Electromagnetic Methods in Surface Geophysics

A.A. KAUFMAN

Colorado School of Mines
Department of Geophysics
Golden, CO, USA

D. ALEKSEEV

Nord-West Ltd.,
Shirshov Institute of Oceanology
Russian Academy of Sciences
Moscow, Russia

M. ORISTAGLIO

Stonewall Ridge
Newtown, CT, USA

ELSEVIER

AMSTERDAM • BOSTON • HEIDELBERG • LONDON • NEW YORK • OXFORD
PARIS • SAN DIEGO • SAN FRANCISCO • SINGAPORE • SYDNEY • TOKYO

Elsevier
Radarweg 29, PO Box 211, 1000 AE Amsterdam, The Netherlands
The Boulevard, Langford Lane, Kidlington, Oxford OX5 1GB, UK
225 Wyman Street, Waltham, MA 02451, USA

British Library Cataloguing in Publication Data
A catalogue record for this book is available from the British Library

Library of Congress Cataloging-in-Publication Data
A catalog record for this book is available from the Library of Congress

ISBN: 978-0-444-53829-1
ISSN: 0076-6895

CONTENTS

PART 2 Propagation and Diffusion of Electromagnetic Fields

PART 4 Electromagnetic Soundings in a Horizontally Layered Medium

INTRODUCTION

This book is devoted to the theory of electromagnetic methods of applied geophysics. It is intended for students and lecturers in geophysical departments, as well as for engineers and scientists involved in the use of electromagnetic fields in resource exploration and crustal geophysics. The content is divided into several parts. The first part is an introduction to the properties of static electric and magnetic fields in conducting and polarizable media. (A static field is one that has no variation in time; the terms steady, stationary, or time invariant are also used.) The full theory and application of steady fields is described in more detail in our previous books. We decided nevertheless to include this theory here, in a shortened form, because the basic physical laws governing the behavior of static fields—the laws of Coulomb and Ampere—remain fully valid for quasistationary electromagnetic fields, which are prominent in applied geophysical methods. The inclusion of this chapter at the start also enables us to avoid later use of the expressions "it can be shown" and "well known" that are a source of justifiable frustration for readers.

The second part of the book, "The propagation and diffusion of electromagnetic fields", treats the full set of physical laws that govern the behavior of the electromagnetic field. In this part, we present the equations that describe the behavior of the field in conducting and polarizable media and provide simple examples showing the effect of conduction and displacement currents and of electric charges that vary in time. We describe in detail sinusoidal and nonsinusoidal (transient) fields in a homogeneous environment that is characterized by specific values of conductivity and dielectric constant. In particular, we describe the concepts of wavelength, velocity of propagation, and attenuation of sinusoidal waves in a medium where conduction currents are much larger than displacement currents.

Our treatment pays special attention to the relationships between the fields, the parameters of the medium and the distance from the source at different intervals of time. These intervals correspond to the time of arrival of the signal, to the interval when conduction and displacement currents significantly affect the field and, finally, to later time intervals when the effects of displacement currents are negligible. We emphasize that even in these later time intervals the field has wavelike properties. In other words, no matter how large the measurement time or how low the frequency, the electromagnetic field (signal) appears as a wave. This observation does not contradict the fact that the electromagnetic field in later time intervals, or at low frequencies, is described with high accuracy by the diffusion equation. Given that most of the well-known books on electromagnetics do not discuss in detail the behavior of the fields in conducting media, it is quite possible that the second part of the book, as well as several subsequent chapters, will be interesting for students of physics.

The third part of the book, "Quasistationary field in a horizontally layered medium", treats the electromagnetic field generated by a vertical or horizontal magnetic dipole when displacement currents in a conducting medium can be neglected. First, we discuss in detail the phenomenon of the skin effect, which forces induced currents to be concentrated near the surface at the part of the conductor that is closest to the source. We discuss in great detail the field of a dipole in a homogeneous conducting medium, because this simple model reveals interesting and useful features present in more complex models. In particular, it is shown that measuring the quadrature and in-phase components of the field at sufficiently low frequencies we observe a very different depth of investigation. This part of the book then fully treats the electromagnetic field in a horizontally layered medium. We derive the main formulas that describe the field and study their asymptotic behaviors in the near field and wave zones, and at early and late times. We consider the characteristics of the electric and magnetic fields in both frequency and time, and show their relationships to the geoelectric parameters of the medium.

The reader who has followed the development to this point will be ready to study the theory and interpretation of "Electromagnetic soundings", the fourth part of the book. This part describes the three main types of electromagnetic sounding of the Earth: magnetotelluric sounding, frequency soundings in the far field, and transient soundings. We first present the history of development of magnetotelluric sounding, which uses natural electromagnetic fields in the atmosphere as the primary or source field for probing the Earth. We discuss the concepts of the wave zone, the subsurface behavior of the field, and the surface impedance and its relationship to the parameters of a horizontally layered medium within different frequency ranges. We emphasize the mathematical basis for methods of interpretation, including concepts such as uniqueness theorems, stable and unstable parameters, and regularization. These concepts are relevant of course to all types of electromagnetic soundings (as well as to other methods of geophysical remote sensing). We describe step by step how the process for interpretation of magnetotelluric soundings has been developed. Finally we describe the theory of frequency and transient soundings with controlled (artificial) sources, since these methods are widely used in many quite different areas of applied geophysics. We first treat the physical and mathematical foundations of frequency soundings in the far field; we then describe in detail the transition to transient soundings at an arbitrary distance from the source.

The final part of the book deals primarily with the principles of electromagnetic methods of mining geophysics. But we also cover briefly the influence of localized inhomogeneities in Earth properties on the interpretation of magnetotelluric sounding. The combination of these two different subjects is not accidental. Concepts such as the galvanic and the vortex parts of the electromagnetic field were first introduced in mining geophysics. In addition, the influence of the field of alternating charges that arise at boundaries between homogeneous regions (and distort magnetotelluric sounding curves) was also explored first in mining geophysics. We begin with a thorough analysis of the frequency and transient responses of magnetic fields generated by currents induced in finite-sized conductors surrounded by a nonconductive medium. We emphasize the relationship between frequency and transient fields, and study the influence of geological noise on both fields. We describe how the parameters of the surrounding medium influence the depth of investigation of sounding methods using the frequency and transient fields.

This last part of the book describes the main electromagnetic methods used to explore for highly conductive bodies (such as mineral deposits), as well as some methods of inductive profiling used in different geological applications. Finally, we return to magnetotelluric sounding and consider two- and three-dimensional Earth models, in each case providing physical explanations for the shapes of sounding curves in the presence of lateral inhomogeneity.

In conclusion, we would like to acknowledge the many prominent scientists, geophysicists, engineers, and practitioners who have worked to develop electromagnetic methods of surface geophysics during its more than 100-year history. We hope the reader will recognize and appreciate the strong influence of these personalities in this book. We would also like to personally thank Mr Paul Prasad Chandramohan, who spent so much effort, preparing the manuscript for the publication.

ACKNOWLEDGMENTS

The subject of applied electromagnetism for geophysical applications is vast. And in attempting to write comprehensively on its different topics, the authors have benefited enormously from the generous contributions of others, including many colleagues and friends. In particular, P. Andrieux, A. Becker, C. Stoyer, Y. Ogawa, S. Terentyev, and V. Mogilatov have all supplied specific information on technology and methods that rendered more vivid and thereby greatly improved different sections of the book. More generally, E. Fainberg helped to clarify the treatment throughout the volume with his clear, patient discussions with the authors. Two key appendices were kindly contributed by others: R. Smith and A. Volkovitsky wrote Appendix 1 on airborne electromagnetic methods; E. Aleksanova and M. Blinova provided data for Appendix 3; A. Petrov contributed Appendix 4. Finally, M. Bosner read carefully and made essential improvements to the sections describing the physical principles of measuring electromagnetic fields in frequency- and time-domain methods of electromagnetic surveying. As with all such contributions, the best parts are the fruits of the contributors, while any errors that may have crept in are the faults of the authors.

PART ONE

The Constant Electric and Magnetic Fields

CHAPTER ONE

The System of Equations of the Constant Electric and Magnetic Fields

INTRODUCTION

Before we begin to study the time-varying electromagnetic fields, it is necessary to focus our attention on the constant electric and magnetic fields. It is related to the fact that Coulomb's and Biot–Savart laws, governing the behavior of these fields, still play the fundamental role for understanding the so-called quasi-stationary fields used in the most electromagnetic methods of the surface and borehole geophysics [1–3]. First, we remind the main features of the constant electric field, which were described in detail in Ref. [2].

1.1. EQUATIONS OF THE CONSTANT ELECTRIC FIELD IN A CONDUCTING AND POLARIZABLE MEDIUM

As was shown in Ref. [2] for the constant electric field, we have three forms of the system of equations at regular points:

$$\text{curl}\,\mathbf{E} = 0 \quad \text{div}\,\mathbf{E} = \frac{\delta}{\varepsilon_0} \tag{1.1}$$

or

$$\text{curl}\,\mathbf{E} = 0 \quad \text{div}\,\mathbf{D} = \delta_0 \tag{1.2}$$

or

$$\text{curl}\,\mathbf{E} = 0 \quad \text{div}\,\mathbf{j} = 0 \tag{1.3}$$

Here $\mathbf{E}$ is the electric field; $\mathbf{D}$ is the vector of electric induction and $\mathbf{j}$ is the current density of free charges; $\mathbf{D} = \boldsymbol{\varepsilon}\mathbf{E}$ and $\boldsymbol{\varepsilon}$ is the dielectric constant of a medium. In accordance with Ohm's law,

$$\mathbf{j} = \gamma\mathbf{E}, \tag{1.4}$$

Methods in Geochemistry and Geophysics, Volume 45
ISSN 0076-6895,
http://dx.doi.org/10.1016/B978-0-444-53829-1.00001-0

where γ is the conductivity and

$$\mathbf{E} = \mathbf{E}^{c} + \mathbf{E}^{ext}$$

Here, $\mathbf{E}^{c}$ and $\mathbf{E}^{ext}$ are the Coulomb's and extraneous (non-Coulomb) electric fields, respectively, and $\mathbf{j}$ is the vector of current density characterizing an ordered movement of free charges, that is, conduction current. The total density δ is the sum as follows:

$$\delta = \delta_0 + \delta_b \tag{1.5}$$

Here, δ_0 and δ_b are the density of the free and bound charges, respectively. Let us make the several comments:

1. Equations 1.1–1.3 are written at regular points where the field has derivatives and it is related to the following. By definition, for any vector field $\mathbf{M}$, we have

$$\text{div}\,\mathbf{M} = \lim \frac{\oint_S \mathbf{M}\cdot d\mathbf{S}}{\Delta V}, \quad \text{if } \Delta V \to 0 \tag{1.6}$$

that is, the divergence of the field characterizes the flux of the field through a closed surface, surrounding an elementary volume which tends to zero. This equation is valid everywhere; however, it is not convenient for calculations because it requires an integration of a surface integral. Taking into account that the surface $\mathbf{S}$ is very small, it is possible to replace integration by differentiation that is much simpler. Inasmuch as such form of divergence contains derivatives, it is valid only at regular points.

Also, by definition, we have for curl of any field $\mathbf{M}$,

$$\text{curl}\,\mathbf{M} = \lim \frac{\oint_L \mathbf{M}\cdot d\mathbf{l}}{\Delta S} \quad \text{if } \Delta S \to 0 \tag{1.7}$$

Here $\mathbf{n}$ is the unit vector perpendicular to the elementary area ΔS, and L is the closed path surrounding this area. Both vectors $d\mathbf{l}$ and $\mathbf{n}$ obey the right-hand rule. It is essential that an area in Eq. (1.7) is oriented in such a way that the numerator has a maximal value. Again the calculation of Eq. (1.7) is cumbersome, but it is possible to replace integration by differentiation since the contour L is very small. As before, the presence of derivatives allows us to use this new expression for curl only at regular points where derivatives exist.

2. At interfaces between media with different electric parameters in place of Eqs (1.1)–(1.3), we have the surface analogy of these equations:

$$E_{2t} - E_{1t} = 0 \quad E_{2n} - E_{1n} = \frac{\sigma}{\varepsilon_0},$$

or

$$E_{2t} - E_{1t} = 0 \quad D_{2n} - D_{1n} = \sigma_0, \tag{1.8}$$

or

$$E_{2t} - E_{1t} = 0 \quad \gamma_2 E_{2n} - \gamma_1 E_{1n} = 0,$$

where E_{1t}, E_{1n} and E_{2t}, E_{2n} are the tangential and normal components of the electric field at the back and front sides of an interface, respectively, and the normal is directed toward the front side; that is into the medium with index "2".

3. The conductivity of a medium can be expressed as

$$\gamma = \delta_0^+ u^+ + \left|\delta_0^-\right| u^- . \tag{1.9}$$

Here u^+ and u^- are the mobility of the positive and negative charges, respectively, which are extremely small numbers. This is the reason why the velocity of free charges, involved in an ordered motion, is usually very small and it does not exceed 10^{-6} m/s. Note that at the same time, these hardly moving charges may create very strong magnetic field.

4. Unlike free charge which can move at distances exceeding molecule size, the bound charges may change their position only within a molecule.
5. As follows from Eq. (1.3), the constant electric field in a conducting and polarizable medium is independent on dielectric constant. In other words, distribution of bound charges does not make influence on the electric field even though these charges are present. Such remarkable feature of the field behavior is also observed for the so-called quasi-stationary fields which vary with time. This means that the density of total charge coincides with that of free charges δ_0 in a nonpolarizable medium where $\varepsilon = \varepsilon_0$.
6. The volume and surface density of these charges are related with the field and conductivity of a medium as

$$\delta(p) = -\varepsilon_0 \frac{\mathbf{E} \cdot \mathrm{grad}\, \gamma}{\gamma}, \quad \sigma(p) = 2\varepsilon_0 K_{12} E_n^{av} \tag{1.10}$$

Here,

$$K_{12} = \frac{\rho_2 - \rho_1}{\rho_2 + \rho_1},$$

and E_n^{av} is the mean value of the normal component of electric field at some point p of the boundary between media with resistivity ρ_1 and ρ_2; the normal $\mathbf{n}$ is directed from medium 1 to medium 2. Besides, it is assumed that an extraneous force is absent in the vicinity of point p. The physical meaning of $E_n^{av}(p)$ is very simple: this is the normal component of the field caused by all charges except that in the vicinity of point p. Note that these charges do not form the current, and they arise at places where conductivity of a medium varies.

7. The second equation of the system for the electric field at regular points is

$$\operatorname{div}\mathbf{E} = \frac{\delta}{\varepsilon_0} \quad \text{or} \quad \operatorname{div}\mathbf{D} = \delta_0$$

and their surface analogy remain valid for the time-varying fields and each of them represents the third Maxwell's equation. One can say that taking into account effect of polarization of a medium and appearance of bounded charges, it became possible from Coulomb's law to obtain the third Maxwell's equation.

Bearing in mind that constant magnetic fields were not discussed in Ref. [2], it is necessary to consider this subject in some detail following [3] and, in particular, derive the system of equations of this field. Importance of this subject is emphasized by the fact that one of these equations is the fourth Maxwell's equation. Moreover, as was pointed out, relationship between the conduction currents and the quasi-stationary magnetic field is the same as in the case of the time-invariant field.

1.2. INTERACTION OF CURRENTS, BIOT–SAVART LAW AND MAGNETIC FIELD

1.2.1. Ampere's Law and Interaction of Currents

Numerous experiments performed in eighteenth century demonstrated that currents in two circuits interact with each other; that is, mechanical forces act at every element of a current circuit. It turns out that this force depends on the current magnitude, the direction of charge movement, the shape and dimension of the current circuit, as well as a distance and mutual orientation of the circuits with respect to each other. This list of factors clearly shows that the mathematical formulation of this phenomenon should be a much more complicated task than that for the electric field. In spite of this fact, Ampere was able to find very quickly an expression for the force of interaction of two elementary currents in a relatively simple form:

$$d\mathbf{F}(p) = \frac{\mu_0}{4\pi} I_1 I_2 \frac{d\mathbf{l}_1(p) \times \left[d\mathbf{l}_2(q) \times \mathbf{L}_{qp}\right]}{L_{qp}^3} \tag{1.11}$$

where I_1 and I_2 are the magnitudes of currents in the linear elements $\mathbf{dl}_1$ and $\mathbf{dl}_2$, respectively, and their direction coincides with that of the current density; L_{qp} is the distance between these elements and $\mathbf{L}_{qp}$ is directed from point q to point p, which are located at the center of current elements; and finally μ_0 is a constant equal to

$$\mu_0 = 4\mu \cdot 10^{-7}\mathrm{H/m},$$

and it is sometimes improperly called the *magnetic permeability of free space*, in spite of the fact that the latter does not have any physical properties. Applying Ampere's law, it is essential to note that the distance between current elements must be much greater than their length:

$$L_{qp} \gg dl_1, \quad L_{qp} \gg dl_2$$

Examples, illustrating an interaction of elementary currents, are given in Figure 1.1.

Making use of the principle of superposition, the force of interaction between two arbitrary closed current circuits is defined as

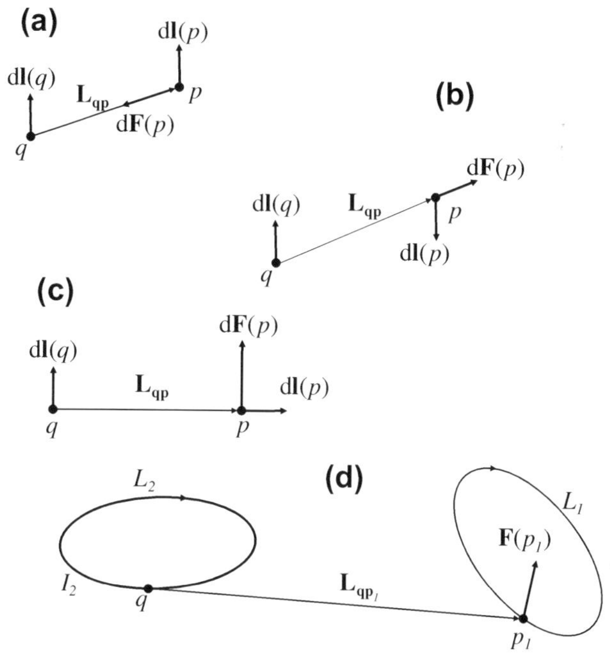

Figure 1.1 (a) Interaction of currents having the same direction. (b) Interaction of currents having opposite direction. (c) Interaction of current elements perpendicular to each other. (d) Interaction of two current loops.

$$\mathbf{F} = \frac{\mu_0}{4\pi} I_1 I_2 \oint_{L_1} \oint_{L_2} \frac{\mathbf{dl}_1 \times (\mathbf{dl}_2 \times \mathbf{L}_{qp})}{L_{qp}^3} \tag{1.12}$$

where L_1 and L_2 are the current lines along which integration is performed and $p \neq q$.

The resultant force $\mathbf{F}$ is the sum of forces acting on different elements of the contour and due to them different types of motion may occur. It is appropriate to notice that in SI units, $\mathbf{F}$ is measured in newtons.

1.2.2. Magnetic Field and Biot–Savart Law

Inasmuch as there is interaction between currents, it is natural, by analogy with the electric field, to assume that current in a contour creates a field and due to the existence of this field, other currents experience the action of the force $\mathbf{F}$. This field is called the *magnetic field*, and it is introduced from Ampere's law as

$$\mathrm{d}\mathbf{F}(p) = I(p)\mathbf{dl}(p) \times \mathrm{d}\mathbf{B}(p) \tag{1.13}$$

Here,

$$\mathrm{d}\mathbf{B}(p) = \frac{\mu_0}{4\pi} I(q) \frac{\mathbf{dl}(q) \times \mathbf{L}_{qp}}{L_{qp}^3} \tag{1.14}$$

and $I(q)$ is the current of the element $\mathbf{dl}(q)$. Equation 1.14 is called the Biot–Savart law and it describes the relationship between the elementary linear current and magnetic field $\mathrm{d}\mathbf{B}$. By definition, the magnitude of the magnetic field caused by the elementary current is

$$\mathrm{d}B(p) = \frac{\mu_0}{4\pi} I(q) \frac{\mathrm{d}l}{L_{qp}^2} \sin(\mathbf{L}_{qp}, \mathbf{dl}) \tag{1.15}$$

Here, $(\mathbf{L}_{qp}, \mathbf{dl})$ is the angle between the vectors $\mathbf{L}_{qp}$ and $\mathbf{dl}$; the vector $\mathrm{d}\mathbf{B}$ is perpendicular to these vectors as is shown in Figure 1.2(a), and these three vectors obey the right-hand rule. The unit vector $\mathbf{b}_0$, characterizing the direction of the field, is defined as

$$\mathbf{b}_0 = \frac{\mathbf{dl} \times \mathbf{L}_{qp}}{|\mathbf{dl} \times \mathbf{L}_{qp}|}$$

In SI units, the magnetic field is measured in teslas and it is related to other units such as gauss and gamma in the following way:

$$1 \text{ tesla} = 10^9 \text{nT} = 10^4 \text{gauss} = 10^9 \text{gamma}$$

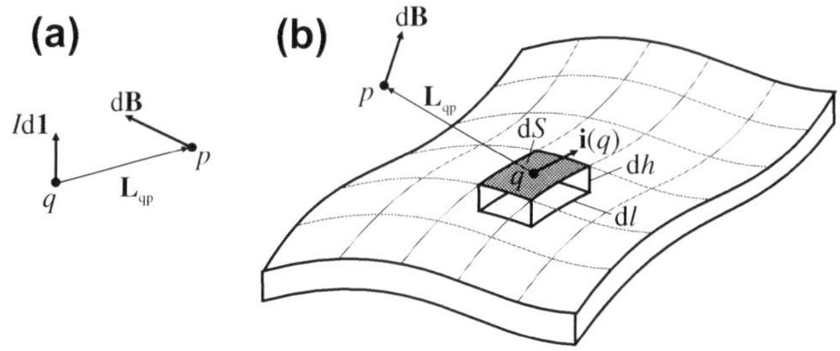

Figure 1.2 (a) Magnetic field of a current element. (b) Magnetic field of the surface current.

Now we will generalize Eq. (1.14) assuming that along with linear currents, there are also volume and surface currents. First let us represent the product $I\mathbf{dl}$ as

$$I\mathrm{d}\mathbf{l} = j\,\mathrm{d}S\,\mathrm{d}\mathbf{l} = \mathbf{j}\,\mathrm{d}S\,\mathrm{d}l = \mathbf{j}\,\mathrm{d}V, \tag{1.16}$$

where $\mathrm{d}S$ is the cross-section of the elementary current tube, $\mathbf{dl}$ is oriented along this tube and $\mathbf{j}$ is the volume current density. If the current is concentrated in a relatively thin layer with thickness $\mathrm{d}h$, which is small with respect to the distance to an observation point, it is convenient to replace this layer by a current surface. As is seen from Figure 1.2(b), the product $I\mathbf{dl}$ can be modified in the following way:

$$I\,\mathrm{d}\mathbf{l} = \mathbf{j}\,\mathrm{d}V = \mathbf{j}\,\mathrm{d}h\,\mathrm{d}S = \mathbf{i}\,\mathrm{d}S \tag{1.17}$$

Here $\mathrm{d}S$ is the surface element and

$$\mathbf{i} = \mathbf{j}\,\mathrm{d}h$$

is the surface density of current. The resultant force $\mathbf{F}$ is the sum of forces acting on different elements of the contour. Applying the principle of superposition for all three types of currents (volume, surface, and linear) and making use of Eqs (1.14), (1.16), and (1.17), we obtain the generalized form of the Biot–Savart law,

$$\mathbf{B}(p) = \frac{\mu_0}{4\pi}\left[\int\limits_V \frac{\mathbf{j}\times\mathbf{L}_{qp}}{L_{qp}^3}\,\mathrm{d}V + \int\limits_S \frac{\mathbf{i}\times\mathbf{L}_{qp}}{L_{qp}^3}\,\mathrm{d}S + \sum_n I_n \oint \frac{\mathrm{d}\mathbf{l}\times\mathbf{L}_{qp}}{L_{qp}^3}\right] \tag{1.18}$$

Now it is appropriate to make several comments.

1. This equation allows us to calculate the magnetic field everywhere inside and outside of volume currents.
2. In general, the currents arise due to a motion of free charges and magnetization of a magnetic medium. Correspondingly, the current density is the sum of the following:

$$\mathbf{j} = \mathbf{j}_c + \mathbf{j}_m$$

where $\mathbf{j}_c$ and $\mathbf{j}_m$ are the volume density of the conduction and magnetization currents, respectively, and a magnetic field of both these currents obey the Biot–Savart law. In most part of this monograph, it is assumed that a magnetization is absent.

3. In accordance with Eq. (1.18), the magnetic field caused by a given distribution of currents depends on the coordinates of the observation point p only, and it is independent of the presence of other currents. The right-hand side of Eq. (1.18) does not contain any terms that characterize the physical properties of the medium, and therefore the field $\mathbf{B}$ at point p, generated by a given distribution of currents, remains the same if a free space is replaced by a nonuniform conducting and polarizable medium. For instance, if the given current circuit is placed inside a magnetic medium, the field $\mathbf{B}$ caused by this current is the same as if it were in free space. Of course, as is well known, the presence of such a medium results in a change of the magnetic field $\mathbf{B}$, but this means that inside of a medium along with a given current, there are other currents which also produce the magnetic field and they are the magnetization currents. This conclusion directly follows from Eq. (1.18) which states that any change of the field $\mathbf{B}$ can happen only due to a change of the current distribution. Later in one special case (quasi-stationary field of a magnetic sphere), we will take into account the influence of currents in a magnetic medium but it is rather exception.
4. Unlike the volume distribution of currents, linear and surface analogies are only the mathematical models of real distribution of current, which are usually introduced to simplify calculations of the field and study its behavior. For this reason, the equation

$$\mathbf{B}(p) = \frac{\mu_0}{4\pi} \int_V \frac{\mathbf{j}(q) \times \mathbf{L}_{qp}}{L_{qp}^3} \, dV \tag{1.19}$$

in essence comprises all possible cases of the current distribution.

5. As will be shown later, the Biot–Savart law, Eq. (1.18), is also valid for the time-varying magnetic field when it is possible to neglect by so-called displacement currents. This means that this law plays an important role in the theory of the frequency and transient soundings, the surface and airborne electromagnetic profiling, in all modifications of the induction logging. In other words, the Biot–Savart law correctly describes the time-varying fields when an influence of diffusion is dominant.
6. All experiments that allowed Ampere to derive Eq. (1.11) were carried out with the closed circuits. At the same time, Eq. (1.11), as well as Eq. (1.14), is written for the element **dl**, where current cannot exist if this element does not constitute a part of the closed circuit. In other words, Eqs (1.11) and (1.14) cannot be proved experimentally but the interaction between closed current circuits takes place in such manner as if the magnetic field **B**, caused by the current element I**dl**, was described by Eq. (1.14).
7. In accordance with the Biot–Savart law, the current is the sole generator of the constant magnetic field, and the distribution of this generator is characterized by the magnitude and direction of the current density vector **j**. As is well known, the vector lines of **j** are always closed. This means that the magnetic field is caused by generators of the vortex type, and unlike the Coulomb's electric field, we are dealing with a vortex field **B**.

1.2.3. Lorentz Force and Electromotive Force at the Moving Circuit

As follows from Eqs (1.13) and (1.16), the current in the elementary volume, placed in the magnetic field **B**, is subjected to the action of force:

$$\mathbf{F} = (\mathbf{j} \times \mathbf{B})\, dV \qquad (1.20)$$

The latter allows us to find force acting on a single electron or ion moving with velocity v.

By definition, the current density **j** can be represented as

$$\mathbf{j} = ne\mathbf{v}$$

where n is the number of particles in the unit volume, and e is the charge of electron or ion. Therefore, the force of the magnetic field **B** acting on all particles is

$$\mathbf{F}_B = ne(\mathbf{v} \times \mathbf{B})\, dV$$

and correspondingly every moving particle, for example, electron, is subjected to a force equal to

$$\mathbf{F}_B = e(\mathbf{v} \times \mathbf{B}) \tag{1.21}$$

Thus, this elementary charge is subjected to the total force equal to

$$\mathbf{F} = \mathbf{F}_\mathrm{e} + \mathbf{F}_\mathrm{m} = e\mathbf{E}_\mathrm{c} + e(\mathbf{v} \times \mathbf{B}), \tag{1.22}$$

and it is called the Lorentz force. Here,

$$\mathbf{F}_\mathrm{e} = e\mathbf{E}_\mathrm{c} \quad \text{and} \quad \mathbf{F}_\mathrm{m} = e(\mathbf{v} \times \mathbf{B})$$

are the forces caused by the electric and magnetic fields, respectively. By analogy with Coulomb's law, let us introduce the electric field of a non-Coulomb's origin as

$$\mathbf{E}_\mathrm{m} = \mathbf{v} \times \mathbf{B}, \tag{1.23}$$

and it acts on the moving charge in the presence of the magnetic field. By definition, this field is perpendicular to both the velocity and the magnetic field, and it reaches a maximum when the angle between these two vectors is equal to $\frac{\pi}{2}$. As in the case of Coulomb's electric field, the voltage of this electric field along an elementary and arbitrary path is

$$\Delta V = \mathbf{E}_\mathrm{m} \cdot \mathbf{dl} = (\mathbf{v} \times \mathbf{B}) \cdot \mathbf{dl} \quad V = \int_L (\mathbf{v} \times \mathbf{B}) \cdot \mathbf{dl} \tag{1.24}$$

In particular, the electromotive force caused by the field $\mathbf{E}_\mathrm{m}$ is

$$\Xi = \oint_L (\mathbf{v} \times \mathbf{B}) \cdot \mathbf{dl} \tag{1.25}$$

Unlike the voltage of the Coulomb's electric field, the second part of Eq. (1.24) is path dependent, and in general, the electromotive force of this field does not vanish. Now we consider several examples, but before it is proper to emphasize that the existence of this field directly follows from Ampere's law which was derived for constant currents.

Example 1

Suppose that the current circuit does not move and it is placed into magnetic field **B** (Figure 1.3(a)). It is clear that moving electrons along the circuit are subjected to the action of the field $\mathbf{E}_\mathrm{m}$, which is usually very small, since the electron velocity is around 10^{-6} m/s. By definition, this field is perpendicular to the Coulomb's field and may cause only insignificant shift of

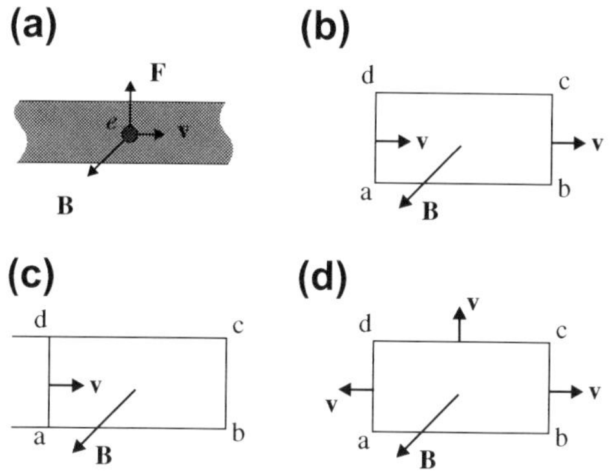

Figure 1.3 (a) Magnetic force acting on a charge which moves with velocity *v*. (b) Rectangular circuit moving with velocity *v* in the magnetic field. (c) Movement of one side of the circuit with velocity *v*. (d) Movement and deformation of a contour in the magnetic field.

charges toward the surface of the circuit, where the positive and negative surface charges tend to appear, but their electric field prevents this shift. For this reason, our attention will be paid to the cases when either whole circuit or its part moves in the magnetic field.

Example 2

Consider the rectangular and conducting circuit *abcd* (Figure 1.3(b)), which moves with the constant velocity **v** along the *x*-axis. The uniform magnetic field **B** is perpendicular to the circuit. Taking into account that direction of currents along paths *ad* and *cb* are opposite to each other, the voltages

$$\Delta V_{ad}, \quad \Delta V_{cb}$$

differ by sign only. The voltages along lines *ab* and *cd* are equal to zero. Indeed, according to vector algebra for the voltage along an arbitrary element of the circuit $\mathbf{dl}$, we have

$$(\mathbf{v} \times \mathbf{B}) \cdot \mathrm{d}\mathbf{l} = (\mathrm{d}\mathbf{l} \times \mathbf{v}) \cdot \mathbf{B} \tag{1.25a}$$

Since in case of lines *ab* and *cd* vectors $\mathrm{d}\mathbf{l}$ and **v** have the same or opposite direction, we conclude that the voltage along these elements of the circuit is zero and, therefore, the electromotive force is also zero. As is seen from Figure 1.3(b), the flux of the magnetic field Φ through the area, surrounded by the path, remains constant. Thus, we have

$$\Xi = 0, \quad \text{and} \quad \frac{\mathrm{d}\Phi}{\mathrm{d}t} = 0$$

Example 3

Now suppose that only the side *ad* slides with the velocity **v**, while the other part of the circuit is at rest (Figure 1.3(c)). Then the electromotive force coincides with the voltage ΔV_{ad}:

$$\Xi = \pm v_x B_z ad, \tag{1.26}$$

where the sign depends on the orientation of the current in this line. It is clear that the product $v_x ad$ means the rate of a change of the area, enclosed by the circuit and, therefore Eq. (1.26) can be also written as

$$\Xi = \pm\frac{d\Phi}{dt}$$

that is, the electromotive force arising in the circuit is defined by the rate of a change of the flux of the magnetic field through the area surrounded by the circuit. By definition, the flux is equal to

$$\Phi = \int_S \mathbf{B}\cdot d\mathbf{S}$$

If the direction along the circuit and the vector d**S** obey the right-hand rule, we have

$$\Xi = -\frac{d\Phi}{dt} \tag{1.27}$$

Example 4

In this case, the magnetic field is aligned in the plane of the circuit which moves with velocity **v**. Then the voltage along an arbitrary element of the circuit is equal to

$$(\mathbf{v}\times\mathbf{B})\cdot d\mathbf{l} = 0$$

and, therefore, the electromotive force is absent in spite of a motion of the circuit and the presence of the magnetic field. Inasmuch as the field **B** is tangential to the circuit, its flux is also equal to zero, and we can again write

$$\Xi = -\frac{d\Phi}{dt} = 0.$$

This shows that only the normal component of the magnetic field makes an influence on the moving charge.

Example 5

Suppose that an arbitrary conducting circuit is located in some plane and each of its element moves with velocity $\mathbf{v}$ which may change from point to point (Figure 1.3(d)). This means that the circuit experiences both a motion and deformation. The component of the magnetic field normal to this plane may also vary. Consider again the elementary voltage along the element $\mathbf{dl}$ which can be written as

$$\Delta V = (\mathbf{v} \times \mathbf{B}) \cdot \mathbf{dl} = (\mathbf{dl} \times \mathbf{v}) \cdot \mathbf{B}$$

The magnitude of the cross product: $\mathbf{dl} \times \mathbf{v}$ is equal to the area covered by the element $\mathbf{dl}$ during 1 s, and correspondingly ΔV is equal to the rate of a change of the elementary flux of the field $\mathbf{B}$. Integrating along the circuit and using again the right-hand rule, we obtain

$$\Xi = -\frac{d\Phi}{dt} \tag{1.28}$$

Later we will describe the Faraday's law which has exactly the same form. However from physical point of view, it has one fundamental difference, namely, it shows that an electromotive force may arise not only because of a movement and deformation of the circuit but also due to a rate of a change of the magnetic field with time when circuit does not move. Besides, Faraday's law is applied to any closed path which can be, for example, an insulator.

1.3. THE VECTOR POTENTIAL OF THE MAGNETIC FIELD

1.3.1. Relation between Magnetic Field and Vector Potential

Although calculation of the magnetic field, making use of the Biot–Savart law is not a very complicated procedure, it is still reasonable to find a simpler way of determining field. With this purpose in mind, by analogy with the scalar potential of the electric field, we will introduce a new function. Besides there is another reason to consider this function, namely, it allows one to derive sufficiently simply the system of equations of the magnetic field. Let us proceed from Biot–Savart law:

$$\mathbf{B}(p) = \frac{\mu_0}{4\pi} \int_V \frac{\mathbf{j}(q) \times \mathbf{L}_{qp}}{L_{qp}^3} dV \tag{1.29}$$

Taking into account that

$$\frac{\mathbf{L}_{qp}}{L_{qp}^3} = \overset{q}{\nabla}\frac{1}{L_{qp}} = -\overset{p}{\nabla}\frac{1}{L_{qp}} \tag{1.30}$$

and substituting Eq. (1.30) into Eq. (1.29), we obtain

$$\mathbf{B}(p) = \frac{\mu_0}{4\pi}\int_V \mathbf{j}(q) \times \overset{q}{\nabla}\frac{1}{L_{qp}}\,\mathrm{d}V = \frac{\mu_0}{4\pi}\int_V \overset{p}{\nabla}\frac{1}{L_{qp}} \times \mathbf{j}(q)\,\mathrm{d}V \tag{1.31}$$

Here, letters q and p indicate that derivatives are taken with respect to coordinates of the point q and p. For instance, in the Cartesian system of coordinates, we have

$$\overset{q}{\nabla}\frac{1}{L_{qp}} = \overset{q}{\mathrm{grad}}\,\frac{1}{L_{qp}} = \frac{\partial}{\partial x_q}\frac{1}{L_{qp}}\mathbf{1}_x + \frac{\partial}{\partial y_q}\frac{1}{L_{qp}}\mathbf{1}_y + \frac{\partial}{\partial z_q}\frac{1}{L_{qp}}\mathbf{1}_z,$$

where $\mathbf{1}_x$, $\mathbf{1}_y$, and $\mathbf{1}_z$ are the orthogonal unit vectors.

Now we will make use of the equality

$$\overset{p}{\nabla} \times \frac{\mathbf{j}}{L_{qp}} = \overset{p}{\nabla}\frac{1}{L_{qp}} \times \mathbf{j} + \frac{\overset{p}{\nabla} \times \mathbf{j}}{L_{qp}},$$

which follows from the vector identity:

$$\nabla \times (\varphi\mathbf{a}) = \nabla\varphi \times \mathbf{a} + \varphi\nabla \times \mathbf{a} \tag{1.32}$$

Applying Eq. (1.32), we can rewrite Eq. (1.31) as

$$\mathbf{B}(p) = \frac{\mu_0}{4\pi}\int_V \overset{p}{\nabla} \times \frac{\mathbf{j}}{L_{qp}}\,\mathrm{d}V - \frac{\mu_0}{4\pi}\int_V \frac{\overset{p}{\nabla} \times \mathbf{j}}{L_{qp}}\,\mathrm{d}V. \tag{1.33}$$

The current density is a function of the point q and does not depend on the location of the observation point p. Therefore, the integrand of the second integral is zero and

$$\mathbf{B}(p) = \frac{\mu_0}{4\pi}\int_V \overset{p}{\mathrm{curl}}\,\frac{\mathbf{j}(q)}{L_{qp}}\,\mathrm{d}V \tag{1.34}$$

Inasmuch as the integration and differentiation in Eq. (1.34) are carried out with respect to different points q and p, we can interchange the order of operations that gives

$$\mathbf{B}(p) = \overset{p}{\text{curl}}\, \frac{\mu_0}{4\pi} \int\limits_V \frac{\mathbf{j}(q)}{L_{qp}}\, \mathrm{d}V$$

or

$$\mathbf{B}(p) = \text{curl}\, \mathbf{A}, \qquad (1.35)$$

where

$$\mathbf{A}(p) = \frac{\mu_0}{4\pi} \int\limits_V \frac{\mathbf{j}(q)}{L_{\text{qp}}}\, \mathrm{d}V \qquad (1.36)$$

Thus, the magnetic field $\mathbf{B}$ caused by constant currents can be expressed through the vector potential $\mathbf{A}$ defined by Eq. (1.36). Comparing Eqs (1.29) and (1.36), we see that the function $\mathbf{A}$ is related to the distribution of currents in a much simpler way than the magnetic field is and, therefore, one reason for introducing this function is already demonstrated. In accordance with Eq. (1.36), $\mathbf{A}$ is a vector, unlike the potential of an electric field, and its magnitude and direction depend essentially on the current distribution. Now let us derive expressions for the vector potential $\mathbf{A}$, caused by surface and linear currents. Making use of Eq. (1.17),

$$\mathbf{j}\, \mathrm{d}V = \mathbf{i}\, \mathrm{d}S,$$

and from Eq. (1.36), it follows that

$$\mathbf{A} = \frac{\mu_0}{4\pi} \int\limits_S \frac{\mathbf{i}\, \mathrm{d}S}{L_{qp}} \quad \text{and} \quad \mathbf{A} = \frac{\mu_0 I}{4\pi} \oint\limits_L \frac{\mathrm{d}\mathbf{l}}{L_{qp}} \qquad (1.37)$$

In general case, when there are volume, surface, and linear currents, we have

$$\mathbf{A} = \frac{\mu_0}{4\pi} \left[\int\limits_V \frac{\mathbf{j}\, \mathrm{d}V}{L_{qp}} + \int\limits_S \frac{\mathbf{i}\, \mathrm{d}S}{L_{qp}} + \sum_{i=1} I_i \oint \frac{\mathrm{d}\mathbf{l}}{L_{qp}} \right] \qquad (1.38)$$

The components of the vector potential can be derived directly from this equation. For instance, in Cartesian coordinates, we have

$$A_x = \frac{\mu_0}{4\pi}\left[\int\limits_V \frac{j_x\,\mathrm{d}V}{L_{qp}} + \int\limits_S \frac{i_x\,\mathrm{d}S}{L_{qp}} + \sum_{i=1} I_i \oint \frac{\mathrm{d}l_x}{L_{qp}}\right]$$

$$A_y = \frac{\mu_0}{4\pi}\left[\int\limits_V \frac{j_y\,\mathrm{d}V}{L_{qp}} + \int\limits_S \frac{i_y\,\mathrm{d}S}{L_{qp}} + \sum_{i=1} I_i \oint \frac{\mathrm{d}l_y}{L_{qp}}\right] \qquad (1.39)$$

$$A_z = \frac{\mu_0}{4\pi}\left[\int\limits_V \frac{j_z\,\mathrm{d}V}{L_{qp}} + \int\limits_S \frac{i_z\,\mathrm{d}S}{L_{qp}} + \sum_{i=1} I_i \oint \frac{\mathrm{d}l_z}{L_{qp}}\right]$$

Similar expressions can be written for the vector potential components in other systems of coordinates. As is seen from Eq. (1.38), if a current flows along a single straight line, the vector potential has only one component, which is parallel to this line. It is also obvious that if currents are situated in a single plane, then the vector potential **A** at every point is parallel to this plane. Later we will consider several examples illustrating the behavior of the vector potential and magnetic field, but now let us derive two useful relations for the function **A**, which simplify to a great extent the task of deriving the system of magnetic field equations.

1.3.2. Divergence and Laplacian of Vector Potential A

First, we will determine the divergence of the vector potential **A**. As follows from Eq. (1.36), we have

$$\overset{p}{\operatorname{div}}\,\mathbf{A} = \overset{p}{\operatorname{div}}\,\frac{\mu_0}{4\pi}\int\limits_V \frac{\mathbf{j}(q)}{L_{qp}}\,\mathrm{d}V$$

Since differentiation and integration in this expression are performed with respect to different points, we can change the order of operations and this gives

$$\overset{p}{\operatorname{div}}\,\mathbf{A} = \frac{\mu_0}{4\pi}\int\limits_V \overset{p}{\operatorname{div}}\,\frac{\mathbf{j}(q)}{L_{qp}}\,\mathrm{d}V \qquad (1.40)$$

The volume over which the integration is carried out includes all currents and therefore it can be enclosed by a surface S such that outside of it currents are absent. Correspondingly, the normal component of the current density at this surface equals zero:

$$j_{\mathrm{n}} = 0 \tag{1.41}$$

The integrand in Eq. (1.40) can be represented as

$$\overset{p}{\nabla}\cdot\frac{\mathbf{j}}{L_{qp}} = \frac{\overset{p}{\nabla}\cdot\mathbf{j}}{L_{qp}} + \mathbf{j}\cdot\overset{p}{\nabla}\frac{1}{L_{qp}} = \mathbf{j}\cdot\overset{p}{\nabla}\frac{1}{L_{qp}},$$

because the current density does not depend on the observation point and

$$\overset{p}{\operatorname{div}}\,\mathbf{j}(q) = 0$$

Then, we have

$$\mathbf{j}\cdot\overset{p}{\nabla}\frac{1}{L_{qp}} = -\mathbf{j}\cdot\overset{q}{\nabla}\frac{1}{L_{qp}} = -\overset{q}{\nabla}\cdot\frac{\mathbf{j}}{L_{qp}} + \frac{\overset{q}{\nabla}\cdot\mathbf{j}}{L_{qp}}$$

As follows from the principle of charge conservation,

$$\overset{q}{\operatorname{div}}\,\mathbf{j} = 0$$

and therefore

$$\mathbf{j}\cdot\overset{p}{\nabla}\frac{1}{L_{qp}} = -\overset{q}{\operatorname{div}}\frac{\mathbf{j}}{L_{qp}}$$

Correspondingly, Eq. (1.40) can be written as

$$\operatorname{div}\mathbf{A} = -\frac{\mu_0}{4\pi}\int\limits_V \overset{q}{\operatorname{div}}\,\frac{\mathbf{j}}{L_{qp}}\,\mathrm{d}V$$

Unlike Eq. (1.40) on the right-hand side of this equation, both integration and differentiation are performed with respect to the same point q so that now we can apply Gauss' theorem:

$$\int\limits_V \operatorname{div}\mathbf{M}\,\mathrm{d}V = \oint\limits_S \mathbf{M}\cdot\mathrm{d}\mathbf{S}$$

Thus, we have

$$\operatorname{div}\mathbf{A} = -\frac{\mu_0}{4\pi}\int\limits_V \overset{q}{\operatorname{div}}\,\frac{\mathbf{j}}{L_{qp}}\,\mathrm{d}V = -\frac{\mu_0}{4\pi}\oint\limits_S \frac{\mathbf{j}\cdot\mathrm{d}\mathbf{S}}{L_{qp}} = -\frac{\mu_0}{4\pi}\oint\limits_S \frac{j_{\mathrm{n}}\,\mathrm{d}S}{L_{qp}}$$

Taking into account the fact that the normal component of the current density j_n vanishes at the surface S which surrounds all currents (Eq. (1.41)), we obtain

$$\operatorname{div} \mathbf{A} = 0 \tag{1.42}$$

This is the first relation that is useful for deriving the system of field equations. Let us note that in accordance with Eq. (1.42), the vector lines of the field **A** are always closed. Next we will obtain one more equation describing this function. As is well known [2], the potential of the electric field U satisfies Poisson's equation

$$\nabla^2 U = -\frac{\delta}{\varepsilon_0},$$

and its solution is

$$U = \frac{1}{4\pi\varepsilon_0} \int\limits_V \frac{\delta \, \mathrm{d}V}{L_{qp}}$$

As follows from Eq. (1.39), every component of the vector potential has the same form as the potential U and, therefore by analogy, it also satisfies the Poisson's equation; that is,

$$\nabla^2 A_x = -\mu_0 j_x, \quad \nabla^2 A_x = -\mu_0 j_y, \quad \nabla^2 A_z = -\mu_0 j_z$$

Multiplying each of these equations by the corresponding unit vector $\mathbf{1}_x$, $\mathbf{1}_y$, and $\mathbf{1}_z$ and performing the summation, we arrive at the Poisson's equation for the vector potential:

$$\nabla^2 \mathbf{A} = -\mu_0 \mathbf{j} \tag{1.43}$$

1.4. SYSTEM OF EQUATIONS OF THE CONSTANT MAGNETIC FIELD

Now we are ready to derive the system of equations of the constant magnetic field. First, making use of Eq. (1.35), we discover that divergence of the field **B** vanishes. In fact, we have

$$\operatorname{div} \mathbf{B} = \operatorname{div} (\operatorname{curl} \mathbf{A}). \tag{1.44}$$

As is well known from vector Calculus, the right-hand term of Eq. (1.44) is identically zero. Therefore,

$$\operatorname{div} \mathbf{B} = 0. \tag{1.45}$$

This means that the magnetic field does not have sources, like charges, and, correspondingly, the vector lines of the magnetic field $\mathbf{B}$ are always closed. Applying Gauss' theorem, we obtain the integral form of this equation

$$\oint_S \mathbf{B}\cdot d\mathbf{S} = 0. \tag{1.46}$$

That is, the total flux of the field $\mathbf{B}$ through any closed surface is always equal to zero. Next we will derive the surface analogy of Eq. (1.45) and with this purpose in mind, consider the flux through an elementary cylindrical surface (Figure 1.4(a)). It is equal to

$$\mathbf{B}^{(2)}\cdot d\mathbf{S}_2 + \mathbf{B}^{(1)}\cdot d\mathbf{S}_1 + \mathbf{B}\cdot d\mathbf{S}_* = 0 \tag{1.47}$$

Here,

$$d\mathbf{S}_2 = dS\mathbf{n}, \quad d\mathbf{S}_1 = -dS\mathbf{n}$$

and $d\mathbf{S}_*$ is the lateral surface of the cylinder. Then reducing the height of the cylinder to zero in place of Eq. (1.47), we obtain

$$B_n^{(2)}\, dS - B_n^{(1)}\, dS = 0 \quad \text{or} \quad B_n^{(2)} = B_n^{(1)} \tag{1.48}$$

Thus, the normal component of the magnetic field $\mathbf{B}$ is always a continuous function of the spatial variables. We have three forms of the first equation that describe the magnetic field caused by constant currents:

$$\oint_S \mathbf{B}\cdot d\mathbf{S} = 0; \;\; \operatorname{div} \mathbf{B} = 0; \;\; B_n^{(2)} - B_n^{(1)} = 0. \tag{1.49}$$

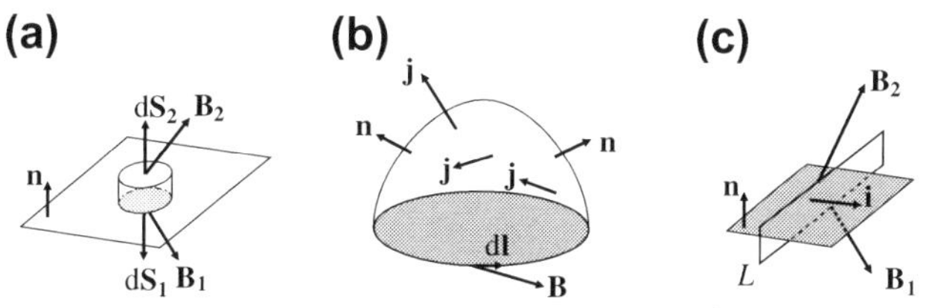

Figure 1.4 (a) Surface analogy of Eq. (1.45). (b) Illustration of Eq. (1.52). (c) Surface analogy of Eq. (1.50).

Each of them expresses the same fact, namely, absence of magnetic charges. Let us make two comments.

1. Equations (1.49) have been derived assuming that the field **B** is caused by conduction currents. However, they remain valid in the presence of magnetic medium when the field is also generated by magnetization currents.
2. These equations were obtained from Biot–Savart law for direct currents, but actually they are still valid for time-varying magnetic fields and in effect represent the fourth Maxwell's equation.

Let us develop the second equation for the magnetic field. Making use of Eq. (1.35) and the identity

$$\text{curl}\,\mathbf{M} = \text{grad}\ \text{div}\,\mathbf{M} - \nabla^2\mathbf{M}$$

we have

$$\text{curl}\,\mathbf{B} = \text{grad}\ \text{div}\,\mathbf{A} - \nabla^2\mathbf{A}.$$

Considering the fact that

$$\text{div}\,\mathbf{A} = 0$$

and taking into account Eq. (1.43), we obtain

$$\text{curl}\,\mathbf{B} = -\nabla^2\mathbf{A} = \mu_0\mathbf{j}.$$

Thus, the second equation for the magnetic field at regular points is

$$\text{curl}\,\mathbf{B} = \mu_0\mathbf{j}. \tag{1.50}$$

Consequently outside of currents, we have

$$\text{curl}\,\mathbf{B} = 0. \tag{1.51}$$

It is clear that Eq. (1.50) expresses the fact that currents are generators of the vortex type and they create the magnetic field. Applying Stokes theorem,

$$\oint_L \mathbf{M}\cdot d\mathbf{l} = \int_S \text{curl}\,\mathbf{M}\cdot d\mathbf{S}$$

where S is the surface bounded by the contour L, we obtain the integral form of the second equation:

$$\oint_L \mathbf{B}\cdot d\mathbf{l} = \int_S \text{curl}\,\mathbf{B}\cdot d\mathbf{S} = \mu_0\int_S \mathbf{j}\cdot d\mathbf{S}$$

or

$$\oint_L \mathbf{B}\cdot d\mathbf{l} = \mu_0 I \tag{1.52}$$

Here, I is the current flowing through the surface S bounded by the path L (Figure 1.4(b)). It is proper to notice that the mutual orientation of vectors $d\mathbf{l}$ and $d\mathbf{S}$ is not arbitrary but obeys the right-hand rule. Thus, the circulation of the magnetic field is defined by the value of current I piercing the surface surrounded by the contour L, and it does not depend on currents located outside the perimeter of this area. Of course, this path L can pass through the media with different physical properties.

In order to derive the surface analogy of Eq. (1.52), consider a closed contour surrounding an element of surface current with density $\mathbf{i}(p)$ (Figure 1.4(c)). Applying Eq. (1.52) to such a path and neglecting by contribution of elements perpendicular to the surface current, we obtain

$$B_t^{(2)} - B_t^{(1)} = \mu_0 i_l \quad \text{or} \quad \mathbf{n}\times\left(\mathbf{B}^{(2)} - \mathbf{B}^{(1)}\right) = \mu_0 \mathbf{i} \tag{1.53}$$

where t and l represent the two mutually perpendicular directions, both tangent to the surface.

Thus, the tangential component of the magnetic field is a discontinuous function at the points where the density of surface current differs from zero. We have derived three forms of the second equation of the field **B**:

$$\oint_L \mathbf{B}\cdot d\mathbf{l} = \mu_0 I, \quad \text{curl}\,\mathbf{B} = \mu_0 \mathbf{j}, \quad \mathbf{n}\times\left(\mathbf{B}^{(2)} - \mathbf{B}^{(1)}\right) = \mu_0 \mathbf{i} \tag{1.54}$$

Here, **i** is the vector of density of surface currents. It is interesting to notice that the last of these equations is valid for any time-varying magnetic field and it is usually regarded as the surface analogy of Maxwell's second equation. Also, as was pointed out earlier, the first two parts of Eq. (1.54) remain valid for quasi-stationary fields which are widely used in the most electromagnetic methods of the surface and borehole geophysics.

Now let us summarize these results and present the system of equations of the magnetic field, caused by conduction currents in a differential form:

1. The system, shown below, has been derived from the Biot–Savart law in the same way that the system of equations for the electric field was derived from Coulomb's law.

2. The Biot–Savart law and Eq. (1.55) contain the same information about the magnetic field. This field is the classical example of the vortex field, and its generators are currents characterized by the current density vector **j**.

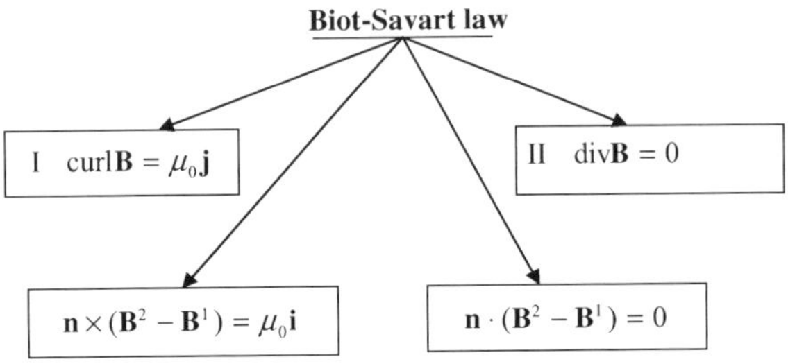

(1.55)

3. At surfaces where the current density **i** equals zero, both the normal and tangential components of the magnetic field are continuous functions.
4. The system Eq. (1.55) describes the behavior of the field in free space as well as in any nonmagnetic conducting medium. Moreover, it turns out that Eq. (1.55) is still valid in the presence of a medium that has influence on the field (magnetic material), provided that the right-hand side of the first equation

$$\mathrm{curl}\,\mathbf{B} = \mu_0\mathbf{j}$$

includes also the magnetization currents.
5. As will be shown later, this system correctly defines the time-varying magnetic field if it is possible to neglect by an effect of propagation of electromagnetic waves.

1.5. BEHAVIOR OF THE MAGNETIC FIELD

Now we consider several examples illustrating the behavior of the magnetic field and its vector potential.

Example 1: Magnetic Field of the Current Filament

Taking into account the axial symmetry of the problem, we choose a cylindrical system of coordinates r,φ, and z, with its origin situated on the

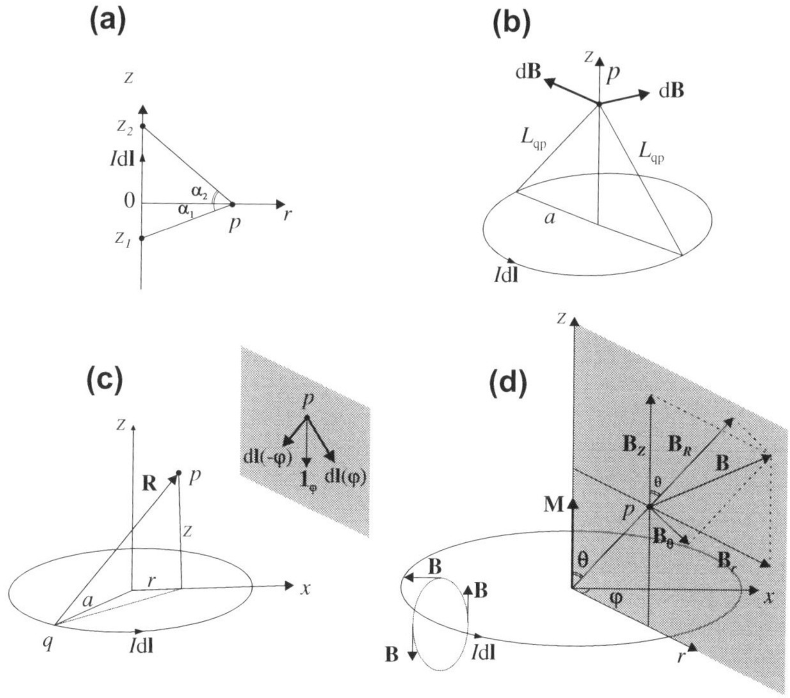

Figure 1.5 (a) Magnetic field of a current line. (b) Magnetic field at the axis of a current loop. (c) Magnetic field of the current loop at an arbitrary point. (d) Magnetic field of the magnetic dipole in spherical and cylindrical coordinate systems.

current-carrying line (Figure 1.5(a)). Starting from the Biot–Savart law, we can say that the magnetic field has only the component B_φ, which is independent of the coordinate φ. From the principle of superposition, it follows that the total field is the sum of fields contributed by the current elements $I\mathrm{d}\mathbf{z}$. Then, we have

$$\mathbf{B}_\varphi = \frac{\mu_0 I}{4\pi} \int_{z_1}^{z_2} \frac{\mathrm{d}\mathbf{z} \times \mathbf{L}_{qp}}{L_{qp}^3} \tag{1.56}$$

where $L_{qp} = \left(r^2 + z^2\right)^{1/2}$ and z is the coordinate of the element $\mathrm{d}z$. The coordinates of the observation point are r and $z = 0$, and z_1 and z_2 are the coordinates of terminal points of the current line. It is clear that the absolute value of the cross-product is

$$\left|\mathrm{d}\mathbf{z} \times \mathbf{L}_{qp}\right| = \mathrm{d}z\, L_{qp} \sin\left(\mathrm{d}\mathbf{z}, \mathbf{L}_{qp}\right) = \mathrm{d}z\, L_{qp} \sin\beta = \mathrm{d}z\, L_{qp} \cos\alpha$$

Thus,

$$B_\varphi = \frac{\mu_0 I}{4\pi} \int_{z_1}^{z_2} \frac{dz}{L_{qp}^2} \cos\alpha \tag{1.57}$$

Inasmuch as $z = r\tan\alpha$, we have

$$dz = r\sec^2\alpha\, d\alpha \quad \text{and} \quad L_{qp}^2 = r^2\left(1+\tan^2\alpha\right) = r^2\sec^2\alpha$$

Substituting these expressions into Eq. (1.57), we obtain

$$B_\varphi = \frac{\mu_0 I}{4\pi r} \int_{\alpha_1}^{\alpha_2} \cos\alpha\, d\alpha$$

Thus, the expression for the magnetic field caused by the current flowing along a straight line has the form

$$B_\varphi(p) = \frac{\mu_0 I}{4\pi r}(\sin\alpha_2 - \sin\alpha_1) \tag{1.58}$$

Here, α_2 and α_1 are the angles, as is shown in Figure 1.5(a). First, suppose that the current-carrying line is infinitely long so that the two angles α_2 and α_1 have values $\frac{\pi}{2}$ and $-\frac{\pi}{2}$, respectively. Then

$$B_\varphi(p) = \frac{\mu_0 I}{2\pi r} \tag{1.59}$$

In the case of a line that is only semi-infinite, $\alpha_1 = 0$ and $\alpha_2 = \frac{\pi}{2}$, we have

$$B_\varphi(p) = \frac{\mu_0 I}{4\pi r} \tag{1.60}$$

Now we will assume that $\alpha_2 = \alpha$ and $\alpha_1 = -\alpha$. Then, in accordance with Eq. (1.58), we obtain

$$B_\varphi(p) = \frac{\mu_0 I}{2\pi r}\sin\alpha = \frac{\mu_0 I}{2\pi r}\frac{l}{\left(l^2+r^2\right)^{1/2}}, \tag{1.61}$$

where $2l$ is the length of the current-carrying line. If l is significantly greater than the distance r, the right-hand side of Eq. (1.61) can be expanded in a series in terms of $(r/l)^2$. This gives

$$B_\varphi = \frac{\mu_0 I}{2\pi r}\left(1 + \frac{r^2}{l^2}\right)^{-1/2} \approx \frac{\mu_0 I}{2\pi r}\left(1 - \frac{1}{2}\frac{r^2}{l^2} + \frac{3}{8}\frac{r^4}{l^4} - \cdots\right)$$

We see that if the length of the current line $2l$ is 4–5 times greater than the separation r, the field is practically the same as that due to infinitely long current-carrying line. It is necessary to make one comment: Eq. (1.58) is used in electromagnetic methods for calculating the primary magnetic field caused by the closed current line of an arbitrary shape.

Example 2: The Vector Potential A and the Magnetic Field B of a Current in a Circular Loop

First assume that the observation point is situated on the axis of a loop with radius a, as is shown in Figure 1.5(b). Then in accordance with Eq. (1.36),

$$\mathbf{A} = \frac{\mu_0 I}{4\pi} \oint_L \frac{\mathbf{dl}}{L_{qp}}$$

Inasmuch as the distance L_{qp} is the same for all points on the loop, we have

$$\mathbf{A} = \frac{\mu_0 I}{4\pi L_{qp}} \oint_L \mathbf{dl}$$

By definition, the sum of the elementary vectors $\mathbf{dl}$ along any closed path is zero. Therefore, the vector potential $\mathbf{A}$ at the z-axis of a circular current loop vanishes. Now we calculate the magnetic field on the z-axis. From the Biot–Savart law (Eq. (1.14)), it can be seen that in a cylindrical system of coordinates, each current element $I\mathbf{dl}$ creates two field components dB_z and dB_r. However, it is always possible to find two current elements $I\mathbf{dl}$ that contribute the same horizontal components at any point of the z-axis but with opposite signs. Therefore, the magnetic field has only a vertical component along the z-axis. According to Figure 1.5(b), we have

$$dB_z = |\mathbf{dB}|\frac{a}{L_{qp}} = \frac{\mu_0 I}{4\pi}\frac{dl}{L_{qp}^2}\frac{a}{L_{qp}} = \frac{\mu_0 Ia}{4\pi}\frac{dl}{L_{qp}^3},$$

since $\left|\mathbf{dl} \times \mathbf{L}_{qp}\right| = L_{qp}\,dl$. Having integrated along the closed path of the loop, we obtain

$$B_z = \frac{\mu_0 I a 2\pi a}{4\pi(a^2+z^2)^{3/2}} = \frac{\mu_0 I a^2}{2(a^2+z^2)^{3/2}} = \frac{\mu_0 M}{2\pi(a^2+z^2)^{3/2}}, \tag{1.62}$$

where

$$M = I\pi a^2 = IS$$

and S being the area enclosed by the loop. When the distance z is much greater than the radius of the loop a, we arrive at an expression for the magnetic field, which plays a very important role in the study of the magnetic and electromagnetic fields applied in geophysics. Neglecting a in comparison with z, we have

$$B_z = \frac{\mu_0 M}{2\pi z^3} \quad \text{if} \quad z \gg a. \tag{1.63}$$

When the intensity of the field does not separately depend on the current or the loop radius, but it is defined by the product $M = IS$, we call this the field of the magnetic dipole. Thus, a relatively small current loop with radius a creates the same magnetic field as a magnetic dipole having the magnitude of the moment equal to $M = \pi a^2 I$. It can also be seen from Eq. (1.62) that when the distance z is at least four times greater than the radius a, the treatment of the loop as the magnetic dipole situated at the center of the loop results in an error of no more than 10%. Now let us return to Eq. (1.62) and find how the field at the loop axis depends on its radius. It is clear that when $a \ll z$, the field B_z increases almost directly proportional toa^2, then it starts to grow slower and reaches the maximum and after it decreases as $1/a$. The maximum takes place when

$$a = 2^{1/2} z$$

This is a useful relation which allows one to find the radius of the loop providing the maximal primary field at a given depth. So far we have considered the vector potential and the magnetic field only along the z-axis. Now we will investigate a general case, and first of all, calculate the vector potential at any point p. Due to symmetry, the vector potential does not depend on the coordinate φ. For simplicity, we can then choose the point p in the x–z plane, where $\varphi = 0$. As can be seen from Figure 1.5(c), every pair of current elements, equally distant from the point p and having coordinates φ and $-\varphi$, creates a vector potential $d\mathbf{A}$ perpendicular to the x–z plane, since each element $I d\mathbf{l}$ creates vector potential of the same orientation as $d\mathbf{l}$. Inasmuch as the whole loop can be represented as the sum of such pairs, we conclude that the vector potential **A** caused by the current-carrying loop has only the component A_φ. Therefore, from Eq. (1.36), it follows that

$$A_\varphi = \frac{\mu_0 I}{4\pi} \oint \frac{dl_\varphi}{R} = \frac{\mu_0 I}{2\pi} \int\limits_0^\pi \frac{a \cos\varphi \, d\varphi}{(a^2 + r^2 - 2ar\cos\varphi + z^2)^{1/2}} \tag{1.64}$$

where dl_φ is the component of $\mathbf{dl}$ along the coordinate line φ and

$$dl_\varphi = dl \cos\varphi, \quad \text{and} \quad L_{qp} = \left(a^2 + r^2 - 2ar\cos\varphi + z^2\right)^{1/2}$$

Letting $\varphi = \pi + 2\alpha$, we have

$$d\varphi = 2\, d\alpha, \quad \cos\varphi = 2\sin^2\alpha - 1$$

and therefore

$$A_\varphi = \frac{aI\mu_0}{\pi} \int\limits_0^{\pi/2} \frac{\left(2\sin^2\alpha - 1\right) d\alpha}{\left[(a+r)^2 + z^2 - 4ar\sin^2\alpha\right]^{1/2}}$$

Introducing new variable,

$$k^2 = \frac{4ar}{(a+r)^2 + z^2}$$

and carrying out some fairly simple algebraic operations, we obtain

$$A_\varphi = \frac{kI\mu_0}{2\pi}\left(\frac{a}{r}\right)^{1/2} \left[\left(\frac{2}{k^2} - 1\right) \int\limits_0^{\pi/2} \frac{d\alpha}{\left(1 - k^2\sin^2\alpha\right)^{1/2}} - \frac{2}{k^2} \int\limits_0^{\pi/2} \left(1 - k^2\sin^2\alpha\right)^{1/2} d\alpha\right] \tag{1.65}$$

$$= \frac{I\mu_0}{\pi k}\left(\frac{a}{r}\right)^{1/2} \left[\left(1 - \frac{k^2}{2}\right)K - E\right]$$

where K and E are the complete elliptical integrals of the first and second kind:

$$K(k) = \int\limits_0^{\pi/2} \frac{d\alpha}{\sqrt{1 - k^2\sin^2\alpha}}, \quad E(k) = \int\limits_0^{\pi/2} \sqrt{1 - k^2\sin^2\alpha}\, d\alpha \tag{1.66}$$

These functions have been studied in detail and there are standard algorithms for their calculations. Using the relationship between the vector potential and magnetic field, as is given by Eq. (1.35), we have in a cylindrical coordinate system

$$B_r = -\frac{\partial A_\varphi}{\partial z}, \quad B_\varphi = 0, \quad B_z = \frac{1}{r}\frac{\partial}{\partial r}\left(rA_\varphi\right) \tag{1.67}$$

As is known for elliptical integrals,

$$\frac{\mathrm{d}K}{\mathrm{d}k} = \frac{E}{k(1-k^2)} - \frac{K}{k}, \quad \frac{\mathrm{d}E}{\mathrm{d}k} = \frac{E}{k} - \frac{K}{k}$$

and

$$\frac{\partial k}{\partial z} = -\frac{zk^3}{4ar}, \quad \frac{\partial k}{\partial r} = \frac{k}{2r} - \frac{k^3}{4r} - \frac{k^3}{4a}$$

Therefore, after differentiation, we obtain

$$B_r = \frac{\mu_0 I}{2\pi}\frac{z}{r\left[(a+r)^2 + z^2\right]^{1/2}}\left[-K + \frac{a^2 + r^2 + z^2}{(a-r)^2 + z^2}E\right]$$

and

$$B_z = \frac{\mu_0 I}{2\pi}\frac{1}{\left[(a+r)^2 + z^2\right]^{1/2}}\left[K + \frac{a^2 - r^2 - z^2}{(a-r)^2 + z^2}E\right] \tag{1.68}$$

Thus, in general, the magnetic field caused by the current in a circular loop can be expressed in terms of elliptical integrals.

Example 3: Magnetic Field of the Magnetic Dipole and its Moment

Suppose that the distance from the center of the current-carrying loop to the observation point R is considerably greater than the loop radius, that is

$$R = \left(r^2 + z^2\right)^{1/2} \gg a$$

Then Eq. (1.64) can be simplified, so that we have

$$\begin{aligned} A_\varphi &= \frac{\mu_0 Ia}{2\pi}\int\limits_0^\pi \frac{\cos\varphi\, d\varphi}{(R^2 - 2ar\cos\varphi)^{1/2}} \\ &= \frac{\mu_0 Ia}{2\pi R}\int\limits_0^\pi \frac{\cos\varphi\, d\varphi}{\left[1-\left(\frac{2ar}{R^2}\right)\cos\varphi\right]^{1/2}} \approx \frac{\mu_0 Ia}{2\pi R}\int\limits_0^\pi\left(1+\frac{ar}{R^2}\cos\varphi\right)\cos\varphi\, d\varphi \\ &= \frac{\mu_0 Ia}{2\pi R}\left[\int\limits_0^\pi \cos\varphi\, d\varphi + \frac{ar}{R^2}\int\limits_0^\pi \cos^2\varphi\, d\varphi\right] \end{aligned} \tag{1.69}$$

where the relation

$$(1+x)^{-n} \approx 1 - nx$$

has been used assuming that $nx \ll 1$. The first integral in Eq. (1.69) vanishes, so that we obtain

$$A_\varphi = \frac{\mu_0 Ia^2 r}{4R^3} \quad \text{or} \quad \mathbf{A} = A_\varphi \mathbf{1}_\varphi = \frac{\mu_0 ISr}{4\pi R^3}\mathbf{1}_\varphi, \tag{1.70}$$

where S is the area of the loop.

Now we make use of the spherical system of coordinates, R, θ, and φ, with the origin at the loop center (Figure 1.5(d)); that is, from points of this axis as $z > 0$, the direction of the current is seen counterclockwise. Then Eq. (1.70) can be written as

$$\mathbf{A} = \mathbf{1}_\varphi \frac{\mu_0 IS}{4\pi R^2}\sin\theta \tag{1.71}$$

Next, we introduce the moment of the small loop as a vector directed along the z-axis, whose magnitude is equal to the product of the current in the loop and its area; that is

$$\mathbf{M} = IS\mathbf{1}_z = M\mathbf{1}_z, \tag{1.72}$$

where $M = IS$. It is essential to note that the moment $\mathbf{M}$ and the direction of the current form the right-hand system. Thus, instead of Eq. (1.71), we can write

$$\mathbf{A} = \frac{\mu_0 \mathbf{M} \times \mathbf{R}}{4\pi R^3}, \tag{1.73}$$

since

$$\mathbf{M} \times \mathbf{R} = \mathbf{1}_{\varphi} MR \sin\theta$$

Now proceeding from Eqs (1.35) and (1.73) and taking into account the fact that

$$A_R = A_\theta = 0$$

we obtain the following expressions for the magnetic field in a spherical system of coordinates,

$$B_R = \frac{1}{R\sin\theta}\frac{\partial(\sin\theta\, A_\varphi)}{\partial\theta}, \quad B_\theta = -\frac{1}{R}\frac{\partial(RA_\varphi)}{\partial R}, \quad B_\varphi = 0$$

hence,

$$B_R = \frac{2\mu_0 M}{4\pi R^3}\cos\theta, \quad B_\theta = \frac{\mu_0 M}{4\pi R^3}\sin\theta, \quad B_\varphi = 0 \qquad (1.74)$$

These equations describe the behavior of the magnetic field of a relatively small current loop; that is, its radius is much smaller than the distance from the loop center to the observation point. This is the most important condition to apply (Eq. (1.74)) while the values of the loop radius and the distance R are not essential. We call the magnetic field, described by Eq. (1.74), that of a magnetic dipole with the moment **M**. Here, it is appropriate to make several comments:

1. In the case of the electric field, a "dipole" means a combination of equal charges having opposite signs, when the field is determined at distances essentially exceeding the separation between these charges. At the same time, the notion of a "magnetic dipole" does not imply the existence of magnetic charges but simply describes the behavior of the magnetic field due to the current in a relatively small loop.
2. The magnetic field of any current system, regardless of its shape and dimensions, is equivalent to that of the magnetic dipole when the field is defined at distances much greater than the size of this system. For instance, a distribution of conduction currents within the upper part of the earth's core is very complicated and it changes with time. However, at the earth's surface, relatively far away from the core, the magnetic field of these currents is very close to that of the magnetic dipole.
3. In the most electromagnetic methods, applied in the surface and borehole geophysics, including the frequency and transient soundings, electromagnetic profiling and induction logging, the current source can be often treated as the magnetic dipole. As was pointed out earlier, Eq. (1.74),

derived for the constant current, is also applied for the quasi-stationary fields and this is the main reason why we consider this example in detail. The main features of the field of the magnetic dipole directly follow from Eq. (1.74) and they are the following:

a. At points of the dipole axis z, the field of the magnetic dipole has only one component B_z directed along this axis and it decreases inversely proportional to z^3:

$$B_z = \frac{\mu_0 M}{2\pi z^3} \tag{1.75}$$

It is proper to note that this component is positive at all points of this axis.

b. At the equatorial plane $\theta = \frac{\pi}{2}$, the radial component B_R vanishes, and the field has the direction opposite to that of the magnetic moment **M**,

$$B_z = -\frac{\mu_0 M}{4\pi r^3} \tag{1.76}$$

Here, r is the distance from the dipole to an observation point.

c. Along any radius ($\theta =$ constant), both components of the field—B_R and B_θ—decrease inversely proportional to R^3. At the same time, the ratio of these components, as well as an orientation of the total vector **B** with respect to the radius **R**, does not change. In fact, in accordance to Eq. (1.74), we have

$$\frac{B_\theta}{B_R} = \frac{1}{2} \tan \theta \tag{1.77}$$

Note that this fact has very important application for paleomagnetic studies. It is also useful to consider the components of the field in the cylindrical system. As follows from Figure 1.5(d), we have

$$B_r(r, z) = B_R \sin\theta + B_\theta \cos\theta \quad \text{and} \quad B_z(r, z) = B_R \cos\theta - B_\theta \sin\theta,$$

where

$$R = \left(r^2 + z^2\right)^{1/2}$$

Taking into account Eq. (1.74), we obtain

$$B_r(r, z) = \frac{3\mu_0 M}{4\pi R^3} \sin\theta \cos\theta, \quad B_z(r, z) = \frac{\mu_0 M}{4\pi R^3}\left(2\cos^2\theta - \sin^2\theta\right)$$

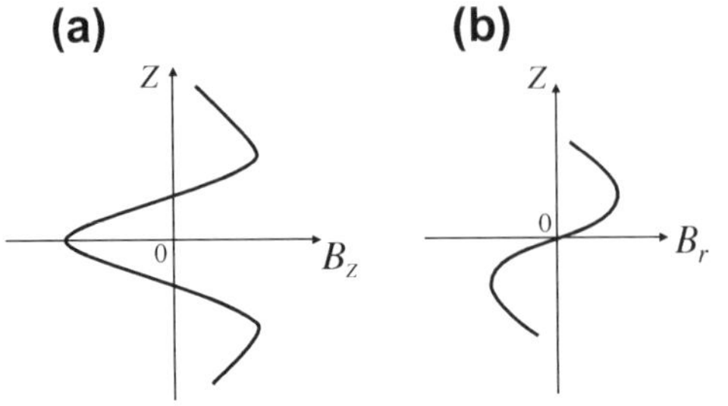

Figure 1.6 Behavior of components B_Z (a) and B_r (b) as functions of z.

or

$$B_r(r,z) = \frac{3\mu_0 Mrz}{4\pi(r^2+z^2)^{5/2}}, \quad B_z = \frac{\mu_0 M}{4\pi(r^2+z^2)^{5/2}}\left(2z^2 - r^2\right) \tag{1.78}$$

If we assume that r is constant, then Eq. (1.78) allows us to study the behavior of the field components parallel to the dipole moment as a function of z (Figure 1.6(a,b)). First of all, it is clear that the radial component B_r is an odd function of z and it changes the sign in the equatorial plane of the dipole. At the same time, the vertical component is an even function of z and it changes sign at points

$$z = \pm\frac{r}{\sqrt{2}}$$

Example 4: Magnetic Field due to a Current in a Cylindrical Conductor

Consider an infinitely long and homogenous cylindrical conductor, as is shown in Figure 1.7(a), with the radius a and current I. In this case, the current density $\mathbf{j}$ is uniformly distributed over the cross-section S and everywhere has only the z-component which is constant:

$$j = j_z = \text{constant} \tag{1.79}$$

In the cylindrical system of coordinates r, φ, z where the z-axis is directed along the conductor, the magnetic field can be characterized by the three components B_r, B_φ, and B_z. However, it turns out that two components are equal to zero. As follows from the Biot–Savart law, the magnetic field caused by the current element is perpendicular to the current density $\mathbf{j}$ and therefore the vertical component B_z equals zero. Next, consider

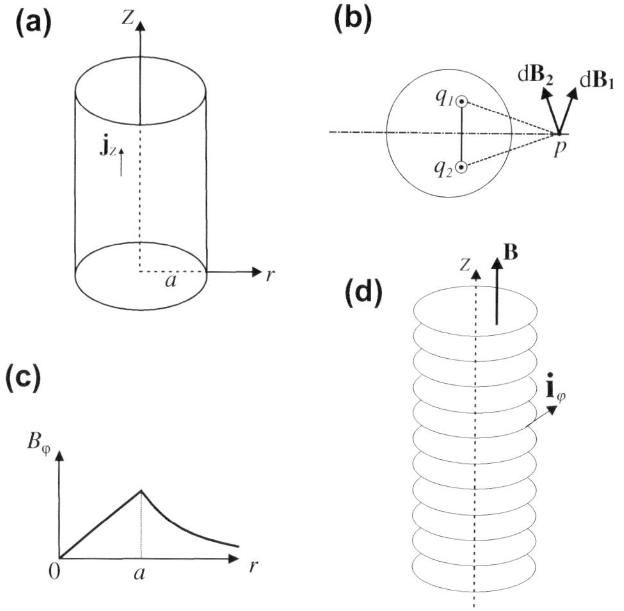

Figure 1.7 (a) Cylindrical conductor with current. (b) Summation of radial components of the magnetic field of cylindrical conductor. (c) Behavior of the magnetic field. (d) Infinitely long solenoid.

two current elements located symmetrically with respect to the half-plane, $\varphi =$ constant (Figure 1.7(b)). It is clear that the sum of radial components of the field is equal to zero. Since the entire conductor can be represented as a combination of such pairs of current elements, we can say that the total magnetic field does not have the radial component, $B_r = 0$. Thus, we demonstrated that

$$\mathbf{B} = \left(0, B_\varphi, 0\right)$$

Taking into account the symmetry of the distribution of currents, we see that the vector lines of the magnetic field are circles located in horizontal planes and their centers are located on the z-axis. In order to determine the component B_φ, we take one such line and apply the first part of Eq. (1.54) in the integral form. This gives

$$\oint_L \mathbf{B} \cdot \mathrm{d}\mathbf{l} = \oint_L B_\varphi \, \mathrm{d}l = B_\varphi \oint_L \mathrm{d}l = 2\pi r B_\varphi = \mu_0 I_S$$

Here, I_S is the current passing through any area bounded by the magnetic line. In deriving this equality, we took into account the fact that the magnitude of the field does not vary along this circle and both vectors: **B** and **dl** are parallel to each other. Thus, the field outside and inside of the current is

$$B_\varphi^e = \frac{\mu_0 I}{2\pi r} \quad \text{if} \quad r \geq a \tag{1.80}$$

and

$$B_\varphi^i = \frac{\mu_0 j}{2} r \quad \text{if} \quad r \leq a \tag{1.81}$$

since $I_S = \pi r^2 j$. In accordance with Eqs (1.80) and (1.81), the magnetic field is equal to zero at the z-axis and increases linearly inside. At the surface of the conductor, it reaches maximum, equal to

$$B_\varphi(a) = \frac{\mu_0 j}{2} a \tag{1.82}$$

and then the field decreases inversely proportional to the distance r (Figure 1.7(c)). In this light, let us notice the following. Considering the magnetic field of the linear current, we found out that the field tends to infinity when an observation point approaches the surface of the current line. As was pointed out earlier, it is a result of replacement of real distribution of currents by its fictitious model. As is seen from Eq. (1.82), at the surface of a conductor, the field has a finite value which is usually rather small.

Example 5: Magnetic Field of Infinitely Long Solenoid

Suppose that at each point of the cylindrical surface S, a distribution of currents is characterized by the density $\mathbf{i}_\varphi$, and it has everywhere the same magnitude (Figure 1.7(d)). Inasmuch as the current has only a component in the $\varphi-$ direction, we have $B_\varphi = 0$. It is a simple matter to show that the radial component also vanishes. In fact, consider two elementary current circuits located symmetrically with respect to plane where an observation point is located (Figure 1.8(a)). We can see that the sum of radial components is equal to zero. Taking into account the fact that the solenoid is infinitely long, one can always find such a pair of current loops, and

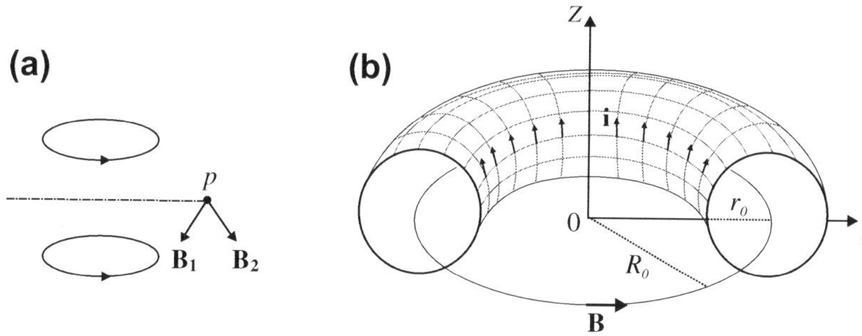

Figure 1.8 (a) Summation of radial components. (b) Magnetic field of a toroid.

therefore, the resultant radial component of the field is always equal to zero. Thus, the total field has only the z-component:

$$\mathbf{B} = (0, 0, B_z) \tag{1.83}$$

This result greatly simplifies algebra because we have to focus on one component only. In principle, it can be evaluated by an integration of the fields caused by elementary current circles with the same radius a, but this is rather cumbersome. For this reason, we make use of a different approach based on the Poisson's equation for the vector potential:

$$\Delta\mathbf{A} = -\mu_0\mathbf{j} \tag{1.84}$$

Taking into account symmetry and the fact that the vector potential has the same component as the current density, we have

$$\mathbf{A} = A_\varphi \mathbf{i}_\varphi \tag{1.85}$$

Outside of the currents, this function satisfies Laplace's equation:

$$\Delta\mathbf{A} = \Delta\left(A_\varphi \mathbf{1}_\varphi\right) = 0.$$

According to vector calculus, we have

$$\Delta uv = v\Delta u + u\Delta v + 2(\text{grad } u \cdot \text{grad } v),$$

where u and v are the arbitrary scalar functions. In our case, A_φ and i_φ depend only on r and φ, respectively. The latter follows from the fact that

$$\mathbf{1}_\varphi = -\mathbf{1}_x \sin\varphi + \mathbf{1}_y \cos\varphi,$$

where $\mathbf{1}_x$ and $\mathbf{1}_y$ are the unit vectors in Cartesian system of coordinates, and they are independent of the coordinates of a point. Therefore, the dot product term vanishes because both gradients involved are orthogonal to each other. Finally, we arrive at the equation,

$$\mathbf{\Delta A} = \Delta\left(A_\varphi \mathbf{1}_\varphi\right) = \mathbf{1}_\varphi \Delta A_\varphi + A_\varphi \mathbf{\Delta 1}_\varphi = 0. \tag{1.86}$$

In a cylindrical system of coordinates, the operator $\mathbf{\Delta}$ is

$$\Delta = \frac{1}{r}\frac{\partial}{\partial r}\left(r\frac{\partial}{\partial r}\right) + \frac{1}{r^2}\frac{\partial^2}{\partial \varphi^2} + \frac{\partial^2}{\partial z^2} \tag{1.87}$$

First, we will find an expression for $\mathbf{\Delta 1}_\varphi$. It is clear that derivatives with respect to r and z are equal to zero and

$$\frac{\partial}{\partial \varphi}\mathbf{1}_\varphi = -\mathbf{1}_x \cos\varphi - \mathbf{1}_y \sin\varphi$$

Thus,

$$\frac{\partial^2}{\partial \varphi^2}\mathbf{1}_\varphi = \mathbf{1}_x \sin\varphi - \mathbf{1}_y \cos\varphi = -\mathbf{1}_\varphi$$

Substitution of the latter into Eq. (1.86) gives Laplace's equation with respect to a scalar component A_φ that greatly simplifies a solution:

$$\frac{\mathrm{d}}{\mathrm{d}r}\left[r\frac{\mathrm{d}A_\varphi(r)}{\mathrm{d}r}\right] - \frac{A_\varphi(r)}{r} = 0 \tag{1.88}$$

It is an ordinary differential equation of the second order and its solution is

$$A_\varphi(r) = Cr + Dr^{-1} \tag{1.89}$$

Taking into account the fact that the magnetic field has to have a finite value and tends to zero at infinity, we represent the vector potential inside and outside of the solenoid as

$$A_\varphi^{(i)} = Cr, \quad A_\varphi^{(e)} = Dr^{-1}, \tag{1.90}$$

where C and D are unknown coefficients. By definition,

$$\mathbf{B} = \text{curl}\,\mathbf{A} = \frac{1}{r}\begin{vmatrix} \mathbf{1}_r & r\mathbf{1}_\varphi & \mathbf{1}_z \\ \dfrac{\partial}{\partial r} & \dfrac{\partial}{\partial_\varphi} & \dfrac{\partial}{\partial z} \\ 0 & rA_\varphi & 0 \end{vmatrix}. \tag{1.91}$$

whence,

$$B_r = 0, \quad B_\varphi = 0, \quad B_z = \frac{1}{r}\frac{\partial}{\partial r}\left(rA_\varphi\right) \tag{1.92}$$

Substitution of $A_\varphi^{(e)}$ into Eq. (1.92) yields

$$B_z^{(e)} = 0 \quad \text{if} \quad r > a$$

and we have proved that the surface currents of the solenoid do not create a magnetic field outside the solenoid. In the same manner for the field inside of the solenoid, we obtain

$$B_z^{(i)} = 2C, \quad \text{if} \quad r \le a \tag{1.93}$$

that is, this magnetic field is constant. In order to determine C, we recall that the difference of tangential components at both sides of the solenoid is

$$2C = \mu_0 i_\varphi \quad \text{or} \quad B_z^{(i)} = \mu_0 i_\varphi$$

Thus, for the field **B** caused by currents in the solenoid, we have

$$B_z^{(i)} = \mu_0 i_\varphi, \quad \text{if} \quad r \le a \quad \text{and} \quad B_z^{(e)} = 0, \quad \text{if} \quad r > a \tag{1.94}$$

We may say that the magnetic field is concentrated only inside the solenoid. Certainly it is a very simple behavior but such result is hardly obvious. First, it is difficult to predict that the field inside, $B_z^{(i)}$,is uniform over the cross-section since the field due to a single current loop varies greatly at its vicinity. Also it is not obvious before calculations that the field outside of the solenoid is zero; that is, the fields caused by all current loops compensate each other. Consider a plane $z =$ constant where an observation point is situated. Current circuits located relatively close to this plane generate a negative component along the z-axis, while current loops situated far away give at the same point a positive contribution, $r > a$. Correspondingly, the field outside is a result of subtraction of elementary fields, and it turns out that in the case of infinitely long solenoid, this difference is equal to zero. Note that inside the solenoid, all terms are positive. Of course, if a solenoid has a finite extension along the z-axis, the field outside is not zero and it is characterized everywhere by both components B_r and B_z, and the latter prevails at its central part.

Example 6: Magnetic Field of a Current Toroid

Consider a toroid with the current density $\mathbf{i}$ shown in Figure 1.8(b) and introduce a cylindrical system of coordinates with the z-axis perpendicular to the toroid. Taking into account the axial symmetry, we see that the vector potential and magnetic field are independent of the coordinate φ. Also imagine two current loops of the toroid located symmetrically with respect to the vertical plane, where a point of observation is located. As can be seen, the sum of vector potentials due to these elementary currents does not have the $\varphi-$ component. Thus, for the vector potential, we have

$$\mathbf{A} = (A_r, 0, A_z) \tag{1.95}$$

Taking into account the fact that

$$\mathbf{B} = \frac{1}{r}\begin{vmatrix} \mathbf{1}_r & r\mathbf{1}_\varphi & \mathbf{1}_z \\ \dfrac{\partial}{\partial r} & \dfrac{\partial}{\partial \varphi} & \dfrac{\partial}{\partial z} \\ A_r & 0 & A_z \end{vmatrix}$$

we obtain

$$B_r = \frac{1}{r}\frac{\partial A_z}{\partial \varphi} = 0, \quad B_\varphi = \left(\frac{\partial A_r}{\partial z} - \frac{\partial A_z}{\partial r}\right), \quad B_z = -\frac{1}{r}\frac{\partial A_r}{\partial \varphi} = 0 \tag{1.96}$$

Thus, the magnetic field has only one component B_φ, but it cannot be calculated from Eq. (1.96), since neither component of the vector potential is known. However, this problem can be easily solved by using the first equation of the field in the integral form:

$$\oint_L \mathbf{B}\cdot d\mathbf{l} = \mu_0 I_S$$

Taking into account the axial symmetry and the fact that $\mathbf{B}$ and $\mathbf{dl}$ have the same direction, this equality is greatly simplified and gives

$$B_\varphi 2\pi r = \mu_0 I_S, \tag{1.97}$$

where L is a circular path of radius r located in the horizontal plane with the center situated at the toroid axis, and I_S is the current passing through a surface S surrounded by this path L. First consider a point p, located outside the toroid. In such a case, either the current does not intersect the surface S or its total value is equal to zero. This means that $B_\varphi = 0$ and, therefore, the magnetic field is absent outside the toroid as in the case of the solenoid:

$$B_\varphi^{(e)} = 0 \tag{1.98}$$

Next consider the magnetic field inside of the toroid. As follows from Eq. (1.97), the field $B_\varphi^{(i)}$ is not uniform and equal:

$$B_\varphi^{(i)} = \frac{\mu_0 I_S}{2\pi r} \tag{1.99}$$

In this case, the path of integration is inside the toroid. Suppose that it is located in the plane $z = 0$ and a change of its radius does not change the flux of the current density. Therefore, within the range,

$$R_0 - r_0 < r < R_0 + r_0$$

where, an increase of r results in a decrease of the field inversely proportional to r. If we consider circular paths in planes with $z \neq 0$, then the current I_S becomes smaller with increase of z. Thus, we observe a nonuniform magnetic field inside the toroid. It is natural to expect that with an increase of the ratio of the toroid radius R_0 to that of its cross-section r_0, the field inside becomes more uniform. Note that if the toroid has an arbitrary but constant cross-section and current density is independent on the coordinate φ, we can still apply Eq. (1.99) and conclude that the field $\mathbf{B}$ is equal to zero outside the toroid. Of course, if the current density is not constant in examples 5 and 6, the magnetic field appears outside too, and $\mathbf{B}^{(e)} \neq 0$.

So far we described examples of the magnetic field caused by linear or surface currents, but now consider a volume distribution of currents.

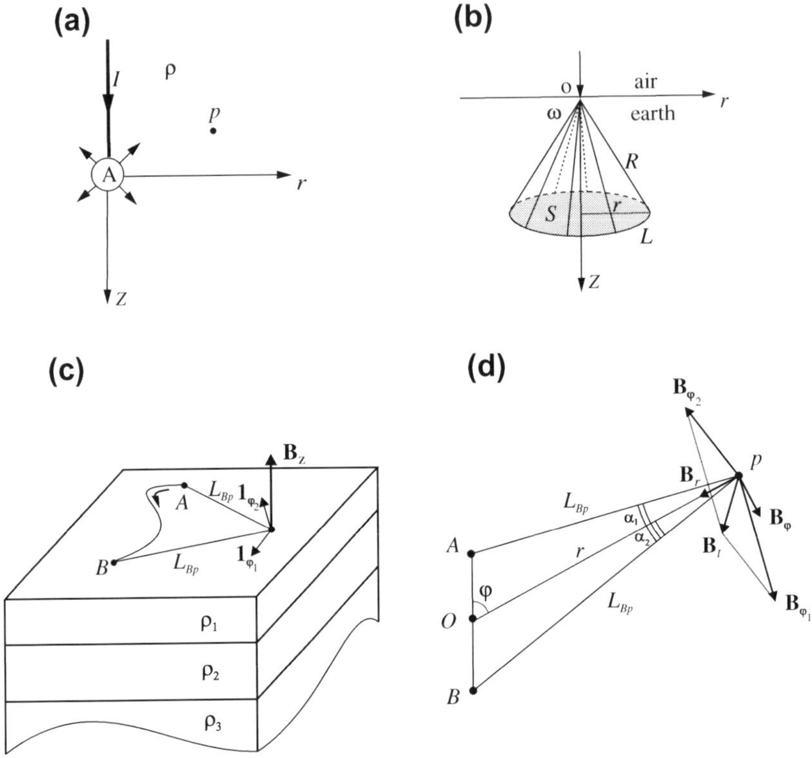

Figure 1.9 (a) Current electrode and semi-infinitely long wire in a uniform medium. (b) Illustration of the solid angle and the field at the surface of a uniform half space. (c) Current line grounded on the surface of a horizontally layered medium. (d) Horizontal magnetic field of the electric dipole.

Example 7: Magnetic Field of Current Electrode in a Uniform Medium

Suppose that the current arrives at the electrode through the wire as is shown in Figure 1.9(a), and therefore, we may imagine that the vector lines of the current density, starting at the electrode, are closed through the wire somewhere at "infinity". Correspondingly, it is possible to apply equations of the field, for example, the first equation in the integral form:

$$\oint_L \mathbf{B}\cdot d\mathbf{l} = \mu_0 I_S \tag{1.100}$$

With this purpose in mind, we introduce the cylindrical system of coordinates r, φ, z such that the origin 0 coincides with the center of the electrode (Figure 1.9(b)). Then it is a simple matter to calculate the current I_S passing through the area surrounded by the closed path L, which is equal to

$$I_S = \frac{\omega}{4\pi} I_0, \tag{1.101}$$

since the current in a medium is distributed uniformly in all directions. Here, I_0 is the current in the wire, and ω is the solid angle subtended by the surface S and viewed from point 0:

$$\omega = 2\pi\left[1 - \frac{|z|}{(r^2+z^2)^{1/2}}\right]$$

Taking into account that

$$\oint_L \mathbf{B}\cdot d\mathbf{l} = 2\pi r B_\varphi, \tag{1.102}$$

we have for the azimuthal component,

$$B_\varphi = \frac{\mu_0 I_0}{4\pi r}\left[1 - \frac{z}{(r^2+z^2)^{1/2}}\right], \quad \text{if} \quad z \geq 0 \tag{1.103}$$

and

$$B_\varphi = \frac{\mu_0}{2\pi r}(I_0 - I_s) = \frac{\mu_0 I_0}{2\pi r}\left(1 - \frac{\omega}{4\pi}\right)$$

or

$$B_\varphi = \frac{\mu_0 I_0}{4\pi r}\left(1 + \frac{|z|}{(r^2+z^2)^{1/2}}\right) \quad \text{if} \quad z \leq 0 \tag{1.104}$$

In the last case, the flux of the current density in the wire and in a medium has opposite signs. As follows from Eq. (1.103) with an increase of the distance r, the field tends to that of the half infinitely long wire (Eq. (1.60)). It is understandable because the flux of the current approaches to $\frac{I_0}{2}$. At the same time when $z \to \infty$, the field becomes infinitely small since this flux tends to zero. Different asymptotic behavior is observed when $z < 0$. If $|z| \to \infty$, the field approaches to that of the current of an infinitely long wire and it happens because an influence of the flux of volume currents becomes small. If the distance r increases unlimitedly, then the field tends to that of a half infinitely long current line, and it occurs since the flux of the volume current density becomes equal to $-\frac{I_0}{2}$. In contrast, when $r \to 0$, the field tends to that of an infinitely long wire (Eq. (1.59)) with the current I_0. Finally at the plane $z = 0$, both expressions for the field give the same value of the half infinitely long wire, because the flux of volume currents is equal to zero. Let us also notice that if we considered the magnetic field of the single current electrode in a uniform medium, then it would be impossible to apply the field equations since the current lines are not closed.

Example 8: Current Electrode on the Surface of a Horizontally Layered Medium

First, we study the case when the electrode is located on the surface of a uniform half space assuming that the current arrives at the electrode through a vertical wire (Figure 1.9(b)). As is known [1], the electric field $\mathbf{E}$ and the vector of current density $\mathbf{j}$ at every point of the conducting medium are

$$\mathbf{E} = \frac{\rho I}{2\pi R^2}\mathbf{R}_0, \quad \mathbf{j} = \frac{I}{2\pi R^2}\mathbf{R}_0, \tag{1.105}$$

where $\mathbf{R}_0$ is the unit vector along $\mathbf{R}$. Thus, the magnetic field $\mathbf{B}$ is caused by the linear and volume currents, and due to the axial symmetry of their distribution in the cylindrical system of coordinates, it is natural to expect that

$$B_r = B_z = 0, \qquad \frac{\partial B_\varphi}{\partial \varphi} = 0.$$

Therefore, it is convenient again to make use of the first field equation in the integral form, Eq. (1.100). It is significant to emphasize that the system of field equations can be applied in this example because the current lines are closed. Then taking into account the axial symmetry of the field, Eq. (1.100) is drastically simplified and we have

$$2\pi r B_\varphi = \mu_0 I_S \quad \text{or} \quad B_\varphi = \frac{\mu_0 I_S}{2\pi r}, \tag{1.106}$$

where L is the circle with radius r located in the horizontal plane and its center is situated on the z-axis (Figure 1.9(b)). As follows from this equation, at the earth's surface and above, we have

$$B_\varphi = \frac{\mu_0 I}{2\pi r} \quad \text{if} \quad z \le 0 \tag{1.107}$$

This expression coincides with that for an infinitely long current-carrying line. According to Eq. (1.60), the current in a half infinitely long line perpendicular to the earth's surface creates at this surface, $z=0$, a magnetic field:

$$B_\varphi = \frac{\mu_0 I}{4\pi r}$$

Therefore, the currents in a conducting medium that flow from the electrode also generate the same field if $z=0$:

$$B_\varphi(r) = \frac{\mu_0 I}{2\pi r} - \frac{\mu_0 I}{4\pi r} = \frac{\mu_0 I}{4\pi r} \quad \text{if} \quad z = 0 \tag{1.108}$$

It is interesting to notice that these volume currents create the field equal to that of a half infinitely long current-carrying wire directed vertically downward with the current I. Also it is important to emphasize that the magnetic field at the earth's surface, caused by currents flowing from the electrode into a horizontally layered medium, does not depend on the sequence of layer's conductivity. This follows directly from the axial symmetry which allows us to apply Eq. (1.107). Next we study the magnetic field in a uniform half space beneath the earth's surface, and with this purpose in mind, it is appropriate to make use of Eq. (1.106). Since the current density vector has a radial component j_R only, the current I_S that passes through the spherical surface S and bounded by the circle L with radius r (Figure 1.9(b)) is

$$I_S = j_R S$$

Then taking into account that

$$j_R = \gamma E_R = \frac{I}{2\pi R^2} \quad \text{and} \quad S = \omega R^2$$

we have

$$I_S = \frac{\omega I}{2\pi}, \tag{1.109}$$

where ω is the solid angle subtended by the surface S as viewed from the electrode. As we already know, the solid angle is

$$\omega = 2\pi(1 - \cos\alpha), \tag{1.110}$$

where

$$\alpha = \cos^{-1}\frac{z}{R}$$

Substitution of Eqs (1.109) and (1.110) into Eq. (1.106) gives

$$B_\varphi = \frac{\mu_0 I}{2\pi r}\left[1 - \frac{z}{\sqrt{r^2 + z^2}}\right] \tag{1.111}$$

This equation describes the magnetic field caused by volume currents as well as the current in the vertical line.

Example 9: The Current Flowing in the Wire Grounded at the Surface of a Horizontally Layered Medium

Using the principle of superposition, this pattern of current flow can be represented as follows:

1. Currents flowing from electrode A into a conducting medium.
2. Currents flowing from the conducting medium into electrode B.
3. Current I in the wire that connects the electrodes.

In accordance with Eq. (1.108), the magnetic field at the earth's surface caused by currents in the conducting medium is

$$\mathbf{B}_t(p) = \frac{\mu_0 I}{4\pi}\left(\frac{\mathbf{1}_{\varphi 1}}{r_{Ap}} + \frac{\mathbf{1}_{\varphi 2}}{r_{Bp}}\right) \tag{1.112}$$

Where r_{Ap} and r_{Bp} are the distances from the electrodes A and B to the observation point p, respectively, and $\mathbf{1}_{\varphi 1}$ and $\mathbf{1}_{\varphi 2}$ are the unit vectors as shown in Figure 1.9(c). It is clear that the field $\mathbf{B}_t$ is tangential to the earth's surface and it is independent of the parameters of the horizontally layered medium. This is why the magnetic field of constant currents is not used to study geoelectric parameters of such medium. Next consider the third element of the system: the current line connecting electrodes and situated at the earth's surface. According to the Biot–Savart law, this current causes only the vertical component of the magnetic field at points of the horizontal earth's surface:

$$\mathbf{B}_z = \frac{\mu_0 I}{4\pi}\mathbf{1}_z \int\limits_B^A \frac{dl}{L_{qp}^2}\sin(d\mathbf{l}, \mathbf{L}_{qp}), \tag{1.113}$$

where $\mathbf{dl}$ is the element of the current-carrying line, and L_{qp} is the distance from any element $\mathbf{dl}$ to the observation point. Equations (1.112) and (1.113) completely describe the magnetic field of a grounded current-carrying line located at the earth's surface when the conducting medium is laterally uniform.

Example 10: Magnetic Field of the Electric Dipole at the Earth's Surface

Now consider a special case when the field $\mathbf{B}$ is studied at distances considerably greater than the separation between electrodes A and B, and current path from B to A is a straight line (Figure 1.9(d)). Then the system of charges arising at the surface of electrodes $\pm 2\varepsilon_0 \rho I$ can be considered as an electric dipole with the moment,

$$\mathbf{p} = 2\varepsilon_0 \rho I AB \mathbf{1}_{AB} \tag{1.114}$$

where $\mathbf{1}_{AB}$ is the unit vector directed along the line BA. To derive expressions for the magnetic field from Eq. (1.106), we will use the following notations: r is the distance from the middle of the dipole to an observation point, and φ is the angle between the dipole moment and the radius vector $\mathbf{r}$. As may be seen from Figure 1.9(d), the following relations hold

$$r_{Ap} = \left[r^2 + \left(\frac{AB}{2} \right)^2 - ABr \cos \varphi \right]^{1/2},$$

$$r_{Bp} = \left[r^2 + \left(\frac{AB}{2} \right)^2 + ABr \cos \varphi \right]^{1/2}$$

$$\frac{AB}{2r_{Ap}} = \frac{\sin \alpha_1}{\sin \varphi}, \quad \frac{AB}{2r_{Bp}} = \frac{\sin \alpha_2}{\sin \varphi}$$

Assuming $r \gg AB$, we can write

$$\left[r^2 + \left(\frac{AB}{2} \right)^2 \pm ABr \cos \varphi \right]^{1/2} = \left[r^2 \left(1 + \left(\frac{AB}{2r} \right)^2 \pm \frac{AB}{r} \cos \varphi \right) \right]^{1/2}$$

$$\approx \left[r^2 \left(1 \pm \frac{AB}{r} \cos \varphi \right) \right]^{1/2},$$

and, since $\sqrt{1 \pm x} \approx 1 \pm \frac{x}{2}$ provided that $|x| \ll 1$, we have

$$r_{Ap} \approx r - \frac{AB}{2}\cos\varphi, \quad r_{Bp} \approx r + \frac{AB}{2}\cos\varphi,$$

$$\cos\alpha_1 \approx \frac{r - \frac{AB}{2}\cos\varphi}{r_{Ap}} \approx 1, \quad \cos\alpha_2 \approx \frac{r + \frac{AB}{2}\cos\varphi}{r_{Bp}} \approx 1$$

and

$$\sin\alpha_1 \approx \frac{AB\sin\varphi}{2\left(r - \frac{AB}{2}\cos\varphi\right)}, \quad \sin\alpha_2 \approx \frac{AB\sin\varphi}{2\left(r + \frac{AB}{2}\cos\varphi\right)} \tag{1.115}$$

or

$$\sin\alpha_1 \approx \sin\alpha_2 \approx \frac{AB}{2r}\sin\varphi.$$

Then making use of Eq. (1.115), we obtain expressions for the tangential components of the field in a cylindrical system of coordinates:

$$B_r = -\frac{\mu_0 I}{4\pi}\left(\frac{\sin\alpha_1}{r_{Ap}} + \frac{\sin\alpha_2}{r_{Bp}}\right) = -\frac{\mu_0 IAB}{4\pi r^2}\sin\varphi,$$

since both φ_1- and φ_2- components corresponding to A and B must be taken with negative signs to produce r-component, and

$$B_\varphi = \frac{\mu_0 I}{4\pi}\left(\frac{\cos\alpha_1}{r_{Ap}} - \frac{\cos\alpha_2}{r_{Bp}}\right) = \frac{\mu_0 IAB}{4\pi r^2}\cos\varphi,$$

because in this case, φ_1- and φ_2- components must be summed with opposite signs.

From Eq. (1.113), we can easily see that z-component of the magnetic field is

$$B_z = \frac{\mu_0 I}{4\pi}\int_B^A \frac{\sin\varphi\,\mathbf{dl}}{r^2} = \frac{\mu_0 IAB}{4\pi r^2}\sin\varphi.$$

Summarizing we see that magnetic field of the electric dipole at the earth's surface is described by the following expressions:

$$B_r = -\frac{\mu_0 IAB}{4\pi r^2}\sin\varphi, \quad B_\varphi = \frac{\mu_0 IAB}{4\pi r^2}\cos\varphi, \quad B_z = \frac{\mu_0 IAB}{4\pi r^2}\sin\varphi \tag{1.116}$$

It is interesting to note that magnitudes of field components $B_r$and B_z are equal to each other and at any point of the earth's surface,

$$B_r + B_z = 0$$

Equation (1.116) clearly demonstrates that measurements of the magnetic field of the dipole at the earth's surface do not contain any information about the distribution of resistivity in a horizontally layered medium. However, it is obvious that this conclusion does not hold when the magnetic field is measured beneath the earth's surface. At the same time, measurements of the magnetic field at the earth's surface can be useful in detecting nonhorizontal structures that are frequently of particular interest in prospecting.

Again making use of the principle of superposition, the magnetic field caused by a current-carrying line with a finite length and arbitrary shape, grounded at the earth's surface, can be represented as a sum of fields described by Eq. (1.116). Note that the current line is the source of the alternating primary electric and magnetic fields in the "infinitely long cable" method, also called "controlled source magnetotelluric method". By investigating the time-varying electromagnetic fields caused by grounded current-carrying line, we will show that Eq. (1.116) plays a useful role when measurements are performed at distances from the source which are smaller than the skin depth; that is, an influence of electromagnetic induction is relatively small.

1.6. THE SYSTEM OF EQUATIONS OF THE CONSTANT ELECTROMAGNETIC FIELD

Now we make the first step toward the Maxwell equations and write down equations of the constant field. As was shown earlier at regular points, we have

$$\text{curl}\, \mathbf{E} = 0, \quad \text{div}\, \mathbf{D} = \delta_0$$

and

$$\text{curl}\, \mathbf{B} = \mu_0 \mathbf{j}_c \quad \text{div}\, \mathbf{B} = 0, \tag{1.117}$$

while at points of interfaces,

$$\mathbf{n} \times (\mathbf{E}_2 - \mathbf{E}_1) = 0 \quad \mathbf{n} \cdot (\mathbf{D}_2 - \mathbf{D}_1) = \sigma_0$$

and

$$\mathbf{n} \times (\mathbf{B}_2 - \mathbf{B}_1) = \mathbf{i}_c \quad \mathbf{n} \cdot (\mathbf{B}_2 - \mathbf{B}_1) = 0 \tag{1.118}$$

Here, $\mathbf{D} = \varepsilon\mathbf{E}$ and $\mathbf{j}_c = \gamma\mathbf{E}$ is the density of conduction current. This system was derived from the Coulomb's law, Biot–Savart law, Ohm's law, and the principle of charge conservation for the constant field. Now we start to consider the time-varying electromagnetic fields and first describe the phenomenon of electromagnetic induction.

REFERENCES

[1] L.M. Alpin, Theory of the Field, Nedra, Moscow, 1966.
[2] A.A. Kaufman, B.I. Anderson, Principles of Electric Methods in Surface and Borehole Geophysics, Elsevier, 2010.
[3] A.A. Kaufman, R.O. Hansen, R.L. Kleinberg, Principles of Magnetic Methods in Geophysics, Elsevier, 2009.

PART TWO

Propagation and Diffusion of Electromagnetic Fields

CHAPTER TWO

Physical Laws and Maxwell's Equations

INTRODUCTION

System of equations of a constant electric field in a conducting and polarizable medium, derived from Coulomb's and Ohm's law, as well as the principle of charge conservation, was described in detail [2]. In the same manner, proceeding from the physical laws of the constant and time-varying electric and magnetic fields, we will arrive at Maxwell's equations. From the historical point of view, such approach is natural since the most of these laws, as well as Faraday's concept about a field, were known before the system of equations of the electromagnetic field was derived.

2.1. FARADAY'S LAW

Experimental investigations performed by Faraday show that a change of flux Φ of the magnetic field $\mathbf{B}$ with time through any surface S bounded by contour L (Figure 2.1(a)) creates an electromotive force Ξ along this contour, which is given by

$$\Xi = -\frac{\partial \Phi}{\partial t} \tag{2.1}$$

where

$$\Phi = \int_S \mathbf{B} \cdot d\mathbf{S}$$

is the flux of the magnetic field, and $\partial\Phi/\partial t$ is its derivative with respect to time. The contour L can have any shape, and it is a closed path passing in general through media having different electric properties including

Methods in Geochemistry and Geophysics, Volume 45
ISSN 0076-6895,
http://dx.doi.org/10.1016/B978-0-444-53829-1.00002-2

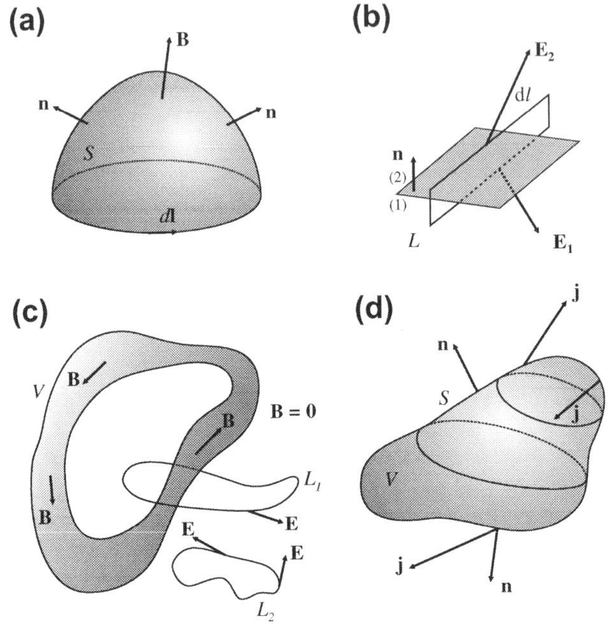

Figure 2.1 (a) Flux of the field $\dot{\mathbf{B}}$. (b) Electric field near an interface. (c) Faraday's law. (d) Flux of the current density point of this contour and changes along it in both magnitude and direction (c). However, the electromotive force in this case is equal to zero.

insulators. Of course, this path L usually does not coincide with current lines. By definition, the electromotive force is

$$\Xi = \oint_L \mathbf{E} \cdot \mathbf{dl}, \tag{2.2}$$

where $\mathbf{E}$ is the electric field at each point of the contour L. In particular, this means that the electromotive force may exist only if there is an electric field. Consequently, in place of Eq. (2.1), we have

$$\oint_L \mathbf{E} \cdot \mathbf{dl} = -\frac{\partial \Phi}{\partial t}, \tag{2.3}$$

and it can be interpreted in the following way. A change of the magnetic flux $\mathbf{\Phi}$ with time gives rise to an electric field. In other words, a change of the magnetic field $\mathbf{B}$ with time is a generator of an electric field as well as electric charges does. This phenomenon was first observed and reported by Faraday and has been called *electromagnetic induction*. The relationship between the electric field and the rate of the change of the magnetic flux, as described by Eq. (2.3), is one of the most fundamental laws of physics. By convention, the electric field which appears due to the electromagnetic induction is called the *inductive electric field* $\mathbf{E}^{\text{ind}}$ to emphasize its origin. Thus, we can rewrite Eq. (2.3) in the form

$$\oint_L \mathbf{E}^{\text{ind}} \cdot d\mathbf{l} = -\frac{\partial \Phi}{\partial t}, \tag{2.4}$$

that is, a change of the magnetic flux with time gives rise to the electric field $\mathbf{E}^{\text{ind}}$. It is essential to note that determination of this field requires additional information even if the function $\partial\Phi/\partial t$ is known. It is related with the fact that, unlike Coulomb's law, the field $\mathbf{E}^{\text{ind}}$ is not given in the explicit form in Eq. (2.3) but rather plays the role of integrand. Thus, in general, there are two generators of the electric field, namely, charges and a change of the magnetic field with time, as shown below.

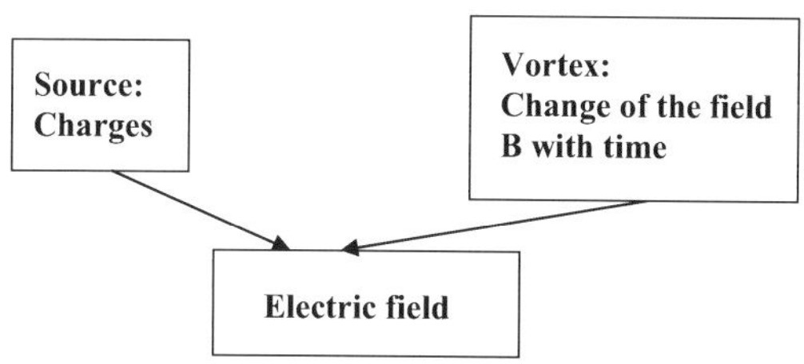

At the same time, we can readily think of particular cases in which one of these generators does not exist, such as the following:

1. A constant field in which derivatives with respect to time are zero and the electric field arises only due to the presence of the electric charges.
2. An alternating electromagnetic field in which the electric field has only a tangential component at interfaces between media with different but constant conductivities. In such a case, charges are absent, and the electric field has an inductive or vortex origin since it appears only due to a change of the magnetic field with time. Now suppose that the total electric field $\mathbf{E}$ can be represented as a sum

$$\mathbf{E} = \mathbf{E}^{\text{c}} + \mathbf{E}^{\text{ind}}, \tag{2.5}$$

where $\mathbf{E}^{\text{c}}$ is the electric field generated by charges and at every instant obeys Coulomb's law, while $\mathbf{E}^{\text{ind}}$ is the part of the electric field which arises due to a change of the magnetic field with time. Combining Eqs (2.4) and (2.5), we have

$$\oint_L \mathbf{E}\cdot d\mathbf{l} - \oint_L \mathbf{E}^c\cdot d\mathbf{l} = -\frac{\partial \Phi}{\partial t}$$

As is well known, the circulation of the Coulomb's electric field is equal to zero, and therefore,

$$\oint_L \mathbf{E}^{ind}\cdot d\mathbf{l} = \oint_L \mathbf{E}\cdot d\mathbf{l} = -\frac{\partial \Phi}{\partial t} \tag{2.6}$$

Sometimes this result leads to some confusion about the role played by charges in forming an electromagnetic field. Actually Eq. (2.6) shows that the electromotive force due to the Coulomb's electric field vanishes. But this conclusion, as in the case of a constant field, cannot be extended to the electric field $\mathbf{E}^c$ itself. In fact, this field has an influence on the distribution of currents in a conducting medium, which in turn creates an alternating magnetic field. Therefore, in general, both parts of the field $\mathbf{E}^c$ and $\mathbf{E}^{ind}$ are related to each other. Next we describe different forms of Faraday's law. First using the definition of the magnetic flux, we have

$$\oint_L \mathbf{E}\cdot d\mathbf{l} = -\frac{\partial}{\partial t}\int_S \mathbf{B}\cdot d\mathbf{S}$$

In general, a change of the flux may happen due to a change of the magnetic field with time, as well as a change of a position of the path L. The influence of the last factor was studied earlier (Lorenz force), but from now we assume that the path L along which the electromotive force is calculated does not move. Then the last equation can be rewritten as

$$\oint_L \mathbf{E}\cdot d\mathbf{l} = -\int_S \dot{\mathbf{B}}\cdot d\mathbf{S} \tag{2.7}$$

where

$$\dot{\mathbf{B}} = \frac{\partial \mathbf{B}}{\partial t}.$$

Equation (2.7) is a formulation of Faraday's law and, as will be seen later, it is the first Maxwell's equation in the integral form. In this equation, the $\mathbf{dl}$

is the elementary displacement and it indicates the direction in which integration is carried along the contour L, while the vector $\mathbf{dS}$ has the direction of the unit vector $\mathbf{n}$ normal to the surface S, bounded by the contour L (Figure 2.1(a)). It is clear that there should be a special relationship between the vectors $\mathbf{dl}$ and $\mathbf{dS}$. Otherwise, in changing the direction of one of these vectors, the same rate of the flux change $\partial\Phi/\partial t$ would create an electromotive force having the same magnitude but the opposite sign. In order to retain the physical meaning of Faraday's law, the vectors $\mathbf{dl}$ and $\mathbf{dS}$ are chosen according to the right-hand rule. This means that an observer facing in the direction of the vector $-\mathbf{dS}$ sees that $\mathbf{dl}$ indicates a direction along the path L counterclockwise. Only in this case, Eq. (2.7) correctly describes the electromagnetic induction phenomenon. Now, making use of Stokes theorem,

$$\oint_L \mathbf{M}\cdot\mathbf{dl} = \int_S \text{curl}\,\mathbf{M}\cdot\mathbf{dS},$$

we obtain the differential form of Eq. (2.7)

$$\oint_L \mathbf{E}\cdot\mathbf{dl} = \int_S \text{curl}\,\mathbf{E}\cdot\mathbf{dS} = -\int_S \frac{\partial\mathbf{B}}{\partial t}\cdot\mathbf{dS}$$

Since the latter is valid for any surface S, from equality of surface integrals, it follows that

$$\text{curl}\,\mathbf{E} = -\frac{\partial\mathbf{B}}{\partial t}, \tag{2.8}$$

where $\mathbf{E}$ and $\mathbf{B}$ are considered in the vicinity of the same point. Both Eq. (2.7) and Eq. (2.8) describe the same physical phenomenon but the differential form, Eq. (2.8), applies only at regular points in which all components of the electric field are continuous functions of spatial variables. Considering that in many problems we must examine electromagnetic fields in media with discontinuous changes in physical properties (interfaces), it is desirable to derive a surface analogy of Eq. (2.8). It is clear that Eq. (2.8) cannot be used for points of the interface between media having different values of dielectric permittivity and conductivity, since the normal component of the electric field is usually a discontinuous function. For this reason, we proceed further with Eq. (2.7) applying it along the elementary path shown in Figure 2.1(b). Taking into account the fact that the flux through the area surrounded by this contour tends to zero, we obtain

$$E_{2t} - E_{1t} = 0, \tag{2.9}$$

where t indicates an arbitrary direction tangential to the interface. The vector form of this equation is

$$\mathbf{n} \times (\mathbf{E}_2 - \mathbf{E}_1) = 0$$

Here, $\mathbf{E}_1$ and $\mathbf{E}_2$ are the electric fields at the back and front sides of the surface, respectively. In accordance with this equation, we can say that the tangential components of the time-varying electric field are continuous functions at an interface as would be the case when the field is caused by constant electric charges. Thus, we derived three forms of Faraday's law:

$$\oint_L \mathbf{E}\cdot d\mathbf{l} = -\int_S \dot{\mathbf{B}}\cdot d\mathbf{S}, \quad \text{curl}\,\mathbf{E} = -\frac{\partial \mathbf{B}}{\partial t}, \quad \mathbf{n} \times (\mathbf{E}_2 - \mathbf{E}_1) = 0, \tag{2.10}$$

and they describe the first Maxwell equation in different forms. Later we will consider numerous examples illustrating the electromagnetic induction and application of Eq. (2.10), but now let us make several comments:

1. Suppose that a change of the magnetic field with time occurs within some volume V, but outside of it the field $\mathbf{B}$ is absent. As follows from Eq. (2.3), the electromotive force along the contour L_1 surrounding this volume (Figure 2.1(c)) is not equal to zero, regardless of how far from the volume V, the path L_1 is located. This means that a change of the field $\mathbf{B}$ with time in some region of space results in an appearance of the electric field $\mathbf{E}^{\text{ind}}$ at any distance from its generator. In this sense, the analogy with the Coulomb's electric field caused by charges is obvious. One can say that neither Coulomb's nor Faraday's laws explain how the field reaches an observation point. Later we will discuss this subject in detail.
2. Consider an arbitrary closed path L_2 which does not enclose the volume where vortices $\partial\mathbf{B}/\partial t$ are located. In general, the electric field caused by these generators exists at each point of this contour and changes along it in both magnitude and direction (Figure 2.1(c)). However, the electromotive force in this case is equal to zero.
3. This analysis shows that at every point of a closed contour (loop), the inductive electric field can be represented as the sum of two fields. One of them is caused by vortices which intersect the area surrounded by the loop, while the other is generated by vortices which do not cross the area of the loop, more precisely; their flux is equal to zero. It is essential that in

accordance with Faraday's law, the electromotive force Ξ is defined by only the first part of the electric field. This fact allows us to use a relatively small loop for measuring the rate of change of the magnetic field $\partial \mathbf{B}/\partial t$ at the points surrounded by the loop.

4. As is well known, the voltage of the Coulomb's electric field between two points is path independent. In contrast, taking into account Eq. (2.4), we have to conclude that in general the voltage of the inductive electric field

$$\int_a^b \mathbf{E}^{\text{ind}} \cdot d\mathbf{l}$$

depends on the path of integration.

Now we take into account Faraday's law and modify the system of equations derived for constant field at regular points in the following way:

$$\text{curl}\,\mathbf{E} = -\frac{\partial \mathbf{B}}{\partial t} \quad \text{div}\,\mathbf{D} = \delta_0$$
$$\text{curl}\,\mathbf{B} = \mu_0 \mathbf{j}_c \quad \text{div}\,\mathbf{B} = 0 \tag{2.11}$$

It seems that this set of equations describes the time-varying electromagnetic field since it takes into account the electromagnetic induction phenomenon, as well as the Coulomb's and Biot–Savart laws. In fact, this system characterizes fields with a high accuracy in a so-called quasi-stationary approximation (diffusion), which plays the dominant role in most electromagnetic methods in geophysics. However, Eq. (2.11) suggests that the field instantly appears at any point of the space regardless of the distance from its generators. Suppose that conduction current was changed at some instant t. Then in accordance with the equation,

$$\oint_L \mathbf{B} \cdot d\mathbf{l} = \mu_0 I$$

the magnetic field synchronously changes at any observation point. The same is true for the electric field. There is another fact which implies that there is a problem with Eq. (2.11). Indeed, from the equation

$$\text{curl}\,\mathbf{B} = \mu_0 \mathbf{j}_c$$

we have

$$\operatorname{div} \mathbf{j}_c = 0, \tag{2.12}$$

since from vector calculus it follows that

$$\operatorname{div} \operatorname{curl} \mathbf{M} = 0$$

Equation (2.12) describes the principle of charge conservation for constant electric field [2], but as we will see, it contradicts with the same principle for the time-varying fields. This discrepancy is of a great importance because its understanding helped one to discover the effect of propagation of electromagnetic fields and arrive at Maxwell equations. Taking into account this fundamental fact, it is proper to describe the general case of the principle of charge conservation.

2.2. THE PRINCIPLE OF CHARGE CONSERVATION

In general, this principle is written in the form

$$\oint_S \mathbf{j}\cdot d\mathbf{S} = -\frac{\partial e_0}{\partial t}, \tag{2.13}$$

where S is an arbitrary closed surface, e_0 is a free charge in the volume V, surrounded by the surface S, and $d\mathbf{S}$ is directed outside the volume (Figure 2.1(d)). Also

$$\mathbf{j} = \mathbf{j}_c$$

is the conduction current only. In accordance with Eq. (2.13), a flux of the current density through S defines the rate of a change of charge over time inside the volume. If for instance the flux is positive, the charge e_0 decreases and, in contrast when the flux is negative, more charges arrive than leave this volume. At the same time, experiments show that in the absence of the current, it is impossible to observe an appearance or a disappearance of charges in macroscopic quantities. Thus, any change of the charge inside the volume V can be made only due to the flux of free charges through the surface S. Now applying Gauss's theorem for regular points of a medium, we obtain

$$\oint_S \mathbf{j}\cdot d\mathbf{S} = \int_V \operatorname{div} \mathbf{j}\, dV = -\frac{\partial}{\partial t}\int_V \delta_0\, dV, \tag{2.14}$$

where δ_0 is the volume density of free charges. Assuming that the volume does not change its shape with time, we have

$$\int_V \operatorname{div} \mathbf{j} \mathrm{d}V = -\int_V \frac{\partial \delta_0}{\partial t} \mathrm{d}V$$

or

$$\operatorname{div} \mathbf{j} = -\frac{\partial \delta_0}{\partial t} = -\dot{\delta}_0 \tag{2.15}$$

Next consider the surface analogy of Eq. (2.15). With this purpose in mind, let us determine the flux of the current density through the surface of an elementary cylinder. Then, making use of Eq. (2.14), we have

$$\mathbf{j}_2 \cdot \mathrm{d}\mathbf{S}_2 + \mathbf{j}_1 \cdot \mathrm{d}\mathbf{S}_1 + \int_{S_l} \mathbf{j} \cdot \mathrm{d}\mathbf{S} = -\dot{\sigma}_0 \mathrm{d}\mathbf{S},$$

where $\mathbf{j}_2$ and $\mathbf{j}_1$ are the current density at the front and back sides of the surface, respectively, S_l is the lateral surface of the cylinder,

$$\mathrm{d}\mathbf{S}_2 = \mathbf{n}\mathrm{d}S, \quad \mathrm{d}\mathbf{S}_1 = -\mathbf{n}\mathrm{d}S$$

and σ_0 is the surface density of free charges. In the limit when the cylinder height tends to zero, we have

$$j_{2n} - j_{1n} = -\dot{\sigma}_0, \tag{2.16}$$

that is, the difference of normal components of the current density defines the rate of a change of the surface density of charges. Thus, we have derived three forms of equations describing the principle of charge conservation:

$$\oint_S \mathbf{j} \cdot \mathrm{d}\mathbf{S} = -\frac{\partial e_0}{\partial t}, \quad \operatorname{div} \mathbf{j} = -\dot{\delta}_0, \quad j_{2n} - j_{1n} = -\dot{\sigma}_0 \tag{2.17}$$

These equations are always valid for a constant and any time-varying electromagnetic field, and they show that a change of a charge in one place is always accompanied by such a change of charge in other places that a total charge remains the same. This is the reason why this phenomenon is called the *principle of charge conservation*. At the same time, there are at least three cases when it is possible to neglect by a change of the charge with time.

Case 1: The constant field.
By definition, the right-hand side of Eq. (2.13) vanishes, and the flux of the current density vector is equal to zero through any closed surface. This means that charges may exist but they do not change with time.

Case 2: Nonconducting medium
In such medium, the conduction current is absent everywhere and free charges remain constant in spite of the presence of arbitrary time-varying fields. This means that these charges cause only the constant electric field.

Case 3: Quasi-stationary electromagnetic field
Suppose that a medium is conductive and a time-varying field changes relatively slowly. Then it turns out that under certain conditions, the right-hand side of Eq. (2.17) can be again neglected. In other words, it is assumed that at each instance, amount of charges arriving at volume is equal to that which leaves it and, as will be shown later, such approximation gives sufficiently correct result when displacement currents are much smaller than the conduction ones. As was mentioned earlier, this case is of a great importance in the most but not all applications of electromagnetic fields in the surface and borehole geophysics. Thus in these special cases, in place of Eq. (2.17), we have

$$\oint_S \mathbf{j}\cdot d\mathbf{S}\approx 0, \quad \operatorname{div}\mathbf{j}\approx 0, \quad j_{2n}\approx j_{1n} \tag{2.18}$$

Returning back to the principle of charge conservation for the time-varying field, it is natural to expect that Eq. (2.17) should follow from the system of Eq. (2.11), describing an electromagnetic field. However, it turns out that in general, we cannot obtain these equations from Eq. (2.11) because the third equation of this set does not take into account one more source of the magnetic field.

2.3. DISTRIBUTION OF ELECTRIC CHARGES

Now proceeding from the second equation of Eq. (2.11),

$$\oint_S \boldsymbol{\varepsilon}\mathbf{E}\cdot d\mathbf{S} = e_0 \quad \text{or} \quad \operatorname{div}\boldsymbol{\varepsilon}\mathbf{E} = \delta_0 \tag{2.19}$$

and Eq. (2.17), we will study a distribution of charges in a conducting and polarizable medium with conductivity γ and dielectric permittivity $\boldsymbol{\varepsilon}$. In this light, it is proper to note that Eq. (2.19) was derived from the

Coulomb's law. This may create impression that our results are applied only for the stationary (time invariant) field. However, as will be shown later, these equations are applied for the time-varying electromagnetic fields, too.

2.3.1. Equation for Volume Density of Charges

First consider regular points of a medium where equations

$$\operatorname{div}\mathbf{j} = -\frac{\partial \delta_0}{\partial t}, \quad \text{and} \quad \operatorname{div}\boldsymbol{\varepsilon}\mathbf{E} = \delta_0 \tag{2.20}$$

are valid. Taking into account Ohm's law, we have

$$\operatorname{div}\mathbf{j} = \operatorname{div}\gamma\mathbf{E} = \operatorname{div}\frac{\gamma}{\varepsilon}\boldsymbol{\varepsilon}\mathbf{E} = \frac{\gamma}{\varepsilon}\operatorname{div}\boldsymbol{\varepsilon}\mathbf{E} + \boldsymbol{\varepsilon}\mathbf{E}\cdot\operatorname{grad}\frac{\gamma}{\varepsilon} = -\frac{\partial \delta_0}{\partial t}$$

or

$$\frac{\gamma}{\varepsilon}\delta_0 + \boldsymbol{\varepsilon}\mathbf{E}\cdot\operatorname{grad}\frac{\gamma}{\varepsilon} = -\frac{\partial \delta_0}{\partial t}$$

Thus, we have arrived at the following differential equation for the volume density of free charges:

$$\frac{\partial \delta_0}{\partial t} + \frac{1}{\tau_0}\delta_0 = -\boldsymbol{\varepsilon}\mathbf{E}\cdot\operatorname{grad}\frac{1}{\tau_0}, \tag{2.21}$$

where

$$\tau_0 = \frac{\varepsilon}{\gamma} = \varepsilon\rho \tag{2.22}$$

is often called the *time constant of the medium.*

2.3.2. Uniform Medium

At the beginning, suppose that in the vicinity of some point the parameter τ_0 does not change or the field $\mathbf{E}$ and grad $\frac{1}{\tau_0}$ are perpendicular to each other,

$$\mathbf{E}\cdot\nabla\frac{1}{\tau_0} = 0 \tag{2.23}$$

Then Eq. (2.21) is simplified and we have

$$\frac{\partial \delta_0}{\partial t} + \frac{1}{\tau_0}\delta_0 = 0 \tag{2.24}$$

The solution of this equation is

$$\delta_0(t) = C \exp\left(-\frac{t}{\tau_0}\right), \tag{2.25}$$

where C is the density of free charge at the initial instant. In a conducting and polarizable medium, the parameter τ_0 is usually very small. For example, if $\rho = 100$ Ohm·m and $\varepsilon = 10\varepsilon_0$, then

$$\tau_0 = 100 \cdot 10 \cdot (36\pi)^{-1} 10^{-9}\text{s} < 10^{-8}\text{s}$$

Thus, a free charge placed inside a conducting medium disappears very quickly. If we are concerned only with charges which exist at times greater than τ_0 ($t \gg \tau_0$) and described by Eq. (2.24), we can assume that they are practically absent. In addition, it is appropriate to notice that the initial volume charge is usually equal to zero inside the conducting medium, that is $C = 0$. Therefore, we can conclude that at points where the medium is uniform with respect to τ_0 or condition (2.23) is met, there are no electric charges and this gives

$$\text{div}\, \varepsilon\mathbf{E} = 0 \tag{2.26}$$

Earlier in Chapter 1, it was mentioned that free charges are accompanied by bound ones:

$$\text{div}\, \mathbf{E} = \frac{\delta_0 + \delta_b}{\varepsilon_0}, \quad \text{and} \quad \text{div}\, \mathbf{E} = \frac{\delta_0}{\varepsilon},$$

since grad $\varepsilon = 0$. Whence,

$$\frac{\delta_0 + \delta_\text{b}}{\varepsilon_0} = \frac{\delta_0}{\varepsilon} \quad \text{and} \quad \delta_\text{b} = \left(\frac{\varepsilon_0}{\varepsilon} - 1\right)\delta_0 \tag{2.27}$$

Therefore, the bound and free charges, located in the vicinity of some point where $\nabla\tau_0 = 0$, decay in the same manner and correspondingly

$$\text{div}\, \mathbf{E} = 0 \quad \text{if} \quad t \gg \tau_0 \tag{2.28}$$

It is also appropriate to notice that the total density of decaying charge, as in the case of the constant field, is ε_r times smaller than the free charge.

$$\delta = \frac{\delta_0(t)}{\varepsilon_\text{r}}, \tag{2.29}$$

where $\varepsilon_\text{r} = \varepsilon/\varepsilon_0$.

2.3.3. Nonuniform Medium

Until now, we have studied the behavior of the charge in the vicinity of points where either the medium is uniform or the condition (2.23) is met, and it was established that charge decays very rapidly. It is essential that such behavior is observed regardless of the presence of the electromagnetic field caused by generators located in other places of the medium. A much more different situation occurs when the medium is not uniform and either $\mathbf{E}\cdot\nabla\gamma\neq 0$ or $\mathbf{E}\cdot\nabla\varepsilon\neq 0$. In this case, the right-hand side of Eq. (2.21) does not vanish, and we have an inhomogeneous differential equation of the first order:

$$\frac{dy}{dt}+\frac{1}{\tau_0}y=f(t), \tag{2.30}$$

where $y=\delta_0(t)$ and $f(t)=-\boldsymbol{\varepsilon}\mathbf{E}\cdot\nabla\frac{1}{\tau_0}$. The general solution of Eq. (2.30) is well known and has the form

$$y(t)=y_0\exp\left(-\frac{t}{\tau_0}\right)+\exp\left(-\frac{t}{\tau_0}\right)\int\limits_0^t\exp\left(\frac{x}{\tau_0}\right)f(x)dx, \tag{2.31}$$

where y_0 is the value of the function $y(t)$ at the instant $t=0$. In accordance with Eq. (2.31),

$$\delta_0(t)=C\exp\left(-\frac{t}{\tau_0}\right)-\exp\left(-\frac{t}{\tau_0}\right)\varepsilon\int\limits_0^t\exp\left(\frac{x}{\tau_0}\right)\mathbf{E}\cdot\nabla\frac{1}{\tau_0}dx \tag{2.32}$$

If the direction of the electric field does not change with time, the last equation can be rewritten as

$$\delta_0(t)=C\exp\left(-\frac{t}{\tau_0}\right)-\exp\left(-\frac{t}{\tau_0}\right)\boldsymbol{\varepsilon}\int\limits_0^t\exp\left(\frac{x}{\tau_0}\right)E(x)dx\left(\mathbf{e}_0\cdot\nabla\frac{1}{\tau_0}\right), \tag{2.33}$$

where

$$\mathbf{E}(t)=E(t)\mathbf{e}_0$$

In general, we can recognize two types of charges whose behavior is quite different as a function of time:

$$\delta_0(t)=\delta_{01}(t)+\delta_{02}(t) \tag{2.34}$$

Here,

$$\delta_{01}(t) = C\exp\left(-\frac{t}{\tau_0}\right)$$

and

$$\delta_{02}(t) = -\exp\left(-\frac{t}{\tau_0}\right)\varepsilon\int_0^t \exp\left(\frac{x}{\tau_0}\right)E(x)\mathrm{d}x\left(\mathbf{e}_0\cdot\nabla\frac{1}{\tau_0}\right). \tag{2.35}$$

The behavior of the function $\delta_{01}(t)$ is the same as in the case of a uniform medium. According to Eq. (2.35), a free charge $\delta_{02}(t)$ arises in the neighborhood of any point where a medium is not uniform, provided that the field $\mathbf{E}(t)$ is not perpendicular to the direction of the gradient of τ_0. In general, the density $\delta_{02}(t)$ depends on the resistivity and dielectric permittivity of the medium as well as on the magnitude and direction of the electric field.

2.3.4. Quasi-Stationary Field

Now we consider one special case which is of a great practical interest for many geophysical applications. Suppose the following inequality holds,

$$\frac{\partial\delta_0}{\partial t} \ll \frac{\delta_0}{\tau_0} \tag{2.36}$$

Then instead of Eq. (2.21), we obtain an approximate equation

$$\gamma\frac{\delta_0(t)}{\varepsilon} + \varepsilon\mathbf{E}(t)\cdot\nabla\frac{1}{\tau_0} = 0 \tag{2.37}$$

Correspondingly, the density of free charge is

$$\delta_0(t) = -\varepsilon^2\rho\mathbf{E}(t)\cdot\nabla\frac{1}{\tau_0}$$

or

$$\delta_0(t) = \varepsilon\mathbf{E}(t)\cdot\frac{\nabla\rho}{\rho} + \mathbf{E}(t)\cdot\nabla\varepsilon, \tag{2.38}$$

since

$$\nabla\frac{1}{\tau_0} = \nabla\frac{1}{\varepsilon\rho} = -\frac{\nabla\varepsilon}{\varepsilon^2}\frac{1}{\rho} - \frac{\nabla\rho}{\rho^2}\frac{1}{\varepsilon}.$$

Therefore, free charges arise in the vicinity of points where either conductivity or dielectric permittivity changes. Of course, this happens only if the electric field is not perpendicular to the direction of the maximal rate of change of these parameters. Note that in the frequency domain, the inequality Eq. (2.36) is equivalent to the following:

$$\frac{\omega\varepsilon}{\gamma} \ll 1$$

As we already know, the free charges are usually accompanied by bound charges and their density is

$$\delta_{\mathrm{b}} = \delta - \delta_0 = \operatorname{div}[(\boldsymbol{\varepsilon}_0 - \boldsymbol{\varepsilon})\mathbf{E}] = \operatorname{div}[(\boldsymbol{\varepsilon}_0 - \boldsymbol{\varepsilon})\mathbf{j}\rho].$$

Making the use of the equation,

$$\operatorname{div}\mathbf{E} = \frac{\delta_0 + \delta_{\mathrm{b}}}{\varepsilon_0}$$

and Eq. (2.38), it can be shown that

$$\delta_{\mathrm{b}}(t) = -\mathbf{E}(t)\cdot\nabla\varepsilon + (\varepsilon_0 - \varepsilon)\mathbf{E}(t)\cdot\frac{\nabla\rho}{\rho}. \tag{2.39}$$

From Eqs (2.38) and (2.39), it follows that the total charge is

$$\delta(t) = \varepsilon_0\mathbf{E}(t)\cdot\frac{\nabla\rho}{\rho}, \tag{2.40}$$

that is, at points where only $\boldsymbol{\varepsilon}$ varies the total charge is equal to zero. This means that in such places, the free and bound charges compensate for each other. At the same time, at points where both parameters change the total charge is the same as if the polarization were absent.

2.3.5. Behavior of Charge Density δ_{02}

Now we return back to a general case, Eq. (2.35), and consider two examples which illustrate the behavior of the charge $\delta_{02}(t)$ when its change with time is taken into account, Eq. (2.21).

Example 1

Let us assume that the electric field varies with time as

$$\mathbf{E}(t) = E_0 \exp\left(-\frac{t}{\tau}\right)\mathbf{e}_0, \tag{2.41}$$

and τ being the parameter characterizing the rate of the field change. Then, Eq. (2.35) becomes

$$\delta_{02}(t) = -\varepsilon \exp\left(-\frac{t}{\tau_0}\right)\left(\mathbf{e}_0 \cdot \nabla \frac{1}{\tau_0}\right)\int\limits_0^t \exp\left[\left(\frac{1}{\tau_0} - \frac{1}{\tau}\right)x\right] dx E_0$$

Carrying out integration, we obtain

$$\delta_{02}(t) = -\frac{\varepsilon \tau_0 E_0 \exp\left(-\frac{t}{\tau_0}\right)}{1 - \frac{\tau_0}{\tau}}\left\{\exp\left[t\left(\frac{1}{\tau_0} - \frac{1}{\tau}\right)\right] - 1\right\}\left(\mathbf{e}_0 \cdot \nabla \frac{1}{\tau_0}\right) \quad (2.42)$$

As follows from this equation, the charge is absent at the instant $t = 0$. Then it increases, reaches a maximum and at greater times exponentially decays. Thus, in general, the dependence of this charge density and the electric field $\mathbf{E}(t)$ on time differs from each other. Assuming that the electric field decays relatively slowly ($\tau\tau_0$) and measurements are performed at sufficiently large times ($t \gg \tau_0$) in place of Eq. (2.42), we obtain

$$\delta_{02}(t) = -\varepsilon \tau_0 \exp\left(-\frac{t}{\tau}\right) E_0 \left(\mathbf{e}_0 \cdot \nabla \frac{1}{\tau_0}\right) \quad (2.43)$$

Correspondingly, the volume density of free charge and the electric field decay in the same manner. For instance, when the time constant of the field τ is 1 s, the function $\delta_{02}(t)$ also decreases with a time constant 1 s regardless of the conductivity and dielectric permittivity of the medium.

Example 2

Now suppose that the electric field varies as a sinusoidal function:

$$\mathbf{E}(t) = E_0 \mathbf{e}_0 \sin \omega t$$

Substituting this expression into Eq. (2.35) and integrating, we have

$$\delta_{02}(t) = -\frac{\varepsilon E_0 \tau_0}{1 + \omega^2 \tau_0^2}\left[\omega \tau_0 \exp\left(-\frac{t}{\tau_0}\right) + (\sin \omega t - \omega \tau_0 \cos \omega t)\right] \times \left(\mathbf{e}_0 \cdot \nabla \frac{1}{\tau_0}\right) \quad (2.44)$$

In particular, assuming that

$$t \gg \tau_0 \quad \text{and} \quad T \gg \tau_0, \quad (2.45)$$

where T is the period of oscillations, we have

$$\delta_{02}(t) = -\varepsilon\tau_0 E_0 \sin \omega t\left(\mathbf{e}_0 \cdot \nabla\frac{1}{\tau_0}\right) \qquad (2.46)$$

It is significant to notice that conditions (2.36) and (2.45) have the same meaning.

2.3.6. Surface Distribution of Charges

So far, we have investigated the distribution of only volume density of charges. Now consider time-varying free charges which arise at interfaces between media with different electric properties. Applying equations,

$$j_{2\mathrm{n}} - j_{1\mathrm{n}} = -\frac{\partial\sigma_0}{\partial t} \qquad D_{2\mathrm{n}} - D_{1\mathrm{n}} = \sigma_0$$

Here,

$$D_{1\mathrm{n}} = \varepsilon_{\mathrm{i}} E_{1\mathrm{n}}, \quad i = 1, 2$$

we have

$$\frac{\gamma_2}{\varepsilon_2}D_{2\mathrm{n}} - \frac{\gamma_1}{\varepsilon_1}D_{1\mathrm{n}} = \frac{1}{2}\left[\left(\frac{\gamma_2}{\varepsilon_2}+\frac{\gamma_1}{\varepsilon_1}\right)(D_{2\mathrm{n}} - D_{1\mathrm{n}}) + \left(\frac{\gamma_2}{\varepsilon_2}-\frac{\gamma_1}{\varepsilon_1}\right)(D_{2\mathrm{n}} + D_{1\mathrm{n}})\right] = -\frac{\partial\sigma_0}{\partial t}$$

or

$$\alpha^{\mathrm{av}}\sigma_0 + (\alpha_2 - \alpha_1)D_{\mathrm{n}}^{\mathrm{av}} = -\frac{\partial\sigma_0}{\partial t}$$

Whence,

$$\frac{\partial\sigma_0(t)}{\partial t} + \frac{1}{\tau_{0s}}\sigma_0(t) = (\alpha_1 - \alpha_2)D_{\mathrm{n}}^{\mathrm{av}} \qquad (2.47)$$

where

$$\tau_{0s} = \frac{1}{\alpha^{\mathrm{av}}} = \frac{2}{1/\tau_{01} + 1/\tau_{02}}$$

is the relaxation time for surface charges, and

$$\alpha_1 - \alpha_2 = \frac{1}{\tau_{01}} - \frac{1}{\tau_{02}}, \qquad D_{\mathrm{n}}^{\mathrm{av}} = \frac{\varepsilon_1 E_{1\mathrm{n}} + \varepsilon_2 E_{2\mathrm{n}}}{2},$$

$$\tau_{01} = \varepsilon_1 \rho_1, \quad \tau_{02} = \varepsilon_2 \rho_2, \quad \alpha_1 = \frac{1}{\tau_{01}}, \quad \alpha_2 = \frac{1}{\tau_{02}}$$

Thus, the equation for the surface density of free charges is a differential equation of the first order similar to that for the volume density. In accordance with Eq. (2.35), the solution of Eq. (2.47) is

$$\sigma_0(t) = C \exp\left(-\frac{t}{\tau_{0s}}\right) + \exp\left(-\frac{t}{\tau_{0s}}\right) \times (\alpha_1 - \alpha_2) \int\limits_0^t D_n^{av}(x) \exp\left(\frac{x}{\tau_{0s}}\right) dx \tag{2.48}$$

that is,

$$\sigma_0 = \sigma_{01} + \sigma_{02}$$

where

$$\sigma_{01} = C \exp\left(-\frac{t}{\tau_{0s}}\right),$$

$$\sigma_{02} = (\alpha_1 - \alpha_2) \exp\left(-\frac{t}{\tau_{0s}}\right) \int\limits_0^t D_n^{av}(x) \exp\left(\frac{x}{\tau_{0s}}\right) dx \tag{2.49}$$

Respectively, there are two types of surface charges. The first type σ_{01} corresponds to the case in which some free charge with density C is placed at the interface. As follows from Eq. (2.49), it decays exponentially with time constant τ_{0s}. In other words, its behavior is controlled by the conductivity and dielectric permittivity of the medium from both sides of the interface, and it is independent of the electric field caused by other generators. Inasmuch as the relaxation time τ_{0s} is usually very small and measurements are usually performed at times much greater than τ_{0s}, again in most cases, we can ignore the presence of this charge. Moreover, with the exception of very special cases, such as the piezoelectric effect and the grounding electrodes, charges are not placed at interfaces. Correspondingly, let us concentrate on the surface of the second type. Of course, as in the case of volume density, the surface charges consist of the free and bound charges and they are related to each other. In fact, from the equations

$$E_{2n} - E_{1n} = \frac{\sigma}{\varepsilon_0}, \quad \text{and} \quad \varepsilon_2 E_{2n} - \varepsilon_1 E_{1n} = \sigma_0$$

we have

$$\sigma_0 = \frac{1}{2}[(\varepsilon_2 + \varepsilon_1)(E_{2n} - E_{1n}) + (\varepsilon_2 - \varepsilon_1)(E_{2n} + E_{1n})]$$

or

$$\sigma_0 = \frac{\varepsilon^{av}}{\varepsilon_0}\sigma + (\varepsilon_2 - \varepsilon_1)E_n^{av}$$

Here,

$$\sigma = \sigma_0 + \sigma_b, \quad \varepsilon^{av} = \frac{\varepsilon_1 + \varepsilon_2}{2}, \quad E_n^{av} = \frac{E_{2n} + E_{1n}}{2}$$

Hence,

$$\sigma_b = \frac{\varepsilon_0 - \varepsilon^{av}}{\varepsilon^{av}}\sigma_0 - \frac{\varepsilon_2 - \varepsilon_1}{\varepsilon^{av}}\varepsilon_0 E_n^{av} \tag{2.50}$$

2.3.7. Case of Relatively Slow Varying Field (Quasi-Stationary Field)

Now let us assume again that the time constant τ_{0s} is very small with respect to time of measurements:

$$t \gg \tau_{0s}$$

or the period of sinusoidal oscillations of the field T is much greater than τ_{0s}: $T \gg \tau_{0s}$. Then it is appropriate to replace the right-hand side of Eq. (2.49) by a series in the parameter τ_{0s}. Carrying out this expansion using integration by parts and discarding all terms except the first one, we obtain

$$\sigma_{02}(t) = (\alpha_1 - \alpha_2)\tau_{0s}D_n^{av}(t) \tag{2.51}$$

It is obvious that the same result follows from Eq. (2.47) if we neglect by the derivative $\partial\sigma_{02}(t)/\partial t$ in comparison with the term σ_{02}/τ_{0s}:

$$\frac{\partial\sigma_{02}}{\partial t} \ll \frac{\sigma_{0s}}{\tau_{0s}} \tag{2.52}$$

As in general case, the free charges are accompanied by bounded ones, but it turns out that the density of the total charge σ does not depend on the dielectric permittivity. In other words, the total charge σ coincides with that of free charges if the medium is not polarizable, provided that the condition (2.52) is met. Correspondingly, letting $\varepsilon_1 = \varepsilon_2 = \varepsilon_0$, Eq. (2.51) can be written as

$$(\gamma_1 + \gamma_2)\frac{\sigma}{2\varepsilon_0} + (\gamma_2 - \gamma_1)E_n^{av} = 0, \tag{2.53}$$

where E_n^{av} is the average magnitude of the normal component of the electric field at point p, located at the interface. Therefore, we arrive at the following expression for the surface density of the total charge:

$$\sigma(p, t) = 2\varepsilon_0 K_{12} E_n^{av}(p, t) \tag{2.54}$$

where

$$K_{12} = \frac{\rho_2 - \rho_1}{\rho_2 + \rho_1} \tag{2.55}$$

As we already know, the same equation describes the density of charges when the field is time invariant. Thus, Eq. (2.54) shows that if the condition (2.52) holds, the density of time-varying charges is related to the electric field and resistivity of the medium as the density of constant charges. Note that Eq. (2.54) plays the fundamental role for understanding the so-called galvanic part of the field. In fact, anomalies observed in some geophysical methods are often caused by mainly electrical charges arising at the surface of an inhomogeneity and surrounding medium. Besides, the galvanic part of the field is rather frequent cause of distortion of the magnetotelluric curves that often makes the quantitative interpretation complicated and ambiguous. It is useful to represent the normal component of the field **E** on either side of the interface as

$$E_{1n}(p, t) = E_n(p, t) - \frac{\sigma(p, t)}{2\varepsilon_0}$$

and

$$E_{2n}(p, t) = E_n(p, t) + \frac{\sigma(p, t)}{2\varepsilon_0} \tag{2.56}$$

Here, $E_n(p,t)$ is the normal component of the field at the point p contributed by all generators except the charge at this point. As is shown [2], this surface charge creates in its vicinity the field

$$\pm\frac{\sigma(p, t)}{2\varepsilon_0},$$

and in accordance with Eq. (2.56), we have to conclude that

$$E_n^{av}(p, t) = E_n(p, t), \tag{2.57}$$

where the normal **n** is directed from the back side "1" to the front side, "2" of the interface. Therefore, the function $E_n^{av}(p,t)$ describes the normal component of the field caused by all generators of the field $\mathbf{E}(p,t)$ except the charge in the vicinity of point p. For this reason, the second term of Eq. (2.53) can be interpreted as the flux of the current density **j** through a closed surface of an elementary cylinder with unit cross-section and an infinitely small height caused by all generators which are located outside this surface. In other words, this flux characterizes the difference between amount of charges, which arrive and leave this volume each second, and this motion of charges is caused by external generators only. The term

$$(\gamma_1 + \gamma_2)\frac{\sigma_0}{2\varepsilon_0}$$

defines the flux of the current density through the same closed surface caused by the electric field of the charge inside the elementary cylinder. Thus in our approximation, Eq. (2.52), the flux of the current density due to the external generators, such as charges and a change of the magnetic field with time, is compensated by the flux caused by the charge in the vicinity of the point p.

2.4. DISPLACEMENT CURRENTS

2.4.1. The Second Generator of Magnetic Field

Next we demonstrate that Eq. (2.11) is not in agreement with the principle of charge conservation for the time-varying fields. With this purpose in mind, let us show that the second equation of this system derived from the Biot–Savart law

$$\text{curl}\,\mathbf{B} = \mu_0 \mathbf{j}_c \tag{2.58}$$

in general, contradicts the principle of charge conservation when an electromagnetic field changes with time. In fact, taking the divergence of both sides of Eq. (2.58), we have

$$\text{div curl}\,\mathbf{B} = \mu_0 \text{div}\,\mathbf{j}_c \quad \text{or} \quad \text{div}\,\mathbf{j}_c = 0,$$

while as follows from Eq. (2.15), it should be equal to a rate of decrease of charge density with time. To remove this contradiction, we first assume that on the right-hand side of Eq. (2.58), there is an additional term **X** which disappears in the case of a time-invariant field. Then, Eq. (2.58) becomes

$$\text{curl}\,\mathbf{B} = \mu_0(\mathbf{j}_c + \mathbf{X}) \tag{2.59}$$

Now, we choose the vector $\mathbf{X}$ in such a way that the principle of charge conservation will be satisfied. Forming the divergence on both sides of Eq. (2.59), we obtain

$$0 = \text{div}\,\mathbf{j}_c + \text{div}\,\mathbf{X}$$

or, in accordance with Eq. (2.15),

$$\text{div}\,\mathbf{X} = \dot{\delta}_0 \tag{2.60}$$

It is a differential equation in partial derivatives with respect to unknown vector $\mathbf{X}$, and it is not clear how to solve it. However, the problem is greatly simplified if take into account the third equation of Eq. (2.11):

$$\text{div}\,\mathbf{D} = \delta_0$$

Assuming that this equation is valid for time-varying fields and taking derivative with respect to time, we have

$$\text{div}\,\dot{\mathbf{D}} = \dot{\delta}_0 \tag{2.61}$$

Comparison of Eqs (2.60 and 2.61) gives

$$\text{div}\,\dot{\mathbf{D}} = \text{div}\,\mathbf{X} \quad \text{or} \quad \text{div}\left(\dot{\mathbf{D}} - \mathbf{X}\right) = 0$$

It is obvious that an infinite number of vectors $\mathbf{X}$ satisfy this equation and they may differ from each other by

$$\text{curl}\,\mathbf{M}$$

where $\mathbf{M}$ is an arbitrary vector since

$$\text{div}\ \ \text{curl}\,\mathbf{M} = 0.$$

However, Maxwell assumed the simplest solution of this equation and let

$$\mathbf{X} = \dot{\mathbf{D}} = \frac{\partial \varepsilon \mathbf{E}}{\partial t}. \tag{2.62}$$

Numerous experimental studies performed during almost two centuries have shown the validity of this assumption, and the vector $\partial \mathbf{D}/\partial t$ is called the *density of displacement current*:

$$\mathbf{j}_d = \frac{\partial \mathbf{D}}{\partial t} \tag{2.63}$$

or

$$\mathbf{j}_{\mathrm{d}} = \varepsilon \frac{\partial \mathbf{E}}{\partial t}, \tag{2.64}$$

if we assume that the dielectric permittivity does not change with time. Consequently, instead of Eq. (2.59), we have

$$\operatorname{curl} \mathbf{B} = \mu_0 \left(\mathbf{j}_{\mathrm{c}} + \varepsilon \frac{\partial \mathbf{E}}{\partial t} \right) \tag{2.65}$$

Thus, the time-varying magnetic field is caused by two types of generators in a nonmagnetic medium, namely, the conduction and displacement currents as illustrated below. Applying Stoke's theorem, we obtain the integral form of Eq. (2.65)

$$\oint_L \mathbf{B} \cdot \mathrm{d}\mathbf{l} = \mu_0 \int_S \left(\mathbf{j}_{\mathrm{c}} + \varepsilon \frac{\partial \mathbf{E}}{\partial t} \right) \cdot \mathrm{d}\mathbf{S} \tag{2.66}$$

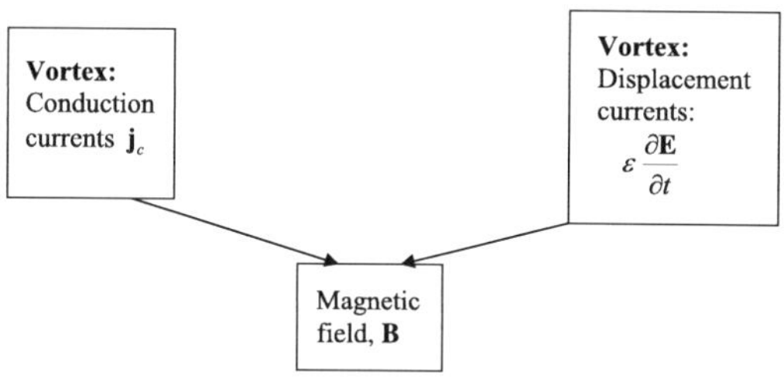

Thus, the circulation of the magnetic field along any contour L is determined by the total current passing through any surface S bound by this contour (Figure 2.2(a)). Now suppose that the path of integration L is an elementary contour as is shown in Figure 2.2(b). Then taking into account

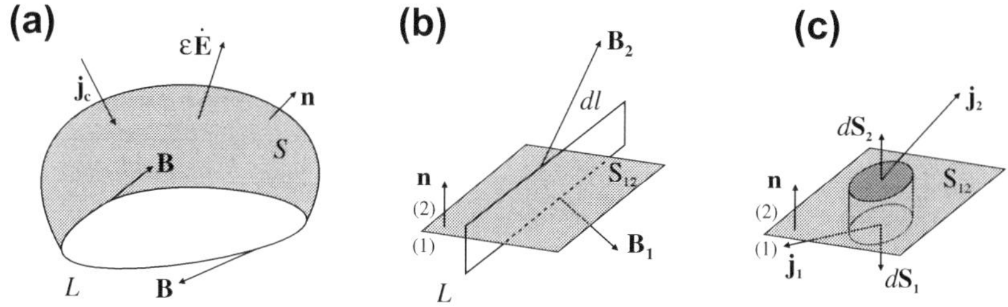

Figure 2.2 (a) Flux of conduction and displacement currents. (b) Field **B** near interface. (c) Continuity of normal component of the total current near interface.

the fact that in the limit when the area surrounded by the path L tends to zero, the flux of both the conduction and displacement currents vanishes, we obtain

$$\mathbf{n} \times (\mathbf{B}_2 - \mathbf{B}_1) = 0 \tag{2.67}$$

Therefore, the tangential component of the magnetic field, as in the case of the time-invariant field, is a continuous function at the interface, if the surface density of conduction currents $\mathbf{i}_c$ is absent. However, sometimes it is convenient to assume that $\mathbf{i}_c \neq 0$; then in place of Eq. (2.67), we have

$$\mathbf{n} \times (\mathbf{B}_2 - \mathbf{B}_1) = \mu_0 \mathbf{i}_c \tag{2.68}$$

2.4.2. The Total Current and the Principle of Charge Conservation

Having introduced the displacement currents, let us represent the principle of charge conservation in a different form. Since

$$\operatorname{div} \mathbf{j}_c = -\dot{\delta}_0 \text{ and } \quad \operatorname{div} \mathbf{D} = \delta_0$$

we have

$$\operatorname{div} \mathbf{j}_c + \operatorname{div} \dot{\mathbf{D}} = 0 \text{ or } \quad \operatorname{div} \mathbf{j} = 0 \tag{2.69}$$

where

$$\mathbf{j} = \mathbf{j}_c + \varepsilon \frac{\partial \mathbf{E}}{\partial t} \tag{2.70}$$

is the density of the total current. In accordance with Eq. (2.69), the current lines of the field $\mathbf{j}$ are always closed, and therefore, the $\mathbf{j}$ is the vortex field. Applying Gauss's theorem, we obtain the integral form of Eq. (2.69)

$$\oint_S \mathbf{j} \cdot d\mathbf{S} = 0, \tag{2.71}$$

that is, the flux of the vector of the total current density through any closed surface is always equal to zero. Considering again an elementary cylinder (Figure 2.2(c)) and calculating the flux of $\mathbf{j}$ through this closed surface, we have

$$j_{1n} + \frac{\partial D_{1n}}{\partial t} = j_{2n} + \frac{\partial D_{2n}}{\partial t} \tag{2.72}$$

Thus, the normal component of the vector **j** is a continuous function at an interface. Let us write down equations for the total current, describing the principle of charge conservation:

$$\oint_S \mathbf{j}\cdot d\mathbf{S} = 0, \quad \operatorname{div}\mathbf{j} = 0, \quad j_{1n} = j_{2n} \tag{2.73}$$

Now it is appropriate to make several comments:

1. Equation (2.65) can be rewritten as

$$\operatorname{curl}\mathbf{B} = \mu_0\mathbf{j}, \tag{2.74}$$

where **j** is the vector of the total current density. The similarity of Eqs (2.58) and (2.74) is obvious. However, it does not mean that in general case, the time-varying magnetic field obeys the Biot–Savart law. At the same time, if the influence of displacement currents is negligible, the magnetic field $\mathbf{B}(t)$ behaves practically in accordance with this law, and we deal with the quasi-stationary field.

2. The displacement currents depend on both the dielectric permittivity and electric field. In particular, in a nonpolarizable medium ($\varepsilon = \varepsilon_0$), it is defined only by the rate of change of the electric field with time.
3. The relationship between the conduction and displacement currents depends on different factors. For instance, in an insulator, only displacement currents can exist, while in a conducting medium, conduction currents usually prevail. Of course, with an increase in the rate of change of the field with time, the relative contribution of displacement currents becomes stronger.
4. Unlike the electromagnetic induction which was discovered by experimental studies, the introduction of displacement currents was a very bold assumption made by Maxwell and only later confirmed by experiments.
5. The quantity

$$\varepsilon\frac{\partial \mathbf{E}}{\partial t}$$

is called the density of displacement currents, though it is not related to a motion of free charges as in the case of conduction currents. In spite of this fundamental difference between the conduction and displacement currents, the latter is also called the current in order to emphasize that both of them generate the magnetic field.

6. The principle of charge conservation has two forms:

$$\operatorname{div} \mathbf{j}_c + \frac{\partial \delta_0}{\partial t} = 0, \quad \text{and} \quad \operatorname{div} \mathbf{j} = 0$$

In the case of the quasi-stationary field, we neglect by the rate of a change of the charge with time, and it is equivalent to assumption that an influence of displacement currents is negligible.

7. Among the different and numerous phenomena based on the existence of displacement currents, we will note only two:

a. Propagation of electromagnetic waves with a finite velocity.

b. Presence of the alternating current in a circuit with a capacitor.

In fact, Eq. (2.58) was derived from the Biot–Savart law which implies that the magnetic field **B** instantly appears at any point, regardless of its distance from conduction currents. In other words, the velocity of propagation of the field is infinitely high. However, this conclusion contradicts all experimental observations which show that the field propagates with finite velocity. For instance, in a nonpolarizable and nonmagnetic medium, the velocity coincides with that of light:

$$c = 3 \times 10^8 \text{m/s}$$

Later we will demonstrate that the propagation of the electromagnetic field is impossible without displacement currents. Now consider the first example illustrating the effect of displacement currents.

2.4.3. Currents in the Circuit with a Capacitor

Suppose that the circuit consists of a conducting part (wire) and an insulator bound by two conducting plates, parallel to each other (Figure 2.3). The last part of the circuit is called a *capacitor*. At the beginning, assume that there is only a conducting current I_c in the wire, while displacement currents do not exist anywhere. As is well known, there is a magnetic field around this circuit. Then applying the equation

$$\oint_L \mathbf{B} \cdot d\mathbf{l} = \mu_0 I_c$$

to the closed contour L (Figure 2.3), we discover the paradox. Indeed, if the surface S_1 intersects the conducting part of the circuit, the circulation of the magnetic field remains the same, and it is equal to $\mu_0 I_c$. However, if the surface S_1 passes through the capacitor, this circulation becomes equal to

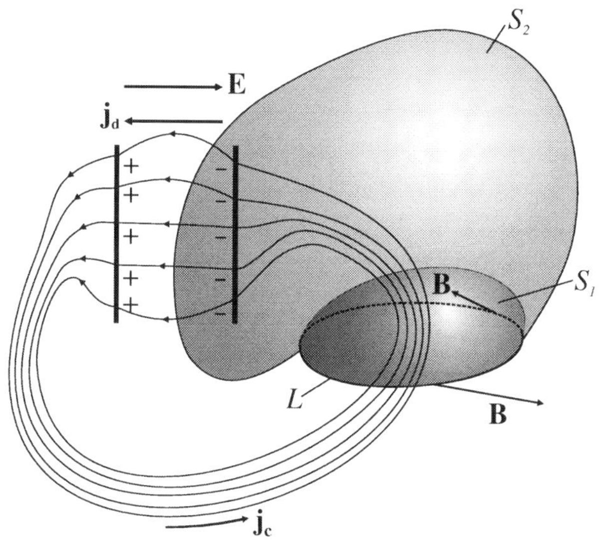

Figure 2.3 Distribution of displacement and conduction currents in a circuit.

zero, since the conduction currents inside the capacitor is absent. This ambiguity indicates that our assumption was incorrect and that in reality, there is a displacement current inside the capacitor. Moreover, this current has to be equal to the conduction current: $I_d = I_c$. Then, applying Eq. (2.66), we see that the circulation of the magnetic field is independent of the place where the surface S_l intersects the circuit.

Later we consider many examples illustrating the role of displacement currents, but now let us study the behavior of the current in the circuit with a capacitor (Figure 2.3). First assume that at some instance, two charges with equal magnitude and opposite sign are placed on the capacitor plates. To facilitate this analysis, we make several assumptions:

1. The inductive electric field caused by a change of the magnetic field with time can be neglected. Therefore, the electric field $\mathbf{E}(t)$ is caused by charges only and it obeys Coulomb's law. In particular, the field $\mathbf{E}(t)$ inside the capacitor is mainly caused by surface charges located on the conducting plates.
2. The distance between capacitor plates is very small compared to their dimensions.
3. At any given instant of time, the current density has the same value at all points of the circuit.

It is clear that charges located on plates create an electric field everywhere including the conducting part of the circuit. As a result, charges appear on the lateral surface of the wire. Due to the electric field of these charges, a conduction current with the density **j** arises and, correspondingly, a decrease of the plate charges is observed. Note that inside the capacitor there can be bound charges. It is obvious that the electric field **E**(t) in the capacitor is directed from the positive to negative charges, as is shown in Figure 2.3. Taking into account the fact that the field **E**(t) decreases, the displacement current has a direction which is opposite that of the electric field. Thus, the conduction current in the wire and displacement current in the capacitor have the same direction. Certainly, displacement currents appear inside the conducting part of the circuit, but they are usually relatively small ($j_c \gg j_d$). In addition, displacement currents exist around the circuit, but we assume that their influence is negligible. In this approximation, Eq. (2.72) can be rewritten as

$$j_c = \varepsilon \frac{\partial E}{\partial t} = j, \tag{2.75}$$

and in our approximation, charges on the plates are located only at points where wire is connected with the capacitor. Thus, we demonstrated that the displacement current in the capacitor represents a continuation of the conduction current in the wire, and in accordance with the principle of charge conservation, the vector lines of the current density **j** are closed. Now, we consider the behavior of both currents in some detail. Suppose that at some instant t, charges with density $\sigma_0(t)$ and $-\sigma_0(t)$ are located on the capacitor plates. Then, as follows from Eq. (2.16), the normal component of density of conduction current is related to the free charge on the plate as

$$j_c = -\frac{\partial \sigma_0}{\partial t} \tag{2.76}$$

Respectively, the current in the wire is equal to

$$I = -\frac{\partial Q_0(t)}{\partial t}, \tag{2.77}$$

where $Q_0(t)$ is the amount of free charge at each plate. By definition, the voltage of the electric field caused by charges is

$$\int_{+}^{-} \mathbf{E} \cdot d\mathbf{l} = U_+ - U_- = IR, \tag{2.78}$$

where U_+ and U_- are the potentials of the plates with positive and negative charges, and R is the resistance of the wire. It is appropriate to emphasize that the use of the potential $U(t)$ is justified since the vortex part of the electric field is neglected. Next, we will find an expression for the same potential difference in terms of capacitor parameters. As is well known, the electric field between plates is directly proportional to density σ_0 and

$$E(t) = \frac{\sigma_0(t)}{\varepsilon} \tag{2.79}$$

At the same time, the free charge on the plate and difference of potentials are related as

$$Q_0(t) = C[U_+(t) - U_-(t)], \tag{2.80}$$

where C is a coefficient of proportionality called the *capacitance*, and it shows the amount of the charge on the plate when the difference of potentials equals unity. In particular in the SI units, the capacitance is measured in farads.

$$1\ \text{F} = 1\ \text{Coulomb}/1\text{V} = 10^9\text{pF}.$$

Assuming that the influence of plate edges is very small and the medium between plates is uniform, it is easy to determine the capacitance C. In fact, from Eq. (2.79), we have

$$U_+ - U_- = \frac{\sigma_0}{\varepsilon} d = \frac{Q_0 d}{\varepsilon S} = \frac{Q_0}{C},$$

where S is the plate area and d is the distance between plates. Thus, the capacitance in this case is

$$C = \frac{\varepsilon S}{d} \tag{2.81}$$

At this point, we are ready to obtain the differential equation which describes the behavior of the charge Q_0 and both currents. From Eqs (2.77), (2.78), and 2.80, we have

$$\frac{dQ_0(t)}{dt} + \frac{Q_0}{CR} = 0 \tag{2.82}$$

Therefore, the charge decays exponentially with time

$$Q_0(t) = Q_0^0 \exp\left(-\frac{t}{CR}\right) \tag{2.83}$$

and correspondingly for the conduction and displacement currents, we have

$$I_c(t) = I_d(t) = \frac{Q_0^0}{CR} \exp\left(-\frac{t}{CR}\right) \tag{2.84}$$

From the physical point of view, it is clear that with an increase of resistance and interaction of charges at opposite plates (capacitance), both currents decay slower. As was pointed out earlier, this is only the simplest example showing the role of displacement currents.

2.5. MAXWELL EQUATIONS OF THE ELECTROMAGNETIC FIELD

2.5.1. Introduction

In the previous sections, we have introduced two new generators of the electromagnetic field, namely, the rate of a change of the electric and magnetic fields with time:

$$\varepsilon\frac{\partial \mathbf{E}}{\partial t}, \quad \text{and} \quad \frac{\partial \mathbf{B}}{\partial t}$$

Together with charges and conduction currents, they form the complete set of generators of the electromagnetic field, shown in Table 2.1.

Now, let us point out several facts concerning the relationship between the electromagnetic field and its generators:

1. In general, the electric field is caused by both sources (charges) and vortices $\partial\mathbf{B}/\partial t$. However, the magnetic field does not have sources; it is generated in a nonmagnetic medium by two types of vortices—the conduction and displacement currents.
2. As is seen from Table 2.1, generators of the magnetic field are defined by the electric field, while one of generators of the electric field is caused by a change of the magnetic field **B** with time. This means that the electric and magnetic fields usually depend on each other and it is in general impossible to separately determine either of them.
3. In studying the time-invariant electric and magnetic fields in the presence of a medium, we shown that usually there is a portion of the generators (charges and currents) which cannot be determined if the field is unknown. This happens because these generators depend on the field,

Table 2.1 Generators of the Electromagnetic Field

Generators	Electric Field	Magnetic Field
Sources	δ	–
Vortexes	$\frac{\partial \mathbf{B}}{\partial t}$	$\mathbf{j}_c, \varepsilon\frac{\partial \mathbf{E}}{\partial t}$

and thus in such cases we could not use, for example, either Coulomb's, or Biot–Savart laws for the field calculations. For this reason, we were compelled to derive a system of equations and formulate boundary value problems.

4. In the case of the time-varying electromagnetic fields, we also have to proceed from the system of field equations. It is related with several factors and among them are the following:
   a. The influence of a medium as in the case of a stationary field.
   b. Interaction between the electric and magnetic fields even in a free space, where an influence of a medium is absent.
   c. In the case of the time-invariant electric field Coulomb's law:

$$\mathbf{E}(p) = \frac{1}{4\pi\varepsilon_0} \int\limits_V \frac{\delta(q)\mathbf{L}_{\mathrm{qp}}}{L_{\mathrm{qp}}^3} \mathrm{d}V$$

allows one to determine the field $\mathbf{E}(p)$ at any point if the charge distribution is known. The same is valid for the Biot–Savart law, and the magnetic field can be calculated as soon as the conduction current $\mathbf{j}_c$ is specified. At the same time, Faraday's law

$$\oint\limits_L \mathbf{E} \cdot \mathrm{d}\mathbf{l} = -\int\limits_S \dot{\mathbf{B}} \cdot \mathrm{d}\mathbf{S}$$

establishes only the linkage between the flux of the vector $\dot{\mathbf{B}}$ and the circulation of the electric field along some line L, where $\mathbf{E}(t)$ usually changes from point to point. This means that even if the magnetic field is known, the field $\mathbf{E}(t)$ cannot be determined without additional information. Of course, this conclusion also applies to the relationship between the magnetic field and the conduction and displacement currents. Therefore, in order to determine the electromagnetic field and study its behavior, we have to proceed from a system of field equations.

As is well known from the vector analysis, the system of field equation for any vector field $\mathbf{M}(p)$ at regular points consists of two equations

$$\operatorname{curl} \mathbf{M}(p) = \mathbf{W}(p) \quad \operatorname{div} \mathbf{M}(p) = \omega(p),$$

where functions $\mathbf{W}(p)$ and $\omega(p)$ describe the distribution of vortices and sources, respectively. Inasmuch as the electric and magnetic fields are related to each other and we have to consider these two vector fields simultaneously, the system of equations of the electromagnetic field includes four equations.

2.5.2. Maxwell's Equations

Before writing down this system, it is appropriate to note again that we restrict ourselves to the study of fields in a piecewise uniform and nonmagnetic medium, since in most cases, this model properly describes a distribution of the conductivity and dielectric permittivity of a real medium, and it is widely used in the theory of electromagnetic methods. Then, taking into Eqns (2.8) and (2.65) and assuming that equations,

$$\text{div}\,\boldsymbol{\varepsilon}\mathbf{E} = \delta_0, \quad \text{div}\,\mathbf{B} = 0$$

derived for the stationary field, remain valid for time-varying fields, we obtain the system of equations of the electromagnetic fields at regular points

$$\begin{array}{llll} \text{I} & \text{curl}\,\mathbf{E} = -\frac{\partial \mathbf{B}}{\partial t} & \text{III} & \text{div}\,\boldsymbol{\varepsilon}\mathbf{E} = \delta_0 \\ \text{II} & \text{curl}\,\mathbf{B} = \mu_0\left(\mathbf{j}_c + \boldsymbol{\varepsilon}\frac{\partial \mathbf{E}}{\partial t}\right) & \text{IV} & \text{div}\,\mathbf{B} = 0, \end{array} \tag{2.85}$$

while its surface analogy is

$$\begin{array}{llll} \text{I} & \mathbf{n} \times (\mathbf{E}_2 - \mathbf{E}_1) = 0 & \text{III} & \mathbf{n}\cdot(\varepsilon_2\mathbf{E}_2 - \varepsilon_1\mathbf{E}_1) = \sigma_0 \\ \text{II} & \mathbf{n} \times (\mathbf{B}_2 - \mathbf{B}_1) = \mu_0\mathbf{i}_c & \text{IV} & \mathbf{n}\cdot(\mathbf{B}_2 - \mathbf{B}_1) = 0, \end{array} \tag{2.86}$$

where $\mathbf{E}_1$, $\mathbf{B}_1$ and $\mathbf{E}_2$, $\mathbf{B}_2$ are the electric and magnetic fields at the back and front sides of the interface, respectively. Equations (2.85) and (2.86) are called Maxwell's equations in differential form. It is clear that the first Maxwell's equation describes Faraday's law, while the second equation is the result of generalization of the Biot–Savart law, which takes into account both the conduction and displacement currents. The third equation was derived from Coulomb's law and it is based on the assumption that it is valid for the time-varying fields. Finally, the last equation of the system follows from Biot–Savart law and also implies that it is always correct; that is, the magnetic field does not have sources (magnetic charges). In this light, it is appropriate to note that Maxwell derived the system of field equations by proceeding from the experimental laws and his assumption that the magnetic field is also caused by displacement currents.

Each equation of this system describes some features of the field behavior. However, only all four equations together characterize such fundamental phenomenon as the propagation of electromagnetic fields which cannot follow from any of them separately.

Often, it is useful to represent Maxwell's equations in integral form, which is valid everywhere including regular points and interfaces. Applying Stoke's and Gauss theorems, we have from Eq. (2.85)

$$\oint_L \mathbf{E}\cdot d\mathbf{l} = -\oint_S \frac{\partial \mathbf{B}}{\partial t}\cdot d\mathbf{S}, \qquad \oint_S \varepsilon\mathbf{E}\cdot d\mathbf{S} = e_0$$
$$\oint_L \mathbf{B}\cdot d\mathbf{l} = \mu_0 \int_S \left(\mathbf{j}_c + \varepsilon\frac{\partial \mathbf{E}}{\partial t}\right)\cdot d\mathbf{S} \qquad \oint_S \mathbf{B}\cdot d\mathbf{S} = 0 \tag{2.87}$$

As follows from Eq. (2.86), tangential components of the electric field are continuous functions. At the same time, there are cases when it is convenient to imagine the presence of the double layer. Then, the tangential components of the electric field $\mathbf{E}(t)$ can have a discontinuity at the layer surface. The tangential component of the magnetic field also has a discontinuity only in those cases when the real distribution of currents near interfaces is replaced by that of surface current. Finally, due to the absence of magnetic charges, the normal component of the field $\mathbf{B}(t)$ is always a continuous function, while the discontinuity of the normal component of the electric field is defined by the surface density of charges.

2.5.3. The Second Form of Maxwell Equations

Now we demonstrate that the equations of the set (2.85), which characterize the divergence of the fields $\mathbf{E}$ and $\mathbf{B}$, can be derived from the first two equations of this system and the principle of charge conservation. In fact, taking the divergence of both sides of the equations,

$$\text{curl}\,\mathbf{E} = -\frac{\partial \mathbf{B}}{\partial t}, \quad \text{curl}\,\mathbf{B} = \mu_0\left(\mathbf{j}_c + \varepsilon\frac{\partial \mathbf{E}}{\partial t}\right) \tag{2.88}$$

we obtain

$$\frac{\partial}{\partial t}\text{div}\,\mathbf{B} = 0 \quad \text{and} \quad \frac{\partial}{\partial t}(-\delta_0 + \varepsilon\text{div}\,\mathbf{E}) = 0,$$

since

$$\text{div}\,\mathbf{j}_c = -\frac{\partial \delta_0}{\partial t}$$

Therefore,

$$\text{div}\,\mathbf{B} = C_1 \quad \text{and} \quad \text{div}\,\mathbf{D} = \delta_0 + C_2,$$

where C_1 and C_2 are independent of time. It is natural to assume that at some time in the past, the fields **E** and **B**, as well as charges, were absent and therefore constants should be equal to zero. Thus, we again obtain the second pair of Maxwell's equations at regular points:

$$\text{div}\,\boldsymbol{\varepsilon}\mathbf{E} = \delta_0 \quad \text{div}\,\mathbf{B} = 0 \tag{2.89}$$

Next let us show that surface analogy of Eq. (2.89) also follows from the first two equations of the set (2.87). For simplicity, assume that the surface density of currents is equal to zero. Then, applying these equations to any elementary closed contour (Figure 2.4(a)), intersecting an interface, it is simple matter to see that tangential components of the electric and magnetic fields are continuous functions. Next imagine two elementary closed paths on both sides of the interface (Figure 2.4(b)) and make use of the equation

$$\oint_L \mathbf{E}\cdot d\mathbf{l} = -\int_S \dot{\mathbf{B}}\cdot d\mathbf{S} = -\int_S \dot{B}_n d\mathbf{S}$$

Since the tangential component of the electric field is continuous across the interface, the left-hand side of this equation has equal values for both elementary paths. Therefore,

$$\dot{B}_{1n}dS - \dot{B}_{2n}dS = 0$$

from which it follows the continuity of the normal components of the field **B**, that coincides with the last relationship of the set (2.86). Finally, applying the equation

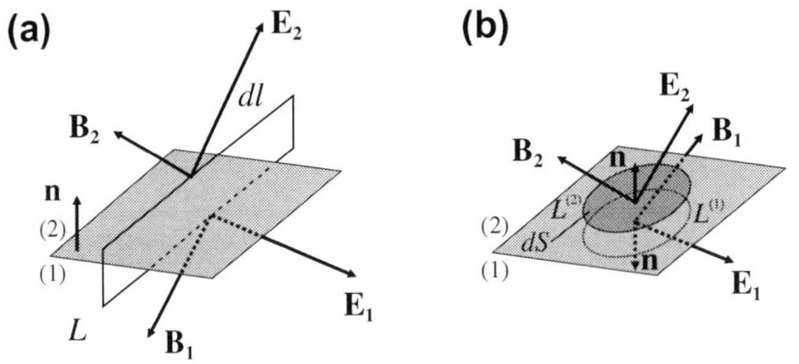

Figure 2.4 (a) Illustration of the proof of the continuity of tangential components of the electric and magnetic fields across the interface. (b) Illustration of continuity of the normal components of the magnetic field and current density vector across the interface.

$$\oint_L \mathbf{B}\cdot d\mathbf{l} = \mu_0 \int_S \left(\mathbf{j}_c + \varepsilon\dot{\mathbf{E}}\right)\cdot d\mathbf{S}$$

to the same closed paths, we find that the normal component of the total current density is continuous:

$$j_{1n} + \varepsilon_1 \dot{E}_{1n} = j_{2n} + \varepsilon_2 \dot{E}_{2n},$$

where

$$j_{1n} = \gamma_1 E_{1n}, \quad j_{2n} = \gamma_2 E_{2n}$$

Integrating both sides of this equality over time and taking into account the fact that

$$j_{2n} - j_{1n} = -\dot{\sigma}$$

we again obtain the third equation of the set (2.86). In other words, Eq. (2.89) and their integral form are valid for any time-varying electromagnetic field. This analysis allows us to represent the system of Maxwell's equations in a different form:

$$\text{curl}\,\mathbf{E} = -\frac{\partial \mathbf{B}}{\partial t}, \quad \text{curl}\,\mathbf{B} = \mu_0\left(\mathbf{j}_c + \varepsilon\frac{\partial \mathbf{E}}{\partial t}\right)$$

and

$$\mathbf{n}\times(\mathbf{E}_2 - \mathbf{E}_1) = 0 \quad \mathbf{n}\times(\mathbf{B}_2 - \mathbf{B}_1) = 0, \quad \text{if } \mathbf{i}_c = 0 \tag{2.90}$$

Also it is possible to write down the integral form of these equations:

$$\oint_L \mathbf{E}\cdot d\mathbf{l} = -\int_S \frac{\partial \mathbf{B}}{\partial t}\cdot d\mathbf{S}, \quad \oint_L \mathbf{B}\cdot d\mathbf{l} = \mu_0 \int_S \left(\mathbf{j}_c + \varepsilon\frac{\partial \mathbf{E}}{\partial t}\right)\cdot d\mathbf{S} \tag{2.91}$$

where $\mathbf{j}_c = \gamma\mathbf{E}$.

Let us emphasize again that in deriving Maxwell's equations, we proceeded from the following physical laws:

1. Coulomb's law
2. Biot–Savart law
3. Faraday's law
4. Principle of charge conservation
5. Ohm's law
6. Maxwell's concept of displacement currents.

2.5.4. Maxwell's Equations in a Piece-Wise Uniform Medium

Earlier it was mentioned that the theory of electromagnetic methods in geophysics is mainly based on the assumption that we deal with a piece-wise uniform medium. Then, as was shown earlier, the volume density of charges is equal to zero and in place of Eq. (2.85), we have at regular points

$$\begin{array}{llll} \text{I} & \operatorname{curl} \mathbf{E} = -\frac{\partial \mathbf{B}}{\partial t} & \text{III} & \operatorname{div} \mathbf{E} = 0 \\ \text{II} & \operatorname{curl} \mathbf{B} = \mu_0\left(\mathbf{j}_c + \varepsilon \frac{\partial \mathbf{E}}{\partial t}\right) & \text{IV} & \operatorname{div} \mathbf{B} = 0 \end{array} \tag{2.92}$$

Here, let us make several comments.

1. As follows from the third equation of this set, the volume charges are absent, but they can be present at interfaces between media with different electric properties.
2. By definition, Eq. (2.92) represents the system of eight scalar partial differential equations of the first order with six unknown components of the electric and magnetic fields. Certainly, in general, it is a very complicated system and it is usually difficult to discover the most important features of the field using directly this set of equations. For this reason, it is natural to replace Eq. (2.92) by much simpler equations and this is the subject of the next section.

2.6. EQUATIONS FOR THE FIELDS E AND B

Now, we replace Maxwell's equations by two equations which contain either the field **E** or the field **B**. Taking the curl of both sides of the first equation of the set (2.92), we have

$$\operatorname{curl}\operatorname{curl} \mathbf{E} = -\operatorname{curl} \dot{\mathbf{B}}$$

or

$$\operatorname{grad}\ \operatorname{div} \mathbf{E} - \nabla^2 \mathbf{E} = -\frac{\partial}{\partial t} \operatorname{curl} \mathbf{B}$$

Making use of the second and third equations of the same set,

$$\operatorname{curl} \mathbf{B} = \mu_0\left(\mathbf{j}_c + \varepsilon \dot{\mathbf{E}}\right) \quad \text{and} \quad \operatorname{div} \mathbf{E} = 0,$$

we obtain

$$-\nabla^2 \mathbf{E} = -\frac{\partial}{\partial t}\left(\gamma \mu_0 \mathbf{E} + \varepsilon \mu_0 \frac{\partial \mathbf{E}}{\partial t}\right)$$

or

$$\nabla^2 \mathbf{E} - \gamma\mu_0 \frac{\partial \mathbf{E}}{\partial t} - \varepsilon\mu_0 \frac{\partial^2 \mathbf{E}}{\partial t^2} = 0$$

By analogy, taking the curl of the second of Maxwell's equations and using the first and fourth equations of the system (2.92), we have

$$\text{curl curl } \mathbf{B} = \text{curl}\left(\gamma\mu_0 \mathbf{E} + \varepsilon\mu_0 \frac{\partial \mathbf{E}}{\partial t}\right)$$

or

$$\text{grad div } \mathbf{B} - \nabla^2 \mathbf{B} = -\gamma\mu_0 \frac{\partial \mathbf{B}}{\partial t} - \varepsilon\mu_0 \frac{\partial^2 \mathbf{B}}{\partial t^2}$$

and

$$\nabla^2 \mathbf{B} - \gamma\mu_0 \frac{\partial \mathbf{B}}{\partial t} - \varepsilon\mu_0 \frac{\partial^2 \mathbf{B}}{\partial t^2} = 0$$

Thus, instead of the system of differential equations of the first order with respect to two fields, we have derived one differential equation in partial derivatives of the second order for each field, **E** and **B**, which is valid at regular points of the conducting and polarizable medium:

$$\nabla^2 \mathbf{E} - \gamma\mu_0 \frac{\partial \mathbf{E}}{\partial t} - \varepsilon\mu_0 \frac{\partial^2 \mathbf{E}}{\partial t^2} = 0$$

and

$$\nabla^2 \mathbf{B} - \gamma\mu_0 \frac{\partial \mathbf{B}}{\partial t} - \varepsilon\mu_0 \frac{\partial^2 \mathbf{B}}{\partial t^2} = 0 \tag{2.93}$$

Then, the electromagnetic fields can be described at regular points and at interfaces with the help of two groups of equations. For the electric field, we have

$$\nabla^2 \mathbf{E} - \gamma\mu_0 \frac{\partial \mathbf{E}}{\partial t} - \varepsilon\mu_0 \frac{\partial^2 \mathbf{E}}{\partial t^2} = 0$$

and

$$\mathrm{n} \times (\mathbf{E}_2 - \mathbf{E}_1) = 0, \quad \gamma_1 \mathbf{E}_{1\mathrm{n}} + \varepsilon_1 \frac{\partial \mathbf{E}_{1\mathrm{n}}}{\partial t} = \gamma_2 \mathbf{E}_{2\mathrm{n}} + \varepsilon_2 \frac{\partial \mathbf{E}_{2\mathrm{n}}}{\partial t}, \tag{2.94}$$

while for the magnetic field,

$$\nabla^2 \mathbf{B} - \gamma\mu_0 \frac{\partial \mathbf{B}}{\partial t} - \varepsilon\mu_0 \frac{\partial^2 \mathbf{B}}{\partial t^2} = 0$$

and

$$\mathbf{n} \times (\mathbf{B}_2 - \mathbf{B}_1) = 0, \quad \mathbf{n} \cdot (\mathbf{B}_2 - \mathbf{B}_1) = 0 \tag{2.95}$$

Here, let us notice the following:

1. The electric and magnetic fields, defined from these equations, are related with each other since they obey the set (2.92).
2. Differential equations for the fields **E** and **B** have one remarkable feature, namely, unlike Maxwell equations, they allow almost instantly to discover two fundamental features of all electromagnetic fields. In fact, first suppose that the last term in Eq. (2.93) is much greater than the second one; that is,

$$\varepsilon\mu_0 \frac{\partial^2 \mathbf{E}}{\partial t^2} \gg \gamma\mu_0 \frac{\partial \mathbf{E}}{\partial t} \quad \text{or} \quad \varepsilon\mu_0 \frac{\partial \mathbf{E}}{\partial t} \gg \gamma\mu_0 \mathbf{E}$$

This means that displacement currents greatly exceeds the conduction currents and we have

$$\nabla^2 \mathbf{E} = \varepsilon\mu_0 \frac{\partial^2 \mathbf{E}}{\partial t^2}, \quad \text{and} \quad \nabla^2 \mathbf{B} = \varepsilon\mu_0 \frac{\partial^2 \mathbf{B}}{\partial t^2} \tag{2.96}$$

As is well known, these are wave equations which show that the electromagnetic field propagates as a wave through a medium with the velocity

$$c = (\varepsilon\mu_0)^{-1/2}$$

This is the first fundamental property of all electromagnetic fields. Next consider the opposite case when the conduction currents prevails and then in place of Eq. (2.93), we obtain

$$\nabla^2 \mathbf{E} = \gamma\mu_0 \frac{\partial \mathbf{E}}{\partial t}, \quad \nabla^2 \mathbf{B} = \gamma\mu_0 \frac{\partial \mathbf{B}}{\partial t} \tag{2.97}$$

The latter are the diffusion equations and they describe completely different process, namely, diffusion. Thus, in accordance with Eq. (2.93), the electromagnetic fields always display two fundamental features: propagation and diffusion. For instance, in a very resistive medium, the influence of diffusion may be insignificant and mainly propagation is observed. In contrast, in a relatively conductive medium, the diffusion usually prevails, but propagation is always present. Later we will discuss this subject in detail. Of course, we can arrive at the same conclusion proceeding from the Maxwell's equations.

2.7. ELECTROMAGNETIC POTENTIALS

Another approach which often facilitates the solution of Maxwell's equations is related to the introduction of scalar and vector potentials. It turns out that in many cases, it is possible to describe the fields **E** and **B** with two or even one component of the vector potential, and this fact allows us to simplify the field determination to a great extent. To introduce potentials, we make use of two of the Maxwell's equations,

$$\text{div}\,\mathbf{E} = 0, \quad \text{and} \quad \text{div}\,\mathbf{B} = 0 \tag{2.98}$$

Inasmuch as the divergence of both fields is zero at regular points, they can be represented as

$$\mathbf{E} = \text{curl}\,\mathbf{A}_{\text{m}}, \quad \text{and} \quad \mathbf{B} = \text{curl}\,\mathbf{A}_{\text{e}}, \tag{2.99}$$

where $\mathbf{A}_{\text{m}}$ and $\mathbf{A}_{\text{e}}$ are called the vector potentials of the magnetic and electric types, respectively. It is clear that an infinite number of vector potentials describe the same electromagnetic field. For instance, adding functions grad φ_{m} and grad φ_{e} to the vector potentials $\mathbf{A}_{\text{m}}$ and $\mathbf{A}_{\text{e}}$, new vector potentials

$$\mathbf{A}_{\text{m}2} = \mathbf{A}_{\text{m}1} + \text{grad}\,\varphi_{\text{m}}, \quad \text{and} \quad \mathbf{A}_{\text{e}2} = \mathbf{A}_{1\text{e}} + \text{grad}\,\varphi_{\text{e}}$$

also describe the same field since

$$\text{curl}(\text{grad}\,\varphi_{\text{m}}) = \text{curl}(\text{grad}\,\varphi_{\text{e}}) = 0$$

In other words, Eq. (2.99) defines the vector potentials up to the gradient of some functions φ_{m} and φ_{e} which are called scalar potentials of the electromagnetic field. This ambiguity in the determination of $\mathbf{A}_{\text{m}}$ and $\mathbf{A}_{\text{e}}$ will be used to our advantage in simplifying the equations for the vector potentials. Let us start with the function $\mathbf{A}_{\text{m}}$. Substituting

$$\mathbf{E} = \text{curl}\,\mathbf{A}_{\text{m}}$$

into the second equation of Eq. (2.92), we have

$$\text{curl}\,\mathbf{B} = \gamma\mu_0\text{curl}\,\mathbf{A}_{\text{m}} + \varepsilon\mu_0\text{curl}\,\dot{\mathbf{A}}_{\text{m}}$$

or

$$\text{curl}\left(\mathbf{B} - \gamma\mu_0\mathbf{A}_{\text{m}} - \varepsilon\mu_0\dot{\mathbf{A}}_{\text{m}}\right) = 0$$

whence,

$$\mathbf{B} - \gamma\mu_0\mathbf{A}_{\text{m}} - \varepsilon\mu_0\dot{\mathbf{A}}_{\text{m}} = \text{grad}\,\varphi_{\text{m}} \tag{2.100}$$

Here, φ_{m} is the scalar potential of the magnetic field and, as in the case of the vector potential, there is infinite number of these functions which describe the same electromagnetic field. Substituting expressions for the fields **E** and **B** in terms of potentials into the first Maxwell's equation, we obtain

$$\mathrm{curl}\ \mathrm{curl}\ \mathbf{A}_{\mathrm{m}} = -\gamma\mu_0\dot{\mathbf{A}}_{\mathrm{m}} - \varepsilon\mu_0\ddot{\mathbf{A}}_{\mathrm{m}} - \mathrm{grad}\ \dot{\varphi}_{\mathrm{m}}$$

or

$$\mathrm{grad}\ \mathrm{div}\ \mathbf{A}_{\mathrm{m}} - \nabla^2\mathbf{A}_{\mathrm{m}} = -\gamma\mu_0\dot{\mathbf{A}}_{\mathrm{m}} - \varepsilon\mu_0\ddot{\mathbf{A}}_{\mathrm{m}} - \mathrm{grad}\ \dot{\varphi}_{\mathrm{m}} \qquad (2.101)$$

where

$$\dot{\mathbf{A}}_{\mathrm{m}} = \frac{\partial\mathbf{A}_{\mathrm{m}}}{\partial t}, \quad \ddot{\mathbf{A}}_{\mathrm{m}} = \frac{\partial^2\mathbf{A}_{\mathrm{m}}}{\partial t^2} \text{ and } \quad \dot{\varphi}_{\mathrm{m}} = \frac{\partial\varphi_{\mathrm{m}}}{\partial t}$$

Taking into account the fact that an infinite number of pairs $\mathbf{A}_{\mathrm{m}}$ and φ_{m} describe the same field, we choose a pair of them such that Eq. (2.101) becomes much simpler. Letting

$$\mathrm{div}\ \mathbf{A}_{\mathrm{m}} = -\frac{\partial\varphi_{\mathrm{m}}}{\partial t} \qquad (2.102)$$

we obtain for the vector potential $\mathbf{A}_{\mathrm{m}}$ exactly the same equation as for the electromagnetic field:

$$\nabla^2\mathbf{A}_{\mathrm{m}} - \gamma\mu_0\frac{\partial\mathbf{A}_{\mathrm{m}}}{\partial t} - \varepsilon\mu_0\frac{\partial^2\mathbf{A}_{\mathrm{m}}}{\partial t^2} = 0 \qquad (2.103)$$

Now making use of the gauge condition, Eq. (2.102), both fields **E** and **B** can be expressed in terms of the vector potential $\mathbf{A}_{\mathrm{m}}$ only. In fact from Eqs (2.99) and (2.100), we have

$$\mathbf{E} = \mathrm{curl}\ \mathbf{A}_{\mathrm{m}}$$

and

$$\dot{\mathbf{B}} = \gamma\mu_0\dot{\mathbf{A}}_{\mathrm{m}} + \varepsilon\mu_0\ddot{\mathbf{A}}_{\mathrm{m}} + \mathrm{grad}\ \mathrm{div}\ \mathbf{A}_{\mathrm{m}} \qquad (2.104)$$

It is necessary to note that taking the divergence of both sides of Eq. (2.103) and integrating over time, we find that the scalar potential φ_{m} also satisfies the same equation as $\mathbf{A}_{\mathrm{m}}$:

$$\nabla^2\varphi_{\mathrm{m}} - \gamma\mu_0\frac{\partial\varphi_{\mathrm{m}}}{\partial t} - \varepsilon\mu_0\frac{\partial^2\varphi_{\mathrm{m}}}{\partial t^2} = 0 \qquad (2.105)$$

Next we derive an equation for the vector potential of the electric type. Substituting the equation

$$\mathbf{B} = \text{curl}\,\mathbf{A}_\text{e}$$

into the first of Maxwell's equation, we obtain

$$\text{curl}\,\mathbf{E} = -\text{curl}\,\frac{\partial \mathbf{A}_\text{e}}{\partial t}$$

or

$$\mathbf{E} = -\frac{\partial \mathbf{A}_\text{e}}{\partial t} + \text{grad}\,\varphi_\text{e}, \tag{2.106}$$

where φ_e is the scalar potential of the electric field. This equation suggests that in general the electric field is caused by a change of the magnetic field with time and electric charges. In other words, there are two parts of this field: the vortex and galvanic one. Replacing the fields **E** and **B** in the second equation of the set (2.92), we have

$$\text{curl curl}\,\mathbf{A}_\text{e} = \mu_0\left(-\gamma\frac{\partial \mathbf{A}_\text{e}}{\partial t} + \gamma\text{grad}\,\varphi_\text{e} - \varepsilon\frac{\partial^2 \mathbf{A}_\text{e}}{\partial t^2} + \varepsilon\text{grad}\,\frac{\partial \varphi_\text{e}}{\partial t}\right)$$

or

$$\text{grad div}\,\mathbf{A}_\text{e} - \nabla^2\mathbf{A}_\text{e} = -\gamma\mu_0\frac{\partial \mathbf{A}_\text{e}}{\partial t} - \varepsilon\mu_0\frac{\partial^2 \mathbf{A}_\text{e}}{\partial t^2} + \text{grad}\left(\gamma\mu_0\varphi_\text{e} + \varepsilon\mu_0\frac{\partial \varphi_\text{e}}{\partial t}\right) \tag{2.107}$$

Assuming that a pair of the vector and scalar potentials obeys the condition,

$$\text{div}\,\mathbf{A}_\text{e} = \gamma\mu_0\varphi_\text{e} + \varepsilon\mu_0\frac{\partial \varphi_\text{e}}{\partial t} \tag{2.108}$$

we obtain for the vector potential $\mathbf{A}_\text{e}$ the same equation as that for the function $\mathbf{A}_\text{m}$.

$$\nabla^2\mathbf{A}_\text{e} - \gamma\mu_0\frac{\partial \mathbf{A}_\text{e}}{\partial t} - \varepsilon\mu_0\frac{\partial^2 \mathbf{A}_\text{e}}{\partial t^2} = 0 \tag{2.109}$$

In this case, the electromagnetic field cannot, in general, be expressed in terms of the vector potential only and we have

$$\mathbf{B} = \text{curl}\,\mathbf{A}_\text{e}, \quad \mathbf{E} = -\frac{\partial \mathbf{A}_\text{e}}{\partial t} + \text{grad}\,\varphi_\text{e} \tag{2.110}$$

At the same time in the absence of electric charges, the latter is greatly simplified and the last equality becomes

$$\mathrm{E} = -\frac{\partial \mathbf{A}_{\mathrm{e}}}{\partial t}$$

It is worth to make one comment here. One should not be confused by the fact that equations for $\mathbf{A}_{\mathrm{m}}$ and $\mathbf{A}_{\mathrm{e}}$ (Eqs (2.103) and (2.109)) are exactly the same. The appropriate boundary value problems for these equations involve different boundary (and initial) conditions leading to different solutions.

2.8. MAXWELL'S EQUATIONS FOR SINUSOIDAL FIELDS

Until now, we have not made any assumptions about the dependence of the electromagnetic field on time. Let us examine the special and very important case, in which the field varies as a sinusoidal function of time. This leads to some important simplifications. First consider the scalar function

$$M = M_0 \sin(\omega t + \varphi), \tag{2.111}$$

where M_0 is the amplitude of the oscillation, φ is the initial phase, and ω is the radial frequency ($\omega = 2\pi f = 2\pi/T$) with T being the period of oscillations. Making use of Euler's formula,

$$e^{-i(\omega t+\varphi)} = \cos(\omega t + \varphi) - i \sin(\omega t + \varphi)$$

we can represent the right-hand side of Eq. (2.111) as the imaginary part of the exponential function

$$M_0 \sin(\omega t + \varphi) = -\mathrm{Im} M^* e^{-i\omega t} \tag{2.112}$$

Here, M^* is the complex amplitude given by

$$M^* = M_0 e^{-i\varphi} \tag{2.113}$$

Therefore, we have

$$M^* e^{-i\omega t} = M_0 e^{-i\varphi} e^{-i\omega t} = M_0 e^{-i(\omega t+\varphi)}$$

and

$$\begin{aligned} -\mathrm{Im} M^* e^{-i\omega t} &= -\mathrm{Im}[M_0 \cos(\omega t + \varphi) - i M_0 \sin(\omega t + \varphi)] \\ &= M_0 \sin(\omega t + \varphi) \end{aligned}$$

Similarly, a cosine function can be represented by the real part of the complex function:

$$M_0 \cos(\omega t + \varphi) = \mathrm{Re} M^* e^{-i\omega t}$$

whereas before $M^* = M_0 e^{-i\varphi}$. It is essential to point out that the complex amplitude M^* is defined by the amplitude of oscillation M_0 and the initial phase φ:

$$|M^*| = M_0, \text{ and } \mathrm{Arg} M^* = \varphi$$

In other words, the complex amplitude contains all information about the corresponding sinusoidal function. Suppose that functions $M_0 \sin(\omega t + \varphi)$ and $M_0 \cos(\omega t + \varphi)$, describing any component of the electromagnetic field, are the solutions of Maxwell's equations. Then, taking into account the fact that these equations are linear, the sum of functions

$$M^* e^{-i\omega t} = M_0 \cos(\omega t + \varphi) - iM_0 \sin(\omega t + \varphi)$$

is also a solution of this system. Therefore, at the beginning, we can represent any component of the electric and magnetic field as

$$M^* e^{-i\omega t},$$

but then, after solving equations for these fields, either the imaginary or real parts of the solution should be only considered. This form of a solution, $M^* exp(-i\omega t)$, has one remarkable feature, namely, it is a product of two functions. One of them is the complex amplitude M^* which depends on the geometric and physical parameters of the medium, the position of an observation point and a frequency but is independent on time. The second function $\exp(-i\omega t)$ is related to time in extremely simple manner so that after differentiation it still remains an exponent. This single fact permits us to write equations of the field in a form which does not contain the argument t that drastically facilitates the solution. It is also appropriate to note that sinusoidal functions have infinite duration in time, so that there is no need to study the field behavior at the initial moment when the electromagnetic fields arise. Now we generalize this result for the vector function. Suppose that

$$\mathbf{M} = M_{01} \sin(\omega t + \varphi_1)\mathbf{1}_x + M_{02} \sin(\omega t + \varphi_2)\mathbf{1}_y + M_{03} \sin(\omega t + \varphi_3)\mathbf{1}_z,$$

where $\mathbf{1}_x$, $\mathbf{1}_y$, and $\mathbf{1}_z$ are the unit vectors along the coordinate axes. The latter can be rewritten as

$$\mathbf{M} = -\mathrm{Im}\left\{ \left[M_{01} e^{-i\varphi_1} \mathbf{1}_x + M_{02} e^{-i\varphi_2} \mathbf{1}_y + M_{03} e^{-i\varphi_3} \mathbf{1}_z \right] e^{-i\omega t} \right\}$$

or

$$\mathbf{M} = -\mathrm{Im} M^* e^{-i\omega t} \tag{2.114}$$

where

$$\mathbf{M}^* = M_{01} e^{-i\varphi_1} \mathbf{1}_x + M_{02} e^{-i\varphi_2} \mathbf{1}_y + M_{03} e^{-i\varphi_3} \mathbf{1}_z \tag{2.115}$$

is the complex amplitude of the sinusoidal vector function **M** which is described by the complex vector. Then representing the field in the form

$$\mathbf{E} = -\mathrm{Im}\left(\mathbf{E}^* e^{-i\omega t}\right), \quad \mathbf{B} = -\mathrm{Im}\left(\mathbf{B}^* e^{-i\omega t}\right)$$

and substituting these expressions into the first two Maxwell's equations, we obtain

$$\mathrm{curl}\, \mathbf{E}^* = i\omega \mathbf{B}^* \quad \mathrm{curl}\, \mathbf{B}^* = \mu_0(\gamma - i\omega\varepsilon)\mathbf{E}^*, \tag{2.116}$$

since

$$\frac{\partial}{\partial t} e^{-i\omega t} = -i\omega e^{-i\omega t}$$

It is obvious that the conditions at the interfaces for the complex amplitudes are the same as those for the field and

$$\mathbf{n} \times \left(\mathbf{E}_2^* - \mathbf{E}_1^*\right) = 0, \quad \mathbf{n} \times \left(\mathbf{B}_2^* - \mathbf{B}_1^*\right) = 0 \tag{2.117}$$

Correspondingly, instead of Eq. (2.93), we have

$$\nabla^2 \mathbf{E}^* + k^2 \mathbf{E}^* = 0, \quad \nabla^2 \mathbf{B}^* + k^2 \mathbf{B}^* = 0, \tag{2.118}$$

where

$$k^2 = i\gamma\mu_0\omega + \omega^2 \varepsilon\mu_0, \tag{2.119}$$

and k is usually called the *wave number*.

It is obvious that complex amplitudes of potentials of the electromagnetic field also satisfy the same equations:

$$\nabla^2 \mathbf{A}_{\mathrm{m}}^* + k^2 \mathbf{A}_{\mathrm{m}}^* = 0, \quad \nabla^2 \mathbf{A}_{\mathrm{e}}^* + k^2 \mathbf{A}_{\mathrm{e}}^* = 0 \tag{2.120}$$

It may be proper to note that in place of the field itself, these equations allow us to determine its complex amplitudes and in order to find the electric and magnetic fields, we have to multiply these functions by $\exp(-i\omega t)$ and take either the imaginary or real part of this product. This consideration shows at least the following two important merits of the use of sinusoidal oscillations:

1. The system of Maxwell's equations with respect to complex amplitudes of the field, as well as Eqs (2.118) and (2.120), does not contain functions of time.

2. A medium with parameters which are independent on time does not change the frequency of oscillations. In other words, if some generator creates a sinusoidal oscillation, then in the presence of a medium, where the conduction and displacement currents appear, the electromagnetic field still remains sinusoidal with the same frequency as the primary field. This is a very remarkable and important feature of the sinusoidal field when the medium can be treated as a linear system. However, in general, when we deal with a nonsinusoidal behavior, the primary and total fields are very different from each other as functions of time. It is proper to note that making use of Fourier transform, the primary field (input) with almost arbitrary dependence on time can be represented as a combination of sinusoidal oscillations and, correspondingly, the field in a medium (output) is also described in terms of sinusoids having different amplitudes and phases. Finally, the study of sinusoidal fields is of a great practical interest since in applications of the surface and borehole geophysics, currents and charges caused by generators of the primary field often vary as sinusoids.

2.9. ELECTROMAGNETIC ENERGY AND POYNTING VECTOR

2.9.1. Principle of Energy Conservation

Until now, we focus on equations describing the electric and magnetic field as a functions of coordinates of a point and time. Also it is useful to characterize a behavior of these fields in terms of its energy. Suppose that a distribution of energy of electromagnetic field is characterized by its density $u(p,t)$. Then the amount of energy inside of some volume V is

$$W = \int_V u(p,t)\mathrm{d}V \tag{2.121}$$

There are several reasons why this energy may change. First of all in the presence of generators of the electromagnetic field, they create every second the energy with density $P(p,t)$. Here p is any point inside the volume. The second reason is related to a motion of charges (current) through a medium. In a fact, in order to create the current, the electromagnetic field performs the work and correspondingly the electromagnetic energy decreases every second in unit volume by some amount $Q(p,t)$ which is transformed into heat. Considering equations for the electric and magnetic field, we emphasized that the electromagnetic field can display both propagation and diffusion; that it is associated with some kind of motion. Then it is natural to assume that energy also moves from one volume to another. In other words, let us suppose that there is a flux of the electromagnetic energy which also causes a change of energy. By definition, we can write for the flux as,

$$\oint_S \mathbf{Y}\cdot d\mathbf{S} \tag{2.122}$$

Here, S is the surface surrounding the volume V. The dot product $\mathbf{Y}\cdot d\mathbf{S}$ characterizes the flux of energy passing the surface dS during 1 s. By definition, $\mathbf{Y}$ shows the direction of energy movement and its magnitude is equal to amount of energy passing during 1 s a unit area oriented perpendicular to this vector. Thus, the principle of conservation of energy can be written in the form

$$\frac{\partial W}{\partial t} = \int_V P(p,t)dV - \int_V Q(p,t)dV - \oint_S \mathbf{Y}(p,t)\cdot d\mathbf{S} \tag{2.123}$$

or making use of Gauss formula,

$$\oint_S \mathbf{Y}\cdot d\mathbf{S} = \int_V \operatorname{div} \mathbf{Y} dV$$

we obtain

$$\frac{\partial W}{\partial t} = \int_V P(p,t)dV - \int_V Q(p,t)dV - \int_V \operatorname{div} \mathbf{Y} dV \tag{2.124}$$

Note that the last two terms are taken with minus since transformation into heat and the positive flux of energy cause a decrease of the energy in the volume V.

2.9.2. Joule's Law

Next, making use of Ohm's law and Maxwell's equations, we will express both sides of Eq. (2.124) in terms of the electromagnetic field. As was pointed out earlier, the electromagnetic field causes a motion of charges (current), and the force acting on the moving charge δ in unit volume is

$$\mathbf{F} = \delta(\mathbf{E}_t + \mathbf{v} \times \mathbf{B}) \tag{2.125}$$

and the work performed by this force per second in unit volume is

$$A = \delta(\mathbf{E}_t + \mathbf{v} \times \mathbf{B}) \cdot \mathbf{v} = \delta \mathbf{E}_t \cdot \mathbf{v} \tag{2.126}$$

since

$$(\mathbf{v} \times \mathbf{B}) \cdot \mathbf{v} = 0$$

Here, $\mathbf{v}$ is the charge velocity, and $\mathbf{E}_t$ is the total electric field; that is, a sum of fields caused by charges, a change of the magnetic field with time, and extraneous forces. By definition, δ is the charge density. Therefore, the product $\delta\mathbf{v}$ is equal to the current density $\mathbf{j}$ and we have

$$A = \mathbf{E}_t \cdot \mathbf{j}$$

or

$$A = \mathbf{j} \cdot \mathbf{E} + \mathbf{j} \cdot \mathbf{E}^{ext} = \mathbf{j} \cdot \mathbf{E} + P \tag{2.127}$$

Here, $\mathbf{E}^{ext}$ is the extraneous field caused by generators inside the elementary volume and

$$\mathbf{j} \cdot \mathbf{E} = Q, \tag{2.128}$$

that is, the work performed by the electromagnetic field in a conducting medium is transformed into heat (Joule's law).

2.9.3. Expressions for the Energy Density and Poynting Vector

Taking into account Eqs (2.121) and (2.128), we have from Eq. (2.124)

$$\mathbf{E} \cdot \mathbf{j} = P - \frac{\partial u}{\partial t} - \nabla \cdot \mathbf{Y} \tag{2.129}$$

The latter describes a distribution of energy in the unit volume of a conducting medium in the presence of extraneous source and, as we will see, it allows us to find formulas for the energy density u and vector $\mathbf{Y}$ in terms of the electric and magnetic fields. With this purpose in mind, we express the left-hand side in terms of the fields $\mathbf{E}$ and $\mathbf{B}$ only and at the beginning, assume that the extraneous source is absent. From the second Maxwell's equation, we have

$$\mathbf{E} \cdot \mathbf{j} = \frac{1}{\mu_0} \mathbf{E} \cdot (\nabla \times \mathbf{B}) - \varepsilon \mathbf{E} \cdot \frac{\partial \mathbf{E}}{\partial t} \tag{2.130}$$

As follows from vector analysis,

$$\nabla\cdot(\mathbf{E}\times\mathbf{B})=\mathbf{B}\cdot(\nabla\times\mathbf{E})-\mathbf{E}\cdot(\nabla\times\mathbf{B})$$

Its substitution into Eq. (2.130) gives

$$\mathbf{E}\cdot\mathbf{j}=-\frac{1}{\mu_0}\nabla\cdot(\mathbf{E}\times\mathbf{B})+\frac{1}{\mu_0}\mathbf{B}\cdot(\nabla\times\mathbf{E})-\varepsilon\mathbf{E}\cdot\frac{\partial\mathbf{E}}{\partial t}$$

Applying the first Maxwell's equation, we obtain

$$\mathbf{E}\cdot\mathbf{j}=-\frac{1}{\mu_0}\nabla\cdot(\mathbf{E}\times\mathbf{B})-\frac{1}{\mu_0}\mathbf{B}\cdot\frac{\partial\mathbf{B}}{\partial t}-\varepsilon\mathbf{E}\cdot\frac{\partial\mathbf{E}}{\partial t}$$

or

$$\mathbf{E}\cdot\mathbf{j}=-\frac{1}{\mu_0}\nabla\cdot(\mathbf{E}\times\mathbf{B})-\frac{\partial}{\partial t}\frac{1}{2}\left(\frac{1}{\mu_0}\mathbf{B}^2+\varepsilon\mathbf{E}^2\right)\tag{2.131}$$

Introducing notations for the energy density and density of flux,

$$u=\frac{1}{2}\left(\varepsilon E^2+\frac{1}{\mu_0}B^2\right)\tag{2.132}$$

and

$$\mathbf{Y}=\frac{1}{\mu_0}(\mathbf{E}\times\mathbf{B})\tag{2.133}$$

we arrive at the principle of conservation of energy in the unit volume where the extraneous force is absent:

$$\frac{\partial u}{\partial t}=-Q-\operatorname{div}\mathbf{Y}\tag{2.134}$$

or in more general case ($P\neq 0$),

$$\frac{\partial u}{\partial t}=P-Q-\operatorname{div}\mathbf{Y}\tag{2.135}$$

Performing integration over an arbitrary volume and using again Gauss's formula, we obtain Eq. (2.123), which shows that the flux of energy through closed surface is equal to

$$\oint\mathbf{Y}\cdot d\mathbf{S}.\tag{2.136}$$

where **Y** is called the Poynting vector.

2.9.4. The Direct Current and Poynting Vector

In accordance with Eq. (2.123) in the case when the energy does not change with time, we have

$$\int_V P\mathrm{d}V = \int_V Q\mathrm{d}V + \oint_S \mathbf{Y}\cdot \mathrm{d}\mathbf{S} \tag{2.137}$$

that is, one part of an energy of the generators inside the volume is transformed into heat while the other part forms the flux of energy intersecting the surface S surrounding this volume. If the extraneous force is absent, we have

$$\oint_S \mathbf{Y}\cdot \mathrm{d}\mathbf{S} = -Q, \tag{2.138}$$

that is, an electromagnetic energy arrives at the volume (flux is negative) where it is transformed into heat. Inasmuch as a movement of energy cannot take place without that of the field, we see that even the time-invariant field is based on propagation. To illustrate a movement of the electromagnetic energy and Poynting vector, consider two examples.

2.9.5. Example 1: Current Circuit

As is shown in Figure 2.5(a), inside the internal part of the circuit, the electric field of the coulomb's origin and current have opposite directions while in the external part, both vectors have the same direction. By definition of the cross-product, the Poynting vector is directed outside and inside within the internal and external parts of the circuit, respectively. This means that electromagnetic energy travels away from the internal part into surrounding

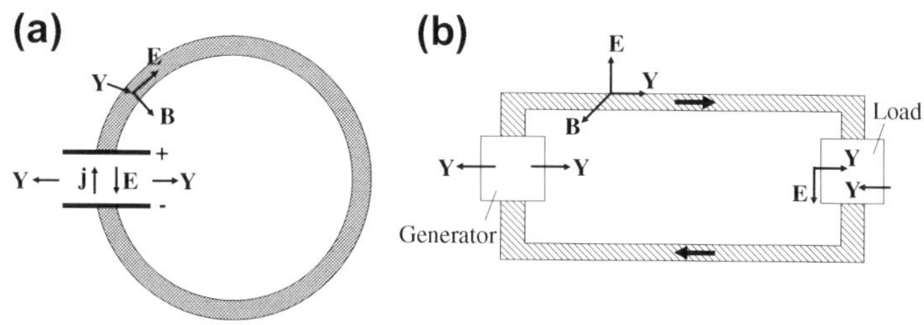

Figure 2.5 (a) The flux energy around a current circuit. (b) Poynting vector in the vicinity of transmission line.

medium and then returns back into circuit in the external part. It may be proper to note the following.

1. This description is rather approximate since the surface charges arise at the lateral surface of the contour and they create the normal component of the electric field. As a result, at points of this surface, the Poynting vector has both the tangential and normal components.

2.9.6. Example 2: Transmission Line

Suppose we have the system, Figure 2.5(b), which consists of the following three parts:

1. The internal part where extraneous forces produce the work that results in appearance of the electromagnetic energy.
2. The long and conductive transmission line.
3. The relatively resistive load.

As was demonstrated earlier, the electromagnetic energy leaves the internal part of the circuit and travels through the surrounding medium. Now we consider the behavior of the field and Poynting vector in the vicinity of the transmission line and the load. Inasmuch as the line has very low resistivity, the tangential component of the electric field **E** is very small inside the line. In fact from Ohm's law, we have

$$E_t = \rho j$$

Due to continuity of the tangential component, it is also small on the external side of the conductor. At the same time, surface charges create outside the line a normal component of the field E_n which is much greater than the tangential component E_t:

$$E_n \gg E_t$$

Then, as is seen from Figure 2.5(b), the Poynting vector is practically tangential to the transmission line. This means that the electromagnetic energy travels along the transmission line; that is, the line plays the role of a guide defining the direction of movement of the energy toward the load. Otherwise, it would travel in all directions from the source of the extraneous force. Of course, due to the presence of the tangential component of the electric field, a small amount of the electromagnetic energy moves into the transmission line and transforms into a heat. This is a pure loss which reduces the amount of energy arriving to the load. Unlike the transmission line, the load is relatively resistive, and correspondingly, the tangential component of the electric field prevails over the normal component: $E_t \gg E_n$. Therefore, the Poynting vector is mainly directed inward and the electromagnetic energy transforms there into heat or other types of energy.

2.10. THEOREM OF UNIQUENESS OF A SOLUTION OF THE FORWARD PROBLEM

2.10.1. The Proof of the Theorem of Uniqueness

The theory of electromagnetic methods of the surface and borehole geophysics is based on a solution of the forward problems that allow one to establish a relationship between the field and the electric and geometric parameters of a medium. From physical and mathematical points of view, it is clear that Maxwell's equations, as well as partial differential equations of the second order for the electric and magnetic fields or vector potentials, have an infinite number of solutions. In other words, they alone do not permit one to determine the field and it is necessary to have additional information which is able to find the field uniquely. Unlike the time-invariant field, these conditions consist of two groups: the initial and boundary ones. It is convenient to consider separately two cases and begin from the simplest one.

Case 1

Suppose that starting from the instant $t = t_0$, the field is considered in some volume V where all points are regular and this volume is surrounded by the surface S. Therefore, at each point, the fields obey the equations

$$\nabla^2\mathbf{E} - \gamma\mu_0\frac{\partial \mathbf{E}}{\partial t} - \varepsilon\mu_0\frac{\partial^2\mathbf{E}}{\partial t^2} = 0, \quad \nabla^2\mathbf{B} - \gamma\mu_0\frac{\partial \mathbf{B}}{\partial t} - \varepsilon\mu_0\frac{\partial^2\mathbf{E}}{\partial t^2} = 0,$$

Also, we assume that at the initial instant ($t = t_0$), the electric and magnetic field is known at every point of the volume:

$$\mathbf{E}(p, t_0) = \mathbf{N}(p), \quad \mathbf{B}(p, t_0) = \mathbf{M}(p) \tag{2.139}$$

In addition, we have to formulate boundary conditions at the surface S, as it was done for the time-invariant electric field. In that case, Gauss's formula served as "bridge" between values of the field inside of the volume and the surface S. In our case, the similar role is played by Eq. (2.123), and this is one of reasons why we studied the principle of conservation of energy. At the beginning, it is assumed that the extraneous generators of the field are absent, $P(p,t) = 0$. Consider two solutions of equations for both fields,

$$\mathbf{E}_1, \mathbf{B}_1, \quad \mathbf{E}_2, \mathbf{B}_2$$

which at the initial moment have the same values inside the volume. Taking into account the fact that these equations are linear, the differences of these solutions

$$\mathbf{E}_3 = \mathbf{E}_2 - \mathbf{E}_1, \quad \mathbf{B}_3 = \mathbf{B}_2 - \mathbf{B}_1 \tag{2.140}$$

also obey the same equations, while the initial condition has the form

$$\mathbf{E}_3(p, t_0) = 0 \quad \text{and} \quad \mathbf{B}_3(p, t_0) = 0 \tag{2.141}$$

To establish the boundary conditions, we will write down the principle of charge conservation of energy for this field as

$$\frac{\partial}{\partial t}\int_V \left(\frac{\varepsilon}{2}E_3^2 + \frac{B_3^2}{2\mu_0}\right)\mathrm{d}V = -\int_V \rho j_3^2 \mathrm{d}V - \frac{1}{\mu_0}\oint_S (\mathbf{E}_3 \times \mathbf{B}_3)\cdot \mathrm{d}\mathbf{S} \tag{2.142}$$

Suppose that either tangential components of $\mathbf{E}_1$ and $\mathbf{E}_2$ or tangential components of $\mathbf{B}_1$ and $\mathbf{B}_2$ coincide with each other at points of the surface S for $t \gg t_0$. For instance, in the case of electric field, this means that the tangential component of $\mathbf{E}_3$ is equal to zero:

$$\mathbf{n} \times \mathbf{E}_3 = 0, \tag{2.143}$$

where $\mathbf{n}$ is the unit vector normal to the surface S. Since

$$\oint_S (\mathbf{E}_3 \times \mathbf{B}_3)\cdot \mathrm{d}\mathbf{S} = \oint_S (\mathbf{E}_3 \times \mathbf{B}_3)\cdot \mathbf{n}\mathrm{d}S = \oint_S (\mathbf{n} \times \mathbf{E}_3)\cdot \mathbf{B}_3 \mathrm{d}S = 0,$$

the surface integral in Eq. (2.142) vanishes and we have

$$\frac{\partial}{\partial t}\int_V \left(\frac{\varepsilon}{2}E_3^2 + \frac{1}{2\mu_0}B_3^2\right)\mathrm{d}V = -\int_V \rho j_3^2 \mathrm{d}V \tag{2.144}$$

Here, $\mathbf{j}_3$ is the vector of current density caused by the field $\mathbf{E}_3$. The right-hand side can be either negative or equals to zero. At the same time, the left-hand side is either equal to zero or positive. This statement requires some explanation. In fact, in general, the derivative of the positive energy with respect to time can be either positive or negative. However, in our case, when at the initial moment the energy is zero (Eq. (2.141)), it is either zero or positive when $t \geq t_0$. Otherwise, the energy would be negative. Thus, the equality Eq. (2.144) takes place only when both right and left sides are equal to zero and this means that for all times ($t > t_0$),

$$\mathbf{E}_3 = \mathbf{E}_2 - \mathbf{E}_1 = 0, \quad \mathbf{B}_3 = \mathbf{B}_2 - \mathbf{B}_1 = 0$$

Therefore, the electromagnetic field within the volume V is uniquely defined for all instances $t > t_0$ by initial values of the electric and magnetic vectors inside the volume V and values of the tangential component either the electric or magnetic field at the surface S, surrounding this volume for instances $t \geq t_0$. Now it is proper to make several comments:

1. The proof of the theorem of uniqueness remains the same if the volume V is surrounded by several surfaces.
2. If the surface S tends to infinity, we can assume that a medium has a finite conductivity and due to transformation of energy into heat, the surface integral in Eq. (2.142) still tends to zero.
3. We have assumed that extraneous forces are absent inside the volume V. At the same time, the initial and boundary conditions remain sufficient to provide uniqueness of the solution if fields $\mathbf{E}_1$, $\mathbf{B}_1$ and $\mathbf{E}_2$, $\mathbf{B}_2$ have the same extraneous sources.

Case 2

Now we consider slightly more complicated case when inside the volume there is an interfaceS_{12} between media with different electrical properties. Inasmuch as the differential equations for the field are not applied at points of the interface, we surround S_{12} by the surface S^*. Correspondingly, at the right-hand side of Eq. (2.142), we have an additional integral:

$$-\frac{1}{\mu_0} \oint_{S^*} (\mathbf{E}_3 \times \mathbf{B}_3) \cdot d\mathbf{S}$$

In approaching S^* to S_{12}, we have to integrate at both sides of the interfaces that gives

$$\frac{1}{\mu_0} \int_{S_{12}} \left[\left(\mathbf{E}_3' \times \mathbf{B}_3'\right)\right] - \left(\mathbf{E}_3'' \times \mathbf{B}_3''\right) \cdot d\mathbf{S} \qquad (2.145)$$

where $\mathbf{E}_3', \mathbf{B}_3'$ and $\mathbf{E}_3'', \mathbf{B}_3''$ are the electric and magnetic field at the back and front sides of the interface, respectively. In accordance with Maxwell's equations, the tangential component of the electric field $\mathbf{E}$ and magnetic field $\mathbf{B}$ are continuous functions at the interface. Thus, integrand in Eq. (2.145) can be represented as

$$\left[\left(\mathbf{E}_3' - \mathbf{E}_3''\right) \times \mathbf{B}_3\right] \cdot d\mathbf{S} = \left[\left(\mathbf{E}_{3n}' - \mathbf{E}_{3n}''\right)(\mathbf{n} \times \mathbf{B}_3) \cdot \mathbf{n} dS\right.$$
$$= \left(\mathbf{E}_{3n}' - \mathbf{E}_{3n}''\right)(\mathbf{n} \times \mathbf{n})\mathbf{B}_3 dS = 0$$

Therefore, the integral over the interface vanishes if solutions satisfy Maxwell's equations at this surface.

2.10.2. Formulation of the Boundary Value Problem

Now, we are ready to formulate the boundary value problem for the electromagnetic field which uniquely defines the field. As follows from the theorem of uniqueness, these conditions are

1. At regular points of the volume V, the field should obey equations

$$\nabla^2 \mathbf{E} - \gamma \mu_0 \frac{\partial \mathbf{E}}{\partial t} - \varepsilon \mu_0 \frac{\partial^2 \mathbf{E}}{\partial t^2} = 0 \quad \nabla^2 \mathbf{B} - \gamma \mu_0 \frac{\partial \mathbf{B}}{\partial t} - \varepsilon \mu_0 \frac{\partial^2 \mathbf{E}}{\partial t^2} = 0$$

2. At the initial moment $t = t_0$, the field

$$\mathbf{E}(p, t_0), \text{ and } \quad \mathbf{B}(p, t_0)$$

is given at each regular point of the volume.

3. At the surface S surrounding the volume V, the tangential component of either the electric or magnetic field $\mathbf{n} \times \mathbf{E}$ or $\mathbf{n} \times \mathbf{B}$ is given at all instances $t \geq t_0$.
4. At interfaces inside the volume, a solution should obey the surface analogy of Maxwell's equations.

As we know now, there is only one electromagnetic field, which obeys all these conditions. Since the field can be expressed in terms of the vector potentials, the value boundary problem is formulated in terms of these functions, too.

REFERENCES AND FURTHER READING

[1] L.M. Alpin, Theory of the Field, Nedra, Moscow, 1966.
[2] V.R. Bursian, Theory of Electromagnetic Fields Applied in Electrical Methods, Nedra, Moscow, 1972.

CHAPTER THREE

Propagation and Quasi-Stationary Field in a Nonconducting Medium

INTRODUCTION

In several methods of the surface and borehole geophysics, measurements are often performed relatively close to the primary sources of the field where it has a sufficiently simple behavior. To understand the origin of its behavior, we first consider the simplest case of wave propagation and then introduce the concept of quasi-stationary field and illustrate it by several examples.

3.1. PLANE WAVE IN A UNIFORM MEDIUM

Suppose that a nonconducting medium with parameters $\boldsymbol{\varepsilon}$ and μ_0 is uniform and the electric and magnetic fields depend on the z-coordinate only; that is, on any plane perpendicular to the z-axis, the field does not change. Also, we assume that the electric field has a single component along the x-axis:

$$\mathbf{E} = E_{0x} e(z, t) \mathbf{1}_x, \tag{3.1}$$

where E_{0x} is a constant, $e_{0x}(z,t)$ is a function which depends on the coordinate z and time t, and $\mathbf{1}_x$ is the unit vector directed along the x-axis. Inasmuch as the field is independent of the x- and y-coordinates, the first equation of the set (Eq. (2.93)) is greatly simplified and we have

$$\frac{\partial^2 e}{\partial z^2} - \varepsilon\mu_0 \frac{\partial^2 e}{\partial t^2} = 0, \tag{3.2}$$

since the medium is nonconductive and

$$\frac{\partial^2 e}{\partial x^2} = \frac{\partial^2 e}{\partial y^2} = 0$$

Methods in Geochemistry and Geophysics, Volume 45
ISSN 0076-6895,
http://dx.doi.org/10.1016/B978-0-444-53829-1.00003-4

3.1.1. Solution of Eq. (3.2)

Equation (3.2) is the well-known partial differential equation of the second order. Applying the trial and error method, D'Alembert found its solution in the following form:

$$e(z,t) = Af\left[a\left(t - \sqrt{\varepsilon\mu_0}z\right)\right] + Bg\left[a\left(t + \sqrt{\varepsilon\mu_0}z\right)\right] \tag{3.3}$$

Here, A and B some constants; f and g are the functions having the first and second derivatives. The constant a has the dimension s^{-1}, since argument of any function is dimensionless. It is a simple matter to show that function $e(z,t)$ obeys Eq. (3.2). In fact, introducing the variable

$$u = a\left(t - \sqrt{\varepsilon\mu_0}z\right)$$

we have for derivatives of the function $f(z,t)$,

$$\frac{\partial f}{\partial z} = \frac{\partial f}{\partial u}\frac{\partial u}{\partial z} = -a\sqrt{\varepsilon\mu_0}f'_u$$

where f'_u is the first derivative with respect to the argument u. Therefore,

$$\frac{\partial^2 f}{\partial z^2} = a^2\varepsilon\mu_0 f''_{uu} \tag{3.4}$$

Also,

$$\frac{\partial f}{\partial t} = \frac{\partial f}{\partial u}\frac{\partial u}{\partial t} = af'_u, \quad \text{and} \quad \frac{\partial^2 f}{\partial t^2} = a^2 f''_{uu} \tag{3.5}$$

From the last two equations, we see that the function

$$f\left[a\left(t - \sqrt{\varepsilon\mu_0}z\right)\right]$$

satisfies Eq. (3.2). Of course, it is also true for the function

$$g\left[a\left(t + \sqrt{\varepsilon\mu_0}z\right)\right]$$

This procedure clearly indicates that any function can be a solution of Eq. (3.2), provided that its argument u is related to z and t by

$$u = a\left(t \pm \sqrt{\varepsilon\mu_0}z\right) \tag{3.6}$$

It is proper to notice that if the argument of a function has the form

$$b\left(z \pm \frac{t}{\sqrt{\varepsilon\mu_0}}\right),$$

Then, it is also a solution of Eq. (3.2). One can say that D'Alembert found such a relationship between arguments z and t of the functions that they obey Eq. (3.2), and this is essence of his solution. Next, let us assume that the electric field is described by the function $f[a(t - \sqrt{\varepsilon\mu_0})]$ only,

$$\mathbf{E} = E_{0x} f\left[a\left(t - \sqrt{\varepsilon\mu_0} z\right)\right] \mathbf{1}_x \tag{3.7}$$

and consider its physical meaning.

3.1.2. Velocity of Propagation of Plane Wave

First of all, a study of the argument u (Eq. (3.6)) allows us to discover the following features of the field behavior:

1. At any point with coordinate z, the field $\mathbf{E}$, in general, changes with time, while at some instant t, it can have different values at points with different values of coordinate z.
2. The electric field $\mathbf{E}$ has the same value at different points and different time if the argument

$$u = a\left(t - \sqrt{\varepsilon\mu_0} z\right)$$

remains the same. As follows from the definition of this argument, with increase of the distance z, the same value of the field is observed at greater times. Let us mentally imagine a system of parallel planes, $z = \text{const}$, which are perpendicular to the z-axis. Let us think of these planes as the surfaces where the field $\mathbf{E}$ has the same value as the time changes (Figure 3.1(a)).

Each of these planes corresponds to some instant of time and the relationship between z and t is

$$t_2 - t_1 = \sqrt{\varepsilon\mu_0}(z_2 - z_1), \quad t_3 - t_1 = \sqrt{\varepsilon\mu_0}(z_3 - z_1),$$

and

$$t_n - t_i = \sqrt{\varepsilon\mu_0}(z_n - z_i)$$

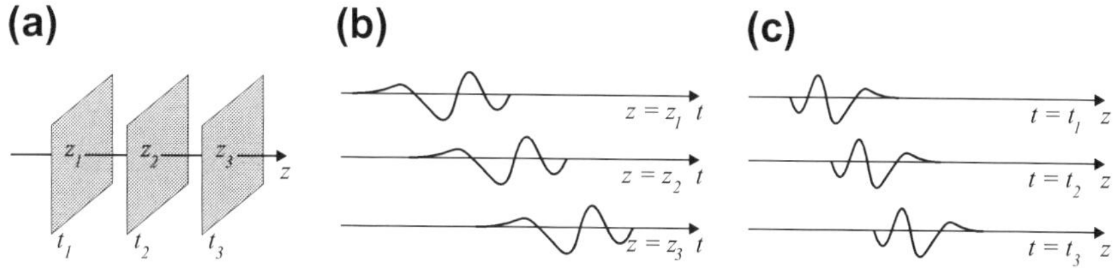

Figure 3.1 (a) A motion of wave. (b) Wave behavior at different distances. (c) A change of wave with time.

since

$$u_i = t_i - \sqrt{\varepsilon\mu_0}z_i = \text{constant}.$$

We can interpret this infinite series of parallel planes as a movement of only one plane, characterized by the same argument, with velocity

$$v = \frac{1}{\sqrt{\varepsilon\mu_0}} = \frac{c}{\varepsilon_r^{1/2}}, \tag{3.8}$$

where c is the speed of light

$$c = 3 \times 10^8 \text{ m/s}$$

In particular, in a free space $v = c$, but, for example, in water

$$v \approx 0.33 \times 10^8 \text{ m/s}$$

In other words, we observe a motion of the field **E** along the z-axis and this phenomenon is called *propagation of the field* or *wave motion* or, even simpler, *a wave*. This is the reason why Eq. (3.2) is called the *wave equation*. It is proper to remind that at all planes, the electric field has the same value which is independent on coordinates x and y, and it is natural that such motion of the field is called the *plane wave*. For illustration, the distribution of the wave as a function of t and z is given in Figure 3.1(b) and (c), which represents the wave distribution along the z-axis and the change of the field with time, respectively.

3.1.3. Magnetic Field of the Plane Wave

Now, we consider a magnetic field which accompanies the field **E**. With this purpose in mind, it is useful to represent Maxwell's equations (assuming conduction currents are absent) in Cartesian system of coordinates.

Then using the expression for curl, we obtain

$$\begin{vmatrix} \mathbf{1}_x & \mathbf{1}_y & \mathbf{1}_z \\ \dfrac{\partial}{\partial x} & \dfrac{\partial}{\partial y} & \dfrac{\partial}{\partial z} \\ E_x & 0 & 0 \end{vmatrix} = -\frac{\partial \mathbf{B}}{\partial t}, \quad \begin{vmatrix} \mathbf{1}_x & \mathbf{1}_y & \mathbf{1}_z \\ \dfrac{\partial}{\partial x} & \dfrac{\partial}{\partial y} & \dfrac{\partial}{\partial z} \\ B_x & B_y & B_z \end{vmatrix} = \varepsilon\mu_0\frac{\partial \mathbf{E}}{\partial t}$$

Equating corresponding components of the fields from both sides of these equations, we have

$$\frac{\partial B_x}{\partial t} = 0, \qquad \frac{\partial E_x}{\partial z} = -\frac{\partial B_y}{\partial t}, \qquad \frac{\partial B_z}{\partial t} = 0$$

and

$$\frac{\partial B_y}{\partial z} = -\varepsilon\mu_0 \frac{\partial E_x}{\partial t}, \qquad \frac{\partial B_x}{\partial z} = 0$$

In deriving these equations, we used the fact that the electric field has only the component E_x and both the electric and magnetic fields are independent of the x- and y-coordinates. Inasmuch as we are interested in a time-varying field, it is natural to assume that the magnetic field has only the component B_y, since $\frac{\partial B_x}{\partial t} = 0$ and $\frac{\partial B_z}{\partial t} = 0$. Therefore, we have

$$\frac{\partial E_x}{\partial z} = -\frac{\partial B_y}{\partial t}, \qquad \frac{\partial B_y}{\partial z} = -\varepsilon\mu_0 \frac{\partial E_x}{\partial t} \tag{3.9}$$

It is easy to see that the equation for the magnetic field derived in Chapter 2 directly follows from Eq. (3.9), which gives again

$$\frac{\partial^2 B_y}{\partial z^2} - \varepsilon\mu_0 \frac{\partial^2 B_y}{\partial t^2} = 0 \tag{3.10}$$

Substituting Eq. (3.7) into Eq. (3.9) and taking into account Eq. (3.8), we obtain

$$\frac{\partial B_y}{\partial t} = \frac{1}{v} E_{0x} a f_u' \left[a\left(t - \frac{z}{v}\right)\right] \quad \text{and} \quad \frac{\partial B_y}{\partial z} = -\varepsilon\mu_0 E_{0x} a f_u' \left[a\left(t - \frac{z}{v}\right)\right]$$

It is obvious that the function

$$B_y(z,t) = B_{0y} f\left[a\left(t - \frac{z}{v}\right)\right] \tag{3.11}$$

satisfies both equations of the set (Eq. (3.9)), provided that

$$B_{0y} = \sqrt{\varepsilon\mu_0} E_{0x} \quad \text{or} \quad E_{0x} = v B_{0y} \tag{3.12}$$

Thus, we have demonstrated that the electromagnetic field propagates along the z-axis with the velocity v and it is described by two vectors:

$$\mathbf{E}(z,t) = E_{0x} e(z,t) \mathbf{1}_x, \quad \text{and} \quad \mathbf{B}(z,t) = B_{0y} b(z,t) \mathbf{1}_y, \tag{3.13}$$

where the coefficients E_{0x} and B_{0y} are related to each other by Eq. (3.12) and

$$e = b = f\left[a\left(t - \frac{z}{v}\right)\right] \tag{3.14}$$

3.1.4. Electromagnetic Plane Wave

Equations derived above allow us to note the following: (1) Both fields are perpendicular to each other and to direction of their propagation. Such an electromagnetic field is called the *transversal plane wave*. By definition, the Poynting vector, that is, the vector of density of the energy flux is

$$\mathbf{Y} = \frac{1}{\mu_0}(\mathbf{E} \times \mathbf{B})$$

Taking into account Eqs (3.12)–(3.14), we have

$$\mathbf{Y} = \sqrt{\frac{\varepsilon}{\mu_0}} E_{ox}^2 e^2(z, t)\mathbf{1}_z \tag{3.15}$$

and it shows a direction of the wave motion. (2) As follows from Eq. (3.9), the electric and magnetic fields support each other at every point of space. In fact when the magnetic field changes with time, in accordance with Faraday's law,

$$\frac{\partial E_x}{\partial z} = -\frac{\partial B_y}{\partial t}$$

it creates the electric field while a change of the field **E** with time; that is, the displacement currents generates the magnetic field:

$$\frac{\partial B_y}{\partial t} = -\varepsilon\mu_0 \frac{\partial E_x}{\partial t}$$

Thus supporting each other the magnetic and electric fields form the electromagnetic wave, which propagates through a nonconducting medium with velocity v. It is proper to note that these generators of the field:

$$\frac{\partial \mathbf{B}}{\partial t}, \quad \text{and} \quad \varepsilon \frac{\partial \mathbf{E}}{\partial t}$$

are vital in forming the electromagnetic waves, and if one of them is neglected, the effect of propagation disappears. For instance, suppose that displacement currents do not have any influence on the field. From the mathematical point of view, this means that the dielectric permittivity ε is equal to zero. Therefore, in accordance with Eq. (3.8), the velocity of propagation becomes infinitely large, and respectively, the fields **E** and **B** would arrive instantly at all points of a medium. This certainly contradicts the concept of propagation. Also in Chapter 2, we described the behavior of displacement currents inside a capacitor, assuming that the change of the magnetic fields of these currents with time is negligible. In other words, we did not take into account the inductive electric field. Correspondingly, inductive effect vanishes and the field **E** between the capacitor plates behaves as a Coulomb's field caused by surface charges. In other words, propagation is absent.

3.1.5. Primary Source of the Plane Wave

In many cases for studying electromagnetic fields, it is convenient to use a model of plane wave; however, it is not realistic one since it requires an infinitely large energy. In order to imagine such source, suppose that two plates of infinite dimension are located at the plane XOY, and charges of equal magnitude but opposite sign are uniformly distributed on them. At some instant $t = 0$, one of plates starts to move, forming a current with the surface density $\mathbf{i}_x$ which is independent of the x- and y-coordinates (Figure 3.2(a)):

$$i_x = \begin{cases} 0 & t < 0 \\ I_0\psi(at) & t \geq 0 \end{cases} \tag{3.16}$$

Due to this current, a magnetic field arises in the vicinity of the plane $z = 0$ and, applying the integral form of the second Maxwell's equation to

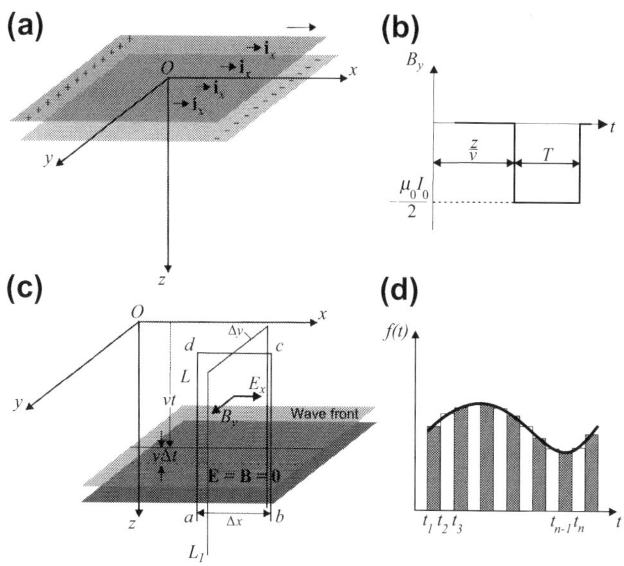

Figure 3.2 (a) Moving plate with surface current density i_x. (b) Propagation of impulse. (c) Illustration of relation between fields **E** and **B**. (d) Propagation when current of the source is arbitrary function of time.

the elementary path around this surface current and located in the plane perpendicular to the x-axis, we obtain

$$B_{2y} - B_{1y} = -\mu_0 i_x \quad \text{if} \quad z = 0$$

where B_{2y} and B_{1y} are the components of the magnetic field at the front and back sides of the plate, respectively. Since the magnetic field B_y is an odd function with respect to the coordinate z, we have

$$B_{2y} = -B_{1y} = -\frac{\mu_0 i_x}{2} \quad \text{if } z = 0 \tag{3.17}$$

Thus, in accordance with Eq. (3.16), the magnetic field at the front side of the current plate behaves as

$$B_y(0,t) = \begin{cases} 0 & t < 0 \\ -\dfrac{\mu_0 I_0}{2}\psi(at) & t \geq 0 \end{cases} \tag{3.18}$$

At the same time, as follows from Eq. (3.11), the dependence of the field on the coordinate z and time t has to be defined by the same function for all points. It is also essential that as follows from Eq. (3.18), this function is equal to zero if the argument is negative. Therefore, we can represent the magnetic field at every point, regardless of the distance from the current plate, in the form

$$B_y(z,t) = \begin{cases} 0 & z > vt \\ -\dfrac{\mu_0 I_0}{2}\psi\left[a\left(t - \dfrac{z}{v}\right)\right] & z \leq vt \end{cases} \tag{3.19}$$

Thus in the vicinity of the plate, the magnetic field varies almost synchronously with the current density of the source, but at some distance z, the same behavior of the field is observed with time delay z/v. In accordance with Eq. (3.12), the electric field is

$$E_x(z,t) = \begin{cases} 0 & z > vt \\ -\dfrac{I_0}{2}\left(\dfrac{\mu_0}{\varepsilon}\right)^{1/2}\psi\left[a\left(t - \dfrac{z}{v}\right)\right] & z \geq vt \end{cases} \tag{3.20}$$

Let us notice that the coefficient $(\mu_0/\varepsilon)^{1/2}$ can be represented as

$$\sqrt{\frac{\mu_0}{\varepsilon}} = \frac{120\pi}{(\varepsilon_r)^{1/2}} \text{ Ohm} \tag{3.21}$$

Now it is appropriate to make several comments.

1. In deriving Eq. (3.17), we took into account the fact that the flux of displacement currents through an infinitely small area surrounded by the elementary path is equal to zero, and this fact allowed us to establish relation between the magnetic field in the vicinity of the source and its current.
2. In essence, Eq. (3.17) is the boundary condition near the source and later it will be derived in a similar manner in more complicated cases.
3. The electric and magnetic field of the plane wave have the same dependence on the distance and time. Such behavior is an exception and it is not observed when the waves are caused by real sources.
4. Electromagnetic field of the plane wave is described by the same function, regardless of the distance from the current source. This is another exception, which does not take place in a general case.
5. The argument $a(t - z/v)$ is usually called the *phase of the wave*, while the plane with constant value of the phase is referred to as the *phase surface*. Thus, the propagation of the plane wave can be treated as a movement of the phase surface with the velocity $\boldsymbol{v}$, and the Poynting vector is perpendicular to this plane. To illustrate the effect of propagation, we consider several examples.

Example 1

Suppose that the current in the plate differs from zero only during some time interval T:

$$i_x = \begin{cases} I_0\psi(at) & 0 \le t \le T \\ 0 & t < 0, t > T \end{cases}$$

Then at an observation point located at the distance z, the field appears when

$$\frac{z}{v} \le t \le \frac{z}{v} + T$$

Hence if the distance z is such that

$$T < \frac{z}{v}$$

the field will be observed in the absence of the conduction current in the plate. This fact vividly demonstrates that the electromagnetic field propagates due to the interaction of the electric and magnetic fields and, in this sense, the wave is not related to the current source.

Example 2

Now consider in detail the propagation of a square waveform, shown in Figure 3.2(b). Then, in accordance with Eq. (3.19), we have

$$B_y(z,t) = \begin{cases} 0 & \text{if} \quad t < \dfrac{z}{v} \\ -\dfrac{\mu_0 I_0}{2} & \text{if} \quad \dfrac{z}{v} \le t \le \dfrac{z}{v} + T \\ 0 & \text{if} \quad t \ge \dfrac{z}{v} + T \end{cases} \tag{3.22}$$

and

$$E_x = vB_y$$

Taking into account the simplicity of the field behavior, let us study the relationship between the electric and magnetic fields proceeding directly from the integral form of Maxwell's equations. With this purpose in mind, we imagine a closed rectangular contour L, situated in any plane which is parallel to the plane XOZ (Figure 3.2(c)). Also we assume that the wave front is located somewhere inside the contour while the back side of the plane wave has not yet reached its side $cd = \Delta x$. Then at any moment t, the flux of the magnetic field $\mathbf{\Phi}$ intersecting the contour is equal to

$$\Phi(t) = z^* \Delta x B_y,$$

where $z^* \Delta x = vt\Delta x$ characterizes the area of the loop where the field B is not zero. Inasmuch as the electromagnetic field moves along the z-axis, this area as well as the flux increases. In particular, at the instant $t + \Delta t$, the flux is

$$\Phi(t + \Delta t) = \Delta x(z^* + v\Delta t)B_y$$

Therefore,

$$\frac{\mathrm{d}\Phi}{\mathrm{d}t} = B_y v \Delta x \tag{3.23}$$

and in accordance with Faraday's law,

$$\oint_L \mathbf{E} \cdot \mathbf{dl} = -\frac{\mathrm{d}\Phi}{\mathrm{d}t},$$

an electromotive force appears in the contour. Integrals along paths which are parallel to the z-axis vanish because the dot product of two perpendicular vectors

$$\mathbf{E} = E_x \mathbf{1}_x, \quad \mathrm{d}\mathbf{l} = \mathrm{d}l \mathbf{1}_z$$

equals zero. At the same time, integral along the path *ab* is also zero since the field has not yet arrived at this side of the contour. Respectively, the electromotive force is defined by the voltage along the path *cd* and is equal to $-E_x \Delta x$; that is,

$$\oint_L \mathbf{E} \cdot \mathrm{d}\mathbf{l} = -E_x \Delta x \tag{3.24}$$

Therefore, due to a movement of the magnetic field, an inductive electric field arises and as follows from Eqs (3.23) and (3.24),

$$E_x = vB_y$$

Next we make use of the second Maxwell's equation,

$$\oint \mathbf{B} \cdot \mathrm{d}\mathbf{l} = \varepsilon\mu_0 \int \dot{\mathbf{E}} \cdot \mathrm{d}\mathbf{S}$$

and apply it to the rectangular contour L_1 located in the plane YOZ. It is obvious that using the same approach as before, we can show that the rate of change of the flux of displacement currents through an area enclosed by the contour L_1 is

$$\varepsilon\mu_0 E_x v \Delta y$$

At the same time, the circulation of the magnetic field along L_1 is

$$B_y \Delta y$$

Thus, displacement currents generate a field $\mathbf{B}$ which is equal to

$$B_y = \varepsilon\mu_0 v E_x = \frac{1}{v} E_x,$$

that, of course, coincides with the relationship between the vectors $\mathbf{E}$ and $\mathbf{B}$ derived before.

Example 3

Now consider a more general case when the current in the source $I_0\psi(at)$ is an arbitrary function of time. It is obvious that this function can be represented as a system of rectangular pulses with different magnitudes arising at different times (Figure 3.2(d)), and with a decrease of time duration of each impulse (its width), this approximation becomes more accurate. Therefore, we can say that at the instant t, the observed field is caused by the current impulse which appears earlier at the instant

$$t_1 = t - \frac{z}{v}$$

For example, if the current in the source I_0 remains constant when $t > 0$, then the time-invariant field is observed at any point with coordinate z, provided $z < vt$. This occurs because the current pulses are identical, and they follow each other continuously. In other words, the front and back of neighbor impulses arise at the same time. Since they are characterized by opposite directions of the field, we observe only the constant field. Thus, the time-invariant electric and magnetic fields arise due to wave propagation. This subject will be considered later in some detail.

Example 4

Another example of the function $\psi(z,t)$ is a sinusoidal oscillation which can also be treated as a system of pulses with different magnitudes and signs. In accordance with Eqs (3.19) and (3.20), if

$$i_x = I_0 \sin \omega t,$$

then

$$B_y(z,t) = \begin{cases} -\dfrac{\mu_0 I_0}{2} \sin \omega\left(t - \dfrac{z}{v}\right) & z < vt \\ 0 & z > vt \end{cases}$$

and (3.25)

$$E_x(z,t) = \begin{cases} -\dfrac{I_0}{2}\left(\dfrac{\mu_0}{\varepsilon}\right)^{1/2} \sin \omega\left(t - \dfrac{z}{v}\right) & z < vt \\ 0 & z > vt \end{cases}$$

Taking into account that $\omega = 2\pi/T$, we can rewrite Eq. (3.25) as

$$B_y = -\frac{\mu_0 I_0}{2} \sin(\omega t - \varphi) \quad \text{and} \quad E_x = vB_y$$

Here,

$$\varphi = \frac{2\pi z}{\lambda} \quad \text{and} \quad \lambda = vT \tag{3.26}$$

The parameter λ is called the *wave length* and it characterizes the distance passed by every elementary impulse during one period. At the same time, φ is the phase shift between the electromagnetic field and current of the source $i_x(0,t)$, and it is defined by the distance from this source expressed in units of the wave length λ. It may be appropriate to note that it is not so easy to visualize the propagation effect of sinusoidal waves as well as any periodical function, because we do not see the front and back of the wave. At the same time, the wave character of the field can be discovered considering the behavior of the phase at different times and distances from the source. We have studied the propagation of a plane wave in a nonconducting medium and demonstrated that this field is generated by two types of vortices at regular points:

$$\frac{\partial \mathbf{B}}{\partial t}, \quad \varepsilon\frac{\partial \mathbf{E}}{\partial t}$$

However, in those cases when the field has a discontinuity, there is also a surface distribution of vortices. For instance, if the wave is represented by the rectangular impulse, these vortices are located at the front and back sides of the impulse. The simplest example of the plane wave in a uniform and nonconducting medium can be treated as some introduction to the more complicated cases when wave propagates through a conducting medium and caused by real current sources, but first let us describe one approximation of the field that is widely used in many geophysical applications.

3.2. QUASI-STATIONARY FIELD IN A NONCONDUCTING MEDIUM

Now it is clear that a field, arising somewhere in a space, cannot reach any place instantly but rather always requires some time which is defined by two parameters, namely, the distance and the velocity of propagation. Later it will be demonstrated that this phenomenon occurs in any medium regardless of its conductivity and dielectric permittivity, and both the electromagnetic induction and displacement currents are vital for field propagation. For instance, letting the parameter ε equals zero, that is, neglecting the displacement currents,

$$\mathbf{j}_{\mathrm{d}} = \varepsilon\frac{\partial \mathbf{E}}{\partial t} \rightarrow 0$$

we arrive at an infinite velocity of propagation of the electromagnetic field. This assumption, which drastically simplifies the study of the field, does not in reality mean that the velocity becomes infinitely large. Quite opposite, it always has a finite value even though is very large; in other words, always without exception, the propagation effect takes place. However, in spite of its presence, there are conditions when with a given accuracy of measurements, it is practically impossible to observe the wave phenomenon. In such cases, this field is called the quasi-stationary one. At the beginning, let us study this field in a nonconducting medium but later influence on conductivity will be investigated in detail. Suppose as before that the field is caused by conduction currents distributed uniformly in the plane as is shown in Figure 3.2(a). Then we have for the magnetic field,

$$B_y(z,t) = B_0\psi\left[a\left(t-\frac{z}{v}\right)\right], \tag{3.27}$$

where $\psi(0,t)$ is defined by the behavior of the conduction currents of the source, and the ratio z/v characterizes the time which is necessary for the field to travel from the source to an observation point. Neglecting in these expression by displacement currents, that is, assuming that

$$\frac{z}{v}=0,$$

we come to conclusion that at every point, the magnetic field and source current vary synchronously. From a mathematical point of view, this means that the velocity of propagation is equal to infinity and this fact can be interpreted in the following way. Suppose that in some area of investigation, the field is studied at an instant t which is much greater than the delay time z/v:

$$t \gg \frac{z}{v} \tag{3.28}$$

Thus, we can say that both the source current and the quasi-stationary field **B** are practically described by the same function $\psi(at)$. From the physical point of view, this study shows that the magnetic field obeys the Biot–Savart law, and this is one of the most essential features of the quasi-stationary magnetic field. It is also useful to represent the inequality (Eq. (3.28)) in a different form which corresponds to the sinusoidal electromagnetic fields. Multiplying both sides of this relationship by the frequency ω, we obtain

$$\omega t \gg \frac{\omega z}{v} = \frac{2\pi z}{\lambda} \tag{3.29}$$

Thus, the field caused by the sinusoidal current source is quasi-stationary if the distance between this source and the observation point is much smaller than the wave length λ:

$$\frac{z}{\lambda} \ll 1 \tag{3.30}$$

Now we consider several examples which illustrate the behavior of the quasi-stationary field in a nonconducting medium.

Example 1: Inductive electric field of the solenoid

Suppose that a magnetic field arises as consequence of an alternating current flowing in an infinitely long cylindrical solenoid as shown in Figure 3.3(a). As we already know in the quasi-stationary approximation, the magnetic field coincides with the time-invariant magnetic field caused by the current at this instant; that is, the Biot–Savart law is valid for such time-varying field. For this reason, using results of Chapter 2, we can say that inside the solenoid, the field is uniform and directed along its axis, while outside the field **B** vanishes. Since the magnetic field changes with time, an inductive electric field arises. Taking into account the fact that both vectors **B** and $\partial\mathbf{B}/\partial t$ have the vertical component only, the electrical field is tangential to the horizontal planes (Figure 3.3(a)). Moreover, due to axial symmetry, the vector lines of **E** are circles with centers located on the solenoid axis. Therefore, the electric field has only component E_φ which is a function of distance r. Making use of Faraday's law,

$$\Xi = -\frac{\partial\Phi}{\partial t}$$

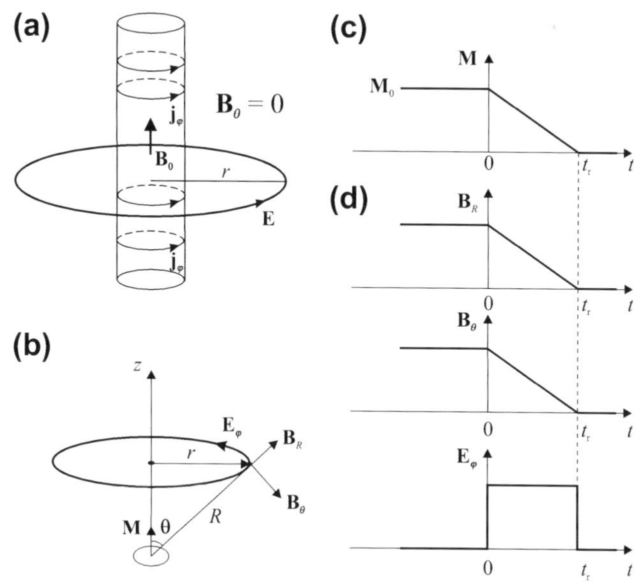

Figure 3.3 (a) Vortex field of solenoid. (b) Quasi-stationary field of magnetic dipole in a nonconducting medium. (c) Behavior of dipole moment with time. (d) Behavior of magnetic and electric field with time.

as well as the axial symmetry, for any circle with radius r located in a horizontal plane, we obtain

$$\oint \mathbf{E}\cdot d\mathbf{l} = 2\pi r E_\varphi = -\frac{\partial \Phi}{\partial t}$$

or

$$E_\varphi = -\frac{1}{2\pi r}\frac{\partial \Phi}{\partial t}, \tag{3.31}$$

where $\partial\Phi/\partial t$ is the rate of change of the magnetic flux within the area bounded by the circle with radius r. Bearing in mind that the magnetic field inside the solenoid is uniform,

$$B = B_0 f(t),$$

we have for the electric field inside the solenoid

$$E^i_\varphi = -\frac{\pi r^2}{2\pi r} B_0 f'(t) = -\frac{B_0 r}{2} f'(t) \quad \text{if } r \leq a \tag{3.32}$$

where a is the radius of the solenoid. Thus, the electric field inside the solenoid increases linearly with the distance from solenoid axis. For all horizontal circles, with radii r exceeding the solenoid radius a, the flux $\mathbf{\Phi}$, as well as the derivative $\partial\Phi/\partial t$, remains the same at any given instant of time and it is equal to

$$\mathbf{\Phi} = \pi a^2 B_0 f(t), \quad \text{and} \quad \mathbf{\Phi}' = B_0 \pi a^2 f'(t)$$

Therefore, the voltage (electromotive force) along any of these circles does not change with further increase of their radius and in accordance with Eq. (3.31), we have

$$E^e_\varphi = -\frac{B_0}{2\pi r}\pi a^2 f'(t) = -\frac{B_0 a^2}{2r} f'(t) \quad \text{if } r \geq a \tag{3.33}$$

As follows from this equation, the vortex electric field outside the solenoid is inversely proportional to the radius r. It is proper to note that this example vividly demonstrates a case when a vortex electric field in the quasi-stationary approximation is nonzero at points where the magnetic field is absent. In reality, due to a change of the magnetic field with time, displacement currents arise everywhere and they also generate the magnetic field. In our approximation, this effect is negligible but it provides a propagation of the wave regardless how small a rate of a change of the current in the solenoid is.

Taking into account that outside the solenoid

$$\text{curl}\,\mathbf{E} = 0$$

the inductive electric field can be expressed in terms of the potential, and a similar behavior takes place in the case of a toroid.

Example 2: The quasi-stationary field of a magnetic dipole in a nonconducting medium

Consider a magnetic dipole with the moment $\mathbf{M}(t)$ directed along the z-axis and situated at the origin of a spherical system of coordinates (Figure 3.3(b)). We again assume that regardless of the distance, the magnetic field at any instant t is defined by the magnitude of the dipole current at the same moment (quasi-stationary approximation). Then making use of the expressions for the magnetic dipole with constant current (Chapter 1), we obtain

$$B_R(t) = \frac{2\mu_0 M(t)}{4\pi R^3}\cos\theta, \quad B_\theta(t) = \frac{\mu_0 M(t)}{4\pi R^3}\sin\theta, \quad B_\varphi = 0 \tag{3.34}$$

The magnetic field is located in longitudinal planes of the spherical system of coordinates and it possesses by the axial symmetry. In this case, as follows from Maxwell's equations, the inductive electric field which arises due to a change of the field $\mathbf{B}$ with time has only a single component $E_\varphi(t)$. Therefore, vector lines of the electric field are circles and their centers are located at the z-axis. As in the previous example, we can write

$$E_\varphi = -\frac{1}{2\pi r}\dot{\Phi}, \tag{3.35}$$

where Φ is the flux piercing the area bounded by a circle with radius r (Figure 3.3(b)). Taking into account the fact that the vector $\mathbf{dS}$ is parallel to the z-axis, we have the following expression for the flux Φ:

$$\Phi = \int_S \mathbf{B}\cdot \mathrm{d}\mathbf{S} = 2\pi \int_0^r rB_z \mathrm{d}r, \tag{3.36}$$

since $\mathrm{d}S = 2\pi r\mathrm{d}r$ and B_z is the vertical component of the magnetic field. As is seen from Figure 3.3(b),

$$B_z(t) = B_R\cos\theta - B_\theta\sin\theta,$$

and considering Eq. (3.34), we obtain

$$B_z(t) = \frac{\mu_0 M(t)}{4\pi R^3}\left(3\cos^2\theta - 1\right) \tag{3.37}$$

Substituting this expression into Eq. (3.36) and then differentiating, we have

$$\dot{\Phi}(t) = \frac{\mathrm{d}\phi}{\mathrm{d}t} = \frac{\mu_0}{2}\frac{\dot{M}(t)}{R^3}r^2, \tag{3.38}$$

where $R = (r^2 + z^2)^{1/2}$ and $\dot{M}(t) = \mathrm{d}M/\mathrm{d}t$. Therefore, the vortex electric field is

$$E_\varphi(t) = -\mu_0 \frac{\dot{M}(t)}{4\pi R^2} \sin\theta \tag{3.39}$$

Thus in the quasi-stationary approximation, when the instantaneous magnitude of the dipole moment defines the magnetic field at the same instant, the expressions for the electromagnetic fields are

$$B_R(t) = \frac{2\mu_0 M(t)}{4\pi R^3}\cos\theta, \quad B_\theta(t) = \frac{\mu_0 M(t)}{4\pi R^3}\sin\theta, \quad E_\varphi(t) = -\mu_0\frac{\dot{M}(t)}{4\pi R^2}\sin\theta \tag{3.40}$$

It should be expected that the electric field is zero on the z-axis ($\theta = 0, \pi$), since the flux through a surface bounded by a circle of radius r tends to zero when the radius decreases.

At the same time as the radius increases, the magnetic vector lines begin to intersect the surface twice. In other words, the component B_z could have opposite sign at different points of the surface. For this reason, if r is sufficiently large, the flux $\mathbf{\Phi}$ gradually decreases in spite of unlimited increase of the surface. Thus, the flux $\mathbf{\Phi}$ as a function of r has a maximum whose position depends on the coordinate z and with its increase, the maximum is observed at greater distances from the z-axis. As follows from Eq. (3.40), at every point of a medium, the magnetic field is accompanied by an inductive electric field. It is natural to expect that if a medium is conductive, this electric field gives rise to a current. The field described by Eq. (3.40) is caused by the current of the magnetic dipole only, and correspondingly, it is often called the primary electromagnetic field. Now let us consider this field when the dipole moment varies with time in relatively simple manner.

Case 1

Suppose that the current in the dipole changes as a sinusoidal function, that is,

$$M = M_0 \sin\omega t, \tag{3.41}$$

where M_0 is the moment amplitude and $\omega = 2\pi f$ is the angular frequency with T being the period of oscillations. Then in accordance with Eqs (3.40) and (3.41), we have for the quasi-stationary field

$$B_R(t) = \frac{2\mu_0 M_0}{4\pi R^3}\cos\theta \sin\omega t, \quad B_\theta(t) = \frac{\mu_0 M_0}{4\pi R^3}\sin\theta \sin\omega t$$

and (3.42)

$$E_\varphi(t) = \frac{\omega\mu_0 M_0}{4\pi R^2}\sin\theta \sin\left(\omega t - \frac{\pi}{2}\right)$$

Thus, one can say that the primary electric field exhibits a phase shift of -90° with respect to the current flowing in the dipole or to the primary magnetic field. Note that at low-frequency methods of the surface and borehole geophysics, based on the use of the magnetic dipole as the primary source, Eq. (3.42) is often very useful for understanding the theory of these methods.

Case 2

Next assume that dipole moment varies with time as is shown in Figure 3.3(c):

$$M(t) = \begin{cases} M_0 & \text{if } \ t \le 0 \\ M_0 - at & \text{if } \ 0 \le t \le t_r \\ 0 & \text{if } \ t \ge t_r \end{cases} \tag{3.43}$$

where $a = M_0/t_r$. As follows from Eq. (3.40), the primary magnetic field is constant if $t < 0$, then it decreases linearly within the interval $0 < t < t_r$ and equals zero when $t > t_r$. Respectively, the primary electric field of the vortex origin exists only within the time interval where the magnetic field changes $(0 \le t \le t_r)$, and in view of its linear dependence on time, the electric field is constant. Thus, we have

$$B_R = \frac{2\mu_0 M_0}{4\pi R^3}\cos\theta, \quad B_\theta = \frac{\mu_0 M_0}{4\pi R^3}\sin\theta, \quad E_\varphi = 0 \quad \text{if } t \le 0$$

and

$$B_R(t) = \frac{2\mu_0 M(t)}{4\pi R^3}\cos\theta, \quad B_\theta(t) = \frac{\mu_0 M(t)}{4\pi R^3}\sin\theta,$$

$$E_\phi(t) = \frac{\mu_0 M_0}{4\pi R^2 t_r}\sin\theta \quad \text{if } 0 < t \le t_r \tag{3.44}$$

and

$$B_R = B_\theta = 0, \quad E_\varphi = 0 \quad \text{if } t > t_r$$

The curves shown in Figure 3.3(d) illustrate the behavior of the magnetic and electric fields as functions of time. Of course, our considerations do not take into account the propagation of the electromagnetic field, and in this approximation, the electric field exists only within the time interval where the dipole moment changes with time.

3.3. INDUCTION CURRENT IN A THIN CONDUCTING RING PLACED IN A TIME-VARYING FIELD

3.3.1. Equation for Induced Current in the Ring

Consider now an example which will be later used for explanation of the skin effect and a behavior of the fields in a conducting medium. As before, assume instead of the real field its quasi-stationary approximation and suppose that a thin conducting ring is placed into the primary field $\mathbf{B}_0$ (Figure 3.4(a)). The appearance of currents in a conducting ring can be described as follows. The time-varying primary magnetic field is accompanied by the inductive electric field. For simplicity, we assume that this electric field has a single component $E_{0\varphi}$, which is tangential to the ring surface. This field is the primary cause of the conduction current in the ring. In turn, these currents generate a secondary electromagnetic field. It should be obvious that the density of the induced current in the ring is defined by both the primary and secondary electric fields. According to Ohm's law, we have

$$j_\varphi = \gamma\left(E_{0\varphi} + E_{s\varphi}\right), \tag{3.45}$$

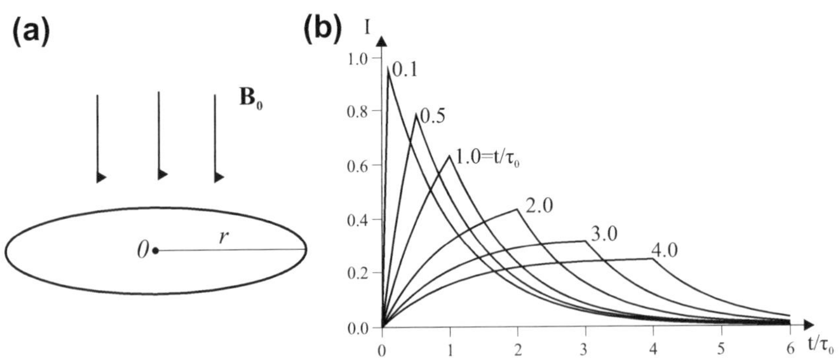

Figure 3.4 (a) Conducting ring in magnetic field. (b) Transient responses of the current.

where j_φ is the current density, γ is the ring conductivity, and $E_{0\varphi}$ and $E_{s\varphi}$ are the primary and secondary electric fields, respectively. To determine the current in the ring, we make use of Faraday's law

$$\Xi = -\frac{d\Phi}{dt} \tag{3.46}$$

The flux Φ through the area bounded by the ring can be written as

$$\Phi = \Phi_0 + \Phi_s \tag{3.47}$$

Here Φ_0 is the flux of the primary magnetic field caused by a given current source, while Φ_s is the flux of the magnetic field generated by the induction current in the ring. Correspondingly, Eq. (3.46) can be written as

$$\Xi = -\frac{d\Phi_0}{dt} - \frac{d\Phi_s}{dt} \tag{3.48}$$

In this equation, only the term $d\Phi_0/dt$ is known while the electromotive force Ξ and the rate of a change of the secondary flux $d\Phi_s/dt$ are unknown. Our objective is to determine the current I flowing in the ring, and so we attempt to express both unknowns in terms of this function. First of all, making use of Ohm's law in the integral form, we have

$$\Xi = IR, \tag{3.49}$$

where R is the ring resistance given by

$$R = \rho\frac{l}{S} \quad \text{if } r \gg r_0 \tag{3.50}$$

Here, ρ is the resistivity of the ring; l is its circumference, and the area of the ring cross-section is $S = \pi r_0^2$. As follows from the Biot–Savart law, the magnetic flux Φ_s caused by the current in ring is directly proportional to I and it can represented as

$$F_s = LI \tag{3.51}$$

Here, L is a coefficient of proportionality known as the inductance of the ring. According to Eq. (3.51), we can say that the ring inductance is the ratio of the secondary magnetic flux through the ring and the current which creates this flux:

$$L = \frac{F_s}{I}$$

In other words, numerically, the inductance is the flux caused by unit current. It is obvious that the inductance is determined by the geometrical parameters of the ring. In general, determination of the inductance involves rather complicated calculations based on the use of Biot–Savart law. But in some special cases including a very thin circular ring, this task is relatively simple, and we have the following well-known expression for the total inductance of such ring in a free space:

$$L = \mu_0 r \left(\ln \frac{8r}{r_0} - 1.75 \right) \tag{3.52}$$

Inductance is measured in henrys in SI units. If instead of one ring we have a coil (solenoid) with n rings per unit length, the inductance increases as the square of number of turns n of the solenoid:

$$L = \mu_0 n^2 r \left(\ln \frac{8r}{r_0} - 1.75 \right), \tag{3.53}$$

where r_0 is the radius of the cross-section of the ring. Thus, the simple form of the conductor and the assumption that the current density is uniformly distributed over the cross-section of the ring have allowed us to find simple expressions for the coefficient of proportionality between the electromotive force Ξ and the secondary flux Φ_s on the one hand and the induced current in the ring on the other hand. Substituting Eqs 3.49 and 3.51 into Eq. 3.48, we arrive at a differential equation with respect to the current I:

$$L\frac{dI}{dt} + RI = -\frac{d\Phi_0}{dt}, \quad \text{or} \quad \frac{dI}{dt} + \frac{1}{\tau_0} I = f(t), \tag{3.54}$$

Here,

$$\tau_0 = \frac{L}{R} \quad \text{and} \quad f(t) = -\frac{1}{L}\frac{d\Phi_0}{dt} \tag{3.55}$$

are given. As is well known, the solution of this ordinary differential equation of the first order is

$$I(t) = I_0 \exp\left(-\frac{t}{\tau_0}\right) - \exp\left(-\frac{t}{\tau_0}\right) \frac{1}{L} \int_0^t \exp\left(\frac{x}{\tau_0}\right) \frac{d\Phi_0(x)}{dx} dx, \tag{3.56}$$

where I_0 is the current at the instant $t = 0$. Now, we study the behavior of induced currents in two cases.

3.3.2. Transient Responses of Induced Current

First suppose that the primary magnetic field varies with time in a similar way as shown in Figure 3.3(c), so that,

$$\frac{d\Phi_0}{dt} = \begin{cases} 0 & \text{if } t < 0 \\ -\dfrac{\Phi_0}{t_r} & \text{if } 0 \le t < t_r \\ 0 & \text{if } t \ge t_r \end{cases} \tag{3.57}$$

During the time interval over which the primary magnetic flux does not change with time ($t < 0$), there are no induced currents in the ring, that is

$$I(t) = 0 \quad \text{if } t < 0$$

Within the ramp time, the primary flux Φ_0 varies linearly with time and therefore an induced current arises. Its magnitude is defined by the rate of a change of the primary magnetic field as well as by two parameters of the ring R and L. When the primary field disappears ($t > t_r$), the behavior of the induced current is controlled by the time constant τ_0 only. In fact, Eq. (3.54) is simplified in this case and we have

$$\frac{dI}{dt} + \frac{1}{\tau_0} I = 0 \quad \text{if } t \ge \tau_r \tag{3.58}$$

and its solution is

$$I(t) = C \exp\left(-\frac{t}{\tau_0}\right) \quad \text{if } t \ge \tau_r \tag{3.59}$$

In order to determine the constant C, we investigate the behavior of the induced currents during the ramp time. In accordance with Eqs (3.56) and (3.57), we obtain

$$I(t) = I_0 \exp\left(-\frac{t}{\tau_0}\right) + \frac{\tau_0}{t_r}\frac{\Phi_0}{L}\left[1 - \exp\left(-\frac{t}{\tau_0}\right)\right] \quad \text{if } 0 \le t \le t_r \tag{3.60}$$

Inasmuch as the induced current is absent at the instant $t = 0$, that is $I_0 = 0$, we have

$$I(t) = \frac{\tau_0}{t_r}\frac{\Phi_0}{L}\left[1 - \exp\left(-\frac{t}{\tau_0}\right)\right] \quad \text{if } 0 \le t \le t_r \tag{3.61}$$

The constant C is readily found from Eqs (3.59) and (3.61). In fact, letting in both equations $t = t_r$, we obtain

$$I(t_r) = C\exp\left(-\frac{t_r}{\tau_0}\right) = \frac{\tau_0}{t_r}\frac{\Phi_0}{L}\left[1 - \exp\left(-\frac{t_r}{\tau_0}\right)\right]$$

Thus,

$$C = \frac{\tau_0}{t_r}\frac{\Phi_0}{L}\left[\exp\left(\frac{t_r}{\tau_0}\right) - 1\right] \tag{3.62}$$

Correspondingly, the equations describing the induced current in the ring are

$$I(t) = \begin{cases} 0 & t < 0 \\ \dfrac{\tau_0}{t_r}\dfrac{\Phi_0}{L}\left[1 - \exp\left(-\dfrac{t}{\tau_0}\right)\right] & 0 < t \le t_r \\ \dfrac{\tau_0}{t_r}\dfrac{\Phi_0}{L}\left[\exp\left(\dfrac{t_r}{\tau_0}\right) - 1\right]\exp\left(-\dfrac{t}{\tau_0}\right) & t \ge t_r \end{cases} \tag{3.63}$$

As follows from Eq. (3.63), the induced current gradually increases during the ramp time, reaches a maximum at the moment $t = t_r$ and then decreases exponentially. Suppose that the ramp time t_r is much less than the time constant τ_0: $t_r \ll \tau_0$. Then expanding the exponential terms in Eq. (3.63) in power series and discarding all terms but those of the first and second order, we obtain

$$I(t) = \begin{cases} 0 & \text{if} \quad t < 0 \\ \dfrac{t}{t_r}\dfrac{\Phi_0}{L} & \text{if} \quad 0 < t \le t_r \\ \dfrac{\Phi_0}{L}\exp\left(-\dfrac{t}{\tau_0}\right) & \text{if} \quad t \ge t_r \end{cases} \tag{3.64}$$

In this case, the induced current increases linearly during the ramp time, and outside of this range ($t \ge t_r$), the current magnitude is independent on the parameter t_r. It is obvious that the magnetic field caused by this current has the same features. As will be shown later, a similar behavior is observed in a more general case of induced currents in volume conductors. It is easy to see that in the opposite case when $t_r \gg \tau_0$, the current I increases linearly at the beginning ($t \ll t_r$) and then slowly approaches a maximum equal to

$$\frac{\tau_0}{t_r}\frac{\Phi_0}{L} \ll \frac{\Phi_0}{L} \quad \text{if } t = t_r$$

Of course, at greater times, the current decays exponentially. Curves illustrating the behavior of induced current are shown in Figure 3.4(b).

3.3.3. Primary Magnetic Field is the Step-Function

Now we investigate the induced current in the ring when the primary magnetic flux changes as the step-function. It might already be obvious that the current I in this case is described by the last equation of the set (Eq. (3.64)) when the ramp time approaches zero. Thus, we have

$$I(t) = \frac{\Phi_0}{L}\exp\left(-\frac{t}{\tau_0}\right) \quad \text{if } t > 0 \text{ and } t_r \to 0 \tag{3.65}$$

that is, the initial value of the induced current does not depend on the ring resistance but rather is determined by the primary flux Φ_0 and the inductance L. Since under real conditions there is always a nonzero ramp time, the initial value of the current Φ_0/L should be treated as being its value at the instant $t = t_r$, provided that t_r is much less than τ_0. At the same time, the current at the initial moment ($t = 0$) is equal to zero. It is still interesting and very useful in understanding skin effect to derive the same result directly from Eq. (3.54). Integrating both parts of this equation within the ramp time interval, we have

$$R\int_0^{t_r} I(t)\mathrm{d}t + L\int_0^{t_r} \frac{\mathrm{d}I(t)}{\mathrm{d}t}\mathrm{d}t = -\int_0^{t_r} \frac{\mathrm{d}\Phi_0}{\mathrm{d}t}\mathrm{d}t$$

whence

$$R\int_0^{t_r} I(t)\mathrm{d}t + L[I(t_r) - I(0)] = \Phi_0(0) - \Phi_0(t_r) \tag{3.66}$$

Inasmuch as at the initial moment,

$$\Phi_0(0) = \Phi_0, \quad I(0) = 0$$

but at the instant $t = t_r$, the primary flux disappears, and Eq. (3.66) can be written as

$$R\int_0^{t_r} I(t)\mathrm{d}t + LI(t_r) = \Phi_0 \tag{3.67}$$

By definition, the integrand $I(t)\mathrm{d}t$ indicates the amount of charge passing through the ring cross-section during the time interval $\mathrm{d}t$. It is obvious that with decreasing the ramp time, the total amount of charge tends to zero. Therefore, in the limit when the primary flux varies as a step function, we have

$$LI(0) = \Phi_0, \quad \text{if } t_r = 0 \tag{3.68}$$

that is, the initial current is

$$I(0) = \frac{\Phi_0}{L} \tag{3.69}$$

It is natural that Eqs (3.65) and (3.68) result in the same magnitude for the initial current. As follows from Eq. (3.63), errors caused by discarding the integral in Eq. (3.67) become smaller with a decrease of the ratio t_r/τ_0. In other words, with an increase of the inductance L or a decrease of the resistance R, Eq. (3.68) defines the initial current more accurately. It is important to emphasize that Eq. (3.68) characterizes the essential feature of the electromagnetic induction and later will be generalized and applied to more complicated models of a medium. In fact, its left-hand side defines the magnetic flux through the area of the ring caused by induced current at the instant $t = 0$ when the primary flux disappears.

Thus, the induced current arises at the ring of such magnitude $I(0)$ that at the first instant magnetic flux, $LI(0)$ is exactly equal to the primary flux Φ_0. In other words, this induced current is trying to preserve the flux due to the primary field. If for example, the primary magnetic field instantly arises at the moment $t = 0$, then the induced current has such direction and magnitude that the total flux Φ through the area bounded by the ring is equal to zero at this instant. In essence, we observe the fundamental phenomenon of the inertia of the magnetic flux which is vital for understanding the skin effect. This study clearly shows that there are two factors which govern the behavior of induced current. One of them is the inertia of the magnetic flux Φ which tends to keep the current unchanged. The second factor is a transformation of the electromagnetic energy into heat which results in decrease of current with time, and with an increase of the resistivity R, this process becomes more rapid.

3.3.4. Primary Magnetic Field is a Sinusoidal Function of Time

Next suppose that the primary magnetic field varies as a sinusoidal function,

$$A \sin \omega t$$

In contrast to the previous case, we are going to examine the primary and secondary fields which have been already established. In order to determine the induced current, we make use of Eq. (3.56). Since the primary flux can be written as

$$\Phi_0 \sin \omega t$$

we have

$$I(t) = I_0 \exp\left(-\frac{t}{\tau_0}\right) - \frac{\omega\Phi_0}{L}\exp\left(-\frac{t}{\tau_0}\right)\int_0^t \exp\left(\frac{x}{\tau_0}\right)\cos \omega x dx$$

Taking into account that

$$\int \exp(\alpha x)\cos \beta x dx = \frac{\exp \alpha x}{\alpha^2 + \beta^2}(\alpha \cos \beta x + \alpha \sin \beta x),$$

we obtain for the induced current in the ring:

$$I(t) = I_0 \exp\left(-\frac{t}{\tau_0}\right) - \frac{\Phi_0}{L}\frac{\omega\tau_0 \cos \omega t}{1+(\omega\tau_0)^2} - \frac{\Phi_0}{L}\frac{(\omega\tau_0)^2 \sin \omega t}{1+(\omega\tau_0)^2}$$
$$+\frac{\Phi_0}{L}\frac{\omega\tau_0}{1+(\omega\tau_0)^2}\exp\left(-\frac{t}{\tau_0}\right)$$

Inasmuch as the initial value of the current is equal to zero ($I_0 = 0$) and we are interested in the behavior of the induced currents when the sinusoidal process is established ($t \gg \tau_0$), we have

$$I(t) = -\frac{\Phi_0}{L}\frac{\omega\tau_0 \cos \omega t}{1+(\omega\tau_0)^2} - \frac{\Phi_0}{L}\frac{(\omega\tau_0)^2 \sin \omega t}{1+(\omega\tau_0)^2} \tag{3.70}$$

Let us introduce notations

$$a(\omega) = -\frac{\Phi_0}{L}\frac{(\omega\tau_0)^2}{1+(\omega\tau_0)^2}, \quad b(\omega) = -\frac{\Phi_0}{L}\frac{\omega\tau_0}{1+(\omega\tau_0)^2} \tag{3.71}$$

Then, we have

$$I(t) = a \sin \omega t + b \cos \omega t, \tag{3.72}$$

that is, the induced current is represented as a sum of two oscillations. One of them

$$a \sin \omega t,$$

and it changes synchronically with the primary magnetic field. Because of this, it is usually called the in-phase component of the current:

$$InI = a \sin \omega t$$

The second oscillation

$$b \cos \omega t$$

is shifted in phase by 90° with respect to the primary magnetic field and is called the *quadrature component* of the current

$$QI = b \cos \omega t$$

Equation (3.72) suggests that we can treat the induced current in the ring as being the sum of two currents shifted in phase by 90°, namely, the in-phase and quadrature components. There is another approach. In fact let us represent the magnitudes of these components in the following form:

$$a = A \cos \varphi, \quad b = A \sin \varphi \tag{3.73}$$

Then the induced current is written as

$$I(t) = A(\cos \varphi \sin \omega t + \sin \varphi \cos \omega t) = A \sin (\omega t + \varphi) \tag{3.74}$$

Therefore, we can say that induced current $I(t)$ is the single sinusoidal oscillations with the same frequency ω as the primary magnetic field $\mathbf{B}_0(t)$. In general, there is a phase shift φ between them. As follows from Eqs (3.71) and (3.73), we have for the amplitude and phase of the current

$$A = \left(a^2 + b^2\right)^{1/2} = \frac{\Phi_0}{L} \frac{\omega \tau_0}{\left[1 + (\omega \tau_0)^2\right]^{1/2}} \qquad \varphi = \tan^{-1} \frac{1}{\omega \tau_0} \tag{3.75}$$

Curves of frequency responses of magnitudes of the quadrature and in-phase components, as well as the amplitude and phase responses, are given in Figure 3.5. Let us note that we can either think that there are two induced currents shifted by phase in 90° with respect to each other or the single

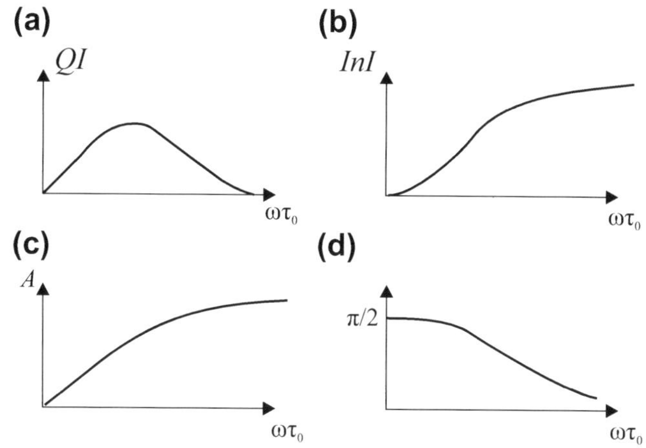

Figure 3.5 (a) Quadrature component of the current. (b) In-phase component of the current. (c) Amplitude of the current. (d) Phase of the current.

current, both approaches are equivalent and choice of them depends on the quantity which is measured. It is important to emphasize that in spite of the apparent simplicity of the model of the very thin circular ring, the frequency responses of the induced current contain some general features which are inherent for much more complicated models of conducting bodies of finite dimensions.

3.3.5. The Range of Small Parameter $\omega\tau_0$ or the Low-Frequency Spectrum of the Induced Current and Its Magnetic Field

Assuming that $\omega\tau_0 < 1$, we can expand the right-hand side of Eq. (3.71) in a series. This gives

$$a(\omega) = \frac{\Phi_0}{L}\left[-(\omega\tau_0)^2 + (\omega\tau_0)^4 - (\omega\tau_0)^6 + (\omega\tau_0)^8 - \cdots\right]$$

and

$$b(\omega) = \frac{\Phi_0}{L}\left[-\omega\tau_0 + (\omega\tau_0)^3 - (\omega\tau_0)^5 + (\omega\tau_0)^7 - \cdots\right] \qquad (3.76)$$

Thus in the low-frequency part of the spectrum, the quadrature and in-phase components of the induced current can be represented as a series containing either odd or even powers of ω. It is interesting to note that this feature of the spectrum remains valid for induced currents arising in any confined conductor surrounded by an insulator and even in some special cases of a medium with an unlimited dimension. It is clear that both series converge if the condition

$$\omega\tau_0 < 1 \tag{3.77}$$

is met. In other words, the radius of convergence of these series is

$$\omega = \frac{1}{\tau_0} \tag{3.78}$$

As follows from the theory of complex variables, the radius of convergence of the power series is the distance from the origin ($\omega = 0$) to the nearest singularity of the functions $a(\omega)$ and $b(\omega)$. To determine the location of this singularity, we have to treat these functions as functions of complex variable ω and consider the denominator in Eq. (3.71). It becomes equal to zero when $\omega = \pm i/\tau_0$. That is, the spectrum has two poles located on the imaginary axis of ω. It is essential that the radius of convergence of the series describing the low-frequency part of the spectrum is expressed in terms of the time constant of the ring. This fact reflects one of the most important aspects of the relationship between the low-frequency part of the spectrum and the late stage of the transient response of currents induced in confined conductors. Now suppose that the frequency is so low that we can consider only the first term in series. Then, we have

$$a(\omega) \approx -\frac{\Phi_0}{L}(\omega\tau_0)^2, \quad \text{and} \quad b(\omega) = -\frac{\Phi_0}{L}\omega\tau_0$$

and

$$QI(\omega) \approx -\frac{\Phi_0}{L}(\omega\tau_0)\cos\omega t, \quad InI(\omega) \approx -\frac{\Phi_0}{L}(\omega\tau_0)^2\sin\omega t \tag{3.79}$$

$$\text{if } \omega\tau_0 \ll 1$$

It is clear that at this part of spectrum, the quadrature component is dominant and directly proportional to the conductivity of the ring and to the frequency, while it does not depend on the inductance L since $\tau_0 = L/R$. Such behavior can be explained readily as follows. If we neglect the flux caused by induced current, that is, $\Phi_s \ll \Phi_0$, then the total flux is practically equal to the primary one:

$$\Phi \approx \Phi_0 \sin\omega t$$

Respectively, the electromotive force in the ring is

$$\Xi = -\frac{d\Phi}{dt} = -\omega\Phi_0 \cos \omega t$$

Applying Ohm's law, we obtain for the induced current

$$QI(\omega) = -\frac{\omega\Phi_0}{R}\cos \omega t = -\frac{\Phi_0}{L}(\omega\tau_0)\cos \omega t$$

Thus, the first term of the series of the quadrature component describes the current which arises due to the primary flux only. This fact is very useful for understanding some methods of the surface geophysics, applied for mapping as well as in the induction logging. In contrast, the in-phase component is caused by a secondary flux. In our approximation, $\omega\tau_0 \ll 1$, the flux generated by the quadrature component of the current is

$$\Phi_1 = L(QI) = -\frac{\omega\Phi_0}{R}L \cos \omega t = -\omega\tau_0\Phi_0 \cos \omega t$$

Therefore,

$$\frac{d\Phi_1}{dt} = \omega^2\tau_0\Phi_0 \sin \omega t,$$

and for the in-phase component of the current induced in the ring, we have

$$InI(\omega) = -\frac{\omega^2\tau_0\Phi_0}{R}\sin \omega t = -\frac{\Phi_0}{L}(\omega\tau_0)^2 \sin \omega t,$$

that is identical to the first term of the series of the in-phase component (Eq. (3.76)). It is obvious that applying the same approach, we can obtain subsequent terms of these series. Note that the name "the low-frequency part of the spectrum" is sometimes confusing. In fact, it does not mean that these equations are valid only when frequencies are small. For instance, if the parameter τ_0 is very small, frequencies can be very large, and they increase with an increase of resistance R and decrease of conductance L.

3.3.6. The Range of Large Parameter $\omega\tau_0$ or the High-Frequency Part of the Spectrum

Next we briefly study the high-frequency part of the spectrum of the induced currents when

$$\omega\tau_0 \gg 1 \tag{3.80}$$

As follows from Eq. (3.71), the in-phase component dominates in this range, and with an increase of the frequency, it approaches a constant which is determined by the primary magnetic flux and geometric parameters of the ring:

$$b(\omega) \to 0, \quad a(\omega) \to -\frac{\Phi_0}{L}, \quad \text{if } \omega\tau_0 \to \infty \tag{3.81}$$

Comparing Eqs (3.69) and (3.81), we see that the magnitudes of the induced current at the early stage of the transient response ($t \ll \tau_0$) and at the high-frequency part of the spectrum coincide. This result is not accidental, and it is valid for an arbitrary conductive medium. At the same time, let us make one comment. The way in which the current in the ring approaches its asymptotic value at high frequency and at the early stage differs from that in the general case. This is related to the fact that the cross-section of the ring has been assumed to be infinitesimal. Except for this detail, the behavior of the frequency and transient responses, given by Eqs (3.69) and (3.81), is similar to those for induced currents in confined conductors with an arbitrary cross-section.

As follows from the Biot–Savart law, the quadrature and in-phase components of the secondary magnetic field are generated by the corresponding components of the induced current. Therefore, the frequency responses of the corresponding components of the magnetic field and those of induced currents are similar. Of course, the transient responses of the induced current and its magnetic field are also similar to each other.

3.3.7. Electromagnetic Induction and Measurements of the Electric and Magnetic Fields

As is well known, coils are often used for measuring time-varying magnetic fields and for this reason, it is proper to consider this subject in some detail. Suppose that a conducting loop, shown in Figure 3.6, is placed in the magnetic field $\mathbf{B}(t)$. In general, the field $\mathbf{E}$ is arbitrary oriented with respect to the loop, and the voltmeter connected in series with this receiver measures the voltage along the path L between its terminal points, b and c:

$$V = \int_c^b \mathbf{E}\cdot d\mathbf{l} = \int_c^b E_l dl, \tag{3.82}$$

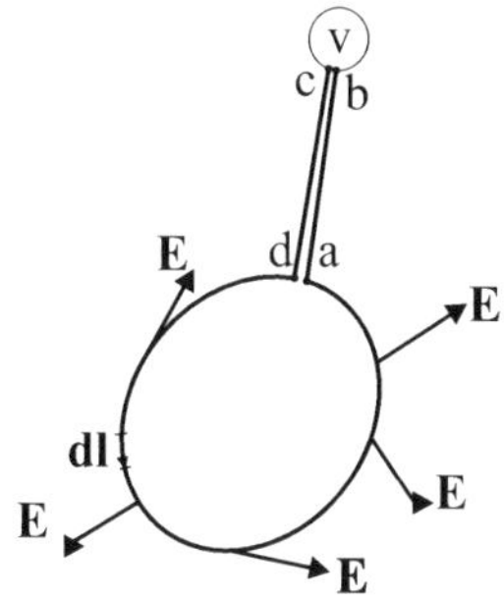

Figure 3.6 Electric field along receiver loop.

where E_l is the field component which is tangential to the loop. We also assume that the radius of the loop cross-section is much smaller than its length and, therefore, the voltage is practically independent of the position of the path L inside the loop. It is clear that if the receiver consists of n loops, then the voltage is

$$V = n \int_c^b E_l dl$$

Since the circuit intervals ab and cd are usually placed very close to each other, we can write

$$\int_a^b \mathbf{E} \cdot \mathbf{dl} = -\int_c^d \mathbf{E} \cdot \mathbf{dl} \quad \text{or} \quad \int_a^b \mathbf{E} \cdot \mathbf{dl} + \int_c^d \mathbf{E} \cdot \mathbf{dl} = 0 \tag{3.83}$$

Taking into account Eq. (3.83) and the fact that points a and d almost coincide, Eq. (3.82) can be rewritten as

$$V = \int_d^a \mathbf{E} \cdot \mathbf{dl} = \oint_L \mathbf{E} \cdot \mathbf{dl} = \oint_L E_l dl = \Xi \tag{3.84}$$

Thus, the voltmeter measures, in essence, the electromotive force along the receiver loop L. As a rule, the internal resistance of the voltmeter is very high and therefore the current in the loop is extremely small. This means we can neglect the influence of its magnetic field, and correspondingly, the electromotive force is defined by only the external electric field $\mathbf{E}(t)$. Inasmuch as the voltage is path dependent, a change of the position, size, and shape of the loop results in a change of the electromotive force even though terminal points of the voltmeter remain in the same place. As is seen from Figure 3.6, in general, regardless how small the loop is, it is impossible to determine the electric field from the electromotive force Ξ. However, there is one exception in which measurements with the loop allow us to calculate $\mathbf{E}(t)$. In fact, suppose that this field is tangential to the loop surface and its magnitude is constant at all its points. Then, in accordance with Eq. (3.84), we have

$$E_l = \frac{\Xi}{L},$$

where L is the loop circumference. It is appropriate to note if there is also a Coulomb's electric field $\mathbf{E}^c$ caused by charges, it does not have any influence on the electromotive force since

$$\oint \mathbf{E}^c \cdot d\mathbf{l} = 0$$

In particular, these charges are often located on the surface of the loop. Moreover, in the quasi-stationary approximation, the field of charges is described by Coulomb's law, and due to the electrostatic induction, the Coulomb's electric field is equal to zero inside the receiver loop.

Next let us discuss the measurements of the magnetic field with the help of the loop. In essence, this type of receiver measures the electromotive force which is a function of the electric field. However, as follows the Faraday's law,

$$\frac{d\Phi}{dt} = \int_S \dot{\mathbf{B}} \cdot d\mathbf{S} = \int_S \dot{B}_n dS = -\Xi \tag{3.85}$$

we can derive some information about the field $\mathbf{B}$. In general, Eq. (3.85) allows us to determine only the rate of change of the flux of the magnetic field through the area bounded by the loop. But if we assume that the loop size is so small that the function $\partial B_n / \partial t$ is uniform within the area enclosed by the loop, then from Eq. (3.85), it follows that

$$\dot{B}_n = -\Xi / S$$

where S is the loop area and B_n is the component of the field normal to the loop. Therefore, in this case, the derivative of the normal component of magnetic field with respect to time $\partial B_n / \partial t$ can be calculated knowing the electromotive force. Then it is not difficult to determine the magnetic field $B(t)$. By definition, we have

$$B_n(t) = B_n(t_*) + \int_{t_*}^{t} \dot{B}_n \mathrm{d}x$$

or

$$B_n(t) = B_n(t_*) - \frac{1}{S}\int_{t_*}^{t} \Xi(x)\mathrm{d}x \tag{3.86}$$

Here, t_* is the time when the field is known. In particular, if $B_n(t_*) = 0$, then

$$B_n(t) = -\frac{1}{S}\int_{t_*}^{t} \Xi(x)\mathrm{d}x$$

It is proper to mention that if the behavior of the field is either sinusoidal or exponential, then the normal component $B_n(t)$ can be calculated directly from a measurement of the electromotive force.

REFERENCES AND FURTHER READING

[1] A.A. Kaufman, Geophysical Field Theory and Methods, Part B, Academic Press, New York, NY, 1994.
[2] I.E. Tamm, Electricity Theory Foundations, Nauka, Moscow, 1987.

CHAPTER FOUR

Propagation and Diffusion in a Uniform Medium

Before we describe the theory of electromagnetic methods based on the use of the quasistationary fields, it is useful to study the propagation in a conducting and polarizable medium and formulate conditions when it is possible to apply the quasistationary approximation. With this purpose in mind, consider several cases and, as mentioned in Chapter 3, start from the simplest model of the plane wave.

4.1. SINUSOIDAL PLANE WAVE IN A UNIFORM MEDIUM

4.1.1. Expressions for the Field

Suppose again that in the plane XOY there is a current source with the density $i_x = i_0 f(at)$, which is independent on the coordinate y. As in the case of a nonconducting medium, we assume that the electromagnetic field is independent of coordinates x and y; and it has only two nonzero components: E_x and B_y. From Eq. (2.93), it follows that

$$\frac{\partial^2 E_x}{\partial z^2} - \gamma\mu_0 \frac{\partial E_x}{\partial t} - \varepsilon\mu_0 \frac{\partial^2 E_x}{\partial t^2} = 0$$

and (4.1)

$$\frac{\partial^2 B_y}{\partial z^2} - \gamma\mu_0 \frac{\partial B_y}{\partial t} - \varepsilon\mu_0 \frac{\partial^2 B_y}{\partial t^2} = 0$$

When the second term is equal to zero (nonconducting medium), we arrive at the wave equation derived earlier and, applying D'Alambert method, obtained its solution for practically arbitrary function $f(at)$ which characterizes a primary source of the field. The most essential feature of this method is the fact that an argument of the solution must have the form $t \pm z/v$ or $z = \pm vt$, where v is the velocity of wave propagation. However, this approach cannot be applied to Eq. (4.1) due to the presence of the second term with the first derivative with respect

Methods in Geochemistry and Geophysics, Volume 45
ISSN 0076-6895,
http://dx.doi.org/10.1016/B978-0-444-53829-1.00004-6

to time. As a result, their solution in general is not known except some special cases, for instance, when the source changes as a sinusoidal function. It turns out that in this case, the field is also sinusoidal function with the same frequency that can be expected since a medium can be treated as a linear system. This is a very important fact since if a solution for sinusoidal function is known, then the use of Fourier integral allows us to obtain a solution when the primary current behaves as an arbitrary function of time. Bearing this fact in mind, suppose that this current varies with time as

$$i_x(t) = i_0 \cos \omega t, \tag{4.2}$$

and therefore, the electromagnetic field is also sinusoidal function of time:

$$E_x(z,t) = E_{0x}(z) \cos(\omega t - \varphi), \quad B_y(z,t) = B_{0y}(z) \cos(\omega t - \psi),$$

or, as was shown in Chapter 2, it can be represented in the form

$$E_x(z,t) = \text{Re } E^*_{0x} \exp(-i\omega t), \quad B_y(z,t) = \text{Re } B^*_{0y} \exp(-i\omega t), \tag{4.3}$$

where the complex amplitudes are

$$E^*_{0x} = E_{0x} \exp(i\varphi) \quad B^*_{0y} = B_{0y} \exp(i\psi), \tag{4.4}$$

and these functions contain information about the amplitude and phase of the field components. As was demonstrated earlier, representation of the field by Eq. (4.3) greatly simplifies a solution of Eq. (4.1). In fact, substituting Eq. (4.3) into Eq. (4.1), we obtain for the complex amplitude of the fields:

$$\text{Re}\left\{\frac{\partial^2 E^*_{0x}}{\partial z^2} + i\gamma\mu_0\omega E^*_{0x} + \omega^2\varepsilon\mu_0 E^*_{0x}\right\} = 0$$

and

$$\text{Re}\left\{\frac{\partial^2 B^*_{0y}}{\partial z^2} + i\gamma\mu_0\omega B^*_{0y} + \omega^2\varepsilon\mu_0 B^*_{0y}\right\} = 0$$

These equations are satisfied if we require that the complex amplitudes are solutions of the Helmholtz equation in one-dimensional case:

$$\frac{\partial^2 E^*_{0x}}{\partial z^2} + k^2 E^*_{0x} = 0, \quad \frac{\partial^2 B^*_{0y}}{\partial z^2} + k^2 B^*_{0y} = 0 \tag{4.5}$$

where

$$k^2 = i\gamma\mu_0\omega + \omega^2\varepsilon\mu_0 \tag{4.6}$$

is the square of the wave number k. The solutions of Eq. (4.5) are well known, and they are

$$E^*_{0x} = C_1 \exp(ikz) + C_2 \exp(-ikz)$$

and (4.7)

$$B^*_{0y} = D_1 \exp(ikz) + D_2 \exp(-ikz),$$

where C and D are some constants. It is obvious that the wave number is complex:

$$k = a + ib, \tag{4.8}$$

and, correspondingly, in place of Eq. (4.7), we have

$$E^*_{0x} = C_1 \exp(-bz) \exp(iaz) + C_2 \exp(bz) \exp(-iaz)$$

and (4.9)

$$B^*_{0y} = D_1 \exp(-bz) \exp(iaz) + D_2 \exp(bz) \exp(-iaz)$$

In addition, it is assumed that $a > 0$ and $b > 0$. From the physical point of view, it is clear that with an increase of the distance from the source, the plane wave has to decay in a uniform medium because the electromagnetic energy is transformed into heat. The second term in these equations does not satisfy this behavior, and for this reason, they have to be discarded. In other words, the factors C_2 and D_2 have to be zeros in order to satisfy this condition. It gives

$$E^*_{0x} = C_1 \exp(-bz) \exp(iaz), \quad B^*_{0y} = D_1 \exp(-bz) \exp(iaz) \tag{4.10}$$

Now we will find out the relationship between constants C_1 and D_1, and with this purpose in mind, substitute Eq. (4.3) into the first Maxwell's equation

$$\text{curl}\,\mathbf{E} = -\frac{\partial \mathbf{B}}{\partial t}$$

Taking into account independence of the field from coordinates x and y, we have

$$\frac{\partial E^*_{0x}}{\partial z} = i\omega B^*_{0y},$$

and this gives

$$C_1 = \frac{\omega}{k} D_1 \tag{4.11}$$

Taking into account that k is complex number, we conclude that there is a phase shift between the electric and magnetic field which depends on conductivity, dielectric constant, and frequency. In particular, in a nonconducting medium, it is equal to zero. In contrast, when conduction currents are dominant, the phase difference is equal to $-\pi/4$. Note that applying Biot–Savart law near the source current, it is a simple matter, as in the case of nonconducting medium, to express the constant D_1, which is real, in terms of the conduction current of the source i_x. Taking into account this fact, we can say that the electric and magnetic fields, given by Eq. (4.10), satisfy all conditions of the theorem of uniqueness, and therefore, an assumption about sinusoidal behavior of the field was correct.

4.1.2. Behavior of the Plane Wave as a Function of Time and Distance

To illustrate a field behavior, consider as an example the function $B_y(z, t)$ which can be represented as

$$B_y = D_1 \exp(-bz) \cos\left[\omega\left(t - \frac{a}{\omega} z\right)\right], \tag{4.12}$$

At each observation point, the magnetic field is a sinusoidal function of time, but its dependence on the distance z is different and it is described by the product of the sinusoidal and exponential functions. One of them is a sinusoidal wave, while the other shows a decay of the field caused by its transformation into heat. Of course, the electrical field has the similar behavior. As follows from Eq. (4.12), the velocity of propagation is

$$v = \frac{\omega}{a} \tag{4.13}$$

By definition, during the period T, the phase plane of sinusoidal wave moves at the distance, equal to the wavelength λ:

$$\lambda = vT, \quad \text{or} \quad \lambda = \frac{v}{f}$$

Thus,

$$\lambda = \frac{2\pi}{a}, \tag{4.14}$$

that is, the wavelength is inversely proportional to the real part of the wave number. Let us introduce the parameter

$$b\lambda = 2\pi\frac{b}{a}, \tag{4.15}$$

which characterizes an attenuation of the wave within the distance equal to wavelength.

Intuitively, it is clear if this parameter is large, we hardly observe propagation in spite of its presence. In other words, considering the field as a function of z, we may not see its sinusoidal behavior. For instance, if $a = b$ at the distance equal, λ field decays almost in 500 times. Of course, in such cases, certainly it is necessary to have a great imagination to see wave propagation. This may allow us to assume that if

$$b\lambda > 1, \tag{4.16}$$

we rather observe diffusion than propagation (Figure 4.1(a), solid line). Correspondingly, in place of Eq. (4.1), the field magnitude is approximately described by diffusion equations

$$\frac{\partial^2 E_x}{\partial z^2} - \gamma\mu_0\frac{\partial E_x}{\partial t} = 0, \qquad \frac{\partial^2 B_y}{\partial z^2} - \gamma\mu_0\frac{\partial B_y}{\partial t} = 0$$

In other words, in such a case, the field behaves as quasistationary one; that is, the Coulomb's and Biot–Savart laws almost correctly describe the field. In contrast, when this product is small

$$b\lambda \ll 1, \tag{4.17}$$

the propagation effect becomes more visible. This means that we can see a sinusoidal behavior of the field and only at large distances greatly exceeding the wavelength, an attenuation effect becomes noticeable (Figure 4.1(a), dotted line). This consideration allows us qualitatively to distinguish three possible situations where we either clearly see the propagation effect, or the field displays features of diffusion; that is, one can use the quasistationary approximation. Of course, there is a transitional case where the field displays equally features of diffusion and propagation. Now let us make two comments: (1) Regardless how small the frequency ω is, the field reaches an observation point due to wave phenomenon. To understand this fact, it is sufficient to recall that a sinusoidal current of the source can be represented as a system of impulses following one after another. (2) Due to transformation of electromagnetic energy into heat, there is always attenuation of

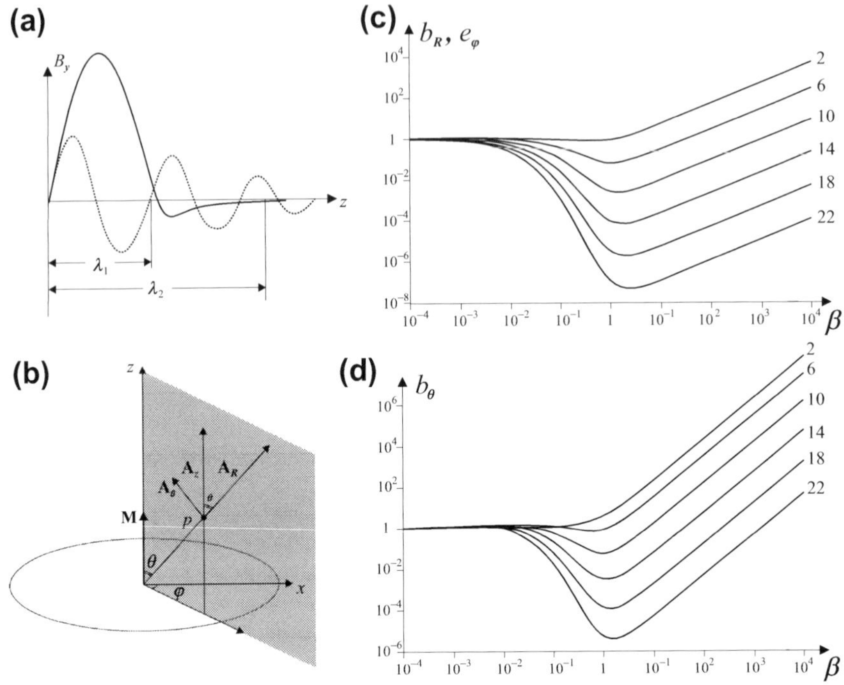

Figure 4.1 (a) Sinusoidal wave when $b\lambda > 1$ (solid line) and $b\lambda \ll 1$ (dotted line). (b) Magnetic dipole and vector potential in spherical coordinates. (c) Frequency responses of the fields b_R and e_φ in a uniform medium; curve index is x. (d) Frequency responses of the field b_θ in a uniform medium; curve index is x.

the field with an increase of the distance from the primary source. Until now we used rather qualitative approach, but next let us discuss this subject in some detail. In order to investigate the behaviors of the plane wave in a uniform medium, consider first dependence of the attenuation $b = \text{Im}\, k$, velocity v, and wavelength λ on the frequency and electric parameters of a medium.

4.1.3. Attenuation, Velocity of Propagation, and Wavelength

At the beginning, we find the real and imaginary parts of the wave number:

$$k = a + ib = \left(i\gamma\mu_0\omega + \omega^2\varepsilon\mu_0\right)^{1/2} \tag{4.18}$$

As follows from Eq. (4.12), the real and imaginary parts of the wave number define a change of the phase and a decrease of the field magnitude per unit distance, respectively. Taking square from both parts of this equality, we obtain the system of equations with respect to a and b:

$$\omega^2 \varepsilon \mu_0 = a^2 - b^2, \quad \text{and} \quad \gamma \mu_0 \omega = 2ab \tag{4.19}$$

Its solution gives

$$a = k_0 \left\{ 2^{-1/2} \left[\left(1 + \beta^{-2}\right)^{1/2} + 1 \right]^{1/2} \right\} \quad b = k_0 \left\{ 2^{-1/2} \left[\left(1 + \beta^{-2}\right)^{1/2} - 1 \right]^{1/2} \right\} \tag{4.20}$$

Here, $k_0 = \omega(\varepsilon\mu_0)^{1/2}$ and $\beta = \omega\varepsilon/\gamma$ are the ratio of the displacement and conduction currents, respectively.

CASE 1: THE HIGH-FREQUENCY SPECTRUM OR THE RANGE OF LARGE PARAMETER β, ($\beta > 1$)

First suppose that parameter β is large; that is, displacement currents prevail. Expanding the right-hand side of Eq. (4.20) in a series, we obtain approximate expressions for the real and imaginary parts of the wave number:

$$a \approx a_\varepsilon \left(1 + \frac{1}{8}\beta^{-2}\right), \quad \text{and} \quad b \approx b_\varepsilon \left(1 - \frac{1}{8}\beta^{-2}\right), \tag{4.21}$$

where

$$a_\varepsilon = \omega(\varepsilon\mu_0)^{1/2}, \quad b_\varepsilon = \frac{\gamma}{2}\left(\frac{\mu_0}{\varepsilon}\right)^{1/2} \tag{4.22}$$

Therefore, at the high-frequency part of the spectrum ($\beta \geq 1$), the real part of the wave number is practically independent of the conductivity and directly proportional to the frequency. As follows from Eq. (4.13) in this frequency range, the velocity of propagation is defined by only the dielectric constant; that is independent on the frequency and conductivity, and it almost coincides with that in a nonconducting medium:

$$v = \frac{1}{(\varepsilon\mu_0)^{1/2}} = \frac{c}{(\varepsilon_r)^{1/2}}, \quad \text{if } \beta \geq 1 \tag{4.23}$$

where $c = 3 \cdot 10^8 \text{m/s}$. The imaginary part b, characterizing the rate of decrease of the field with distance, is governed by the term b_ε, which is directly proportional to conductivity regardless of how strong an influence of displacement currents, and it is practically independent on the frequency. As follows from Eq. 4.22,

$$b_\varepsilon \lambda \approx \frac{\pi}{\beta} \ll 1$$

Correspondingly, we may say that with an increase of the frequency, the effect of the field decay over the interval equal to wavelength decreases and the wave phenomenon becomes more noticeable. Of course, it happens because the wavelength becomes smaller. At the same time, it is clear that the field is not able to propagate far away from the source because of a strong attenuation. Note that the ratio of a_ε and b_ε is

$$\frac{b_\varepsilon}{a_\varepsilon} = \frac{1}{2\beta} < 1 \tag{4.24}$$

As may be seen from Eq. (4.21), even when the value of β becomes close to unity, the real and imaginary parts of the wave number k differ only slightly from the limiting values of a_ε and b_ε. This means that at the high-frequency spectrum, sinusoidal waves decay almost in the same rate, and they have practically equal velocity (ω/a) which slightly increases with an increase of frequency.

CASE 2: THE LOW-FREQUENCY SPECTRUM OR THE RANGE OF SMALL PARAMETER β, (β < 1)

In contrast, when the conduction currents prevail, Eq. (4.20) gives

$$a \approx a_\gamma \left(1 + \frac{\beta}{2}\right), \quad \text{and} \quad b = b_\gamma \left(1 - \frac{\beta}{2}\right) \quad \text{if} \ \beta < 1 \tag{4.25}$$

Here,

$$a_\gamma = b_\gamma = \left(\frac{\gamma \mu_0 \omega}{2}\right)^{1/2} = \frac{1}{\delta} \tag{4.26}$$

The parameter δ is called the *skin depth*:

$$\delta = \left(\frac{2}{\gamma \mu_0 \omega}\right)^{1/2} \tag{4.27}$$

Thus at this range of β, the attenuation is described by Eq. (4.26), and it becomes smaller with a decrease of conductivity and frequency. As follows from Eq. (4.12), the skin depth is equal to distance where the field magnitude of the plane wave decreases in e times; that is, approximately 2.7 times. It is proper to note that in real conditions when field depends on coordinates y and z, the decrease of the field is even stronger. For the velocity of propagation, we have

$$v = \frac{\omega}{a} = \left(\frac{2\omega\rho}{\mu_0}\right)^{1/2} \quad \text{or} \quad v = (10\rho f)^{1/2}\ \text{km/s}, \tag{4.28}$$

and it becomes smaller with a decrease of frequency and resistivity of a medium, and in the limit tends to zero. The latter can be also represented as

$$v = (2\beta)^{1/2}\frac{c}{(\varepsilon_r)^{1/2}}$$

that is at this part of the spectrum, the velocity is much smaller than at the high frequencies. As follows from Eq. (4.28), the wavelength is

$$\lambda = 2\pi\delta, \quad \text{or} \quad \lambda = (10\rho T)^{1/2}\ \text{km} \tag{4.29}$$

and it increases with a decrease of frequency. From Eqs (4.26) and (4.27), we have

$$b_\gamma \lambda = 2\pi,$$

that is, the field attenuation within the interval equal the wavelength is extremely strong, $\exp(-2\pi)$, which means that the field decreases almost by 500 times. Of course, within this range, it is not simple to see wave phenomenon, but only here the field can propagate at relatively large distances. With an increase of frequency, attenuation increases in the same manner as the wavelength becomes smaller so that their product remains constant. Behavior of functions $a(\beta)$ and $b(\beta)$ normalized by their limiting values are shown in Figure 4.2a,b as well as functions, $b\lambda$ (Figure 4.2(c)).

It should be emphasized that we assume that conductivity, as well as dielectric permittivity, is independent of the frequency, while in some rocks, this behavior may not be always be guaranteed. Now let us make two comments: (1) Regardless how low the frequency is, there is always a propagation of waves through a conducting medium. Otherwise, the field would instantly appear at any point of a medium regardless how far this point is located from the primary source. (2) Influence of conductivity comes out in two ways. First of all, due to transformation of an electromagnetic energy into heat, the field decays with the distance from the source. Second, there is in general a frequency dispersion of the wave velocity.

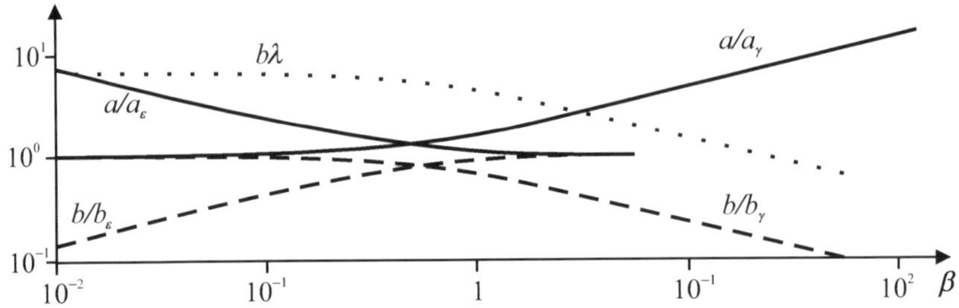

Figure 4.2 (a). Functions $b/_{b_\varepsilon}(\beta)$ and $\nu/_{\nu_\varepsilon}(\beta)$. (b) Function $b\lambda(\beta)$.

4.2. FIELD OF THE MAGNETIC DIPOLE IN A UNIFORM MEDIUM (FREQUENCY DOMAIN)

4.2.1. Introduction

Next we investigate the frequency responses of the magnetic and electric field caused by the magnetic dipole (small current loop), located in a uniform conducting and polarizable medium. The moment of the dipole is

$$\mathbf{M} = \text{Re}\, M_0 \exp\left(-\, i\omega t\right)\mathbf{1}_z \tag{4.30}$$

Here, $M_0 = I_0 n S$ is the magnitude of the dipole moment, I_0 is the amplitude of the sinusoidal current, n is the number of turns in the loop, and S is the area enclosed by one turn of the loop. The dipole moment $\mathbf{M}$ is directed along the z-axis (Figure 4.1(c)), and $\mathbf{1}_z$ is the corresponding unit vector. Before we formulate the boundary value problem and find its solution, it is useful to recall the behavior of the quasistationary electric field of the magnetic dipole in a nonconducting medium (Chapter 3). In the spherical system of coordinates (Figure 4.1(c)), this field has only one component E_φ. Since $M = M_0 \cos \omega t$, Eq. (3.40) gives

$$E_\varphi(\omega) = \frac{\mu_0 \omega M_0 \sin \omega t}{4\pi R^2} \cos\theta, \quad \text{or} \quad E_\varphi^*(\omega) = \frac{i\omega\mu_0 M_0}{4\pi R^2} \cos\theta \tag{4.31}$$

Here, E_φ^* is the complex amplitude of the electric field caused by the change with time of the primary magnetic field of the dipole. Inasmuch as the electric field has only one component located in horizontal planes, it is natural to assume that conduction and displacement currents arising in a conducting and polarizable medium, which is isotropic one, also have the same component j_φ^*. As was shown earlier, an inductive electric field can be represented in terms of vector potential

$$\mathbf{E}^* = \text{curl}\, \mathbf{A}^*, \quad \text{or} \quad \mathbf{E}^* = \nabla \times \mathbf{A}^*, \tag{4.32}$$

where $\mathbf{A}^*$ is the complex amplitude of the vector potential of the magnetic type. As was demonstrated in Chapter 2, the function $\mathbf{A}^*$ satisfies the Helmholtz equation, and now we are ready to solve the corresponding boundary value problem.

4.2.2. Solution of Helmholtz Equation

Taking into account the axial symmetry of the field, we suppose that the vector potential has the single z-component and in the spherical coordinates, it depends on the coordinate R only; that is,

$$\mathbf{A}^* = A_z^*(k, R)\mathbf{1}_z \tag{4.33}$$

Then Helmholtz's equation is simplified:

$$\frac{1}{R^2}\frac{\mathrm{d}}{\mathrm{d}R}\left(R^2\frac{\mathrm{d}A_z^*}{\mathrm{d}R}\right) + k^2 A_z^* = 0 \tag{4.34}$$

To solve this equation, we introduce a new function

$$W = A_z^* R$$

Performing differentiation, we have

$$\frac{\mathrm{d}A_z^*}{\mathrm{d}R} = -R^{-2}W + R^{-1}\frac{\mathrm{d}W}{\mathrm{d}R}, \quad \text{and} \quad R^2\frac{\mathrm{d}A_z^*}{\mathrm{d}R} = -W + R\frac{\mathrm{d}W}{\mathrm{d}R}$$

Whence

$$\frac{\mathrm{d}}{\mathrm{d}R}\left(R^2\frac{\mathrm{d}A_z^*}{\mathrm{d}R}\right) = R\frac{\mathrm{d}^2 W}{\mathrm{d}R^2}$$

Therefore, Eq. (4.34) becomes

$$\frac{\mathrm{d}^2 W}{\mathrm{d}R^2} + k^2 W = 0,$$

which was already used in the previous section, and its solutions are exponential functions $\exp(\pm ikR)$. Now it becomes clear why we introduced new function W. Thus, the expression for the z-component of the vector potential is

$$A_z^* = C\,\frac{\exp(ikR)}{R} + D\,\frac{\exp(-\,ikR)}{R} \tag{4.35}$$

Inasmuch as

$$-\,ikR = -i(a + ib)R = -iaR + bR$$

we conclude that with increase of the distance, the second term of Eq. (4.35) unlimitedly increases and therefore it has to be discarded. Otherwise, condition at infinity would not be satisfied. Thus, we have

$$A_z^*(k,R) = C\frac{\exp(ikR)}{R}, \tag{4.36}$$

which obeys the Helmholtz equation and condition at infinity. In order to satisfy the condition near source (magnetic dipole), we have to determine the unknown C, and with this purpose in mind, let us find expressions for the field components.

4.2.3. Expressions for the Complex Amplitudes of the Electric and Magnetic Fields

Inasmuch as we expect that the electric field has only the φ^- component, it is proper to use the spherical system of coordinates. Then in the spherical coordinates, Eq. (4.32) gives

$$E_\varphi^* = \frac{1}{R}\left[\frac{\partial}{\partial R}\left(RA_\theta^*\right) - \frac{\partial A_R^*}{\partial \theta}\right] \tag{4.37}$$

As is seen from Figure 4.1(c),

$$A_R^* = A_z^* \cos\theta, \quad \text{and} \quad A_\theta^* = -A_z^* \sin\theta$$

Substituting these expressions into Eq. (4.37), and performing simple algebra operations, we obtain

$$E_\varphi^* = \frac{C}{R^2}(1 - ikR)\exp(ikR)\sin\theta \tag{4.38}$$

In approaching the magnetic dipole, we have

$$E_\varphi^* = \frac{C}{R^2}\sin\theta \tag{4.39}$$

From the physical point of view, it is clear in the vicinity of the dipole that the electromagnetic field tends to that of the field caused by the current of the dipole only. In particular, in this case, the electrical field is generated by the change of the primary magnetic field only (Eq. (4.31)). In essence, we formulated the boundary condition near the source. This allows us to determine unknown constant C. In fact, comparing Eqs (4.31) and (4.39), we obtain

$$C = \frac{i\omega\mu_0 M_0}{4\pi} \tag{4.40}$$

Thus, the complex amplitude of the vector potential is

$$\mathbf{A}^* = \frac{i\omega\mu_0 M_0}{4\pi R} \exp(ikR)\mathbf{1}_z, \tag{4.41}$$

and this function satisfies both boundary conditions and Helmholtz equation; that is, we solved the boundary value problem. In particular, this means that all assumptions about behavior of the vector potential are correct. From Eqs (4.38) and (4.40), we have for the complex amplitude of the electric field:

$$E_\varphi^* = \frac{i\omega\mu_0 M_0}{4\pi R^2}(1 - ikR)\exp(ikR)\sin\theta \tag{4.42}$$

To determine the complex amplitudes of the magnetic field, we make use of the relation derived in Chapter 2 (Eq. (2.104)):

$$i\omega\mathbf{B}^* = k^2\mathbf{A}^* + \text{grad div }\mathbf{A}^* \tag{4.43}$$

Inasmuch as the vector potential is directed along the z-axis and the field is independent on the φ - coordinate, the component B_φ is absent and we have to determine B_R and B_θ only. First let us calculate div $\mathbf{A}^*$. Taking into account that divergence is independent of the system of coordinates, it is convenient to perform this operation in the Cartesian system where it has very simple form:

$$\text{div }\mathbf{A}^* = \frac{dA_z^*}{dz}$$

Carrying out differentiation, we obtain

$$\text{div }\mathbf{A}^* = -\frac{i\omega\mu_0 M_0}{4\pi R^2}\exp(ikR)(1 - ikR)\cos\theta, \tag{4.44}$$

since

$$\frac{dR}{dz} = \cos\theta$$

Taking into account the expression for grad in spherical coordinates, we have

$$i\omega B_R^* = k^2 A_R^* + \frac{\partial}{\partial R}\text{div }\mathbf{A}^*, \quad \text{and} \quad i\omega B_\theta^* = k^2 A_\theta^* + \frac{1}{R}\frac{\partial}{\partial\theta}\text{div }\mathbf{A}^*$$

Differentiating and taking into account that

$$A_R^* = A_z^*\cos\theta, \quad \text{and} \quad A_\theta^* = -A_z^*\sin\theta,$$

we obtain

$$B_R^* = \frac{2\mu_0 M_0}{4\pi R^3} \exp(ikR)(1 - ikR) \cos\theta,$$
$$B_\theta^* = \frac{\mu_0 M_0}{4\pi R^3} \exp(ikR)\left(1 - ikR - k^2R^2\right) \sin\theta \tag{4.45}$$

Thus, Eqs (4.42) and (4.45) describe the complex amplitudes of the field of the magnetic dipole in any point of a conducting and polarizable medium. This field is caused by the dipole current, as well as conduction and displacement currents arising in a medium; correspondingly it depends on several factors, such as the dipole moment, frequency, product kR, and coordinates of observation point. It is convenient to express the field in terms of the primary one generated by only the dipole current. Then we have

$$e_\varphi^* = \frac{E_\varphi^*}{E_{0\varphi}^*} = (1 - ikR)\exp(ikR), \quad b_R^* = \frac{B_R^*}{B_{0R}^*} = (1 - ikR)\exp(ikR)$$

and

$$b_\theta^* = \frac{B_\theta^*}{B_{0\theta}^*} = \left(1 - ikR - k^2R^2\right)\exp(ikR) \tag{4.46}$$

These expressions depend on only parameter kR, and it strongly simplifies the study of the field behavior. At the same time, the primary field components $E_{0\varphi}^*$, B_{0R}^*, and $B_{0\theta}^*$ are described by very simple formulas given earlier. By definition, the amplitude and phase of the sinusoidal field, as well as its quadrature and in-phase components, can be expressed in terms of the complex amplitudes. For instance, for the amplitude and phase of the electric field, we have

$$\left|e_\varphi^*\right| = \exp(-bR)\left[(1+bR)^2 + a^2R^2\right]^{1/2}$$

and

$$\psi_\varphi = aR - \tan^{-1}\frac{aR}{1+bR} \tag{4.47}$$

Its in-phase and quadrature components are

$$Ine_\varphi = Inb_R = \left|e_\varphi^*\right| \cos\psi_\varphi \quad \text{and} \quad Qe_\varphi = Qb_R = \left|e_\varphi^*\right| \sin\psi_\varphi \tag{4.48}$$

Similarly, for the azimuthal component of the magnetic field, we have

$$\left|b_\theta^*\right| = \exp(-bR)\left[\left(1 + bR + b^2R^2 - a^2R^2\right)^2 + \left(aR + 2abR^2\right)^2\right]^{1/2}$$

and

$$\psi_\theta = aR - \frac{aR + 2abR^2}{1 + bR + b^2R^2 - a^2R^2} \tag{4.49}$$

For the in-phase and quadrature components, we have

$$Inb_\theta = \left|b_\theta^*\right| \cos\theta, \quad \text{and} \quad Qb_\theta = \left|b_\theta^*\right| \sin\theta \tag{4.50}$$

In Eqs (4.47) and (4.49), a and b are real and imaginary parts of the complex number k.

4.2.4. The Frequency Responses of the Field

To study the frequency responses of the field amplitudes, it is convenient to introduce two parameters, namely a characteristic length R_0 and characteristic frequency ω_0 The parameter R_0 is defined from the relation $b_\varepsilon R_0 = 1$ or

$$R_0 = \frac{2}{\gamma(\mu_0/\varepsilon)^{1/2}} = \frac{\rho(\varepsilon_r)^{1/2}}{188.5} \tag{4.51}$$

The characteristic frequency is defined as being the frequency at which the displacement and conduction currents are equal

$$\frac{\omega_0\varepsilon}{\gamma} = 1 \tag{4.52}$$

The complex amplitudes b_R^*, b_θ^*, and e_φ^* can be treated as functions of the dimensionless parameters $\beta = \omega/_{\omega_0}$ and $X = R/_{R_0}$, where R is the distance from the dipole to the location at which the field is being evaluated. A set of typical curves illustrating the dependence of the function b_R^* on the frequency parameter β is given in Figure 4.1(d). These curves show that with an increase in frequency, the electromagnetic field first decreases in a conducting medium, and then near the characteristic frequency, the amplitude of the field passes through a minimum value and with further increase of the frequency grows without limit, as soon as it is assumed that both the conductivity and dielectric permittivity are independent of the frequency. With a decrease of the characteristic length R_0, that is, with a decrease of resistivity and ε, the minimum value becomes smaller and shifts toward higher frequencies. If the frequencies are lower than ω_0 for all practical purposes, the field is almost independent of dielectric permittivity and this part of the spectrum is defined as being the quasistationary field. For frequencies higher than the characteristic frequency, the field strength becomes greater and can be in orders larger than the primary field caused by the dipole current. Certainly this result is not obvious and it is difficult to predict. Over this part of the spectrum, the field depends on both conductivity and dielectric permittivity, but the imaginary part of the wave number, b, becomes independent on the frequency. In particular, measuring the magnitude of the field and its phase in this part of the spectrum, we obtain information about the conductivity and dielectric permittivity, respectively. Note that we study the field magnitudes normalized by the primary magnetic field which decreases with the distance inversely proportional to the cube of the distance R.

4.2.5. Dependence of the Field on the Distance from the Magnetic Dipole

Consider a field behavior as a function of the separation R. As follows from Eq. (4.46) and taking into account that

$$k = a + ib = |k| \exp(i\xi) \quad \text{and} \quad B_R = \mathrm{Re}\left[B_R^* \exp(-i\omega t)\right]$$

we obtain

$$B_R = \frac{2\mu_0 M_0}{4\pi} \exp(-bR)\left[\frac{\cos(\omega t - aR)}{R^3} - \frac{|k|}{R^2}\sin(\omega t - aR - \xi)\right]\cos\theta$$

or

$$B_R = \frac{2\mu_0 M_0}{4\pi R^3} \exp(-bR)[\cos(\omega t - aR) - |k|R\sin(\omega t - aR - \xi)]\cos\theta,$$

$$B_\theta = \frac{\mu_0 M_0}{4\pi} \exp(-bR)\left[\frac{\cos(\omega t - aR)}{R^3} - \frac{|k|}{R^2}\sin(\omega t - aR - \xi) - \frac{|k^2|}{R}\cos(\omega t - aR - 2\xi)\right]\cdot\sin\theta$$

or

$$B_\theta = \frac{\mu_0 M_0}{4\pi R^3} \exp(-bR)\left[\cos(\omega t - aR) - |k|R\sin(\omega t - aR - \xi) - |k^2|R^2\cos(\omega t - aR - 2\xi)\right]\cdot\sin\theta$$

and

$$E_\varphi = \frac{\mu_0 \omega M_0}{4\pi} \exp(-bR)\left[\frac{\sin(\omega t - aR)}{R^2} - \frac{|k|}{R}\cos(\omega t - aR - \xi)\right]\sin\theta$$

or

$$E_\varphi = \frac{\mu_0 \omega M_0}{4\pi R^2} \exp(-bR)[\sin(\omega t - aR) - |k|R\cos(\omega t - aR - \xi)]\sin\theta \tag{4.53}$$

As we see, the field behavior depends on the parameters $|k|R$ and R, and regardless of their values, it is a combination of sinusoidal waves with the same frequency decaying differently with the distance. There are two factors which results in a decrease of the field: one of them is attenuation caused by transformation electromagnetic energy into heat. The term $\exp(-bR)$ appears due to this factor. The second reason is geometry of the source; that is, the wave moves in all directions, and the density of its energy decreases with the distance even in a nonconducting medium. As follows from Eq (4.53), the field behavior changes with the distance. At relatively small separations, when $|k|R \ll 1$, the magnetic field almost synchronously varies with the dipole current and rapidly decreases with distance as $1/R^3$. This range of distances is usually called the *near zone*. For instance, if conduction currents prevail $(\omega < \omega_0)$, we have

$$k \approx (i\omega\mu_0\gamma)^{1/2} = \frac{(2i)^{1/2}}{\delta} = \frac{1+i}{\delta}$$

and (4.54)

$$\delta = \left(\frac{2}{\gamma\mu_0\omega}\right)^{1/2} \quad \text{or} \quad \delta = \frac{10^3}{2\pi}\left(\frac{10\rho}{f}\right)^{1/2} \text{ m}$$

Correspondingly, the near zone takes place when

$$\frac{R}{\lambda} \ll 1 \tag{4.55}$$

where $\lambda = 2\pi\delta$. For illustration, consider two numerical examples. First, suppose that $\rho = 10$ Ohm·m, and $f = 10$ kHz.

Then $\delta \approx 16$ m and this distance approximately defines the boundary of the near zone. If the frequency is strongly increased, $f = 10^6$ Hz, the first term of Eq. (3.49) describes the field behavior at distances which are smaller than 1 m. As was pointed out in those cases when it is possible to neglect by the phase aR, we observe the quasistatic field; that is, both the current and magnetic fields change simultaneously. In the intermediate zone, all terms play an important role and with an increase of the distance, their contribution varies. At the beginning, the field decays almost in the same manner as in the near zone and then it starts to decrease slower. Also the phase shift between the field and the dipole current changes with the distance. This change is caused by superposition of terms differently decreasing with the distance. Finally, when $R > \lambda$, we observe the wave zone where

$$B_R = -\frac{2\mu_0 M_0}{4\pi} \exp(-bR)\frac{|k|}{R^2} \sin(\omega t - aR - \xi)\cos\theta$$

$$B_\theta = -\frac{\mu_0 M_0}{4\pi} \exp(-bR)\frac{|k^2|}{R} \cos(\omega t - aR - 2\xi)\sin\theta \tag{4.56}$$

$$E_\varphi = -\frac{\mu_0 \omega M_0}{4\pi} \exp(-bR)\frac{|k|}{R} \cos(\omega t - aR - \xi)\sin\theta$$

The latter shows that in the wave zone, the field in general decays relatively slower and $B_\theta > B_R$, provided that the field is not considered in the vicinity of the z-axis. In conclusion, let us notice the following. (1) Regardless of the frequency, we always observe three zones where a field behavior differs from each other. (2) A change of frequency causes a change of boundaries of these zones. (3) At each zone there are three effects, namely, propagation of sinusoidal waves, attenuation due to transformation into heat, and geometric spreading, and they display at different degree. (4) At the high-frequency spectrum $\omega > \omega_0$, the wave propagates with almost the same velocity as in a nonconducting medium and attenuation is directly proportional to conductivity. In contrast, at the low-frequency spectrum $\omega < \omega_0$, the velocity of propagation is defined by only conductivity and frequency

$$v = \frac{\omega}{a} = \left(\frac{2\omega}{\gamma\mu_0}\right)^{1/2}, \quad \text{and} \quad \lambda = 2\pi\delta = 2\pi\left(\frac{2}{\gamma\mu_0\omega}\right)^{1/2},$$

and field behavior can be derived from the diffusion equation which implies that displacement currents are neglected. The latter is equivalent to the assumption that the field instantly appears at all points of a medium regardless of the distance from the source.

4.3. EQUATIONS FOR TRANSIENT FIELD OF THE MAGNETIC DIPOLE IN A UNIFORM CONDUCTING AND POLARIZABLE MEDIUM

4.3.1. Expression for the Vector Potential

For better understanding of an influence of propagation and diffusion on the field behavior, we also consider a nonstationary field caused by the magnetic dipole located in a uniform medium. First, suppose that dipole current arises instantaneously, so that it can be described as being a step-function of time:

$$M_0 = \begin{cases} 0 & t < 0 \\ M_0 & t \geq 0 \end{cases} \tag{4.57}$$

In order to derive expressions of the field, we will proceed from the Fourier integrals:

$$F(t) = \frac{1}{2\pi}\int_{-\infty}^{\infty} S(\omega)\exp(-i\omega t)\,d\omega \quad \text{and} \quad S(\omega) = \int_{-\infty}^{\infty} F(t)\exp(i\omega t)\,dt \tag{4.58}$$

The first equation shows that the function $F(t)$ can be represented as a sum of infinite number of sinusoids (harmonics), and their amplitude and phase are characterized by the spectrum (complex amplitude) $S(\omega)$, given by the second equation. In accordance with Eq. (4.58), the spectrum of the step function is

$$S(\omega) = M_0\int_{0}^{\infty}\exp(i\omega t)\,dt,$$

since the current is absent when $t < 0$. To calculate this integral, we consider slightly different integral which converges

$$\int_{0}^{\infty}\exp[(i\omega - p)t]\,dt,$$

where p is the small positive number and then find its limit when $p \to 0$. Performing this simple procedure, we obtain the spectrum of the step function when the dipole current is turned on

$$S(\omega) = -\frac{M_0}{i\omega} \tag{4.59}$$

Thus the step-function can be represented as a sum of sinusoidal functions which have the same phase, but their amplitude decreases with an increase of frequency. In other words, the maximum of energy of the dipole is concentrated at the low-frequency part of the spectrum, which relatively slow decays with the distance from the source and, correspondingly has greater depth of investigation. As we know, the vertical component of the complex amplitude A_z^* is equal to

$$A_z^* = \frac{i\omega\mu_0 M_0}{4\pi}\frac{\exp(ikR)}{R},$$

If the moment is described as $-M_0/_{i\omega}$, then for the vector potential at each frequency, we have

$$-\frac{\mu_0 M_0}{4\pi}\frac{\exp(ikR)}{R}$$

Therefore, applying the Fourier integral, we obtain the vector potential in the case when the dipole moment behaves as the step function

$$A_z = -\frac{\mu_0 M_0}{8\pi^2 R}\int\limits_{-\infty}^{\infty} \exp\, i(kR - \omega t)\, \mathrm{d}\omega \tag{4.60}$$

Here, $k = (i\gamma\omega\mu_0 + \varepsilon\mu_0\omega^2)^{1/2}$, and A_z is the vector potential of the magnetic type. It turns out that this integral is tabular and it gives

$$A_z = \begin{cases} 0 & t < \tau_0 \\ -\frac{\mu_0 M_0}{4\pi R}\left\{\exp(-q\tau_0)\delta(t-\tau_0) + q\tau_0 \exp(-qt)\frac{I_1\left[q\left(t^2-\tau_0^2\right)^{\frac{1}{2}}\right]}{\left(t^2-\tau_0^2\right)^{\frac{1}{2}}}\right\} & t \geq \tau_0 \end{cases} \tag{4.61}$$

Here,

$$q = \frac{1}{2}\frac{\gamma}{\varepsilon} \quad \tau_0 = (\varepsilon\mu_0)^{1/2}R, \tag{4.62}$$

and R is the distance from the dipole to the observation point, M_0 is the magnitude of the magnetic dipole, $I_1\left[q(t^2 - \tau_0^2)^{1/2}\right]$ is a modified Bessel's function of the first order, $\delta(t - \tau_0)$ is the Dirac (or delta) function defined mathematically as

$$\int\limits_{b_1}^{b_2} f(x')\delta^{(n)}(x' - x)\, \mathrm{d}x' = \begin{cases} (-1)^n f^{(n)}(x) & (b_1 \leq x \leq b_2) \\ 0 & (x < b_1, x > b_2) \end{cases} \tag{4.63}$$

and t is the time measured from the instant when the dipole current is turned on. As follows from Eq. (4.61), the electromagnetic field can be observed at a point of a medium only after instant

$$\tau_0 = (\varepsilon\mu_0)^{1/2}R, \tag{4.64}$$

and with an increase of the distance from the dipole, the appearance of signal is observed at later times. Thus, the first arrival of the signal propagates with the phase velocity of the high-frequency sinusoids, equal to

$$v = \frac{1}{(\varepsilon\mu_0)^{1/2}} = \frac{c}{(\varepsilon_r)^{1/2}} \tag{4.65}$$

4.3.2. Expressions for the Field Components

Making use of Eq. (4.61), it is relatively simple matter to derive expressions for the components of the electromagnetic field. By analogy with the frequency domain, we have

$$E_\varphi = -\frac{\partial A_z}{\partial R}\sin\theta,$$

and omitting some algebraic operations obtain

$$E_\varphi = 0, \quad \text{if } t < \tau_0$$

and (4.66)

$$E_\varphi = E_\varphi^{(1)} + E_\varphi^{(2)} \quad \text{if } t \geq \tau_0$$

where

$$E_\varphi^{(1)} = -\frac{\mu_0 M_0}{4\pi R^2}\left[\left(1 + q\tau_0 + \frac{q^2\tau_0^2}{2}\right)\delta(t-\tau_0)\right.$$

$$\left. + \tau_0\delta'(t-\tau_0)\right]\exp(-q\tau_0)\sin\theta \text{ if } t \geq \tau_0,$$

and

$$E_\varphi^{(2)} = -\frac{\mu_0 M_0}{4\pi R^2}q^2\tau_0^3\exp(-qt)\frac{I_2\left[q\left(t^2-\tau_0^2\right)^{1/2}\right]}{t^2-\tau_0^2}\sin\theta \quad \text{if } t > \tau_0, \tag{4.67}$$

since

$$I_2(x) = I_0(x) - \frac{2I_1(x)}{x}$$

To determine the magnetic field, we will make use of the first Maxwell equation

$$\text{curl}\,\mathbf{E} = -\partial\mathbf{B}/\partial t,$$

that gives

$$\dot{B}_R = -\frac{1}{R\sin\theta}\frac{\partial}{\partial\theta}E_\varphi\sin\theta, \quad \text{and} \quad \dot{B}_\theta = \frac{1}{R}\frac{\partial}{\partial R}RE_\varphi$$

It turns out that equations for B_θ component look fairly complicated and in the subsequent analysis we will consider B_R and E_φ components only.

For the time derivative of the radial component of the magnetic field, we have

$$\dot{B}_R = \dot{B}_R^{(1)} + \dot{B}_R^{(2)},$$

where each term is equal to zero if $t \le \tau_0$ and taking into account Eq. 4.67,

$$\dot{B}_R^{(1)} = \frac{\mu_0 M_0}{2\pi R^3}\left[\left(1 + q\tau_0 + \frac{q^2\tau_0^2}{2}\right)\delta(t-\tau_0) + \tau_0\delta'(t-\tau_0)\right]\exp(-q\tau_0)\cos\theta;$$

and

$$\dot{B}_R^{(2)} = \frac{\mu_0 M_0}{2\pi R^3}q^2\tau_0^3\exp(-qt)\frac{I_2\left[q\left(t^2-\tau_0^2\right)^{1/2}\right]}{t^2-\tau_0^2}\cos\theta \tag{4.68}$$

Performing integration with respect to time, we obtain

$$B_R = 0, \quad \text{and} \quad B_\theta = 0 \quad \text{if} \quad t < \tau_0 \tag{4.69}$$

$$B_R^{(1)} = \frac{\mu_0 M_0}{2\pi R^3}[(1+q\tau_0)h(t-\tau_0) + \tau_0\delta(t-\tau_0)]\exp(-q\tau_0)\cos\theta \tag{4.70}$$

Finally,

$$B_R^{(2)} = \frac{\mu_0 M_0}{2\pi R^3}q^2\tau_0^3\int_{\tau_0}^{t}\exp(-qx)\frac{I_2\left[q\left(x^2-\tau_0^2\right)^{1/2}\right]}{x^2-\tau_0^2}dx\cos\theta \tag{4.71}$$

where I_2 and I_3 are modified Bessel functions of the second and third order, respectively, while $h(t - \tau_0)$ is the step function

$$h(t-\tau_0) = \begin{cases} 0 & t < \tau_0 \\ 1 & t > \tau_0 \end{cases}$$

As follows from Eqs (4.69)–(4.71), the electromagnetic field depends on the following parameters: (1) The distance R from the dipole and the angle θ. (2) The velocity of wave propagation of the high-frequency waves,

$$v = \frac{1}{(\varepsilon\mu_0)^{1/2}}$$

(3) The parameter q equal to $\gamma/2\varepsilon$ which has dimension of t^{-1}, and it characterizes the decay of the high-frequency waves in a conducting medium. (4) The time t of the field measurement.

In the same way as we did it in the frequency domain, it is also convenient to represent the field in units of the magnetic field when the dipole current is constant:

$$B_R = \frac{2\mu_0 M_0}{4\pi R^3} b_R \cos\theta$$

4.4. BEHAVIOR OF THE FIELD IN A NONCONDUCTING MEDIUM

4.4.1. Expressions for the Field

At the beginning, we study the field in a nonconducting medium. In accordance with Eqns (4.67)–(4.70), since $q = 0$, we have

$$b_R^{(1)} = h(t-\tau_0) + \tau_0\delta(t-\tau_0)$$
$$e_\varphi^{(1)} = -\delta(t-\tau_0) - \tau_0\delta'(t-\tau_0), \tag{4.72}$$

where

$$E_\varphi = \frac{\mu_0 M_0}{4\pi R^2} e_\varphi \sin\theta \tag{4.73}$$

Thus, the magnetic field in an insulator, as well as the dipole moment, is described by a step-function if $t \neq \tau_0$, and it is natural here to distinguish three successive stages. Until the moment $t = \tau_0 = R/v$, the field is absent at the observation point located at the distance R from the dipole. Due to the instantaneous change of the dipole moment, the wave front is described by a step-function and its derivative with respect to time. The electric field is nonzero only in the vicinity of this front ($t = \tau_0$). After the arrival of the signal ($t > \tau_0$), the magnetic field instantly becomes constant in the same way as the dipole moment, while the electric field vanishes. It is obvious that any field sensor measures its mean value within some small time interval. Correspondingly, in the case of electric field, we have

$$E(\tau_0) = \frac{1}{\Delta t} \int\limits_{\tau_0 - \Delta t/2}^{\tau_0 + \Delta t/2} E(t)\, dt \tag{4.74}$$

Taking into account Eqs (4.72) and (4.73), we obtain

$$E(\tau_0) = -\frac{\mu_0 M_0}{4\pi R^2} \sin\theta \tag{4.75}$$

Note that the step function can be represented as a sum of rectangular impulses of equal magnitude following each other. As a result, the electric field of neighbor impulses is compensated, except the instant $t = \tau_0$.

4.4.2. Duhamel's Integral

We have considered the field behavior when the current source changes as a step function that of course is impossible to realize. To study more realistic cases, it is convenient to apply Duhamel's integral. First suppose that the dipole moment $M(t)$ is an arbitrary function of time,

$$M(t) = \begin{cases} 0 & t < 0 \\ M(t) & t \geq 0 \end{cases}$$

As is seen from Figure 4.3, the dipole moment $M(t)$ can be represented as a sum of subsequent elementary step-functions $h(t-\tau)$ with an amplitude $M'(\tau)\,\mathrm{d}\tau$, and it gives Duhamel's integral:

$$M(t) = M(0) + \int_0^t \frac{\mathrm{d}M}{\mathrm{d}\tau} h(t-\tau)\,\mathrm{d}\tau$$

Before we continue, let us notice that function $M(t)$ shown in Figure 4.3 increases monotonically, but of course it can be an arbitrary function

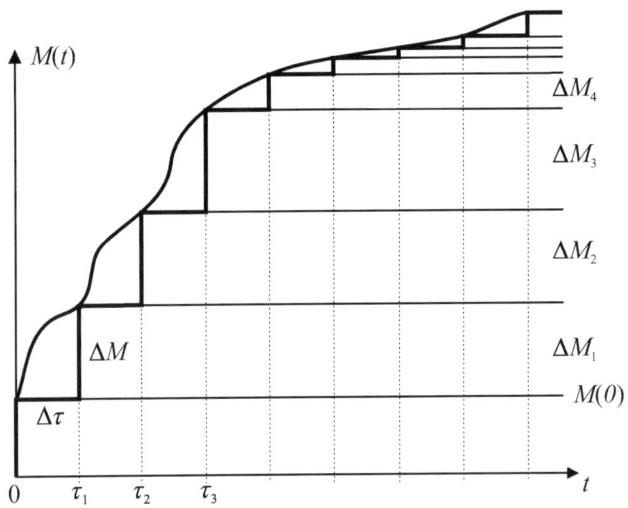

Figure 4.3 Representation of an arbitrary function as a sum of step functions.

including periodical one. Applying the principle of superposition and taking into account that electromagnetic field responses (A_E andA_B), generated by a single-step function are known, we obtain expressions for the field for an arbitrary function $M(t)$

$$B(t) = M(0)A_B(t) + \int_0^t \frac{dM(\tau)}{d\tau} A_B(t-\tau)\, d\tau$$

and

$$E(t) = M(0)A_E(t) + \int_0^t \frac{dM(\tau)}{d\tau} A_E(t-\tau)\, d\tau \qquad (4.76)$$

For instance, in accordance with Eq. (4.72),

$$b_R^{(1)} = \int_0^t \frac{dM}{d\tau}\left[h(t-\tau_0-\tau) + \tau_0\delta(t-\tau_0-\tau)\right] d\tau,$$

since $M(0) = 0$. Taking into account that the step function is equal to unity for positive values of argument $0 < \tau < t - \tau_0$, we have

$$\int_0^t \frac{dM}{d\tau} h(t-\tau_0-\tau)\, d\tau = \int_0^{t-\tau_0} M'(t-\tau_0-\tau)\, d\tau = M(t-\tau_0)$$

Also applying Eq. (4.63), we obtain

$$b_R^{(1)}(t) = M(t-\tau_0) + \tau_0 M'(t-\tau_0) \qquad (4.77)$$

By analogy,

$$e_\varphi^{(1)}(t) = -M'(t-\tau_0) - \tau_0 M''(t-\tau_0) \qquad (4.78)$$

Thus, the electromagnetic field in a nonconducting medium as a function of time is expressed in terms of the moment $M(t)$ and it has the first and second derivatives.

4.4.3. Behavior of the Field of the Magnetic Dipole in a Nonconducting Medium

For illustration, consider two examples.

Example 1

Suppose that

$$M(t) = \begin{cases} 0 & t < 0 \\ kt & 0 \le t \le T, k = \frac{1}{T} \\ 1 & t > T \end{cases}$$

Applying Eq. (4.77), we have for the radial component of the field

$$b_R^{(1)}(t) = \begin{cases} 0 & t < \tau_0 \\ kt & \tau_0 \le t \le \tau_0 + T, \\ 1 & t > \tau_0 + T \end{cases}$$

and it is shown in Figure 4.4. The same behavior has the azimuthal component of the magnetic field. Thus when the moment is a linear function within the initial interval, the magnetic field changes in the similar way, except the instant $t = \tau_0 + T$ where it changes abruptly, and with an increase of the distance from the source, it becomes more noticeable. This discontinuity appears because of the discontinuity of $M'(t)$ at instant $t = \tau_0 + T$, although dipole moment is a continuous function of time. At the same time, the electric field is

$$e_\varphi^{(1)} = \begin{cases} 0 & t < \tau_0 \\ 1/T & \tau_0 \le t \le \tau_0 + T \\ 0 & t > \tau_0 + T \end{cases}$$

that is, it is everywhere zero except the initial interval where it is constant.

Example 2

Consider the case when the moment varies as

$$M(t) = \begin{cases} 0 & t < 0 \\ \sin \omega t & 0 < t < C \\ 0 & t > C \end{cases}$$

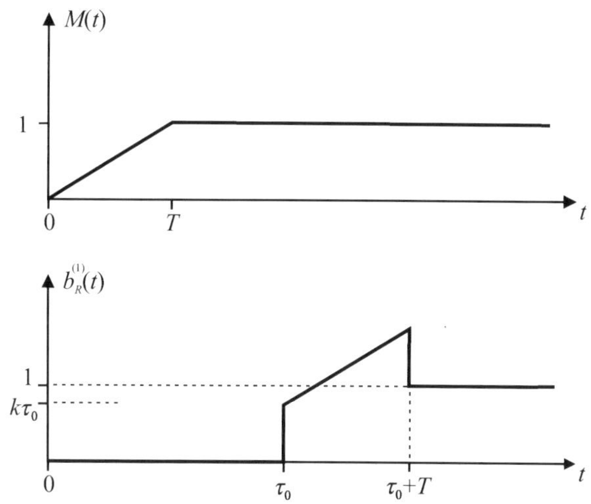

Figure 4.4 Behavior of the field when the moment is linear function of time.

Then applying again Eqs (4.77) and (4.78), we see that each component can be represented as a sum of two sinusoidal functions with different amplitudes and shifted in phase by 90°. For instance, the radial component is

$$b_R(t) = \begin{cases} 0 & t < \tau_0 \\ \sin(\omega(t-\tau_0)) + \omega\tau_0 \cos(\omega(t-\tau_0)) & \tau_0 \le t \le \tau_0 + C \\ 0 & t > \tau_0 + C \end{cases}$$

Of course, if $T \gg \tau_0$ and $t \gg \tau_0$, the field changes practically synchronously with the dipole moment, and we deal with the quasistatic field.

4.4.4. The Second Form of Duhamel's Integral and Representation of the Field as a Sum of Impulses

It is also useful to represent the dipole moment $M(t)$ in a different form, and with this purpose in mind, we will proceed from equality derived earlier

$$M(t) = M(0) + \int_0^t \frac{dM}{d\tau} h(t-\tau)\, d\tau$$

or

$$M(t) = M(0) + \int_0^t h(t-\tau)\, dM(\tau)$$

Integrating by parts, we obtain

$$M(t) = M(0) + M(x)\, h(t-x)|_0^t - \int_0^t M(\tau)\, \delta(t-\tau)\, \mathrm{d}\tau \tag{4.79}$$

Applying the mathematical approach, it was shown that any function can be represented as a system of impulses following one after another. For instance, it could be constant or sinusoidal function. From the physical point of view, it is also obvious because a source current produces within each small time interval some impulse having the front and back side. This approach again suggests that we always deal with propagation even in the case of the constant field or very low-frequency oscillations.

4.5. BEHAVIOR OF THE TRANSIENT FIELD IN A CONDUCTING MEDIUM

Next we investigate the field behavior in a more general case of a conducting medium and first let us study dependence of electric field on time.

4.5.1. The Electric Field at the First Arrival in Conducting Medium

As in the case of a nonconducting medium, the first arrival travels with the velocity $v = c/(\varepsilon_r)^{1/2}$ and until the moment $t = \tau_0$, the field magnitude is zero. As was shown earlier, the intensity $e_\varphi^{(1)}$ of the signal at this moment essentially depends on the parameter $q\tau_0$, which can be represented in the form:

$$q\tau_0 = \frac{1}{2}\gamma\left(\frac{\mu_0}{\varepsilon}\right)^{1/2} \cdot R = b_\varepsilon \cdot R = \frac{R}{R_0} = m, \tag{4.80}$$

where $R_0 = 1/b_\varepsilon$ is the characteristic length which was introduced in the previous section, and b_ε coincides with the limiting value of the imaginary part of the wave number k for the high frequency of the spectrum. Taking into account Eq. (4.51), we have

$$m = \frac{188.5}{\rho(\varepsilon_r)^{1/2}} R \tag{4.81}$$

and for the most practical cases of surface electroprospecting, it is very large number, and one can assume that the first arrival of this part of the field is practically equal to zero due to the very small values of exponential term.

4.5.2. The Dependence of the Field $e_\varphi^{(2)}$ on Time ($t \geq \tau_0$)

It is convenient to represent the function $E_\varphi^{(2)}$ as

$$E_\varphi^{(2)} = -\frac{M\rho}{2\pi R^4} e_\varphi^{(2)} \sin\theta \tag{4.82}$$

$$e_\varphi^{(2)} = m^3 \exp(-mn) \frac{I_2\left[m(n^2-1)^{1/2}\right]}{n^2-1}$$

and

$$n = \frac{t}{\tau_0} \geq 1 \tag{4.83}$$

First consider the field at the moment when it arrives, $n=1$. Applying the expansion of the function $I_2(z)$ in series in z and using only the leading term $I_2(z) \approx \frac{z^2}{8}$, we obtain

$$e_\varphi^{(2)} \approx \frac{1}{8} m^5 \exp(-m) \tag{4.84}$$

Function $e_\varphi^{(2)}$ has a maximum when $m=5$. For this reason, if the distance R from the dipole does not exceed $5R_0$, then with an increase of conductivity, the field also increases. The same behavior is observed with an increase of the distance if $R < 5R_0$. As was pointed out for large values of m, the field is extremely small and it can be treated as zero. Thus, one may say that in most cases, the field is equal to zero at the first arrival. Next we investigate the behavior of the electric field behind the wave front when $t > \tau_0$. Suppose that the argument of the function I_2 in Eq. (4.83) is large,

$$m(n^2-1) \gg 1$$

Then replacing this function by its asymptotic expression,

$$I_2(z) \approx \frac{\exp(z)}{(2\pi z)^{1/2}}, \quad \text{if } z \gg 1$$

we obtain

$$e_\varphi^{(2)} = \frac{1}{(2\pi)^{1/2}} \frac{m^{5/2}}{(n^2-1)^{5/4}} \exp m\left[(n^2-1)^{1/2} - n\right] \tag{4.85}$$

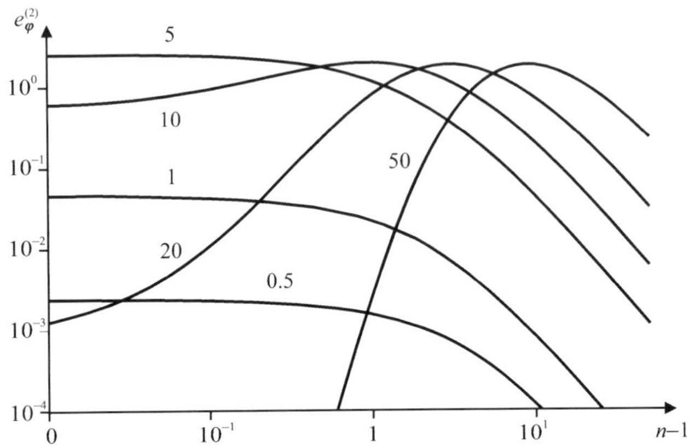

Figure 4.5 Transient responses of the electric field, $e_\varphi(n)$. Index of curves is m.

In particular, this equation is applied when measurements are performedat times which essentially exceed τ_0, $(n \gg 1)$. Then it can be rewritten as

$$e_\varphi = \frac{1}{(2\pi)^{1/2}} \left(\frac{m}{n}\right)^{5/2} \exp\left(-\frac{m}{2n}\right), \tag{4.86}$$

and it describes the quasistationary field which is independent on dielectric permittivity. Curves of the function $e_\varphi(n)$ are shown in Figure 4.5. The index of curves is the parameter $m = R/R_0$. If the distance from the dipole does not exceed $5R_0$, the electric field decreases monotonically with time. However, with an increase of the parameter m (an increase of conductivity or distance, or a decrease of dielectric permittivity), the maximum which appears on curves of e_φ is shifted to larger times. Therefore, in observing the field in a conducting and polarizable medium, we can distinguish the following stages of the transient response: (1) The field is equal to zero until the moment $\tau_0 = \frac{R}{c}(\varepsilon_r)^{1/2}$. For instance, if measurements are performed at distances from the dipole around 1 m, the time of arrival τ_0 is on the order of nanoseconds. (2) At the instant τ_0, the field intensity is a function of the distance from the dipole and electric parameters of the medium. In a conducting medium after the wave front passes through an observation point, the electric field does not disappear instantly. At the beginning, when the time of observation is close to τ_0, both the conduction and displacement currents generate the magnetic field. In other words, the rate of change of the electric field with time cannot be neglected at the early stage of the transient response when $\tau_0 < t < 3\tau_0$. With an increase of resistivity medium, the time interval where displacement currents play an essential role becomes wider. (3) In the last stage of the transient response when the electric field varies with time relatively slowly, the displacement currents can be neglected. In this last interval, the field behaves as a quasistationary field and, as will be shown later, demonstrates features which are typical for diffusion. These ranges of the transient field are observed in any conducting medium, regardless how small its conductivity. However, the moment of transition to a quasistationary field becomes more noticeable earlier with an increase of conductivity.

4.5.3. Dependence of the Electric Field with Distance

Next, consider the electric field as a function of the distance from the dipole. With this purpose in mind, let us make use of Eq. (4.67) and represent it as

$$E_{\varphi}^{(2)} = -\frac{M_0 \rho}{2\pi} \frac{(qt)^3}{(\nu t)^4} F_E \sin\theta$$

$$F_E = x \exp(-qt) \frac{I_2\left[qt(1-x^2)^{1/2}\right]}{1-x^2} \quad \text{and} \quad x = \frac{R}{\nu t}$$

Graphs illustrating the distribution of the electric field in a medium as a function of the distance are shown in Figure 4.6. The index of the curves is the parameter qt As is seen from these curves for small values of qt, the maximum of the field intensity takes place near the wave front and on approaching the dipole, the field decreases almost linearly. In the opposite

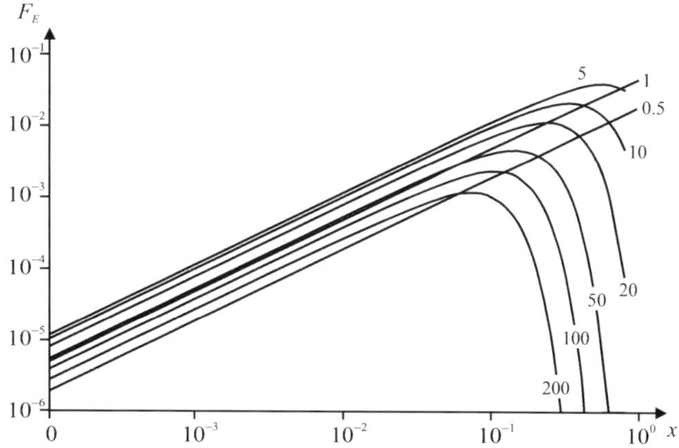

Figure 4.6 Behavior of function F_E. Index of curves, qt.

case of large values of qt, a different behavior is observed, in particular, a maximum moves away from the wave front.

4.6. PROPAGATION AND DIFFUSION

In studying an electromagnetic field of the magnetic dipole in a uniform medium, we considered two cases. In one of them, the dipole current changes as a step function, while in the other, it is a sinusoidal function. In the first case, several time intervals were distinguished, namely:

1. The absence of the field if $t < \tau_0$.
2. The moment of the first arrival $t = \tau_0$ when the field in a conducting medium is usually extremely small.
3. The time interval $\tau_0 \leq t \leq n\tau_0$, $n > 1$. Within this interval, the electric field changes with time relatively quickly and both displacement and conduction currents generate the field, which usually becomes much greater than that of the first arrival.
4. Within the last time interval $(t > n\tau_0)$, the influence of displacement currents can be neglected but conduction currents play the dominant role. Their behavior, that is the magnitude, distribution in space, dependence on time, interaction between them with a high accuracy can be described by the diffusion equation. The latter can be treated as an approximation of more general equation which takes into account influence of both currents. As we already know, within this range of time, the behavior of the field can be explained with a high accuracy by Biot–Savart and Coulomb's laws which imply the instant propagation of the field. Of course, in reality the field travels with a finite velocity but it turns out that a contribution of this factor is insignificant. Thus, the propagation is present at all times, but within the last stage it can be neglected and for this reason we can deal with only diffusion; that is, a quasistationary field. It is obvious that the use of the diffusion equation does not allow us to study the first arrival as well as the next stage of the transient response. Also let us notice the following. In the case of sinusoidal oscillations, the quasistationary approximation (diffusion equation) shows that the field is described by the sinusoidal wave which strongly decays with the distance and its velocity depends on a frequency. Inasmuch as the first arrival is absent, an accuracy of the quasistationary approximation is defined only by the ratio between the displacement and conduction currents. Imagine that instead of the step function, the dipole

current behaves as the rectangular impulse. It is obvious that it can be represented as a difference of two step functions arising at different times. Considering propagation of such impulse, we again observe its attenuation and distinguish four time intervals. Intuitively, it is clear that the impulse width makes influence on a contribution of each time interval. If it is very small and comparable with τ_0, then the field within the impulse is subjected to influence of both the displacement and conduction currents. In other words, this field propagates as the high-frequency waves. At the same time behind the impulse, the field is relatively weak because the fields caused by each step function almost cancel each other, and correspondingly the quasistationary stage is hardly noticeable. Correspondingly, the system of such alternating impulses following one after another approximately represents the high-frequency wave, where we observe only propagation and transformation into heat. With an increase of the impulse width, the last (quasistationary) stage appears inside of the impulse as well as outside because cancellation effect becomes weaker. The same takes place in the case of a system of alternating impulses; that is, the diffusion effect begins to play the dominant role. As was already pointed out, the sinusoidal field can also be represented a system of the alternating impulses. In particular, these impulses can have a sinusoidal shape with width equal of a half period. This consideration also suggests that the effect of propagation is present regardless of how low the frequency is.

REFERENCES AND FURTHER READING

[1] R. Gabbilard, Reflexions sur le problem de le propagation d'une onde electron netique dans le sol, Rev. Ins. Fr. Ret. 28 (9) (1963).
[2] A.A. Kaufman, Geophysical Theory and Methods, Part B, Academic Press, New York, NY, 1994.
[3] J.A. Stratton, Electromagnetic Theory, McGraw-Hill, 1941.

PART THREE

Quasi-Stationary Field in a Horizontally Layered Medium

CHAPTER FIVE

The External and Internal Skin Effect, Diffusion

INTRODUCTION

Now we focus on the behavior of quasi-stationary field in a horizontally layered medium that will allow us to describe the principles of the frequency and transient soundings, in particular, the magnetotelluric soundings. As we already know, neglecting by displacement currents, the system of equations of this field at regular points of a piecewise uniform medium is

$$\operatorname{curl} \mathbf{E} = -\frac{\partial \mathbf{B}}{\partial t}, \quad \operatorname{div} \mathbf{E} = 0$$

$$\operatorname{curl} \mathbf{B} = \mu_0 \mathbf{j}, \quad \operatorname{div} \mathbf{B} = 0,$$

and at points of interfaces,

$$\mathbf{n} \times (\mathbf{E}_2 - \mathbf{E}_1) = 0, \quad \mathbf{n} \times (\mathbf{B}_2 - \mathbf{B}_1) = \mu_0 \mathbf{i},$$

where $\mathbf{j}$ and $\mathbf{i}$ are the volume and surface density of the conduction current, respectively. At the same time, the principle of charge conservation is written as

$$\operatorname{div} \mathbf{j} = 0, \quad j_{2n} - j_{1n} = 0$$

It is proper to notice that these equations do not take into account the influence of extraneous forces. Applying Stokes and Gauss's theorems, we obtain the integral form of Maxwell's equations

$$\oint \mathbf{E} \cdot d\mathbf{l} = -\frac{\partial \Phi}{\partial t} \quad \oint \mathbf{E} \cdot d\mathbf{S} = 0$$

and

$$\oint \mathbf{B} \cdot d\mathbf{l} = \mu_0 I \quad \oint \mathbf{B} \cdot d\mathbf{S} = 0,$$

Methods in Geochemistry and Geophysics, Volume 45
ISSN 0076-6895,
http://dx.doi.org/10.1016/B978-0-444-53829-1.00005-8

where Φ and I are the flux of the magnetic field and current correspondingly. These equations are valid when

$$\varepsilon \frac{\partial \mathbf{E}}{\partial t} \ll \gamma \mathbf{E} = \mathbf{j}$$

From mathematical point of view, this is equivalent to assumption that the dielectric permittivity of a medium is zero: $\boldsymbol{\varepsilon}_{\mathrm{r}} = 0$. Taking into account the fact that maximal velocity of wave propagation occurs at the high-frequency spectrum and it is equal to

$$v = \frac{c}{(\varepsilon_{\mathrm{r}})^{1/2}},$$

we conclude that the quasi-stationary field implies that this velocity tends to infinity; that is, an electromagnetic field instantly propagates through the medium. Correspondingly, as was mentioned earlier, we can use Biot–Savart and Coulomb's laws as in the case of time-invariant field. In other words, the relationship between the electric field and charges, as well as the magnetic field and currents, is the same as in the constant field. In this sense, we see some analogy between the quasi-stationary and stationary fields. However, there is also an essential difference due to the electromagnetic induction which is absent in the case of the time-invariant field. This is the reason why solutions of the equations, given above, are called quasi (almost)-stationary fields. Also, let us remind the following. As was shown earlier, we distinguished two time intervals after the arrival of the signal at the observation point. In the first one, both the displacement and conduction currents play an important role, while in the second interval (tail of the signal), the influence of the dielectric permittivity is negligible; that is, we deal with the quasi-stationary field. In other words, when displacement currents are neglected, we are not able to study the wave arrival and the first time interval. Bearing in mind that field equations at regular points are equivalent to diffusion equation, we see that the behavior of the transient quasi-stationary field has features of diffusion. It may be proper to emphasize again that in reality, the signal arrives at the observation point due to propagation of electromagnetic waves, but the quasi-stationary approximation allows us with high accuracy to describe only the last stage of this phenomenon. As usual, we will start to study electromagnetic fields in the simplest models of a medium, but at the beginning, it is important to describe in detail the phenomenon of the skin effect.

5.1. THE SKIN EFFECT

5.1.1. Equations for Currents in a System of Conducting Circuits

First we assume that the primary magnetic field changes in the simplest possible way, namely, as a step function. However in such a case, the right-hand side of the first equation

$$\operatorname{curl} \mathbf{E} = -\frac{\partial \mathbf{B}}{\partial t}$$

tends to infinity at the instant $t = t_0$, because the current source either abruptly arises or disappears. The similar situation happens if the system of equal amount of positive and negative charges suddenly vanishes. For this reason, we will use a special approach which allows us to find the field behavior at the initial moment. It is obvious that any volume distribution of currents in a confined conductor can be mentally represented as a system of current tubes, and then assume that under certain conditions, it may be replaced by linear currents with unknown magnitudes and orientation. Therefore, instead of an original model of a conducting medium, we obtain a system of N linear currents placed in the primary magnetic field $\mathbf{B}_0(t)$. Certainly, it is only approximation which becomes more accurate with an increase of number of rings. Earlier in Chapter 3, we derived an equation of the induced current in the single circuit. Now it is a simple matter to generalize this result for an arbitrary number of rings. Applying again Faraday's law for the induced current in k ring, we have

$$R_k I_k = -\frac{\partial \Phi_{0k}}{\partial t} - \frac{\partial \Phi_{sk}}{\partial t} \tag{5.1}$$

Here, R_k and I_k are the resistance and the current in the k circuit, respectively. Φ_{0k} and Φ_{sk} are the magnetic fluxes of the primary and secondary fields. It is obvious that the flux Φ_{sk} caused by induced currents in all rings can be written in the form

$$\Phi_{sk} = M_{1k}I_1 + M_{2k}I_2 + \cdots + M_{ik}I_i + L_k I_k + \cdots + M_{Nk}I_N \tag{5.2}$$

Here, L_k is the self-inductance of the k in ring and M_{ik} is the mutual inductance between circuits "i" and "k". In other words, M_{ik} characterizes the flux through the area of the k the circuit, caused by the unit current in the "i"th ring:

$$\Phi_{ik} = M_{ik} I_i \tag{5.3}$$

Respectively, Eq. (5.1) can be written as

$$R_k I_k + L_k \frac{\partial I_k}{\partial k} + \sum_{n=1}^{N} M_{nk} \frac{\partial I_n}{\partial t} = -\frac{\partial \Phi_{0k}}{\partial t} \tag{5.4}$$

Note that now and later this summation does not include term for which $n = k$. Having written this equation for all rings, we arrive at the system of equations with respect to induced currents provided that parameters of rings are known. Such situation takes place only when geometry of the current rings is known. However, our goal is not a solution of the forward problem, and for this reason it is no matter whether these parameters are known or not. First assume that the primary flux through kth ring Φ_{0k} caused by the primary source starts to decrease at some instant 0 from value $\Phi_k^{(0)}$ to zero as is shown in Figure 3.3(c).

5.1.2. The Law of Inertia for the Magnetic Flux

Then integrating Eq. (5.4) within the ramp time: $0 \le t \le t_r$, we obtain

$$R_k \int_0^{t_r} I_k \mathrm{d}t + L_k \int_0^{t_r} \frac{\partial I_k}{\partial t} \mathrm{d}t + \sum_{n=1}^{N} M_{nk} \int_0^{t_r} \frac{\partial I_n}{\partial t} \mathrm{d}t = \Phi_k^{(0)}, \tag{5.5}$$

since

$$\Phi_{0k}(t_r) = 0$$

Taking into account the fact that induced currents are absent at the instant $t = 0$ and the ramp time t_r is so small that during this time the amount of charges moving through any cross-section of the circuit is negligible, we have

$$\int_0^{t_r} I_k \mathrm{d}t \to 0$$

and Eq. (5.5) can be approximately represented in the form:

$$L_k I_k(t_r) + \sum_{n=1}^{N} M_{nk} I_n(t_r) \approx \Phi_k^{(0)}$$

The left-hand side of this equality characterizes the flux of the magnetic field through the k circuit, caused by induced currents in all circuits at the moment t_r when the primary field $\mathbf{B}_0$ vanishes. At the same time, the right-hand side describes the flux of the primary field through the same ring at the initial moment $t = 0$. Thus in the limit when t_r tends to zero and the primary field becomes the step function, we as in the case of the single circuit again observe the fundamental feature of the electromagnetic induction. In a fact, at the first instant induced currents are distributed in such a way that the flux of the field $\mathbf{B}$, caused by these currents through any circuit remains equal to the flux which was present previously. Omitting index k, we have

$$\Phi_s = \Phi_0, \quad \text{if } t = 0 \tag{5.6}$$

that is, if the primary field suddenly vanishes (turn off) the secondary flux through any circuit is equal to the primary one when it existed. In contrast, if the primary flux Φ_0 appears (turn on), then the flux of induced currents Φ_s compensates the Φ_0 in the conducting medium:

$$\Phi_0 + \Phi_s = 0 \quad \text{if } t = 0 \tag{5.7}$$

Unlike the previous case, both fluxes exist at the instant t_r which tends to zero. The last two equations clearly demonstrate the inertia of the magnetic flux which is one of the most remarkable features of its behavior. Taking into account this amazing behavior of the flux, let us derive again the last two equations in a slightly different manner. As before, we will proceed from Faraday's law

$$\oint\limits_l \mathbf{E}\cdot d\mathbf{l} = -\frac{\partial \Phi}{\partial t} \tag{5.8}$$

Suppose that l describes the central line of an elementary current tube. Then applying Ohm's law, we have

$$I\oint\limits_l \frac{\rho\mathbf{j}\cdot d\mathbf{l}}{I} = -\frac{\partial \Phi}{\partial t} \quad \text{or} \quad RI = -\frac{\partial \Phi_0}{\partial t} - \frac{\partial \Phi_s}{\partial t}, \tag{5.9}$$

where R is the resistance of the current tube, Φ_0 is the flux of the primary magnetic field, and Φ_s is the secondary flux caused by all induced currents including the current I. It is proper to notice that unlike the first approach, we do not replace the real distribution of volume currents by a system of linear currents. Now integrating Eq. (5.9) within the ramp time, where the primary field changes and becomes equal to zero, we obtain

$$R\int_0^{t_r} I\mathrm{d}t = \Phi_0 - \Phi_s(t_r) \tag{5.10}$$

Here the integral defines the amount of charge which passes through any cross-section within the time interval t_r. As is well known, the movement of these charges is accompanied by transformation of the electromagnetic energy into the heat, and this process becomes more rapid with an increase of resistivity of a medium. It is obvious that the left-hand side of Eq. (5.10) is exactly equal to zero at least in two cases.

1. The primary field changes as a step function and the integral vanishes since $t_r = 0$.
2. The medium is an ideal conductor and therefore $R = 0$. In this case, regardless of the behavior of the primary field as a function of time, we observe the principle of inertia. In other words, the flux of the magnetic field cannot change inside an ideal conductor. It is clear that in a real conducting medium with a decrease of either resistivity ρ or time interval t_r, the integral in Eq. (5.9) becomes smaller and in the limit we again observe the principle of inertia. Now let us make several comments.

1. Equation (5.6) or (5.7) is exact if the primary flux changes as a step function, while it is only approximation if a finite change of Φ_0 requires some time.
2. The instantaneous change of the primary flux leads to an instantaneous change of the secondary flux at the same moment.
3. In deriving Eq. (5.6) which describes the principle of inertia, we implied that the path l corresponds to the current tube with the current I and resistance R. However, this assumption is not necessary. In fact, applying Ohm's law in the differential form, we have

$$\oint_l \rho\mathbf{j}\cdot\mathrm{d}\mathbf{l} = -\frac{\partial\Phi}{\partial t} = -\frac{\partial\Phi_0}{\partial t} - \frac{\partial\Phi_s}{\partial t}, \tag{5.11}$$

where l is an arbitrary contour inside the conducting medium. Performing integration of Eq. (5.11) within elementary time interval $\Delta t = t_r$ and taking into account that both functions ρ and $\mathbf{j}$ have finite values, we again arrive at the principle of inertia of the magnetic flux Φ. Let us formulate it in the following way: in the limit when the primary flux Φ_0

behaves as a step function at some instant t_0, the flux Φ through an area bounded by any contour l, which is located inside a conducting medium, does not change at this moment.

5.1.3. Behavior of the Magnetic Field at the Initial Moment $t = t_0$

Now we are prepared to understand one of the most important features of current distribution. First suppose that the generators of the primary magnetic field are located outside the conducting medium and at the instant $t = t_0$, they arise instantly. Correspondingly, the primary flux behaves as

$$\Phi_0(p,t) = \begin{cases} 0 & t < t_0 \\ \Phi_0(p) & t \geq t_0 \end{cases}$$

Then in accordance with the principle of inertia (Eq. (5.7)), the flux of the magnetic field,

$$\Phi = \int_S \mathbf{B} \cdot d\mathbf{S}$$

inside the conducting medium remains equal to zero at the moment $t = t_0$. What is surprising is the fact that it happens in spite of the appearance of the primary field $\mathbf{B}_0$. Therefore, induced currents must arise in order to cancel the primary flux at each point of this medium

$$\Phi(p) = \Phi_0(p) + \Phi_s(p) = 0 \quad \text{if } t = t_0 \tag{5.12}$$

Certainly, without these currents the principle of inertia would be invalid. The natural question arises: where are induced currents located at this moment, which provide the principle of inertia? To find answer, let us represent Eq. (5.12) as

$$\int_S \mathbf{B} \cdot d\mathbf{S} = 0, \quad \text{if } t = t_0 \tag{5.13}$$

where $\mathbf{B}$ is the total magnetic field and S is an arbitrary area inside the conducting medium. Inasmuch as Eq. (5.13) is valid for any surface S, regardless of its shape, orientation, and dimensions, we have to conclude that the magnetic field $\mathbf{B}$ inside the medium is also equal to zero at the instant when the primary field arises; that is,

$$\mathbf{B}(p) = \mathbf{B}_0(p) + \mathbf{B}_s(p) = 0 \quad \text{if } t = t_0 \tag{5.14}$$

Certainly, this is a remarkable result which shows that the secondary field, $\mathbf{B}_s$, caused by induced currents compensates the primary field $\mathbf{B}_0$ at each point of the conducting medium. In other words, these currents are distributed in such "clever" way that at each point of a conducting medium without exception, the magnetic field is zero at the initial moment.

5.1.4. Location of Induced Currents at the Initial Instant

Now it is a simple matter to determine the position of the induced currents. In fact, taking into account Eq. (5.14) and the second Maxwell's equation for the quasi-stationary field,

$$\text{curl}\,\mathbf{B} = \mu_0 \mathbf{j},$$

we find that at the initial moment, the volume density of currents inside a conducting medium is equal to zero:

$$\mathbf{j}(p) = 0 \quad \text{if } t = t_0 \tag{5.15}$$

Thus, only one place of a conductor is left which was not considered, and we have to conclude that at this instant, the induced currents appear only on the surface of a conducting medium. And they are distributed in such a manner that the total magnetic field inside a conducting medium is still equal to zero at $t = t_0$. It is essential that this behavior of induced currents and magnetic field is independent on a resistivity of a medium. For instance, Eqs (5.14) and (5.15) are valid in a very good conductor (metal) as well as in extremely resistive rocks like granite. Let us emphasize that the secondary magnetic field, caused by surface currents, does not compensate the primary field $\mathbf{B}_0$ outside a conducting medium and, in this sense, there is some analogy with the electrostatic induction. Next suppose that the primary field $\mathbf{B}_0$ instantly disappears at the instant $t = t_0$. Then, as follows from Eq. (5.6), the flux of the magnetic field inside the conducting medium remains the same at this moment in spite of the fact that the primary field is absent:

$$\int_S \mathbf{B}_s \cdot d\mathbf{S} = \Phi_0 \quad \text{if } t = t_0 \tag{5.16}$$

or

$$\int_S \mathbf{B}_s \cdot d\mathbf{S} = \int_S \mathbf{B}_0 \cdot d\mathbf{S}$$

Bearing in mind that the latter is valid for any surface S, we conclude that at every point of this medium the magnetic field is equal to the primary field:

$$\mathbf{B}_s = \mathbf{B}_0 \quad \text{if } t = t_0 \tag{5.17}$$

Inasmuch as a time-invariant field $\mathbf{B}_0$ cannot generate conduction currents, Ampere's law for any contour l inside a conducting medium has a form

$$\oint_l \mathbf{B}_0 \cdot \mathbf{dl} = 0 \tag{5.18}$$

Then, taking into account Eq. (5.17), we see that at the instant when the primary field vanishes the circulation of the secondary field is also equal to zero:

$$\oint_l \mathbf{B}_s \cdot \mathbf{dl} = 0 \quad \text{if } t = t_0$$

Thus, currents are absent inside the conducting medium while surface currents arise; and due to these surface currents, the field in a conductor remains equal to $\mathbf{B}_0(p)$ at the moment $t = t_0$. For example, if the primary field is caused by a single current and it is turned off, then the surface currents usually have the same direction as the primary current. In contrast, when the primary source is turned on, the induced currents have opposite direction.

5.1.5. The External and Internal Skin Effect

The phenomenon of appearance of induced currents on the conductor surface at the moment when the primary source either instantly appears or disappears is called *skin effect*. Assuming that the primary source behaves as a step function, let us describe the skin effect in some detail. Suppose that the single current source of the field $\mathbf{B}_0$, located outside a conductor begins to work instantly at the moment t_0. Then the electromagnetic energy propagates from the generator through a nonconducting medium and reaches the conducting surface. In the quasi-stationary approximation, it is assumed that the field arrives instantly at the conductor surface where induced currents appear. And as we know, the distribution of these currents is such that the total magnetic field remains zero inside the conductor at the instant $t = t_0$. This result is independent of the conductivity which can be arbitrary function of a point. In particular, the conductor can be piecewise uniform medium with several interfaces inside. This discussion indicates that at the initial moment induced currents appear only on the external surface where the primary electromagnetic field arrives at the first place. As the primary source, for example a current loop, approaches this surface, the induced currents are mainly concentrated near the source. This happens because the primary electric field is maximal within this area. Of course, if there are several sources of the primary field which are located outside the conductor, the distribution of surface currents becomes much more complicated. However, in all possible cases, the magnetic field of these currents must satisfy Eq. (5.14). This consideration allows us to distinguish two essential features characterizing the skin effect.

1. The appearance of only surface currents at the initial moment.
2. The generators of the secondary field arise at the external surface between a conductor and an insulator where the current sources of the primary field are located. In this case, we speak about the external skin effect. In contrast, suppose that the generators of the primary field are located somewhere inside the conducting medium. In this case, induced currents appear at the internal surface of the conductor which surrounds the generators, and we deal with the internal skin effect. Certainly, the magnetic field of these currents, as in the previous case, satisfies Eq. (5.14). If the volume where primary generators are placed becomes smaller, then in the limit, the internal surface coincides with the lateral surface of the primary current tube which creates the field $\mathbf{B}_0$. In such case, the magnitude of the secondary and primary currents is equal to each other, while the mutual direction is defined by the behavior of the field $\mathbf{B}_0$ at the initial moment. Thus, in principle, the internal and external skin effects do not differ from each other. In fact, in both cases, induced currents arise at the instant $t = t_0$ (quasi-stationary approximation) at the surface which is the closest to the generators of the primary field. If these current sources are situated both outside and inside the conductor, then the skin effect is displaced on its external surface as well as on the corresponding surfaces inside the medium. Of course, this phenomenon can occur either simultaneously or at different times, and it is defined by the behavior of the generators of the primary field.

Example

To illustrate the skin effect, let us consider a conducting shell and assume that as before, the generators of the primary field change with time as the step functions. If these generators are surrounded by the shell, then induced currents appear at the instant $t = t_0$ only on the internal surface of the shell. It is proper to note that these currents preserve the magnetic field at $t = t_0$ not only within the shell but also in an external nonconducting medium surrounding this shell. Certainly, such behavior is not observed inside the shell where the primary generators are located. Next suppose that these generators are located outside this shell. Correspondingly, at the initial instant, induced currents arise only on the external surface, and their magnetic field obeys Eq. (5.14) in the volume surrounded by this surface which also includes the internal nonconducting part of the medium.

When the primary electromagnetic field is caused by the current source, it is natural to distinguish two cases. In the first one, the primary vortex electric field does not intersect the external surface of the conducting medium, and therefore charges are absent at the moment when the primary field changes as the step function. In the second case, which is more general one, this field has a normal component on the conductor surface. For this reason, at the instant $t = t_0$, the skin effect manifests itself in the appearance of both surface currents and charges. Because of the presence of these charges, conduction currents always remain tangential to the interface between a conducting medium and an insulator. In conclusion of this section, let us make two comments. (1) At the beginning, it was assumed a model of a confined conductor, but the same skin effect takes place also in the case of a medium of infinite extension. (2) Neglecting by the displacement currents, we assume that the primary field instantly reaches all points of the conductor surface. Inasmuch as the velocity of propagation of electromagnetic field is extremely high, this assumption does not result in significant errors for many geophysical applications, but there may be some exceptions.

5.2. DIFFUSION OF INDUCED CURRENTS

The study of the skin effect allowed us to see that if the primary magnetic field behaves as a step function of time, then at the initial moment, the induced currents appear at the surface of a conductor. With an increase of time, this current decreases, and this occurs because the electromagnetic energy transforms into heat. A change of the surface currents cause a change of their magnetic field with time and therefore the induced electric field arises inside of a conducting medium. As follows from Ohm's law, the induced currents appear, and our goal is to find a distribution of these currents and the magnetic and electric field associated with them. Assuming that displacement currents are equal to zero; that is, the effect of propagation is neglected, we deal only with the quasi-stationary field which obeys diffusion equations

$$\nabla^2 \mathbf{E} = \gamma\mu_0 \frac{\partial \mathbf{E}}{\partial t}, \quad \nabla^2 \mathbf{B} = \gamma\mu_0 \frac{\partial \mathbf{B}}{\partial t}, \quad \nabla^2 \mathbf{j} = \gamma\mu_0 \frac{\partial \mathbf{j}}{\partial t}, \tag{5.19}$$

5.2.1. One-Dimensional Diffusion Equation

To illustrate the phenomenon of the skin effect and diffusion, let us consider the simplest model of a medium and the field where its study does not require sufficiently complicated derivations. Suppose that the primary magnetic field $\mathbf{B}_0$ is caused by currents which are uniformly distributed on a horizontal surface (Figure 5.1), and a uniform half space with conductivity γ and magnetic permeability μ_0 is located beneath these currents. We choose the Cartesian system of coordinates. The surface current density of the primary source has only the component $\mathbf{i}_{0y}$ ($\mathbf{i}_{0y} = \text{constant}$). As follows from the Biot–Savart law, such a distribution of currents generates above and beneath the plane $z = -H$ a uniform magnetic field that has only the component:

$$B_{0x} = \frac{\mu_0 i_{0y}}{2} \quad \text{if } z > -H \tag{5.20}$$

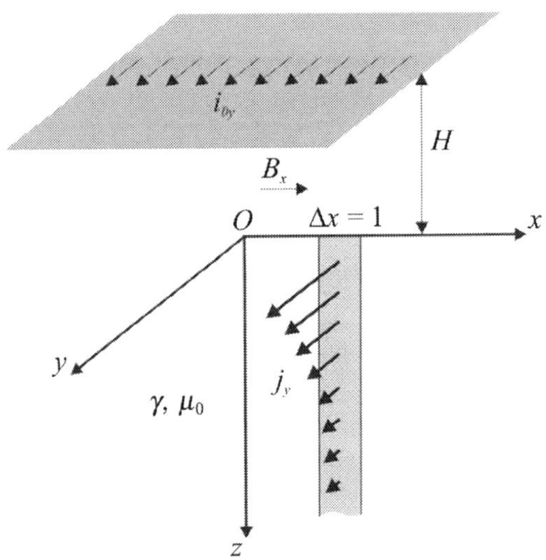

Figure 5.1 Model of a medium and a vertical strip with unit width.

Assuming that the primary field changes as a step function,

$$B_{0x}(t) = \begin{cases} B_{0x} & t \le 0 \\ 0 & t > 0 \end{cases} \tag{5.21}$$

we already know that at the initial moment $t=0$, induced currents arise at the surface of a conducting half space, and their magnetic field inside the medium coincides with the primary field B_{0x} when $t=0$. We also assume that the induced currents have only the component j_y if $t \geq 0$. This implies that the electric and magnetic fields have components E_y and B_x, and in place of Eq. (5.19), we have

$$\frac{\partial^2 E_y}{\partial z^2} = \gamma\mu_0 \frac{\partial E_y}{\partial t}, \quad \frac{\partial^2 B_x}{\partial z^2} = \gamma\mu_0 \frac{\partial B_x}{\partial t} \quad \frac{\partial^2 j_y}{\partial z^2} = \gamma\mu_0 \frac{\partial j_y}{\partial t}, \tag{5.22}$$

since fields depend on coordinate z only. Equation (5.22) is the partial differential equations of the second order and in deriving them, we have used three physical laws, namely, Ampere's, Faraday's, and Ohm's laws. It is obvious that there are an unlimited number of models of a conducting medium and generators of the primary field, when the different quasi-stationary fields in a medium depend only on the coordinate z and time t. Moreover, equations of this type may describe completely different physical phenomenon, such as diffusion of ions or penetration of heat. Also, the magnetic field and current density have different behavior, but both of them satisfy the same equation. All these facts illustrate again that Eq. (5.22) has an infinite number of solutions. First, we investigate the distribution of induced currents.

5.2.2. Expression for the Current Density

In order to find the current density, we recall that in the frequency domain, the complex amplitude of the current obeys the equation

$$\frac{d^2 j_y^*}{dz^2} + k^2 j_y^* = 0,$$

and its solution, obeying the condition at infinity, is described by the function $\exp(ikz)$. Then taking into account the spectrum of the step function and applying inverse Fourier transform, we obtain for the current density

$$j_y(z,t) = C \frac{\exp\left(-\gamma\mu_0 z^2/4t\right)}{t^{1/2}} \tag{5.23}$$

Here C is some constant. As in the case of wave equation, it may be proper to show that the function given by Eq. (5.23) obeys the diffusion equation. Taking derivatives with respect to z, we have

$$\frac{\partial j_y}{\partial z} = -\frac{C}{t^{1/2}}\left(\frac{\gamma\mu_0 z}{2t}\right)\exp\left(-\frac{\gamma\mu_0 z^2}{4t}\right) = -\frac{C(\gamma\mu_0 z)}{2t^{3/2}}\exp\left(-\frac{\gamma\mu_0 z^2}{4t}\right)$$

and

$$\frac{\partial^2 j_y}{\partial z^2} = -\frac{\gamma\mu_0 C}{2t^{3/2}}\exp\left(-\frac{\gamma\mu_0 z^2}{4t}\right)\left[1-\frac{\gamma\mu_0 z^2}{2t}\right] \tag{5.24}$$

Next, forming the derivative with respect to time, we obtain

$$\frac{\partial j_y}{\partial t} = \frac{-C\exp\left(-\gamma\mu_0 z^2/4t\right)}{2t^{3/2}}\left[1-\frac{\gamma\mu_0 z^2}{2t}\right]$$

or

$$\gamma\mu_0\frac{\partial j_y}{\partial t} = -\frac{C\gamma\mu_0}{2t^{3/2}}\exp\left(-\frac{\gamma\mu_0 z^2}{4t}\right)\left[1-\frac{\gamma\mu_0 z^2}{2t}\right] \tag{5.25}$$

Comparing Eqs (5.24) and (5.25), we confirm that function given by Eq. (5.23) is a solution of the diffusion equation. It is immediately clear that the behavior of the current density does not contradict the phenomenon of the skin effect. In fact, as follows from Eq. (5.23), the volume density vanishes at the instant $t=0$ at any regular point of the conducting medium ($z \neq 0$). Before we begin to study the current distribution, let us determine the constant C. With this purpose in mind, it is useful to remember that the primary magnetic field $\mathbf{B}_{0x}$ is caused by currents which are uniformly distributed on the horizontal plane $z=-H$, and this field is independent of the coordinate z. In other words, this field is constant within a conducting medium. Certainly, it is unrealistic case and for this reason, it is natural to expect some peculiar features of the field behavior. Taking into account that at the initial moment $t=0$, the secondary field is equal to the primary one and making use of Eq. (5.20), we conclude that the density of the induced surface currents coincides with that of the primary currents:

$$i_y = i_{0y} \quad \text{if } t=0 \quad \text{and } z=0 \tag{5.26}$$

5.2.3. Determination of Constant C

To calculate the constant C, we consider a vertical strip having unit width in a conducting medium which is perpendicular to the y-axis and calculate the current I passing through this strip (Figure 5.1). Making use of Eq. (5.23), the current through this strip is

$$I = \frac{C}{t^{1/2}} \int_0^\infty \exp(-pz^2)\,dz, \tag{5.27}$$

where

$$p = \frac{\gamma\mu_0}{4t}$$

The integral on the right-hand side of Eq. (5.27) is well known and it can be represented as

$$\int_0^\infty \exp(-pz^2)\,dz = \frac{1}{2}\left(\frac{\pi}{p}\right)^{1/2} = \left(\frac{\pi t}{\gamma\mu_0}\right)^{1/2} \tag{5.28}$$

Thus, we have

$$I = C\left(\frac{\pi}{\gamma\mu_0}\right)^{1/2} \tag{5.29}$$

As follows from this equation, the current I, unlike the density j_y, does not depend on time. In particular, at the instant $t = 0$, this current is located only on the surface of a conducting medium and, in accordance with Eq. (5.21), it is equal to i_{0y}:

$$I = i_{0y} \tag{5.30}$$

Therefore for the constant C, we have

$$C = \left(\frac{\gamma\mu_0}{\pi}\right)^{1/2} i_{0y} \tag{5.31}$$

and

$$j_y = i_{0y}\left(\frac{\gamma\mu_0}{\pi t}\right)^{1/2} \exp\left(-\frac{\gamma\mu_0 z^2}{4t}\right) \tag{5.32}$$

It seems that Eqs (5.30) and (5.32) contradict with each other. In fact, as follows from Eq. (5.32), the current density tends to zero with an increase of time, and it is understandable because an electromagnetic energy transforms into heat. At the same time, the total current I (Eq. (5.30)) does not decrease with time. Such contradiction is one of the peculiar features of the plane-wave field and it does not take place in real conditions.

5.2.4. Dependence of Induced Currents on Time

Now we are ready to describe in detail the most essential features of the current distribution in the conducting medium which characterizes diffusion. First of all, at the instant $t=0$ currents are concentrated on the surface and they are absent inside the conductor. Indeed, from Eq. (5.32), it follows that if $z \neq 0$, the volume density vanishes at the moment $t=0$ since the exponential term decreases more rapidly than the increase of the function $t^{-1/2}$. Thus, as mentioned above, we observe the skin effect. In particular, in the case of the ideal conductor, $\gamma \to \infty$, the currents remain on the surface regardless of time and they do not appear inside the medium. This behavior is obvious and it can be explained in the following way. Inasmuch as resistivity is equal to zero, the electromagnetic energy does not transform into heat. Respectively, currents on the surface as well as their magnetic field remain constant. For this reason, the inductive electric field is equal to zero, and therefore, currents do not arise inside a perfect conductor. However, in real media which have a finite value of conductivity, we observe a completely different behavior of currents. First, consider an arbitrary point beneath the surface with coordinate z. As follows from Eq. (5.32), the current density at the beginning increases very rapidly with time, reaches a maximum and then relatively slowly, as $t^{-1/2}$, decreases. It is proper to note that the current density vector does not change direction, which coincides with that of the surface currents. To describe the time dependence of j_y, it is useful to rewrite Eq. (5.32) as

$$j_y(t,z) = \frac{2i_{0y}}{z}\left(\frac{\tau}{\pi t}\right)^{1/2} \exp(-\tau/t), \tag{5.33}$$

where τ can be called the *time constant* equal to

$$\tau = \frac{\gamma\mu_0 z^2}{4} \tag{5.34}$$

It is directly proportional to the conductivity and the distance from the conductor surface squared; that is, this time constant strongly depends on the position of the observation point with respect to the surface $z=0$. It is natural to distinguish three stages of the transient response of the current density j_y:

(1) The early stage, $t \ll \tau$. (2) The intermediate stage when time of observation is comparable with τ. (3) The late stage when the time t is greater than τ. These stages are observed at any point inside the medium, regardless of the distance from the surface $z=0$. But with an increase of the distance, each stage manifests itself at greater times. As follows from Eq. (5.32) at the early stage, the current density j_y grows very rapidly with an increase of time. In the intermediate stage, it reaches a maximum, and it occurs when

$$t_{\max} = 2\tau = \frac{\gamma\mu_0 z^2}{2} \tag{5.35}$$

It is obvious that at different distances from the conductor surface, the maximal value of j_y is observed at different times and these times become greater with increasing z-coordinate. Also, with a decrease of conductivity, $t_{\max}$ becomes smaller. From Eqs (5.32) and (5.35), it follows that the maximal magnitude of the current density is

$$j_{\max} = \left(\frac{2}{\pi e}\right)^{1/2} \frac{i_{0y}}{z} \tag{5.36}$$

and therefore, it is inversely proportional to z. Next consider the late stage when $\tau/t \ll 1$. In accordance with Eq. (5.32) at relatively large times, the current density is

$$j_y \approx i_{0y}\left(\frac{\gamma\mu_0}{\pi t}\right)^{1/2}, \tag{5.37}$$

that is, it is uniformly distributed along the z-axis and decreases with time as $t^{-1/2}$. Representing $\exp\left(-\frac{\tau}{t}\right)$ as

$$\exp\left(-\frac{\tau}{t}\right) = \sum_{n=0}^{\infty} \frac{(-1)^n}{n!}\left(\frac{\tau}{t}\right)^n$$

we have for the current density

$$j_y(t) = \frac{2i_{0y}}{\pi^{1/2} z} \sum_{n=0}^{\infty} \frac{(-1)^n}{n!}\left(\frac{\tau}{t}\right)^{\left(n+\frac{1}{2}\right)} \tag{5.38}$$

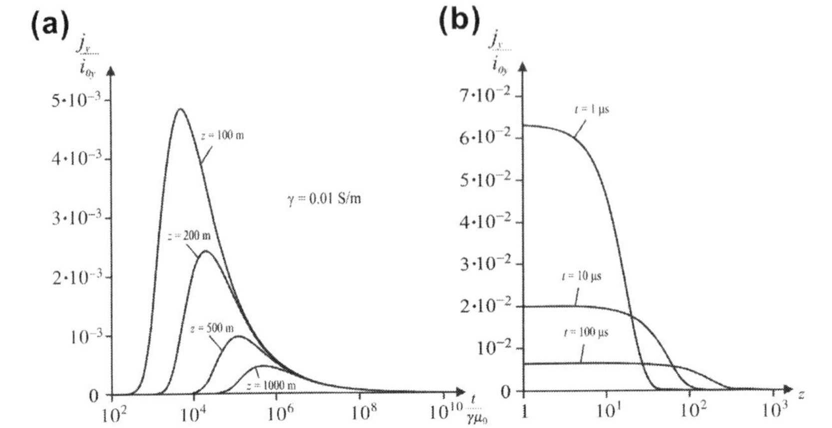

Figure 5.2 Behavior of the current density as a function of (a) Time and (b) Distance.

This form is very convenient to describe the late stage since the series converges very quickly when $t > \tau$. It is appropriate to note that a similar representation, which contains the integer and fractional powers of the parameter τ/t, describes the late stage of the transient response of the current and field in much more complicated models of a layered medium with different primary sources. Results of calculation of transient responses of the current density (Eq. (5.32)) are shown in Figure 5.2(a). Index of these curves is the parameter τ, which characterizes the distance from the conductor surface. It is clearly seen that with an increase of the distance from this surface, the transient responses of the current density change. For instance, their maximum becomes smaller and as follows from Eq. (5.32) for every given ratio τ/t, the magnitude of j_y is inversely proportional to the distance z.

5.2.5. Dependence of the Current Density on the Distance *z*

It is also useful to discuss the dependence of the current density on the distance from the conductor surface. With this purpose in mind, we distinguish three ranges of distances which correspond to the small, intermediate, and large values of the parameter τ/t. For instance, at relatively small distances from the surface $z = 0$ when $t \gg \tau$, the current is almost uniformly distributed and decreases at rate inversely proportional to square root of time. Then, with an increase of z, the density j_y starts to decrease (Figure 5.2(b)). The index of curves is the value of $(\gamma\mu_0/\pi t)^{1/2}$. These curves vividly demonstrate the process of penetration of currents into a conducting medium. At the beginning, they are located at the vicinity of the conductor surface when the skin effect manifests itself. Then with an increase of time, they appear at greater distances. It is essential that the character of the current distribution varies with time. In particular, at the first instant, the volume density is equal to zero, while in the opposite case of relatively large times, the currents are almost uniformly distributed at a large part of the conducting medium.

5.2.6. About Diffusion of Currents

We have described the behavior of currents in a conducting medium as a function of time and of the distance from the surface, where they appear at the initial moment. This process of penetration of currents into the conducting medium is called *diffusion*. From the mathematical point of view, it is related to the fact that Eq. (5.19), describing the quasi-stationary field, is the diffusion. As was mentioned earlier, these equations also describe the penetration of the heat and movement of ions in solutions where the concentration of ions changes. Certainly, there is some similarity in the behavior of functions describing these completely different phenomena. For instance, the magnitude of currents, the amount of heat, as well as number of ions change in such a way that their maximum appears at different places at different times. However, it is appropriate to make one comment. In our case, the current density has only one component j_y. This means that charges located at some depth $z = z_0$ remain during diffusion always at the same distance from the conductor's surface. In other words, the charges which form the current are not involved in the movement along the z-axis. The same behavior is observed in many other cases where we also observe a diffusion of the current magnitude only. Of course, in relatively complicated models of a conducting medium and primary source, charges may move in all directions at different times, but this fact is not essential feature of the current diffusion. In accordance with Eq. (5.32), the magnitude of the current density differs from zero at any distance from the surface of a conducting medium as soon as time t is not zero. This feature of current clearly demonstrates the fundamental difference between propagation and diffusion since it implies that induced currents appear instantly at all points of a medium. This apparent paradox only emphasizes the fact that the quasi-stationary field correctly characterizes the field at each point only when time of observation is greater than the time of wave arrival at this point.

5.3. DIFFUSION OF THE MAGNETIC FIELD

5.3.1. Equation for the Magnetic Field

Inasmuch as the behavior of the electric field,

$$E_y = \rho j_y = i_{0y}\left(\frac{\rho\mu_0}{\pi t}\right)^{1/2} \exp\left(-\frac{\gamma\mu_0 z^2}{4t}\right) \tag{5.39}$$

as a function of the distance z and time t is the same as that of the current density, we only pay attention to the magnetic field. With this purpose in mind, let us make use of the second Maxwell equation

$$\text{curl}\,\mathbf{B} = \mu_0 \mathbf{j},$$

which in the case of the plane-wave field has the form

$$\frac{\partial B_x}{\partial z} = \mu_0 j_y \tag{5.40}$$

5.3.2. Magnetic Field at Infinity

In order to determine the field B_x from this differential equation, it is necessary to know the field at some point of the medium. For this reason, it is useful to study the field behavior at infinitely large distances from the conductor's surface. First of all, consider again the initial moment $(t=0)$ when the primary field vanishes. Correspondingly, the induced currents, which arise at this moment at the plane $z=0$, create the same magnetic field B_{0x} everywhere in the conducting medium including infinity (skin effect). Moreover, taking into account the antisymmetry of the field B_x caused by these currents with respect to the plane $z=0$, we have to conclude that above the conducting medium, the magnetic field is

$$B_x = -B_{0x} \tag{5.41}$$

if $t=0$ and $z=0$.

A natural question arises. What happens to the field B_x at infinity $(z \to \infty)$ and above the conductor when the currents penetrate into the medium? To solve this task, let us recall that the total current I passing through the elementary vertical strip shown in Figure 5.1 is independent on time (Eq. (5.29)). Also it is useful to mentally represent a current distribution as a system of elementary horizontal layers with density $j_y(z,t)$. Then taking into account the fact that the magnetic field on either side of each current layer is uniform and the total current of all layers is constant, we can conclude that at infinity ($z \to \infty$), the field B_x caused by induced currents remains equal to the primary field. At the same time above the conducting medium, this field has opposite direction but the same magnitude. Thus, regardless of time, we obtain

$$B_x = \begin{cases} B_{0x} & z \to \infty \\ -B_{0x} & z < 0 \end{cases} \tag{5.42}$$

In addition, let us make several comments: (1) In deriving Eq. (5.42), we used the fact that the field depends only on one coordinate (one-dimensional model). (2) This remarkable boundary condition also remains valid in the medium when conductivity changes only along the z-axis, provided that the primary magnetic field is uniform and

$$B_0 = (B_{0x}, 0, 0)$$

(3) It is a simple matter to see that in the case when the primary field B_{0x} arises at some instant $t = 0$, we have

$$B_x = \begin{cases} 0 & z \to \infty \\ 2B_{0x} & -H < z < 0 \end{cases} \tag{5.43}$$

Thus, the magnetic field above the conductor surface, in particular in its vicinity, is two times greater than the primary field B_{0x}. (4) The independence of the quasi-stationary magnetic field of the conductivity and time when it is considered above the medium, in particular at its vicinity, is a fundamental feature of one-dimensional electromagnetic field. Applying the principle of superposition, it is easy to show that this result is still valid when such primary field is an arbitrary function of time.

5.3.3. Expression for the Magnetic Field

Now consider again Eq. (5.40). Integrating this equation from z to infinity and taking into account Eq. (5.42), we have

$$B_{0x} - B_x(z,t) = \mu_0 \int_z^\infty j_y \mathrm{d}z$$

or

$$B_x(z,t) = B_{0x} - \mu_0 \int_z^\infty j_y \mathrm{d}z \tag{5.44}$$

Let us introduce a new variable:

$$u = p^{1/2} z \quad \text{and} \quad p = \frac{\gamma \mu_0}{4t}$$

Then, we can represent Eq. (5.44) as

$$B_x(z,t) = B_{0x} - \frac{2\mu_0 i_{0y}}{\pi^{1/2}} \int\limits_u^\infty \exp(-u^2)\,\mathrm{d}u$$

Inasmuch as

$$\int\limits_0^\infty \exp(-u^2)\,\mathrm{d}u = \frac{\pi^{1/2}}{2}$$

and

$$F(u) = \frac{2}{\sqrt{\pi}} \int\limits_0^u \exp(-x^2)\,\mathrm{d}x \tag{5.45}$$

is the probability integral, we have

$$B_x(z,t) = B_{0x}\{1 - 2[1 - F(u)]\}$$

or

$$B_x(z,t) = B_{0x}[2F(u) - 1] \tag{5.46}$$

where

$$u = \left(\frac{\gamma \mu_0}{t}\right)^{1/2} \frac{z}{2} = \left(\frac{\tau}{t}\right)^{1/2} \tag{5.47}$$

Thus, we expressed the field **B** in terms of the probability integral which is very well-known function and its behavior is simple (Figure 5.3). In particular, when the argument u is small, the function $F(u)$ can be represented as a power series:

$$F(u) = \frac{2}{\pi^{1/2}} \sum_{k=0}^{\infty} \frac{(-1)^k u^{2k+1}}{k!(2k+1)} \tag{5.48}$$

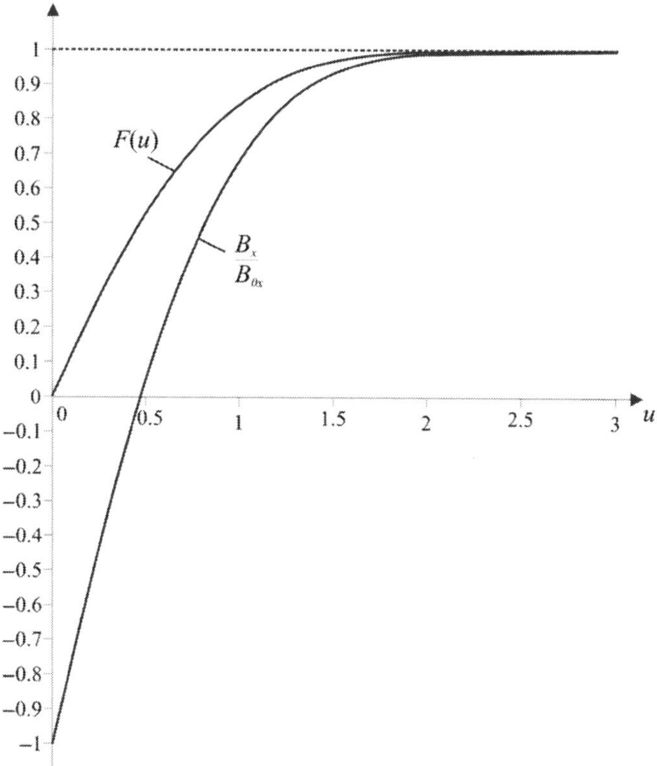

Figure 5.3 Behavior of the function $F(u)$ and the transient response of the magnetic field.

In the opposite case of large values of u, the probability integral is described by the asymptotical series

$$F(u) \approx \frac{\exp(-u^2)}{\pi^{1/2}u}\left[1 + \sum_{k=1}^{\infty}(-1)^k \frac{1\cdot 3\cdot 5\cdot 7\cdots(2k-1)}{2^k u^{2k}}\right] \qquad (5.49)$$

5.3.4. Behavior of the Magnetic Field

Now we are prepared to study the behavior of the magnetic field. Taking into account Eqs (5.46)–(5.49), the field B_x at the early stage is

$$B_x(t,z) \approx B_{0x}\left[1 - 2\left(\frac{t}{\tau\pi}\right)^{1/2}\exp(-\tau/t)\right] \quad \text{if } \frac{t}{\tau} \ll 1 \qquad (5.50)$$

As is seen from this equation, the behavior of the field B_x clearly demonstrates the principle of inertia of the magnetic flux. In fact, we have

$$B_x(t, z) \rightarrow B_{0x}, \quad \text{if } t \rightarrow 0$$

and at greater distances from the conductor surface, this approximate equality holds for greater times. It is interesting to notice that within the early stage, the difference between the secondary and primary fields is defined by a very small exponential term. In other words, at the early stage, the field B_x is practically equal to B_{0x}. Next consider the opposite case; that is the late stage. Since the parameter u is small, it is appropriate to make use of the series (Eq. (5.48)). Then, we obtain

$$B_x(t, z) = -B_{0x} + 2B_{0x} \sum_{k=0}^{\infty} \frac{(-1)^k}{k!(2k+1)} \left(\frac{\tau}{t}\right)^{k+\frac{1}{2}} \tag{5.51}$$

Thus, at the late stage, the magnetic field also consists of the following two parts:

1. A constant which describes the field above the conductor surface and is equal to $-B_{0x}$.
2. A second part which is a series containing fractional powers of the ratio τ/t. Discarding all terms of this series except the first three, we have

$$B_x(t) \approx -B_{0x}\left[1 - 2\left(\frac{\tau}{t}\right)^{1/2} + \frac{2}{3}\left(\frac{\tau}{t}\right)^{3/2} - \frac{1}{3}\left(\frac{\tau}{t}\right)^{5/2} + \cdots\right] \quad \text{if } t \gg \tau \tag{5.52}$$

The behavior of the transient responses of the magnetic field is shown in Figure 5.3. Here it is proper to note that in spite of the absence of the primary magnetic field when $t > 0$, the leading term of the series describing the late stage is a constant and it differs from the primary field B_{0x} only by sign. As will be shown later, this behavior is an exception and related to the fact that in this unrealistic model, the total current I is independent of time.

REFERENCE AND FURTHER READING

[1] A.A. Kaufman, Geophysical Theory and Methods, Part B, Academic Press, New York, NY, 1994.

CHAPTER SIX

Quasi-Stationary Field of the Magnetic Dipole in a Uniform Medium

INTRODUCTION

In developing the theory of electromagnetic methods, we are mainly concerned with the behavior of fields observed on the Earth's surface or above, or in the borehole, as well as at the sea bottom. However, in order to understand better their behavior, it is useful at the beginning to investigate in detail the field and currents in uniform medium. This approach will allow us to obtain some insight into physical principles on which many electromagnetic methods are based, even though the effect of boundaries between various media cannot be considered. In this chapter, we assume that the quasi-stationary field is caused by a magnetic dipole, and this choice is related to the fact that an inductive excitation of the primary field is very often but not always used in many electromagnetic methods.

6.1. QUASI-STATIONARY FIELD OF THE MAGNETIC DIPOLE (FREQUENCY DOMAIN)

6.1.1. Expressions for the Field

First of all, let us discuss the mechanism which causes the secondary electromagnetic field. When the magnetic dipole with sinusoidal current is placed in a uniform conducting medium, a change of the primary magnetic field causes the primary vortex electric field, and the latter gives rise to the induced currents. These currents interact with each other, and the secondary magnetic and electric fields appear. It turns out that this interaction does not change the current direction, and in the spherical system of coordinates, they have only the φ component. Inasmuch as the medium can be treated as a linear system, the secondary field is also the sinusoidal function of time which has the same frequency as the primary one. In Chapter 4, we derived equations for the electromagnetic field of the magnetic dipole in a uniform

Methods in Geochemistry and Geophysics, Volume 45
ISSN 0076-6895,
http://dx.doi.org/10.1016/B978-0-444-53829-1.00006-X

medium when both the conduction and displacement currents are present. Taking into account Eqs (4.42) and (4.45), we have the following for the complex amplitudes of the field:

$$E_\varphi^* = \frac{i\mu_0\omega M_0}{4\pi R^2}(1 - ikR)\exp(ikR)\sin\theta$$
$$B_R^* = \frac{\mu_0 M_0}{2\pi R^3}(1 - ikR)\exp(ikR)\cos\theta \tag{6.1}$$
$$B_\theta^* = \frac{\mu_0 M_0}{4\pi R^3}\left(1 - ikR - k^2R^2\right)\exp(ikR)\sin\theta$$

Since we study the quasi-stationary field, the wave number has the form

$$k = \frac{1+i}{\delta}, \quad \delta = \left(\frac{2}{\gamma\mu_0\omega}\right)^{1/2} = \frac{10^3}{2\pi}(10\rho T)^{1/2}\ m \tag{6.2}$$

Here as before, δ is the skin depth. Let us remind that the dipole moment varies as

$$M = M_0 \cos \omega t, \tag{6.3}$$

and in accordance with Biot–Savart law, its current creates the primary magnetic field $B_R^{(0)}$ and $B_\theta^{(0)}$, equal to

$$B_R^{(0)} = \frac{\mu_0 M_0}{2\pi R^3}\cos\theta\cos\omega t, \quad \text{and} \quad B_\theta^{(0)} = \frac{\mu_0 M_0}{4\pi R^3}\sin\theta\cos\omega t, \tag{6.4}$$

that is, this field is located in the meridian planes, and it synchronously changes with the dipole current. Earlier we called this field as the quasi-stationary field of the magnetic dipole in a nonconducting medium. Its variation with time causes the vortex electric field (Chapter 4), and its complex amplitude is equal to

$$E_\varphi^{(0)^*} = \frac{i\omega\mu_0 M_0}{4\pi R^2}\sin\theta \tag{6.5}$$

This means that this field is

$$E_\varphi^{(0)} = \mathrm{Re}\left[\frac{i\omega\mu_0 M_0}{4\pi R^2}\sin\theta\exp(-i\omega t)\right] = \frac{\omega\mu_0 M_0}{4\pi R^2}\sin\theta\sin\omega t, \tag{6.6}$$

which is located in horizontal planes, and it exists at any point regardless of whether this medium is conductive or nonconductive. It is obvious that the primary electric and magnetic fields are shifted with respect to each other in phase by 90°. As in general case (Chapter 4), it is convenient to express the complex amplitudes of the field in the corresponding units of the primary field; that is,

$$b_R^* = (1 - ikR)\exp(ikR), \quad b_\theta^* = \left(1 - ikR - k^2R^2\right)\exp(ikR),$$

and

$$e_\varphi^* = (1 - ikR)\exp(ikR) \tag{6.7}$$

6.1.2. Two Forms of Field Presentation

Inasmuch as the right-hand side of Eq. (6.7) is complex, we can say that there is a phase shift between the field and the dipole current. For instance, in the case of the radial component of the magnetic field, we have

$$B_R = \frac{\mu_0 M_0}{2\pi R^3}\cos\theta \mathrm{Re}\left[(c_R + id_R)\exp(-i\omega t)\right]$$

or

$$B_R = \frac{\mu_0 M_0}{2\pi R^3}\cos\theta\left[c_R\cos\omega t + d_R\sin\omega t\right] \tag{6.8}$$

Here,

$$b_R^* = c_R + id_R$$

By analogy,

$$B_\theta = \frac{\mu_0 M_0}{4\pi R^3}\sin\theta\left[c_\theta\cos\omega t + d_\theta\sin\omega t\right] \tag{6.9}$$

where

$$b_\theta^* = c_\theta + id_\theta$$

As we know, Eqs (6.8) and (6.9) can be interpreted in two ways. One of them suggests that at each point the magnetic field changes with time as a sinusoidal function and there is a phase shift between this field and the dipole current. In essence, this approach represents the field as a sinusoidal wave which relatively rapidly decays with the distance from the dipole. Also these equations allow us to imagine that at every point, the magnetic field, normalized by the primary field, is a sum of two sinusoids called the *in-phase* and *quadrature components*:

$$Inb_R = c_R \cos \omega t \quad Qb_R = d_R \sin \omega t,$$

and

$$Inb_\theta = c_\theta \cos \omega t \quad Qb_\theta = d_\theta \sin \omega t \tag{6.10}$$

By definition, the real and imaginary parts of complex amplitude are the amplitudes of the in-phase and quadrature components, respectively. The in-phase component changes synchronously with the primary field, while the quadrature component is shifted by phase in 90°. In general, these oscillations have different amplitudes. Both approaches are equivalent to each other and a choice between them is defined by the quantity which is measured. Taking into account that in several conventional methods of the surface and borehole geophysics the quadrature and in-phase components are measured and our purpose is to understand physical principles of these methods, we will use the second approach in this section. In a similar manner, the electric field as well as the current density can be represented as a sum of the quadrature and in-phase components. In accordance with the Biot–Savart law, the quadrature component of the magnetic field arises due to currents which are shifted in the phase by 90° with respect to the current in the dipole, while the in-phase component of the field is the algebraic sum of the primary and secondary fields. The in-phase component of the secondary field is contributed by induction currents in the medium shifted by either 180° or 0° with respect to the dipole current. Thus we assume that at each point of the medium, there are quadrature and in-phase components of the current, and this representation is very useful for understanding electromagnetic methods when these components of the magnetic field are measured. Now we are prepared to discuss the field behavior regardless of a position of an observation point with respect to the dipole. At the same time, it is natural to distinguish two special cases. In one of them, $\theta = 0$ and only the radial component is measured. In the other, $\theta = \pi/2$ and the component B_θ is the subject of measurements.

6.1.3. An Asymptotic Behavior of the Field

First, consider the range of small parameter p, which is often called the *low-frequency spectrum* of the field. Expanding exp (ikR) in the series

$$\exp(ikR) = \sum_{n=0}^{\infty} \frac{(ikR)^n}{n!}$$

and substituting this into Eq. (6.7) after some simple algebra, we have

$$b_R^* = 1 + \sum_{n=2}^{\infty} \frac{1-n}{n!} 2^{n/2} p^n \exp\left(\frac{i3\pi n}{4}\right) \tag{6.11}$$

Here,

$$p = \left(\frac{\gamma\mu_0\omega}{2}\right)^{1/2} R = \frac{R}{\delta} \tag{6.12}$$

is the parameter characterizing the distance between the dipole and the observation point, expressed in units of the skin-depth δ. Sometimes, this parameter is called the *induction number*. Taking into account Eq. (6.12), we see that the series describing the low-frequency spectrum, as in the case of plane field, contains integer and fractional powers of ω. As follows from Eq. (6.11),

$$\text{Im}\, b_R^* = d_R \approx p^2 - \frac{2}{3}p^3, \quad \text{and} \quad \text{Re}\, b_R^* = c_R \approx 1 - \frac{2}{3}p^3 \quad \text{if } p \ll 1 \tag{6.13}$$

As is seen from Eq. (6.13) at this range of the parameter p, the quadrature and in-phase components are related to the frequency, conductivity, and the distance from the dipole in a completely different manner. In fact, from Eqs (6.12) and (6.13), we obtain for both components

$$\text{Im}\, B_R^* \approx \frac{\mu_0 M_0}{2\pi R^3} \cos\theta \left\{ \frac{\gamma\mu_0 R^2}{2}\omega - \frac{(\gamma\mu_0 R^2)^{3/2}}{3(2)^{1/2}}\omega^{3/2} + \ldots \right\} \tag{6.14}$$

and

$$\text{Re}\, B_R^* \approx \frac{\mu_0 M_0}{2\pi R^3} \cos\theta \left\{ 1 - \frac{(\gamma\mu_0 R^2)^{3/2}}{3(2)^{1/2}}\omega^{3/2} \ldots \right\} \tag{6.15}$$

It is clear that the first term at the right-hand side of Eq. (6.15) characterizes the primary field, caused by only the dipole current, while the next term describes the in-phase component of the secondary magnetic field which arises due to the induced currents in a conductive medium. At the same time, all the terms in the right-hand side of the expression for quadrature component (Eq. (6.14)) correspond to the secondary magnetic field. Comparison of last two equations shows that the in-phase component of the secondary field is more sensitive to a change of conductivity, and in this low-frequency approximation (range of small parameter p), it is independent of the separation. As follows from Biot–Savart law, this interesting feature indicates that the in-phase component may have a much greater depth of investigation than the quadrature component within this part of the spectrum, ($p \ll 1$). In a similar manner, we have for azimuthal component of the field:

$$\text{Im}\, b_\theta^* \approx -p^2 + \frac{4}{3}p^3, \quad \text{and} \quad \text{Re}\, b_\theta^* \approx 1 + \frac{4}{3}p^3 \tag{6.16}$$

As follows from Eq. (6.7) in the opposite case, that is at high frequencies or more precisely when $p \gg 1$, the in-phase and quadrature components of the magnetic field tend to be zero:

$$\text{Re}\, \mathbf{b}^* \to 0 \ \text{ or } \ \text{Re}\, \mathbf{b}^{s^*} = -\mathbf{b}^0 \quad \text{and} \quad \text{Im}\, \mathbf{b}^* \to 0$$

In other words, in this part of the spectrum, we observe the internal skin effect when induced currents concentrate in the vicinity of the dipole. Correspondingly, the secondary in-phase component differs from the primary field by sign only.

6.1.4. Behavior of the Field as a Function of Parameter *p*

Inasmuch as the radial and azimuthal components have the similar behavior, let us focus on the radial component which can be represented as

$$\text{Im}\, b_R^* = \exp(-p)[(1+p)\sin p - p\cos p]$$

and

$$\text{Re}\, b_R^* = \exp(-p)[(1+p)\cos p + p\sin p] \tag{6.17}$$

Now we consider curves illustrating the dependence of the amplitudes of the quadrature and in-phase components of this field on the parameter p (Figure 6.1). With an increase in the induction number, the quadrature

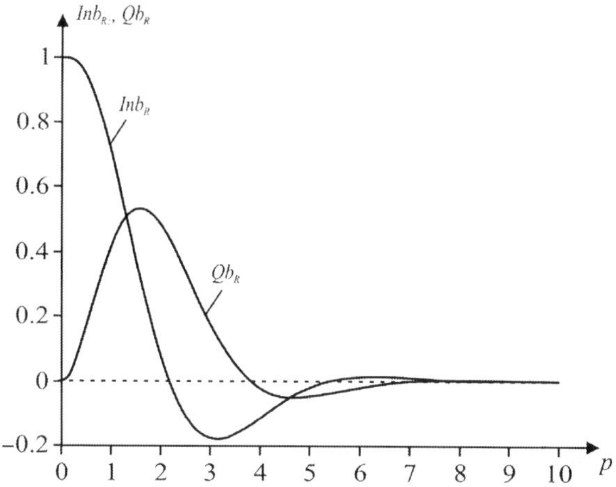

Figure 6.1 The quadrature and in-phase components of the magnetic field.

component (Im b_R^*) increases, reaches a maximum, and after it tends to zero. In contrast, the in-phase component of the field decreases and then in oscillating manner, like the quadrature component, approaches zero. Let us examine the low-frequency part of the spectrum in detail. According to Eqs (6.14) and (6.15), the amplitude of the quadrature component prevails over that of the in-phase component Inb_R^s, and we have

$$QB_R = \frac{\mu_0 M_0}{4\pi R} \gamma \mu_0 \omega \cos\theta \sin\omega t, \quad p \ll 1 \tag{6.18}$$

Hence, in the range of small parameter values, the quadrature component is directly proportional to the conductivity and the frequency and inversely proportional to the distance from the magnetic dipole. As will be shown later, some of these features of the field also remain valid in a nonuniform conducting medium. From Eq. (6.17), we see that over this part of the spectrum, the secondary field is much smaller than the primary field:

$$QB_R \ll B_R^{(0)}, \tag{6.19}$$

and this is the reason why measurements of QB_R require compensation of the primary field with a high accuracy when $p \ll 1$.

6.1.5. Expression for Induced Currents

At this point, it is very useful to explain the behavior of the field in terms of distribution of induced currents. Making use of the first equation of the set (Eq. (6.1)) and Ohm's law in the differential form,

$$\mathbf{j} = \gamma \mathbf{E},$$

we have the following expression for the current density at every point in a uniform medium:

$$j_{\varphi}^{*} = \frac{i\gamma\mu_0\omega M_0}{4\pi R^2} \exp(ikR)(1 - ikR)\sin\theta \tag{6.20}$$

As in the case of the magnetic field, we represent the current density as a sum of the quadrature and in-phase components, and in accordance with Eq. (6.20), we obtain

$$\operatorname{Im} j_{\varphi}^{*} = \frac{\gamma\mu_0\omega M_0 r}{4\pi R^3} \exp(-p)[(1+p)\cos p + p\sin p]$$

and

$$\operatorname{Re} j_{\varphi}^{*} = -\frac{\gamma\mu_0\omega M_0 r}{4\pi R^3} \exp(-p)[(1+p)\sin p - p\cos p] \tag{6.21}$$

Here,

$$\sin\theta = \frac{r}{R}$$

It is obvious that the current density field can be imagined as a system of current toroids or rings which have a common axis with that of the dipole, and they are located in planes perpendicular to this axis (Figure 6.2(b)).

6.1.6. The Induced Current $j_{\varphi}^{(0)}$

First of all, we determine the induced currents which arise due to the primary electric field only. As follows from Eq. (6.5), the density of these currents is

$$j_{\varphi}^{(0)^*} = \gamma E_{\varphi}^{(0)^*} = \frac{i\gamma\mu_0\omega M_0 r}{4\pi R^3}, \tag{6.22}$$

and it is shifted in phase by 90° with respect to the dipole current. If the influence of the magnetic field caused by the induced currents in the medium is negligible, the distribution of currents would be defined by Eq. (6.22). In such a case, the current density at any point in the medium is a function which is described by the product of two terms. One of them depends on the dipole moment, frequency, and conductivity at the point where the density is calculated, whereas the other is the function of the point

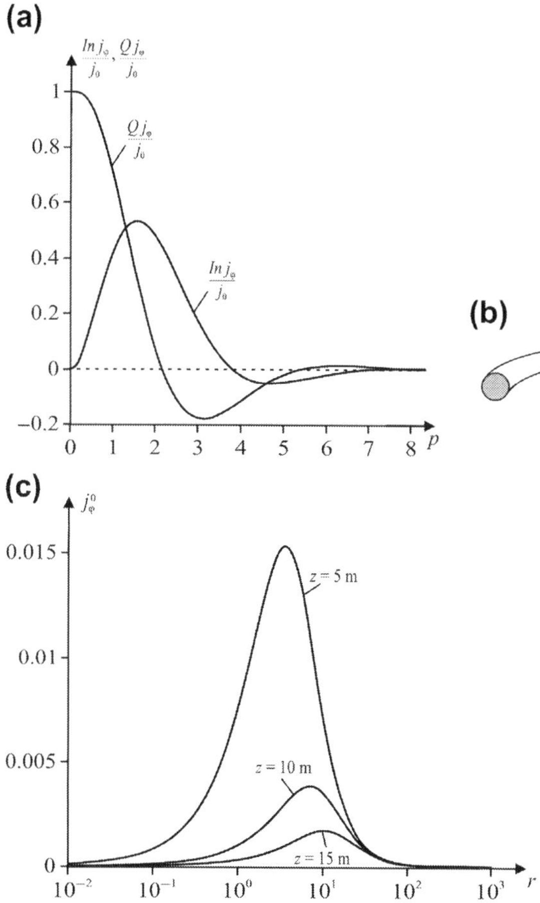

Figure 6.2 (a) Quadrature and in-phase components of the current density. (b) Geometry of current tubes. (c) Current j_φ^0.

coordinates. It is remarkable that if we neglect by interaction of induced currents, determination of current distribution is an elementary task. Correspondingly, it is a simple matter, applying Biot–Savart law, to determine the magnetic field of these currents even in a nonuniform medium, unless the primary electric field does not intersect boundaries. This last condition is very important because appearance of the electric charges causes a change of a direction of the current density. In other words, geometry of current tubes becomes unknown and it is impossible to use Biot–Savart law. Later we will demonstrate that the approximation, based on the use of Eq. (6.22), is very useful in some methods of the surface and borehole geophysics. The behavior of the current $j_\varphi^{(0)}$ in planes perpendicular to the dipole axis is shown in Figure 6.2(c). It can be seen that with increasing z, the distance from the z-axis to the ring with the maximal density also increases.

6.1.7. Behavior of the Quadrature and In-Phase Components of the Current Density

Let us introduce the notation

$$j_0 = \frac{\gamma\mu_0\omega M_0}{4\pi}\frac{r}{R^3},$$

and rewrite Eq. (6.21) as

$$Q\,j_\varphi = j_0 \exp(-p)[(1+p)\cos p + p\sin p]\sin\omega t$$

and

$$Inj_\varphi = j_0 \exp(-p)[(1+p)\sin p - p\cos p]\cos\omega t \qquad (6.23)$$

An analysis of these functions permits us to explore how the actual current density j_φ differs from j_0 for various values of the parameter p, especially for different distances from the dipole. Curves for the quadrature and in-phase components of the current density, normalized by j_0, are shown in Figure 6.2(a). For small values of the parameter p, the quadrature component of the current density is essentially the same as the current density j_0; that is, the interaction between currents is negligible in this case. With an increase in the parameter p, the ratio Qj_φ/j_0 decreases, passes through zero and for larger values of the parameter p approaches zero in oscillating manner. The curve for the ratio of the in-phase component of the current density to j_0 has a completely different character. At small values of p, the ratio Inj_φ/j_0 approaches zero, then increases to maximum when p is about 1.5, and for larger values of p it tends to zero again in oscillating manner. Therefore, the actual distribution of currents, in contrast to the behavior of j_0, is determined by both geometric factors and interaction of currents. This last factor is taken into account in the case of a uniform medium by the parameter p. Comparing the curves in Figure 6.2(a), we can see that for small values of p, the quadrature component of the current density dominates. However, there is a range of values of p over which the in-phase component is significantly larger. The curves in Figure 6.2(a) can be analyzed from two points of view. If the conductivity and frequency are held constant, one of the curves shows a change in the quadrature component of the current density related to j_0 when the distance from the dipole to an observation point increases. On the other hand, if the position of this point is fixed, then this curve illustrates the frequency responses of the ratio Qj_φ/j_0 or its behavior when the conductivity changes. This approach allows us to explain the main features of the magnetic field proceeding from the distribution of the corresponding components of the current density. As can be seen from Figure 6.2(a), for relatively small values of p, the current density Qj_φ does not differ significantly from j_0. If the frequency is low enough and the medium has a relatively high resistivity, the range of distances where the actual current density Qj_φ is almost equal to j_0 can be sufficiently large. This means that the magnetic field $Q\mathbf{B}$ at the observation point is entirely defined by currents in this region. In other words, the depth of investigation does not exceed its dimensions; that is, it does not increase with a decrease of frequency in spite of the fact that depth of penetration of the electromagnetic field becomes greater. Inasmuch as the current density Qj_φ in this region of the medium is directly proportional to the frequency (Eq. (6.22)), the magnetic field caused by these currents is also proportional to the frequency. Within some range of the parameter p, the dimensions of this volume remain much greater than the distance from the dipole to an observation point. With an increase of the value of the parameter p (for example, as would be caused by an increase of the frequency), the size of this volume where currents are practically equal to j_0 becomes smaller. Correspondingly, currents outside this volume where the density is less than j_0 begin to have an effect on the magnetic field at the observation point. This influence results in decrease of the rate of the growth of the field $Q\mathbf{B}$. With a further increase in the frequency, as a consequence of the rapid decrease of the quadrature component of the current density, the growth of the quadrature component of the magnetic field ceases and then it starts to decrease. By analogy, the behavior of the in-phase component of the magnetic field can also be explained with the use of the in-phase component of the current. As we will see later, the general features of the behavior of the quadrature and in-phase components of the field remain valid for a nonuniform medium as well.

In summary, let us enumerate some aspects of the behavior of the quasi-stationary field which are of practical interest. First, the quadrature component of the magnetic field is mainly determined by currents flowing in the volume surrounding the dipole and an observation point. The interaction between currents becomes significant if the dimension of the zone, in which the quadrature component Qj_{φ} almost coincides with j_0, is less than the distance between the dipole and an observation point. From the behavior of the quadrature component of the current density, it is hard to recognize a diffusion phenomenon. In contrast, the behavior of the in-phase component of the current density clearly shows a diffusion process. For instance, a maximum of Inj_{φ} moves away from the dipole with a decrease of frequency and this means that the corresponding component of the magnetic field can be more sensitive to currents induced at the distant parts of a medium than the quadrature component. With decreasing frequency, the depth of investigation along with a depth of penetration, when the in-phase component is measured, gradually increases regardless of the distance between the dipole and an observation point. This feature of the in-phase component manifests primarily when the separation between the dipole and receiver is comparable or less than the skin depth. Similar behavior is observed in measuring the transient field. Note, that at the high-frequency spectrum with an increase of the frequency, both components rapidly decay and oscillate.

As was pointed out earlier, the behavior of the quadrature component of the magnetic field is defined by the quadrature component of the current Qj_{φ}. On the other hand, the function Qj_{φ}/j_0 coincides with the function describing the behavior of the in-phase component of the magnetic field. Therefore, from the physical point of view, it becomes obvious that there is a relationship between both components of the magnetic field. Of course, we arrive at the same conclusion in considering the in-phase component of the magnetic field. Later we will investigate this subject in some detail.

With an increase in the distance between the dipole and the observation point, the induction number p also increases and therefore the interaction of currents manifests more strongly. Certainly, the position of the receiver does not change the character of the current distribution in the conducting medium. The increase of influence of the skin effect in this case can be explained as follows. When the separation between the dipole and receiver is small, the depth of investigation in measuring **QB** is also small. Therefore, the electromotive force in the receiver is mainly defined by currents which are approximately equal to j_0. In other words, currents subjected to the influence of the skin effect (interaction) for given values of γ and ω are situated at distances which exceed the range of investigation. As the separation between the dipole and the observation point increases, the depth of investigation of the dipole-receiver array also becomes greater. Correspondingly, the relative contribution of distant currents which undergo an influence of their interaction becomes more significant. In this case when the in-phase component is measured, we observe a completely different behavior. Inasmuch as the density of the current Inj_φ around the dipole is very small, the field component $In\mathbf{B}$ is mainly defined by currents located relatively far away. For this reason, a change of sufficiently small separations between the dipole and receiver does not practically affect the field. However, with further increase of this separation, an observation point approaches closer to currents causing this component of the field and the distance makes stronger influence.

6.2. TRANSIENT FIELD OF THE MAGNETIC DIPOLE IN UNIFORM MEDIUM

6.2.1. Expressions of the Field

In Chapter 4, we have derived equations describing the transient field of the magnetic dipole in a general case when both the conduction and displacement currents are taken into account, and the moment of the dipole changes as a step function. Of course, these formulas allow us to obtain expressions for the quasi-stationary field, assuming that displacement currents are relatively small and time of observation greatly exceeds the time of the signal arrival. At the same time, it is convenient to obtain these formulas independently proceeding from the expression for the vector potential of the magnetic type (Eq. (4.41)):

$$A_z(t) = -\frac{\mu_0 M_0}{8\pi^2 R} \int\limits_{-\infty}^{\infty} \exp(-ikR)\exp(-i\omega t)d\omega, \qquad (6.24)$$

providing that the wave number is equal to $k = (i\gamma\mu_0\omega)^{1/2}$. This integral is tabular and it gives

$$A_z = -\frac{\mu_0 M_0}{4\pi(2\pi)^{1/2}} \frac{1}{R} \frac{u}{t} \exp\left(-\frac{u^2}{2}\right), \qquad (6.25)$$

where

$$u = \frac{2\pi R}{\tau} \quad \text{and} \quad \tau = \left(2\pi\rho t 10^7\right)^{1/2} \qquad (6.26)$$

By definition,

$$\mathbf{E} = \text{curl}\,\mathbf{A}$$

Then, taking into account that

$$A_R = A_z \cos\theta \quad \text{and} \quad A_\theta = -A_z \sin\theta$$

we find

$$E_\varphi = -\frac{M_0\rho}{4\pi R^4} e_\varphi \sin\theta \quad \text{and} \quad e_\varphi = \left(\frac{2}{\pi}\right)^{1/2} u^5 \exp\left(-\frac{u^2}{2}\right)$$

or

$$E_\varphi = -\left(\frac{2}{\pi}\right)^{1/2} \frac{M_0\rho}{4\pi R^4} u^5 \exp\left(-\frac{u^2}{2}\right)\sin\theta \qquad (6.27)$$

In order to find the magnetic field, we first make use the Maxwell's equation

$$\text{curl}\,\mathbf{E} = -\frac{\partial\mathbf{B}}{\partial t},$$

which allows us to determine the derivatives of the magnetic field with time: $\partial B_R/\partial t$ and $\partial B_\theta/\partial t$. Their integration with respect to time gives

$$B_R = \frac{2\mu_0 M_0}{4\pi R^3} b_R \cos\theta = \frac{2\mu_0 M_0}{4\pi R^3}\left[1 - \Phi(u) + \left(\frac{2}{\pi}\right)^{1/2} u \exp\left(-\frac{u^2}{2}\right)\right]\cos\theta,$$

and

$$\begin{aligned} B_\theta &= \frac{\mu_0 M_0}{4\pi R^3} b_\theta \sin\theta \\ &= \frac{\mu_0 M_0}{4\pi R^3}\left[1 - \Phi(u) + \left(\frac{2}{\pi}\right)^{1/2} u(1+u^2)\exp\left(-\frac{u^2}{2}\right)\right]\sin\theta \end{aligned} \qquad (6.28)$$

Here,

$$\Phi(u) = \left(\frac{2}{\pi}\right)^{1/2} \int_0^u \exp\left(-\frac{x^2}{2}\right) dx$$

is the probability integral. Let us remind that Eqs (6.27) and (6.28) are valid when displacement currents are negligible with respect to conduction currents, that is, the field is measured at times significantly exceeding τ_0, which is the time required for the signal to arrive at the observation point. In this light, it may be proper to point out again that we always deal with the propagation, but the quasi-stationary field indicates that we focus on the last stage of propagation in a conducting medium (diffusion).

6.2.2. Transient Responses of the Field

The behavior of the functions b_R, b_θ, and e_φ, when the dipole current is turned on at the moment $t=0$, are shown in Figure 6.3. Note that the left part of curves corresponds to the late stage, whereas the right part is the early stage. Since the source is turned on with an increase of time, the magnetic field tends to that of a direct current, while the electric field vanishes:

$$b \to b_0, \quad e_\varphi \to 0 \quad \text{if } t \to \infty \text{ or } u \to 0$$

First, let us consider the early stage of the transient response. In accordance with the electromagnetic induction law at the first moment, the field is absent in a conducting medium. Such behavior takes place because induced currents arise in the vicinity of the dipole, and their field compensates the primary field of the dipole current. It is appropriate to notice that in a general case (Chapter 4), when displacement currents are taken into

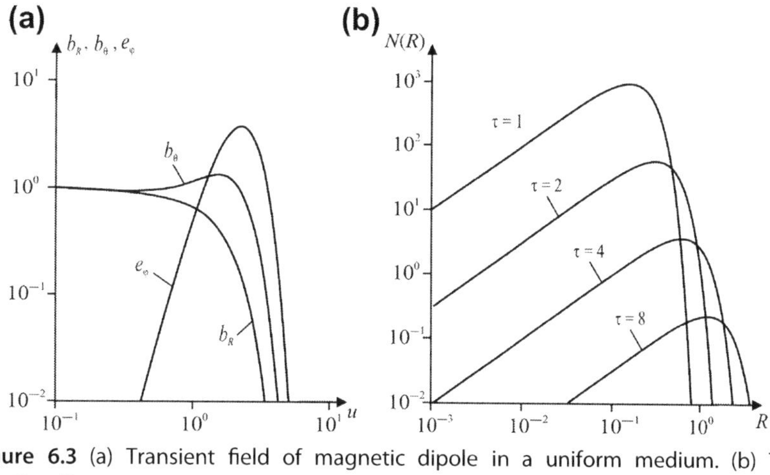

Figure 6.3 (a) Transient field of magnetic dipole in a uniform medium. (b) The function *N*.

account at the initial instant, the field is also absent at all points of medium. As is seen from Figure 6.3(a) with an increase of time, the electric and magnetic fields rapidly increase. Next consider the late stage of the transient response. Applying the known expansion of the probability integral in a series of small parameter u, which corresponds to relatively large times or small distances from the dipole to an observation point, or sufficiently small conductivity, we obtain approximate formulas for the secondary field:

$$B_R \approx \frac{\mu_0 M_0}{6\pi R^3}\left(\frac{2}{\pi}\right)^{1/2} u^3\left(1-\frac{3}{10}u^2\right)\cos\theta$$

$$B_\theta \approx \frac{\mu_0 M_0}{6\pi R^3}\left(\frac{2}{\pi}\right)^{1/2} u^3\left(1-\frac{3}{5}u^2\right)\sin\theta \quad u \ll 1 \tag{6.29}$$

$$E_\varphi \approx -\frac{M_0\rho}{4\pi R^4}\left(\frac{2}{\pi}\right)^{1/2} u^5\left(1-u^2\right)\sin\theta$$

Restricting ourselves to the first terms, we have

$$B_R \approx \frac{\mu_0 M_0}{12\pi(\pi)^{1/2}}\frac{\mu_0^{3/2}\gamma^{3/2}}{t^{3/2}}\cos\theta$$

$$B_\theta \approx \frac{\mu_0 M_0}{12\pi(\pi)^{1/2}}\frac{\mu_0^{3/2}\gamma^{3/2}}{t^{3/2}}\sin\theta \tag{6.30}$$

$$E_\varphi \approx -\frac{M_0}{16\pi(\pi)^{1/2}}\frac{\mu_0^{5/2}\gamma^{3/2}}{t^{5/2}}R\sin\theta$$

These formulas describe the field with sufficient accuracy for values of the parameter $u < 0.2$. Thus, at the late stage of the transient response, the magnetic field is practically independent of the distance from the dipole, and it is related to the conductivity of the medium in the same way as the in-phase component of the secondary magnetic field in the low-frequency spectrum. Also the same behavior is displaced by the second term of the quadrature component of the field at this part of the spectrum. The fact that the distance from the dipole to an observation point does not have an influence on the field at the late stage suggests the following. In this case, induced currents that generate the magnetic field are located from the dipole at distances which are essentially greater than the separation between the dipole and receiver. As follows from Eq. (6.27), the current density in the medium is

$$j_\varphi = -\left(\frac{2}{\pi}\right)^{1/2}\frac{M_0}{4\pi}\frac{\sin\theta}{R^4}u^5\exp\left(-\frac{u^2}{2}\right) \qquad (6.31)$$

To illustrate its behavior, the graphs of the function

$$N(R) = \frac{1}{R^4}u^5\exp\left(-\frac{u^2}{2}\right)$$

are shown in Figure 6.3(b). The index of curves is the parameter $\tau = (2\pi\rho t 10^7)^{1/2}$. It is vividly seen that with an increase of time, the curve maximum is shifted toward greater distances, and correspondingly, the magnetic field and electromotive force measured in the receiver become more sensitive to the distant parts of a medium. In the conclusion of this chapter, it may be proper to emphasize again that the quasi-stationary field obeys the diffusion equation and correspondingly has the basic features of diffusion phenomenon. It is specially seen when we consider the distribution of the current density of the transient field or its in-phase component in the frequency domain, in particular, a movement of the current maximum with time. At the same time, charges which form the current remain inside of the same current tube and practically do not move.

REFERENCES AND FURTHER READING

[1] A.A. Kaufman, Theory of the Induction Logging, Nauka, Novosibirsk, 1965.
[2] A.A. Kaufman, V.P. Sokolov, Theory of the Transient Induction Logging, Nauka, Novosibirsk, 1972.

CHAPTER SEVEN

The Hilbert and Fourier Transforms

INTRODUCTION

In the forthcoming chapters, we consider the frequency and transient soundings based on measuring the amplitude and phase of the field, as well as the transient responses. In order to understand better these methods and linkage between them, it is useful to consider two types of well-known relationships, namely, the Hilbert and Fourier transforms. Earlier, we already studied some general relationships between various responses of the electromagnetic field, but now let us study this subject in some detail.

7.1. HILBERT TRANSFORM

Suppose that **M** represents the complex vector of either the electric or magnetic fields and

$$\mathbf{M} = In\mathbf{M} + iQ\mathbf{M}, \tag{7.1}$$

where $In\mathbf{M}$ and $Q\mathbf{M}$ are the in-phase and quadrature components, respectively. Assuming that time dependence is described by the function $\exp(-i\omega t)$ and substituting Eq. (7.1) into Helmholtz's equation,

$$\nabla^2\mathbf{M} + k^2\mathbf{M} = 0, \tag{7.2}$$

we have

$$\nabla^2(In\mathbf{M} + iQ\mathbf{M}) + \left(i\gamma\mu_0\omega + \omega^2\varepsilon\mu_0\right)(In\mathbf{M} + iQ\mathbf{M}) = 0$$

or

$$\nabla^2 In\mathbf{M} + \omega^2\varepsilon\mu_0 In\mathbf{M} = \gamma\mu_0\omega Q\mathbf{M}$$

and

$$\nabla^2 Q\mathbf{M} + \omega^2\varepsilon\mu Q\mathbf{M} = -\gamma\mu_0\omega In\mathbf{M} \tag{7.3}$$

Methods in Geochemistry and Geophysics, Volume 45
ISSN 0076-6895,
http://dx.doi.org/10.1016/B978-0-444-53829-1.00007-1

Thus, there is a linkage between the in-phase and quadrature components of the spectrum.

Now, we find this relationship in the explicit form which does not require a solution of a differential equation in the partial derivatives. Let us assume that a solution of Maxwell's equations for sinusoidal oscillations has the form

$$\mathrm{Re}[M \exp(-i\omega t)],$$

where M is the complex amplitude of the field component in an arbitrary direction. Of course, in order to obtain an actual sinusoidal field, it is necessary to take the real part of the product:

$$\mathrm{Re}[M \exp(-i\omega t)] = M_0 \cos(\omega t + \varphi),$$

since

$$M = M_0 \exp(-i\varphi)$$

As we know, if the solution contains a complex amplitude term, from the physical point of view this means that there is a phase shift between the field and, for example, the current which generates the primary field. Correspondingly, the field can be represented as being the sum of the quadrature (Q) and the in-phase (In) components, and we have

$$M = InM + iQM = M_0 \cos\varphi - iM_0 \sin\varphi \tag{7.4}$$

7.1.1. Cauchy Formula

Making use of the conventional symbols for representing a complex function and its argument, we can write M as

$$M(z) = U(z) + iV(z), \tag{7.5}$$

where $U(z)$ and $V(z)$ are the real and imaginary parts of the function $M(z)$, respectively, and z is an argument defined as

$$z = x + iy$$

Here, x and y are the coordinates on the complex plane z. In our case, the complex variable z is the frequency

$$\omega = \mathrm{Re}\omega + i\mathrm{Im}\omega,$$

that is,

$$x = \mathrm{Re}\omega, \quad \text{and} \quad y = \mathrm{Im}\omega$$

Of course, measurements of the field are always performed at real frequencies. However, in order to study a field behavior and, in particular, to establish relationship between the quadrature and in-phase component, as well as between the amplitude and phase, it is important to consider the field behavior on the complex plane of ω. First, assume that the complex amplitude $M(\omega)$ is an analytical function of frequency in the upper part of $\omega-$plane $\mathrm{Im}\,\omega > 0$. This means that the function does not have singularities, such as poles or branch points above the real axis of ω. Later we consider the case when singularities are present. As is well known, the Cauchy–Riemann conditions

$$\frac{\partial U}{\partial x} = \frac{\partial V}{\partial y} \quad \text{and} \quad \frac{\partial U}{\partial y} = -\frac{\partial V}{\partial x} \tag{7.6}$$

are necessary and sufficient conditions for analyticity of a function. These conditions express in the differential form the relationship that exists between the real and imaginary parts of the analytical function in the vicinity of the point z. Our first goal is to describe the relationship between the quadrature and in-phase components of the field for real values of ω, because, as was pointed out, the electromagnetic field is observed only for real frequencies. Moreover, unlike Cauchy–Riemann conditions, we will establish such relationships that both components will be represented in an explicit form. With this purpose in mind, let us make use of Cauchy formula. It shows that if the function $M(z)$ is analytic inside the closed contour C, as well as along this contour, and a is any point in the z plane, then

$$\oint\limits_C \frac{M(z)}{z-a}\mathrm{d}z = 2\pi i M(a) \times \begin{cases} 1, & \text{if } a \text{ is inside } C \\ 1/2, & \text{if } a \text{ is on } C \\ 0, & \text{if } a \text{ is outside } C \end{cases} \tag{7.7}$$

It is clear that the Cauchy formula allows us to evaluate $M(a)$ at any point inside the contour C when the values of $M(z)$ are known along this contour. This relationship is a consequence of the close connection which exists between values of an analytic function on the complex z plane. Consider a path consisting of the x-axis and a semicircle with an infinitely large radius. Its center is located at the coordinate origin (Figure 7.1(a)).

The internal area of this contour includes the upper half plane. We will attempt to find the quadrature component of the function $M = U + iV$ by assuming that the in-phase component U is known along the x-axis or vice versa. Applying the Cauchy formula, we have

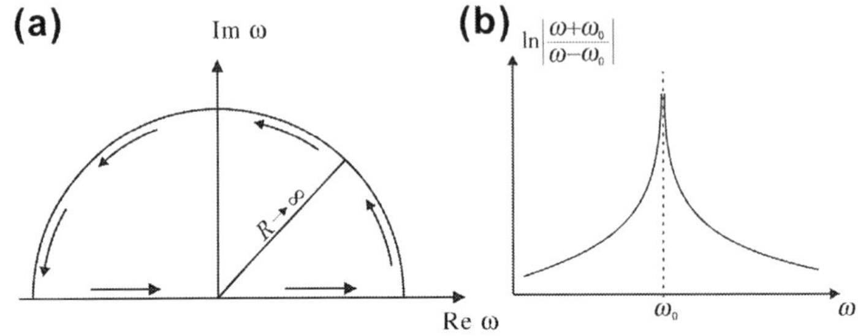

Figure 7.1 (a) Path of integration used in Cauchy formula. (b) Weighing factor.

$$M(\xi) = \frac{1}{i\pi} P \oint_C \frac{M(z)}{z - \xi} dz \tag{7.8}$$

The point $\xi = \varepsilon + i\eta$ is located on the path of integration, and the symbol P denotes the principle value of the integral. Inasmuch as the path of integration coincides with the x-axis, $\eta = 0$, we obtain

$$M(\varepsilon, 0) = \frac{1}{i\pi} P \int_{-\infty}^{\infty} \frac{M(x, 0)}{x - \varepsilon} dx \tag{7.9}$$

In deriving Eq. (7.9), it has been assumed that the value for the integral along the semicircular part vanishes when the radius of the circle increases without limit. This conclusion follows from the fact that the function $M(z)$ representing the field is analytic at infinity and with an increase of z, the ratio $\frac{M(z)}{z}$ tends to zero. Because

$$M(\varepsilon, 0) = U(\varepsilon, 0) + iV(\varepsilon, 0) \quad \text{and} \quad M(x, 0) = U(x, 0) + iV(x, 0)$$

Equation (7.9) gives

$$U(\varepsilon, 0) = \frac{1}{\pi} P \int_{-\infty}^{\infty} \frac{V(x, 0)}{x - \varepsilon} dx \quad V(\varepsilon, 0) = -\frac{1}{\pi} P \int_{-\infty}^{\infty} \frac{U(x, 0)}{x - \varepsilon} dx \tag{7.10}$$

7.1.2. Hilbert Transform

The integrands in these expressions are characterized by a singularity at $x = \varepsilon$, which can be removed by making use of the identity

$$P\int_{-\infty}^{\infty}\frac{dx}{x-\varepsilon}=0$$

Then we can rewrite Eq. (7.10) in the form

$$U(\varepsilon,0)=\frac{1}{\pi}\int_{-\infty}^{\infty}\frac{V(x,0)-V(\varepsilon,0)}{x-\varepsilon}dx \tag{7.11}$$

and

$$V(\varepsilon,0)=-\frac{1}{\pi}\int_{-\infty}^{\infty}\frac{U(x,0)-U(\varepsilon,0)}{x-\varepsilon}dx, \tag{7.12}$$

since

$$V(\varepsilon,0)\int_{-\infty}^{\infty}\frac{dx}{x-\varepsilon}=U(\varepsilon,0)\int_{-\infty}^{\infty}\frac{dx}{x-\varepsilon}=0$$

Expanding the numerator in Eqs (7.11) and (7.12) in the power series, we can see that the integrand does not have singularity, and therefore, the letter P in front of integrals is omitted. Unlike Cauchy–Riemann conditions, these equations establish the relationship between the real and imaginary parts of the analytical function in the explicit form.

7.1.3. Relationships between the Amplitude and Phase of the Spectrum

Let us return to consideration of the complex amplitude of the field

$$M(\omega)=InM(\omega)+iQM(\omega)$$

In accordance with Eqs (7.11) and (7.12), the relationships between the in-phase and quadrature components are

$$InM(\omega_0)=\frac{1}{\pi}\int_{-\infty}^{\infty}\frac{QM(\omega)-QM(\omega_0)}{\omega-\omega_0}d\omega$$

and

$$QM(\omega_0)=-\frac{1}{\pi}\int_{-\infty}^{\infty}\frac{InM(\omega)-InM(\omega_0)}{\omega-\omega_0}d\omega \tag{7.13}$$

Thus, when the spectrum of one of the components is known, the other component of the field can be calculated by making use of Eq. (7.13). It is now a simple matter to find the relationship between the amplitude and phase responses of the field component. Taking the logarithm of the complex amplitude $M = M_0 \exp(-i\varphi)$, we have

$$\ln M = \ln M_0 - i\varphi$$

From this equation, we see that the relationship between the amplitude and phase responses is the same as that for the quadrature and in-phase components, provided that $M(\omega) \neq 0$ at all points of the half plane $\mathrm{Im}\,\omega > 0$. For instance, for the phase we have

$$\varphi(\omega_0) = \frac{1}{\pi} \int_{-\infty}^{\infty} \frac{\ln M_0(\omega) - \ln M_0(\omega_0)}{\omega - \omega_0} d\omega \tag{7.14}$$

Often, it is preferable to express the right-hand side of this equation in another form. After some algebraic operations, we obtain

$$\varphi(\omega_0) = -\frac{1}{\pi} \int_{-\infty}^{\infty} \frac{\partial L}{\partial u} \left|\ln \coth \frac{u}{2}\right| du, \tag{7.15}$$

where

$$L = \ln M_0, \quad \text{and} \quad u = \ln \frac{\omega}{\omega_0}$$

It can be seen from Eq. (7.15) that the phase response depends on the slope of the amplitude response curve plotted on a logarithmic scale. Inasmuch as the integration is carried out over the entire frequency range, the phase at any particular frequency ω_0 depends on the slope of the amplitude response curve at all frequencies. However, the relative importance of the slope over various portions of the spectrum is controlled by a weighing factor $\left|\ln \coth \frac{u}{2}\right|$ which, can also be written as

$$\ln \frac{\omega + \omega_0}{\omega - \omega_0},$$

and its behavior is shown in Figure 7.1(b). This factor increases as the frequency approaches ω_0 and is logarithmically infinite at that point. Therefore, the slope of the amplitude response near the frequency for which the phase is calculated is much more important than the slope of the amplitude response at more distant frequencies.

From the geophysical point of view, Eqs (7.13) and (7.14) lead us to the following conclusions. First of all, measurements of the phase response do not provide additional information about parameters of the medium when the amplitude response is already known. However, it may happen that the shape of the phase response curve more clearly reflects some diagnostic features of the resistivity section of a medium than the amplitude response curve. It is also important to stress that while there is in essence a unique relationship between the quadrature and in-phase responses, as well as between the phase and amplitude responses, this does not mean that there is a point-by-point relationship between them. In fact, measurements of both the amplitude and phase at one or a few frequencies may provide different information characterizing the geoelectric section in a different manner. The same conclusion can be reached for measurements of the quadrature and in-phase components of the spectrum.

7.1.4. Zeroes of the Spectrum on the Upper Part of the ω–Plane

Equations (7.14) and (7.15) imply that at all points above the real axis of ω, the spectrum $M(\omega)$ differs from zero; otherwise $\ln M(\omega)$ is not an analytical function and Cauchy formula cannot be used. Now consider more general case and suppose at the beginning that at some point ω_1 above the real axis of ω,

$$\omega_1 = \mathrm{Re}\,\omega_1 + i\mathrm{Im}\,\omega_1 \tag{7.16}$$

the spectrum is equal to zero:

$$M(\omega_1) = 0$$

Inasmuch as we cannot apply Cauchy formula to function $M(\omega)$ and find the relationship between its amplitude and phase, let us present this function as follows:

$$M(\omega) = M_1(\omega)\frac{\omega - \omega_1}{\omega - \omega_1^*}, \tag{7.17}$$

where ω_1^* is conjugate of ω,

$$\omega_1^* = \mathrm{Re}\,\omega_1 - i\mathrm{Im}\,\omega_1 \tag{7.18}$$

and M_1 does not have zero values. It is clear that both functions $M(\omega)$ and $M_1(\omega)$ have equal amplitudes and their phases are related as

$$\varphi(\omega) = \varphi_1(\omega) - 2\tan^{-1}\frac{\text{Im}\,\omega_1}{\omega - \text{Re}\,\omega_1} \tag{7.19}$$

In fact,

$$\frac{\omega - \omega_1}{\omega - \omega_1^*} = \exp(-2i\alpha),$$

where

$$\alpha = \tan^{-1}\frac{\text{Im}\,\omega_1}{\omega - \text{Re}\,\omega_1},$$

since ω is a real number. Inasmuch as $M_1(\omega)$ does not have zero, we can apply Hilbert transform; that is,

$$\varphi(\omega_0) = \frac{1}{\pi}\int_{-\infty}^{\infty}\frac{\ln|M_1(\omega)| - \ln|M_1(\omega_0)|}{\omega - \omega_0}d\omega - 2\tan^{-1}\frac{\text{Im}\,\omega_1}{\omega_0 - \omega_1} \tag{7.20}$$

It is a simple matter to generalize this equality when spectrum $M(z)$ is equal to zero at several points. In such case, we can write

$$M(\omega) = M_1 \cdot \Pi\frac{\omega - \omega_i}{\omega - \omega_i^*}$$

Here, Π is the product of fractions shown above. In accordance with Eq. (7.20) in order to calculate the phase response, it is necessary to know both the amplitude response and coordinates of the points at the complex plane of ω where the spectrum is equal to zero. Thus, we see that Eq. (7.13) is valid if the complex amplitude of the field as a function of frequency is analytic in the upper half plane of ω. At the same time in order to use Eqs (7.14) and (7.15), that is to calculate the phase from the field amplitude, it is necessary to satisfy one more condition, namely, $M(\omega) \neq 0$ if $\text{Im}\,\omega > 0$. This analysis shows that in those cases when the measured and calculated phases do not coincide with each other, we deal with a geoelectric section where the complex amplitude either is nonanalytic or has zero values at some points of complex plane of the frequency, and such case should not cause surprise.

7.2. FOURIER INTEGRALS

7.2.1. Different Forms of Fourier Transform

Next we describe some relationships between the frequency and transient responses which follows from Fourier transform:

$$M(t) = \frac{1}{2\pi} \int_{-\infty}^{\infty} M(\omega) \exp(-i\omega t) d\omega$$

and

$$M(\omega) = \int_{-\infty}^{\infty} M(t) \exp(i\omega t) dt, \tag{7.21}$$

where $M(t)$ and $M(\omega)$ are the transient response and its spectrum, respectively. Assuming that the transient response appears at some instant $t = 0$, we can rewrite the last equation of the set (7.21) as

$$\begin{aligned} \mathrm{Re}\, M(\omega) + i\mathrm{Im}\, M(\omega) &= \int_0^{\infty} M(t) \exp(i\omega t) dt \\ &= \int_0^{\infty} M(t) \cos \omega t dt + i \int_0^{\infty} M(t) \sin \omega t dt \end{aligned}$$

or

$$\mathrm{Re}\, M(\omega) = \int_0^{\infty} M(t) \cos \omega t dt, \quad \text{and} \quad \mathrm{Im}\, M(\omega) = \int_0^{\infty} M(t) \sin \omega t dt \tag{7.22}$$

Therefore, the real and imaginary parts of the complex spectrum are even and odd functions of ω, respectively:

$$\mathrm{Re} M(-\omega) = \mathrm{Re} M(\omega), \quad \text{and} \quad \mathrm{Im} M(-\omega) = -\mathrm{Im} M(\omega), \tag{7.23}$$

and such behavior is one of the most remarkable features of the spectrum. Let us notice that Eq. (7.22) allows us to calculate the in-phase and quadrature components of the spectrum as well as their derivatives with respect to ω, when the transient response is known. This procedure is useful for reduction of different types of noise.

7.2.2. The Step Function of Excitation

In most cases considered in the monograph, it is assumed that a transient electromagnetic field is excited by a step function current. Correspondingly, the primary magnetic field accompanying this current behaves likewise

$$B_0(t) = M_0(t) = \begin{cases} B_0 & t < 0 \\ 0 & t \geq 0 \end{cases} \tag{7.24}$$

Now we demonstrate that the transient response can be expressed in terms of either the quadrature or in-phase component of the spectrum. Similar results can be derived for the electric field too. As was found earlier and as it follows from Eqs (7.21) and (7.24), the spectrum for the primary magnetic field is

$$B_0(\omega) = \frac{B_0}{i\omega} \tag{7.25}$$

The amplitude of this spectrum decreases at a rate inversely proportional to frequency while the phase remains constant. Inasmuch as the energy of the primary field is mainly concentrated at the low-frequency part of the spectrum, when a step function excitation is applied, its use is often preferable in the practice of geophysical methods. As was mentioned earlier, this is related to the fact that with a decrease in the frequency, the field penetrates to a greater depth. In accordance with the Fourier transform formula, the primary field can be written as

$$B_0(t) = \frac{B_0}{2\pi} \int\limits_{-\infty}^{\infty} \frac{\exp(-i\omega t)}{i\omega} \mathrm{d}\omega, \tag{7.26}$$

where the path of integration is not permitted to pass through the point $\omega = 0$. Let us write the integral as a sum as follows:

$$\begin{aligned} \frac{1}{2\pi} \int\limits_{-\infty}^{\infty} \frac{\exp(-i\omega t)}{i\omega} \mathrm{d}\omega = & \frac{1}{2\pi} \int\limits_{-\infty}^{\varepsilon} \frac{\exp(-i\omega t)}{i\omega} \mathrm{d}\omega \\ & + \frac{1}{2\pi} \int\limits_{-\varepsilon}^{\varepsilon} \frac{\exp(-i\omega t)}{i\omega} \mathrm{d}\omega \\ & + \frac{1}{2\pi} \int\limits_{\varepsilon}^{\infty} \frac{\exp(-i\omega t)}{i\omega} \mathrm{d}\omega \end{aligned}$$

Here, ε is very small number. We select a semicircular path of integration around the origin whose radius tends to be zero. In calculating the middle integral, it is convenient to introduce a new variable φ:

$$\omega = r \exp(i\varphi), \quad \text{and} \quad d\omega = ir \exp(i\varphi)d\varphi$$

Thus,

$$\frac{1}{2\pi}\int_{-\varepsilon}^{\varepsilon}\frac{\exp(-i\omega t)}{i\omega}d\omega = \frac{1}{2\pi i}\int_{\pi}^{2\pi}\frac{ir\exp(i\varphi)}{r\exp(i\varphi)}d\varphi = \frac{1}{2}$$

since

$$\exp(-i\omega t) \to 1$$

Respectively, an expression for the primary field when the variable of integration ω takes only real nonzero values is

$$B_0(t) = \frac{B_0}{2} + \frac{B_0}{2\pi}\int_{-\infty}^{\infty}\frac{\exp(-i\omega t)}{i\omega}d\omega \tag{7.27}$$

Making use of the principle of superposition, we obtain the following expressions for a nonstationary field:

$$B(t) = \frac{1}{2\pi i}\int_{-\infty}^{\infty}\frac{B(\omega)}{\omega}\exp(-i\omega t)d\omega \tag{7.28}$$

or

$$B(t) = \frac{B_0}{2} + \frac{1}{2\pi i}\int_{-\infty}^{\infty}\frac{B(\omega)}{\omega}\exp(-i\omega t)d\omega, \tag{7.29}$$

where

$$B(\omega) = InB(\omega) + iQB(\omega)$$

is the response function. In other words, $B(\omega)$ characterizes the field in a medium when the primary field varies as the sinusoidal function $B_0 \sin \omega t$. As is well known, the derivative of a step function with respect to time is a Dirac delta function, and in accordance with Eq. (7.29), we have for such case

$$B(t) = -\frac{1}{2\pi}\int_{-\infty}^{\infty} B(\omega)\exp(-i\omega t)\mathrm{d}\omega, \tag{7.30}$$

where $B(t)$ is the impulse response caused by delta function excitation.

Let us represent Eq. (7.29) in the form

$$B(t) = \frac{B_0}{2} + \frac{1}{2\pi}\int_{-\infty}^{\infty} \frac{QB(\omega)\cos\omega t - InB(\omega)\sin\omega t}{\omega}\mathrm{d}\omega$$
$$-\frac{i}{2\pi}\int_{-\infty}^{\infty} \frac{QB(\omega)\sin\omega t + InB(\omega)\cos\omega t}{\omega}\mathrm{d}\omega \tag{7.31}$$

From Eq. (7.23), it follows that the second integral in Eq. (7.31) is equal to zero since the numerator of the integrand is even function while denominator is odd. It is consistent with the physical point of view which suggests that the field $B(t)$ cannot be a complex-valued function of time. Thus, we obtain

$$B(t) = \frac{B_0}{2} + \frac{1}{\pi}\int_{0}^{\infty} \frac{QB(\omega)\cos\omega t - InB(\omega)\sin\omega t}{\omega}\mathrm{d}\omega \tag{7.32}$$

If time t is positive, then $-t$ is negative and from Eq. (7.32), we have

$$B(-t) = \frac{B_0}{2} + \frac{1}{\pi}\int_{0}^{\infty} \frac{QB(\omega)\cos\omega t + InB(\omega)\sin\omega t}{\omega}\mathrm{d}\omega$$

On the other hand, $B(-t) = B_0$ and therefore

$$B_0 = \frac{B_0}{2} + \frac{1}{\pi}\int_{0}^{\infty} \frac{QB(\omega)\cos\omega t + InB(\omega)\sin\omega t}{\omega}\mathrm{d}\omega$$

or

$$0 = -\frac{B_0}{2} + \frac{1}{\pi}\int_{0}^{\infty} \frac{QB(\omega)\cos\omega t + InB(\omega)\sin\omega t}{\omega}\mathrm{d}\omega \tag{7.33}$$

It is proper to note that in these last expressions, the time t is taken as positive.

Combining Eqs (7.32) and (7.33), we obtain

$$B(t) = \frac{2}{\pi}\int_0^\infty \frac{QB(\omega)}{\omega}\cos\omega t d\omega \quad \text{or} \quad B(t) = B_0 - \frac{2}{\pi}\int_0^\infty \frac{InB(\omega)}{\omega}\sin\omega t d\omega \tag{7.34}$$

Equation (7.34) permits us to calculate the transient response when either the quadrature or in-phase components of the spectrum are known. Of course, making use of Eq. (7.34), it is a simple matter to express derivatives of the field with respect to time in terms of the spectrum. In particular, we have

$$\frac{\partial B}{\partial t} = -\frac{2}{\pi}\int_0^\infty QB(\omega)\sin\omega t d\omega, \quad \text{or} \quad \frac{\partial B}{\partial t} = \frac{2}{\pi}\int_0^\infty InB(\omega)\sin\omega t d\omega \tag{7.35}$$

REFERENCES AND FURTHER READING

[1] P.M. Morse, H. Feshbach, Methods of Theoretical Physics, McGraw-Hill, New York, 1953.

[2] K. Aki, P. Richards, Quantitative Seismology, Theory and Method, Freeman and Co, 1980.

[3] A.A. Kaufman, A.L. Levchin, K.L. Larner, Acoustic and Elastic Wave Fields in Geophysics, Elsevier, 2002.

CHAPTER EIGHT

Vertical Magnetic Dipole in the Presence of Uniform Half Space

8.1. FORMULATION OF BOUNDARY VALUE PROBLEM

First suppose that a vertical magnetic dipole is located in a nonconducting medium at a height h above the surface, as is shown in Figure 8.1(a). It is proper to remind that a horizontal loop behaves practically as a vertical magnetic dipole if the distance from the loop center to an observation point is in several times greater than the loop radius. Let us introduce the cylindrical system of coordinates with the origin at the Earth's surface and with the z-axis perpendicular to this surface and directed downward. The conductivity and magnetic permeability of the medium beneath the surface are γ and μ_0, respectively. We solve the boundary value problem for the field in the frequency domain and then, applying a Fourier transform, determine its transient response. It is appropriate to note that in a conducting medium, the displacement currents are neglected; that is, the period of oscillations is much

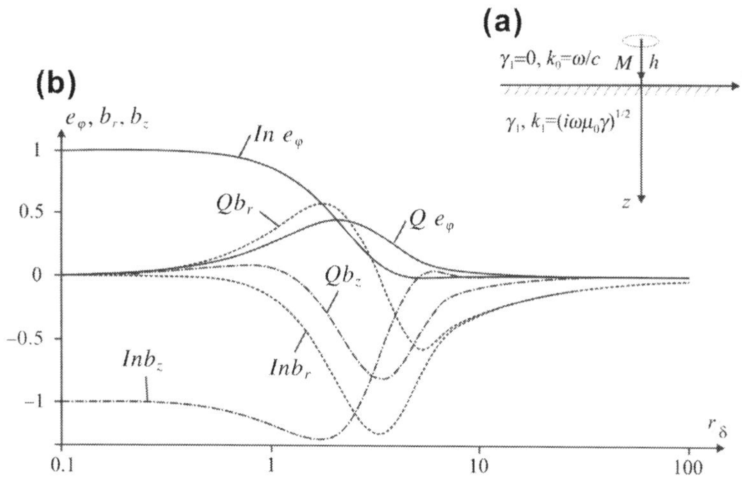

Figure 8.1 (a) Vertical magnetic dipole above the uniform half space. (b) The quadrature and in-phase components of e^*_φ, b^*_r, and b^*_z.

Methods in Geochemistry and Geophysics, Volume 45
ISSN 0076-6895,
http://dx.doi.org/10.1016/B978-0-444-53829-1.00008-3

greater than the time of arrival of the signal, τ_0, at any point beneath the Earth's surface. As was demonstrated earlier, the primary electric field arising due to a change of the primary magnetic field with time has only the φ-component. Thus, induced currents generated in a conducting medium also have a tangential component j_φ. It is natural to expect that interaction between currents should not change their direction. Of course, it is essential that the primary electric field is tangent to the surface $z = 0$, and correspondingly, the surface charges are absent. Therefore, the induced currents are located in horizontal planes and their current lines are circles with the centers on the z-axis. As a consequence of these assumptions, one can expect that the secondary magnetic field has two components B_r and B_z, but only one component of the electric field E_φ. To simplify a solution of the boundary value problem, we again introduce the vector potential of the magnetic type. As we know, in the frequency domain,

$$\mathbf{E}^* = \text{curl } \mathbf{A}^*, \quad i\omega\mathbf{B}^* = k^2\mathbf{A}^* + \text{grad div } \mathbf{A}^* \tag{8.1}$$

Taking into account the orientation of the electric field, we assume that the complex vector $\mathbf{A}^*$ has only one component A_z^* and it is independent of the coordinate φ; that is,

$$A_r^* = A_\varphi^* = 0 \quad \text{and} \quad \mathbf{A}_z^* = A_z^*(r, z)\mathbf{1}_z, \tag{8.2}$$

where $\mathbf{1}_z$ is the unit vector along the z-axis. Correspondingly, the complex amplitudes of the electromagnetic field are expressed in terms of A_z^* as

$$E_\varphi^* = -\frac{\partial A_z^*}{\partial r},$$

and

$$i\omega B_r^* = \frac{\partial^2 A_z^*}{\partial r \partial z}, \quad B_\varphi^* = 0, \quad i\omega B_z^* = k^2 A_z^* + \frac{\partial^2 A_z^*}{\partial z^2} \tag{8.3}$$

At regular points, the component A_z^* satisfies Helmholtz's equation; in the vicinity of the magnetic dipole, located at the point $r = 0$ and $z = -h$, it approaches to that of the primary field, A_0^*, and at infinity, it tends to zero. Thus,

$$\nabla^2 A_z^* + k^2 A_z^* = 0,$$

$$A_z^* \to A_0^* = \frac{i\omega\mu_0 M_0}{4\pi R_1}\exp(ik_0 R_1) \quad \text{if } R_1 \to 0 \tag{8.4}$$

and

$$A_z^* \to 0 \quad \text{if} \quad R_1 \to \infty$$

Here, $R_1 = \left[(z+h)^2 + r^2\right]^{1/2}$ and $k_0 = \omega/c$ is the wave number of a nonconducting half space, and $c = 10^8$ m/s is the speed of light.

Inasmuch as tangential components of the electromagnetic field are continuous functions at the boundary, the function A_z^* has to satisfy the following conditions:

$$\frac{\partial A_{1z}^*}{\partial r} = \frac{\partial A_{2z}^*}{\partial r} \quad \text{and} \quad \frac{\partial^2 A_{1z}^*}{\partial r \partial z} = \frac{\partial^2 A_{2z}^*}{\partial r \partial z}, \quad \text{if } z = 0 \tag{8.5}$$

where A_{1z}^* and A_{2z}^* are the z-components of the complex amplitude of the vector potential in the upper space and the conducting medium, respectively. Suppose that some function is continuous at the surface S; that is, it has the same value on both sides of this surface. Then it is obvious that the derivative of this function in a direction which is tangential to the surface is also continuous function. For this reason, Eq. (8.5) can be greatly simplified, namely

$$A_{1z}^* = A_{2z}^*, \quad \frac{\partial A_{1z}^*}{\partial z} = \frac{\partial A_{2z}^*}{\partial z} \tag{8.6}$$

It is convenient to represent the vector potential in the upper space as a sum as follows:

$$A_{1z}^* = A_0^* + A_s^*, \tag{8.7}$$

where A_s^* is the z-component of the vector potential describing the secondary field. Now we are ready to formulate the boundary value problem in terms of the vector potential. In accordance with Eqs (8.4), (8.6) and (8.7), this function must satisfy the following conditions:

1. At regular points,

$$\nabla^2 A_{1z}^* + k_0^2 A_{1z}^* = 0 \quad \text{if } z < 0$$

and

$$\nabla^2 A_{2z}^* + k_1^2 A_{2z}^* = 0 \quad \text{if } z > 0$$

2. In the vicinity of the magnetic dipole,

$$A_{1z}^*(r, z) \to \frac{i\omega\mu_0 M_0}{4\pi R_1} \exp\left(ik_0 R_1\right) \quad \text{if } R_1 \to 0 \tag{8.8}$$

3. At the surface, $z=0$:

$$A^*_{1z} = A^*_{2z}, \qquad \frac{\partial A^*_{1z}}{\partial z} = \frac{\partial A^*_{2z}}{\partial z}$$

4. At infinity,

$$A^*_{1z} \to 0, \quad A^*_{2z} \to 0 \quad \text{if } R_1 \to \infty$$

As follows from the uniqueness theorem, these conditions uniquely define the complex amplitude of the vector potential and, therefore, the electromagnetic field.

8.2. SOLUTION OF HELMHOLTZ EQUATIONS

First of all, we find a solution of Helmholtz's equation, which in cylindrical system of coordinates r, φ, and z has the form

$$\frac{\partial^2 A^*_z}{\partial r^2} + \frac{1}{r}\frac{\partial A^*_z}{\partial r} + \frac{\partial^2 A^*_z}{\partial z^2} + k^2 A^*_z = 0, \tag{8.9}$$

since

$$\frac{\partial A^*_z}{\partial \varphi} = 0$$

Applying the method of separation of variables, we represent the function A^*_z as

$$A^*_z = U(r)V(z),$$

and substituting this into Eq. (8.9), we obtain two ordinary differential equations:

$$\frac{\partial^2 U}{\partial r^2} + \frac{1}{r}\frac{\partial U}{\partial r} + m^2 U = 0 \tag{8.10}$$

and

$$\frac{\partial^2 V}{\partial z^2} - \left(m^2 - k^2\right) V = 0, \tag{8.11}$$

where m is the variable of separation. As is well known, the first equation is Bessel's equation, and its solution are Bessel functions of the first and second kinds, $J_0(mr)$ and $Y_0(mr)$. The function $Y_0(mr)$ has an infinite value along the z-axis, where $r = 0$, and therefore, it cannot be used to describe the electromagnetic field. It is a simple matter to see that the solution of Eq. (8.11) is

$$V = C\exp\left[-\left(m^2 - k^2\right)^{1/2} z\right] + D\exp\left[\left(m^2 - k^2\right)^{1/2} z\right],$$

where C and D are constants. Inasmuch as partial solution of Eq. (8.9) is the product of functions U and V for given m, its general solution can be written as

$$A_z^* = \frac{i\omega\mu_0 M_0}{4\pi}\int_0^\infty \left[C_m \exp\left(-m_i z\right) + D_m \exp\left(m_i z\right)\right] J_0(mr)\mathrm{d}m, \tag{8.12}$$

where

$$m_i = \left(m^2 - k_i^2\right)^{1/2}, \quad i = 0, 1$$

We further consider the function A_z^* provided that

$$\mathrm{Re}\, m_i = \mathrm{Re}\left(m^2 - k_i^2\right)^{1/2} > 0,$$

Then, taking into account the fact that the field vanishes at infinity ($z \to \infty$), the integrand for A_{2z}^* in Eq. (8.12) is simplified and we have

$$A_{2z}^* = \frac{i\omega\mu_0 M_0}{4\pi}\int_0^\infty C_m \exp\left(-m_1 z\right) J_0(mr)\mathrm{d}m \quad \text{if} \quad z \geq 0 \tag{8.13}$$

Here

$$m_1 = \left(m^2 - k_1^2\right)^{1/2}, \quad \text{and} \quad k_1^2 = i\gamma\mu_0\omega$$

In the upper half space, the field is a sum of the primary and secondary one and we obtain the following expression for A_{1z}^*:

$$A_{1z}^* = \frac{i\omega\mu_0 M_0}{4\pi}\left[\frac{1}{R_1}\exp\left(ik_0 R_1\right) + \int_0^\infty D_m \exp\left(m_0 z\right) J_0(mr)\mathrm{d}m\right] \text{if} \quad z < 0 \tag{8.14}$$

It is obvious that the function A_{1z}^* satisfies the condition near the dipole.

8.3. EXPRESSIONS FOR THE VECTOR POTENTIAL

In order to satisfy the boundary conditions at the Earth's surface, it is convenient to make use of Sommerfeld identity

$$\frac{\exp(ik_0R_1)}{R_1} = \int_0^\infty \frac{m}{m_0} \exp\left(-m_0|z+h|\right)J_0(mr)\mathrm{d}m \qquad (8.15)$$

Then Eq. (8.14) becomes

$$A^*_{1z} = \frac{i\omega\mu_0 M_0}{4\pi}\int_0^\infty \left\{\frac{m}{m_0}\exp\left[m_0(-|z+h|)\right] + D_m \exp\left(m_0 z\right)\right\}J_0(mr)\mathrm{d}m \qquad (8.16)$$

Substituting these expressions for the vector potential into boundary conditions (Eq. (8.6)), and making use of the orthogonality of the Bessel functions, we can determine the unknown coefficients D_m and C_m. In doing so, we obtain the system of equations with two unknowns:

$$\frac{m}{m_0}\exp\left(-m_0 h\right) + D_m = C_m$$

$$-m\exp\left(-m_0 h\right) + m_0 D_m = -m_1 C_m$$

Solution of this system gives

$$C_m = \frac{2m}{m_0+m_1}\exp(-m_0 h), \quad \text{and} \quad D_m = \frac{m}{m_0}\frac{m_0-m_1}{m_0+m_1}\exp\left(-m_0 h\right) \qquad (8.17)$$

Whence,

$$A^*_{1z} = \frac{i\omega\mu_0 M_0}{4\pi}$$

$$\times\int_0^\infty \frac{m}{m_0}\left\{\exp(-|z+h|m_0) + \frac{m_0-m_1}{m_0+m_1}\exp\left[m_0(z-h)\right]\right\}J_0(mr)\mathrm{d}m$$

and

$$A^*_{2z} = \frac{i\omega\mu_0 M_0}{4\pi}\int_0^\infty \frac{2m}{m_0+m_1}\exp\left(-m_0 h\right)\exp\left(-m_1 z\right)J_0(mr)\mathrm{d}m \qquad (8.18)$$

Note that since these equations satisfy all conditions of the boundary value problem, our assumptions about the symmetry of the field were correct. Now we focus our attention on the quasi-stationary field in a conducting half space and mainly at its surface, where measurements are usually performed.

8.4. THE FIELD OF THE MAGNETIC DIPOLE IN A CONDUCTING MEDIUM PROVIDED THAT $h=0$, $k_0=0$

If the vertical magnetic dipole is situated on the surface ($h=0$) and letting $k_0=0$, that is, neglecting by propagation in a free space, $m_0=m$, these equations are slightly simplified, and we have

$$A^*_{1z}=\frac{i\omega\mu_0 M_0}{4\pi}\int\limits_0^\infty\left[\exp\left(-m|z|\right)+\frac{m-m_1}{m+m_1}\exp\left(mz\right)\right]J_0(mr)dm$$

and

$$A^*_{2z}=\frac{i\omega\mu_0 M_0}{4\pi}\int\limits_0^\infty\frac{2m}{m+m_1}\exp\left(-m_1 z\right)J_0(mr)dm \tag{8.19}$$

From Eq. (8.3), we obtain the following expressions for the electromagnetic field components within the conducting medium:

$$E^*_\varphi=\frac{i\omega\mu_0 M_0}{2\pi}\int\limits_0^\infty\frac{m^2}{m+m_1}\exp\left(-m_1 z\right)J_1(mr)dm$$

$$B^*_r=\frac{\mu_0 M_0}{2\pi}\int\limits_0^\infty\frac{m_1 m^2}{m+m_1}\exp\left(-m_1 z\right)J_1(mr)dm \tag{8.20}$$

$$B^*_z=\frac{\mu_0 M_0}{2\pi}\int\limits_0^\infty\frac{m^3}{m+m_1}\exp\left(-m_1 z\right)J_0(mr)dm,$$

where $J_1(mr)$ is the Bessel function of the first kind and order one and is related to the Bessel function of zero order as

$$J_1(x)=-J_0'(x)$$

Similar expressions can be derived for the field in the upper half space. When both the vertical magnetic dipole and the observation point are located at the Earth's surface ($z = 0$), we have

$$A_z^* = \frac{i\omega\mu_0 M_0}{2\pi} \int_0^\infty \frac{m}{m + m_1} J_0(mr)\mathrm{d}m \tag{8.21}$$

$$E_\varphi^* = \frac{i\omega\mu_0 M_0}{2\pi} \int_0^\infty \frac{m^2}{m + m_1} J_1(mr)\mathrm{d}m \tag{8.22}$$

$$B_r^* = \frac{\mu_0 M_0}{2\pi} \int_0^\infty \frac{m^2 m_1}{m + m_1} J_1(mr)\mathrm{d}m \tag{8.23}$$

and

$$B_z^* = \frac{\mu_0 M_0}{2\pi} \int_0^\infty \frac{m^3}{m + m_1} J_0(mr)\mathrm{d}m \tag{8.24}$$

We also consider the function $\partial A_z^*/\partial z$ on the Earth's surface. From Eq. (8.19), it follows that

$$\frac{\partial A_z^*}{\partial z} = -\frac{i\omega\mu_0 M_0}{2\pi} \int_0^\infty \frac{m m_1}{m + m_1} J_0(mr)\mathrm{d}m \quad \text{if } z = 0 \tag{8.25}$$

Perhaps, it is proper to make one comment. In deriving equations of the field, we assumed that displacement currents are absent in a free space. Of course, in a reality, there is a direct wave propagating from the source through this medium with the speed of light. Our assumption simply means that the distance between the magnetic dipole and an observation point is much smaller than the wave length in the upper space; that is,

$$r \ll \frac{c}{f}$$

8.5. THE FIELD EXPRESSIONS AT THE EARTH'S SURFACE

It is proper to note that Eqs (8.22)–(8.25) cannot be used directly for calculation of the field since the integrals in these equations cannot be taken.

However, we will show that electromagnetic field at the Earth's surface can be still expressed in terms of elementary and Bessel functions if $h=0$, and they of course have finite values. With this purpose in mind, we make use of three relationships, namely,

$$\frac{\exp(ik_1R)}{R} = \int_0^\infty \frac{m}{m_1} \exp(-m_1|z|)J_0(mr)\mathrm{d}m \tag{8.26}$$

$$I_0\left[\frac{ik_1(R-z)}{2}\right] K_0\left[\frac{ik_1(R+z)}{2}\right] = \int_0^\infty \frac{\exp(-m_1|z|)}{m_1} J_0(mr)\mathrm{d}m, \tag{8.27}$$

where

$$R = \left(r^2+z^2\right)^{1/2}$$

and

$$I_\nu\left(\frac{ik_1r}{2}\right) K_\nu\left(\frac{ik_1r}{2}\right) = \int_0^\infty \left(\frac{m_1-m}{m_1+m}\right)^\nu \frac{J_0(mr)}{m_1}\mathrm{d}m, \tag{8.28}$$

where I_ν and K_ν are modified Bessel functions.

Let us start with function A_z^*. Rewriting the integral in Eq. (8.21) in the form

$$\int_0^\infty \frac{m}{m+m_1} J_0(mr)\mathrm{d}m = \int_0^\infty \frac{m(m-m_1)}{m^2-m_1^2} J_0(mr)\mathrm{d}m$$
$$= \frac{1}{k_1^2}\left[\int_0^\infty m^2 J_0(mr)\mathrm{d}m - \int_0^\infty mm_1 J_0(mr)\mathrm{d}m\right],$$

and applying Eq. (8.26), we have

$$\int_0^\infty m^2 \exp(-m|z|)J_0(mr)\mathrm{d}m = \frac{\partial^2}{\partial z^2}\frac{1}{R} \tag{8.29}$$

$$\int_0^\infty mm_1 \exp(-m_1|z|)J_0(mr)\mathrm{d}m = \frac{\partial^2}{\partial z^2}\frac{\exp(ik_1R)}{R} \tag{8.30}$$

Taking the second derivative and letting $z=0$, we obtain

$$\int_0^\infty m^2 J_0(mr)\mathrm{d}m = -\frac{1}{r^3} \tag{8.31}$$

$$\int_0^\infty m m_1 J_0(mr)\mathrm{d}m = -\frac{1}{r^3}(1 - ik_1 r)\exp(ik_1 r) \tag{8.32}$$

Therefore, the expression for the vertical component of the vector potential can be represented in the form

$$A_z^* = -\frac{i\omega\mu_0 M_0}{2\pi k_1^2 r^3}[1 - (1 - ik_1 r)\exp(ik_1 r)], \tag{8.33}$$

if $z=h=0$.

Then, as follows from Eq. (8.3), we have

$$E_\varphi^* = -\frac{M\rho_1}{2\pi r^4}\left[3 - \left(3 - 3ik_1 r - k_1^2 r^2\right)\exp(ik_1 r)\right], \tag{8.34}$$

where ρ_1 is the resistivity of the conducting half space. Carrying out similar procedures on the right-hand side of Eq. (8.24), we obtain

$$B_z^* = \frac{9\mu_0 M_0}{2\pi k_1^2 r^5}\left[1 - \left(1 - ik_1 r - \frac{4}{9}k_1^2 r^2 + \frac{1}{9}ik_1^3 r^3\right)\exp(ik_1 r)\right] \tag{8.35}$$

In order to find an expression for B_r^*, we have to perform some algebraic operations with the integrand at the right-hand side of Eq. (8.19) describing A_{2z}^*. This gives

$$\frac{2m}{m + m_1} = \frac{m}{m_1} + F,$$

where

$$F = \frac{2m}{m + m_1} - \frac{m}{m_1} = \frac{m}{m_1}\frac{m_1 - m}{m_1 + m} = -k_1^2\frac{m}{m_1}\frac{1}{(m + m_1)^2}$$

Therefore,

$$A_{2z}^* = \frac{i\omega\mu_0 M_0}{4\pi}\left[\frac{\exp(ik_1 R)}{R} - k_1^2\int_0^\infty \frac{m}{m_1}\frac{\exp(-m_1 z)}{(m + m_1)^2}J_0(mr)\mathrm{d}m\right]$$

Since

$$\frac{\partial}{\partial z}\frac{\exp(ik_1R)}{R} = 0 \quad \text{if} \ z = 0$$

we obtain at the Earth's surface

$$\frac{\partial A_{2z}^*}{\partial z} = \frac{i\omega\mu_0 M_0}{4\pi} k_1^2 \int_0^\infty \frac{mJ_0(mr)\mathrm{d}m}{(m+m_1)^2} \quad \text{if} \ z = 0$$

The term

$$\frac{m}{(m+m_1)^2}$$

can be represented in the form

$$\frac{m}{(m+m_1)^2} = \frac{1}{4m_1}\left[1 - \left(\frac{m_1 - m}{m_1 + m}\right)^2\right]$$

Then making use of Eqs (8.27) and (8.28), we have

$$\frac{\partial A_{2z}^*}{\partial z} = \frac{i\omega\mu_0 M_0 k_1^2}{16\pi}\left[I_0\left(\frac{ik_1r}{2}\right)K_0\left(\frac{ik_1r}{2}\right) - I_2\left(\frac{ik_1r}{2}\right)K_2\left(\frac{ik_1r}{2}\right)\right] \tag{8.36}$$

Taking the derivative of $\partial A_{2z}^*/\partial z$ with respect to r and applying the following relationships between modified Bessel functions,

$$I_{n-1}(x) - I_{n+1}(x) = \frac{2n}{x} I_n(x)$$

$$K_{n-1}(x) - K_{n+1}(x) = -\frac{2n}{x} K_n(x)$$

$$I_{n-1}(x) + I_{n+1}(x) = 2I_n'(x)$$

$$K_{n-1}(x) + K_{n+1}(x) = -2K_n'(x),$$

we arrive at the expression for the horizontal component of the magnetic field,

$$B_r^* = -\frac{\mu_0 M_0 k_1^2}{4\pi r}\left[I_1\left(\frac{ik_1r}{2}\right)K_1\left(\frac{ik_1r}{2}\right) - I_2\left(\frac{ik_1r}{2}\right)K_2\left(\frac{ik_1r}{2}\right)\right] \tag{8.37}$$

Thus, expressions for the electromagnetic field components when both the vertical magnetic dipole and an observation point are located on the Earth's surface and $k_0 = 0$ can be written as

$$B_z^* = B_z^0 b_z^*, \quad B_r^* = B_z^0 b_r^*, \quad E_\varphi^* = E_\varphi^{0^*} e_\varphi^*, \tag{8.38}$$

where B_z^0 and $E_\varphi^{0^*}$ describe the magnetic and electric fields of the magnetic dipole current (primary field) on the plane $z = 0$, respectively

$$B_z^0 = \frac{\mu_0 M_0}{4\pi r^3} \quad \text{and} \quad E_\varphi^{0^*} = \frac{i\omega\mu_0 M_0}{4\pi r^2}, \tag{8.39}$$

while

$$b_z^* = \frac{18}{k_1^2 r^2}\left[1 - \left(1 - ik_1 r - \frac{4}{9}k_1^2 r^2 + \frac{1}{9}ik_1^3 r^3\right)\exp\left(ik_1 r\right)\right],$$

$$b_r^* = -k_1^2 r^2\left[I_1\left(\frac{ik_1 r}{2}\right)K_1\left(\frac{ik_1 r}{2}\right) - I_2\left(\frac{ik_1 r}{2}\right)K_2\left(\frac{ik_1 r}{2}\right)\right] \tag{8.40}$$

$$e_\varphi^* = -\frac{2}{k_1^2 r^2}\left[3 - \left(3 - 3ik_1 r - k_1^2 r^2\right)\exp\left(ik_1 r\right)\right]$$

Let us remind again that we were forced to perform sometimes rather cumbersome transformations since the integral representations of the field on the Earth's surface cannot be used to calculate its frequency responses. In accordance with Eq. (8.40), the components of the electromagnetic field, when they are normalized by the primary field, are functions only of the parameter p:

$$p = \frac{r}{\delta} = 2\pi\frac{r}{\lambda}$$

As was pointed out earlier, this parameter is sometimes called the *induction number*, and it characterizes the distance between the dipole and an observation point on the Earth's surface expressed in terms of the skin depth δ or wave length λ. First we consider the asymptotic behavior of the field when the parameter p is either small or large.

8.6. THE RANGE OF SMALL PARAMETER *P* OR NEAR ZONE

Suppose that the parameter is small,

$$p = \frac{r}{\delta} < 1 \quad \text{or} \quad r \ll \lambda \tag{8.41}$$

The zone, represented by this range of distances, is usually called the *inductive or near zone* since a separation is much smaller than the wave length. Any such name has some shortcomings. For instance, the inductive effect, that is interaction between currents, is inherent regardless of the value of r/δ. Also it may happen that condition (Eq. (8.41)) can be applied to very large distances from the dipole, so that making use of the term near zone can be misleading too. However, we will use the latter following conventional terminology in electrodynamics. In accordance with Eq. (8.41), small induction numbers are always observed at the vicinity of the dipole, but with increasing resistivity of the medium or with a decrease of the frequency, the dimensions of the zone of small parameter becomes larger. Regardless of the distance from the dipole to the observation point, there is always a frequency below which the field practically coincides with that in the zone with small values for induction number p. It should also be clear that if the distance r becomes smaller, the field behavior, corresponding to the near zone, is observed at higher frequencies. It is convenient to write expressions for the field components in the form of series. Expanding the exponential in a power series and performing some algebra, we obtain

$$b_z^* = -1 - 2\sum_{n=4}^{\infty} \frac{(n-1)(n-3)^2}{n!}(ik_1 r)^{n-2} \tag{8.42}$$

Taking into account the fact that

$$ik_1 r = p(i-1)$$

we have

$$Inb_z^* \approx -1 - \frac{8}{15}p^3 + \frac{1}{2}p^4 - \frac{16}{105}p^5 + \cdots$$

$$= -1 - \frac{2^{3/2}}{15}\left(\gamma_1\mu_0\omega r^2\right)^{3/2} + \frac{1}{8}\left(\gamma_1\mu_0\omega r^2\right)^2 - \frac{2^{3/2}}{105}\left(\gamma_1\mu_0\omega r^2\right)^{5/2} + \cdots$$

and

$$Qb_z^* \approx \frac{1}{2}p^2 - \frac{8}{15}p^3 + \frac{16}{105}p^5 - \cdots$$
$$= \frac{\gamma_1\mu_0\omega r^2}{4} - \frac{2^{3/2}}{15}\left(\gamma_1\mu_0\omega r^2\right)^{3/2} + \frac{2^{3/2}}{105}\left(\gamma_1\mu_0\omega r^2\right)^{5/2} - \cdots \tag{8.43}$$

Carrying out the same steps with the expression for the electric field, we have

$$e_\varphi^* \approx 1 + 2\sum_{n=4}^{\infty}\frac{(n-1)(3-n)}{n!}(ik_1r)^{n-2} \tag{8.44}$$

From this, we have

$$Ine_\varphi^* \approx 1 - \frac{4}{15}p^3 + \frac{1}{6}p^4 - \frac{4}{105}p^5 + \cdots$$

and

$$Qe_\varphi^* \approx \frac{1}{2}p^2 - \frac{4}{15}p^3 + \frac{4}{105}p^5 - \cdots \tag{8.45}$$

It should be clear that the electric field E_φ is shifted in phase by $\pi/2$ with respect to the function e_φ^*. In order to find an asymptotic expression for the horizontal component b_r^*, we make use of the following representation for Bessel functions when their argument is small:

$$I_1(x) \approx \frac{x}{2} + \frac{x^3}{16} + \cdots, \quad I_2(x) \approx \frac{x^2}{8} + \frac{x^4}{96} + \cdots$$

$$K_1(x) \approx \frac{1}{x} + \left(\frac{x}{2} + \frac{x^3}{16}\right)\ln \alpha x + \cdots \tag{8.46}$$

$$K_2(x) \approx \frac{2}{x^2} - \frac{1}{2} - \left(\frac{x^2}{8} + \frac{x^4}{96}\right)\ln \alpha x + \cdots,$$

where $\alpha \approx 0.8905362$. Substitution of Eq. (8.46) into Eq. (8.40) gives

$$b_r^* \approx -\frac{ip^2}{2} - \frac{1}{4}p^4\ln\frac{p}{2^{1/2}}$$

or

$$Inb_r^* \approx -\frac{p^4}{4}\ln\frac{p}{2^{1/2}} = -\frac{\left(\gamma_1\mu_0\omega r^2\right)^2}{16}\ln\frac{\left(\gamma_1\mu_0\omega r^2\right)^{1/2}}{2} \tag{8.47}$$

and

$$Qb_r^* \approx -\frac{\gamma_1\mu_0\omega r^2}{4} - \frac{3\pi}{64}\left(\gamma_1\mu_0\omega r^2\right)^2$$

Equations (8.43), (8.45), and (8.47) describe the behavior of a quasi-stationary field for small induction numbers r/δ. First of all, we notice that the quadrature component of the magnetic field is significantly larger than the in-phase component of the secondary field. With decreasing p, the quadrature components of b_z^* and b_r^* approach each other

$$\left|Qb_z^*\right| \to \left|Qb_r^*\right|, \tag{8.48}$$

and they are mainly defined by the first terms of the series (Eqs (8.43) and (8.47)). In this case, the quadrature component of the magnetic field is

$$QB_z^{(1)} = -QB_r^{(1)} = \frac{\mu_0 M_0}{16\pi r}\gamma_1\mu_0\omega, \tag{8.49}$$

where index "1" means that only the first term of the series is considered. In accordance with Eq. (8.49), this part of the field is directly proportional to the conductivity of the medium, as well as the frequency and inversely proportional to the distance from the dipole. Considering a uniform medium, we have shown that such behavior holds when the interaction between induced currents can be neglected. In other words, the induced currents are defined by the primary electric field and with a decrease of frequency, the volume, where such behavior takes place, increases. Inasmuch as $p \ll 1$, we conclude that the quadrature component of the magnetic field is much less than the primary field B_z^0. With increasing the distance r, the function QB_z decreases, but the ratio between the quadrature component and the primary field becomes larger. From comparison of the field in a uniform medium with that at the Earth's surface, we see that the quadrature component $QB_z^{(1)}$ is twice smaller. This is easily explained since the dipole and observation point are located at the surface $z=0$ and in the upper space induced currents, caused by the primary field E_φ^0, are absent.

It is obvious that at the Earth's surface, the horizontal component of the primary magnetic field is zero and the observed component B_r is generated only by induced currents in the conducting half space. Next consider the following term in Eq. (8.43) for the quadrature component of B_z. As follows from this equation, we have

$$QB_z^{(2)} = -\frac{\mu_0 M_0}{4\pi r^3}\frac{2^{3/2}}{15}\left(\gamma_1\mu_0\omega r^2\right)^{3/2} = -\frac{\mu_0 M_0 2^{1/2}}{30\pi}(\gamma_1\mu_0\omega)^{3/2} \quad (8.50)$$

Unlike the first term $QB_z^{(1)}$, this function is more sensitive to conductivity and does not depend on the distance r. The latter means that the currents which are responsible for this term are located at relatively large distances in comparison with the distance r. The second term in the series, describing the quadrature component of B_r, differs markedly from $QB_z^{(2)}$. From Eq. (8.47), we have

$$QB_r^{(2)} = -\frac{3\mu_0 M_0}{256} r(\gamma_1\mu_0\omega)^2 \quad (8.51)$$

In contrast to the case with $QB_z^{(2)}$, this function is directly proportional to the distance r and even more sensitive conductivity. From Eqs (8.50) and (8.51), we see that

$$QB_r^{(2)} \ll QB_z^{(2)}, \quad p \ll 1$$

It can be readily seen that different terms of series describing the quadrature and in-phase components of the field are related in a different manner to the distance r and conductivity γ_1. Also all these terms, except the first one for the quadrature component of the magnetic field, arise as the result of interaction between induced currents. Now consider the behavior of the in-phase component of B_z and B_r for the small induction numbers. It is clear that the in-phase component of the secondary field B_z^s is significantly smaller than the primary field B_z^0 as well as the quadrature component QB_z; that is,

$$\left|InB_z^s\right| \ll |QB_z| \ll \left|B_z^0\right| \quad \text{if } p \ll 1$$

This contributes considerable difficulty to field measurements of the in-phase component of $\mathbf{B}^s$ at low induction number. In accordance with Eq. (8.43), the first term in the series for the in-phase component of B_z^s is

$$InB_z^{s(1)} = -\frac{\mu_0 M_0 2^{1/2}}{30\pi}(\gamma_1\mu_0\omega)^{3/2} \quad (8.52)$$

This expression is the same as that for $QB_z^{(2)}$. Thus the in-phase component $InB_z^{s(1)}$ is independent of the separation r and is more sensitive to conductivity than the leading term of the series describing the component

QB_z. In particular, it implies that in the range of low induction number, the in-phase component of the field has a greater depth of investigation than the quadrature component $QB_z^{(1)}$. Comparing the leading terms in the series for the in-phase components, we see that

$$\left|InB_r^{(1)}\right| < \left|InB_z^{s(1)}\right|, \quad \text{if } p < 1$$

and the in-phase component $InB_r^{(1)}$ is more sensitive to conductivity. Here, it is appropriate to make several comments concerning the electric field. As follows from Eq. (8.40), the leading term of the series (Eq. (8.45)) corresponding to the quadrature component of the electric field is

$$QE_\varphi^{(1)} = \frac{\omega\mu_0 M_0}{4\pi r^2} \tag{8.53}$$

This part of the electric field is contributed only by the primary magnetic field B_z^0, and therefore, it is not function of conductivity. As follows from Eq. (8.40), the leading term of the series describing the in-phase component is

$$InE_\varphi^{(1)} = -\frac{M_0}{16\pi}\gamma_1\mu_0^2\omega^2 \tag{8.54}$$

This part of the field can be easily explained. It arises due to the change with time of the vertical component of the magnetic field $QB_z^{(1)}$, given by Eq. (8.49). For this reason, both the field quantities $QB_z^{(1)}$ and $InE_\varphi^{(1)}$ are characterized by the same dependence on conductivity. Since the magnetic field $QB_z^{(1)}$ is inversely proportional to the distance r, the flux of this field through a circle lying in a horizontal plane $z = 0$ is directly proportional to r. Therefore, the electromotive force along such a circle is also proportional to the radius, and correspondingly, the in-phase component of the electric field $InE_\varphi^{(1)}$ does not depend on r. The subsequent terms in the series (Eq. (8.44)) arise from the change with time of the successive terms of in the series (Eq. (8.43)) describing the field B_z. In fact comparing the series for the functions b_z and e_φ, one can say that the corresponding terms in each series are related to one another as

$$e_\varphi^{(n)} = \frac{1}{n-3} b_z^{(n)} \quad \text{if } n \geq 4$$

In conclusion, let us make several comments: (1) As will be shown later, the electromagnetic field in the near zone can be in general represented on the surface of a horizontally layered medium as a series containing whole and fractional powers of ω as well as logarithmic terms $\ln \omega$. However, in the case of a uniform half space, the series representing the electric field and the vertical component of the magnetic field on the Earth's surface do not contain logarithmic terms. (2) As in the case of a uniform half space in the vicinity of the dipole, there is a region where induced currents are mainly defined by the primary electric field, and correspondingly, they are directly proportional to conductivity and frequency. Inasmuch as they are shifted by phase in $90°$, these currents cause the quadrature component of the magnetic field at an observation point. With a decrease of the frequency, the volume occupied by such currents becomes greater and may happen that it gives the main contribution to this component of the field. This means that in such case, the interaction between currents (diffusion) does not practically make influence on the quadrature component of the field measured at an observation point. In contrast, outside this volume, the diffusion starts to play an important role. (3) Inside of such volume, the in-phase component of induced currents is negligible and it appears outside due to interaction of currents, where at some distance from the source, it has a maximum and then due to transformation into the heat it tends to zero, as well as the quadrature component does. With decrease in the frequency, this maximum is observed at greater distances from the dipole and it becomes smaller. Certainly, a behavior of the in-phase component of induced currents shows all features of diffusion. Of course, the quadrature component also demonstrates diffusion but it is masked by the presence of currents due to the primary electric field. (4) Bearing in mind that every elementary current generates the magnetic field according to Biot–Savart law, we should conclude that the quadrature and in-phase components of the field have different depth of investigation. For instance, the quadrature component is usually subjected to influence of currents around the dipole, while the in-phase component can be mainly cased by remote parts of a medium, regardless of separation between the dipole and an observation point. (5) Considering the low-frequency part of the spectrum, we focus our attention on such questions as dependence of the field from conductivity and separation because it is directly related to a resolution and depth of investigation of the electromagnetic methods. (6) As follows from Eqs (8.34), (8.35), and (8.37), the field components can be represented as a sum of the decaying sinusoidal waves with different amplitudes and phases propagating with the same velocity from the source. In the near zone, when the distance r is much smaller than the wave length, $r \ll \lambda$, the wave field is practically defined by the leading terms of these sums, and the magnetic field changes synchronously with the dipole current. Of course, it is difficult to see the wave character of the sinusoidal field in the near zone.

8.7. THE RANGE OF LARGE PARAMETERS p OR WAVE ZONE

8.7.1. The Dipole and Observation Point are at the Earth's Surface

Next, we consider the behavior of the electromagnetic field for large values of the parameter r/δ:

$$\frac{r}{\delta} \gg 1 \tag{8.55}$$

When the distance r from the dipole to an observation point becomes significantly larger than the skin depth δ, the exponential term on the right-hand side of expressions for b_z^* and e_φ^* is small and can be neglected. Then, we obtain asymptotic expressions for b_z^* and e_φ^* at the Earth's surface when the parameter p is large:

$$b_z^* = \frac{18}{k_1^2 r^2} = \frac{18}{i\gamma_1 \mu_0 \omega r^2}, \quad \text{and} \quad e_\varphi^* = -\frac{6}{k_1^2 r^2} = -\frac{6}{i\gamma_1 \mu_0 \omega r^2} \tag{8.56}$$

The modified Bessel functions can be represented as follows for large values of their argument:

$$\begin{aligned}
I_1(x) &\approx \frac{\exp(x)}{(2\pi x)^{1/2}}\left(1 - \frac{3}{8x} - \frac{15}{128x^2}\cdots\right) \\
I_2(x) &\approx \frac{\exp(x)}{(2\pi x)^{1/2}}\left(1 - \frac{15}{8x} + \frac{105}{128x^2} - \cdots\right) \\
K_1(x) &\approx \exp(-x)\left(\frac{\pi}{2x}\right)^{1/2}\left(1 + \frac{3}{8x} - \frac{105}{128x^2}\cdots\right) \\
K_2(x) &\approx \exp(-x)\left(\frac{\pi}{2x}\right)^{1/2}\left(1 + \frac{15}{8x} + \frac{105}{128x^2} - \cdots\right)
\end{aligned} \tag{8.57}$$

Substituting Eq. (8.57) into Eq. (8.40) for b_r^*, we obtain

$$b_r^* = -k_1^2 r^2 \left[\frac{3}{2\left(ik_1 r/2\right)^2} \right] \frac{1}{ik_1 r} = \frac{6}{ik_1 r} \tag{8.58}$$

Thus, the components of the field are

$$E_\varphi^* = -\frac{6M_0\rho_1}{4\pi r^4}, \quad B_z^* = \frac{18\rho_1 M_0}{4\pi\omega r^5} \exp\left(-i\pi/2\right)$$

and

$$B_r^* = \frac{6\mu_0 M_0}{4\pi i k_1 r^4} = \frac{6\mu_0 M_0}{4\pi(\gamma_1\mu_0\omega)^{1/2} r^4} \exp\left(-i3\pi/4\right) \quad \text{if } p \gg 1 \tag{8.59}$$

It is clear that approximate Eq. (8.59) becomes more accurate with an increase of the parameter r/δ. As was mentioned previously, the following terms in the expression for E_φ^* and B_z^* are exponentials. However, the asymptotic expression for B_r^* is represented as a sum of terms, each of which is inversely proportional to r/δ. Because of this, the asymptotic for the horizontal component B_r^* becomes valid at larger values of r/δ than in the case for B_z^* and E_φ^*. According to Eq. (8.59), the horizontal component B_r^* is larger than the vertical component B_z^*:

$$\frac{B_z^*}{B_r^*} \approx \frac{3i}{k_1 r} \quad \text{if } p \gg 1 \tag{8.60}$$

With an increase of p, this ratio (Eq. (8.60)) becomes progressively smaller and magnetic field tends to be horizontal. For large values of p, both the electric field and horizontal component of the magnetic field decrease as $1/r^4$, and this ratio remains essentially constant equal to

$$\frac{E_\varphi^*}{B_r^*} = -\frac{ik_1\rho_1}{\mu_0} = -\frac{1}{\mu_0}\left(\frac{\omega\mu_0}{\gamma_1}\right)^{1/2} \exp\left(-i\pi/4\right) = -\frac{Z}{\mu_0}$$

where

$$Z = (\omega\mu_0\rho_1)^{1/2} \exp\left(-i\pi/4\right), \tag{8.61}$$

and it is called the *impedance* of the plane wave introduced earlier. This feature of the behavior of the electromagnetic field is one of the reasons why this range of parameters p is called the *wave zone*. This name is not occasional because within this range, $r>\lambda$, the field on the Earth's surface behaves as a sinusoidal wave propagating mainly downward. As follows from Eq. (8.59), the Poynting vector has two components in the wave zone. One of them is the smallest and horizontal directed along the Earth's surface, while the other one is the vertical component which dominates and is oriented downward. Certainly, there is analogy with power line and a load where the earth in the wave zone plays role of a load.

8.7.2. The Field Beneath the Earth's Surface ($h = 0$)

Now we consider one more feature of the field when the ratio r/δ is large. With this purpose in mind, let us study the electromagnetic field beneath the Earth's surface, provided that $p \gg 1$. Introducing a new variable $x = mr$ on the right-hand side of the second equation of the set (8.19), we obtain the following expression for the vector potential within the conducting medium:

$$A_{2z}^{*} = \frac{i\omega\mu_0 M_0}{2\pi r} \int\limits_0^{\infty} \frac{x}{x + x_1} \exp\left(-x_1 z_0\right) J_0(x) \mathrm{d}x, \qquad (8.62)$$

where

$$x_1 = \left(x^2 - k_1^2 r^2\right)^{1/2}, \quad z_0 = \frac{z}{r}$$

It turns out that if the parameter $k_1 r$ is very large, then the integral is mainly defined by the initial part of domain of integration when $x \ll 1$. Later this question will be discussed in some detail. Correspondingly, we can replace $\exp(-x_1 z_0)$ by $\exp(ik_1 z)$ since

$$\begin{aligned}\exp\left(-x_1 z_0\right) &= \exp\left(-\left(x^2 - k_1^2 r^2\right)^{1/2} \cdot \frac{z}{r}\right) \approx \exp\left(-i\left(k_1^2\right)^{1/2} z\right) \\ &= \exp\left(-ik_1 z\right)\end{aligned}$$

Here, we have chosen $-k_1$ as a square root of k_1^2 in order to satisfy an appropriate condition at infinity (if $z \to \infty$). Thus in place of Eq. (8.62), we obtain

$$A_{2z}^{*} = \frac{i\omega\mu_0 M_0}{2\pi r} \exp\left(ik_1 z\right) \int\limits_0^{\infty} \frac{x}{x + x_1} J_0(x) \mathrm{d}x \quad \text{if } p \gg 1$$

or

$$A_{2z}^* = \frac{i\omega\mu_0 M_0}{2\pi r^3 k_1^2} \exp(ik_1 z)\left[\int_0^\infty x^2 J_0(x)\mathrm{d}x - \int_0^\infty x x_1 J_0(x)\mathrm{d}x\right]$$

$$= \frac{i\omega\mu_0 M_0}{2\pi k_1^2 r^3} \exp(ik_1 z)\left[\frac{\partial^2}{\partial z_1^2}\frac{1}{R} - \frac{\partial^2}{\partial z_1^2}\frac{\exp(ik_1 R)}{R}\right] \quad \text{if } z_1 = 0$$

where

$$R = \left(r^2 + z_1^2\right)^{1/2}$$

As follows from Eqs (8.31) and (8.32),

$$A_{2z}^* \approx -\frac{i\omega\mu_0 M_0}{2\pi k_1^2 r^3} \exp(ik_1 z)[1 - (1 - ik_1 r)\exp(ik_1 r)]$$

Thus, when the ratio r/δ is large,

$$A_{2z}^* = -\frac{i\omega\mu_0 M_0}{2\pi k_1^2 r^3} \exp(ik_1 z) \tag{8.63}$$

From Eq. (8.3), we have

$$E_\varphi^* = \frac{3i\omega\mu_0 M_0}{2\pi k_1^2 r^4} \exp(ik_1 z), \quad B_r^* = -\frac{3\mu_0 M_0}{2\pi i k_1 r^4} \exp(ik_1 z) \tag{8.64}$$

Using similar approach and proceeding from the last expression of Eq. (8.20), we have

$$B_z^* = \frac{9M_0\rho_1}{2\pi i\omega r^5} \exp(ik_1 z) \tag{8.65}$$

Here, we have to notice that we are unable to derive an expression for B_z^* directly from Eq. (8.63) since this approximation is not accurate enough for double differentiation with respect to z.

Thus, for large values of the ratio r/δ, the electromagnetic field observed beneath the surface of the earth decays exponentially with increase of z, just as in the case of a plane wave propagating downward. An important feature of this behavior is that the phase does not change in a horizontal plane. In other words, the horizontal planes are surfaces of equal phase. In contrast to the case with plane wave propagation downward into the earth, surfaces of equal amplitude do not form horizontal planes since the field strength decreases with the distance from the dipole.

As is well known, in carrying out magnetotelluric soundings, the natural electromagnetic fields on the Earth's surface are used. In most cases, these fields are measured at the range of large parameters, regardless of the type of generators of the primary field, and correspondingly, they behave as quasi-plane waves. For this reason, the ratio of tangential components of the field is independent of the distance from these generators; in other words, the impedance coincides with that for the plane wave moving downward.

Also it may be proper to notice that the wave zone does not mean that induced currents are located near the boundary. It happens only when the skin depth tends to zero, and in such case, the induced currents are concentrated near the primary source and at each instant, they are directed opposite to its current. This behavior of currents is observed when the wave zone takes place very close to the dipole. In contrast, at large distances, induced currents can be noticeable even far away from the interface as soon as $z/\delta \leq 1$.

In the same manner, we can derive expressions for the field in the wave zone in the upper half space and demonstrate that the field represents a sum of two waves, $k_0 \neq 0$. One of them is the incident spherical wave caused by the primary source (magnetic dipole), while the other describes the secondary field and depends on the wave number of a conducting medium.

8.8. FREQUENCY RESPONSES OF THE FIELD

We studied in detail the electromagnetic field over two asymptotic ranges of the parameter p. In particular, for given distance r and conductivity γ, they correspond to the low- and high-frequency spectra. The behavior of the quadrature and in-phase components of the field, as well as their amplitudes and phases within the near, intermediate, and wave zones are shown in Figures 8.1(b) and 8.2. It may be proper to note that these functions can be treated as the frequency responses of the field, but they do not show the field dependence on the distance as it is seen from Eqs (8.38) and (8.39). As follows from Hilbert transform, both the quadrature and in-phase components of the field contain the same information. For instance, if the frequency response of the quadrature component is known at some point, the in-phase component can be calculated and vice versa. Since at the upper part of the complex plane of ω the complex amplitude of any component of the field does not become zero, the phase and amplitude contain the same information, and the former

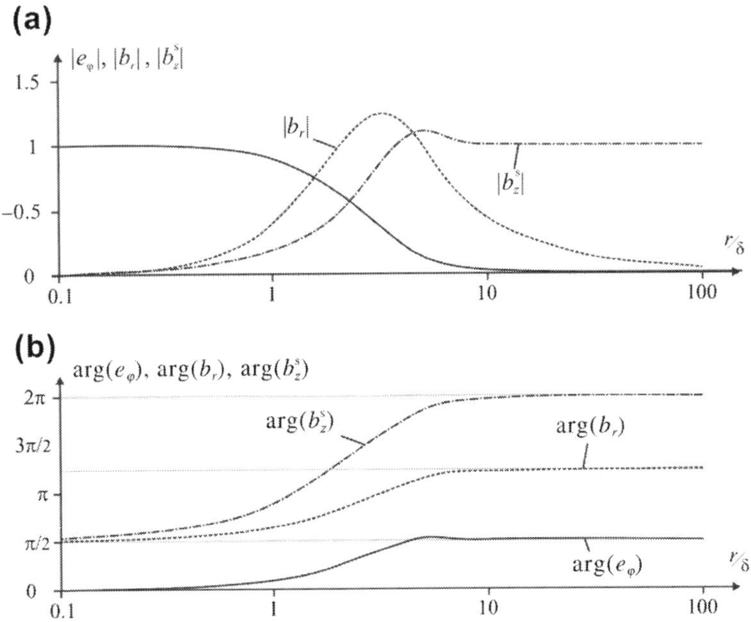

Figure 8.2 (a) Amplitude and (b) Phase responses of e^*_φ, b^*_r, and b^*_z.

can be calculated from the amplitude response. As follows from Maxwell's equation div $\mathbf{B} = 0$, there is a relationship between the vertical and horizontal components of the magnetic field, which shows that if the vertical component is known at the Earth's surface, then it can be used to calculate the horizontal one at the same frequency and vice versa.

8.9. THE VERTICAL MAGNETIC DIPOLE ON THE SURFACE OF A UNIFORM HALF SPACE (TIME DOMAIN)

Until now, we have considered the behavior of the field in the frequency domain. Next let us study the transient responses of the field of the vertical magnetic dipole on the Earth's surface assuming that the dipole moment changes as a step function:

$$M(t) = \begin{cases} 0 & t < 0 \\ M_0 & t \geq 0 \end{cases} \tag{8.66}$$

Note that in most cases, such type of excitation is sufficiently close to that applied in conventional systems of electromagnetic methods. Moreover, any function can be represented as a system of step functions arising at different times.

8.9.1. Expressions for the Field

In deriving the equations describing the transient field, we proceed from Fourier transform, which in the case of the step function excitation has the form

$$\mathbf{B}(t) = -\frac{1}{2\pi i}\int_{-\infty}^{\infty}\frac{\mathbf{B}^*(\omega)}{\omega}\exp(-i\omega t)d\omega$$

and

$$\mathbf{E}(t) = -\frac{1}{2\pi i}\int_{-\infty}^{\infty}\frac{\mathbf{E}^*(\omega)}{\omega}\exp(-i\omega t)d\omega, \tag{8.67}$$

where $\mathbf{B}^*(\omega)$ and $\mathbf{E}^*(\omega)$ are the complex amplitudes of the fields in the frequency domain. First, we obtain expressions for the electric field E_φ and the vertical component of the magnetic field B_z. For this purpose in mind, the following integrals will be used:

$$\frac{1}{2\pi}\int_{-\infty}^{\infty}\frac{\exp(-i\omega t)}{-i\omega}d\omega = \begin{cases} 0 & t<0 \\ 1 & t>0 \end{cases}$$

and

$$\frac{1}{2\pi}\int_{-\infty}^{\infty}\exp(ik_1 r)\frac{\exp(-i\omega t)}{-i\omega}d\omega = \begin{cases} 0 & t<0 \\ 1-\Phi(u) & t>0 \end{cases} \tag{8.68}$$

where $\Phi(u)$ is the probability integral introduced earlier and

$$\Phi(u) = \left(\frac{2}{\pi}\right)^{1/2}\int_0^u \exp\left(-x^2/2\right)dx,$$

and

$$u = \frac{2\pi r}{\tau}, \quad \tau = \left(2\pi\rho_1 t 10^7\right)^{1/2}, \quad \mu = \mu_0$$

Differentiating the second equality in Eq. (8.68) successively with respect to r, we obtain three useful identities:

$$\frac{1}{2\pi}\int_{-\infty}^{\infty} ik_1 r \exp\left(ik_1 r\right)\frac{\exp\left(-i\omega t\right)}{-i\omega}\mathrm{d}\omega = -\left(\frac{2}{\pi}\right)^{1/2} u \exp\left(-\frac{u^2}{2}\right) \quad (8.69)$$

$$\frac{1}{2\pi}\int_{-\infty}^{\infty} \left(ik_1 r\right)^2 \exp\left(ik_1 r\right)\frac{\exp\left(-i\omega t\right)}{-i\omega}\mathrm{d}\omega = \left(\frac{2}{\pi}\right)^{1/2} u^3 \exp\left(-\frac{u^2}{2}\right) \quad (8.70)$$

$$\frac{1}{2\pi}\int_{-\infty}^{\infty} \left(ik_1 r\right)^3 \exp\left(ik_1 r\right)\frac{\exp\left(-i\omega t\right)}{-i\omega}\mathrm{d}\omega = -\left(\frac{2}{\pi}\right)^{1/2} u^3\left(u^2-1\right)\exp\left(-\frac{u^2}{2}\right) \quad (8.71)$$

Taking into account Eq. (8.34) and making use of the appropriate substitutions in the expression for the electric field in Eq. (8.67), we obtain

$$E_\varphi = -\frac{3M\rho_1}{2\pi r^4}\left[\Phi(u) - \left(\frac{2}{\pi}\right)^{1/2}\exp\left(-\frac{u^2}{2}\right)u\left(1+\frac{u^2}{3}\right)\right] \quad (8.72)$$

Applying the same approach to Eq. (8.35) and making use of Eq. (8.67), it follows that

$$\frac{\partial B_z}{\partial t} = -\frac{9M\rho_1}{2\pi r^5}\left[\Phi(u) - \left(\frac{2}{\pi}\right)^{1/2}\exp\left(-\frac{u^2}{2}\right)u\left(1+\frac{u^2}{3}+\frac{u^4}{9}\right)\right] \quad (8.73)$$

and

$$B_z(t) = \int_0^t \frac{\partial B_z}{\partial t}\mathrm{d}t \quad (8.74)$$

Considering that

$$\int_0^u \Phi(x)\mathrm{d}x = u\Phi(u) + \left(\frac{2}{\pi}\right)^{1/2}\left[\exp\left(-\frac{u^2}{2}\right) - 1\right]$$

and integrating each term in Eq. (8.73) by parts, we obtain

$$B_z = -\frac{\mu_0 M}{4\pi r^3}\left[1 - \left(1 - \frac{9}{u^2}\right)\Phi(u) - \left(\frac{2}{\pi}\right)^{1/2}\exp\left(-u^2/2\right)\left(\frac{9}{u} + 2u\right)\right] \tag{8.75}$$

In order to determine the horizontal component of the magnetic field, we first consider the function $\partial A_{2z}/\partial z$. As was shown earlier, it can be expressed in terms of the function,

$$F = \frac{\mu_0 M_0}{4\pi a^2}\int_0^\infty \frac{mJ_0(mr)}{(m+m_1)^2}\mathrm{d}m \tag{8.76}$$

where

$$a^2 = \frac{1}{\gamma_1 \mu_0}$$

To begin with, we calculate the second derivative with respect to time of Fourier transform of function F:

$$L = \frac{M\mu_0}{8\pi^2 a^2}\int_0^\infty \left[\int_{-\infty}^{\infty} \frac{(i\omega)^2 \exp\,(-\,i\omega t)}{(m+m_1)^2}\mathrm{d}\omega\right]\cdot mJ_0(mr)\mathrm{d}m \tag{8.77}$$

where

$$L = \frac{1}{2\pi}\frac{\partial^2}{\partial t^2}\int_{-\infty}^{\infty} F\,\exp\,(-\,i\omega t)\mathrm{d}\omega$$

Consider for the moment the integral C defined by

$$C = \frac{1}{8\pi^2 a^2}\int_{-\infty}^{\infty} \frac{(i\omega)^2 \exp\,(-\,i\omega t)}{(m+m_1)^2}\mathrm{d}\omega \tag{8.78}$$

where

$$m_1 = \left(m^2 - i\frac{\omega}{a^2}\right)^{1/2}$$

The integrand in Eq. (8.78) has a branch point when $m_1 = 0$; that is,

$$\left(m^2 - i\frac{\omega}{a^2}\right)^{1/2} = 0, \quad \text{or} \quad \omega = -im^2 a^2$$

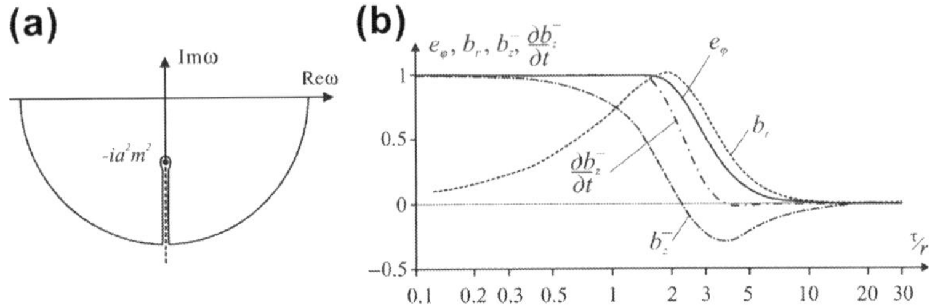

Figure 8.3 (a) Path of integration in Eq. (8.79). (b) Transient responses of the field.

Making use of Cauchy theorem and integrating along the negative imaginary axis instead of along the real values of ω (Figure 8.3(a)), and considering that ω changes within the interval:

$$-i\infty < \omega < -ia^2m^2$$

and that on either side of the path the radical m_1 has opposite sign, we obtain

$$C + \int_{-i\infty}^{-ia^2m^2} + \int_{-ia^2m^2}^{-i\infty} = 0$$

or

$$C = \frac{1}{8\pi^2a^2}\left[\int_{-i\infty}^{-ia^2m^2} \frac{(i\omega)^2\exp(-i\omega t)}{(m+m_1)^2}d\omega + \int_{-ia^2m^2}^{-i\infty} \frac{(i\omega)^2\exp(-i\omega t)}{(m-m_1)^2}d\omega\right] \tag{8.79}$$

Introducing new variable $m_1 = -i\nu$, we have

$$i\omega = a^2\left(m^2+\nu^2\right), \quad d\omega = -2ia^2\nu d\nu$$

and

$$C = \frac{a^4 i}{4\pi^2}\left[\int_{\infty}^{0} \frac{\left(m^2+\nu^2\right)^2}{(m-i\nu)^2}\exp\left[-a^2t\left(m^2+\nu^2\right)\right]\nu d\nu + \int_{0}^{\infty} \frac{\left(m^2+\nu^2\right)^2}{(m+i\nu)^2}\exp\left[-a^2t\left(m^2+\nu^2\right)\right]\nu d\nu\right]$$

or

$$C = -\frac{a^4 i}{4\pi^2}\int_0^\infty \left(m^2+\nu^2\right)^2 \exp\left[-a^2 t\left(m^2+\nu^2\right)\right]\nu\left[\frac{1}{(m+i\nu)^2}-\frac{1}{(m-i\nu)^2}\right]d\nu$$

$$= -\frac{a^4 m \exp\left(-a^2 m^2 t\right)}{\pi}\int_0^\infty \nu^2 \exp\left(-a^2 t\nu^2\right)d\nu \tag{8.80}$$

The integral on the right-hand side of Eq. (8.80) is known:

$$\int_0^\infty \nu^2 \exp\left(-a^2 t\nu^2\right)d\nu = \frac{\pi^{1/2}}{4a^3}\frac{1}{t^{3/2}}$$

Thus,

$$C = \frac{\pi^{1/2} am}{4\pi^2 t^{3/2}}\exp\left(-a^2 t m^2\right) \tag{8.81}$$

Certainly, it was worth to carry out these rather cumbersome derivations since we obtain very simple expression for C instead of Eq. (8.78). Substitution of Eq. (8.81) into Eq. (8.77) gives

$$\frac{\partial^2 F}{\partial t^2} = \frac{\mu_0 M_0 \pi^{1/2} a}{4\pi^2 t^{3/2}}\int_0^\infty m^2 \exp\left(-a^2 m^2 t\right)J_0(mr)dm \tag{8.82}$$

or

$$\frac{\partial^2 F}{\partial t^2} = -\frac{\mu_0 M_0 \pi^{1/2}}{4\pi^2 a t^{3/2}}\frac{\partial}{\partial t}\int_0^\infty \exp\left(-a^2 m^2 t\right)J_0(mr)dm$$

Making use of tabulated integral

$$\int_0^\infty \exp\left(-a^2 m^2 t\right)J_0(mr)dm = \frac{\pi^{1/2}}{2at^{1/2}}\exp\left(-\frac{r^2}{8a^2 t}\right)I_0\left(\frac{r^2}{8a^2 t}\right)$$

we have

$$\frac{\partial^2 F}{\partial t^2}=\frac{\mu_0 M_0}{8\pi a^2 t^2}\frac{\partial}{\partial t}\left[\exp\left(-r^2/8a^2 t\right) I_0\left(\frac{r^2}{8a^2 t}\right)\right] \tag{8.83}$$

The argument of the Bessel function can also be written as

$$\frac{r^2}{8a^2 t}=\frac{u^2}{4}, \quad \text{and} \quad u=\frac{2\pi r}{\tau}$$

Then after the indicated differentiation and integration, we arrive at the following equations:

$$\frac{\partial B_r}{\partial t}=-\frac{\mu_0 M_0 \rho_1}{\pi r^5} u^4 \exp\left(-u^2/4\right) \times\left[\left(1+\frac{u^2}{2}\right) I_0\left(\frac{u^2}{4}\right)-\left(2+\frac{u^2}{2}+\frac{8}{u^2}\right) I_1\left(\frac{u^2}{4}\right)\right] \tag{8.84}$$

and after integration, we have

$$B_r=-\frac{\mu_0 M_0}{4\pi r^3} b_r,$$

where

$$b_r=4\exp\left(-u^2/4\right)\left[\left(2+\frac{u^2}{4}\right) I_1\left(\frac{u^2}{4}\right)-\frac{u^2}{4} I_0\left(\frac{u^2}{4}\right)\right] \tag{8.85}$$

In summary, let us write down expressions for the quasi-stationary transient field observed on the Earth's surface of a uniform half space:

$$E_\varphi=-\frac{3M_0\rho_1}{2\pi r^4} e_\varphi, \quad B_z=-\frac{\mu_0 M_0}{4\pi r^3} b_z, \quad B_r=-\frac{\mu_0 M_0}{4\pi r^3} b_r,$$

where

$$e_\varphi=\Phi(u)-\left(\frac{2}{\pi}\right)^{1/2}\exp\left(-u^2/2\right) u\left(1+\frac{u^2}{3}\right)$$

$$b_z=1-\left(1-\frac{9}{u^2}\right)\Phi(u)-\left(\frac{2}{\pi}\right)^{1/2}\exp\left(-u^2/2\right)\left(\frac{9}{u}+2u\right) \tag{8.86}$$

$$b_r=4\exp\left(-u^2/4\right)\left[\left(2+\frac{u^2}{4}\right) I_1\left(\frac{u^2}{4}\right)-\frac{u^2}{4} I_0\left(\frac{u^2}{4}\right)\right]$$

and

$$\frac{\partial B_z}{\partial t} = -\frac{9M_0\rho_1}{2\pi r^5}\left[\Phi(u) - \left(\frac{2}{\pi}\right)^{1/2}\exp\left(-\frac{u^2}{2}\right)u\left(1+\frac{u^2}{3}+\frac{u^4}{9}\right)\right]$$

$$\frac{\partial B_r}{\partial t} = \frac{M_0\rho_1}{\pi r^5}u^4\exp\left(-\frac{u^2}{4}\right)\left[\left(1+\frac{u^2}{2}\right)I_0\left(\frac{u^2}{4}\right) - \left(2+\frac{u^2}{2}+\frac{8}{u^2}\right)I_1\left(\frac{u^2}{4}\right)\right] \tag{8.87}$$

Equation (8.86) describes the field components when the dipole current is turned on at the instant $t=0$. In order to obtain expressions for the field when the current is turned off, it is necessary to subtract the function b_z from unity, while the functions e_φ and b_r do not change. First, as in the case of frequency domain, we consider the asymptotic behavior of the field.

8.9.2. Early Stage of the Transient Response of the Electric and Magnetic Fields

Making use of Eq. (8.86) and taking into account that

$$\Phi(u) \to 1 \text{ if } u \to \infty,$$

we have

$$E_\varphi = -\frac{3M\rho_1}{2\pi r^4}, \quad B_z = -\frac{9M\rho_1}{2\pi r^5}t$$

and

$$B_r = -\frac{3\mu_0 M}{\pi r^4}\left(\frac{t}{2\gamma_1\mu_0\pi}\right)^{1/2} \tag{8.88}$$

As follows from Eq. (8.88), the dependence of the field on the resistivity of the medium and separation r is the same as that in the frequency domain for large induction numbers. For example, on the Earth's surface, the electric field does not depend on time at the first instant and the magnetic field is zero that of course demonstrates the principle of inertia of the magnetic flux; that is, the skin effect. With an increase of time, both components of the magnetic field grow but at the early stage,

$$|B_z| \ll |B_r|$$

With greater conductivity in the medium or a distance from the dipole, the early-stage behavior is observed at later times. Comparison with the exact solution shows that Eq. (8.88) is valid with a reasonable accuracy for the electric field E_φ and the component B_z when $\tau/r < 2$, but for the horizontal component B_r, only if $\tau/r < 1$. It is clear that at the initial moment, induced currents concentrate near the dipole and with increasing time, their magnitude becomes larger at greater distances from the surface $z = 0$. Note that at the first instant, the induced current flows around the current loop and has an opposite direction. Correspondingly, the total current is equal to zero and it is natural that the electric and magnetic fields are equal to zero. At relatively short separations r, the early stage is observed when currents are located quite close to the Earth's surface. However, at large separations, the field behavior described by Eq. (8.88) still holds when the induced currents are located at sufficiently great depths. As was pointed out, from Eq. (8.88), it follows that at the initial instant, when the current is turned on, the magnetic field is zero at every point of the Earth's surface and beneath but it begins to grow immediately. In other words, regardless of the separation for any small time t, the field is not zero. This means again that the equations describing the quasi-stationary field can only be used when a measurement time is in several times greater than the time which is necessary for propagation of the field from the dipole to an observation point. It is proper to note that as in the in the wave zone of the frequency domain, when $\tau/r < 1$, the Poynting vector has very interesting feature. It is mainly directed downward while the tangential component of this vector is relatively small. Thus in this zone, the electromagnetic energy mainly penetrates into a conducting medium where it transforms into a heat.

8.9.3. Transient Field at the Late Stage

Next, we consider the opposite case, namely, the late stage. Expanding the probability integral and modified Bessel functions for small values of the parameter u,

$$\Phi(u) \approx \left(\frac{2}{\pi}\right)^{1/2}\left(u - \frac{u^3}{6} + \frac{u^5}{40} - \cdots\right),$$

$$I_0(u) \approx 1 + \frac{u^2}{4} + \frac{u^4}{16} + \cdots, \quad I_1(u) \approx \frac{u}{2} + \frac{u^3}{16}$$

we arrive at approximate expressions for the field components when the ratio $\tau/r \gg 1$. They are

$$E_\varphi \approx -\frac{\mu_0^{5/2}\gamma_1^{3/2}M_0}{40\pi(\pi)^{1/2}t^{5/2}}r, \quad B_z^+ \approx -\frac{\mu_0 M_0}{4\pi r^3}\left[1+\frac{2}{15}r^3\frac{(\gamma_1\mu_0)^{3/2}}{(\pi)^{1/2}t^{3/2}}\right],$$

$$B_r \approx \pm\frac{\mu_0 M_0}{4\pi}\frac{r}{32t^2}(\gamma_1\mu_0)^2, \quad B_z^- \approx \frac{\mu_0 M_0}{30\pi(\pi)^{1/2}}\frac{(\gamma_1\mu_0)^{3/2}}{t^{3/2}}, \tag{8.89}$$

where B_z^+ and B_z^- are the vertical components of the magnetic field during the "time on" and "time off," respectively. From Eq. (8.89), it follows, in contrast to the early stage, the vertical component of the magnetic field is greater than the horizontal one $(B_z > B_r)$, and is independent of the separation r. As in the case of a uniform medium, the electromagnetic field is characterized by a relatively high sensitivity to changes in conductivity of the medium:

$$E_\varphi \sim \gamma_1^{3/2}, \quad B_z \sim \gamma_1^{3/2}, \quad B_r \sim \gamma_1^2$$

Equation (8.89) describes the field with reasonable accuracy when the parameter $\tau/r > 16$.

8.9.4. Transient Responses of the Field

The transient responses for functions b_z^-, b_r and its derivatives with respect to time and e_φ are shown in Figure 8.3(b). At the initial moment, when the current is turned off, the vertical component of the magnetic field is equal to the time-invariant field caused only by the dipole current, B_z^0. Then with an increase of time, the field B_z decreases, passes through zero at $\tau/r \approx 2.3$, and at the late stage is inversely proportional to $t^{3/2}$. Such a behavior indicates that the field is caused by two groups of currents and their vertical components differ by sign. With time, a distribution of these currents changes and it turns out that at the beginning, one group of currents gives the main contribution, but at the late stage, the other plays the dominant role. It is natural that there is a moment when the fields of these two groups cancel each other. Let us remind that since the quasi-stationary field is considered, we can apply Biot–Savart law to calculate the magnetic field. As is seen from Figure 8.3(b), the horizontal component has a maximum around $\tau/r = 2$ where its magnitude slightly exceeds that of the vertical component of the primary field B_z^0.

REFERENCES AND FURTHER READING

[1] V.P. Bursian, Theory of Electromagnetic Fields Applied in Electroprospecting, Nedra, Moscow, 1972.
[2] A.A. Kaufman, Geophysical Field Theory and Methods, Part C, Academic Press, New York, NY, 1994.
[3] S.M. Sheinman, On transient electromagnetic fields in the earth, Applied Geophysics 3 (1947).
[4] L.L. Vanyan, Principles of Electromagnetic Soundings, Nedra, Moscow, 1965.
[5] J.R. Wait, Geoelectromagnetism, Academic Press, New York, NY, 1982.

CHAPTER NINE

Quasi-Stationary Field of Vertical Magnetic Dipole on the Surface of a Horizontally Layered Medium

INTRODUCTION

In studying the field of the magnetic dipole in a uniform medium and at the surface of a uniform half space, our main attention was paid to a relationship between the frequency and transient responses and influence of parameters of a medium on the field at different separations between the current source and observation point. Also, we considered the distribution of induced currents in a medium and demonstrated that the quadrature and in-phase components of induced current, in general, have completely different behavior. In particular, it was shown that the behavior of the in-phase component of the current and its transient response resembles diffusion, while the quadrature component in the vicinity of the source is mainly defined by the primary electric field. Also, we analyzed the behavior of the sinusoidal and transient magnetic and electric fields and found that they have completely different behavior at the near, intermediate, and wave zones. The same approach will be used in studying of the field on the surface of a horizontally layered medium that will allow us to formulate the main principles of different types of electromagnetic soundings. At the beginning, we derive equations for the field in the frequency domain and then, applying Fourier's transform, determine its transient responses when the current of the magnetic dipole varies as a step function.

9.1. THE FIELD EXPRESSIONS ON THE SURFACE OF *N*-LAYERED MEDIUM

9.1.1. Formulation of Boundary Value Problem

Suppose that a magnetic dipole with the moment amplitude M_0 is located at the Earth's surface at the origin of a cylindrical system of coordinates (Figure 9.1). Because the primary electric field has only a tangential

Methods in Geochemistry and Geophysics, Volume 45
ISSN 0076-6895,
http://dx.doi.org/10.1016/B978-0-444-53829-1.00009-5

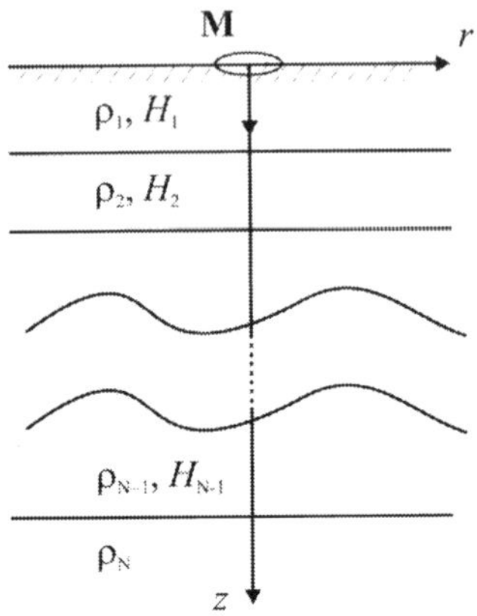

Figure 9.1 Model of a layered medium.

component E^0_φ, which does not cross the interfaces between media with different conductivities, the electric charges are absent. For this reason, the induced currents are situated in horizontal planes and in view of the cylindrical symmetry of the electromagnetic field, just as in the case of a uniform half space, we assume that there are only three nonzero components of the field:

$$\mathbf{E} = \{0, E_\varphi, 0\}, \quad \mathbf{B} = \{B_r, 0, B_z\} \tag{9.1}$$

Let us again, as in the previous cases, describe the field with help of the vector potential of the magnetic type **A** which has only a z-component

$$\mathbf{A} = \{0, 0, A_z\} \tag{9.2}$$

Of course, Eqs (9.1) and (9.2) are based on our assumptions, but as it will follow from the theorem of uniqueness, it turns out that they are correct. As we already know, the complex amplitude of the vector potential $\mathbf{A}^*$ satisfies Helmholtz's equation in each medium:

$$\nabla^2 A^*_{iz} + k_i^2 A^*_{iz} = 0, \tag{9.3}$$

where A^*_{iz} is the z-component of the vector potential in the ith medium. From the continuity of tangential components of the electromagnetic field at the interfaces, it follows that at each boundary between layers,

$$A^*_{iz} = A^*_{i+1,z}, \quad \frac{\partial A^*_i}{\partial z} = \frac{\partial A^*_{i+1,z}}{\partial z}, \quad \text{if } z = h_i \tag{9.4}$$

where h_i is the distance from the Earth's surface to the boundary between the i and $i+1$ media. Equations (9.3) and (9.4) along with conditions near the dipole and at infinity constitute a boundary value problem which we will solve in the frequency domain. At the beginning, let us assume that the wave number of the upper half space k_0 is not equal to zero.

9.1.2. Three-Layered Medium

First consider the three-layered medium. As was shown in the Chapter 8, the vector potential $\mathbf{A}^*$ can be written in the following form for each medium:

$$A^*_{0z} = \frac{i\omega\mu_0 M_0}{4\pi}\int_0^\infty \left[\frac{m}{m_0}\exp(-m_0|z|) + D_0 \exp(m_0 z)\right] J_0(mr)dm \quad \text{if } z \le 0$$

$$A^*_{1z} = \frac{i\omega\mu_0 M_0}{4\pi}\int_0^\infty [C_1 \exp(-m_1 z) + D_1 \exp(m_1 z)] J_0(mr)dm \quad \text{if } 0 \le z \le h_1$$

$$A^*_{2z} = \frac{i\omega\mu_0 M_0}{4\pi}\int_0^\infty [C_2 \exp(-m_2 z) + D_2 \exp(m_2 z)] J_0(mr)dm \quad \text{if } h_1 \le z \le h_2 \tag{9.5}$$

$$A^*_{3z} = \frac{i\omega\mu_0 M_0}{4\pi}\int_0^\infty C_3 \exp(-m_3 z) J_0(mr)dm \quad \text{if } \quad z \ge h_3$$

where $m_1 = (m^2 - k_1^2)^{1/2}$, $m_2 = (m^2 - k_2^2)^{1/2}$, $m_3 = (m^2 - k_3^2)^{1/2}$, and k_1, k_2, and k_3 are the wave numbers for the first, second, and third medium, respectively. Also, $m_0 = (m^2 - k_0^2)^{1/2}$. Note that Eq. (9.5) is written assuming that

$$\text{Re } m_0 > 0 \quad \text{Re } m_1 > 0, \quad \text{Re } m_2 > 0 \text{ and } \text{Re } m_3 > 0,$$

otherwise conditions at infinity ($z \to \pm\infty$) are not satisfied. From boundary conditions, we obtain a system of algebraic equations for determining the unknown coefficients D_0, C_1, D_1, C_2, D_2, and C_3. In fact, making use of the orthogonality of Bessel functions, instead of equality of integrals, we have

$$\frac{m}{m_0} + D_0 = C_1 + D_1, \quad -m + m_0 D_0 = -m_1 C_1 + m_1 D_1$$

$$C_1 \exp(-m_1h_1) + D_1 \exp(m_1h_1) = C_2 \exp(-m_2h_1) + D_2 \exp(m_2h_1) \tag{9.6}$$

$$-m_1C_1 \exp(-m_1h_1) + m_1D_1 \exp(m_1h_1)$$
$$= -m_2C_2 \exp(-m_2h_1) + m_2D_2 \exp(m_2h_1)$$

$$C_2 \exp(-m_2h_2) + D_2 \exp(m_2h_2) = C_3 \exp(-m_3h_2)$$

$$-C_2m_2 \exp(-m_2h_2) + D_2m_2 \exp(m_2h_2) = -m_3C_3 \exp(-m_3h_2)$$

Inasmuch as our interest is to study the field on the Earth's surface, we determine only the coefficient D_0. Eliminating C_3 from the last two equations of this set, we obtain

$$D_2 = -\frac{m_3 - m_2}{m_3 + m_2} \exp(-2m_2h_2)C_2 = m_{23} \exp(-2m_2h_2)C_2$$

where

$$m_{23} = \frac{m_2 - m_3}{m_2 + m_3}$$

Substituting this result in a second pair of Eq. (9.6), we find

$$D_1 = \frac{m_{12} + m_{23} \exp(-2m_2H_2)}{1 + m_{12}m_{23} \exp(-2m_2H_2)} \exp(-2m_1H_1)C_1$$

or

$$D_1 = P \exp(-2m_1H_1)C_1, \tag{9.7}$$

where

$$m_{12} = \frac{m_1 - m_2}{m_1 + m_2}, \quad m_{23} = \frac{m_2 - m_3}{m_2 + m_3}$$

and

$$P = \frac{m_{12} + m_{23} \exp(-2m_2H_2)}{1 + m_{12}m_{23} \exp(-2m_2H_2)} \tag{9.8}$$

Also,

$$H_1 = h_1 \quad \text{and} \quad H_2 = h_2 - h_1$$

Now making use of the first pair of equations in Eq. (9.6), we obtain

$$\frac{m}{m_0} + D_0 = [1 + P\exp(-2m_1H_1)]C_1$$

$$-m + m_0D_0 = -m_1[1 - P\exp(-2m_1H_1)]C_1$$

and

$$D_0 = -\frac{m}{m_0}\frac{m_{10} - P\exp(-2m_1H_1)}{1 - m_{10}P\exp(-2m_1H_1)} \tag{9.9}$$

where

$$m_{10} = \frac{m_1 - m_0}{m_1 + m_0}$$

Let us consider, as a test, several simpler models of a medium.

Case 1: Uniform half space

Here $k_1 = k_2 = k_3$ and, therefore, $P = 0$ and we arrive at the expression derived earlier:

$$D_0 = -m_{10}\frac{m}{m_0} = \frac{m}{m_0}\frac{m_0 - m_1}{m_0 + m_1}$$

Case 2: Two-layered medium ($k_2 = k_3$)

Then, as it follows from Eq. (9.8), $P = m_{12}$ and

$$D_0 = -\frac{m}{m_0}\frac{m_{10} - m_{12}\exp(-2m_1H_1)}{1 - m_{10}m_{12}\exp(-2m_1H_1)} \tag{9.10}$$

Case 3: Two-layered medium ($k_1 = k_2$)

In this case, $m_{12} = 0$ and

$$P = m_{23}\exp(-2m_2H_2)$$

Whence,

$$D_0 = -\frac{m}{m_0}\frac{m_{20} - m_{23}\exp[-2m_2(H_1 + H_2)]}{1 - m_{20}m_{23}\exp[-2m_2(H_1 + H_2)]}$$

This result corresponds to that of a two-layered medium, in which the thickness of the upper layer is equal to $H_1 + H_2$. It has been shown previously that

$$E_\varphi^* = -\frac{\partial A_z^*}{\partial r}, \quad i\omega B_r^* = \frac{\partial}{\partial r}\frac{\partial A_z^*}{\partial z} \quad i\omega B_z^* = \frac{\partial^2 A_z^*}{\partial z^2} + k^2 A_z^*,$$

and therefore, we have at the points of the Earth's surface

$$E_\varphi^* = E_\varphi^{0*} + \frac{i\omega\mu_0 M_0}{4\pi}\int_0^\infty m D_0 J_1(mr)\mathrm{d}m$$

$$B_r^* = -\frac{\mu_0 M_0}{4\pi}\int_0^\infty m m_0 D_0 J_1(mr)\mathrm{d}m \qquad (9.11)$$

$$B_z^* = B_z^0 + \frac{\mu_0 M_0}{4\pi}\int_0^\infty m^2 D_0 J_0(mr)\mathrm{d}m,$$

where E_φ^{0*} and B_z^0 describe the primary electromagnetic field. Deriving expression for B_r^*, we took into account that derivative

$$\frac{\partial}{\partial z}\int_0^\infty \frac{m}{m_0}\exp(-m_0|z|)J_0(mr)\mathrm{d}m$$

is equal to zero if $z = 0$.

9.2. EXPRESSIONS FOR THE FIELD IN *N*-LAYERED MEDIUM

Using the same approach, it is possible to obtain an expression for D_0 when the medium consists of any finite number of layers. As the number of layers increases, this expression becomes more complicated. In order to find new expression for the quantity D_0 and the field components in a much more compact form, we will make use of another approach and illustrate it using the model of three-layered medium. Taking again the ratio of the last two equations in the set Eq. (9.6), we have

$$\frac{C_2\exp(-m_2h_2) + D_2\exp(m_2h_2)}{C_2\exp(-m_2h_2) - D_2\exp(m_2h_2)} = \frac{m_2}{m_3}$$

or

$$\frac{1 + D_2/C_2\exp(2m_2h_2)}{1 - D_2/C_2\exp(2m_2h_2)} = \frac{m_2}{m_3}$$

Inasmuch as

$$\frac{D_2}{C_2} = \exp\left(\ln\frac{D_2}{C_2}\right),$$

we have

$$\frac{1 + D_2/C_2 \exp(2m_2h_2)}{1 - D_2/C_2 \exp(2m_2h_2)} = \frac{1 + \exp\left[2\left(m_2h_2 + ½\ln D_2/C_2\right)\right]}{1 - \exp\left[2\left(m_2h_2 + ½\ln D_2/C_2\right)\right]}$$

$$= -\coth\left[m_2h_2 + ½\ln D_2/C_2\right] = m_2/m_3$$

Therefore

$$m_2h_2 + \frac{1}{2}\ln D_2/C_2 = -\coth^{-1}\left(m_2/m_3\right) \qquad (9.12)$$

Using the next pair of equations of the set Eq. (9.6), which represents the boundary conditions at the interface $z = h_1$, we have

$$\frac{C_1\exp(-m_1h_1) + D_1\exp(m_1h_1)}{C_1\exp(-m_1h_1) - D_1\exp(m_1h_1)} = \frac{m_1}{m_2}\frac{C_2\exp(-m_2h_1) + D_2\exp(m_2h_1)}{C_2\exp(-m_2h_1) - D_2\exp(m_2h_1)}$$

or

$$\coth\left(m_1h_1 + \frac{1}{2}\ln\frac{D_1}{C_1}\right) = \frac{m_1}{m_2}\coth\left(m_2h_1 + \frac{1}{2}\ln\frac{D_2}{C_2}\right) \qquad (9.13)$$

Combining this with Eq. (9.12), we obtain

$$\coth\left(m_1h_1 + \frac{1}{2}\ln\frac{D_1}{C_1}\right) = \frac{m_1}{m_2}\coth\left(m_2h_1 - m_2h_2 - \coth^{-1}\frac{m_2}{m_3}\right)$$

or

$$\coth\left(m_1H_1 + \frac{1}{2}\ln\frac{D_1}{C_1}\right) = -m_1/m_2\coth\left(m_2H_2 + \coth^{-1}\frac{m_2}{m_3}\right) \qquad (9.14)$$

From the first two equations of the set Eq. (9.6), it follows that

$$\frac{m + m_0D_0}{-m + m_0D_0} = -\frac{C_1 + D_1}{C_1 - D_1}\cdot\frac{m_0}{m_1},$$

or

$$\frac{m+m_0D_0}{m-m_0D_0}=-\frac{m_0}{m_1}\coth\frac{1}{2}\ln\frac{D_1}{C_1}$$

Applying Eq. (9.14), it is apparent that

$$\frac{m+m_0D_0}{m-m_0D_0}=\frac{m_0}{m_1}\coth\left\{m_1H_1+\coth^{-1}\left[\frac{m_1}{m_2}\coth\left(m_2H_2+\coth^{-1}\frac{m_2}{m_3}\right)\right]\right\}$$

Whence,

$$\frac{m+m_0D_0}{m-m_0D_0}=\frac{m_0}{m_1}R_3$$

and after simple algebra,

$$\frac{m}{m_0}+D_0=\frac{2m}{m_0+m_1/R_3}, \qquad (9.15)$$

where

$$R_3=\coth\left\{m_1H_1+\coth^{-1}\left[\frac{m_1}{m_2}\coth\left(m_2H_2+\coth^{-1}\frac{m_2}{m_3}\right)\right]\right. \qquad (9.16)$$

Thus, according to the first equation of Eq. (9.5), the following expression holds for the vector potential on the surface of a three-layered medium

$$A^*_{1z}=\frac{i\omega\mu_0M}{2\pi}\int\limits_0^\infty\frac{mJ_0(mr)}{m_0+m_1/R_3}\,\mathrm{d}m \qquad (9.17)$$

As follows from Eq. (9.16) in the case of a uniform half space ($H_1\rightarrow\infty$), we have

$$R_1=1, \qquad (9.18)$$

Since $\coth(\infty)\rightarrow 1$, but for two-layered medium

$$R_2=\coth\left(m_1H_1+\coth^{-1}\frac{m_1}{m_2}\right) \qquad (9.19)$$

Equation (9.16) obviously can be written in the form

$$R_3=\coth\left[m_1H_1+\coth^{-1}\left(\frac{m_1}{m_2}\right)R_2(k_2,k_3,H_2,m)\right] \qquad (9.20)$$

where

$$R_2 = \coth\left(m_2 H_2 + \coth^{-1} m_2/_{m_3}\right) \tag{9.21}$$

Now, applying the method of mathematical induction, the expression for the function R_n of the corresponding n-layered medium can be written as

$$R_n = \coth\left[m_1 H_1 + \coth^{-1}\frac{m_1}{m_2} R_{n-1}\right]$$

$$R_{n-1} = \coth\left[m_2 H_2 + \coth^{-1}\frac{m_2}{m_3} R_{n-2}\right] \tag{9.22}$$

$$R_1 = 1$$

Generalizing Eq. (9.15) for the case of n-layered medium, we have

$$D_0 = \frac{m}{m_0}\frac{m_0 R_n/_{m_1} - 1}{m_0 R_n/_{m_1} + 1}, \qquad \frac{m}{m_0} + D_0 = \frac{2\, m R_n/_{m_1}}{1 + m_0 R_n/_{m_1}},$$

and

$$-m + m_0 D_0 = -\frac{2}{1 + m_0 R_n/_{m_1}} m \tag{9.23}$$

Correspondingly, taking into account Eqs (9.5) and (9.23) for the vector potential A_z^* and field components, we have

$$A_z^* = \frac{i\omega\mu_0 M_0}{2\pi}\int_0^\infty \frac{m J_0(mr)}{m_0 + m_1/R_n} \mathrm{d}m \tag{9.24}$$

$$E_\varphi^* = \frac{i\omega\mu_0 M_0}{2\pi}\int_0^\infty \frac{m^2 J_1(mr)}{m_0 + m_1/R_n} \mathrm{d}m$$

$$B_r^* = \frac{\mu_0 M_0}{2\pi}\int_0^\infty \frac{m_1}{R_n}\frac{m^2}{m_0 + m_1/R_n} J_1(mr)\mathrm{d}m \tag{9.25}$$

$$B_z^* = \frac{\mu_0 M_0}{2\pi}\int_0^\infty \frac{m^3}{m_0 + m_1/R_n} J_0(mr)\mathrm{d}m$$

An important feature of these equations is their relative simplicity, and the fact that they have the same form regardless of the number of layers. Let us introduce a new variable, $x = mr$, so that Eq. (9.25) can be written as

$$E_{\varphi}^{*} = \frac{i\omega\mu_0 M_0}{2\pi r^2}\int_0^{\infty}\frac{x^2}{x_0 + x_1/R_n}J_1(x)\mathrm{d}x = E_{\varphi}^{(0)*}e_{\varphi}^{*}$$

$$B_r^{*} = \frac{\mu_0 M_0}{2\pi r^3}\int_0^{\infty}\frac{x_1}{R_n}\frac{x^2}{x_0 + x_1/R_n}J_1(x)\mathrm{d}x = B_z^{(0)}b_r^{*} \qquad (9.26)$$

$$B_z^{*} = \frac{\mu_0 M_0}{2\pi r^3}\int_0^{\infty}\frac{x^3}{x_0 + x_1/R_n}J_0(x)\mathrm{d}x = B_z^{(0)}b_z^{*},$$

where R_n depends on $x_i = \left(x^2 - k_i^2 r^2\right)^{1/2}$ and the ratio H_i/r $(i = 1, 2, \ldots, n)$. As follows from Eq. (9.26), the field on the surface of a horizontally layered medium, just as in the case of a uniform half space, is written as the product of two functions and one of them is the primary field of the magnetic dipole. The second function $(e_{\varphi}^{*}, b_r^{*}, b_z^{*})$ indicates the influence of induced currents on the electromagnetic field and depends only on such dimensionless parameters as

$$1)\ r/\delta_1 \quad 2)\ r/H_1 \quad 3)\ \rho_2/\rho_1, \rho_3/\rho_1, \ldots \rho_n/\rho_1 \quad 4)\ H_2/H_1, H_3/H_1 \ldots H_n/H_1,$$

where δ_1 is the skin depth in the upper layer, while ρ_i and H_i are the resistivity and thickness of the *i*th layer, respectively. These functions, e_{φ}^{*}, b_r^{*}, or b_z^{*}, are the measure of how much the corresponding component of the field differs from the primary field. In accordance with Eq. (9.25), the field can be interpreted as being an infinite sum of cylindrical harmonics having the form $T_{m,i}J_i(mr)$ where $i = 0$ or 1, and T_m is the complex amplitude of the harmonic characterized by the spatial frequency m. As follows from Eq. (9.22) with an increase of m, the function R_n tends to unity:

$$R_n \to 1, \quad \text{if } m \to \infty$$

and the integrands in Eq. (9.25) approach that for a uniform half space with the resistivity of the uppermost layer. In other words, the cylindrical harmonics with relatively high spatial frequencies m contain practically no information about the geoelectric section except resistivity ρ_1. Let us make one more comment. The electromagnetic field described by Eq. (9.25) is caused by currents flowing in the horizontally layered medium and by the current in the magnetic dipole. However, integral representation of the field due to the source current and currents in a uniform half space cannot be used for numerical calculations because these integrals are numerically divergent. In particular, when $k_0 = 0$ and variable of integration m increases, the functions $mJ_1(mr)$, $m^2J_1(mr)$, and $m^2J_1(mr)$ are oscillatory and increase in amplitude. In order to remove the singularity in the expression of the field, one straightforward approach is to subtract from the total field the part which corresponds to a uniform half space with resistivity ρ_1. At the same time, as was shown earlier, these last integrals are expressed in terms of known functions. Now we begin to study the field at the Earth's surface of a horizontally layered medium, assuming that wave number in the upper half space $k_0 = 0$ and start with analyzing its asymptotic behavior.

9.3. BEHAVIOR OF THE FIELD WHEN INTERACTION BETWEEN INDUCED CURRENTS IS NEGLIGIBLE

9.3.1. The Quadrature Component of Magnetic Field and In-Phase Component of Electric Field

First consider the three-layered medium and suppose that the thickness of layers and the distance of the receiver from the source, measured in units of the skin depth; that is, H_i/δ_i and r/δ_i are very small. Then expanding the functions m_{10}, m_{12}, and m_{23} in the power series in k_i^2/m^2 and discarding all terms but the first which is proportional to ω, we obtain

$$m_{10} = \frac{m - m_1}{m + m_1} = \frac{m - m\left(1 - \frac{k_1^2}{m^2}\right)^{1/2}}{m + m\left(1 - \frac{k_1^2}{m^2}\right)^{1/2}} \approx \frac{1}{4}\frac{k_1^2}{m^2}$$

By analogy,

$$m_{12} \approx \frac{\left(k_2^2 - k_1^2\right)}{4m^2}, \qquad m_{23} \approx \frac{\left(k_3^2 - k_2^2\right)}{4m^2}$$

Substituting these last expressions into Eqs (9.8) and (9.9), one arrives at an approximation to the function

$$D_0 \approx \frac{k_1^2}{4m^2} + \frac{k_2^2 - k_1^2}{4m^2}\exp(-2mH_1) + \frac{k_3^2 - k_2^2}{4m^2}\exp[-2m(H_1 + H_2)] \tag{9.27}$$

Replacing the function D_0 by the right-hand part of Eq. (9.27), and taking into account that

$$\int_0^\infty \exp(-m|z|)J_0(mr)\mathrm{d}m = \frac{1}{(r^2+z^2)^{1/2}}$$

$$\int_0^\infty \exp(-m|z|)J_1(mr)\mathrm{d}m = \frac{(r^2+z^2)^{1/2}-z}{r(r^2+z^2)^{1/2}} \tag{9.28}$$

$$\int_0^\infty \frac{\exp(-m|z|)}{m}J_1(mr)\mathrm{d}m = \frac{(r^2+z^2)^{1/2}-z}{r},$$

we obtain the asymptotic expressions for the quadrature component of the magnetic field and for the in-phase component of the electric field, caused by induced currents in a conducting medium (secondary field):

$$E^s_\varphi = \frac{\rho_1 k_1^2 M_0}{4\pi}\left\{\frac{k_1^2}{4}+\frac{(k_2^2-k_1^2)}{4}\frac{(4H_1^2+r^2)^{1/2}-2H_1}{r}\right.$$
$$\left.+\frac{(k_3^2-k_2^2)}{4}\frac{[r^2+4(H_1+H_2)^2]^{1/2}-2(H_1+H_2)]}{r}\right\}$$

$$B^s_z = \frac{\mu_0 M_0}{4\pi}\left\{\frac{k_1^2}{4r}+\frac{k_2^2-k_1^2}{4}\frac{1}{(4H_1^2+r^2)^{1/2}}+\frac{k_3^2-k_2^2}{4[4(H_1+H_2)^2+r^2]^{1/2}}\right\}$$

$$B^s_r = -\frac{\mu_0 M_0}{4\pi}\left\{\frac{k_1^2}{4r}+\frac{k_2^2-k_1^2}{4r}\frac{(4H_1^2+r^2)^{1/2}-2H_1}{(4H_1^2+r^2)^{1/2}}\right.$$
$$\left.+\frac{k_3^2-k_2^2}{4r}\frac{[4(H_1+H_2)^2+r^2]^{1/2}-2(H_1+H_2)}{[4(H_1+H_2)^2+r^2]^{1/2}}\right\} \tag{9.29}$$

9.3.2. The Role of Initial Part of Integration

Before analyzing the meaning of the expressions in Eq. (9.29), it would be proper to discuss one aspect of derivation. The amazing simplicity of these equations is a consequence of the fact that the radical expressions m_i have been expanded in a series of powers of k_i^2/m^2. This is equivalent to assumption that inequality $k_i^2/m^2 < 1$ holds over the range of integration. This is not strictly correct because there are always some small values of m, for which the ratio k_i^2/m^2 is greater than unity. In other words, this approach cannot be used over the range in m corresponding to the initial part of integration:

$$0 \leq m \leq |k_i|$$

It should be noted that as $|k_i|$ decreases, this range becomes narrower, and it turns out that incorrect representation of the function D_0 over the initial part of integration does not result in an error in determining the leading term of the series for the quadrature component of the magnetic field and the in-phase component of the electric field, when parameters r/δ_i and H_i/δ_i are small. It can readily be seen that an attempt to take successive terms of an expansion of radicals m_1, m_2, and m_3 into account leads to divergent integrals.

9.3.3. Assumption about Interaction of Induced Currents

As follows from Eq. (9.29), these asymptotic expressions for the magnetic field describe only its quadrature component, which is proportional to ω but they do not contain any information about the in-phase component of this field. This means that such asymptotic approach does not take into account the interaction of induced currents. Therefore, the in-phase component of the secondary magnetic field is assumed to be negligible with respect to the quadrature component over this particular range of parameters r/δ_i. To obtain asymptotic expressions for the in-phase component $In\mathbf{B}^s$ and improve the representation of the quadrature component, we describe in the next section another approach, which is based on the correct expression for the function D_0 for small parameters r/δ_i over entire range of variable m. Of course, the in-phase component of the electric field is also derived assuming that interaction of induced currents is absent. To emphasize the physical principles of such approximation, let us discuss the origin of the electric and magnetic fields, described by Eq. (9.29). First of all, as demonstrated earlier, the change of the primary magnetic field with time causes the primary electric field:

$$E_{\varphi}^{0*} = \frac{i\omega\mu_0 M}{4\pi R^2}\sin\theta,$$

where R, θ are spherical coordinates. In accordance with Ohm's law under the action of this field, the induced currents arise

$$j_\varphi^* = \gamma E_\varphi^{0*},$$

which has only the φ component. Thus, we can imagine an infinite system of current elementary tubes with cross-sections dS. The center of all these current tubes is located at the z-axis. As follows from Biot–Savart law, the current dI at each current tube generates the magnetic field, which is described by the quadrature component and it is directly proportional to frequency and conductivity. As is well known, the magnetic field for every current tube is expressed through elliptical integrals, and in order to perform their summation over the whole conducting space, we have to use surface integration. Certainly such procedure is rather cumbersome while the method based on expansion of the integrand is much simpler. In this approximation, the electric field arises due to a change of the quadrature component of the magnetic field, and correspondingly, we obtain only its in-phase component.

9.3.4. Concept of Geometric Factor

Referring again to Eq. (9.29), it can readily be seen that these equations can be written in a more compact form,

$$\begin{aligned} InE_\varphi^* &= \frac{i\omega\mu_0 M}{16\pi} i\omega\mu_0\left[\gamma_1 G_1^E + \gamma_2 G_2^E + \gamma_3 G_3^E\right] \\ QB_z^* &= \frac{\mu_0 M}{16\pi r} i\omega\mu_0\left[\gamma_1 G_1^z + \gamma_2 G_2^z + \gamma_3 G_3^z\right\} \\ QB_r^* &= -\frac{\mu_0 M}{16\pi r} i\omega\mu_0\left[\gamma_1 G_1^r + \gamma_2 G_2^r + \gamma_3 G_3^r\right], \end{aligned} \tag{9.30}$$

where

$$\begin{aligned} G_1^E &= 1 - \frac{\left(4H_1^2 + r^2\right)^{1/2} - 2H_1}{r} \\ G_2^E &= \frac{\left(4H_1^2 + r^2\right)^{1/2} - 2H_1}{r} - \frac{\left[4(H_1 + H_2)^2 + r^2\right]^{1/2} - 2(H_1 + H_2)}{r} \\ G_3^E &= \frac{\left[4(H_1 + H_2)^2 + r^2\right]^{1/2} - 2(H_1 + H_2)}{r} \end{aligned} \tag{9.31}$$

and

$$G_1^z = 1 - \frac{r}{\left(4H_1^2 + r^2\right)^{1/2}}$$

$$G_2^z = \frac{r}{\left(4H_1^2 + r^2\right)^{1/2}} - \frac{r}{\left[4(H_1 + H_2)^2 + r^2\right]^{1/2}} \tag{9.32}$$

$$G_3^z = \frac{r}{\left[4(H_1 + H_2)^2 + r^2\right]^{1/2}}$$

For the horizontal component B_r, we have

$$G_1^r = 1 - \frac{\left(4H_1^2 + r^2\right)^{1/2} - 2H_1}{\left(4H_1^2 + r^2\right)^{1/2}}$$

$$G_2^r = \frac{\left(4H_1^2 + r^2\right)^{1/2} - 2H_1}{\left(4H_1^2 + r^2\right)^{1/2}} - \frac{\left[4(H_1 + H_2)^2 + r^2\right]^{1/2} - 2(H_1 + H_2)}{\left[4(H_1 + H_2)^2 + r^2\right]^{1/2}} \tag{9.33}$$

$$G_3^r = \frac{\left[4(H_1 + H_2)^2 + r^2\right]^{1/2} - 2(H_1 + H_2)}{\left[4(H_1 + H_2)^2 + r^2\right]^{1/2}}$$

These functions G_i are usually called *geometric factors* because they depend only on the distance r and thickness of the layers; that is, entirely on geometric parameters of a medium. Equation (9.30) vividly demonstrates the remarkable simplicity of the behavior of the field when the parameters r/δ_i are very small. In fact, the components of the field $\mathbf{B}$ and in-phase component of the electric field are the sums of three terms, and each of them is a function of the geometric parameters and the conductivity of the appropriate layer. This means that in such approximation, the interaction between induced currents is neglected; that is, induced currents arise only as a consequence of the primary electric field $\mathbf{E}_\varphi^0$. In other words, it is assumed that the magnetic field of these currents is significantly less than the primary field $\mathbf{B}_0$, and for this reason, it has no appreciable effect on the induced currents. Thus, the density of induced currents at any point of a medium is independent of the intensity of currents in nearby parts of a medium. This independence is clearly reflected in terms of geometric factors that describe the field. In this light, it is proper to emphasize that the concept of geometric factor can be introduced when the electric field does not intersect boundaries between media with different conductivities, since in general, the surface density of charges is defined by both resistivity. However, there is one important exception of the Earth's surface when the upper space is an insulator and the field is caused by an arbitrary oriented current loop. Also this consideration indicates that it is impossible to introduce geometric factor when the field arises due to a grounded current line.

9.3.5. Behavior of Geometric Factors

Now consider some features of the geometric factors. First, as follows from Eq. (9.33), the functions G for three-layered medium depend on the ratio:

$$h_1 = \frac{H_1}{r}, \quad h_2 = \frac{H_2}{r},$$

and the sum of geometric factors is always equal to unity:

$$G_1^E + G_2^E + G_3^E = G_1^z + G_2^z + G_3^z = G_1^r + G_2^r + G_3^r = 1 \qquad (9.34)$$

For instance, if the separation r is very small with respect to the layer thickness H_1, geometric factor of this layer for each field component tends to unity, while two other geometric factors become very small. This means that performing measurements of the quadrature component of the magnetic field or the in-phase component of the electric field near the dipole usually yields information about the conductivity of the upper layer only. In contrast, with an increase of the separation r, the geometric factor of the basement with conductivity γ_3 approaches unity. Therefore, the influence of induced currents in the layers with thickness H_1 and H_2 becomes very small, and the field on the Earth's surface is practically defined by conductivity γ_3. This consideration shows that measurements of the field $Q\mathbf{B}$ or $In\mathbf{E}$ in the range of small parameters with different separations allow one to obtain information about the resistivity structure. At the same time, measurements with fixed separation r, that is profiling, permit us to trace lateral changes of conductivity, and this approach is often used in practice. In accordance with Eqs (9.31)–(9.33), geometric factors of different components of the field differ from each other. This indicates that in the range of small parameter r/δ and at a given distance from the current source, these field components have a different depth of investigation. Let us note that unusual simplicity of the expressions for the electromagnetic field greatly facilitates the solution of the forward and inverse problems. In contrast to the electromagnetic methods based on the use of the direct and time-varying currents, we observe a unique situation where the field caused by induced currents in a given layer is independent of the conductivity of all other layers.

9.3.6. Generalization for *N*-Layered Medium

Taking into account this fact, it is obvious that an approximate solution for N-layered medium has the following form:

$$InE_{\varphi}^{*} = \frac{i\omega\mu_0 M}{16\pi} i\omega\mu_0 \sum_{i=1}^{N} \gamma_i G_i^E \quad QB_z^{*} = \frac{\mu_0 M}{16\pi r} i\omega\mu_0 \sum_{i=1}^{N} \gamma_i G_i^z$$

$$QB_r^{*} = -\frac{\mu_0 M}{16\pi r} i\omega\mu_0 \sum_{i=1}^{N} \gamma_i G_i^r, \tag{9.35}$$

which are valid regardless of the distance between the vertical magnetic dipole and an observation point, as far as one can neglect by interaction of currents. In particular, in a uniform half space we have

$$InE_{\varphi}^{*} = \frac{i\omega\mu_0 M}{16\pi} i\gamma\omega\mu_0, \quad QB_z^{*} = \frac{\mu_0 M}{16\pi r} i\omega\mu_0\gamma, \quad QB_r^{*} = -\frac{\mu_0 M}{16\pi r} i\omega\mu_0\gamma,$$

which were derived earlier. As follows from Eqs (9.31) and (9.32) derived for three-layered medium, the geometric factors of the uppermost layer and the lowermost medium are particular cases for the intermediate layer and, correspondingly, the geometric factor of any layer can be written as

$$G^E = \left[\left(4h_t^2 + 1\right)^{1/2} - 2h_t\right] - \left[\left(4h_b^2 + 1\right)^{1/2} - 2h_b\right]$$

$$G^z = \frac{1}{\left(1 + 4h_t^2\right)^{1/2}} - \frac{1}{\left(1 + 4h_b^2\right)^{1/2}} \tag{9.36}$$

$$G^r = \frac{\left(1 + 4h_t^2\right)^{1/2} - 2h_t}{\left(1 + 4h_t^2\right)^{1/2}} - \frac{\left(1 + 4h_b^2\right)^{1/2} - 2h_b}{\left(1 + 4h_b^2\right)^{1/2}}$$

Here, h_t and h_b characterize the distance from the Earth's surface to the top and bottom of the layer, normalized by a separation r. Behavior of these functions for magnetic field is given in Figure 9.2(a,b).

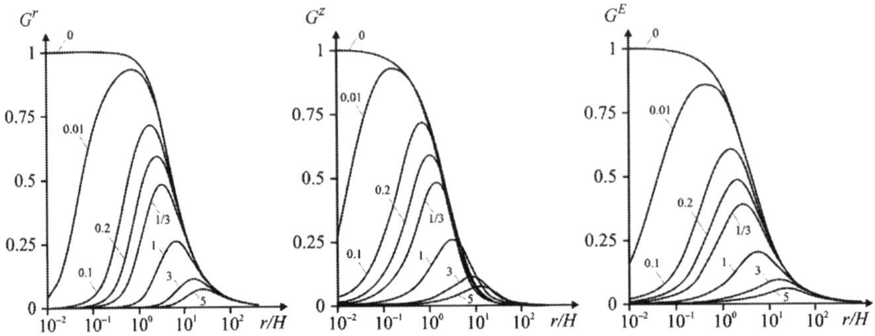

Figure 9.2 Geometrical factors G^r, G^z, G^E. Curve index is $\frac{h_t}{h_b - h_t}$ ratio.

9.3.7. About the Depth of Investigation

The study of geometrical factors, describing the quadrature component of magnetic field at the range of small parameters $(r/\delta_i \ll 1)$, shows that the depth of investigation increases with an increase of the separation between the current source and observation point. In other words, the upper layers become transparent and the field coincides with that of the basement. In contrast, when the separation r is relatively small, the depth of investigation is usually also small. Especially, if the upper layer is relatively conductive, its contribution becomes dominant. Such dependence on the separation is related with the fact that at small distances the main contribution to the magnetic field is given by induced currents near the dipole that follows from Biot–Savart law. With an increase of r, the influence of more remote currents becomes stronger. As follows from Eq. (9.35), a decrease of the frequency does not have any effect on the depth of investigation even though the depth of penetration of the field increases. Bearing in mind that within this range of parameter r/δ_i the in-phase component of the secondary magnetic field is very small, we may say that the amplitude of the secondary or the total fields have the same dependence on the separation and frequency.

In this light, let us imagine that we perform frequency soundings on the Earth's surface measuring the quadrature component of the magnetic field or its amplitude with sufficiently small array ($r < H$), where r is the total thickness of layers. Certainly, with a decrease of frequency, the depth of field penetration increases since an influence of skin effect decreases. However, the measured quadrature component of the magnetic field or the in-phase component of the electric field still remains sensitive only to the upper portion of the geoelectric section. This is one of the reasons why the conventional frequency soundings are carried out with separations greatly exceeding the total thickness of layers situated above the resistive basement.

9.4. THE FIELD OF A VERTICAL MAGNETIC DIPOLE IN THE RANGE OF SMALL PARAMETERS r/δ_i

Now we study in detail the behavior of the field in the range of small parameters r/δ_i, in particular, in the low frequency part of the spectrum. This topic is of a considerable practical interest since this portion of the spectrum contains information about the deepest layers which are being explored. In contrast to the previous section, the interaction between induced currents is taken into account. To obtain an analytic representation of the field in the range of small parameters at the beginning, we make use of the expression for the vector potential on the surface of a two-layered medium caused by induced currents. From Eqs (9.5) and (9.10), we have

$$A^*_{zs} = \frac{i\omega\mu_0 M}{4\pi r} \int_0^\infty D_0\left(x, r/\delta_1, s\right) J_0(x)\mathrm{d}x \tag{9.37}$$

where

$$D_0 = -\frac{x_{10} - x_{12}\exp(-2x_1 h_1)}{1 - x_{10}x_{12}\exp(-2x_1 h_1)} \tag{9.38}$$

and

$$x_{10} = \frac{x_1 - x}{x_1 + x}, \quad x_{21} = -x_{12} = \frac{x_2 - x_1}{x_2 + x_1}, \quad x_1 = \left(x^2 + \alpha^4\right)^{1/2}$$

$$x_2 = \left(x^2 + s\alpha^4\right)^{1/2}, \quad \alpha^4 = -k_1^2 r^2, \quad s = \frac{\gamma_2}{\gamma_1}$$

and

$$h_1 = \frac{H_1}{r}, \quad k_1 = (i\gamma_1\mu_0\omega)^{1/2}$$

Now we find an expression for the vector potential of the secondary field in the form of a power series in terms of the parameter α. This can be accomplished inasmuch as the skin depth in both media is much greater than the distance from the dipole to the observation point. Consider the integral in Eq. (9.37) as a sum of two integrals,

$$\int_0^\infty D_0 J_0(x)\mathrm{d}x = \int_0^{|\alpha|} D_0 J_0(x)\mathrm{d}x + \int_{|\alpha|}^\infty D_0 J_0(x)\mathrm{d}x, \tag{9.39}$$

where $|\alpha|$ is much less than unity and we look for a power series expansion for each of integrals. It should be clear that

$$\alpha^4 = -i\gamma_1\mu_0\omega r^2 = -2i\left(\frac{r}{\delta_1}\right)^2 \tag{9.40}$$

9.4.1. Expansion of the Internal Integral, $0 \le m \le |\alpha^2|$

First consider the internal integral, and at the beginning assume that the lower medium has a finite resistivity, $\gamma_2 \neq 0$, while the limiting case $\rho_2 \to \infty$ will be studied later. This means that the magnitude of the term $x_{10}x_{21}\exp(-2x_1h_1)$ is less than unity over the entire range of integration:

$$|x_{10}x_{21}\exp(-2h_1x_1)| < 1$$

Then, the function D_0 can be written as a series

$$D_0(x) = -\left[\sum_{l=1}^{\infty}(-1)^{l-1}x_{10}^{l}x_{21}^{l-1}\exp[-2x_1h_1(l-1)] + \sum_{l=1}^{\infty}(-1)^{l-1}x_{10}^{l-1}x_{21}^{l}\exp(-2x_1h_1l)\right] \tag{9.41}$$

The series on the right-hand side of this equation alternates and converges rapidly for any value of x. Assuming that over the range of integration $0 \le x \le |\alpha|$, the value of $|2x_1lh_1|$ is less than unity, we expand the functions $\exp[-2x_1h_1(l-1)]$ and $\exp(-2x_1h_1l)$ in power series. Then, again making use of the identity,

$$\frac{1}{1+x} = \sum_{k=0}^{\infty}(-1)^k x^k \quad \text{if } |x| < 1$$

and performing simple algebraic transformations, we obtain

$$D_0(x) = \sum_{p=1}^{\infty} F_p(x), \tag{9.42}$$

where, for instance,

$$F_1 = -\frac{s\alpha^4}{(x_2+x)^2}, \quad F_2 = 2xh_1\frac{(s-1)\alpha^4}{(x_2+x)^2} \tag{9.43}$$

and so on. Then the first integral on the right-hand side of Eq. (9.39) can be written as

$$\sum_{p=1}^{\infty} \int_0^{|\alpha|} F_p(x) J_0(x) \mathrm{d}x \tag{9.44}$$

Because the variable x is very small, the function $J_0(x)$ can be represented as a rapidly converging series:

$$J_0(x) = \sum_{i=0}^{\infty} f_i x^{2i},$$

where f_i are known coefficients. After substitution of this series into Eq. (9.44) and multiplying the denominators of the functions F_p by their complex conjugates, the integrals in Eq. (9.44) reduce to tabulated ones of the form

$$I_n = \int_0^{|\alpha|} x^n \left(x^2 + s\alpha^4\right)^{1/2} \mathrm{d}x$$

Having expressed these integrals in terms of elementary functions, we can write their expansion as a series in α. As an example, the series describing integrals I_0 and I_1 are given below:

$$I_0 = \frac{1}{2}\alpha^2 + \left(\frac{1}{4}s + \frac{1}{2}s\ln 2 - \frac{1}{4}s\ln s\right)\alpha^4 - \frac{1}{2}s\alpha^4\ln\alpha + \frac{1}{16}s^2\alpha^6 - \frac{1}{64}s^3\alpha^8$$
$$+ \frac{5}{32\cdot 24}s^4\alpha^{10} - \frac{7}{63\cdot 32}s^5\alpha^{12} + \cdots$$

and

$$I_1 = \frac{\alpha^3}{3} + \frac{1}{2}s\alpha^5 - \frac{1}{3}s^{3/2}\alpha^6 + \frac{1}{8}s^2\alpha^7 - \frac{1}{48}s^3\alpha^9 + \frac{1}{128}s^4\alpha^{11} - \frac{1}{256}s^5\alpha^{13} + \cdots \tag{9.45}$$

Performing a summation of the series corresponding to different functions F_p (Eq. (9.44)), we obtain an expansion of the integral

$$\int_0^{|\alpha|} D_0(x) J_0(x) \mathrm{d}x$$

It is essential to emphasize that terms of the series contain integer and fractional powers of ω as well as its logarithm $\ln \omega$.

9.4.2. Expansion of the External Integral

Next consider an expansion in a series in α of the second integral on the right-hand side of Eq. (9.39):

$$\int\limits_{|\alpha|}^{\infty} D_0(x)J_0(x)\mathrm{d}x$$

The method used above based on a series expansion of the term $\exp(-2x_1h_1l)$ and $J_0(x)$, cannot be applied for evaluating the outer integral since the variable x increases without limit. Therefore, it is necessary to seek an expansion in a series in α of integrals, in which the integrand is a product of each term of series (9.41) by the factor $J_0(x)$. For instance, for the first three integrals, we have the following expressions:

$$N_0 = -\int\limits_{|\alpha|}^{\infty} \frac{x_1 - x}{x_1 + x} J_0(x)\mathrm{d}x, \quad N_1 = \int\limits_{|\alpha|}^{\infty} \frac{x_2 - x_1}{x_2 + x_1} \exp(-2x_1h_1)J_0(x)\mathrm{d}x$$

and

$$N_2 = \int\limits_{|\alpha|}^{\infty} \left(\frac{x_1 - x}{x_1 + x}\right)^2 \left(\frac{x_2 - x_1}{x_2 + x_1}\right) \exp(-2x_1h_1)J_0(x)\mathrm{d}x \tag{9.46}$$

It turns out that the first term in series expansion of each subsequent integral N_i begins with higher power of α. This does not occur if the expansion of integrals N_i is considered over only the initial range of integration ($0 \le x \le |\alpha|$). In this case, each series starts with a term having the same power in α. It is obvious that for the outer range of integration, the condition

$$x^2 \ge |\alpha^4|, \tag{9.47}$$

is met. Therefore, the radicals $(x^2 + \alpha^4)^{1/2}$ and $(x^2 + s\alpha^4)^{1/2}$, which are present in integrands of N_i, can be expanded in a reasonably rapidly converging series of powers of α^4/x^2. For instance, using such a series, we arrive at the following expansion:

$$\frac{x_2 - x_1}{x_2 + x_1} \approx a_1\frac{\alpha^4}{x^2} + a_2\frac{\alpha^8}{x^4} + a_3\frac{\alpha^{12}}{x^6} + a_4\frac{\alpha^{16}}{x^8} + a_5\frac{\alpha^{20}}{x^{10}} + \cdots$$

where a_i are coefficients depending on s only. Correspondingly, the integrals N_i can be expressed in terms of integrals, which do not contain the parameter α in their integrands. These integrals have the form

$$L_k = \int_{|\alpha|}^{\infty} \frac{\exp(-\beta_i x)}{x^k} J_0(x)\mathrm{d}x, \tag{9.48}$$

where

$$\beta_0 = 0, \quad \beta_1 = 2h_1, \quad \beta_2 = 4h_1, \quad \beta_3 = 6h_1 \cdots$$

Now we introduce coefficients $f_n^{(i)}$ from the expansion

$$\exp(-\beta_i x)J_0(x) = \sum_{n=0}^{\infty} f_n^{(i)} x^n, \tag{9.49}$$

and represent the integrals L_k as series

$$L_k = \int_{|\alpha|}^{\infty} \exp(-\beta_i x)x^{-k} J_0(x)\mathrm{d}x = B_k + \sum a_n^k \alpha^n + a_0^k \ln\alpha \tag{9.50}$$

where n can be negative. Taking a derivative of both sides of Eq. (9.50) with respect to α and setting the coefficients of the same power of α equal, we readily find the coefficients α_n^k in terms of an expansion for L_k. For instance, we have

$$L_0 = B_0 - f_0^{(i)}\alpha - \frac{1}{2}f_1^{(i)}\alpha^2 - \frac{1}{3}f_2^{(i)}\alpha^3 - \frac{1}{4}f_3^{(i)}\alpha^4 - \frac{1}{5}f_4^{(i)}\alpha^5 - \cdots$$

$$L_1 = B_1 - f_0^{(i)}\ln\alpha - f_1^{(i)}\alpha - \frac{1}{2}f_2^{(i)}\alpha^2 - \frac{1}{3}f_3^{(i)}\alpha^3 - \frac{1}{4}f_4^{(i)}\alpha^4 - \cdots \tag{9.51}$$

$$L_2 = B_2 + \frac{f_0^{(i)}}{\alpha} + f_1^{(i)}\ln\alpha - f_2^{(i)}\alpha - \frac{1}{2}f_3^{(i)}\alpha^2 - \frac{1}{3}f_4^{(i)}\alpha^3 -$$

and so on. In these formulas, the coefficients B_k remain unknown. To determine them, the following relationships are used,

$$\frac{\partial L_n}{\partial \beta} = -L_{n-1},$$

and

$$J_{i0}(\alpha) = \int\limits_{|\alpha|}^{\infty} \frac{J_0(u)}{u} \mathrm{d}u, \quad J_{i1}(\alpha) = \int\limits_{|\alpha|}^{\infty} \frac{J_1(u)}{u} \mathrm{d}u, \tag{9.52}$$

where the latter equations are known functions. For the integral L_0, we have

$$L_0 = \int\limits_{|\alpha|}^{\infty} \exp(-\beta_i x) J_0(x) \mathrm{d}x$$

Letting α be zero, we obtain the Lipschitz integral

$$L_0(0) = \int\limits_{0}^{\infty} \exp(-\beta_i x) J_0(x) \mathrm{d}x = \left(1 + \beta_i^2\right)^{1/2}$$

Thus,

$$B_0 = \left(1 + \beta_1^2\right)^{-1/2}$$

The same approach allows us to determine other constants from Eq. (9.51). Thus, making use of the procedures described above, we obtain the expansion of the integral

$$\int\limits_{|\alpha|}^{\infty} D_0(x) J_0(x) \mathrm{d}x$$

as series in α.

9.4.3. Expression for the Vector Potential at the Range of Small Parameters

Collecting the coefficients with the same power in α in each of the series, representing the inner and outer ranges of integration, we arrive at the following expression for the vector potential at the range of small parameters r/δ_i :

$$A_z^{*s} = \frac{i\omega\mu_0 M}{4\pi r}\left[\varphi_0\alpha^2 + \varphi_1\alpha^4 \ln\alpha + \varphi_2\alpha^4 + \varphi_3\alpha^6 + \varphi_4\alpha^8 \ln\alpha + \varphi_5\alpha^8 + \varphi_6\alpha^{10} + \cdots\right], \tag{9.53}$$

where φ_i are coefficients depending on electric and geometric parameters of a medium and separation between the dipole and observation point. It is appropriate to note that though this method of deriving asymptotic expressions is rather cumbersome, it is perhaps the only way to find in the explicit form relationships between the parameters of a medium and the field components in the range of small parameters. Moreover, this asymptotic expression played an important role in developing and understanding the electromagnetic soundings with relatively small separations. Let us choose an arbitrary small value of the variable of integration $x = x_0 < 1$. Then Eq. (9.39) can be written in the form

$$\int_0^\infty D_0(x)J_0(x)\mathrm{d}x = \int_0^{x_0} D_0(x)J_0(x)\mathrm{d}x + \int_{x_0}^\infty D_0(x)J_0(x)\mathrm{d}x \tag{9.54}$$

It is clear that for the outer range of integration $x \geq x_0$, we have

$$x > \left|\alpha^4\right| \quad \text{and} \quad x > \left|s\alpha^4\right|$$

This means that the second integral on the right-hand side of Eq. (9.54) can be expanded in a series in α with powers $4n$. In other words, this expansion contains only integer powers of ω. Correspondingly, the fractional powers of ω, as well as logarithmic terms ln ω, arise in the series (9.53) due to integration over very small values of x. It may be proper to make one comment. We illustrated method of deriving the asymptotic formulas, assuming that a point of observation is located at the Earth's surface but this approach remains the same for arbitrary values of z.

9.4.4. The Expression for the Field at the Range of Small Parameters r/δ_i

Now taking corresponding derivatives of the vector potential (Eq. (9.53)), and discarding all terms except the leading ones, we obtain expressions for the field of a magnetic dipole on the surface of a two-layered medium when the basement has a finite resistivity:

$$QB_z^* = \frac{\mu_0 M_0}{4\pi r^3}\left[\frac{\omega\mu_0 r^2}{4}\left(\gamma_1 G_1^z + \gamma_2 G_2^z\right) - \frac{2^{3/2}}{15}\left(\gamma_2\mu_0\omega r^2\right)^{3/2} + \cdots\right] \tag{9.55}$$

$$InB_z^* = \frac{\mu_0 M_0}{4\pi r^3}\left[1 + \frac{2^{3/2}}{15}\left(\gamma_2\mu_0\omega r^2\right)^{3/2} - \cdots\right]$$

$$QB_r^* = -\frac{\mu_0 M_0}{4\pi r^3}\left[\frac{\omega\mu_0 r^2}{4}\left(\gamma_1 G_1^r + \gamma_2 G_2^r\right) + \frac{3\pi}{64}\left(\gamma_2\mu_0\omega r^2\right)^2 + \cdots\right]$$

$$InB_r^* = -\frac{\mu_0 M_0}{4\pi r^3}\left[\frac{1}{16}\left(\gamma_2\mu_0\omega r^2\right)^2 \ln\frac{\left(\gamma_2\mu_0\omega r^2\right)^{1/2}}{2} + \cdots\right] \tag{9.56}$$

$$QE_\varphi^* = \frac{\omega\mu_0 M_0}{4\pi r^2}\left[1 - \frac{2^{1/2}}{15}\left(\gamma_2\mu_0\omega r^2\right)^{3/2} + \cdots\right] \tag{9.57}$$

$$InE_\varphi^* = -\frac{\omega\mu_0 M_0}{4\pi r^2}\left[\frac{\omega\mu_0}{4}r^2\left(\gamma_1 G_1^E + \gamma_2 G_2^E\right) - \frac{2^{1/2}}{15}\left(\gamma_2\mu_0\omega r^2\right)^{3/2} + \cdots\right]$$

It should be obvious that by including more terms from Eq. (9.53), it is possible to provide a significantly better asymptotic representation of the field.

9.4.5. Behavior of the Field at the Range of Small Parameters r/δ_i

As is seen from Eqs (9.55) and (9.57), different terms of the series are related to the parameters of the medium in various ways. The leading terms of the series for the quadrature component of the magnetic field and the in-phase component of the electric field, which are proportional to conductivity, were in detail considered in the previous section. The analysis of other terms allows us to reveal some fundamental features of the field behavior within the range of small parameters r/δ_i. First of all, the in-phase component of the secondary magnetic field as well as the quadrature component of the electric field approaches the values corresponding to the half space with the conductivity of the basement as the frequency decreases. This occurs regardless of the distance from the dipole to the receiver, that is,

$$InB^{s*} \rightarrow InB^{s*}(\gamma_2), \quad QE^{s*} \rightarrow QE^{s*}(\gamma_2) \quad \text{if} \ r/\delta_i \rightarrow 0 \qquad (9.58)$$

In other words, with a decrease of the frequency, the first layer becomes transparent to the electromagnetic field, and this phenomenon is independent of the conductivity and is observed at any distance from the dipole. Suppose that the separation r is greater than the thickness H_1 of the first layer. Then, the coincidence of functions InB^{s*} and QE^{s*}, observed on the surface of a two layered medium, with corresponding components for a uniform half space with conductivity γ_2, occurs when the parameter r/δ is small in both media. If the separation is less than the thickness of the upper layer, the same coincidence is observed, provided that the parameters H_1/δ_i are small in both media. It is proper to note that this behavior of InB^{s*} and QE^{s*} takes place regardless of the ratio of conductivities. In particular, the conductivity of the upper layer can be many times greater than the conductivity of the underlying medium. This phenomenon was, in essence, explained when we studied the field in a uniform medium and shown that with a decrease of frequency the maximum of the in-phase component of currents in the medium shifts to greater distances from the dipole. Therefore, the influence of induced currents in the vicinity of the dipole becomes smaller. This means that the depth of investigation when measuring the functions QE^{s*} and InB^{s*} increases as frequency decreases, no matter what the separation r is. For example, the distance r can be much smaller than the thickness H_1. Thus with decreasing frequency, there is always some low frequency for which both functions QE^{s*} and InB^{s*} practically coincide with the corresponding components, which would be measured at the surface of a uniform half space with conductivity γ_2. A further decrease in frequency does not change the relationship between the field and geoelectric parameters of the medium. The succeeding terms in the series for the in-phase component of **B** and quadrature component of **E** contain information about the conductivity γ_1 and thickness H_1, as well as the basement conductivity γ_2. These terms describe the interaction between the induced currents in both media. With decreasing the frequency, the in-phase component of the induced currents concentrates principally in the basement but because of the interaction, some currents arise in the upper layer. However, over the range of very small values for r/δ, their effect becomes negligible, and the in-phase component of the magnetic field InB^{s*} depends only γ_2.

9.4.6. Comparison of Both Components of the Field at the Range of Small Parameters

Now let us compare the behavior of the quadrature and in-phase components of the field for small values of parameters r/δ_1 and H_1/δ. We can draw several obvious conclusions from Eqs (9.55)–(9.57).

1. The quadrature component of B_z^* is directly proportional to the frequency ω, but the in-phase component InB_z^{s*} is proportional to $\omega^{3/2}$.
2. The quadrature component of B_z^* is proportional to the conductivity γ_1 and γ_2, while the in-phase component InB_z^{s*} is proportional to $\gamma_2^{3/2}$.
3. The quadrature component of B_z^* is larger than the in-phase component of B_z^{s*}, and with decreasing frequency, the ratio InB_z^{s*}/QB_z^* becomes smaller.
4. When the separation between the dipole and receiver is less than the thickness of the upper layer, the component QB_z^* is mainly determined by its conductivity γ_1. In contrast, the in-phase component of the magnetic field InB_z^{s*} is equal to that for a uniform half space with the conductivity of the basement γ_2 regardless of the separation. In other words, the influence of the separation r on the depth of investigation is completely different for these two components. In the case of the quadrature component, the depth of investigation decreases as the separation r is reduced, no matter how low the frequency is. When the in-phase component of the secondary magnetic field or QE_φ^{s*} is measured, the depth of investigation does not depend on the separation r/δ_1 and it is maximal since both functions InB_z^{s*} and QE_φ^{s*} are practically defined by the conductivity of the basement.
5. With an increase in the separation r, the depth of investigation increases when the quadrature component of **B** is measured. In the limit, it becomes a function of the conductivity γ_2 only, along with the in-phase component of the secondary magnetic field. However, as was mentioned earlier, the in-phase component is more sensitive to change in conductivity than is the quadrature component.
6. The depth of investigation over the range of small parameters depends primarily on the distance between the dipole and an observation point when the quadrature component of the magnetic field is measured.
7. When parameters r/δ_1 and H/δ_1 decrease, the secondary magnetic field is mainly determined by the quadrature component. As was pointed out earlier, the frequency soundings, made with relatively short separations

$\left({}^{r}/_{H_1} < 1\right)$ and based on observations of the amplitude and phase or the quadrature and in-phase components of the total field, are limited in their depth of investigation. Correspondingly, they usually provide only information about the thickness and conductivity of the upper layer, unless $\gamma_1/\gamma_2 \ll 1$.

8. Measurements of the in-phase component of the secondary magnetic field or the quadrature component of the secondary electric field usually represent a difficult task when the parameter r/δ is small. This is related to the fact that this part of the field is much smaller than the function InB_z^{0*} or QE_φ^{0*} generated by the current in the dipole.

In this light, consider the series for the quadrature component of the magnetic field (Eq. (9.55)). The second term in the expansion for the quadrature component of B_z^*, like the first term of the in-phase component, does not depend on the conductivity of the first layer γ_1. Thus, in order to obtain the same depth of investigation as that for the in-phase component at relatively short separations $\left({}^{r}/_{H_1} < 1\right)$, the first term in the series for the quadrature component has to be eliminated. This might be done by making measurements at two frequencies. For instance, the difference

$$QB_z^*(\omega_1) - \frac{\omega_1}{\omega_2} QB_z^*(\omega_2) \tag{9.59}$$

allows us to eliminate the influence of the first term of the series and it depends on γ_2 only, if ${}^{r}/_{\delta_1} \ll 1$ and $H_1/_{\delta_1} \ll 1$. It is interesting to notice that this result remains valid even in the presence of lateral changes of conductivity within the upper layer.

9. As was also the case for the vertical component B_z^*, the leading term of the series (9.56) which describes the in-phase component of B_r^*, is controlled by the conductivity γ_2. In contrast to InB_z^*, this component of B_r^* is more closely related with the conductivity (as $\gamma_2^2 \ln \gamma_2$). However, as follows from Eqs (9.55) and (9.56), we have

$$\left|InB_r^*\right| < \left|InB_z^*\right|, \quad \text{if } {}^{r}/_{\delta_1} \ll 1, \text{ and } H_1/_{\delta_1} \ll 1 \tag{9.60}$$

10. In accordance with Eq. (9.56), the second term in the expansion for the quadrature component B_r^* is proportional to ω^2 and does not depend on the conductivity of the upper layer. As was previously noted, the terms in the various series (Eqs (9.55)–(9.57)) which do not contain the conductivity γ_1 coincide with the corresponding terms of the series, representing the case of a uniform half space with the conductivity γ_2. From Eq. (9.57), it follows that the main features of the behavior of the electric field E_φ^* for small values of the parameters r/δ_1, r/δ_2, and H_1/δ_1 are practically the same as those for the vertical component of the magnetic field B_z^*. Perhaps it is appropriate to point out again that this analysis of asymptotic expressions allowed us to realize that, in principle, it is possible to reach any depth of investigation regardless of the separation between the dipole and an observation site.

11. Also we can say that Eq. (9.53) shows that the electromagnetic field can be represented in terms of a generalized Mac Lauren series at the range of small parameters. These series contain integer and fractional powers of the frequency ω, as well as logarithmic terms ($\ln \omega$), and each field component has the form

$$\sum a_n\omega^n + \sum b_n\omega^{n+1/2} + \sum c_n\omega^n \ln \omega \qquad (9.61)$$

For example, the series for the quadrature component of the magnetic field always starts with a term proportional to ω, while the corresponding leading term of the series representing the in-phase component of the electric field is proportional to ω^2. As was shown earlier, this behavior occurs when interaction between induced currents can be neglected. From Eqs (9.55) and (9.56), it follows that the second term in the series for the quadrature component of the magnetic field, along with the leading term of the series for the in-phase component of the secondary field, is different for B_z^* and B_r^*. They contain either a fractional power of ω or $\ln \omega$. For instance, we have

$$InB_z^{s*} \sim \omega^{3/2} \qquad InB_r^{s*} \sim \omega^2 \ln \omega$$

9.4.7. The Range of Small Parameters on the Surface of *N*-Layered Medium

Derivation of various asymptotic formulas has shown that only the initial portion of the integral in Eq. (9.17) is responsible for the fractional powers in ω and logarithmic terms. Making use of this fact, let us consider the asymptotic behavior of the electromagnetic field on the surface of an N-layered medium, provided that the basement is conductive; that is, $\gamma_N \neq 0$. As follows from Eq. 9.24,

$$A_z = \frac{i\omega\mu_0 M}{2\pi}\int_0^\infty \frac{mJ_0(mr)}{m + m_1/R_N}\,dm,$$

where

$$R_N = \coth\left[m_1 H_1 + \coth^{-1}\frac{m_1}{m_2}R_{N-1}\right] \tag{9.62}$$

Thus, in order to determine terms in the series containing fractional powers of ω or logarithmic terms, we must evaluate the limit of the integrand of Eq. (9.62) when $k_i \to 0$ and $m \to 0$; that is,

$$\lim \frac{m}{m + m_1/R_N} \quad \text{if } m \to 0 \text{ and } k_i \to 0$$

From Eq. (9.22), we have

$$R_2 \to \frac{m_1}{m_2}, \quad R_3 \to \frac{m_1}{m_3}$$

and finally,

$$R_N \to \frac{m_1}{m_N} \quad \text{if } m \to 0 \text{ and } k_i \to 0 \tag{9.63}$$

where

$$m_N = \left(m^2 - i\gamma_N\mu_0\omega\right)^{1/2} \tag{9.64}$$

Substituting Eq. (9.63) into the integrand for A_z^*, we see that

$$\frac{m}{m + m_1/R_N} \to \frac{m}{m + m_N} \quad \text{if } m \to 0,\, k_i \to 0 \tag{9.65}$$

This limit corresponds to the case of a uniform half space with conductivity γ_N; that is, the leading terms of the series which contain fractional powers of ω or $\ln\omega$ coincide with the same terms for a uniform half space with conductivity of the lowermost layer. It should be clear that this result is invalid for the terms proportional to the integer powers of ω, because all spatial harmonics m in Eq. (9.62) contribute to this part of the series. Therefore, taking into account the results derived in this section, we arrive at the following asymptotic expression for the field of the magnetic dipole on the surface of N-layered medium when parameters r/δ_i and H_i/δ_i are small and $\gamma_N \neq 0$:

$$QB_z^* = \frac{\mu_0 M_0}{4\pi r^3}\left[\frac{1}{4}\sum_{i=1}^{N}\left(\gamma_i\mu_0\omega r^2\right)G_i^z - \frac{2^{3/2}}{15}\left(\gamma_N\mu_0\omega r^2\right)^{3/2} + \cdots\right]$$

$$InB_z^* = -\frac{\mu_0 M_0}{4\pi r^3}\left[1 + \frac{2^{3/2}}{15}\left(\gamma_N\mu_0\omega r^2\right)^{3/2} + \cdots\right]$$

$$QB_r^* = -\frac{\mu_0 M_0}{4\pi r^3}\left[\frac{1}{4}\sum_{i=1}^{N}\left(\gamma_i\mu_0\omega r^2\right)G_i^r + \frac{3\pi}{64}\left(\gamma_N\mu_0\omega^2\right)^2 + \cdots\right]$$

$$InB_r^* = -\frac{\mu_0 M_0}{4\pi r^3}\left[\frac{1}{16}\left(\gamma_N\mu_0\omega r^2\right)^2 \ln\frac{\left(\gamma_N\mu_0\omega r^2\right)^{1/2}}{2} + \cdots\right] \qquad (9.66)$$

$$QE_\varphi^* = \frac{\omega\mu_0 M_0}{4\pi r^2}\left[1 - \frac{2^{1/2}}{15}\left(\gamma_N\mu_0\omega r^2\right)^{3/2} + \cdots\right]$$

$$InE_\varphi^* = -\frac{\omega\mu_0 M_0}{4\pi r^2}\left[\frac{1}{4}\sum_{i=1}^{N}\left(\gamma_i\mu_0\omega r^2\right)G_i^E - \frac{2^{1/2}}{15}\left(\gamma_N\mu_0\omega r^2\right)^{3/2} + \cdots\right]$$

Thus, the in-phase component of the secondary magnetic field, in contrast to the quadrature component, tends to the in-phase component corresponding to a uniform half space with conductivity γ_N regardless of the distance between the dipole and the observation point; the same result applies for the quadrature component of the secondary electric field; that is,

$$InB_r^* \to InB_r^*(\gamma_N), \quad InB_z^{s*} \to InB_z^{s*}(\gamma_N), \quad QE_\varphi^{s*} \to QE_\varphi^s(\gamma_N)$$
$$\text{if } r/\delta_i \to 0,\ H_i/\delta_i \to 0 \qquad (9.67)$$

In other words, over the low-frequency portion of the spectrum, when the in-phase component of the secondary magnetic field or the quadrature component of the electric fields are measured, all the conducting layers become transparent no matter what the separation r; that is, diffusion phenomenon takes place. This conclusion remains valid when the second term in the series expansion of components $Q\mathbf{B}^*$ and $In\mathbf{E}^*$ are measured. As was pointed out earlier, the successive terms in the asymptotic series represent interaction of induced currents in the various layers and, therefore, they usually depend on their thickness and conductivity.

9.5. APPROXIMATE METHOD OF FIELD CALCULATION

Now we apply a different method to study the field at the range of relatively small parameters. Later it will be demonstrated that this approach may follow from the integral equation for the field, and it represents its first approximation. Here we will arrive at the same result in different way, and for illustration, consider two-layered medium assuming the following:

1. The induced currents in the upper layer arise only due to the primary electric field.
2. In the basement, the skin effect manifests itself in the same manner as in a uniform half space with the conductivity γ_2.

Both assumptions imply that interaction between currents in the upper layer, as well as between currents of this layer and those in the basement are absent. This allows us to use concept of the geometric factor G_1.

First, consider a uniform half space with conductivity γ_2. Taking into account that within its upper part with the thickness H_1 the interaction of currents is absent, we can write for the vertical component of the field, caused by currents in the basement:

$$B_z^* = B_z^*(\gamma_2) - \frac{\mu_0 M}{4\pi r^3}\frac{i\omega\mu_0\gamma_2 r^2}{4} G_1^z \tag{9.68}$$

It is obvious that the latter describes the field when the dipole and observation point are located at the height H_1 above a uniform half space with conductivity γ_2. It is essential that Eq. (9.68) to some extent takes into account the skin effect. Now adding the field due to currents in the upper layer, we obtain an approximate expression for the field at the surface of two-layered medium:

$$B_z^*(\gamma_1, \gamma_2) = i\frac{\mu_0 M}{4\pi r^3}\frac{\omega\mu_0 r^2}{4}(\gamma_1 - \gamma_2)G_1^z + B_z^*(\gamma_2) \tag{9.69}$$

Here, $B_z^*(\gamma_2)$ is the vertical component of the field on the surface of a uniform half space which we studied in detail earlier. In the same manner, we have

$$B_r^*(\gamma_1, \gamma_2) = -i\frac{\mu_0 M}{4\pi r^3}\frac{\omega\mu_0 r^2}{4}(\gamma_1 - \gamma_2)G_1^r + B_r^*(\gamma_2) \tag{9.70}$$

$$E_{\varphi}^{*}(\gamma_1, \gamma_2) = -\frac{\omega\mu_0 M}{4\pi r^2}\frac{\omega\mu_0 r^2}{4}(\gamma_1 - \gamma_2)G_1^E + E_{\varphi}^{*}(\gamma_2)$$

By analogy, we can write equations for the field for multilayered medium. It is proper to note that from these equations directly follows that at the range of small parameters, the upper layer becomes transparent, provided that the in-phase component of the secondary magnetic field is measured.

9.6. THE FIELD WITHIN THE RANGE OF SMALL PARAMETERS WHEN BASEMENT IS AN INSULATOR

So far, we have only discussed the behavior of the field for small values of the parameters r/δ_i and H_i/δ_i when underlying medium has a finite resistivity and demonstrated how diffusion manifests itself at the low-frequency spectrum. Next let us consider the field at the surface of a layered medium when the basement is insulator. In reality, the medium always has some nonzero conductivity; however, asymptotic formulas derived above give a satisfactory accuracy only when difference in resistivity is not large. For this reason, it is very useful to obtain formulas for limiting case of a nonconducting basement. The method of expansion of the integrand in Eq. (9.37) in a series which has been used previously cannot be applied in this case because the function $x_{10}x_{21}\exp(-2x_1h_1)$ in Eq. (9.38) tends to unity when $x \to 0$ if $k_2 = 0$, and therefore, the corresponding series diverges. In order to avoid this problem, we describe another approach which is illustrated for the case of a two-layered medium. As before, we write the vector potential as the sum of two integrals:

$$A_z^* = \frac{i\omega\mu_0 M}{2\pi}\left[\int_0^{m_0}\frac{mJ_0(mr)\mathrm{d}m}{m + m_1/R_2} + \int_{m_0}^{\infty}\frac{mJ_0(mr)\mathrm{d}m}{m + m_1/R_2}\right], \tag{9.71}$$

where

$$R_2 = \coth\left(m_1 H_1 + \coth^{-1}\frac{m_1}{m}\right)$$

As we know, the second integral can be represented as a series containing only even powers of k_1; that is, integer powers of ω, since $m_0^2 \gg k_1^2$. Because the range of integration of the inner integral is small, the function m_1/R_2 can be expanded in a Maclaurin series in power of m:

$$\frac{m_1}{R_2} \approx -ik_1\frac{1}{R_2(0)} + \frac{ik_1 m}{R_2^2(0)}\frac{\partial R_2}{\partial m} + \cdots, \tag{9.72}$$

since

$$m_1(0) = \left(-k_1^2\right)^{1/2} = -ik_1$$

Here $R_2(0) = \coth(-ik_1H_1)$ because

$$\lim \coth^{-1}\frac{m_1}{m} = 0 \quad \text{if } m \to 0$$

It should be clear that

$$\frac{\partial R_2}{\partial m} = -\frac{1}{\sinh^2 x}\frac{\partial x}{\partial m},$$

where

$$x = m_1 H_1 + \coth^{-1}\frac{m_1}{m}$$

and

$$\frac{\partial x}{\partial m} = \frac{mH_1}{m_1} + \frac{1}{1 - m_1^2/m^2}\frac{\partial}{\partial m}\frac{m_1}{m} = \frac{mH_1}{m_1} + \frac{m^2}{k_1^2}\frac{\partial}{\partial m}\frac{m_1}{m}$$

and

$$\frac{\partial}{\partial m}\frac{m_1}{m} = \frac{m}{m_1}m^{-1} - \frac{m_1}{m^2} = \frac{m^2 - m_1^2}{m_1 m^2} = \frac{k_1^2}{m_1 m^2}$$

or

$$\frac{\partial x}{\partial m} = \frac{mH_1}{m_1} + \frac{1}{m_1}$$

Inasmuch as

$$\sinh^2 x = \frac{1}{\coth^2 x - 1}$$

we have

$$\frac{\partial R_2(0)}{\partial m} = \frac{1}{ik_1}\left[\coth^2(-ik_1H_1) - 1\right]$$

and

$$\frac{m}{R_2} = -\frac{ik_1}{R_2(0)} + \frac{m}{R_2^2(0)}\left[R_2^2(0) - 1\right]$$

and

$$m + \frac{m_1}{R_2} = \frac{-ik_1}{R_2(0)} + m\left[2 - R_2^{-2}(0)\right]$$

Therefore, the inner range of integration can be written in the form

$$\int_0^{m_0} \frac{mJ_0(mr)\mathrm{d}m}{a + bm}, \tag{9.73}$$

where a and b are coefficients depending on the parameters of the medium and frequency. It should be clear that this representation does not change for N-layered medium. Now expanding the Bessel function $J_0(mr)$ in power series in mr, we represent the integral in Eq. (9.73) as a sum of tabulated integrals, which have the form

$$T_n = \int_0^{m_0} \frac{m^n}{a + bm}\mathrm{d}m$$

Having taken these integrals, it is a simple matter to expand them in a series with respect to the parameter $p = r/\delta_1$, which contains only integer powers of ω and logarithmic terms of ω. Applying the known approach of expansion of the integral from m_0 and infinity, we can form a series which describes the electromagnetic field on the surface of a layered medium when the basement is an insulator. Its essential feature is the fact that it does not contain terms with fractional powers of ω. Then we have

$$\begin{aligned}
InB_z^{s*} &= \frac{\mu_0 M}{4\pi r^3}\left[\mathrm{d}_1 p^4 + \mathrm{d}_2 p^6 + \mathrm{d}_3 p^8 \ln p + \mathrm{d}_4 p^8 + \cdots\right] \\
QB_z^{s*} &= \frac{\mu_0 M}{4\pi r^3}\left[n_1 p^2 + n_2 p^6 \ln p + n_3 p^6 + n^4 p^8 + \cdots\right] \\
InB_r^* &= \frac{\mu_0 M}{4\pi r^3}\left[l_1 p^4 + l_2 p^8 \ln p + \cdots\right] \\
QB_r^* &= \frac{\mu_0 M}{4\pi r^3}\left[f_1 p^2 + f_2 p^6 + f_3 p^8 + \cdots\right]
\end{aligned} \tag{9.74}$$

$$InE_\varphi^* = \frac{\omega\mu_0 M}{4\pi r^2}\left[c_1 p^4 + c_2 p^8 \ln p + c_3 p^8 + \cdots\right]$$

$$QE_\varphi^* = \frac{\omega\mu_0 M}{4\pi r^2}\left[u_1 p^6 + u_2 p^8 + u_3 p^{10} \ln p + \cdots\right],$$

Here, $p = r/\delta_1$, δ_1 is the skin depth in the upper layer, and d_i, n_i, l_i, f_i, u_i are coefficients which depend on the electric properties of a medium and the distance r. For instance, for the quadrature component of the magnetic field QB_z^* and the in-phase component of the electric field on the surface of a two-layered medium, we have

$$QB_z^* = \frac{\mu_0 M}{4\pi r^3}\left[\frac{\omega\mu_0\gamma_1 r^2}{4} - \frac{1}{4}\frac{\omega\mu_0\gamma_1 r^2}{\left(r^2+4H_1^2\right)^{1/2}} + \left(\frac{\omega\mu_0 S_1 r}{2}\right)^3 \ln\frac{\omega\mu_0 S_1 r}{2} + \cdots\right],$$

and

$$InE_\varphi^* = \frac{\mu_0 M}{4\pi r^2}\left\{\frac{\omega\mu_0\gamma_1 r^2}{4} - \frac{\omega\mu_0\gamma_1 r^2}{4}\,\frac{\left[\left(4H_1^2+r^2\right)^{1/2} - 2H_1\right]}{r}\right.$$

$$\left. + \frac{1}{2}\left(\frac{\omega\mu_0 S_1 r}{2}\right)^3 \ln\frac{\omega\mu_0 S_1 r}{2} - \ldots\right\}, \quad \text{if } \gamma_2 = 0, \frac{r}{\delta_1}\to 0, \frac{H_1}{\delta_1}\to 0 \tag{9.75}$$

where $S_1 = \gamma_1 H_1$ is the conductance of the upper layer. It is not difficult to generalize this result for N-layered medium, introducing geometric factors of layers and replacing the conductance S_1 by the total conductance S equal to

$$S = \sum_{i=1}^{N-1} \gamma_i H_i$$

Later it will be shown that the third term in Eq. (9.75) defines the late stage of the transient field, which plays an important role in an interpretation of electromagnetic soundings and profiling since this stage allows one to evaluate the total conductance of a resistivity structure using the asymptotic formulas of the field only.

9.7. THE FIELD ON THE SURFACE OF A LAYERED MEDIUM AT THE WAVE ZONE

Now we suppose that the separation between the magnetic dipole and the observation point r is greater than the wave length,

$$r/\lambda > 1, \tag{9.76}$$

where λ is some parameter, depending on frequency and geoelectric section of a medium.

First of all, let us assume that the layers and the basement are characterized by finite values of resistivity.

9.7.1. Derivation of Asymptotic Formulas (the First Approach)

Let us proceed from the known equations of the field on the Earth's surface,

$$\begin{aligned}
A_z^* &= \frac{i\omega\mu_0 M_0}{2\pi}\int\limits_0^\infty \frac{m}{m+m_1/R_N} J_0(mr)dm, \\
B_z^* &= \frac{\mu_0 M_0}{2\pi}\int\limits_0^\infty \frac{m^3}{m+m_1/R_N} J_0(mr)dm \\
B_r^* &= \frac{\mu_0 M_0}{2\pi}\int\limits_0^\infty \frac{m_1}{R_N}\frac{m^2}{m+m_1/R_N} J_1(mr)dm \\
E_\varphi^* &= \frac{i\omega\mu_0 M_0}{2\pi}\int\limits_0^\infty \frac{m^2}{m+m_1/R_N} J_1(mr)dm,
\end{aligned} \tag{9.77}$$

and pay attention to the asymptotic values of the field when the distance r unlimitedly increases. As follows from the theorem, proved by A. Tikhonov, in this case the integrals in Eq. (9.77) are defined by extremely small values of variable of integration m:

$$\lim \int\limits_0^\infty \Phi(m)J_i(mr)dm \rightarrow \int\limits_o^\infty \Phi(0)J_i(mr)dm \quad \text{if } r\rightarrow\infty \tag{9.78}$$

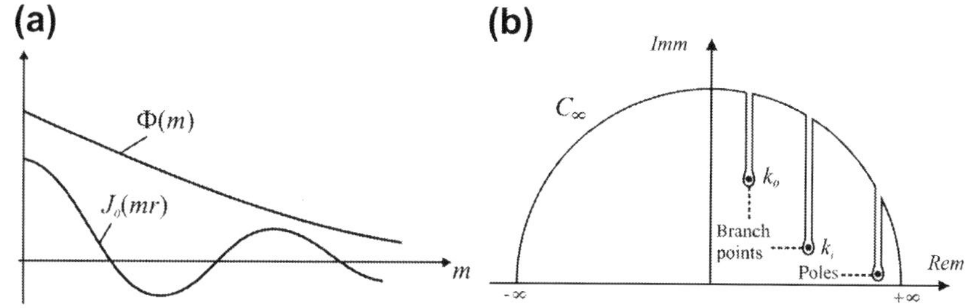

Figure 9.3 (a) Integrand in Eq. (9.78) as a product of two functions. (b) Integration in the complex plane of m.

In fact, in accordance with these expressions, the field can be considered to be the infinite sum of cylindrical harmonics with a continuous spectrum, where m plays a role of the spatial frequency. It is clear that the larger the value of m, the more rapidly the corresponding harmonic of the field will change. For instance, studying the field relatively far away from the source, it is natural to think that the field varies rather slowly. Correspondingly, one can expect that at such distances, the field is mainly defined by small values of m. This expectation also follows from Eq. (9.77) since with an increase of the distance, the integral is mainly determined by small values of m because of the oscillating behavior of Bessel function for large arguments. Indeed, the integrand is the product of two functions $\Phi(m)$ and $J_{0,1}(mr)$: one of them changes relatively slowly while Bessel functions oscillate very rapidly, if $mr > 1$. In this case, as is seen from Figure 9.3(a), integrals over neighboring periods are almost equal by magnitude but have opposite signs, and they cancel each other.

This means that the main contribution to the integral comes from the initial part of integration where $mr < 1$; that is, m tends to zero. Taking into account this fact, we expand the functions $\Phi(m)$ in Eq. (9.77) in a power series in m and keep only the first term in this expansion:

$$m + m_1/R_N \approx m - ik_1/R_N(0) \tag{9.79}$$

Here,

$$R_N(0) = \lim R_N \quad \text{as } m \to 0$$

For example,

$$R_2(0) = \coth\left(-ik_1H_1 + \coth^{-1}\frac{k_1}{k_2}\right)$$

$$= \coth\left[-ik_1H_1 + \coth^{-1}\left(\frac{\rho_2}{\rho_1}\right)^{1/2}\right]$$

and

$$R_3(0) = \coth\left(- ik_1 H_1 + \coth^{-1}\left(\left(\frac{\rho_2}{\rho_1}\right)^{1/2} \coth\left[- ik_2 H_2 + \coth^{-1}\left(\frac{\rho_3}{\rho_2}\right)^{1/2}\right]\right)\right), \tag{9.80}$$

and so on. Whence,

$$\frac{1}{m + m_1/R_N} \approx -\frac{R_N(0)}{ik_1}\frac{1}{1 - R_N(0)\left(m/_{ik_1}\right)} \approx -\frac{R_N(0)}{ik_1}\left(1 + \frac{R_N(0)}{ik_1}m + \cdots\right) \tag{9.81}$$

Note that the equality $m_i = -ik_i$ takes place because the condition $\mathrm{Re}\, m_i > 0$ should always be valid. Substituting this series into Eq. (9.77), we obtain the following expression for the vertical component of the magnetic field:

$$B_z^* = \frac{\mu_0 M_0}{2\pi}\left[-\frac{R_N}{ik_1}\int\limits_0^\infty m^3 J_0(mr)\mathrm{d}m + \frac{R_N^2}{k_1^2}\int\limits_0^\infty m^4 J_0(mr)\mathrm{d}m\right] \tag{9.82}$$

In spite of the fact that both integrals diverge, they can be evaluated from the known equality:

$$\frac{1}{(r^2 + z^2)^{1/2}} = \int\limits_0^\infty \exp(- m|z|)J_0(mr)\mathrm{d}m$$

Differentiating the latter with respect to z and letting $z = 0$, we obtain expressions of integrals in Eq. (9.82). This gives

$$B_z^* = \frac{9\mu_0 M_0 R_N^2(0)}{2\pi r^5 k_1^2} = \frac{9\mu_0 M_0 R_N^2(0)}{2\pi r^3}\frac{1}{k_1^2 r^2} \tag{9.83}$$

By analogy,

$$\begin{aligned} B_r^* &= \frac{\mu_0 M_0}{2\pi}\int\limits_0^\infty \frac{m^2 J_1(mr)}{1 + \left(R_N(0)/_{ik_1}\right)m}\mathrm{d}m \\ &= \frac{\mu_0 M_0}{2\pi}\int\limits_0^\infty m^2 J_1(mr)\mathrm{d}m + \frac{\mu_0 M_0 R_N(0)}{2\pi i k_1}\int\limits_0^\infty m^3 J_1(mr)\mathrm{d}m \\ &= -\frac{\mu_0 M_0 R_N(0)}{2\pi k_1}\frac{\partial}{\partial r}\frac{1}{r^3} = \frac{3\mu_0 M_0 R_N(0)}{2\pi i k_1 r^4} \end{aligned} \tag{9.84}$$

Also for the electric field, we have

$$E_\varphi^* = i\omega\mu_0\frac{M_0}{2\pi}\frac{R_N(0)}{ik_1}\left[\int\limits_0^\infty m^2 J_1(mr)\mathrm{d}m - \frac{R_N}{ik_1}\int\limits_0^\infty m^3 J_1(mr)\mathrm{d}m\right]$$

or

$$E_\varphi^* = -\frac{i\omega\mu_0 M_0}{2\pi k_1^2}\frac{\partial}{\partial r}\int\limits_0^\infty m^2 J_1(mr)\mathrm{d}m = -\frac{3M_0\rho_1}{2\pi r^4}R_N^2(0) \tag{9.85}$$

Thus, the electromagnetic field observed at the surface of a layered medium for large values of the parameter r/λ is described by the expressions

$$\begin{aligned} B_z^* &= \frac{9\mu_0 M_0 R_N^2(0)}{2\pi r^3}\frac{1}{k_1^2 r^2}, \quad B_r^* = \frac{3\mu_0 M_0 R_N(0)}{2\pi i k_1 r^4}, \\ E_\varphi^* &= -\frac{3M_0\rho_1 R_N^2(0)}{2\pi r^4}, \quad \text{if } r/\lambda > 1 \end{aligned} \tag{9.86}$$

where $R_N(0)$ is a function of the geoelectric parameters of the medium and frequency.

Before we continue, let us make several comments. (1) The amazing simplicity of these equations was one of the reasons why the first modification of the frequency soundings was based on measuring the field far away from the current source. (2) In deriving formulas, it was assumed that $k_i > m$. This means that in the limit when r tends to infinity in every layer, we have $r/\lambda_i > 1$. In other words, in the limit, conditions r/λ and r/λ_i are equivalent to each other. (3) Inasmuch as the distance from an observation point to the source exceeds the parameter λ, it is natural to call this range *the wave zone.* (4) With a decrease of a frequency or a conductivity of a medium, the wave zone behavior is observed at greater distances from the source. (5) Also with an increase of the distance r, this asymptotic behavior takes place at greater values of the wavelength. (6) Earlier we demonstrated that the terms of the low-frequency spectrum, which contain fractional powers of ω and $\ln\omega$, are defined by small values of variable of integration, m, as well as the field in the wave zone. As will be shown later, this coincidence has interesting consequences when we consider the behavior of the transient field.

9.7.2. Behavior of the Field when $r/\lambda > 1$ (Wave Zone)

It should be obvious that for a uniform half space ($H_1 \to \infty$), the function $R_N(0)$ is equal to unity and Eq. (9.86) is the same as those for a uniform half space with resistivity ρ_1. The function $R_N(0)$ represents the influence of the layers, and correspondingly, it can be described in terms of the ratio of the field in a horizontally layered medium to that in a uniform half space:

$$\frac{B_z^*}{B_z^*(\rho_1)} = \frac{E_\varphi^*}{E_\varphi^*(\rho_1)} = R_N^2(0) \quad \text{and} \quad \frac{B_r^*}{B_r^*(\rho_1)} = R_N(0) \tag{9.87}$$

We can see that the horizontal component of the field B_r^* is less sensitive to a change in the geoelectric parameters than either of components (E_φ^*, B_z^*) when the parameter r/λ is large. However, it should be noted that this conclusion may not hold for other forms of the field excitation or for other models of a medium. As was also the case for a uniform half space, the electromagnetic field in this range of parameters has some features which are similar to the plane wave traveling downward. In particular, its behavior as function of the depth in the Earth is characterized by the exponential terms $\exp(\pm ikz)$. Moreover, as will be shown later, the ratio of the tangential components of the electric and magnetic fields is equal up to a constant to the impedance of a plane wave field propagating vertically into an N-layered medium. In accordance with Eq. (9.86), the components B_r^* and E_φ^* vary with the distance r in the same manner and their ratio is

$$\frac{E_\varphi^*}{B_r^*} = \frac{Z_N}{\mu_0} = -\frac{ik_1\rho_1}{\mu_0} R_N(0) = -\frac{Z_1}{\mu_0} R_N(0), \tag{9.88}$$

where

$$Z_N = -ik_1\rho_1 R_N(0) \tag{9.89}$$

and

$$R_N(0) = \coth\left[-ik_1H_1 + \coth^{-1}\left(\frac{\rho_2}{\rho_1}\right)^{½} \coth(-ik_2H_2 + \cdots)\right],$$

but $Z_1 = -ik_1\rho_1$ is the impedance at the surface of a uniform half space. In contrast to the behavior of a plane wave propagating vertically into a conducting medium along the z-axis, the field generated by a dipole changes with the distance on the Earth's surface as well as inside the medium; that is, we deal with the quasi-plane wave. Also, the vertical component of the magnetic field is not zero though $|B_z^*| < |.B_r^*|$, and their ratio

$$\frac{B_z^*}{B_r^*} = \frac{3R_N(0)}{k_1 r}$$

decreases with an increase of r.

The behavior of the field over this range of distances is characterized by some essential features of the propagation of the field through free space and a layered medium. When the parameter r/λ is large, absorption of the field in a conducting medium layer can be relatively strong. Therefore, the signal arrives at the observation point mainly through free space, but of course, it depends on parameters of a medium. At the same time at distances from the dipole to the receiver, which are not large enough to meet the conditions for the asymptotic behavior, the Earth's surface may serve to guide the propagation of the electromagnetic field. Here, the tangential component of the Poynting vector may be significant. But at the range of large parameter $r/\lambda \gg 1$, the normal component of the Poynting vector is dominant, and the latter is mainly directed downward. Now let us continue to discuss the field as given by Eq. (9.86). These equations are remarkably simple. In fact, the right-hand side of each expression can be represented as the product of two functions. One of these functions depends on the dipole moment M_0, the distance r, the frequency ω, and the resistivity of the uppermost layer ρ_1. This function describes the corresponding components of the field in a uniform half space, when the parameter r/λ_1 is greater than unity. The second term in this product, $R_N(0)$ or $R_N^2(0)$, is a function which depends only on the parameters of the medium and the frequency but does not depend on the dipole moment or the distance r. Taking into account the fact that some geophysical methods, such as the magnetotelluric soundings, are based on measuring the field in the range of large parameters, let us study in some detail the behavior of the function $R_N(0)$ and with this purpose in mind consider several cases.

Case 1. The parameter H_1/λ_1 is large.

From Eq. (9.89), we see that when the parameter H_1/λ_1 increases, the function $R_N(0)$ tends to unity:

$$\lim R_N(0) \to 1 \text{ if } H_1/\lambda_1 \to \infty \tag{9.90}$$

This indicates that as the skin depth δ_1 decreases, induced currents concentrate mainly in the uppermost layer, and the field corresponds to that of a uniform half space having resistivity ρ_1.

Case 2. The parameters H_i/λ_i are small.

Now consider the opposite case and suppose that the skin depth in each layer increases and therefore the magnitude of $k_i H_i$ goes to zero. Then, from Eq. (9.89), we have

$$\lim R_2(0) = \left(\frac{\rho_2}{\rho_1}\right)^{1/2} = \frac{k_1}{k_2}, \quad \lim R_3(0) = \left(\frac{\rho_3}{\rho_1}\right)^{1/2} = \frac{k_1}{k_3},$$

and in general

$$\lim R_N(0) = \left(\frac{\rho_N}{\rho_1}\right)^{1/2} \quad \text{if } |k_i H_i| \to 0 \tag{9.91}$$

Substituting this last equation into Eq. (9.86), we obtain

$$B_z^* = \frac{9\mu_0 M_0}{2\pi r^3}\frac{1}{k_N^2 r^2}, \quad B_r^* = \frac{3\mu_0 M_0}{2\pi i k_N r^4}, \quad E_\varphi^* = -\frac{3M_0 \rho_N}{2\pi r^4} \tag{9.92}$$

Thus, when the skin depth in each of the layers increases, the electromagnetic field approaches that for a uniform half space with the resistivity of the lowermost medium ρ_N. In this case, all layers above the basement become transparent.

Case 3. The highly resistive basement and skin depth in each layer is greater than its thickness.

First consider the two-layered medium with sufficiently resistive basement $\rho_2 \gg \rho_1$. Then we can write

$$R_2(0) = \coth\left(-ik_1 H_1 + \coth^{-1}\frac{k_1}{k_2}\right) = \frac{1 + \left(k_1/k_2\right)\coth(-ik_1 H_1)}{\coth(-ik_1 H_1) + k_1/k_2}$$

Since the skin depth is greater than the layer thickness, we have

$$R_2(0) \approx \frac{1 - \left(k_1/k_2\right)\frac{1}{ik_1 H_1}}{\frac{k_1}{k_2} - \frac{1}{ik_1 H_1}} = \frac{k_1 - ik_1 k_2 H_1}{k_2 + \omega\mu_0 S_1} \quad \text{if } |k_1 H_1| \ll 1$$

Bearing in mind that $\rho_2 \gg \rho_1$, we finally obtain

$$R_2(0) = \frac{k_1}{k_2 + \omega\mu_0 S} \quad \text{if } \rho_2 \gg \rho_1,\ |k_1 H_1| < 1 \tag{9.93}$$

where $S = \gamma_1 H_1$ is the conductance of the upper layer. Using the same approach, we obtain the following expression for R_3:

$$R_3(0) = \frac{k_1}{k_3 + \omega\mu_0 S} \quad \text{if } |k_i H_i| < 1, \rho_1 \ll \rho_3, \rho_2 \ll \rho_3 \tag{9.94}$$

Here, $S = S_1 + S_2 = \gamma_1 H_1 + \gamma_2 H_2$. This result can be easily generalized to the case of an N-layered medium:

$$R_N(0) = \frac{k_1}{k_N + \omega\mu_0 S}, \quad \text{if } H_i/\lambda_i < 1 \tag{9.95}$$

and the basement is relatively resistive.

Substituting this last result into Eq. (9.86), we obtain asymptotic expressions for the field, describing the low-frequency part of the spectrum in the wave zone:

$$B_z^* = \frac{9\mu_0 M_0}{2\pi r^5} \frac{1}{(\omega\mu_0 S + k_N)^2}, \quad B_r^* = \frac{3i\mu_0 M_0}{2\pi r^4} \frac{1}{(\omega\mu_0 S + k_N)}$$

and

$$E_\varphi^* = -\frac{3i\omega\mu_0 M_0}{2\pi r^4} \frac{1}{(\omega\mu_0 S + k_N)^2} \quad \text{if } H_i/\delta_i < 1 \tag{9.96}$$

and the basement is sufficiently resistive. Here,

$$S = \sum_{i=1}^{N-1} S_i = \sum_{i=1}^{N-1} \gamma_i H_i$$

is the total conductance of the layers. As follows from Eq. (9.96), the field is defined by two parameters of the geoelectric section, namely, the total conductance of the section S and the conductivity of the underlying medium γ_N. It is clear that within the range of large parameters, where Eq. (9.96) correctly describes the field, one can distinguish three intervals and each with a characteristic behavior. If the frequency is not sufficiently low that $\omega\mu_0 S \gg |k_N|$, then the field is defined by the longitudinal conductance S, and

$$B_z^* = \frac{9\mu_0 M_0}{2\pi r^5} \frac{1}{\omega^2\mu_0^2 S^2}, \quad B_r^* = \frac{i3\mu_0 M_0}{2\pi r^4} \frac{1}{\omega\mu_0 S}, \quad E_\varphi^* = -\frac{3iM_0}{2\pi r^4\, \omega\mu_0 S^2} \tag{9.97}$$

It is natural that this range of frequencies is usually called the S-zone. With decreasing frequency, both terms in the dominator of Eq. (9.96) become comparable, and correspondingly, the second interval is controlled by both parameters S and ρ_N. With a further decrease in frequency, the term $\omega\mu_0 S$ can be neglected, and the field approaches the value for a uniform half space with resistivity ρ_N.

Case 4. Very conductive basement and the skin depth in each layer is more than its thickness.

Next we study the case when the lowermost medium is relatively conductive and frequency is sufficiently low. Starting with a two-layered medium and making use of Eq. (9.80), we have

$$R_2(0) \approx \frac{1 - 1/ik_2H_1}{-1/ik_1H_1} = k_1\left(\frac{1}{k_2} - iH_1\right), \tag{9.98}$$

since the basement is more conductive and

$$\left|\frac{k_1}{k_2}\right| \ll \frac{1}{|k_1|H_1}$$

A similar expression is obtained for the function $R_3(0)$:

$$R_3(0) \approx k_1\left[\frac{1}{k_3} - i(H_1 + H_2)\right]$$

$$\text{if} \quad H_1/\delta_1 < 1, \quad H_2/\delta_2 < 1, \quad \rho_3/\rho_2 \ll 1, \quad \rho_3/\rho_1 \ll 1$$

By analogy for N-layered medium, we have

$$R_N \approx k_1\left(\frac{1}{k_N} - iH\right), \tag{9.99}$$

where $H = \sum H_i$ is the total thickness of layers. Substituting this expression for R_N into Eq. (9.86), we arrive at equations for the field, when $r/\lambda_1 > 1$ and the skin depth in each layer is greater than its thickness:

$$B_z^* = -\frac{9\mu_0 M_0}{2\pi r^5}\left(\frac{1}{k_N} - iH\right)^2, \quad B_r^* = \frac{3\mu_0 M_0}{2\pi r^4}\left(\frac{1}{k_N} - iH\right)$$

and

$$E_\varphi^* = \frac{3i\omega\mu_0 M_0}{2\pi r^4}\left(\frac{1}{k_N} - iH\right)^2 \tag{9.100}$$

Thus, over this range, the field is defined by two parameters of the medium, namely, the basement resistivity ρ_N and the total thickness of the layers above the basement H. Again, it is proper to distinguish here three intervals of frequencies. In particular, if $\left|k_N^{-1} < H\right|$, the field is determined by the total thickness of the layers only. This frequency interval is often called the H-zone. With a decrease of frequency, both geoelectric parameters make an influence on the field and, finally, when frequency is sufficiently low, the field coincides in the wave zone with that of a uniform half space with resistivity of the basement.

9.7.3. The Field in the Wave Zone when the Basement is an Insulator $\rho_N \to \infty$

In deriving the formulas for the wave zone when the ratio r/λ is large, we assumed that the underlying medium (basement) has a finite resistivity. Now we investigate the behavior of the electromagnetic field in the wave zone when the basement is insulator. As an example, consider a two-layered medium. By definition,

$$R_2 = \coth\left[m_1 H_1 + \coth^{-1}\frac{m_1}{m}\right],$$

and when $m \to 0$, we have

$$\frac{m_1}{R_2} \approx m_1 \tanh\left[(-ik_1 H_1) + \tanh^{-1}\left(-\frac{m}{ik_1}\right)\right]$$

$$= ik_1 \frac{\tanh(ik_1 H_1) + m/_{ik_1}}{1 + m/_{ik_1} \tanh(ik_1 H_1)} \approx ik_1 \tanh ik_1 H_1 + m\left(1 - \tanh^2 ik_1 H_1\right)$$

Whence,

$$m + \frac{m_1}{R_2} \approx ik_1 \tanh ik_1 H_1 + m\left(2 - \tanh^2 ik_1 H_1\right)$$

and

$$\frac{1}{m + m_1/_{R_2}} \approx \frac{\coth ik_1 H}{ik_1} + \frac{m}{k_1^2}\left(2\coth^2 ik_1 H_1 - 1\right)$$

Repeating the procedure for deriving the field components that was carried out, we obtain

$$B_z^* = \frac{9\mu_0 M_0}{2\pi k_1^2 r^5}\left(2\coth^2 ik_1H_1 - 1\right), \quad B_r^* = \frac{3\mu_0 M_0}{2\pi ikr^4}\left(\coth ik_1H_1\right)$$

and

$$E_\varphi^* = -\frac{3M_0\rho_1}{2\pi r^4}\left(2\coth^2 ik_1H_1 - 1\right) \tag{9.101}$$

At high frequencies $\left(H_1/\delta_1 > 1\right)$, as a result of the skin effect, the electromagnetic field is the same as for a uniform half space with resistivity ρ_1. In the opposite case, for relatively low frequencies $\left(r/\lambda_1 > 1 \text{ and } H_1/\delta_1 < 1\right)$ replacing coth x by the first term in its expansion $1/x$, we have

$$B_z^* = \frac{9\mu_0 M_0}{\pi r^5}\frac{1}{\omega^2\mu_0^2 S^2}, \quad B_r^* = \frac{i3\mu_0 M_0}{2\pi r^4}\frac{1}{\omega\mu_0 S}, \quad E_\varphi^* = -\frac{3iM_0}{\pi r^4\,\omega\mu_0 S^2} \tag{9.102}$$

Comparison with Eq. (9.97) indicates that the field, described by Eq. (9.102), results from the propagation of the field principally along two paths: (1) in free space along the earth's surface and (2) vertically downward into the conducing medium along the z-axis, then horizontally through the insulating basement and vertically upward to the observation point. In fact, in accordance with Eq. (9.102) at the low frequencies $\left(H_1/\delta_1 < 1\right)$, the components B_z and E_φ are twice as large as the corresponding components of the field arriving through free space because attenuation in the conducting layer is negligible when $H_1/\delta_1 < 1$. This analysis allows us to arrive at some conclusion about the field behavior when the basement has a relatively high resistivity ρ_2. It is obvious that if the skin depth in the basement δ_2 is significantly greater than the separation r, the field attenuation in the basement is small, and therefore, the field is well described by Eq. (9.102), which are derived for the limiting case $\rho_2 = \infty$. In other words, these equations characterize the field when the condition

$$\lambda_1 < r < \lambda_2$$

holds. It is clear that with increasing separation r due to attenuation of the field in the resistive basement, the field observed at the Earth's surface approaches to that described by Eq. (9.86).

9.8. THE SECOND APPROACH OF DERIVING THE ASYMPTOTIC FORMULAS FOR WAVE ZONE

Now we apply a different method to derive the asymptotic formulas for the wave zone ($r > \lambda$), which is based on the use of Cauchy's theorem.

9.8.1. Deformation of the Contour Integration

In accordance with this theorem, the integral along a closed path is equal to zero, if inside an area surrounded by the contour C, the function $f(m)$ is analytical; that is singularities, such as branch points and poles, are absent:

$$\oint_C f(m)\mathrm{d}m = 0 \tag{9.103}$$

It is essential that equality (9.103) is independent on the path C, as soon as inside of each such area function remains analytical. In other words, we can deform the closed path as we like, provided that this contour does not intersect singularities. This result can be formulated differently, namely, the integrals between two points along different paths remain the same if between these paths singularities are absent. In accordance with Eqs (9.24) and (9.25), the integration takes place along the real axis of m from zero to infinity. Our first goal is to deform this contour of integration in such a way that allows us to obtain the simple expressions of the field in the wave zone when $|kr| \gg 1$. Here k is parameter, depending on the resistivity structure and frequency. First of all, the kernel function $\Phi(m_i)$ has branch points $m = k_i$, where corresponding radicals are equal to zero and the function $\Phi(m_i)$ is not analytical. For this reason performing deformation of the initial contour of integration, we have to choose such path C, that these points are not situated inside the area enclosed by this closed contour. In order to perform a deformation of the path of integration, it is useful first of all to change the path of integration and make it along the whole real axis of m. With this purpose in mind, the following equality is used:

$$J_0(mr) = \frac{1}{2}\left[H_0^{(1)}(mr) + H_0^{(2)}(mr)\right], \tag{9.104}$$

where $H_0^{(1)}(mr)$ and $H_0^{(2)}(mr)$ are Hankel functions of the first and second kinds. As is known, their asymptotic behavior when $mr > 1$ is

$$H_0^{(1)}(mr) \approx \left(\frac{2}{\pi mr}\right)^{1/2} \exp\left[i\left(mr - \pi/4\right)\right],$$

and

$$H_0^{(2)}(mr) \approx \left(\frac{2}{\pi mr}\right)^{1/2} \exp\left[-i\left(mr - \pi/4\right)\right] \tag{9.105}$$

This means that if $|mr| \to \infty$, these functions tend to zero in the upper ($Im m > 0$) and the lower ($Im m < 0$) parts of the complex plane of m, respectively. Making use of Eq. (9.104), the integral in Eq. (9.24) can be represented as

$$I = \frac{1}{2}\left[\int_0^\infty \Phi(m)H_0^{(1)}(mr)\mathrm{d}m + \int_0^\infty \Phi(m)H_0^{(2)}(mr)\mathrm{d}m\right] \tag{9.106}$$

Bearing in mind that in the first quadrant of the complex plane of m, there are branch points, and the function $H_0^{(1)}(mr)$ tends to zero if $\mathrm{Im}\, m > 0$ and $|m| \to \infty$, we represent Eq. (9.106) as

$$I = \frac{1}{2}\left[\int_0^\infty \Phi(m)H_0^{(1)}(mr)\mathrm{d}m - \int_0^\infty \Phi(m)H_0^{(1)}(-mr)\mathrm{d}m\right], \tag{9.107}$$

since

$$H_0^{(2)}(mr) = -H_0^{(1)}(-mr)$$

Thus,

$$I = \frac{1}{2}\left[\int_0^\infty \Phi(m)H_0^{(1)}(mr)\mathrm{d}m + \int_0^{-\infty} \Phi(-m)H_0^{(1)}(mr)\mathrm{d}m\right]$$

As is seen from Eq. (9.24), function Φ describing the field has the following property:

$$\Phi(m) = -\Phi(-m)$$

Correspondingly, the integral I can be written in the form

$$I = \frac{1}{2}\int_{-\infty}^{\infty} \Phi(m)H_0^{(1)}(mr)\mathrm{d}m, \tag{9.108}$$

and the first task is accomplished. As follows from Cauchy theorem, we can imagine infinite number of paths with the same terminal points $(-\infty, \infty)$, in the upper plane of the complex variable m along which the integral I remains the same, but we choose the path shown in Figure 9.3(b). It consists of the semicircle C_∞ of an infinitely large radius, pairs of hyperbolae $Rem_i = 0$ and semicircles surrounding branch points with infinitely small radii. Besides, in general, there are paths around poles, but their contribution will be discussed later. Due to the behavior of the Hankel's function, the integral along C_∞ is equal to zero (Jordan's lemma). Since the path length in the vicinity of branch points tends to zero, integrals along these circles are infinitely small. Thus, integration is reduced to only the lines, called *branch cuts*. Correspondingly, ignoring at the beginning by influence of poles, we have

$$I = \sum_{i=1}^{n} I_i, \tag{9.109}$$

where

$$I_i = \frac{1}{2} \int_{C_i} \Phi(m) H_0^{(1)}(m) \mathrm{d}m,$$

Each path includes two lines located infinitely close to each other, and integration along them is carried out in the opposite directions. Note that when integration is performed around the branch point, corresponding radical changes sign. In fact, the argument of the function $m^2 - k_i$ continuously varies, and along the circle around a branch point k_i changes by 2π. Therefore, the argument of the radical varies by π, that is, equivalent to a change of sign. For this reason, we have

$$I_i = \frac{1}{2}\left[\int_{i\infty}^{k_i} \Phi(m, m_1 \ldots m_i \ldots m_n) H_0^{(1)}(mr) \mathrm{d}m + \int_{k_i}^{i\infty} \Phi(m, m_1 \ldots - m_i \ldots m_n) H_0^{(1)}(mr) \mathrm{d}m \right]$$

or

$$I_i = \frac{1}{2} \int_{k_i}^{i\infty} [\Phi(\ldots - m_i \ldots) - \Phi(\ldots m_i \ldots)] H_0^{(1)}(mr) \mathrm{d}m \tag{9.110}$$

9.8.2. Evaluation of Integrals along Branch Cuts and Poles

Before we derive an asymptotic expression for the field, associated with the branch points, it is convenient to consider again a uniform medium with the wave number k_0 of the upper space. As we know, the primary field is expressed in terms of the Sommerfeld integral:

$$\int_0^\infty \frac{m}{m_0}\exp(-m_0|z|)J_0(mr)\mathrm{d}m = I_0 = \frac{1}{2}\int_{-\infty}^{\infty} \frac{m}{m_0}\exp(-m_0|z|)H_0^{(1)}(mr)\mathrm{d}m$$

that is,

$$\frac{\exp(ik_0R)}{R} = \frac{1}{2}\int_{C_0} \frac{m}{m_0}\exp(-m_0|z|)H_0^{(1)}(mr)\mathrm{d}m$$

or

$$\frac{\exp(ik_0R)}{R} = \frac{1}{2}\int_{k_0}^{i\infty} -\frac{m}{m_0}[\exp(m_0|z|)+\exp(-m_0|z|)]H_0^{(1)}(mr)\mathrm{d}m$$

Thus,

$$\frac{\partial^2}{\partial z^2}\frac{\exp(ik_0R)}{R} = -\int_{k_0}^{i\infty} mm_0H_0^{(1)}(mr)\mathrm{d}m \quad \text{if } z=0 \tag{9.111}$$

Performing differentiation and letting $z=0$, we obtain

$$\int_{k_0}^{i\infty} mm_0H_0^{(1)}(mr)\mathrm{d}m = \frac{1}{r^3}(1-ik_0r)\exp(ik_0r) \tag{9.112}$$

In particular, in the case of the upper half space, where $k_0=\omega/c$, we have

$$\int_{k_0}^{i\infty} mm_0H_0^{(1)}(mr)\mathrm{d}m = \frac{1}{r^3}, \tag{9.113}$$

provided that the product $k_0 r$ is very small. Now we are ready to derive the asymptotical expressions for the field, and for illustration of this approach, consider the vector potential (Eq. 9.24),

$$A_z^* = \frac{i\omega\mu_0 M_0}{2\pi}\frac{1}{2}\int\limits_{-\infty}^{\infty}\frac{mH_0^{(1)}(mr)}{m_0 + m_1/R_N}\mathrm{d}m \tag{9.114}$$

In accordance with Eq. (9.110), the integral along the path C_0 is

$$I_0 = \frac{1}{2}\int\limits_{k_0}^{i\infty} m\left[\frac{1}{-m_0 + m_1/R_N} - \frac{1}{m_0 + m_1/R_N}\right]H_0^{(1)}(mr)\mathrm{d}m \tag{9.115}$$

or

$$I_0 = \int\limits_{k_0}^{i\infty}\frac{mm_0}{m_1^2/R_N^2 - m_0^2}H_0^{(1)}(mr)\mathrm{d}m, \tag{9.116}$$

since R_N is independent on m_0. Taking into account the asymptotic behavior of Hankel's function (Eq. (9.105)), we can say that for large values of the distance r, the integral is defined by mainly the initial part of integration, located in the vicinity of the branch point, while outside of this interval, integrand exponentially decays. Correspondingly, let us neglect by the second term of the dominator $\left(m_0^2 = 0\right)$, while for the first term assume that m_1 equals k_1. Then taking into account Eq. (9.113), in place of Eq. (9.116), we obtain its approximate expression which is valid for sufficiently large distances:

$$I_0 \approx -\frac{1}{k_1^2 r^3}R_N^2(0) \tag{9.117}$$

Here, $R_N(0)$ is the function where variable m is replaced by zero; that is, we assume that $k_n \gg k_0$. Thus, we found a contribution to the vector potential, given by branch point k_0, provided that the separation between the observation point and the current dipole is sufficiently large and an observation is performed in the near zone of the primary field ($k_0 r < 1$). Of course, making use of Eq. (9.112), we can easily apply an appropriate version of Eq. (9.117) to any values of the product $k_0 r$, respectively. Thus, we have for the vector potential (Eq. (9.114)):

$$A_z^* = -\frac{i\omega\mu_0 M_0}{2\pi}\frac{R_N^2(0)}{k_1^2 r^3} \tag{9.118}$$

Before we continue, let us show one interesting feature of this equation. Suppose that the resistivity of the upper layer becomes very large, and correspondingly, $k_1 \to 0$. Then in accordance with Eq. (9.80), we have

$$R_N(0) \to \frac{k_1}{k_2} R_{N-1} \quad \text{and} \quad A_z^* = -\frac{i\omega\mu_0 M_0}{2\pi}\frac{R_{N-1}^2(0)}{k_2^2 r^3},$$

That is, an influence of the upper layer disappears and the wave zone is observed when $|k_2 r| > 1$. Now performing differentiation by r, the electric field is

$$E_\varphi^* = \frac{3\rho_1 M_0}{2\pi r^4} R_N^2(0),$$

that coincides with Eq. (9.85), providing that an underlying medium is conductive. In the same manner, we can derive again expressions for the magnetic field. Next we demonstrate that integrals along branch cuts, corresponding to wave numbers of layers with finite thickness, are equal to zero. In other words, in order to find asymptotic expression of the field, it is sufficient to consider integrals just along two branch cuts, related to points k_0 and k_N, where the latter is the wave number of the underlying medium. As follows from Eq. (9.110), the integral along the n branch cuts is

$$I_n = \frac{1}{2}\int_{k_i}^{i\infty} m\left[\frac{1}{m_0 + m_1^{(-)}/R_N^{(-)}} - \frac{1}{m_0 + m_1/R_N}\right] H_0^{(1)}(mr)\mathrm{d}m \tag{9.119}$$

Here,

$$\begin{aligned} R_N &= \coth\left[m_1 H_1 + \coth^{-1}\frac{m_1}{m_2} R_{N-1}\right] \\ R_{N-1} &= \coth\left[m_2 H_2 + \coth^{-1}\frac{m_2}{m_3} R_{N-2}\right], \end{aligned} \tag{9.120}$$

and $R_N^{(-)}$ differs from R_N by the presence of $-m_i$ instead of m_i. First of all, consider the case when $i = 1$. It is clear that the function R_{N-1} does not contain m_1, and therefore, R_N differs by sign at the opposite branch cuts around the point $m = k_1$. Also $m_1^{(-)} = -m_1$. Whence,

$$m_1^{(-)}/R_N^{(-)} = m_1/R_N, \tag{9.121}$$

and the integral $I_1 = 0$. As concerns the other integrals, when $i > 1$, it is a simple matter to see that in the set (9.120), the radical m_n appears only in the term:

$$\frac{m_{n-1}}{m_n}\coth\left(m_n H_n + \coth^{-1}\frac{m_n}{m_{n+1}} R_{N-n}\right),$$

where R_{N-n} is independent on m_n; that is, equality (9.121) is again satisfied and

$$I_n = 0$$

At the same time, the last integral around the branch point k_N differs from zero. In the case of a conducting underlying medium, this part of the field exponentially decays and its influence at the wave zone is negligible. Until now we focused out attention to the field, associated with branch points, and next it may be useful to draw analogy with propagation of acoustic medium in a layered medium. As is well known, in general, one can observe different acoustic waves, such as reflected, refracted, and channeling waves. Because of the reflection at different boundaries, the reflected waves appear at the observation point, and its asymptotic behavior is defined by the stationary points of the spectrum; that is, by the integrand of equation like in Eq. (9.24). In contrast, a contribution of the reflected electromagnetic wave at an observation point, located at the earth surface and far away from the source, is negligible and it happens due to a strong attenuation. At the same time, the asymptotic behavior of the refracted waves is related to the branch points of the spatial spectrum. By analogy, we may say that Eq. (9.118) describes at the wave zone the lateral electromagnetic field, propagating along the Earth's surface.

Also under certain conditions besides the reflected and refracted acoustic waves, it is also possible to observe the channel waves, propagating inside of a layer. For instance, it may happen when the wave length is twice smaller than the layer thickness, and wave velocity of the surrounding medium is greater than that of the layer. The behavior of these waves is characterized by poles of the spectrum, which are real ($\mathrm{Im}\, m = 0$), provided that there is no transformation of an acoustic energy into heat. The channel wave is always accompanied by an inhomogeneous wave in a surrounding medium, which relatively rapidly decays with the distance from the layer, even when heat loses are absent. For this reason, its contribution to the signal at an observation point located at the Earth's surface can be neglected. Note that if conditions for appearance of the channel waves are absent, the spectrum has complex poles. For instance, it happens when the wave length of the acoustic wave exceeds the thickness of each layer and the destructive interference of waves, reflected from the top and bottom of every layer, takes place. Again by analogy, one can expect that channel electromagnetic waves may arise in a layered medium. Unlike the acoustic case, these poles are always complex since there is always attenuation of an electromagnetic field into heat. This means that an influence of channeling waves on the field at the Earth's surface certainly can be neglected. In other words, we do not need to take into account the contribution of poles even if they are present inside of the contour of integration. Besides the wavelength of the quasi-stationary field, $\lambda = (10\rho T)^{1/2}\ km$ is usually much greater than the thickness of a layer above a basement, and therefore, constructive interference does not occur. Of course, in general, one can imagine a geoelectric section, when channel waves propagate inside of a nonconductive layer, but still due to decaying, an inhomogeneous wave outside its contribution will be still exponentially small.

In conclusion of this section, it may be proper to emphasize again that the wave zone behavior can be observed at any distance from the source, but only at relatively large distance (the far zone), it is possible to carry out the sounding of a resistivity structure. It happens because in such case within a certain range of frequencies, one can satisfy both conditions, namely, the distance r exceeds the wave length, but the latter may be greater than the total thickness of layers above a basement.

Now in the same manner as in the frequency domain, we describe the transient field of the magnetic dipole on the surface of a layered medium and start from the early stage where a behavior is similar to that at the range of large parameters, $r/\lambda > 1$.

9.9. TRANSIENT FIELD AT THE RANGE OF LARGE PARAMETER r/τ AT THE SURFACE OF A LAYERED MEDIUM (WAVE ZONE)

Taking into account that in most applications, the current of the primary source changes almost as the step function or its combination, suppose that the primary field $\mathbf{B}_0$, caused by the current in the vertical magnetic dipole, varies as

$$B_0(t) = \begin{cases} B_0 & t < 0 \\ 0 & t > 0 \end{cases}$$

Then as in the case of a uniform half space, we can distinguish three stages of the transient response caused by induced currents in the medium, namely, (1) the range of large parameters r/τ (early stage), (2) the intermediate stage, and (3) the range of small parameter r/τ (late stage). Here, the parameter τ plays the same role as λ at the frequency domain. In order to study the asymptotic behavior, we will precede as in general case from the Fourier's transform,

$$F(t) = \frac{1}{2\pi} \int_{-\infty}^{\infty} \frac{F(\omega)}{-i\omega} \exp(-i\omega t) d\omega, \tag{9.122}$$

where $F(\omega)$ is the spectrum of the field components; that is, the field caused by sinusoidal current of the source and currents in a surrounding medium, while the spectrum of the unit step function is $-1/_{i\omega}$.

9.9.1. Comparison of Fields in the Ranges of Large Parameters r/δ and r/τ for Uniform Half Space

First let us compare the frequency and transient responses at the surface of a uniform half space when $r/\delta \gg 1$ and $r/\tau \gg 1$. As was shown earlier,

$$E^*_\varphi(\omega) = \frac{3M_0\rho}{2\pi r^4} \exp(i\pi) \quad E_\varphi(t) = -\frac{3M_0\rho}{2\pi r^4}$$

$$B^*_z(\omega) = \frac{9M_0\rho}{2\pi r^5\,\omega} \exp\left(-i\,\pi/_2\right) \quad B_z(t) = \frac{9M_0\rho}{2\pi r^5} t \tag{9.123}$$

$$B^*_r(\omega) = \frac{3\mu_0 M_0 \exp\left(-i\,3\pi/_4\right)}{2\pi(\gamma\mu_0\omega)^{1/2} r^4} \quad B_r(t) = -\frac{3\mu_0 M_0}{\pi r^4}\left(\frac{t}{\gamma\mu_0\pi}\right)^{1/2}$$

As may be seen from these expressions, the behavior of the field for large induction numbers and large parameter r/τ_1 is almost identical. In particular, the equations for the electric field and the vertical component of the magnetic field when $r/_{\tau_1} \gg 1$ can be obtained by replacing ω by $1/t$ in the corresponding expressions in the frequency domain. This suggests that applying Fourier's transform and making use of only expressions for the field in the range of large parameter r/λ_1, we can derive asymptotic equations that correspond to the range of large parameter r/τ_1. For example, in the case of the electric field in a uniform half space, we have

$$E_\varphi(t) = -\frac{3M_0\rho}{2\pi r^4}\frac{1}{2\pi}\int\limits_{-\infty}^{\infty}\frac{\exp(-i\omega t)}{-i\omega}\mathrm{d}\omega = -\frac{3M_0\rho}{2\pi r^4}$$

Of course, this transition from the frequency to time domain is an approximate one, and it is valid only over a limited range of time since values of the field at the small and intermediate ranges of parameter r/λ_1 are replaced by those which correspond to large values of this parameter.

9.9.2. Derivation of Formulas for the Range of Large Parameters r/τ

Now applying this approach for a layered medium and taking into account Eq. (9.86), we obtain

$$B_z(t) = \frac{9\mu_0 M_0}{2\pi r^5}\frac{1}{2\pi}\int\limits_{-\infty}^{\infty}\frac{R_N^2(0)\exp(-i\omega t)}{-i\omega k_1^2}\mathrm{d}\omega$$

$$B_r(t) = \frac{3\mu_0 M_0}{2\pi r^4}\frac{1}{2\pi}\int\limits_{-\infty}^{\infty}\frac{R_N(0)\exp(-i\omega t)}{\omega k_1}\mathrm{d}\omega \tag{9.124}$$

$$E_\varphi(t) = -\frac{3M_0\rho_1}{2\pi r^4}\frac{1}{2\pi}\int\limits_{-\infty}^{\infty}\frac{R_N^2(0)\exp(-i\omega t)}{-i\omega}\mathrm{d}\omega$$

These equations are valid when the following condition is satisfied:

$$\frac{r}{\tau} > 1. \tag{9.125}$$

Each expression of the set (9.124) can be represented as being the product of two terms; one term depends on the dipole moment and the separation r, while the second is a function of the time and the geoelectric parameters of the medium. Thus, during the early stage of the transient response, as in the wave zone, the relationship between the field components and a distribution of conductivity in a horizontally layered medium is independent of the separation r. Therefore, there is a strong resemblance in the behavior of the field for large parameters r/λ in the frequency domain and that in the time domain when $r/\tau > 1$. In other words, we deal here with the wave zone in the time domain. The integrals on the right-hand side of Eq. (9.122) generally cannot be expressed in terms of known functions, and they must be evaluated numerically.

9.9.3. Behavior of the Field at the Surface of Two-Layered Medium (Early Stage)

At the same time in the case of a two-layered medium, these expressions can be simplified. As we know,

$$R_2(0) = \coth\left[-ik_1H_1 + \coth^{-1}\left(\frac{\rho_2}{\rho_1}\right)^{1/2}\right]$$

Inasmuch as

$$\coth\left(a + \coth^{-1} b\right) = \frac{1 + b\coth a}{b + \coth a} = \frac{1 + \left(b - 1/_{b+1}\right)\exp(-2a)}{1 - \left(b - 1/_{b+1}\right)\exp(-2a)},$$

we have

$$R_2(0) = \frac{1 + Q\exp(-2ik_1H_1)}{1 - Q\exp(-2ik_1H_1)} \quad \text{and} \quad \frac{Q = \rho_2^{1/2} - \rho_1^{1/2}}{\rho_2^{1/2} + \rho_1^{1/2}}$$

Assuming that the basement is not perfect insulator ($Q \neq 1$), we can expand the function

$$\frac{1}{1 - Q\exp(-2ik_1H_1)}$$

in the power series. Then after some algebraic operations, we have

$$R_2(0) = 1 + 2\sum_{n=1}^{\infty} Q^n \exp(-2ik_1H_1n)$$

and

$$R_2^2(0) = 1 + 4\sum_{n=1}^{\infty} nQ^n \exp(-2ik_1H_1n) \tag{9.126}$$

Substituting Eq. (9.126) into Eq. (9.122), we can represent the transient field at large parameters r/τ on the surface of the two-layered medium in terms of the probability integral $\Phi(x)$. As an example, we have for the electric field

$$E_\varphi(t) = -\frac{3M_0\rho_1}{2\pi r^4}\frac{1}{2\pi} \times \int_{-\infty}^{\infty}\frac{1}{-i\omega}\left[1 + 4\sum_{n=1}^{\infty} Q^n n\exp(-2ik_1H_1n)\right]\exp(-i\omega t)\Bigg]d\omega$$

$$= -\frac{3M_0\rho_1}{2\pi r^4}\left[\frac{1}{2\pi}\int_{-\infty}^{\infty}\frac{\exp(-i\omega t)}{-i\omega}d\omega + 4\sum_{n=1}^{\infty} nQ^n\frac{1}{2\pi}\int_{-\infty}^{\infty}\frac{\exp(-2ik_1H_1n)}{-i\omega}d\omega\right]$$

or

$$E_\varphi = -\frac{3M_0\rho_1}{2\pi r^4}\left\{1 + 4\sum_{n=1}^{\infty} nQ^n\left[1 - \Phi\left(\frac{4\pi n}{\tau_1/H_1}\right)\right]\right\} \tag{9.127}$$

Let us examine the behavior of the electric field as a function of the parameter τ_1/H_1. During the early stage, when the parameter τ_1/H_1 tends to zero, the probability integral approaches to unity, and therefore,

$$E_\varphi \to -\frac{3M_0\rho_1}{2\pi r^4}, \quad t \to 0$$

That is, the field is the same as that for a uniform half space with resistivity ρ_1. In contrast, when the parameter τ_1/H_1 increases, the probability integral tends to zero and we have

$$E_\varphi(t) = -\frac{3M_0\rho_1}{2\pi r^4}\left[1 + 4\sum_{n=1}^{\infty} nQ^n\right] \quad \text{if } \tau_1/H_1 > 1$$

Inasmuch as

$$\sum_{n=1}^{\infty} nQ^n = \frac{Q}{(1-Q)^2},$$

we obtain

$$E_\varphi(t) \rightarrow -\frac{3M\rho_2}{2\pi r^4}$$

Thus during the early stage $\tau/r < 1$, when τ_1 is significantly greater than the thickness of the first layer, it becomes transparent to the transient field and the field coincides with that for a uniform half space having resistivity ρ_2, $\left(r/H_1 \gg 1\right)$. The same behavior is shown by the magnetic field and is valid for a medium with any number of layers. As was already mentioned, the use of the notation "the early stage" for this asymptotic is often confusing. For instance, this stage does not always mean that induction currents are located only near the Earth's surface. Suppose that the separation r is less than the thickness H_1, $r/H_1 < 1$. Then during the early stage, we also have $H_1/\tau_1 \gg 1$, and therefore, the field is defined by the resistivity of the upper layer since induction currents are mainly located in the upper layer. In other words, if the separation r is less than or comparable to the thickness H_1, the transient field observed during the early stage does not contain information about the medium beneath the upper layer. However, this field at the early stage can be used in principle to investigate the geoelectric section, provided that the separation r is much greater than the total thickness of layers. In this case, the condition $r/\tau_1 > 1$ is met, even when parameter τ_i/H_i may exceed unity in some layers. Accordingly, at the early stage, induction currents can be distributed over greater distances from the dipole beneath the Earth's surface. The similarity with the wave zone in the frequency domain is obvious. It is clear that with an increase of the distance r or with a decrease in the resistivity of the medium, the early stage persists to greater times. Note that many years ago, the transient soundings at the early stage found a broad application in the petroleum and geothermal geophysics.

9.10. THE LATE STAGE OF THE TRANSIENT FIELD ON THE SURFACE OF A LAYERED MEDIUM

9.10.1. Expansion of the Field in Series by Powers t^{-n}

In this section, we study the field behavior on the Earth's surface when the parameter τ/r in every layer is much greater than unity:

$$\frac{\tau}{r} \gg 1 \tag{9.128}$$

With this purpose in mind, as in the case of the early stage, we proceed from the Fourier integral. For instance, for the secondary magnetic field, we have

$$B_s(t) = \frac{2}{\pi} \int_0^\infty \varphi_1(\omega) \cos \omega t d\omega$$

and

$$B_s(t) = -\frac{2}{\pi} \int_0^\infty \varphi_2(\omega) \sin \omega t d\omega, \tag{9.129}$$

where

$$\varphi_1(\omega) = \frac{QB_s(\omega)}{\omega}, \quad \varphi_2(\omega) = \frac{\mathrm{Re}\, B_s(\omega)}{\omega}, \tag{9.130}$$

and B_s is the complex amplitude of any component of the secondary field caused by currents induced in a horizontally layered medium. Assuming that time increases without limit and integrating Eq. (9.129) by parts, we obtain

$$\begin{aligned} B_s(t) &= \frac{2}{\pi}\left[\frac{\varphi_1 \sin \omega t}{t}\bigg|_0^\infty - \frac{1}{t}\int_0^\infty \varphi_1' \sin \omega t d\omega\right] \\ &= \frac{2}{\pi}\left[\frac{\varphi_1 \sin \omega t}{t}\bigg|_0^\infty + \frac{\varphi_1'}{t^2}\cos \omega t\bigg|_0^\infty - \frac{1}{t^2}\int_0^\infty \varphi_1'' \cos \omega t d\omega\right] \end{aligned}$$

or

$$B_s(t) = \frac{2}{\pi}\left[\frac{\varphi_2 \cos \omega t}{t}\Bigg|_0^\infty - \frac{1}{t}\int_0^\infty \varphi_2'(\omega)\cos \omega t d\omega\right]$$
$$= \frac{2}{\pi}\left[\frac{\varphi_2 \cos \omega t}{t}\Bigg|_0^\infty - \frac{\varphi_2' \sin \omega t}{t^2}\Bigg|_0^\infty + \frac{1}{t^2}\int_0^\infty \varphi_2''(\omega)\sin \omega t d\omega\right] \tag{9.131}$$

Thus, we have obtained a series expansion in powers inversely proportional to t, which can be used in determining the late stage of the transient field. As follows from Eq. (9.131), this stage is controlled by the low-frequency spectrum including its derivatives with respect to frequency, such as

$$\frac{d\varphi_1}{d\omega}, \frac{d\varphi_2}{d\omega}, \frac{d^2\varphi_1}{d\omega^2}, \frac{d^2\varphi_2}{d\omega^2} \ldots$$

and so on. In other words, the intermediate- and high-frequency parts of the spectrum have practically no control over the late stage of the transient field. Depending on the field components and the medium, either the magnitude or its derivatives with respect to frequency control the late stage behavior. If the derivatives with respect to frequency are the dominant factors, the behavior of the field component over the low-frequency part of the spectrum is usually quite different from the behavior of this component at the late stage. The study of the low-frequency spectrum has shown that any component of the field can be written as a sum

$$\sum_{n=1}^{\infty} C_{1n}k^{2n} + \sum_{n=1}^{\infty} C_{2n}k^{2n+1} + \sum_{n=1}^{\infty} C_{3n}k^{2n} \ln k, \tag{9.132}$$

where $k = (i\gamma\mu_0\omega)^{1/2}$ and C_{in} are the coefficients depending on the geoelectric parameters, the separation r and the dipole moment.

9.10.2. Contribution of the First Sum of Eq. (9.132) into the Late Stage

Now we demonstrate that the sum

$$\sum_{n=1}^{\infty} C_{1n}k^{2n} \tag{9.133}$$

has no effect on the late stage of the transient response. Let us write the latter as a sum of the in-phase and quadrature components, respectively,

$$\sum_{n=1}^{\infty} C_{1n}k^{2n} = \sum_{n=1}^{\infty} a_{1n}\omega^{2n} + i\sum_{n=1} b_{1n}\omega^{2n-1} \tag{9.134}$$

Substituting this into the Fourier transform, we obtain two types of integrals, namely,

$$L_n = \int_0^{\infty} \omega^{2n-1} \sin \omega t d\omega \quad \text{and} \quad M_n = \int_0^{\infty} \omega^{2n-2} \cos \omega t d\omega \tag{9.135}$$

Of course, they diverge and in order to evaluate these integrals, we will consider them as being the limiting cases for large t of more general integrals which converge as

$$L_n = \lim \int_0^{\infty} \omega^{2n-1} \exp(-\beta\omega)\sin \omega t d\omega$$

and

$$M_n = \lim \int_0^{\infty} \omega^{2n-2} \exp(-\beta\omega)\cos \omega t d\omega \quad \text{as } \beta \to 0, t \to \infty \tag{9.136}$$

This approach is valid because the introduction of the exponential term $\exp(-\beta\omega)$ does not change the initial part of integration, which defines the value for the integral in Eq. (9.136) when the parameter t tends to infinity. The integrals

$$\int_0^{\infty} \omega^{2n-1} \exp(-\beta\omega)\sin \omega t d\omega \quad \text{and} \quad \int_0^{\infty} \omega^{2n-2} \exp(-\beta\omega)\cos \omega t d\omega$$

are well known and they are expressed in terms of elementary functions. It is essential that their values vanish as the parameter β tends to zero. Therefore, we can conclude that the first sum in Eq. (9.132), which contains the integer powers of ω, makes no contribution to the late stage of the transient field. Correspondingly, only the fractional powers of ω and logarithmic terms are important in determining the late stage behavior. This fact plays a fundamental role in understanding the relationship between the frequency domain and time domain responses of the electromagnetic field in a conducting medium. For example, the quadrature component of the magnetic field at the low frequencies is principally controlled by the leading term in the series representation, being directly proportional to ω. But the following terms, which contain fractional powers of ω and $\ln\,\omega$, have relatively negligible effect. However, in accordance with what was shown above, these less important terms in the frequency domain define or control the behavior of the transient field at the late stage. Usually the relationship of the first term in the series for the quadrature component **QB**, which is directly proportional to ω, with the geoelectric parameters of the medium and with the separation r differs essentially from that of the rest of terms. Therefore, we can readily understand the fundamental difference between the behavior of this component of the magnetic field at the low frequency and the transient response during the late stage. On the other hand, if the lower part of the medium is characterized by a finite resistivity, then as was shown earlier, the leading term in the series expansion for the in-phase component of the magnetic field at the low frequencies contains either a fractional power of ω or $\ln\,\omega$. Correspondingly, one might expect that the behavior of this component of the magnetic field at the low frequencies will be practically the same as the behavior of the transient field at the late stage.

Also let us notice the following. As we know, the complex amplitudes of the electromagnetic field in the frequency domain are expressed in terms of an integral of the type:

$$\int\limits_{0}^{\infty} \Phi(m, k_1, k_2, \ldots k_n) J_i(mr)dm \quad \text{and } i = 0 \text{ or } 1$$

However, it has been demonstrated that the only initial part of the range of integration, $m \to 0$, is responsible for the existence of the fractional powers of ω and logarithmic terms. Therefore, we can say that the late stage of the transient field is controlled by the long-period spatial harmonics characterized by very small values of m.

9.10.3. Derivation of the Asymptotic Formulas

Next we describe the method of deriving asymptotic formulas for the transient field at the late stage. After separation of the real and imaginary parts, the second sum in Eq. (9.132) can be written as

$$\sum_{n=1}^{\infty} C_{2n}k^{2n+1} = \sum_{n=1}^{\infty} a_{2n}\omega^{2n+1/2} + i\sum_{n=1}^{\infty} b_{2n}\omega^{2n+1/2}$$

For illustration, consider the in-phase component of the sum

$$\sum_{n=1}^{\infty} a_{2n}\omega^{2n+1/2} = a_2\omega^{3/2} + a_4\omega^{5/2} + \cdots$$

Substituting this sum into the Fourier integral, we obtain

$$-\frac{2}{\pi}\sum_{n=1}^{\infty} a_{2n}\int_0^{\infty} \omega^{n-1/2}\sin\omega t\,\mathrm{d}\omega \tag{9.137}$$

Letting $n = 1$, we have

$$I_1 = \int_0^{\infty} \omega^{1/2}\sin\omega t\,\mathrm{d}\omega,$$

which can be considered as the limiting case of the integral,

$$\int_0^{\infty} \omega^{1/2}\exp\left(-\beta\omega\right)\sin\omega t\,\mathrm{d}\omega \quad \text{if} \quad \beta\to 0$$

For simplicity, the exponential term below is omitted. Integrating I_1 by parts and taking into account the fact that the quasi-stationary field at the high frequencies tends to zero, we obtain

$$I_1 = -\frac{1}{t}\int_0^{\infty}\omega^{1/2}\mathrm{d}\cos\omega t = -\frac{1}{t}\left[\omega^{1/2}\cos\omega t\Big|_0^{\infty} - \frac{1}{2}\int_0^{\infty}\frac{\cos\omega t}{\omega^{1/2}}\mathrm{d}\omega\right]$$

$$= \frac{1}{2t}\int_0^{\infty}\frac{\cos\omega t}{\omega^{1/2}}\mathrm{d}\omega = \frac{1}{2t^{3/2}}\int_0^{\infty}\frac{\cos x}{x^{1/2}}\mathrm{d}x$$

This last integral is well known:

$$\int_0^{\infty}\frac{\cos x}{x^{1/2}}\mathrm{d}x = \left(\frac{\pi}{2}\right)^{1/2}$$

Thus,

$$I_1 = \frac{\pi^{1/2}}{2^{3/2} t^{3/2}} \tag{9.138}$$

For $n = 2$, we have

$$I_2 = \int_0^\infty \omega^{3/2} \sin \omega t \mathrm{d}\omega$$

Integrating I_2 twice by parts, we obtain

$$I_2 = -\frac{1}{t}\left[\omega^{3/2} \cos \omega t\Big|_0^\infty - \frac{3}{2}\int_0^\infty \omega^{1/2} \cos \omega t \mathrm{d}\omega\right] = \frac{3}{2t}\int_0^\infty \omega^{1/2} \cos \omega t \mathrm{d}\omega$$

$$= \frac{3}{2t^2}\int \omega^{1/2} \mathrm{d}\sin \omega t = \frac{3}{2t^2}\left[\omega^{1/2} \sin \omega t\Big|_0^\infty - \frac{1}{2}\int_0^\infty \frac{\sin \omega t}{\omega^{1/2}} \mathrm{d}\omega\right]$$

$$= -\frac{3}{4t^{5/2}}\int_0^\infty \frac{\sin \omega t}{\omega^{1/2}} \mathrm{d}\omega$$

Inasmuch as

$$\int_0^\infty \frac{\sin x}{x^{1/2}} \mathrm{d}x = \left(\frac{\pi}{2}\right)^{1/2}$$

we have

$$I_2 = -\frac{3}{4}\left(\frac{\pi}{2}\right)^{1/2} \frac{1}{t^{5/2}} \tag{9.139}$$

Making use of the similar approach, one can calculate any integral in the sum (9.137), and we can see that any term proportional to $\omega^{n+1/2}$ generates a term in the time domain proportional to $t^{-n-1/2}$. Therefore, the portion of the spectrum described by the sum

$$\sum_{n=1}^{\infty} a_{2n} \omega^{n+1/2}$$

is responsible for the appearance of the sum of the type

$$\sum_{n=1}^{\infty} a_{2n}^{*} \frac{1}{t^{n+1/2}} \tag{9.140}$$

in the expression for the late stage of the transient response.

The third sum in Eq. (9.132) can be written as

$$\sum_{n=1}^{\infty} C_{3n} k^{2n} \ln k = \ln k \sum_{n=1}^{\infty} C_{3n} k^{2n}$$

$$= \ln\left[(\gamma\mu_0\omega)^{1/2} \exp\left(i\pi/4\right)\right] \sum_{n=1}^{\infty} C_{3n}(\gamma\mu_0\omega)^n \exp\left(in\pi/2\right)$$

$$= \left[\ln(\gamma\mu_0\omega)^{1/2} + i\pi/4\right]$$

$$\times \left[\sum_{n=1}^{\infty} C_{3n}(\gamma\mu_0\omega)^n \cos\frac{\pi}{2} n + i\sum_{n=1}^{\infty} C_{3n}(\gamma\mu_0\omega)^n \sin\frac{\pi}{2} n\right]$$

Letting $n = 2p$ and $n = 2p - 1$ in the first and second sums, respectively, and taking into account the fact that $\cos\pi p = (-1)^p$ and $\sin\left(2p - 1/2\right)\pi = (-1)^{p-1}$, we have the following expressions for the real and imaginary parts of the third sum:

$$\left[\frac{1}{2}\ln(\gamma\mu_0\omega) \sum_{p=1}^{\infty} (-1)^p C_p(\gamma\mu_0\omega)^{2p} - \frac{\pi}{4} \sum_{p=1}^{\infty} (-1)^{p-1} C_p(\gamma\mu_0\omega)^{2p-1}\right]$$

$$i\left[\frac{\pi}{4} \sum_{p=1}^{\infty} (-1)^p C_p(\gamma\mu_0\omega)^{2p} + \frac{1}{2}\ln(\gamma\mu_0\omega) \sum_{p=1}^{\infty} (-1)^{p-1} C_p(\gamma\mu_0\omega)^{2p-1}\right] \tag{9.141}$$

Substituting the real part of this last equation into the Fourier transform, we obtain two types of integrals:

$$A_p = \int_0^{\infty} \omega^{2p-2} \sin \omega t d\omega \quad \text{and} \quad B_p = \int_0^{\infty} \omega^{2p-1} \ln \omega \sin \omega t d\omega \tag{9.142}$$

For example, when $p=1$, we have

$$A_1 = \int_0^\infty \sin \omega t d\omega = \lim \int_0^\infty \exp(-\beta\omega)\sin \omega t d\omega = \frac{1}{t} \quad \text{if } \beta \to 0, t \to \infty$$

and

$$B_1 = \int_0^\infty \omega \ln \omega \sin \omega t d\omega = \int_0^\infty F(\omega)\sin \omega t d\omega,$$

where $F(\omega) = \omega \ln \omega$. Integrating by parts, we obtain

$$B_1 = -\frac{1}{t}\int_0^\infty F(\omega) d\cos \omega t = -\frac{1}{t}\left[F(\omega)\cos \omega t\Big|_0^\infty - \int_0^\infty F'(\omega)\cos \omega t d\omega\right]$$

$$= -\frac{1}{t}\left[F(\omega)\cos \omega t\Big|_0^\infty - \int_0^\infty F'(\omega) d \sin \omega t\right]$$

$$= -\frac{1}{t}F(\omega)\cos \omega t\Big|_0^\infty + \frac{1}{t^2}\left[F'(\omega)\sin \omega t\Big|_0^\infty - \int_0^\infty F''(\omega)\sin \omega t d\omega\right]$$

Since

$$F'(\omega) = 1 + \ln \omega, \quad F'' = \frac{1}{\omega}$$

we have

$$B_1 = -\frac{1}{t^2}\int_0^\infty \frac{\sin \omega t}{\omega} d\omega = -\frac{\pi}{2t^2}$$

Applying the same approach, one can derive integrals A_p and B_p when p is not unity. It is readily seen that the portion of the low-frequency spectrum described by the last sum in Eq. (9.132) gives rise to terms in the representation of the late stage of the transient field, proportional to t^{-n}. Thus, the following sum appears in the expression for the transient field:

$$\sum_{n=1}^{\infty} a_{3n}^{*} \frac{1}{t^n} \tag{9.143}$$

9.10.4. Formulas for the Late Stage when Basement is Not an Insulator

Therefore, in accordance with Eqs (9.140) and (9.143), the late stage of the transient electric and magnetic fields in a horizontally layered medium can be represented as follows:

$$\sum_{n=1}^{\infty} a_{2n}^{*} \frac{1}{t^{n+1/2}} + \sum_{n=1}^{\infty} a_{3n}^{*} \frac{1}{t^{n}},$$

where a_{2n}^{*} and a_{3n}^{*} are related to the coefficients describing the low-frequency part of the spectrum. They are functions of the electrical properties of the medium and separation between the dipole and observation site. Under certain conditions, some of the coefficients are equal to zero. For instance, as well known, the late stage of the transient field, when basement is an insulator, is described by a sum containing only integer powers of t; that is, $a_{2n}^{*} = 0$.

Having completed a general analysis of the field at the late stage, we should now obtain the leading terms of the asymptote for each component of the electromagnetic field. From Eqs (9.55)–(9.57) and (9.129), we have for the vertical component of the magnetic field:

$$B_z(t) = \frac{\mu_0 M_0}{2\pi^2 r^3} \left[\frac{1}{4} \sum_{i=1}^{N} \left(\gamma_i \mu_0 r^2\right) G_i^z \int_0^\infty \cos \omega t d\omega - \frac{2^{3/2}}{15} \left(\gamma_N \mu_0 r^2\right)^{3/2} \int_0^\infty \omega^{1/2} \cos \omega t d\omega \right] \tag{9.144}$$

As was shown earlier, the first integral in Eq. (9.144) is zero, while the second is

$$\int_0^\infty \omega^{1/2} \cos \omega t d\omega = -\frac{1}{2t} \int_0^\infty \frac{\sin \omega t}{\omega^{1/2}} d\omega = -\frac{\pi^{1/2}}{2^{3/2} t^{3/2}}$$

Thus,

$$B_z \to \frac{M_0}{30\pi^{3/2}} \frac{\mu_0^{5/2} \gamma_N^{3/2}}{t^{3/2}} \tag{9.145}$$

Of course, the same result follows if the in-phase component of B_z^* is used. In the case of the horizontal component $B_r(t)$, we use the quadrature component as follows:

$$B_r(t) = -\frac{\mu_0 M_0}{2\pi^2 r^3}\left[\frac{1}{4}\sum_{i=1}^{N}\left(\gamma_i\mu_0 r^2\right)G_i^r\int_0^\infty \cos\omega t d\omega + \frac{3\pi}{64}\left(\gamma_N\mu_0 r^2\right)^2\int_0^\infty \omega\cos\omega t d\omega\right]$$

The first integral is again zero but the second integral can be written as

$$\int_0^\infty \omega\cos\omega t d\omega = \frac{\partial}{\partial t}\int_0^\infty \exp(-\beta t)\sin\omega t d\omega = \frac{1}{t^2} \quad \text{if } \beta\to 0, t\to\infty$$

Thus,

$$B_r(t)\to -\frac{\mu_0 M_0}{128\pi r^3}\frac{\left(\gamma_N\mu_0 r^2\right)^2}{t^2} \tag{9.146}$$

Finally, for the transient electric field, we have

$$E_\varphi(t)\to \frac{\mu_0^{5/2}\gamma_N^{3/2} r M_0}{40\pi^{3/2} t^{5/2}} \tag{9.147}$$

In summary, the late stage of the transient field on the Earth's surface of a horizontally layered medium is

$$B_z\to \frac{M_0}{30\pi^{3/2}}\frac{\mu_0^{5/2}\gamma_N^{3/2}}{t^{5/2}}, \quad B_r(t)\to -\frac{\mu_0 M_0}{128\pi r^3}\frac{\left(\gamma_N\mu_0 r^2\right)^2}{t^2}, \quad E_\varphi(t)\to \frac{\mu_0^{5/2}\gamma_N^{3/2} r M_0}{40\pi^{3/2} t^{5/2}} \tag{9.148}$$

Therefore, at sufficiently late times, induction currents are primarily located in the lowermost medium, and for this reason, expression (9.148) is exactly the same as those for a uniform half space having conductivity γ_N. In other words, all overlying layers regardless of their conductivity and thickness become transparent during the late stage of the transient response. This means that the depth of investigation, obtained with the transient soundings, is in principle controlled only by time, and the separation has no principle significance. As follows from Eq. (9.148), the vertical component of the magnetic field during the late stage does not depend on the distance between the dipole and a point of measurements. This feature of the behavior of B_z is related to the fact that induced currents, generating this component, are mainly located far away from the Earth's surface.

This study also shows that during the late stage, the component $B_z(t)$ is more sensitive to a change of conductivity than the quadrature component observed at small induction numbers or the total field when the parameter r/δ is large. Although the horizontal component B_r is less than B_z at the late stage, it possesses a higher sensitivity to a change of conductivity. It is readily seen that at the late stage, the following relationship holds

$$E_\varphi(t) = -\frac{r}{2}\frac{\partial B_z}{\partial t} \tag{9.149}$$

Of course, the conclusion concerning the depth of investigation during the late stage is also valid for the in-phase component of the magnetic field over the range of small values of induction numbers.

Until now, we have considered only the leading term of the asymptote for each component of the field at the late stage. These expressions can be markedly improved using formulas of the low-frequency spectrum derived earlier. For instance, taking into account the terms of the series expansion for the vector potential in the case of a two-layered medium and making use of the Fourier transform, after simple algebra, we obtain

$$\begin{aligned}
B_z(t) &= \frac{\mu_0 M_0}{4\pi}\left[\frac{2}{15}\frac{(\mu_0\gamma_2)^{3/2}}{\pi^{1/2}t^{3/2}} - \frac{(\mu_0\gamma_1)^2 sH_1}{8t^2}(s-1) - \frac{r^2(\gamma_2\mu_0)^{5/2}}{35\pi^{1/2}t^{5/2}} + \cdots\right] \\
B_r(t) &= \frac{\mu_0 M_0 r}{4\pi}\left[\frac{(\gamma_2\mu_0)^2}{32t^2} + \frac{32}{35}\frac{(\mu_0\gamma_1)^{3/2}H_1 s^{3/2}(1-s)}{\pi^{1/2}t^{5/2}}\right. \\
&\qquad \left. - \frac{(\mu_0\gamma_1)^3}{64t^3}H_1^2(s-1)s(6-5s) + \cdots\right] \\
E_\varphi(t) &= \frac{\mu_0 M_0}{4\pi}\, r\left\{\frac{(\mu_0\gamma_2)^{3/2}}{10\pi^{1/2}t^{5/2}} - \frac{(\mu_0\gamma_1)^2 H_1 s(s-1)}{8t^3}\right. \\
&\qquad \left. - \frac{(\mu_0\gamma_1)^{5/2}}{112\pi^{1/2}t^{7/2}}\left[2r^2s^2 + 4H_1^2 s^{1/2}(1-s)(8s-9)\right] + \cdots\right\}
\end{aligned} \tag{9.150}$$

where

$$s = \frac{\gamma_2}{\gamma_1} \neq 0$$

It is clear that the second and following terms in the asymptotic series (9.150) reflect the influence of induced currents in the upper layer, and these expressions can be used if the corresponding terms decrease rapidly.

9.10.5. Formulas for the Late Stage when Basement is an Insulator

Now applying the same approach, we arrive at the asymptotic formulas for the field at the late stage for N-layered medium when the basement is an insulator:

$$B_z(t) \approx \frac{\mu_0^4 M_0 S^3}{16\pi t^3}, \quad B_r(t) \approx -\frac{3\mu_0^5 M_0 S^4}{64\pi t^4} r, \quad E_\varphi(t) \approx \frac{3\mu_0^4 M_0 S^3}{32\pi t^3} r, \tag{9.151}$$

where S is the total conductance of layers above a basement. Thus, the vertical component of the magnetic field during the late stage does not depend on the separation r and is directly proportional to the cube of the longitudinal conductance S. For this reason, relatively small changes in the thickness of a sedimentary sequence, resting on the basement, can be observed by measuring the transient field, regardless of the separation between the dipole and observation point. As follows from Eq. (9.151), the horizontal component B_r is even more sensitive to a change of conductance S, but it is less in magnitude than the vertical component. It is a simple matter to improve the expressions for the late stage. In particular, in the case of a two-layered medium, we have

$$\begin{aligned} B_z(t) &= \frac{\mu_0^4 M_0 S_1^3}{16\pi t^3}\left(1 - \frac{3\mu_0 S_1 H_1}{t} + \cdots\right) \\ B_r(t) &= -\frac{3\mu_0^5 M_0 S_1^4}{64\pi t^4} r\left(1 - \frac{14}{3}\frac{\mu_0 S_1 H_1}{t} + \cdots\right) \\ E_\varphi(t) &= \frac{3\mu_0^4 M_0 S_1^3}{32\pi t^3} r\left(1 - \frac{4\mu_0 S_1 H_1}{t}\right) \end{aligned} \tag{9.152}$$

9.11. FIELD OF A VERTICAL MAGNETIC DIPOLE IN THE PRESENCE OF A HORIZONTAL CONDUCTING PLANE

In studying the behavior of the frequency and transient electromagnetic fields of a vertical magnetic dipole located on the surface of a layered medium, we paid some attention to one special case when the underlying medium (basement) is an insulator. This analysis indicated that under certain conditions, the field is defined by the total conductance of layers S:

$$S = \sum_{i=1}^{N-1} \gamma_i H_i$$

Such behavior of the field occurs if one of the following conditions takes place:

1. The quadrature component of the magnetic field is measured at the range of small parameters:

$$r/\delta_i \ll 1 \quad \text{and} \quad H_i/\delta_i \ll 1,$$

but the separation r is much greater than the total thickness of layers

$$r \gg \sum_{i=1}^{N-1} H_i$$

2. The in-phase components of the magnetic field $In\mathbf{B}_s^*$ or the quadrature component of the electric field QE_φ^* are measured, and they are caused by induced currents in the range of small parameters regardless of the distance between the dipole and the observation point.
3. The field at the wave zone when

$$r/\lambda > 1, \quad \text{and} \quad H/\lambda_{\min} < 1,$$

where H is the total thickness of layers above an insulator.

4. The transient field at the early stage if

$$r/\tau > 1 \quad \text{but} \quad H/\tau_{\min} < 1$$

5. The transient field at the late stage

$$\tau_i/H_i \gg 1,$$

regardless of the distance r.

Assuming that one of these conditions is met, we can say that a system of layers is equivalent to a thin conducting plane with the conductance S. Taking into account this fact, it is natural to investigate the field generated by conduction currents in the limiting case of the plane with the conductance S.

9.11.1. Formulation of the Boundary Value Problem in the Frequency Domain

With this purpose in mind, suppose that a vertical magnetic dipole with the moment

$$\mathbf{M}^* = M_0 1_z$$

is located in a nonconducting medium at the origin of a cylindrical system of coordinates at the distance h from the plane with the conductance S (Figure 9.4), and it is directed along the z-axis. Such plane is equivalent to an infinitely thin layer with infinitely large conductivity. It is obvious that the

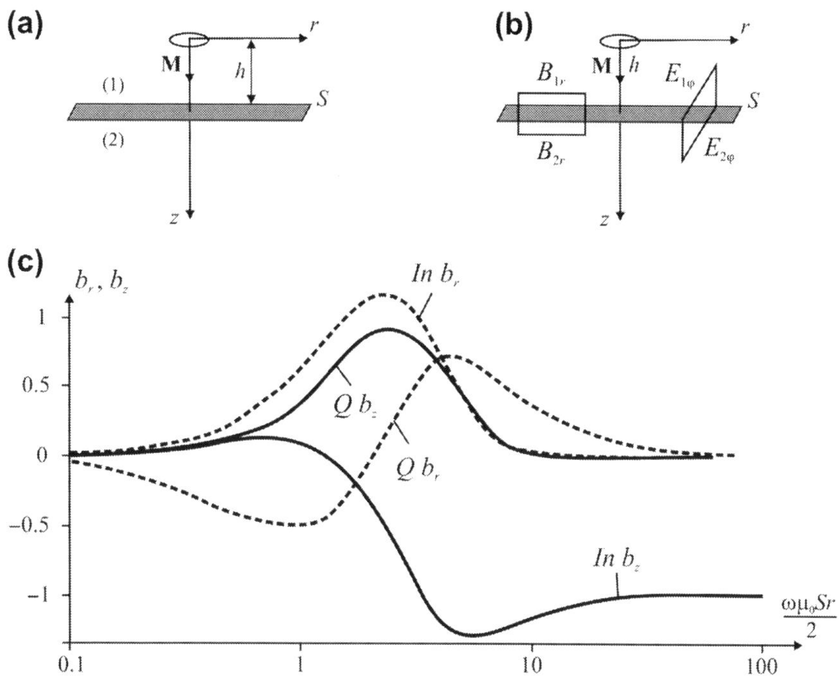

Figure 9.4 (a) Vertical magnetic dipole over the conducting plane. (b) Tangential components of the electromagnetic field near the plane S. (c) Behavior of quadrature and in-phase components of the fields b_z^* and b_r^*.

currents induced in the plane possess an axial symmetry and have only a φ component. Correspondingly, in a nonconducting medium above and beneath the plane, the electromagnetic field has the following components:

$$\mathbf{E} = \left(0, E_\varphi, 0\right) \quad \text{and} \quad \mathbf{B} = (B_r, 0, B_z)$$

As before, we describe the field with the help of the vertical component of the vector potential of the electric type A_z^*, which is related to the field as

$$E_\varphi^* = -\frac{\partial A_z^*}{\partial r}, \quad i\omega B_z = \frac{\partial^2 A_z^*}{\partial z^2}, \quad i\omega B_r = \frac{\partial^2 A_z^*}{\partial r \partial z} \tag{9.153}$$

Making use of the results derived earlier, we have

$$A_{1z}^* = i\omega \frac{\mu_0 M_0}{4\pi} \int_0^\infty \{\exp(-m|z|) + C_m \exp mz\} J_0(mr) dm \quad \text{if } z < h$$

and

$$A_{2z}^* = i\omega \frac{\mu_0 M_0}{4\pi} \int_0^\infty D_m \exp(-mz) J_0(mr) dm \quad \text{if } z > h \tag{9.154}$$

9.11.2. Boundary Conditions at the Plane *S*

Because the conducting sheet is vanishingly thin, instead of the ordinary boundary conditions, we can make use of approximate ones, which do not require determination of the field inside the conducting layer. In fact, writing Maxwell's equations in the integral form, we have

$$\oint_L \mathbf{B} \cdot d\mathbf{l} = \mu_0 \int_S \mathbf{j} \cdot d\mathbf{S} \quad \text{and} \quad \oint_L \mathbf{E} \cdot d\mathbf{l} = \mu_0 \oint_S \dot{\mathbf{B}} \cdot d\mathbf{S} \tag{9.155}$$

Evaluation of the first integral of the set (9.155) along the path shown in Figure 9.4(b) gives

$$B_{2r}^* - B_{1r}^* = \mu_0 i_\varphi^*, \tag{9.156}$$

where B_{2r}^* and B_{1r}^* are the complex amplitudes of tangential components of the magnetic field at each side of the conducting plane, and i_φ^* is the complex amplitude of the surface density of currents in the plane given by

$$i_\varphi^* = S E_\varphi^* \tag{9.157}$$

In deriving Eq. (9.155), we assumed that the component B_z has a finite value that corresponds to the limited value of the current density i_φ. Evaluating the second integral in the set (9.155) and taking into account the fact that the vertical component of the electric field is zero, we obtain the second boundary condition:

$$E^*_{1\varphi} = E^*_{2\varphi} \tag{9.158}$$

Thus, the electromagnetic field on both sides of the conducting plane is related in the following way:

$$B^*_{2r} - B^*_{1r} = \mu_0 i^*_\varphi, \quad E^*_{1\varphi} = E^*_{2\varphi} \quad \text{if } z = h \tag{9.159}$$

That is, the electric field is a continuous function across the conducting sheet, but the tangential component of the magnetic field is discontinuous, and this discontinuity is defined by the current density at a given point. It is essential that these boundary conditions contain only the tangential components of the field for free space that drastically facilitates its determination. The components of the magnetic field on either side of the conducting sheet B_{1r} and B_{2r} are caused by the dipole current and by induced currents in the plane S. The primary magnetic field $\mathbf{B}_0$ is continuous across the sheet. Therefore, the first condition in Eq. (9.159) can be written as

$$B^{s*}_{2r} - B^{s*}_{1r} = \mu_0 S E^*_\varphi, \tag{9.160}$$

where E^*_φ is the total field, but B^{s*}_r is only the secondary magnetic field caused by induction currents in the conducting plane. Consider any point p of the plane. The current in the vicinity of this point generates tangential components of the magnetic field on either side of the sheet which are equal by the magnitude but different in sign. According to Biot–Savart law, the magnetic field caused by currents over the rest of the conducting plane is continuous and, in particular, the horizontal component of this part of the magnetic field is zero at this point p. Therefore, Eq. (9.160) can be represented as

$$2B^{s*}_{2r}(p) = \mu_0 S E^*_\varphi(p), \quad \text{if } z = h$$

since

$$B^{s*}_{1r}(p) = -B^{s*}_{2r}(p)$$

9.11.3. Expressions for the Field Components

Now, taking into account Eq. (9.153), the boundary conditions for the vector potential A_z^* have the form

$$A_{1z}^* = A_{2z}^*$$

and

$$\frac{\partial A_{2z}^*}{\partial z} - \frac{\partial A_{1z}^*}{\partial z} = -i\omega\mu_0 S A_{1z}^* = -i\omega\mu_0 S A_{2z}^* \tag{9.161}$$

Substituting Eq. (9.154) into Eq. (9.159), we arrive at a set of two equations with two unknowns:

$$\exp(-mh) + C_m \exp mh = D_m \exp(-mh)$$

and

$$\begin{aligned} &-mD_m \exp(-mh) + m\exp(-mh) - mC_m \exp mh \\ &= -i\omega\mu_0 S D_m \exp(-mh) \end{aligned} \tag{9.162}$$

Whence,

$$D_m = \frac{2m}{2m - i\omega\mu_0 S} \quad \text{and} \quad C_m = i\omega\mu_0 S \frac{\exp(-2mh)}{2m - i\omega\mu_0 S} \tag{9.163}$$

Thus above the plane, we have

$$A_{1z}^* = i\omega \frac{\mu_0 M_0}{4\pi}\left[\frac{1}{R} + i\omega\mu_0 S \int_0^\infty \frac{\exp(-2mh)\exp mz}{2m - i\omega\mu_0 S} J_0(mr)\mathrm{d}m\right] \tag{9.164}$$

In the case when both the magnetic dipole and the observation point are situated on the conducting plate, Eq. (9.164) is slightly simplified, and it gives

$$A_{1z}^*(p) = i\omega \frac{\mu_0 M_0}{2\pi} \int_0^\infty \frac{mJ_0(mr)\mathrm{d}m}{2m - i\omega\mu_0 S}\mathrm{d}m \tag{9.165}$$

since

$$\frac{1}{r} = \int_0^\infty J_0(mr)\mathrm{d}m$$

Correspondingly,

$$\frac{\partial A_{1z}^*}{\partial z} = -\omega^2 \mu_0^2 \frac{M_0}{4\pi} S \int_0^\infty \frac{m J_0(mr)}{2m - i\omega\mu_0 S} \mathrm{d}m \tag{9.166}$$

Thus,

$$i\omega\mu_0 S A_{1z}^* = 2\frac{\partial A_{1z}^*}{\partial z} \quad \text{if } z = h = 0$$

Then from Eq. (9.153), we obtain

$$E_\varphi^* = \frac{i\omega\mu_0 M_0}{2\pi} \int_0^\infty \frac{m^2 J_1(mr)}{2m - i\omega\mu_0 S} \mathrm{d}m \tag{9.167}$$

$$B_r^* = -i\omega\mu_0 S \frac{M_0 \mu_0}{4\pi} \int_0^\infty \frac{m^2 J_1(mr) \mathrm{d}m}{2m - i\omega\mu_0 S} \tag{9.168}$$

and

$$B_z^* = \frac{\mu_0 M_0}{2\pi} \int_0^\infty \frac{m^3 J_0(mr) \mathrm{d}m}{2m - i\omega\mu_0 S} \quad \text{if } z = h = 0 \tag{9.169}$$

Inasmuch as the horizontal component of the primary magnetic field is zero on the conducting plane ($h = 0$), the tangential components B_r^* and E_φ^* must satisfy the condition

$$B_{2r}^* = \frac{\mu_0 S}{2} E_\varphi^* \tag{9.170}$$

and this also follows from Eqs (9.167) and (9.168). It is convenient to introduce a new variable $x = mr$ in Eqs (9.165)–(9.167). Then we have

$$E_\varphi^* = i\omega\mu_0 \frac{M_0}{4\pi r^2} e_\varphi = E_\varphi^0 e_\varphi \tag{9.171}$$

where

$$e_\varphi = \int_0^\infty \frac{x^2 J_1(x)}{x + q} \mathrm{d}x \tag{9.172}$$

and

$$B_z^* = \frac{\mu_0 M_0}{4\pi r^3} b_z = B_z^0 b_z \tag{9.173}$$

while

$$b_z = \int_0^\infty \frac{x^3 J_0(x)}{x+q} \mathrm{d}x \tag{9.174}$$

and

$$q = -\frac{i\omega\mu_0 Sr}{2} = -ip_s$$

The parameter

$$p_s = \frac{\omega\mu_0 Sr}{2}$$

plays the same role as the ratio between the distance r and the skin depth δ for a conducting half space.

9.11.4. The Range of Small Parameters p_s

First, we study the range of small parameters p_s. Applying the method of deriving asymptotic formulas described above, we obtain

$$b_z \approx -1 + q^2 + q^3 \ln q - a_0 q^3 - q^4 + \ldots \quad \text{if } |q| \ll 1$$

or

$$Qb_z \approx \left(\frac{\omega\mu_0 Sr}{2}\right)^3 \ln\frac{\omega\mu_0 Sr}{2} - a_0\left(\frac{\omega\mu_0 Sr}{2}\right)^3 + \ldots$$

and

$$Inb_z^s \approx -\left(\frac{\omega\mu_0 Sr}{2}\right)^2 + \frac{\pi}{2}\left(\frac{\omega\mu_0 Sr}{2}\right)^3 - \left(\frac{\omega\mu_0 Sr}{2}\right)^4 + \ldots \tag{9.175}$$

where

$$a_0 = \ln 2 - C + 2, \quad C = 0.57721566$$

Therefore, in contrast to the case of a layered medium, the in-phase component of the secondary field B_z dominates over the quadrature component, which is proportional to $\omega^3 \ln \omega$ and the cube of longitudinal conductance S^3. It is obvious that in this range ($p_s < 1$), the component QB_z is practically independent of the distance from the dipole to the receiver. In fact, as follows from Eqs (9.173) and (9.174), we have

$$QB_z \approx \frac{\mu_0 M_0}{4\pi}\left(\frac{\omega\mu_0 S}{2}\right)^3 \ln\omega \quad \text{if } p_s < 1 \tag{9.176}$$

It is proper to note that the vertical component of the transient magnetic field $B_z(t)$ is also proportional to S^3 and is independent of the distance r at the late stage. This particular behavior of the quadrature component QB_z is observed only when both the dipole and the observation site are located on the conducting plane. For example, if the dipole and observation point are placed at a height h above the plane S, the secondary magnetic field B^s_{1z} can be written as

$$B^{s*}_{1z} = \frac{i\omega\mu_0 S}{4\pi}\mu_0 M_0 \int_0^\infty \frac{m^2\exp(-mh)}{2m - i\omega\mu_0 S} J_0(mr)\mathrm{d}m$$

In the limit when the parameter p_s tends to zero, we have

$$QB^*_{1z} \approx \frac{\mu_0 M_0}{8\pi}\omega\mu_0 S \frac{h}{(h^2+r^2)^{3/2}}, \quad \text{if } p_s \to 0 \tag{9.177}$$

that is, the leading term of the quadrature component of the field is directly proportional to ω, and it is not equal to zero. In this light, it is proper to note that in defining the conducting sheet, it was assumed that this model would be equivalent to a relatively thin layer with finite thickness. However, as this analysis shows these models are not equivalent when the quadrature component of B^*_z is considered at the range of small parameter p_s and the dipole and observation point are located at the plane S. It is not surprising because the distance from the array to plane is zero, but such equivalence can be expected when the distance is much greater than the layer thickness. At the same time, the first term of the expression for the quadrature component and the second term for the in-phase component (Eq. (9.175)) are the same as corresponding terms in the series representing the field at the surface of two-layered medium lying above an insulating basement. Applying the same approach for deriving asymptotic formulas, we have the following formulas for the electric field in the range of small parameters p_s:

$$InE^*_\varphi \approx -\frac{\omega\mu_0 M_0}{4\pi r^2}\left[\frac{\omega\mu_0 Sr}{2} + \left(\frac{\omega\mu_0 Sr}{2}\right)^3 \ln\frac{\omega\mu_0 Sr}{2} + b_1\left(\frac{\omega\mu_0 Sr}{2}\right)^3 + \ldots\right]$$

and

$$QE_\varphi^* \approx \frac{\omega\mu_0 M_0}{4\pi r^2}\left[1 - \left(\frac{\omega\mu_0 Sr}{2}\right)^2 + \frac{\pi}{4}\left(\frac{\omega\mu_0 Sr}{2}\right)^3 - \ldots\right] \tag{9.178}$$

where

$$b_1 = \frac{C}{4} - \frac{\ln 2}{4}$$

The first term in the expression for the quadrature component is the primary electric field E_φ^0. At small values of p_s, this field nearly defines the current in the conducting plane:

$$i_\varphi = \frac{\omega\mu_0 M_0}{4\pi r^2} S \tag{9.179}$$

This current gives rise to the magnetic field which is also proportional to ω and S. But, as was shown earlier, the vertical component of this magnetic field is zero if $z = h = 0$.

The first term of the expression for the in-phase component (Eq. (9.178)) is caused by a change of the quadrature component of the magnetic field with time which is proportional to ω^2. The second and following terms in Eq. (9.178) also reflect the interaction between currents. Finally from Eq. (9.170), we have for the horizontal component B_r^*:

$$InB_r^* \approx \frac{\omega\mu_0 S}{8\pi r^2}\mu_0 M_0\left[\frac{\omega\mu_0 Sr}{2} + \left(\frac{\omega\mu_0 Sr}{2}\right)^3 \ln\frac{\omega\mu_0 Sr}{2} + b_1\left(\frac{\omega\mu_0 Sr}{2}\right)^3 + \ldots\right]$$

and

$$QB_r^* \approx \frac{\omega\mu_0 S}{8\pi r^2}\mu_0 M_0\left[-1 + \left(\frac{\omega\mu_0 Sr}{2}\right)^2 - \frac{\pi}{4}\left(\frac{\omega\mu_0 Sr}{2}\right)^3 + \ldots\right] \tag{9.180}$$

Comparing Eqs (9.175) and (9.180), it can be seen that in contrast to the case of B_z, the quadrature component of B_r contains a term which is directly proportional to ω.

9.11.5. The Range of Large Parameters p_s

Next consider the behavior of the field when parameter p_s is large. Representing the fraction $1/_{x+q}$ in Eqs (9.172) and (9.174) as a series,

$$\frac{1}{x+q} = \frac{1}{q}\sum_{n=0}^{\infty}(-1)^n\left(\frac{x}{q}\right)^n,$$

and performing integration, we obtain the following expressions for the leading terms of the series which describe the field in the range of large parameters:

$$b_z \approx \frac{36}{(\omega\mu_0 Sr)^2}, \quad e_\varphi \approx -\frac{12}{(\omega\mu_0 Sr)^2} \quad \text{if } p_s \gg 1 \tag{9.181}$$

Thus in this range, the in-phase component of B_z^*, as well as the quadrature component of the electric field, are dominant and they are inversely proportional to S^2. For instance,

$$InB_z^* \approx \frac{9\mu_0 M_0}{\pi\omega^2\mu_0^2 S^2 r^5} \quad \text{if } p_s \gg 1 \tag{9.182}$$

Taking into account Eq. (9.170), it can be shown that the quadrature component of the horizontal magnetic field B_r is less sensitive to a change in the longitudinal conductance than the in-phase component of B_z. In fact, we have

$$QB_r^* \approx \frac{3\mu_0 M_0}{2\pi r^3}\frac{1}{\omega\mu_0 Sr} \quad \text{if } p_s \gg 1$$

Frequency responses of the field are shown in Figure 9.4. Finally note that as follows from Eq. (9.164), the integrand, describing the secondary field, has one singularity on the upper part of the complex plane of m and it is equal to

$$m = i\frac{\omega\mu_0 S}{2}$$

Applying the residual theorem, one can show that the vector potential is expressed in terms of the Hankel's function. Using the similar approach, we investigate the transient responses of the field.

9.12. TRANSIENT RESPONSES OF CURRENTS IN A CONDUCTING PLANE

9.12.1. Derivation of Formulas

Now we investigate the transient responses of the electromagnetic field caused by currents in a conducting plane. Applying Fourier's response to spectrum of the vector potential of the secondary field normalized by $i\omega$ (Eq. (9.164)), we have

$$A_{1z}(t) = -\frac{i\mu_0^2 SM_0}{16\pi^2}\int\limits_0^\infty \exp m(z-2h)J_0(mr)\mathrm{d}m\cdot \int\limits_{-\infty}^{\infty} \omega\frac{\exp(-i\omega t)}{m - i\omega\big/b}\mathrm{d}\omega, \tag{9.183}$$

where

$$b = \frac{2}{\mu_0 S} \tag{9.184}$$

First consider the integration over frequency:

$$L = \int\limits_{-\infty}^{\infty} \frac{\omega \exp(-i\omega t)}{m - i\omega\big/b}\mathrm{d}\omega \tag{9.185}$$

The integrand has a pole when

$$\omega_0 = -imb, \tag{9.186}$$

but there are no branch points. Placing the path of integration as is shown in Figure 9.5 and applying the residue theorem:

$$\frac{1}{2\pi i}\oint f(\omega)\mathrm{d}\omega = \frac{\varphi_1(\omega_0)}{\varphi_2'(\omega_0)},$$

where

$$f(\omega) = \frac{\varphi_1(\omega)}{\varphi_2(\omega)}$$

we have

$$\varphi_1(\omega_0) = -imb\exp(-mbt), \quad \varphi_2'(\omega_0) = -\frac{i}{b}$$

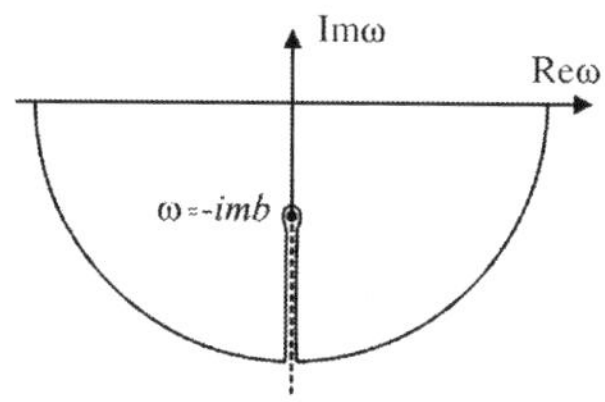

Figure 9.5 Path of integration in Eq. (9.185).

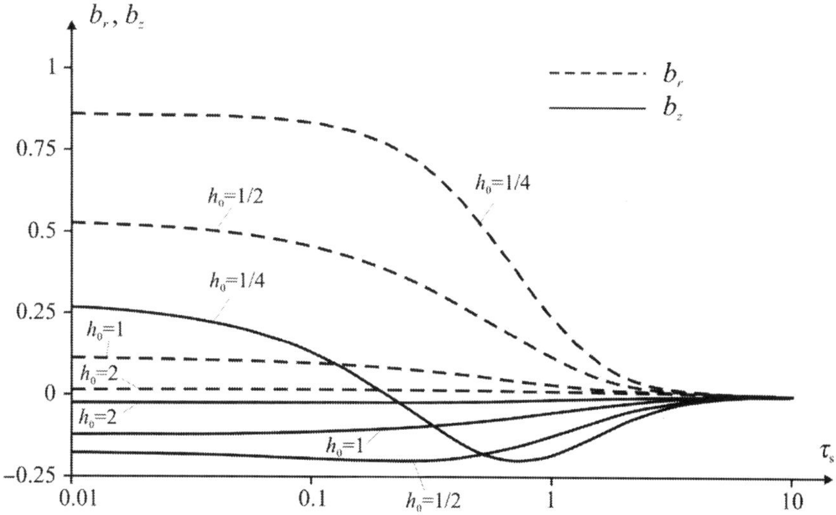

Figure 9.6 Transient responses of the fields b_z and b_r.

Thus,

$$L = 2\pi i b^2 m \exp(-mbt)$$

and correspondingly,

$$A_{1z}(t) = -\frac{\mu_0 M_0 b}{8\pi} \int_0^\infty m \exp[-m(bt + 2h - z)] J_0(mr)\mathrm{d}m \quad \text{if } z < h$$

or

$$A_{1z} = -\frac{\mu_0 M_0 b}{8\pi} \frac{\partial}{\partial z} \left[r^2 + (bt + 2h - z)^2\right]^{-1/2}$$

that is,

$$A_{1z} = -\frac{\mu_0 M_0 b}{8\pi} \frac{(bt + 2h - z)}{\left[r^2 + (bt + 2h - z)^2\right]^{3/2}} \tag{9.187}$$

Respectively, for the electric field, we have

$$E_\varphi = -\frac{\partial A_{1z}}{\partial r} = -\frac{3M_0}{2\pi S} \frac{r(bt + 2h - z)}{\left[r^2 + (bt + 2h - z)^2\right]^{5/2}} \tag{9.188}$$

Applying the same approach for the magnetic field, we obtain

$$B_z(z) = \frac{\mu_0 M_0}{4\pi} \frac{r^2 - 2(bt + 2h - z)^2}{\left[r^2 + (bt + 2h - z)^2\right]^{5/2}} \tag{9.189}$$

$$B_r(t) = \frac{3\mu_0 M_0}{4\pi} \frac{r(bt + 2h - z)}{\left[r^2 + (bt + 2h - z)^2\right]^{5/2}} \tag{9.190}$$

These expressions are remarkably simple and they are useful in an interpretation of transient soundings. It can be seen from Eqs (9.188)–(9.190) that in all cases, there is a single relationship between the electric field and the horizontal component of the magnetic field:

$$B_r(t) = -\frac{\mu_0 S}{2} E_\varphi(t) \tag{9.191}$$

Now we consider several specific cases.

9.12.2. The Dipole and Observation Point are Located at the Same Axis

Letting $r = 0$, we obtain

$$B_z(t) = -\frac{\mu_0 M_0}{2\pi} \frac{1}{(bt + 2h - z)^3} \tag{9.192}$$

and

$$B_r = E_\varphi = 0$$

It is clear that during the early stage of the transient response,

$$B_z(t) \to -\frac{\mu_0 M_0}{2\pi(2h-z)^3} \quad \text{if } bt < 2h - z$$

This expression describes the field of a fictitious magnetic dipole located on the z-axis at the point $z = 2h$. This analogy is easily generalized to a more common case when neither t nor r are zero. The late stage begins when

$$\frac{2t}{\mu_0 S} \gg 2h - z \tag{9.193}$$

and then we have

$$B_z(t) \approx -\frac{M_0 \mu_0^4 S^3}{16\pi t^3} \tag{9.194}$$

We see that during the late stage, the field B_z is independent of distance (either z or h), and it is characterized by a very high sensitivity to the parameter S:

$$B_z \sim S^3$$

Note that from the physical point of view, the condition (9.193) indicates that induction currents are situated relatively far away from the z-axis. As follows from Eq. (9.192), in order to determine the conductance S and the height h, it is sufficient in principle to make measurements at only two times, corresponding, for example, to the intermediate portion of the transient response. If the receiver and transmitter coils are coincident, then from Eq. (9.188), we have

$$B_z(t) = -\frac{\mu_0 M_0}{2\pi(bt+2h)^3},$$

and therefore, a single coil system can be used in determining both parameters S and h.

9.12.3. The Dipole and Observation Point are Situated on the Plane *S*

Assuming $z = h = 0$, we have

$$B_z(t) = \frac{\mu_0 M_0}{4\pi}\frac{r^2 - 2b^2t^2}{(r^2+b^2t^2)^{5/2}} \quad \text{and} \quad B_r(t) = \frac{3\mu_0 M_0}{4\pi} r \frac{bt}{(r^2+b^2t^2)^{5/2}}$$

and

$$E_\varphi(t) = -\frac{3M_0}{2\pi S} r \frac{bt}{(r^2 + b^2 t^2)^{5/2}} \tag{9.195}$$

Let us represent these equations in the form

$$B_z(t) = \frac{\mu_0 M_0}{4\pi r^3} b_z, \quad B_r(t) = \frac{\mu_0 M_0}{4\pi r^3} b_r, \quad E_\varphi(t) = -\frac{M_0}{2\pi S r^3} e_\varphi$$

where

$$b_z = \frac{1 - 2\tau_s^2}{\left(1 + \tau_s^2\right)^{5/2}}, \quad b_r = \frac{3\tau_s}{\left(1 + \tau_s^2\right)^{5/2}}, \quad e_\varphi = b_r \tag{9.196}$$

and

$$\tau_s = \frac{bt}{r} = \frac{2t}{\mu_0 S r} \tag{9.197}$$

During the early stage when τ_s tends to zero, we have

$$b_z \approx 1 - \frac{9}{2}\tau_s^2, \quad b_r \approx 3\tau_s - \frac{15}{2}\tau_s^3, \quad \text{if } \tau_s \ll 1 \tag{9.198}$$

Thus, when the dipole current is turned off, induction currents arise immediately in the vicinity of the dipole creating a vertical component of the magnetic field which is equal to the primary magnetic field ($t \to 0$). At that instant, the horizontal component of the magnetic field along with the electric field is zero at the plane S. This means that at the instant $t = 0$, induced currents are absent everywhere except the lateral surface of the current source. In the other extreme during the late stage, we obtain

$$b_z \to -\frac{2}{\tau_s^3}, \quad b_r \to \frac{3}{\tau_s^4}$$

or

$$B_z \to -\frac{\mu_0^4 M_0 S^3}{16\pi t^3}, \quad B_r \to \frac{3\mu_0 M_0}{64\pi} r \frac{\mu_0^4 S^4}{t^4},$$
$$E_\varphi \to -\frac{3M_0}{32\pi} r \frac{\mu_0^4 S^3}{t^4} \quad \text{if } \tau_s \gg 1 \tag{9.199}$$

Thus, at the late stage of the transient response, the vertical component of the magnetic field is independent of the separation r. Such behavior has already been observed in the case of a horizontally layered medium. At the same time, the horizontal component B_r is less than the vertical component, but it is more sensitive to a change in the conductance:

$$B_r \sim S^4$$

In accordance with Eq. (9.195), the current density at any point in the plate S is

$$i_\varphi = -\frac{3M_0}{2\pi r^3}\frac{\tau_s}{\left(1+\tau_s^2\right)^{5/2}}$$

It is clear that regardless of the distance r ($r \neq 0$), the surface density i_φ increases with time, reaches a maximum, and then tends to zero. The maximal magnitude of the current density is

$$\left(i_\varphi\right)_{\max} \approx 0.3\frac{M_0}{\pi r^3}$$

9.12.4. The Dipole and Observation Point are at the Height *h* above the Plane *S*

Letting in Eqs (9.188)–(9.190) $z = 0$, we obtain

$$B_z = B_z^0 b_z, \quad B_r = B_z^0 b_r \quad E_\varphi = -\frac{M_0}{2\pi S r^3} e_\varphi$$

where

$$b_z = \frac{1-2(\tau_s+2h_0)^2}{\left[1+(\tau_s+2h_0)^2\right]^{5/2}} \quad b_r = \frac{3(\tau_s+2h_0)}{\left[1+(\tau_s+2h_0)^2\right]^{5/2}} \tag{9.200}$$

and

$$e_\varphi = b_r, \quad h_0 = h/r$$

In this case, the early stage occurs if

$$h \gg t/\mu_0 S \tag{9.201}$$

If so we have

$$b_z = \frac{1-8h_0^2}{\left(1+4h_0^2\right)^{5/2}}, \quad b_r = \frac{6h_0}{\left(1+4h_0^2\right)^{5/2}}$$

Comparing Eqs (9.200) and (9.196), we see that they practically coincide when the inequality

$$\tau_s \gg 2h_0 \tag{9.202}$$

is met. In other words, the influence of height vanishes if the time of observation satisfies the condition

$$t > \mu_0 Sh \tag{9.203}$$

It is also obvious that the late stage of the transient field contains no information about the height h and it is described by Eq. (9.199).

REFERENCES AND FURTHER READING

[1] A.A. Kaufman, G.M. Morozova, Theoretical Principles of Transient Soundings in the Near Zone, Nauka, Novosibirsk, 1970.
[2] A.A. Kaufman, G.V. Keller, Frequency and Transient Sounding, Elsevier, Amsterdam, 1983.
[3] L.L. Vanyan, Principles of Electromagnetic Soundings, Nedra, Moscow, 1965.
[4] J.R. Wait, Geoelectromagnetism, Academic Press, New York, NY, 1982.
[5] J.R. Wait, Electromagnetic Waves in Stratified Media, Pergamon Press, New York, NY, 1972.

CHAPTER TEN

Horizontal Magnetic Dipole above the Surface of a Layered Medium

10.1. FORMULATION OF BOUNDARY VALUE PROBLEM FOR VECTOR POTENTIAL

10.1.1. Relation between the Vector Potential and Electromagnetic Field

Suppose that a horizontal magnetic dipole is located above the Earth's surface at the height h_1, and its moment $\mathbf{M}$ is directed along the x-axis as shown in Figure 10.1(a). First we derive expressions for the electromagnetic field in the frequency domain and then consider several cases. It is convenient to introduce both Cartesian and cylindrical systems of coordinates with a common origin where the dipole is situated. As before, we make use of the vector potential of the electric type which is related to the electric and magnetic fields by

$$\mathbf{E}^* = \text{curl}\,\mathbf{A}^*, \quad i\omega\mathbf{B}^* = k^2\mathbf{A}^* + \text{grad div}\,\mathbf{A}^* \tag{10.1}$$

and at regular points satisfies Helmholtz equation

$$\nabla^2\mathbf{A}^* + k^2\mathbf{A}^* = 0, \tag{10.2}$$

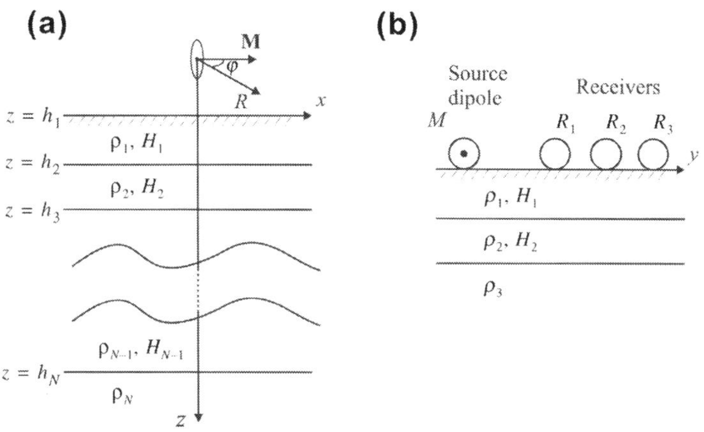

Figure 10.1 The horizontal magnetic dipole above the earth's surface.

Methods in Geochemistry and Geophysics, Volume 45
ISSN 0076-6895,
http://dx.doi.org/10.1016/B978-0-444-53829-1.00010-1

where $\mathbf{A}^*$ is the complex amplitude. Inasmuch as the primary electric field is located in vertical planes, it intersects the Earth's surface, and we have to expect an appearance of electric charges. This means that unlike the previous cases when the field is generated by the vertical magnetic dipole, the electromagnetic field of the horizontal dipole is caused by both the induction currents and surface charges. Correspondingly, the geometry of the field becomes more complicated, and for this reason, we will try to solve the boundary value problem with help of two components of the vector potential A_x^* and A_z^*, assuming that $A_y^* = 0$. It is clear that the complex amplitude of each component of the vector potential obeys the Helmholtz equation:

$$\nabla^2 A_x^* + k^2 A_x^* = 0, \quad \nabla^2 A_z^* + k^2 A_z^* = 0, \tag{10.3}$$

and for components of the electromagnetic field, we have

$$E_x^* = \frac{\partial A_z^*}{\partial y}, \quad E_y^* = \frac{\partial A_x^*}{\partial z} - \frac{\partial A_z^*}{\partial x}, \quad E_z^* = -\frac{\partial A_x^*}{\partial y},$$

and

$$i\omega B_x^* = k^2 A_x^* + \frac{\partial}{\partial x}\text{div } \mathbf{A}^*, \quad i\omega B_y^* = \frac{\partial}{\partial y}\text{div } \mathbf{A}^* \tag{10.4}$$

and

$$i\omega B_z^* = k^2 A_z^* + \frac{\partial}{\partial z}\text{div } \mathbf{A}^*$$

Therefore, in order to provide continuity of the tangential components of the electric and magnetic fields at the ith surface, the vector potential should satisfy the following conditions:

$$\begin{aligned}
&\frac{\partial A_{z,i}^*}{\partial y} = \frac{\partial A_{z,i+1}^*}{\partial y} \\
&\frac{\partial A_{x,i}^*}{\partial z} - \frac{\partial A_{z,i}^*}{\partial x} = \frac{\partial A_{x,i+1}^*}{\partial z} - \frac{\partial A_{z,i+1}^*}{\partial x} \\
&k_i^2 A_{x,i}^* + \frac{\partial}{\partial x}\text{div } \mathbf{A}_i^* = k_{i+1}^2 A_{x,i+1}^* + \frac{\partial}{\partial x}\text{div } \mathbf{A}_{i+1}^* \\
&\frac{\partial}{\partial y}\text{div } \mathbf{A}_i^* = \frac{\partial}{\partial y}\text{div } \mathbf{A}_{i+1}^*
\end{aligned} \tag{10.5}$$

Certainly this system looks like much more complicated than that for the vertical magnetic dipole, but it can be drastically simplified if we represent Eq. (10.5) as two groups of equations, where the first group is

$$k_i^2 A_{x,i}^* = k_{i+1}^2 A_{x,i+1}^*, \quad \text{and} \quad \frac{\partial A_{x,i}^*}{\partial z} = \frac{\partial A_{x,i+1}^*}{\partial z}, \tag{10.6}$$

and the second group

$$A_{z,i}^* = A_{z,i+1}^*, \quad \text{div}\, \mathbf{A}_i^* = \text{div}\, \mathbf{A}_{i+1}^* \tag{10.7}$$

It is obvious that if Eqs (10.6) and (10.7) are satisfied, then the components of the vector potential also obey Eq. (10.5). It is essential that conditions for the component A_x, directed along the dipole moment, do not contain the component A_z. This fact simplifies a solution of the boundary value problem since it allows us to determine first of all the component A_x and then find out the expression for the vertical component A_z. Correspondingly, let us start from formulation of the boundary value problem for the component A_x.

10.1.2. Boundary Value Problem for the Component A_x

In accordance with Eqs (10.3) and (10.6) at regular points and at interfaces, the complex amplitude of the component A_x satisfies the following conditions:

$$\nabla^2 A_x^* + k^2 A_x^* = 0$$

and

$$k_i^2 A_{x,i}^* = k_{i+1}^2 A_{x,i+1}^*, \quad \frac{\partial A_{x,i}^*}{\partial z} = \frac{\partial A_{x,i+1}^*}{\partial z}$$

Near the dipole, the primary field, caused by the dipole current, dominates and we have

$$A_x^* \rightarrow A_{x0}^* = \frac{i\omega\mu_0 M_0}{4\pi R} = \frac{i\omega\mu_0 M_0}{4\pi} \int_0^\infty \exp\left(-m|z|\right) J_0(mr)\mathrm{d}m \quad \text{if } R \rightarrow 0 \tag{10.8}$$

In contrast, at infinity, A_x^* tends to zero:

$$A_x^* \rightarrow 0 \quad \text{if } R \rightarrow \infty \tag{10.9}$$

We formulated the boundary value problem for the vector potential A_x, and in accordance with the theorem of uniqueness, all these conditions define this field uniquely.

10.1.3. Expressions for the Horizontal Component A_x

Taking into account Eq. (10.8) and by analogy with the case of the vertical magnetic dipole, solution of Helmholtz equation for the complex amplitude of the component A_x in each layer has the form

$$A^*_{x,i}(r,z) = \frac{i\omega\mu_0 M_0}{4\pi}\int\limits_0^\infty [C_i \exp(-m_i z) + D_i \exp(m_i z)]J_0(mr)\mathrm{d}m$$

Thus, we have

$$A^*_{x,0} = \frac{i\omega\mu_0 M_0}{4\pi}\int\limits_0^\infty [\exp(-m|z|) + D_0 \exp(mz)]J_0(mr)\mathrm{d}m \quad \text{if} \quad z < h_1$$

$$A^*_{x,1} = \frac{i\omega\mu_0 M_0}{4\pi}\int\limits_0^\infty [C_1 \exp(-m_1 z) + D_1 \exp(m_1 z)]J_0(mr)\mathrm{d}m$$

if $h_1 < z < h_2$

$$A^*_{x,2} = \frac{i\omega\mu_0 M_0}{4\pi}\int\limits_0^\infty [C_2 \exp(-m_2 z) + D_2 \exp(m_2 z)]J_0(mr)\mathrm{d}m$$

if $h_2 < z < h_3$

And so forth. For Nth layer we have:

$$A^*_{x,N} = \frac{i\omega\mu_0 M_0}{4\pi}\int\limits_0^\infty C_N \exp(-m_N z)J_0(mr)\mathrm{d}m \quad \text{if} \quad z > h_N \qquad (10.10)$$

where

$$m_i = \left(m^2 - k_i^2\right)^{1/2} \quad \text{and} \quad k_i^2 = i\gamma_i\mu_0\omega$$

10.1.4. The System of Equations for A_x^* and Its Solution

Now, applying conditions Eq. (10.6) at each surface and assuming that above the earth surface the wave number is equal to zero, we obtain a system of equations for determination of unknown coefficients:

$$k_1^2[C_1 \exp(-m_1h_1) + D_1 \exp(m_1h_1)] = 0,$$

$$m_1[-C_1 \exp(-m_1h_1) + D_1 \exp(m_1h_1)]$$

$$= m[-\exp(-mh_1) + D_0 \exp(mh_1)],$$

$$k_1^2[C_1 \exp(-m_1h_2) + D_1 \exp(m_1h_2)]$$

$$= k_2^2[C_2 \exp(-m_2h_2) + D_2 \exp(m_2h_2)],$$

$$m_1[-C_1 \exp(-m_1h_2) + D_1 \exp(m_1h_2)]$$
$$= m_2[-C_2 \exp(-m_2h_2) + D_2 \exp(m_2h_2)], \ldots$$

$$k_{N-1}^2[C_{N-1} \exp(-m_{N-1}h_N) + D_{N-1} \exp(m_{N-1}h_N)]$$
$$= k_N^2 C_N \exp(-m_N h_N),$$

$$m_{N-1}[-C_{N-1} \exp(-m_{N-1}h_N) + D_{N-1} \exp(m_{N-1}h_N)]$$
$$= -m_N C_N \exp(-m_N h_N).$$

Performing the standard procedure for solving this system, we find that

$$C_i = D_i = 0$$

and (10.11)

$$D_0 = \exp(-2mh_1)$$

Therefore in a conducting medium, the horizontal component of the vector potential is equal to zero:

$$A_x = 0 \quad \text{if} \quad z > h_1 \tag{10.12}$$

10.1.5. Remarkable Behavior of the Vertical Component of Electric Field E_z

In accordance with Eq. (10.1), we have

$$E_z = -\frac{\partial A_x}{\partial y}$$

Thus beneath the Earth's surface, the vertical component of the electric field is equal to zero:

$$E_z = 0 \tag{10.13}$$

Certainly, it is an unexpected result since the primary electric field has the vertical component and it is not easy to predict that the vertical component of the secondary electric field completely compensates at each point of a conducting medium the same component of the primary field, and the total electric field inside the medium has only horizontal component. It turns out that the charges arising on the Earth's surface are distributed in such "clever" way that the vertical component of the secondary electric field at each point of a conducting medium and this component of the primary electric field are equal by magnitude but differs by sign. Correspondingly, in accordance with Ohm's law, the induced currents are located in horizontal planes too.

10.1.6. The Vertical Component of the Electric Field above the Earth's Surface

At the same time, unlike a conducting medium above the Earth's surface, the component A_x differs from zero and it consists of two parts. In fact, as follows from Eqs (10.10) and (10.11), we have

$$A^*_{x,0} = \frac{i\omega\mu_0 M_0}{4\pi} \int\limits_0^\infty \{\exp(-m|z|) + \exp[-m(2h_1 - z)]\} J_0(mr)\mathrm{d}m$$

or

$$A^*_{x,0} = \frac{i\omega\mu_0 M_0}{4\pi}\left(\frac{1}{R} + \frac{1}{R_1}\right) \quad \text{if } z < h_1 \tag{10.14}$$

Here,

$$R = \left(r^2 + z^2\right)^{1/2}, \quad R_1 = \left[r^2 + (2h_1 - z)\right]^{1/2}$$

In particular, when the dipole is located at the Earth's surface, $h_1 = 0$, we have

$$A^*_{x,0} = \frac{i\omega\mu_0 M_0}{2\pi R} \tag{10.15}$$

In upper half space, unlike the conducting medium, the normal component of the electric field is not equal to zero but it is independent on a conductivity of a medium. By definition,

$$E^*_{z,0} = -\frac{\partial A^*_{x,0}}{\partial y} = \frac{i\omega\mu_0 M_0}{4\pi} y \left(\frac{1}{R^3} + \frac{1}{R_1^3}\right) \tag{10.16}$$

The latter allows us to calculate the density of charges on the Earth's surface which is equal to

$$\sigma(r,0) = \varepsilon_0[E_{1z}(r,0) - E_{0z}(r,0)]$$

or

$$\sigma(r,0) = -\varepsilon_0 E_{0z}(r,0)$$

Taking into account Eq. (10.16), we obtain

$$\sigma^*(r,0) = -\varepsilon_0 \frac{i\omega\mu_0 M_0}{2\pi R^3} y \quad \text{if } z = h_1 \tag{10.17}$$

We see that distribution of charges is defined by geometry of the primary electric field. Its vector lines are located in planes perpendicular to the x-axis, and each of them intersects the surface twice. Correspondingly (Eq. (10.17)), charges of opposite sign appear at the same instant at points where $y > 0$ and $y < 0$, and their magnitude decreases rapidly with the distance from the dipole. Of course, in the vicinity of the x-axis, where the primary electric field is zero, charges are absent. Also these charges are absent at all interfaces beneath the Earth's surface.

10.2. THE VERTICAL COMPONENT OF THE VECTOR POTENTIAL A_z^*

10.2.1. Integral Representation of A_z^*

Next we find an analytical expression for the vertical component of the vector potential. First of all, in accordance with Eq. (10.12), the equation for div **A** in the conducting medium is greatly simplified and we have

$$\text{div}\,\mathbf{A} = \frac{\partial A_z}{\partial z} \quad \text{if } z > h_1$$

Therefore, the boundary conditions for the complex amplitude of this component are

$$A^*_{z,i} = A^*_{z,i+1} \quad \text{and} \quad \frac{\partial A^*_{z,i}}{\partial z} = \frac{\partial A^*_{z,i+1}}{\partial z}, \quad \text{if} \;\; i \neq 0 \tag{10.18}$$

and at the Earth's surface,

$$A^*_{z,0} = A^*_{z,1} \quad \text{and} \quad \frac{\partial A^*_{x,0}}{\partial x} + \frac{\partial A^*_{z,0}}{\partial z} = \frac{\partial A^*_{z,1}}{\partial z} \quad \text{if} \;\; z = h_1 \tag{10.19}$$

It is obvious that in order to satisfy the last equality in the set Eq. (10.19), all terms must have the same dependence on coordinates of the point. In other words, this equality should be valid regardless on a position of a point at the Earth's surface. It is clear that

$$\frac{\partial A^*_{x,0}}{\partial x} = \frac{\partial A^*_{x,0}}{\partial r}\frac{\partial r}{\partial x} = \cos\varphi \frac{\partial A^*_{x,0}}{\partial r}$$

Here, φ is the angle between the x-axis along which the dipole moment is directed and radius $\mathbf{r}$ of an observation point. Thus,

$$\frac{\partial A^*_{x,0}}{\partial x} = -\frac{i\omega\mu_0 M_0}{4\pi}\cos\varphi \int\limits_0^\infty m[\exp(-m|z|) + \exp(-m2h_1)\cdot\exp(mz)]J_1(mr)\mathrm{d}m \tag{10.20}$$

since

$$J_0'(x) = -J_1(x)$$

The requirement that the derivative $\partial A^*_z/\partial z$ in Eq. (10.19) has to have the same dependence on r and φ as $\partial A^*_{x,0}/\partial x$ allows us to find an expression for the component $A^*_{z,i}$:

$$A^*_{z,i} = \frac{i\omega\mu_0 M_0}{4\pi}\cos\varphi \int\limits_0^\infty [C^*_i \exp(-m_i z) + F^*_i \exp(m_i z)]J_1(mr)\mathrm{d}m, \tag{10.21}$$

and this function obeys Helmholtz equation:

$$\frac{\partial^2 A^*_{z,i}}{\partial r^2} + \frac{1}{r}\frac{\partial A^*_{z,i}}{\partial r} + \frac{\partial^2 A^*_{z,i}}{\partial z^2} + \frac{1}{r^2}\frac{\partial^2 A^*_{z,i}}{\partial \varphi^2} + k_i^2 A^*_{z,i} = 0$$

In particular, above the Earth's surface, we have

$$A^*_{z,0} = \frac{i\omega\mu_0 M_0}{4\pi}\cos\varphi \int\limits_0^\infty F^*_0 \exp(mz) J_1(mr)\mathrm{d}m \tag{10.22}$$

Comparison of Eqs (10.10) and (10.21) shows that both components of the vector potential have different dependence on coordinates of a point.

10.2.2. The First Form of Coefficient F_0^* for Three-Layered Medium

Now, applying boundary conditions (Eqs (10.18) and (10.19)) at the interfaces, we can determine the unknown coefficients C_i^* and F_i^*. For illustration, consider the case of the three-layered medium. This gives

$$
\begin{aligned}
F_0^* \exp(mh_1) &= C_1^* \exp(-m_1h_1) + F_1^* \exp(m_1h_1) - 2m \exp(-mh_1) \\
&+F_0^* m \exp(mh_1) = -m_1\left[C_1^* \exp(-m_1h_1) - F_1^* \exp(m_1h_1)\right], \\
&C_1^* \exp(-m_1h_2) + F_1^* \exp(m_1h_2) = C_2^* \exp(-m_2h_2) \\
&+F_2^* \exp(m_2h_2) \\
m_1\left[C_1^* \exp(-m_1h_2) - F_1^* \exp(m_1h_2)\right] &= m_2\left[C_2^* \exp(-m_2h_2) - F_2^* \exp(m_2h_2)\right] \\
&C_2^* \exp(-m_2h_3) + F_2^* \exp(m_2h_3) = C_3^* \exp(-m_3h_3) \\
&m_2\left[C_2^* \exp(-m_2h_3) - F_2^* \exp(m_2h_3)\right] = m_3C_3^* \exp(-m_3h_3)
\end{aligned} \tag{10.23}
$$

Eliminating all coefficients except F_0^*, we obtain

$$F_0^* = \frac{2m \exp(-2mh_1)}{m + m_1P_3}, \tag{10.24}$$

where

$$P_3 = \frac{1 - P_2 \exp(-2mH_1)}{1 + P_2 \exp(-2mH_1)}, \quad P_2 = \frac{m_1 - m_2P_1}{m_1 + m_2P_1}$$

and

$$P_1 = \frac{1 - m_{23} \exp(-2m_2H_2)}{1 + m_{23} \exp(-2m_2H_2)}$$

Here

$$m_{23} = \frac{m_2 - m_3}{m_2 + m_3}$$

and H_2 and H_3 are the layer thicknesses. To illustrate Eq. (10.24), consider several cases.

Case 1: Uniform half space with conductivity γ_1
Then $H_1 \to \infty$ and $P_3 = 1$. Thus,

$$F_0^* = \frac{2m \exp(-2mh_1)}{m + m_1} \tag{10.25}$$

Case 2: Uniform half space with conductivity γ_2
Letting $\gamma_1 = \gamma_2$ and $H_2 \to \infty$, we have $P_1 = 1$, and $P_2 = 0$. Therefore, as before $P_3 = 1$ and

$$F_0^* = \frac{2m \exp(-2mh_1)}{m + m_2}$$

Case 3: Two-layered medium, $\gamma_2 = \gamma_3$
Then, $m_{23} = 0$ and $P_1 = 1$. Thus, $P_2 = m_{12}$ and

$$P_3 = \frac{1 - m_{12} \exp(-2m_1 H_1)}{1 + m_{12} \exp(-2mH_1)} \tag{10.26}$$

Case 4: Two-layered medium, $\gamma_1 = \gamma_2$
Then,

$$P_2 = \frac{1 - P_1}{1 + P_1} = m_{13} \exp(-2m_1 H_2)$$

whence,

$$P_3 = \frac{1 - m_{13} \exp[-2m_1(H_1 + H_2)]}{1 + m_{13} \exp[-2m_1(H_1 + H_2)]},$$

and the latter again corresponds to two-layered medium when the thickness of the upper layer is $H_1 + H_2$.

10.2.3. The Second Form for the Coefficient F_0^*

By analogy with the vertical magnetic dipole, we can derive the recursive relation for the function F_0^*, which is applied for *N*-layered medium, and this gives (Eq. (10.23)),

$$F_0^* = \frac{2m \exp(-2mh_1)}{m + m_1/R_3}, \tag{10.27}$$

where R_3 is a well-known function:

$$R_3 = \coth\left\{m_1 H_1 + \coth^{-1}\frac{m_1}{m_2}\coth\left[m_2 H_2 + \coth^{-1}\frac{m_2}{m_3}\right]\right\}$$

Correspondingly, the vertical component of the vector potential in N-layered medium can be represented as

$$A_z^{*0} = \frac{i\omega\mu_0 M_0}{2\pi}\cos\varphi\int_0^\infty \frac{m\exp(-2mh_1)\exp(mz)}{m + m_1/R_N} J_1(mr)dm \qquad (10.28)$$

Note that comparison with Eq. (10.24) shows that $R_3 = P_3^{-1}$.

10.3. THE COMPONENT OF THE MAGNETIC FIELD B_X

In accordance with Eqs (10.14) and (10.22), the components of the vector potential above the Earth's surface are

$$A_{x,0}^* = \frac{i\omega\mu_0 M_0}{4\pi}\left(\frac{1}{R} + \frac{1}{R_1}\right)$$

and

$$A_{z,0}^* = \frac{i\omega\mu_0 M_0}{2\pi}\cos\varphi\int_0^\infty \frac{m\exp(-2mh_1)}{m + m_1 P_3}\exp(mz)J_1(mr)dm$$

10.3.1. Derivation of Expression for B_x

Usually in performing electromagnetic profiling, the horizontal component B_x is measured at the equatorial plane of the dipole ($\varphi = \pi/2$). Taking into account this fact, we focus on this configuration of an array. As follows from Eq. (10.4),

$$i\omega B_x = \frac{\partial}{\partial x}\text{div}\,\mathbf{A}_0, \quad \text{if } z \le h_1$$

Inasmuch as

$$\text{div}\,\mathbf{A}_0 = \frac{\partial A_{x,0}}{\partial x} + \frac{\partial A_{z,0}}{\partial z} = \frac{i\omega\mu_0 M_0}{4\pi}\frac{\partial}{\partial x}\left(\frac{1}{R} + \frac{1}{R_1}\right)$$

$$+\frac{i\omega\mu_0 M_0}{2\pi}\cos\varphi\int_0^\infty \frac{m^2\exp(-2mh_1)}{m + m_1 P_3}\exp(mz)J_1(mr)dm,$$

we have

$$B_x = \frac{\mu_0 M_0}{4\pi}\left[\frac{\partial^2}{\partial x^2}\left(\frac{1}{R}+\frac{1}{R_1}\right) + 2\frac{\partial}{\partial x}\frac{x}{r}\int_0^\infty \frac{m^2 \exp(-2mh_1)\exp(mz)}{m+m_1P_3}J_1(mr)dm\right]$$

Correspondingly, at the y-axis when $x=0$, we obtain

$$B_x^* = \frac{\mu_0 M_0}{4\pi}\left[-\frac{1}{R^3}-\frac{1}{R_1^3}+\frac{2}{r}\int_0^\infty \frac{m^2 \exp(-2mh_1)\exp(mz)}{m+m_1P_3}J_1(mr)dm\right]$$

Next suppose that both the dipole and receiver are located on the Earth's surface. Then bearing in mind that $P_3 = 1/R_3$, we have in the case of three-layered medium

$$B_x^* = \frac{\mu_0 M_0}{2\pi r}\left[-\frac{1}{r^2}+\int_0^\infty \frac{m^2 J_1(mr)}{m+m_1/R_3}\,dm\right] \tag{10.29}$$

10.3.2. Quadrature Component of the Field at the Range of Small Parameters *r*/*δ*

Now we will perform some transformations of Eq. (10.29) assuming that wave numbers in each layer tend to zero. This is related to the fact that the transmitter–receiver separation is usually much smaller than the skin depth in each layer. Taking into account that

$$m_i \approx m - \frac{1}{2}\frac{k_i^2}{m} \quad \text{if } k_i \to 0$$

we have

$$m_{23} \approx \frac{1}{4m^2}\left(k_3^2-k_2^2\right) \quad \text{and} \quad P_1 \approx 1-\frac{1}{2m^2}\left(k_3^2-k_2^2\right)\exp(-2mH_2)$$

$$P_2 \approx \frac{1}{4m^2}\left[\left(k_2^2-k_1^2\right)+\left(k_3^2-k_2^2\right)\exp(-2mH_2)\right]$$

$$P_3 \approx 1-\frac{1}{2m^2}\left[\left(k_2^2-k_1^2\right)+\left(k_3^2-k_2^2\right)\exp(-2mH_2)\right]\exp(-2mH_1)$$

Whence

$$m_1 P_3 = \left(m - \frac{k_1^2}{2m}\right)\left(1 - \frac{\Delta}{2m^2}\right)$$

where

$$\Delta = \left[\left(k_2^2 - k_1^2\right) + \left(k_3^2 - k_2^2\right)\exp\left(-2mH_2\right)\right]\exp\left(-2mH_1\right)$$

and

$$m + m_1 P_3 = 2m - \left[\frac{k_1^2}{2m} + \frac{\Delta}{2m}\right]$$

Substituting the latter into Eq. (10.29), we obtain

$$B_x^* \approx \frac{\mu_0 M_0}{2\pi}\left[-\frac{1}{r^3} + \frac{1}{2r}\int_0^\infty mJ_1(mr)\mathrm{d}m + \frac{k_1^2}{8r}\int_0^\infty \frac{J_1(mr)}{m}\mathrm{d}m\right.$$

$$+\frac{k_2^2 - k_1^2}{8r}\int_0^\infty \frac{\exp\left(-2mH_1\right)}{m}J_1(mr)\mathrm{d}m$$

$$\left.+\frac{k_3^2 - k_2^2}{8r}\int_0^\infty \frac{\exp\left[-2m(H_1 + H_2)\right]}{m}J_1(mr)\mathrm{d}m\right]$$

Note that the first two terms in brackets describe the primary field. Inasmuch as

$$\int_0^\infty \frac{\exp\left(-m|z|\right)}{m}J_1(mr)\mathrm{d}m = \frac{\left(r^2 + z^2\right)^{1/2} - z}{r},$$

we have

$$\int_0^\infty \frac{J_1(mr)}{m}\mathrm{d}m = 1, \quad \int_0^\infty \frac{\exp(-2mH_1)}{m}J_1(mr)\mathrm{d}m = \frac{\left(r^2 + 4H_1^2\right)^{1/2} - 2H_1}{r}$$

and

$$\int_0^\infty \frac{\exp\left[-2m(H_1 + H_2)\right]}{m}J_1(mr)\mathrm{d}m = \frac{\left[r^2 + 4(H_1 + H_2)^2\right]^{1/2} - 2(H_1 + H_2)}{r}$$

Thus, the quadrature component of the field, proportional to ω, is equal to

$$B_x^{s*} = \frac{\mu_0 M_0}{16\pi r}\left\{k_1^2\left[1 - \frac{\left(r^2+4H_1^2\right)^{1/2} - 2H_1}{r}\right] + k_2^2\left[\frac{\left(r^2+4H_1^2\right)^{1/2} - 2H_1}{r} - \frac{\left[r^2+4(H_1+H_2)^2\right]^{1/2} - 2(H_1+H_2)}{r}\right] + k_3^2\left[\frac{\left[r^2+4(H_1+H_2)^2\right]^{1/2} - 2(H_1+H_2)}{r}\right]\right\} \tag{10.30}$$

Here, $r = y$.

10.3.3. Geometric Factors of Layers

This equation clearly demonstrates that at the low-frequency part of the spectrum, the quadrature component of the magnetic field QB_x, as in the case of the vertical magnetic dipole, can be described with the help of geometric factors. In fact, the field QB_x can be represented in the form:

$$QB_x = \frac{\omega\mu_0^2 M_0}{16\pi r}[\gamma_1 G_1 + \gamma_2 G_2 + \gamma_3 G_3] \tag{10.31}$$

where

$$G_1 = 1 - \frac{\left(r^2+4H_1^2\right)^{1/2} - 2H_1}{r}$$

$$G_2 = \frac{\left(r^2+4H_1^2\right)^{1/2} - 2H_1}{r} - \frac{\left[r^2+4(H_1+H_2)^2\right]^{1/2} - 2(H_1+H_2)}{r}$$

$$G_3 = \frac{\left[r^2+4(H_1+H_2)^2\right]^{1/2} - 2(H_1+H_2)}{r} \tag{10.32}$$

are geometric factors for the upper and intermediate layers, as well as for the lowermost medium, respectively. The function G_2 is called the geometric factor of a layer of finite thickness, and it is obvious that functions G_1 and G_3 can be derived from G_2. As follows from Eq. (10.32),

$$G_1 + G_2 + G_3 = 1$$

The behavior of geometric factor of the layer is shown in Figure 10.2.

This analysis shows that though the distribution of the quadrature component of induced currents is different from that when the field is caused by a vertical magnetic dipole, we can still use the concept of the geometric factor. This is related to the fact that surface charges are absent beneath the Earth's surface and at the low-frequency part of the spectrum, the interaction of the quadrature component of induced currents is negligible, and of course, the upper medium is not conductive. Therefore, we can generalize Eq. (10.31) for N-layered medium and this gives

$$QB_x^* = \frac{\omega\mu_0^2 M_0}{16\pi r}\sum_{i=1}^{N}\gamma_i G_i \tag{10.33}$$

The latter is used for interpretation of electromagnetic profiling with different separations between the dipole and the receiver (Figure 10.1(b)). Let us notice that Eq. (10.33) was derived assuming that interaction between induced currents is neglected. In other words, the current at each point is caused by the primary electric field of the vortex origin and the field due to the surface electric charges, which obey Coulomb's law. As in the case of the vertical magnetic dipole, this asymptotic can be greatly improved. With this purpose in mind, suppose that in underlying medium, the skin effect displays in the same manner as in a uniform half space with the same conductivity γ_N. Then we obtain

$$B_x^* = \frac{\omega\mu_0^2 M_0}{16\pi r}\sum_{i=1}^{N-1}(\gamma_i - \gamma_N)G_i + B_x^{*\text{un}} \tag{10.34}$$

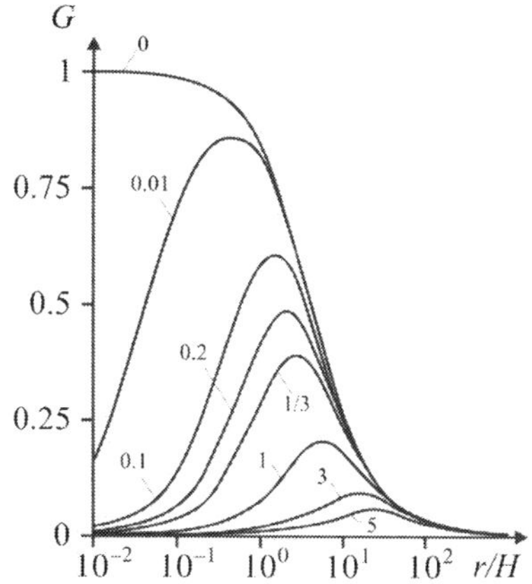

Figure 10.2 Geometric factor of the layer. Curve index is $h_t/h_b - h_t$ ratio, where h_t and h_b are the depths of the layer's top and bottom.

Here B_x^{*un} is the field on the surface of a uniform half space with conductivity γ_N. Unlike Eq. (10.33), the latter allows us to find the quadrature component at higher frequencies, since to some extent it takes into account the skin effect in the underlying medium. Moreover, it shows again that the in-phase component of the secondary field is mainly defined by the conductivity of this medium, while upper layer becomes in such case transparent, and it demonstrates the diffusion character of this component.

REFERENCE AND FURTHER READING

[1] A.A. Kaufman, Geophysical Field Theory and Methods, Part C, Academic Press, New York, NY, 1994.

PART FOUR

Electromagnetic Soundings in a Horizontally Layered Medium

CHAPTER ELEVEN

Principles of Magnetotellurics

INTRODUCTION

The study of electromagnetic fields in the previous chapters allows us to understand the principles of almost all types of the electromagnetic methods used in onshore geophysics, provided that we can treat a medium beneath observation sites as a horizontally layered one. We start with magnetotellurics (MT) based on the measurements of natural electromagnetic field variations. Above the Earth's surface, there are different systems of currents in an ionosphere, which vary with time and generate the primary electromagnetic field. Correspondingly in a conducting medium, induced currents as well as charges arise, and the secondary field appears. At any point, the total field can be observed which contains information about the resistivity structure. The spectrum of the field is extremely wide, but in the applied geophysics, we will be mainly interested by approximately the following range of periods:

$$0.001 \text{ s} < T < 1000 \text{ s}$$

For instance, if we assume that the resistivity of sediments is about 20 Ωm, then the wave length changes as

$$0.4 \text{ km} < \lambda < 500 \text{ km}$$

Intensity of the MT field depends on several factors such as sun activity, time of the year and time of the day, position of an observation point on the Earth's surface, as well as a resistivity structure. For instance, at middle latitude locations, the amplitude of the electric (telluric) field does not usually exceed several millivolts per kilometer while the amplitude of the magnetic (magneto) field has order of 0.1^{-1} nT. At higher latitudes, the field usually becomes stronger. As we already know, any component of the electric and magnetic field depends on an intensity of the primary source, its distance from an observation point, frequency, or time of measurements and resistivity of a medium. It is clear that in the case of natural fields, we know neither the distribution of currents in ionosphere, which generate the field, nor the distance from them. In order to eliminate the influence of these

Methods in Geochemistry and Geophysics, Volume 45
ISSN 0076-6895,
http://dx.doi.org/10.1016/B978-0-444-53829-1.00011-3

factors, it was suggested to measure the horizontal components of the electric and magnetic fields in two mutually perpendicular directions, for instance E_x and B_y, and take their ratio:

$$\frac{E_x}{B_y}$$

It was expected that performing measurements at different frequencies, it would be possible to study resistivity vs depth distribution; in other words, to carry out the electromagnetic soundings which was naturally called *magnetotelluric soundings*, emphasizing the fact that both the magnetic (magneto) and electric (telluric) fields are measured. However, as was shown earlier, this ratio is in general a function of the distance to the primary source (current system in an ionosphere), while the sensitivity to parameters of the medium also depends on the distance from the source. At the same time, there is one important exception, namely, the wave zone where the field behaves as a quasi-plane wave, and its components still change with the distance but the ratio of horizontal components of the electric and magnetic fields on the Earth's surface and beneath is independent on the distance and coincides with that of the plane wave propagating downward into a medium. Fortunately, in most practical cases of the exploration geophysics, these measurements of natural fields are performed in the wave zone. At the time when this method was introduced, the behavior of the electromagnetic field in this zone and its concept were hardly known within geophysical community. Moreover, in developing the theory of the method, it was assumed that the field on the Earth's surface and beneath in a conducting medium behaves as plane wave propagating downward into the Earth. Certainly, such case is far away from the reality, but in spite of this fact, the theory based on such assumption has found a broad application in the applied geophysics. It happened because the impedances of the plane wave and that in the wave zone coincide. In this chapter, we describe the principles of MT for a horizontally layered medium proceeding from this theory but later the influence of the lateral changes of resistivity will be studied in some detail. Also, it may be proper to note the following.

Prior to the development of the MT, two other methods based on the use of the natural electromagnetic field have been proposed for studying the Earth's interior structure. The first method was developed at the end of the nineteenth century and it was based on the separation of the observed magnetic field into so-called internal and external parts. Earlier in studying constant magnetic field, Gauss represented the field observed on the Earth's surface as a series of the spherical harmonics which are the functions of coordinates of a point. Making use of the results of measurements of the magnetic field at different points, Gauss was able to find coefficients of the first harmonics of this series and evaluate the dipole part of the total field and separate the field, caused by currents inside the Earth, from that generated by external sources. Later the same spherical analysis was applied to long-period magnetic variation data to obtain the information about the electrical conductivities and thicknesses of the internal layers including mantle, as well as for studying the current systems in the ionosphere. Certainly, the idea of the separation of the magnetic field into internal and external parts can be useful for studying the Earth's structure on a global scale, but is unlikely to provide sufficiently detailed resistivity structure imaging, which is required in geophysical exploration.

The second approach was the so-called *telluric method* which is much closer to the MT and can be treated as its predecessor. This method was first suggested by Schlumberger and later has been used extensively in Europe and Africa, as well as in Russia. Its modest application in USA was related to the fact that the method was only effective as a reconnaissance tool used in areas where geology was poorly known, and its main goal was mapping the surface of a sufficiently resistive basement. With this purpose in mind, the natural electric field was measured along profiles at different points, and at each observation point, the voltages in two mutually perpendicular directions were recorded. Inasmuch as the frequencies of the measured field are rather low (0.01–0.1 Hz), the skin depth is much greater than the total thickness of sediments above the basement, and correspondingly the distribution of currents with depth does not practically differ from that of the time-constant field. As was well known from the theory, in such case, the electric field is inversely proportional to the total conductance of sediments, S. For this reason, the result of interpretation is primarily a map showing lateral variations in conductance, which allows one to reveal relatively large structures on the basement surface. The field in telluric method is recorded simultaneously at local observation site and some reference site. This procedure permits one to cancel out an influence of time variations of the source, and the ratio between the electric fields or their combination is used for interpretation. This is the classical method of profiling which essentially differs from soundings. Even if we imagine that measurements of the electric field are performed at different frequencies, the accurate and detailed quantitative interpretation is barely possible. For this reason, transition to measuring the tangential components of both the electric and magnetic fields was very important step allowing the quantitative interpretation of MT data. Before we outline the theory of the method, let us briefly discuss the contribution of the scientists to the development of MT and start with Cagniard.

11.1. INVENTION OF THE METHOD

Understanding the shortcomings of the telluric method, Cagniard invented the MT and applied for the patent in 1950, describing the principle of this new method for exploration geophysics. One can say that inability of the telluric method, suggested also in France to perform soundings, motivated Cagniard to develop the new electromagnetic method in the applied geophysics. At the same time, the basic geophysical aspects of this method were published only in 1953. This paper described the physical principles of the method, solution of the forward problem in horizontally layered medium and main features of interpretation. First of all, Cagniard, analyzing the experimental data obtained with telluric method, noticed that within relatively small area on the Earth's surface, both the electric and magnetic fields are almost uniform. Correspondingly, he assumed that on the Earth's surface, that is at a horizontal plane, the field does not practically change and then making use of Maxwell's equations solved the forward problem for the plane wave propagating downward in a conducting medium. As an example, he considered a medium with one, two, and three horizontal boundaries. In order to eliminate an influence of the intensity on the primary plane wave, he suggested considering the ratio of the mutually perpendicular electric and magnetic fields, that is the impedance of the plane wave:

$$\frac{E_x}{B_y} \quad \text{or} \quad \frac{E_y}{B_x}$$

He did not give a name to this ratio (now we refer it as Z_{xy} or Z_{yx} impedance), but it was not quite essential since such notation was already well known and described in all text books on electrodynamics. As a result, Cagniard found a relationship between the impedance of the plane wave on the Earth's surface and resistivity structure of a medium, and then by analogy with the direct current soundings introduced the apparent resistivity as a function of the period of the sinusoidal oscillations of the field which can be represented in the form

$$\frac{\rho_a}{\rho_1} = \left|\frac{Z}{Z(\rho_1)}\right|^2$$

Here, Z is the measured impedance, while $Z(\rho_1)$ is the impedance corresponding to a uniform half space with the resistivity ρ_1. Unlike the telluric method, the MT imply the measurements at different frequencies, and correspondingly, it represents the frequency soundings as the tool for studying the resistivity medium. His paper contains very rigorous analysis of the two- and three-layer apparent resistivity curves, which in essence laid down foundation of the quantitative interpretation of the MT data in horizontally layered medium. Also, Cagniard suggested applying this method in the marine environment. In addition, it is proper to give Cagniard a credit for introduction in the patent of linear relations between horizontal components of the electric and magnetic fields, measured in two mutually perpendicular directions:

$$B_x = a_1 E_x + b_1 E_y \quad \text{and} \quad B_y = a_2 E_x + b_2 E_y$$

Later the coefficients in these equations were called the *admittance tensor elements*. Admittance tensor plays an important role for interpretation of the MT data in nonhorizontally layered medium. Also, Cagniard pointed out that in a medium with lateral changes of resistivity, the method can be still useful but the quantitative interpretation would be hardly possible, and here he was again very close to the truth. Without any doubts, he was a founder of the MT method in the exploration geophysics, which is already applied more than 60 years. In the same year 1950, Japanese researchers Y. Kato and T. Kikuchi as well as Russian scientist A. Tikhonov published their papers in which they described the field of the plane wave on the surface of two-layered medium, provided that a basement is an ideal conductor. These last two papers were practically identical and dedicated to using the MT method for study of the Earth's crust.

The analysis of long-period data, measured in 1930s in Arizona (USA), has shown that the electric and magnetic fields practically do not change over relatively large regions, as it was also noticed by Cagniard but for relatively higher frequencies. Based on these findings, Tikhonov assumed that the field behaves as a plane wave traveling downward. He also supposed that the crust lies on the extremely conductive medium and correspondingly considered two-layered model with an ideal conductor as a basement and found the relationship between the impedance on the Earth's surface and crust's thickness. Exactly, the same assumption was made in two papers published slightly earlier by Kato and Kikuchi. Both of them belonged to the school of the famous geophysicists Prof. Rikitake and during 1930s, Japanese scientists of this group made strong contribution to study of the

Earth using electromagnetic fields. In particular, they suggested measuring the ratio of the horizontal components of the field that is the impedance. Note that this situation is not unique and rather often a new method is introduced almost simultaneously and independently in different countries. Next we start to describe the Cagniard's theory of the MT method in a horizontally layered medium.

11.2. WAVE ZONE, QUASI-PLANE WAVE AND THE IMPEDANCE OF PLANE WAVE

11.2.1. Wave Zone and Quasi-Plane Wave

Assuming that the MT method is applied in the wave zone, first let us make use of formulas derived in Chapter 9 for the vertical magnetic dipole located at the Earth's surface. Of course, the real ionospheric current systems are not dipoles and they are located at some distance from the surface, but as well known, the magnetic field generated by any system of currents of an arbitrary configuration resembles the field of the dipole provided that the distance to the source is sufficiently large. Taking into account this fact, we start from Eq. (9.86)

$$B_z^* = \frac{9\mu_0 M R_N^2(0)}{2\pi r^3}\frac{1}{k_1^2 r^2}, \quad B_r^* = \frac{3\mu_0 M R_N(0)}{2\pi i k_1 r^4}, \quad E_\varphi^* = -\frac{3M\rho_1 R_N^2(0)}{2\pi r^4}, \tag{11.1}$$

where

$$R_N(0) = \coth\left[-ik_1 H_1 + \coth^{-1}\frac{k_1}{k_2}R_{N-1}(0)\right] \tag{11.2}$$

As follows from Eq. (11.1) and it was discussed earlier, the tangential components of the electric and magnetic fields depend on the distance and the source moment in the same manner and their ratio is equal to

$$Z_N^* = \mu_0 \frac{E_\varphi^*}{B_r^*} = -\rho_1 i k_1 R_N(0) \tag{11.3}$$

Here, Z_N^* is called the *impedance* of the plane wave on the surface of N-layered medium. Since $k_1^2 = i\gamma_1\mu_0\omega$, we have

$$Z_N^* = \frac{\omega\mu_0}{k_1} R_N(0) \tag{11.4}$$

As was pointed out in the wave zone, the electromagnetic field in a conducting medium behaves as a quasi-plane wave propagating downward, and this implies that the electric and magnetic fields vary in the horizontal plane. Remarkable feature of this field is the fact that the ratio of these components coincides with the impedance of the ideal plane wave moving downward, and because of this, Cagniard's theory correctly describes the impedance on the surface of a horizontally layered medium. In the Cartesian system of coordinates, we can write

$$Z_{xy}^{*} = \mu_0 \frac{E_x^{*}}{B_y^{*}} = \frac{\omega\mu_0}{k_1} R_N(0) \tag{11.5}$$

In deriving Eq. (11.1), it was assumed that the current source (magnetic dipole) is situated at the Earth's surface, but in reality, currents are located in ionosphere at the height d that is around 70–100 km, which is usually much smaller than the distance from the observation points ($d \ll r$). This is the reason why at such distances r, we still deal with quasi-plane wave. Besides, there is one more theoretical rather than realistic case when this behavior can be observed at very small distances too. It may happen if the observation point is located beneath a uniform distribution of currents which creates a plane field.

Before we continue, let us make several comments. (1) It is assumed that at each frequency, the electric and magnetic fields are oriented in perpendicular directions, that is, the field vectors change their magnitudes (in oscillating manner) with time but maintain their directions. In other words, the linear polarization of the field takes place. (2) The quasi-plane wave propagating downward from the Earth's surface is observed in the wave zone regardless of the type of field excitation. (3) The observations are performed at distances r, which exceed the parameter λ:

$$r > \lambda, \tag{11.6}$$

where λ depends both on frequency and resistivity structure. For instance, in case of a uniform half space,

$$\lambda = (10\rho T)^{1/2}$$

Here, ρ is the resistivity of a medium and T is the period of oscillations. At the same time, when a system of layers, situated above a nonconducting basement, can be replaced by a thin layer with the conductance S, we have

$$\lambda = \left(\omega\mu S 10^3\right)^{-1},$$

where ω is the frequency and S is the total conductance of the layers. (4) It is also assumed that the wavelength above the Earth's surface is much greater than the distance r. (5) Of course, the condition (11.6) is not sufficient; it is also necessary to use the frequencies low enough to provide the sufficient depth of investigation. (6) Orientation of the field vectors in a quasi-plane wave at observation point does not make influence on the impedance in a horizontally layered medium. In particular, this means that Cagniard's theory also applies when the field is elliptically polarized since the latter is a superposition of linearly polarized fields.

11.3. THE IMPEDANCE OF THE PLANE WAVE

Now following Cagniard, we derive the impedance of the plane wave moving downward from the Earth's surface and demonstrate that it coincides with that of the quasi-plane wave. For illustration, consider the three-layered medium and choose the Cartesian system of coordinates with the origin at the Earth's surface and the z-axis directed downward (Figure 11.1). Assume that the electric and magnetic fields of the plane wave are directed along the x and y axes, respectively,

$$\mathbf{E} = (E_x, 0, 0), \quad \text{and} \quad \mathbf{B} = \left(0, B_y, 0\right) \tag{11.7}$$

As was shown earlier, their complex amplitudes obey at regular points of each layer the same equations

$$\frac{d^2 E_x^*}{dz^2} + k^2 E_x^* = 0 \quad \text{and} \quad \frac{d^2 B_y^*}{dz^2} + k^2 B_y^* = 0, \tag{11.8}$$

and at interfaces, both components are continuous functions:

$$E_{x,i}^* = E_{x,i+1}^*, \quad \text{and} \quad B_{y,i}^* = B_{y,i+1}^* \tag{11.9}$$

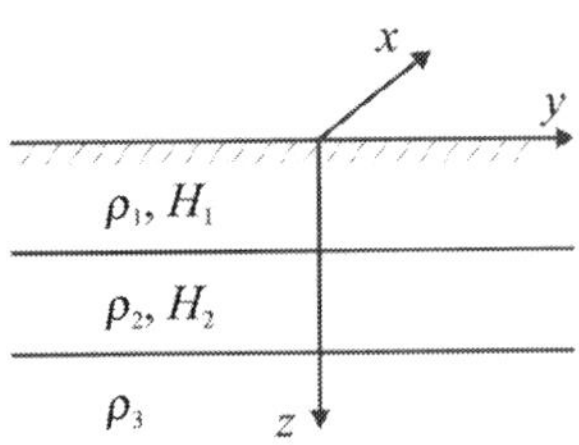

Figure 11.1 Three-layered medium.

From the first Maxwell's equation,

$$\text{curl}\,\mathbf{E}^* = i\omega\mathbf{B}^*$$

follows a very simple relationship between complex amplitudes of the plane wave fields:

$$\frac{\partial E_x^*}{\partial z} = i\omega B_y^* \tag{11.10}$$

For this reason, the boundary value problem can be formulated in terms of the electric field only, as follows:

1. At regular points,

$$\frac{d^2 E_x^*}{dz^2} + k^2 E_x^* = 0$$

2. At interfaces,

$$E_{x,i}^* = E_{x,i+1}^* \quad \text{and} \quad \frac{\partial E_{x,i}^*}{\partial z} = \frac{\partial E_{x,i+1}^*}{\partial z}$$

3. At infinity, the field tends to zero

$$E_x^* \to 0 \quad \text{and} \quad \frac{\partial E_x^*}{\partial z} \to 0 \quad \text{if } z \to \infty$$

Unlike the solution of the boundary value problem when the field is caused by the known source, for instance the magnetic dipole, we suppose that both components of the electromagnetic field are known at the Earth's surface. Such assumption is natural since these quantities are measured. This allows us to avoid a determination of the field above the Earth's surface. Thus, at the Earth's surface, we have

4. $E_x^*(x, y, 0) = E^* \quad B_y^*(x, y, 0) = \frac{1}{i\omega}\frac{\partial E_x^*}{\partial z} = B^*$

where E^* and B^* are known since they are measured. In accordance with Eq. (11.8), the complex amplitude of the electric field has the form

$$E_x^* = C\exp(ikz) + D\exp(-ikz)$$

Correspondingly, taking into account the boundary conditions at interfaces and at infinity, we obtain for three-layered medium the system of equations with respect to unknown coefficients:

$$C_1 + D_1 = E^*$$

$$\frac{k_1}{\omega}(C_1 - D_1) = B^*$$

$$C_1 \exp(ik_1h_1) + D_1 \exp(-ik_1h_1) = C_2 \exp(ik_2h_1) + D_2 \exp(-ik_2h_1) \tag{11.11}$$

$$k_1[C_1 \exp(ik_1h_1) - D_1 \exp(-ik_1h_1)]$$

$$= k_2[C_2 \exp(ik_2h_1) - D_2 \exp(-k_2h_1)]$$

$$C_2 \exp(ik_2h_2) + D_2 \exp(-ik_2h_2) = C_3 \exp(ik_3h_2)$$

$$k_2[C_2 \exp(ik_2h_2) - D_2 \exp(-ik_2h_2)] = k_3C_3 \exp(ik_3h_2)$$

As before from the last two equations, we have

$$\coth\left(ik_2h_2 - \frac{1}{2}\ln\frac{D_2}{C_2}\right) = \frac{k_2}{k_3}$$

or

$$\frac{1}{2}\ln\frac{D_2}{C_2} = ik_2h_2 - \coth^{-1}\frac{k_2}{k_3} \tag{11.12}$$

The next pair of equations of the set (11.11) gives

$$\coth\left[ik_1h_1 - \frac{1}{2}\ln\frac{D_1}{C_1}\right] = \frac{k_1}{k_2}\coth\left[ik_2h_1 - \frac{1}{2}\ln\frac{D_2}{C_2}\right],$$

or substituting Eq. (11.12), we have

$$\coth\left[ik_1h_1 - \frac{1}{2}\ln\frac{D_1}{C_1}\right] = \frac{k_1}{k_2}\coth\left[-ik_2H_2 + \coth^{-1}\frac{k_2}{k_3}\right]$$

or

$$-\frac{1}{2}\ln\frac{D_1}{C_1} = -ik_1H_1 + \coth^{-1}\left[\frac{k_1}{k_2}\coth\left(-ik_2H_2 + \coth^{-1}\frac{k_2}{k_3}\right)\right]$$

where $H_2 = h_2 - h_1$ is the thickness of the second layer. The upper pair of equations gives

$$Z_{xy} = \frac{E_x}{B_y} = \frac{\omega\mu_0}{k_1}\coth\left(-\frac{1}{2}\ln\frac{D_1}{C_1}\right)$$

or

$$Z_{xy} = \frac{\omega\mu_0}{k_1}\coth\left\{-ik_1H_1 + \coth^{-1}\left[\frac{k_1}{k_2}\coth\left(-ik_2H_2 + \coth^{-1}\frac{k_2}{k_3}\right)\right]\right\} \tag{11.13}$$

The latter is the same as that for the quasi-plane wave and, as was pointed out, due to this fact the Cagniard's theory became foundation of the theory of MT. Of course, this coincidence takes place for N-layered medium too. Let us make two comments. (1) Until now, we assumed that the electric and magnetic fields are linearly polarized. As was mentioned earlier, it is not difficult to show that the amplitude and phase of the impedance for the elliptically polarized field propagating downward in a horizontally layered medium do not depend on the orientation of the sensors measuring the electric and magnetic fields. Just as in the case of linear polarization, the generation of a field simultaneously by several sources does cause a change of the impedance with a direction of sensors if the medium is horizontally layered. Of course, the impedance coincides with that for the linear polarization. The situation is completely different if the medium contains lateral inhomogeneities. (2) The method, which did not require the man-made generator, was suggested at the time when computers were at the beginning of their development and the amazing simplicity of the MT forward problem solution even compared to Sclumberger soundings was also its very attractive feature.

11.4. THE APPARENT RESISTIVITY AND ITS BEHAVIOR IN A HORIZONTALLY LAYERED MEDIUM

11.4.1. Impedance of a Uniform Half Space

In accordance with Eq. (11.4) at the surface of a uniform half space, we have

$$Z_1 = \frac{\omega\mu_0}{k_1} = (\omega\mu_0\rho_1)^{1/2}\exp\left(-i\frac{\pi}{4}\right) \quad \text{or} \quad |Z_1| = (\omega\mu_0\rho_1)^{1/2} \tag{11.14}$$

Whence,

$$\rho_1 = \frac{|Z_1|^2}{\omega\mu_0} = \frac{10^7\ T}{8\pi^2}|Z_1|^2, \tag{11.15}$$

where T is the period of the electromagnetic field oscillations. According to this equation at any period, we can calculate ρ_1 for a uniform half space by carrying out the algebraic operations indicated on the right-hand side of the equation. As a matter of reference, the similar approach is used in direct current methods to introduce the apparent resistivity, which allows one to obtain resistivity ρ_1 regardless of the separation between the current and receiver electrodes. Now, making use of Eq. (11.15), we introduce the apparent resistivity in the following way:

$$\rho_T = \frac{|Z_n|^2}{\omega\mu_0} \tag{11.16}$$

In particular, in a uniform medium, the latter instantly gives

$$\rho_T = \rho_1$$

As follows from Eqs (11.15) and (11.16), we have

$$\frac{\rho_T}{\rho_1} = \frac{|Z_n|^2}{|Z_1|^2} \tag{11.17}$$

It is clear that the ratio ρ_T/ρ_1 shows the degree to which the impedance measured at the Earth's surface differs from that for a uniform half space. Note that this is only one reason why the apparent resistivity is introduced in many electric and electromagnetic methods. Now let us make two comments: (1) the definition of the apparent resistivity ρ_T is based only on the use of the modulus of impedance. The phase shift between the electric and magnetic fields is a second parameter, which carries information about the properties of the medium. However, in most cases knowing the frequency response of the apparent resistivity, it is possible to calculate the phase and, thus avoid measurements of the phase. In particular, it is true for a horizontally layered medium. In other words, usually the phase response contains the same information as the apparent resistivity response. (2) Bearing in mind the meaning of the apparent resistivity, the most natural way to introduce it from Eq. (11.17). Now we are going to describe the apparent resistivity curves and the relation between $\rho_T(T)$ and geoelectrical parameters of a medium. Note that first of all, this subject has the historical interest since at the beginning, an interpretation of MT data was based as in the case of Schlumberger soundings on the curve matching. Also, the knowledge which was gained during this stage became foundation for the second stage of interpretation in more complicated models of a medium. For instance, it became very quickly clear that taking into account the presence of different types of a noise, there are parameters of a medium (resistivity and thickness of layers) which cannot be determined with a sufficient accuracy. In contrast, for certain types of sections, it is possible to evaluate some parameters with relatively high accuracy, such as a total conductance of a layered medium or the depth to sufficiently conductive basement. This classification already indicated that an inverse problem is ill posed one and there are two groups of parameters, namely, stable and unstable, so that the result of interpretation was based on the determination of stable parameters (regularization). One can say that the modern interpretation of MT data contains essential elements, which were discovered during the first stage of this procedure performing the curve matching.

11.4.2. Apparent Resistivity for a Two-Layered Medium

We start from the apparent resistivity for the case of a two-layered medium. From Eqs (11.13) and (11.17), we have

$$\rho_T = \rho_1 \left| \coth\left[-ik_1 H_1 + \coth^{-1}\left(\frac{\rho_2}{\rho_1}\right)^{1/2}\right]\right|^2 \tag{11.18}$$

or

$$\rho_T = \rho_1 \left| \tanh\left[-ik_1 H_1 + \tanh^{-1}\left(\frac{\rho_2}{\rho_1}\right)^{1/2}\right]\right|^2$$

First, consider the high-frequency part of the spectrum when $|k_1H_1| \gg 1$. Inasmuch as $\coth x \to 1$ if $x \to \infty$, we have in the limit $\rho_T \to \rho_1$ if $T \to 0$. Thus, the left asymptote of the ρ_T curve when it is plotted as a function of $|k_1H_1|$ takes the value of resistivity equal to that of the first layer. This is a consequence of the skin effect since all currents are concentrated in the first layer when H_1/δ_1 is much less than unity. Now, consider the behavior of the high-frequency part of the spectrum of ρ_T as it approaches the value of ρ_1 asymptotically. Let us introduce the notations

$$\frac{2\pi H_1}{\lambda_1} = x \quad \text{and} \quad \coth^{-1}\left(\frac{\rho_2}{\rho_1}\right)^{1/2} = y$$

Here, the wave length λ_1 is

$$\lambda_1 = 2\pi\delta_1 = 2\pi\left(\frac{2}{\gamma_1\mu_0\omega}\right)^{1/2} = (10\rho_1 T)^{1/2} \text{ km} \tag{11.19}$$

Thus, we have

$$\coth(-ik_1H_1+y)=\coth[-i(1+i)x+y]=\coth[(x+y)-ix]$$
$$=\frac{\coth(x+y)\coth ix-1}{\coth ix-\coth(x+y)} \tag{11.20}$$

Using the identity

$$\coth ix=-i\text{ctg}x,$$

we obtain

$$\coth(-ik_1H_1+y)=\frac{1+i\coth(x+y)\text{ctg}x}{i\text{ctg}x+\coth(x+y)}$$

Whence,

$$\rho_T=\rho_1|R_2|^2=\frac{1+\coth^2(x+y)\text{ctg}^2\left(\frac{2\pi H_1}{\lambda_1}\right)}{\text{ctg}^2\left(\frac{2\pi H_1}{\lambda_1}\right)+\coth^2(x+y)} \tag{11.21}$$

It is clear that when $\text{ctg}^2\left(\frac{2\pi H_1}{\lambda_1}\right)=1$, the apparent resistivity ρ_T is equal to ρ_1. This happens each time that

$$\frac{2\pi H_1}{\lambda_1}=\frac{\pi}{4}(2n+1), \tag{11.22}$$

where $n=0,1,2,3,\ldots$ or

$$\frac{\lambda_1}{H_1}=\frac{8}{(2n+1)} \tag{11.23}$$

Thus, $\rho_T=\rho_1$ when $\lambda_1/H_1=8, 8/3, 8/5$, and so on. If we represent the function $\rho_T/\rho_1\left(\frac{\lambda_1}{H_1}\right)$ as a curve, then we have a series of points where the apparent resistivity curve intersects the line $\rho_T=\rho_1$. The abscissas of these points do not depend on the resistivity of the underlying medium, ρ_2. Thus, the apparent resistivity curve approaches the left asymptote, ρ_1, in an oscillatory manner. However, the amplitudes of these oscillations are relatively small and become progressively smaller with increasing frequency. Next consider the low-frequency part of the apparent resistivity curve.

According to Eq. (11.18), as the period of oscillations increases, $\lambda_1/_{H_1} \to \infty$, we have

$$\rho_T \to \rho_1 \left| \left(\frac{\rho_2}{\rho_1} \right)^{1/2} \right|^2 = \rho_2, \tag{11.24}$$

and the apparent resistivity approaches the value of resistivity of the second layer; that is at low frequencies, the first layer becomes transparent, and the impedance becomes equal to that of a uniform half space with resistivity ρ_2. It is interesting to study the manner in which the apparent resistivity curve approaches the right asymptote over the low-frequency part of the spectrum. As follows from Eq. (11.18),

$$Z_2 = Z_1 \frac{k_2 - k_1 \coth ik_1 H_1}{k_1 - k_2 \coth ik_1 H_1} \tag{11.25}$$

For relatively large periods, we have

$$\coth ik_1 H_1 \approx \frac{1}{ik_1 H_1}$$

and this gives

$$Z_2 \approx Z_1 \frac{k_2 - k_1 \left(\frac{1}{ik_1 H_1} \right)}{k_1 - k_2 \left(\frac{1}{ik_1 H_1} \right)} = Z_1 \frac{ik_2 H_1 - 1}{ik_1^2 H_1 - k_2} k_1 = \omega\mu_0 \frac{1 - ik_2 H_1}{k_2 - ik_1^2 H_1} \tag{11.26}$$

It is also useful to consider separately two cases: (1) the underlying medium has higher resistivity ρ_2, and (2) the second layer is more conductive than the surface layer.

Case 1: Let $|k_1 H_1| \ll 1$ and $\rho_2 \gg \rho_1$, so that $|k_2 H_1| \ll 1$. With these conditions, we can ignore the second term in the numerator of Eq. (11.26) and obtain

$$Z_2 \approx \omega\mu_0 \frac{1}{k_2 - ik_1^2 H_1}$$

or

$$Z_2 \approx \omega\mu_0 \frac{1}{(i\gamma_2 \omega\mu_0)^{1/2} + \omega\mu_0 S_1} = \frac{1}{S_1 + \left(\frac{\gamma_2}{\omega\mu_0} \right)^{1/2} \exp\left(i\frac{\pi}{4}\right)}, \tag{11.27}$$

where S_1 is the conductance of the first layer. First assume that the basement is an insulator, $\gamma_2 = 0$. Then,

$$Z_2 = \frac{1}{S_1} \tag{11.28}$$

The complex amplitude of the impedance is a real number equal to $1/S_1$ and is independent of the period. In other words, the phase shift between the electric and magnetic fields is zero. Thus, in measuring the impedance Z_2 over this range of periods, we obtain information only about the longitudinal conductance of the first layer. From Eq. (11.28), we have

$$\rho_T = \frac{10^7}{8\pi^2}\frac{T}{S_1^2}\frac{1}{} \tag{11.29}$$

Thus, the right-hand asymptote of the apparent resistivity curve increases in direct proportion to the period T with the proportionality coefficient S_1^{-2}. Let us return to Eq. (11.27), which describes the impedance when the second layer has a finite resistivity and $\lambda_1/H_1 > 10$. Over this portion of the spectrum, the apparent resistivity curve can be presented in the below form:

$$\rho_T \approx \frac{10^7 T}{8\pi^2}\left|\frac{1}{S_1 + \left(\frac{\gamma_2}{\omega\mu_0}\right)^{1/2}\exp\left(i\frac{\pi}{4}\right)}\right|^2 \tag{11.30}$$

It is clear that with increasing period, the second term of the denominator becomes larger than the first one and the apparent resistivity ρ_T approaches the value of resistivity characterizing the second layer, ρ_2. Unlike the general case at the low-frequency part of the spectrum $\left(\frac{\lambda_1}{H_1} > 10\right)$, the apparent resistivity is controlled by the conductance of the first layer, the resistivity of the second layer and a period T. From Eq. (11.30), it follows that if

$$S_1 \gg \left(\frac{\gamma_2}{\omega\mu_0}\right)^{1/2},$$

there is a transitional part of the apparent resistivity curve ($\rho_2 \neq \infty$), which can be considered the S-zone, where ρ_T is given by Eq. (11.28). As is seen from Eq. (11.30), with decreasing resistivity of the second layer (basement), the range of periods corresponding to the S-zone becomes narrower.

Case 2: Now we consider the case when the lower medium is much more conductive. Assuming that $|k_2| \gg |ik_1^2 H_1|$, we have from Eq. (11.26)

$$Z_2 \approx \omega\mu_0 \frac{1 - ik_2 H_1}{k_2} = \frac{\omega\mu_0}{k_2} - i\omega\mu_0 H_1 \tag{11.31}$$

The first term corresponds to the impedance of a uniform half space with resistivity ρ_2. When the second layer is perfectly conducting, we have

$$Z_2 = -i\omega\mu_0 H_1 = \frac{8\pi^2 10^{-7}}{T} H_1 \exp\left(-i\frac{\pi}{2}\right) \quad \text{if } \lambda_1/H_1 > 10,\ \rho_2 = 0 \tag{11.32}$$

In this limiting case, the impedance is a function of the thickness of the first layer H_1 and it does not depend on resistivity ρ_1. Hence,

$$\rho_T \approx \frac{8\pi^2 10^{-7}}{T} H_1^2, \quad \frac{\lambda_1}{H_1} > 10,\ \rho_2 = 0 \tag{11.33}$$

Thus over the low-frequency part of the spectrum, the apparent resistivity changes in proportion to H_1^2, and this right-hand asymptote can be used to determine H_1. When the resistivity of the lower medium is not equal to zero, there can be a transitional zone of the apparent resistivity curve over which the apparent resistivity is reasonable well described by Eq. (11.33). This part of the spectrum can be called the *H*-zone, and such behavior occurs when the following conditions are satisfied:

$$\frac{\lambda_1}{H_1} > 10, \quad \frac{\rho_2}{\rho_1} \ll 1 \quad \text{and} \quad \rho_2 < \frac{8\pi^2}{T} 10^{-7} H_1^2$$

It is obvious that with increasing resistivity of the conducting lower medium, the width of the *H*-zone becomes smaller. From Eq. (11.31), we have

$$\rho_T = \frac{8\pi^2 10^{-7}}{T} \left| \frac{1}{k_2} - iH_1 \right|^2 \tag{11.34}$$

As may be seen from this equation, with increasing period T, the term $1/k_2$ becomes dominant and ρ_T approaches ρ_2. Before considering some characteristics of two-layer ρ_T curves, we should note the following. In the general case of a multilayered medium, the apparent resistivity can be expressed as a function

$$\rho_T = \rho_1 F\left(\frac{\lambda_1}{H_1}, \mu_1, \mu_2, \ldots \mu_{n-1}, \nu_1, \nu_2 \ldots \nu_{n-1}\right), \tag{11.35}$$

where ρ_1 is the resistivity of the first layer, H_1 is its thickness, and the wavelength in this layer is

$$\lambda_1 = 2\pi\delta_1 = 2\pi\left(\frac{2}{\gamma_1\mu_0\omega}\right)^{1/2} = (10\rho_1 T)^{1/2}\ \text{km}, \tag{11.36}$$

and

$$\mu_1 = \frac{\rho_2}{\rho_1}, \quad \mu_2 = \frac{\rho_3}{\rho_1}, \quad \mu_{n-1} = \frac{\rho_n}{\rho_1};$$

$$\nu_1 = \frac{H_2}{H_1}, \ldots \nu_{n-1} = \frac{H_n}{H_1}$$

In particular, for two layers,

$$\rho_T = \rho_1 F\left(\frac{\lambda_1}{H_1}, \mu_1\right) \tag{11.37}$$

In this case, the apparent resistivity depends on three parameters:

$$\rho_1, \frac{\lambda_1}{H_1}, \mu_1$$

If the function ρ_T was plotted using the linear scale, the shape of the ρ_T curve would depend on the value of ρ_1. Similarly, the value of the thickness H_1 would have an influence on the shape of the ρ_T curve when it is plotted in the linear graph. As in the case of Schlumberger soundings, in order to avoid this, the ρ_T curve is plotted on a log–log graph scale. This converts the ρ_T curve to the form

$$\log \rho_T = \log \rho_1 + \log F\left(\log \frac{\lambda_1}{H_1}, \mu_1\right)$$

Now the changes in either H_1 or ρ_1 do not change the shape of the curve but cause only a shift of the curve in the horizontal or vertical directions. Log–log scales are used for plotting MT apparent resistivity responses for the same reasons, as that they are used in plotting DC resistivity (Schlumberger) responses. Figure 11.2 shows a set of apparent resistivity curves for the two-layered medium with the ratio ρ_T/ρ_1, plotted as a function of the parameter

$$\frac{\lambda_1}{H_1} = \frac{(10\rho_1 T)^{1/2}}{H_1}$$

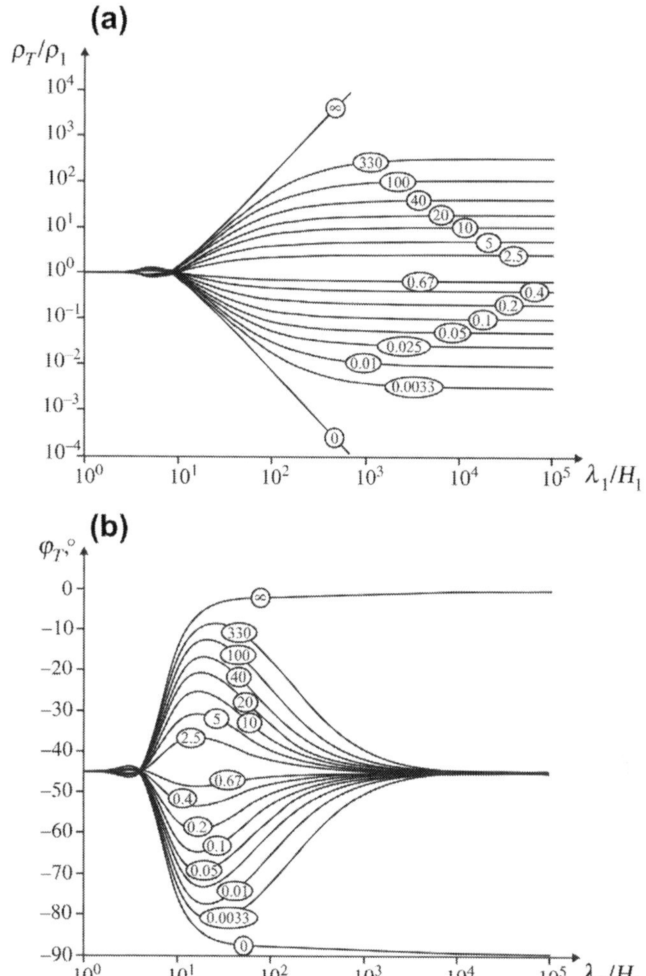

Figure 11.2 Magnetotelluric responses for two-layered medium. (a) Apparent resistivity curves; and (b) phase curves. Curve index is ρ_2/ρ_1 ratio.

Here, the numerator and denominator are expressed in kilometers. The reference point for this set of curves has the coordinates $\lambda_1/H_1 = 8$, and $\rho_T/\rho_1 = 1$, and it is defined in the following way. The abscissa of such point is $\lambda_1/H_1 = 8$, where all curves intersect the line $\rho_T/\rho_1 = 1$ and the parameter for the individual curves in the set is $\mu = \rho_2/\rho_1$. Now we consider some of the general features of these curves. (1) When $\lambda_1/H_1 < 8$ and $(\rho_2 > \rho_1)$, the apparent resistivity ρ_T approaches ρ_1 in an oscillatory manner.

Over this part of the spectrum, the influence of resistivity of the second layer is very small and it is manifested as a variation in values for the maxima and minima of curves. (2) With increasing period T, the apparent resistivity ρ_T increases ($\rho_2 > \rho_1$) and approaches the right-hand asymptote, $\rho_T = \rho_2$. When $\mu = \rho_2/\rho_1$ takes larger values, the approach to the right asymptote requires larger values of T. When the ratio ρ_2/ρ_1 is much larger than unity, there is almost always a transitional zone on the curve, which is nearly coincident with the curve for a nonconducting basement. This range corresponds to the S zone where the impedance is inversely proportional to S. (3) According to Eq. (11.29), the equation for this asymptotic behavior can be written as

$$\rho_T = \frac{10^7 T}{8\pi^2}\frac{\rho_1^2}{H_1^2} \quad \text{or} \quad \frac{\rho_T}{\rho_1} = \frac{1}{8\pi^2}\left(\frac{\lambda_1}{H_1}\right)^2, \tag{11.38}$$

since $\lambda_1 = (10^7 \rho_1 T)^{1/2}$. Thus on a log–log scale, we have

$$\log\frac{\rho_T}{\rho_1} = 2\log\frac{\lambda_1}{H_1} - \log 8\pi^2,$$

and correspondingly in the limiting case of the nonconducting basement, the slope of the right-hand asymptote to the ρ_T curve is equal to $63°23'$. This straight line intersects the horizontal line for $\rho_T = \rho_1$ at a point with abscissa:

$$\left(\frac{\lambda_1}{H_1}\right)_S = 8^{1/2}\pi \approx 8.9 \tag{11.39}$$

This point of intersection is close to the reference point for the curve set: $\lambda_1/H_1 = 8$. (4) When the lower medium is more conductive than the upper layer, ρ_T decreases monotonically for $\lambda_1/H_1 > 8$ and approaches the right-hand asymptote ρ_2. In the limiting case with $\rho_2 = 0$, the right-hand asymptote takes the form of a straight line whose equation can be obtained from Eq. (11.33)

$$\frac{\rho_T}{\rho_1} = \frac{8\pi^2 H_1^2}{10^7 \rho_1 T} = \frac{8\pi^2}{\left(\frac{\lambda_1}{H_1}\right)^2}$$

and

$$\log\left(\frac{\rho_T}{\rho_1}\right) = \log 8\pi^2 - 2\log\left(\frac{\lambda_1}{H_1}\right) \tag{11.40}$$

Thus, the low-frequency asymptote for the limiting case is a straight line making an angle of $63°23'$ with the horizontal and intersecting the line $\rho_T = \rho_1$ at the point,

$$\left(\frac{\lambda_1}{H_1}\right)_H = \left(\frac{\lambda_1}{H_1}\right)_S = 8.9. \tag{11.41}$$

(5) When the resistivity contrast is relatively large and $\rho_2 \ll \rho_1$, there is always an interval where the apparent resistivity curve practically coincides with that for a perfectly conducting second layer. Over this range of periods, the apparent resistivity is determined by the thickness of the first layer (H zone). (6) Curves with reciprocal values for μ, for example, 2 and 1/2, are symmetrical to one another. This feature of the curves follows from the identity

$$\coth\left(a + \coth^{-1} b\right) = \frac{1}{\coth\left[a + \coth^{-1}\left(\frac{1}{b}\right)\right]} \tag{11.42}$$

In fact, consider two cases of layered media with the parameters ρ_1, H_1, ρ_2', and ρ_1, H_1, ρ_2'' with $\rho_2'' = 1/\rho_2'$. Then, $\mu' = 1/\mu''$. From Eq. (11.18), we have the following equations for ρ_T:

$$\rho_T' = \rho_1 \left|\coth\left(-ik_1 H_1 + \coth^{-1}\mu'\right)\right|^2$$

and

$$\rho_T'' = \rho_1 \left|\coth\left(-ik_1 H_1 + \coth^{-1}\mu''\right)\right|^2$$

Using Eq. (11.42), we have

$$\rho_T'' = \rho_1 \frac{1}{\left|\coth(-ik_1 H_1 + \coth^{-1}\mu')\right|^2}$$

Whence,

$$\log\left(\frac{\rho_T'}{\rho_1}\right) = -\log\left(\frac{\rho_T''}{\rho_1'}\right) \tag{11.43}$$

Thus, apparent resistivity curves with parameters μ_1 and $1/\mu_1$ are symmetric to each other. As will be shown later, this behavior is valid for phase response, and for any number of layers.

11.5. DEVELOPMENT OF MAGNETOTELLURIC INVERSE PROBLEM SOLUTION

As is well known, solution of inverse problem of MT in a horizontally layered medium is unique, as well as for Schlumberger method. In other words, it is possible to determine a resistivity and thickness of layers regardless how small this thickness, provided that the field is known with absolute precision. At the same time, due to the presence of different types of noises, such as geological noise, errors of measurements, and limited number of frequencies at which the field is measured, this problem is ill-posed as in other geophysical methods. This means that there are almost always stable and unstable parameters of a medium, and our goal is to determine the first one. The procedure of their determination is carried out by applying the method of subsequent approximations, and in order to facilitate this process, it is often very important to use for the first guess approximate values of parameters of a medium which are reasonably close to real ones. Now, we demonstrate how knowledge of curves of the apparent resistivity can be used to obtain information about values of some parameters of the resistivity model. Various methods are available for interpreting the MT responses, including curve matching, the use of asymptotes, and generalized inversion. In essence, the curve matching is the vital element of the modern inversion procedure which is performed in the form different than graphical comparison of the theoretical and experimental curves.

11.5.1. Determination of Two-Layer Resistivity Model Parameters by Matching

Of course, this section has rather a historical interest, but it turns out that the study of a relation between the behavior of the apparent resistivity curves and parameters of a geoelectric section in many cases is very useful for application of modern methods of interpretation. The simplest method, which was used at the beginning, is a matching a field curve of the apparent resistivity with one of a set of precalculated theoretical curves. The field curve is plotted as the logarithm of values of ρ_T along the ordinate axis with the logarithm of the square root of T along the abscissa, as shown in Figure 11.3(a). The field curve (dashed line) is matched with the curve from the theoretical set which matches most closely. The reference point for the two-layer theoretical set of curves corresponds to the point with coordinates $\left(\rho_0, T_0^{1/2}\right)$ on the field plot, so that $\rho_1 = \rho_0$ and $\rho_2 = \mu\rho_0$, where μ is the ratio

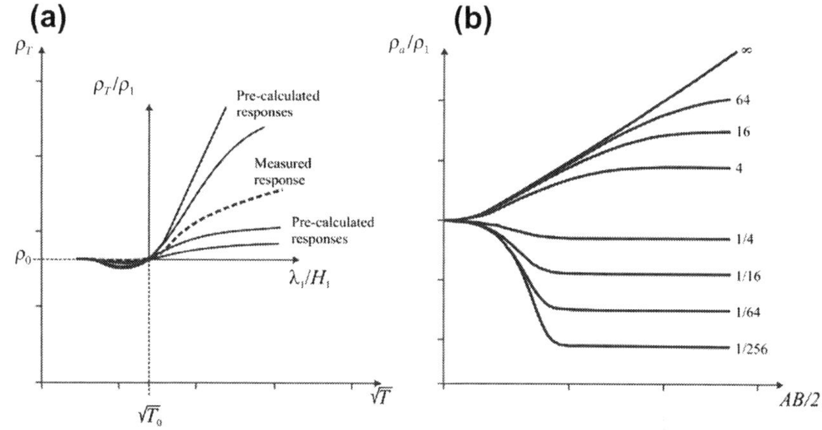

Figure 11.3 (a) Matching the field response data with modeled curves. (b) Schlumberger curves (curve index is ρ_2/ρ_1 ratio).

of resistivity for the theoretical curve which matches. From theoretical curves, it follows that if measurements are performed at sufficiently broad range, then parameters of two-layered medium are defined with the same accuracy as that of the useful signal. In other words, the resistivity and thickness of the first layer as well as the resistivity of the basement are usually stable parameters. Thus in the presence of sufficiently low level of noise, MT response allows us to determine the parameters of a two-layered medium.

11.5.2. The Use of Asymptotic Behavior and Special Points of the Curves

Now suppose that curves of ρ_T are given only within a relatively narrow interval of periods and consider two cases.

Case 1: $\rho_2 \gg \rho_1$

First assume that measured values of the apparent resistivity ρ_T fall within the S zone.

In this case, we are able to determine only the longitudinal conductance S_1. According to Eq. (11.29),

$$\log \rho_T = \log \frac{10^7}{8\pi^2 S_1^2} + \log T \qquad (11.44)$$

Thus, the S asymptote intersects the horizontal line $\rho_T = 1$ at a point given by

$$\frac{10^7\ T}{8\pi^2} = S_1^2$$

Whence,

$$S_1 = \left(\frac{10^3}{2\pi}5^{1/2}\right)T^{1/2} \tag{11.45}$$

If the resistivity of the first layer can be determined, its thickness is calculated as being $H_1 = S_1\rho_1$. It can also be found graphically from the intersection of the horizontal line $\rho_T = \rho_1$ with the S_1 line. From Eq. (11.44), we have

$$\begin{aligned}\log \rho_1 &= \log\left(\frac{10^7\rho_1^2}{8\pi^2 H_1^2}\right) + 2\log T^{1/2}\\ &= 2\log\rho_1 - 2\log H_1 + \log\left(\frac{10^7 T}{8\pi^2}\right),\end{aligned}$$

that is,

$$2\log H_1 = \log\left(\frac{10^7\rho_1 T}{8\pi^2}\right) \quad \text{and} \quad H_1 = 10^3(5\rho_1 T)^{1/2}/2\pi \tag{11.46}$$

It is natural to expect that this result is the same as that obtained from Eq. (11.38).

Next assume that $\lambda_1/_{H_1} > 10$, so that from Eq. (11.27),

$$Y_2 = 1/_{Z_2} = S_1 + \left(\frac{\gamma_2}{\omega\mu_0}\right)^{1/2} \exp\left(i\frac{\pi}{4}\right), \tag{11.47}$$

where Y_2 is called the *admittance*. It is clear that

$$\mathrm{Re}\, Y_2 = S_1 + \left(\frac{\gamma_2}{2\omega\mu_0}\right)^{1/2} = S_1 + \frac{10^3}{4\pi}\left(\frac{10T}{\rho_2}\right)^{1/2} \tag{11.48}$$

and

$$\mathrm{Im}\, Y_2 = \left(\frac{\gamma_2}{2\omega\mu_0}\right)^{1/2} = \frac{10^3}{4\pi}\left(\frac{10T}{\rho_2}\right)^{1/2} \tag{11.49}$$

Therefore, the resistivity of the second layer is

$$\rho_2 = \frac{10^7 T}{(4\pi)^2(\mathrm{Im}\, Y_2)^2} \tag{11.50}$$

Substituting Eq. (11.50) into Eq. (11.48), we obtain a value for the conductance S_1. Also the resistivity for the second layer and the value of S_1

can be obtained from the real part of Y_2 by taking values at two periods, T_1 and T_2. We may obtain still another relationship between ρ_1 and H_1, proceeding from the left part of the curve. When $\rho_2/\rho_1 \gg 1$, the minimum of the apparent resistivity curve corresponds to the argument

$$\frac{\lambda_1}{H_1} = 5.3 \quad \text{or} \quad H_1 = \left(\frac{10^7}{28}\right)\frac{T}{S_1}$$

Knowing the value for S_1 and using Eq. (11.48), we can evaluate ρ_1 and H_1. If the resistivity of the second layer is significantly larger than that of the first, the minimum value of ρ_T is practically independent of the value for ρ_2, and we have

$$(\rho_T)_{\min} \approx 0.77\rho_1$$

From the latter equations, the following relationship between coordinates of the minimum point can be obtained:

$$T_{\min}(\rho_T)_{\min} = 3.6 \cdot 10^{-6} S_1^2 \tag{11.51}$$

or

$$S_1 = 520(T \cdot \rho_T)_{\min}^{1/2} \tag{11.52}$$

Case 2: $\rho_2 \ll \rho_1$

If the apparent resistivity ρ_T falls within the H range, we are able to determine only the thickness of the first layer. From Eq. (11.33), we have

$$\log \rho_T = \log\left(8\pi^2 10^{-7} H_1^2\right) - \log T$$

Hence, the asymptote corresponding to the H zone intersects the horizontal line $\rho_T = 1$ when

$$8\pi^2 10^{-7} H_1^2 = T \quad \text{or} \quad H_1 = \frac{10^3}{2\pi}(5T)^{1/2} \approx 0.36 \cdot 10^3 T^{1/2} \tag{11.53}$$

Comparing this equation with Eq. (11.45), we see that H and S lines intersect the horizontal line for $\rho_T = \rho_1$ at a common point. Now assume that $\lambda_1/H_1 > 10$ and $\rho_2 < \rho_1$. From Eq. (11.31), we have

$$Z_2 = \omega\mu_0\left[\frac{1}{(i\gamma_2\mu_0\omega)^{1/2}} - iH_1\right]$$

Hence,

$$\mathrm{Re}\, Z_2 = \left(\frac{1}{2\gamma_2\mu_0\omega}\right)^{1/2}\omega\mu_0 = 2\pi 10^{-3}\left(\frac{\rho_2}{10T}\right)^{1/2}$$

and

$$\mathrm{Im}\, Z_2 = -2\pi 10^{-3}\left(\frac{\rho_2}{10T}\right)^{1/2} - H_1\left(8\pi^2 10^{-7}\right)/T \tag{11.54}$$

The real part of the impedance depends on ρ_2 but is independent of H_1. Determining ρ_2 from the real part of Z_2 and substituting this in the expression for ImZ_2, we can calculate the value for H_1. Also, as in the previous case, we may use the left part of the apparent resistivity curve. For instance, from the coordinates of the maximum point,

$$\left(\frac{\lambda_1}{H_1}\right)_{\max} \approx 5.3$$

it follows that

$$\rho_1 = \left(28.1 H_1^2 10^{-7}\right)/T \tag{11.55}$$

If the resistivity of the second layer is much lower than that of the first layer, the maximum value for ρ_T is practically independent of ρ_2, and we have

$$(\rho_T)_{\max} = 1.3\rho_1 \quad \text{or} \quad \rho_1 = (\rho_T)_{\max}/1.3 \tag{11.56}$$

From Eqs (11.55) and (11.56), it follows that

$$(\rho_T)_{\max}\cdot T_{\max} = 36.6 H_1^2\cdot 10^{-7}$$

or

$$H_1 = 0.52\cdot 10^3\left[(\rho_T)_{\max} T_{\max}\right]^{1/2} \tag{11.57}$$

Before leaving the subject of interpretation of the two-layered curves, it is useful to compare MT-response curves with DC resistivity responses such as those obtained with the Schlumberger array (Figure 11.3(b)). (1) MT curves approach their left-hand asymptotes in an oscillatory manner, while direct current sounding curves approach the same asymptote in a monotonic manner. MT curves cross the line $\rho_T = \rho_1$ an infinite number of times, while direct current resistivity curves gradually approach this line. (2) The curves for both the DC resistivity method and MT method approach the right-hand asymptotes monotonically. (3) Unlike the behavior of direct current soundings curves, MT curves are symmetrical with respect to the value for the contrast in resistivity between layers.

11.5.3. Apparent Resistivity Curves for Three-Layered Medium

Now consider the behavior of apparent resistivity curves for a three-layered medium as shown in Figures 11.4–11.7. A three-layered sequence is characterized by five parameters: ρ_1, h_1, ρ_2, h_2, and ρ_3. With three values for resistivity, four different sequences of resistivity contrast can be recognized:

A. $\rho_1 < \rho_2 < \rho_3$ H. $\rho_1 > \rho_2 < \rho_3$ K. $\rho_1 < \rho_2 > \rho_3$ Q. $\rho_1 > \rho_2 > \rho_3$

First, we will examine the general features of three-layered apparent resistivity curves. By definition,

$$\rho_T = \rho_1 \left| \coth\left\{ -ik_1H_1 + \coth^{-1}\left(\left(\frac{\rho_2}{\rho_1}\right)^{1/2} \right.\right.\right.$$
$$\left.\left.\left. \times \coth\left[-ik_2H_2 + \coth^{-1}\left(\frac{\rho_3}{\rho_2}\right)^{1/2}\right]\right)\right\}\right|^2 \quad (11.58)$$

At the high frequencies, the first term dominates and

$$\rho_T \to \rho_1 \text{ as } T \to 0$$

With increasing period, the influence of the second layer becomes larger, and the apparent resistivity decreases (type H and Q curves) or increases when $\rho_2 > \rho_1$ (type A and K curves). With further increase in period, the apparent resistivity approaches the resistivity of the lowermost medium, ρ_3. Thus, in principle, the sequence of maximum and minimum values of the MT curve reflects the sequence of real subsurface resistivities. Consider now the behavior of ρ_T curves when penetration (skin depth) is greater than the thickness of the first and second layer. As follows from the expression of impedance, when $|k_1H_1| < 1$ and $|k_2H_2| < 1$, for curves of type A and H, we have

$$Z_3 \approx \frac{1}{S + k_3/\omega\mu_0} = \frac{1}{S + \left(\frac{\gamma_3}{\omega\mu_0}\right)^{1/2} \exp\left(i\frac{\pi}{4}\right)}, \quad (11.59)$$

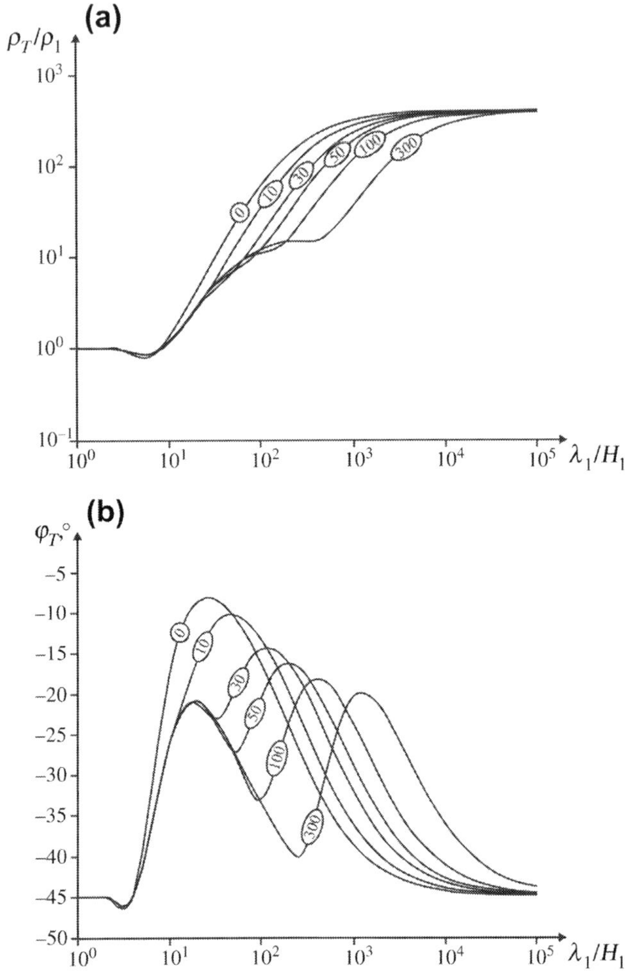

Figure 11.4 Magnetotelluric responses of a type A. (a) Apparent resistivity curves; and (b) phase curves. Curve index is H_2/H_1 ratio.

where $S = S_1 + S_2$ is the total conductance of the first two layers. It is a simple matter to generalize this equation for n-layered medium and it gives

$$Z_n \approx \frac{1}{S + \left(\frac{\gamma_n}{\omega\mu_0}\right)^{1/2} \exp\left(i\frac{\pi}{4}\right)}, \tag{11.60}$$

where S is the total conductance and γ_n is the conductivity of the lowermost layer. Note that maps of a distribution of the total conductance allow one to

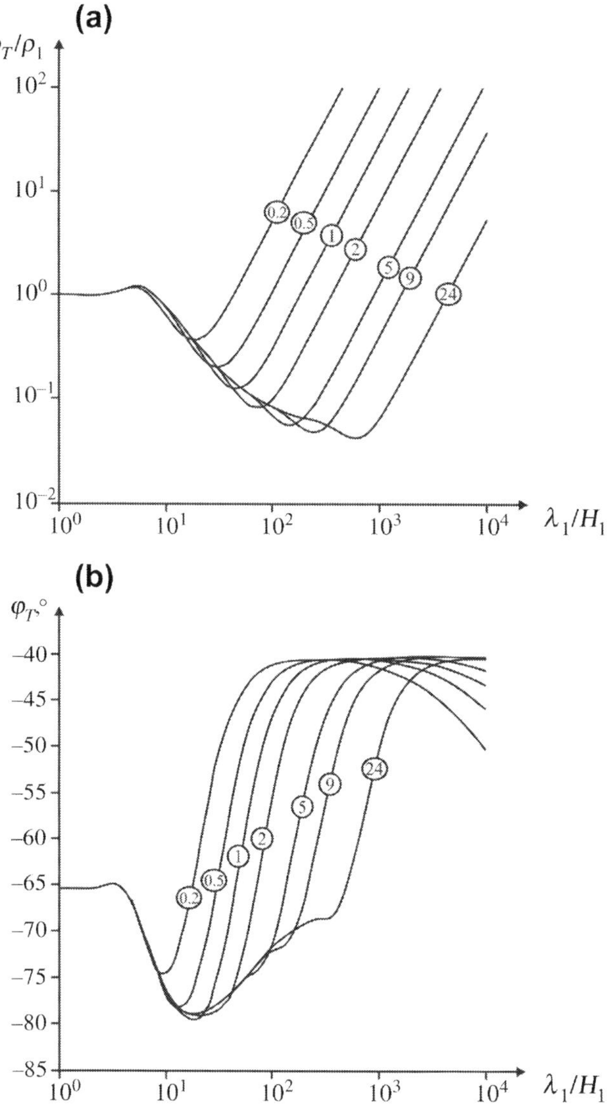

Figure 11.5 Magnetotelluric responses of a type H. (a) Apparent resistivity curves; and (b) phase curves. Curve index is H_2/H_1 ratio.

obtain information about relatively large structures on the basement surface, that especially in past was important for oil exploration. Thus from Eq. (11.59), we have

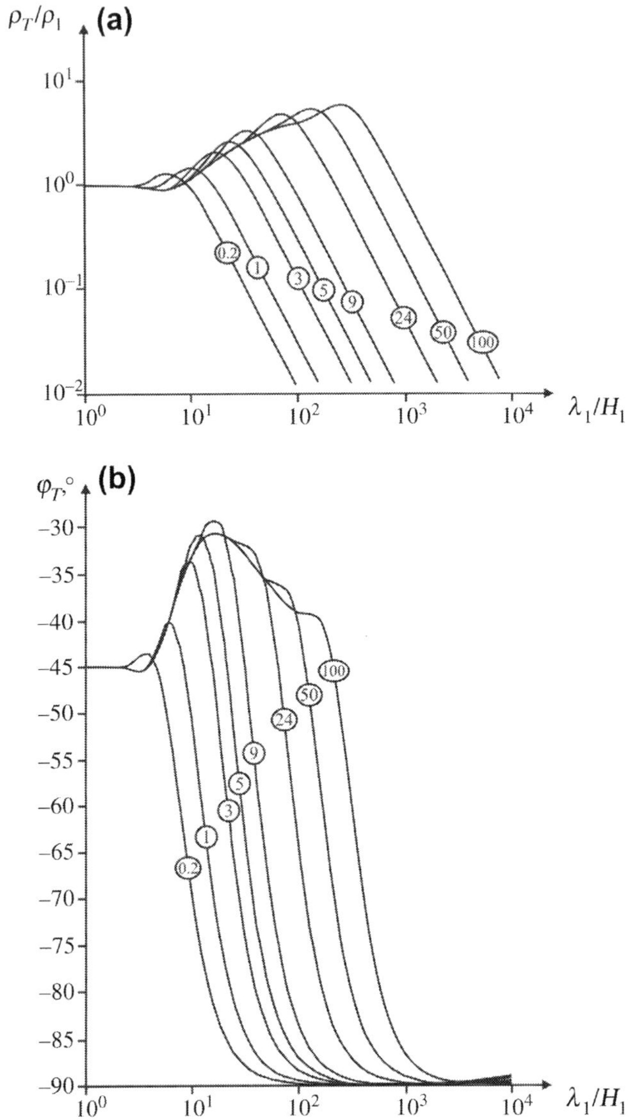

Figure 11.6 Magnetotelluric responses of a type K. (a) Apparent resistivity curves; and (b) phase curves. Curve index is H_2/H_1 ratio.

$$\rho_T = \frac{1}{\omega\mu_0}\left|\frac{1}{S + \left(\frac{\gamma_3}{\omega\mu_0}\right)\exp\left(i\frac{\pi}{4}\right)}\right|^2 \tag{11.61}$$

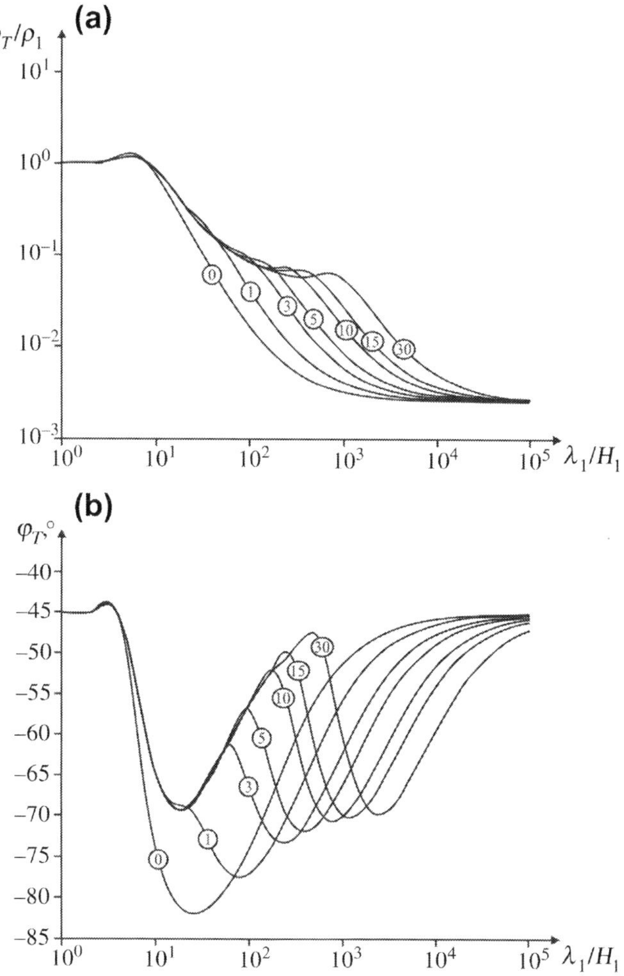

Figure 11.7 Magnetotelluric responses of a type Q. (a) Apparent resistivity curves; and (b) phase curves. Curve index is H_2/H_1 ratio.

This equation describes the right-hand part of either A or type H curves. If

$$S \gg \left(\frac{\gamma_3}{\omega\mu_0}\right)^{1/2} = \frac{10^3}{2\pi}\left(\frac{5T}{\rho_3}\right)^{1/2}$$

we again obtain an expression for the S zone behavior

$$\rho_T \approx \frac{10^7\,T}{8\pi^2 S^2} = \frac{10^7\,T}{8\pi^2 (S_1 + S_2)^2} \tag{11.62}$$

At the same time, the phase difference between the electric and magnetic fields is zero. As in the case for a two-layered medium within the S zone, the impedance is independent of the period. Next consider the alternative cases of Q and K types of resistivity distribution. Then for relatively large wavelength, we have

$$Z_3 \approx \omega\mu_0 \frac{1 - ik_3 H}{k_3} = \frac{8\pi^2 10^{-7}}{T} H \exp\left(-i\frac{\pi}{2}\right) + \left(\frac{\omega\mu_0}{\gamma_3}\right)^{1/2} \exp\left(-i\frac{\pi}{4}\right) \tag{11.63}$$

where $H = H_1 + H_2$ is the total depth to the bottom of the second layer. As in the previous case, this equation can be generalized as follows:

$$Z_n \approx \omega\mu_0 \frac{1 - ik_n H}{k_n} = \frac{8\pi^2 10^{-7}}{T} H \exp\left(-i\frac{\pi}{2}\right) + \left(\frac{\omega\mu_0}{\gamma_n}\right)^{1/2} \exp\left(-i\frac{\pi}{4}\right)$$

Here, H is the total thickness of layers. Over the interval where the condition $H \gg |k_3|^{-1}$ takes place, the impedance Z_3 is determined by the value for H only, and this part of the curve is termed the H zone. According to Eq. (11.63), the expression for ρ_T has the following form:

$$\rho_T = \left|\omega\mu_0 H \exp\left(-i\frac{\pi}{2}\right) + \frac{1}{k_3}\right|^2 = \frac{8\pi^2 10^{-7}}{T}\left|H \exp\left(-\frac{i\pi}{2}\right) + \frac{1}{k_3}\right|^2 \tag{11.64}$$

Within H zone, we have

$$\rho_T = \omega\mu_0 H^2 = \frac{8\pi^2 10^{-7}}{T} H^2 \tag{11.65}$$

Examining the behavior of the ρ_T curves, we can see that the left-hand part of the curve over which the influence of the basement is small may coincide rather closely with the curve for two-layered medium defined by the parameters ρ_1, H_1, and ρ_2. The behavior of the right-hand part of the curves is similar to that for two-layered curves for which the last layer has a resistivity ρ_3. As was pointed out, this analysis of the two- and three-layered curves of the apparent resistivity, which emphasize the relationship between different parts of the curves and resistivity structure parameters, is often very useful for interpretation. At the same time, the analysis of the apparent resistivity curves for three- and more layered medium, as well as the experience of curve matching, shown that unlike the case of two-layered medium, there are model parameters which cannot be determined with a sufficient accuracy since the useful signal is always known with some error. In other words, a relatively broad change of these parameters does not practically change the useful signal. For instance, a resistivity and thickness of the second layer in three-layered medium might be considered unstable parameters. This fact can make influence of an accuracy of evaluation of the other parameters. The knowledge of these features of interpretation suggests that the experience obtained during the curve matching was useful in developing the modern methods in a solution of the inverse problem. Note that application of curve matching with help of master curves was naturally limited to a model of a horizontally layered medium. For more general cases such as two- and three-dimensional models, it was necessary to develop new approaches, but still the idea of comparison of the theoretical and experimental data remains the same.

11.6. SOLUTION OF INVERSE PROBLEM OF THE ELECTROMAGNETIC SOUNDINGS FOR THE HORIZONTALLY LAYERED MEDIUM

Unlike previous sections, this subject is also applied to different types of the soundings over a horizontally layered medium, as well as in a more general case, when we deal with the two- and three-dimensional models of a medium.

11.6.1. The Useful Signal and Noise

First of all, there is analogy between a solution of the inverse problem of the DC resistivity method and electromagnetic soundings, including MT. As was pointed out earlier, in the case of Schlumberger soundings, the solution of inverse problem for horizontally layered for all types of electromagnetic soundings is unique, and this fact is of a great importance. At the same time taking into account the presence of different types of noise, it is impossible to resolve the resistivity model parameters with infinite accuracy since we deal with the so-called *ill-posed problem*. The presence of errors is caused mainly by several factors: (1) Measured field is always affected by the influence of geological noise caused by lateral changes of resistivity, such as topography and a presence of different subsurface inhomogeneities. (2) Measuring devices can only operate to a precision that depends on a design of the instrument, as well as external factors such as variation of temperature. (3) In practice measurements of the electric and magnetic fields are performed for a limited number of periods. Thus, the measured impedance response is a sum as follows:

$$Z = Z_U + Z_N \tag{11.66}$$

Here, Z_U is the true response (useful signal) caused by induced currents in the horizontal layer medium and Z_N is the noise. The total impedance can be measured with a relatively high accuracy but the error of determination of the true response

$$Z_U = Z - Z_N \tag{11.67}$$

depends on the level of noise. It is clear that reduction of noise is one of the most important elements of interpretation in any method, including MT, since determination of parameters of a horizontally layered medium must be based on comparison of a calculated impedance (the solution of the forward problem) not with the measured impedance but with the true response that contains information about the medium. Bearing in mind the fact that we never know the value of noise signal but rather its range of variation, it is appropriate to speak also of an interval of variation of the true response. To emphasize this fact, consider the apparent resistivity curve shown in Figure 11.8. Because of influence of noise, we only know that the value of Z_U is located somewhere inside the interval (bar). Its boundary is obtained from Eq. (11.67) assuming a certain level of noise. If at an observation point the measured field and that due to noise have the same sign, then we have $Z_U < Z$. In contrast, if Z and Z_N have different signs, the value of the useful signal is larger than the measured field. Notice that the boundary of a bar where the useful signal is located is defined approximately and based on some additional information about the accuracy of measurements and the influence of geological noise. From this consideration, it is clear that the accuracy of forward problem solution is can, in practice, be about the same as that of the true response. Because of this, we should never expect to get an exact agreement between the true and modeled impedances. In other words, an attempt to fit measured data with a field calculated to a higher precision than the range of useful signal does not make sense. Recognizing that the useful signal is known with an error that can often exceed several percent and more, let us consider the process of interpretation. First, as was pointed out, the useful part of the electric and magnetic fields at each period

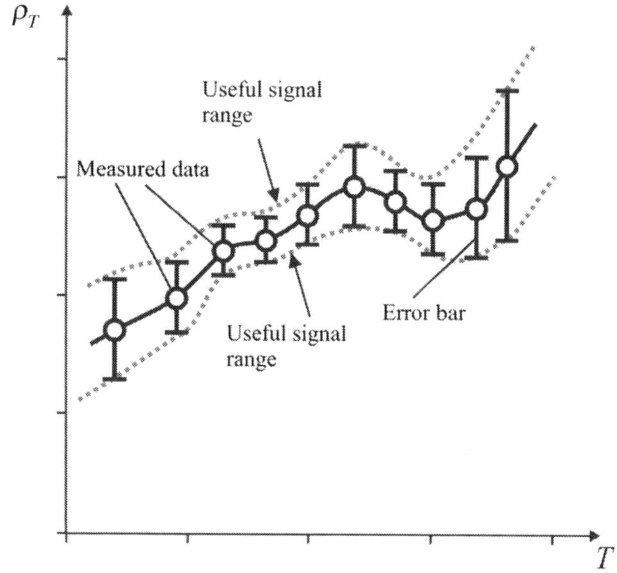

Figure 11.8 Measured response and range of useful signal.

can be represented as a sum of fields caused by currents and charges induced in different layers, and their contribution depends essentially on a period and location of interfaces with respect to the Earth's surface. For instance, we can imagine such changes of resistivity and thickness in some layers that the signal would still remain somewhere inside the range of useful signal (within error bar). In other words, because of the presence of noise, there can be unlimited number of different resistivity models that generate practically the same useful signal, even if the inverse problem for a particular class of models (such as a layered medium) has a unique answer for ideal measurements. Since the useful signal is an integrated effect caused by all induced currents inside the layers, possible changes in some parts of the resistivity model can have a significant effect on the apparent resistivity curve at some observation point, whereas similar changes in other parts of the section will result in much smaller changes. For this reason, performing an interpretation, it is natural to distinguish at least two groups of parameters describing a resistivity model, namely, the stable and unstable parameters.

11.6.2. Stable and Unstable Parameters

1. The stable parameters have a strong effect on the field; that is, relatively small changes of their values produce a detectable change in the useful

signal, for instance, the conductance of a relatively conductive layer even when its thickness is sufficiently small.

2. The unstable parameters could have observable influence on the field only if their values are greatly changed. This simply means that they cannot be determined from the useful signal measured with some error. One of the examples of such parameters can be resistivity and thickness of a sufficiently thin and resistive layer, surrounded by more conductive medium. Therefore, we can say that an interpretation or a solution of the inverse problem consists mainly of determining the parameters of the first group; that is, the resistivity and thickness of layers or their combination that have a significant effect on the measurements, even though they do not completely characterize the resistivity model. It is clear that this stable group of parameters describes a structure that differs to some extent from the actual one but both of them have common parameters.

The error of evaluation of some unstable parameters can be extremely large. In other words, these parameters cannot be practically determined. To illustrate a little more this fact, imagine that we change the value of a single parameter section p_i by the amount Δp_i. Let $\Delta Z_i(T)$ be a change in calculated impedance (the useful signal) at different periods caused by the change in the ith parameter. We can write the following approximate relation between the change of the useful signal and Δp_1:

$$\Delta Z_i(T) = k_i(T)\Delta p_i \quad \text{or} \quad \Delta p_i = k_i^{-1}\Delta Z_i(T) \tag{11.68}$$

Here, $k_i(T)$ is the coefficient of proportionality between the change in the ith parameter of a resistivity model and the change of the useful signal $\Delta Z_i(T)$. The factor $k_i(T)$ is usually called a "sensitivity coefficient" and will be a different function for different parameters. If it is large, then a small change of parameter p_i causes a relatively large change of useful signal. This means that in performing a solution of the inverse problem, we can determine the value of this parameter with sufficiently high accuracy. In contrast, when k_i is relatively small, a large change in parameter p_i will produce only a small change in useful signal ΔZ_i. As it follows from the second equation of the set (11.68), when the sensitivity coefficient k_i gets smaller and smaller, our ability to determine the ith parameter deteriorates, and it might not be possible to determine a given parameter to any reasonable precision. In principle, a lower limit cannot be established for some coefficients k_i. This is the most important feature of an ill-posed problem. In other words, even an unlimited change of some parameters (for instance, the resistivity of a relatively thin and resistive intermediate layer) does not produce a noticeable variation of the field exceeding a bar interval, and as a result of this, it is impossible to resolve these parameters.

Thus, in any practical inverse problem, there are two groups of model parameters: stable and unstable parameters. The goal of interpretation and inversion is to separate them and determine the stable ones. The latter characterizes a new model of a real subsurface structure, and the coefficients k_i for its parameters are relatively large. In this case, we arrive at a well-posed problem in which some parameters of the real model remain unknown because they are not well determined from the measurements. This situation may seem unsatisfactory but in fact reflects the reality of a solution of nearly all problems in geophysics, in particular, the inverse problem of all types of electromagnetic methods. The transition from an ill-posed problem to well-posed one is called the *regularization* of the inverse problem and is of a great practical interest. It is obvious that an interpretation of MT data is useful for practical application if the parameters of a medium, which are the survey targets, can be determined within a narrow enough range of values. Often selection of this group of parameters can be done automatically by making use of the standard algorithm for solving inverse problem employing Eq. (11.68). Of course, with an increase in the number of model parameters, the approximation of a real model can in principle be better. However, the error with which some of these parameters are determined also increases. As in the theoretical case, when the field is known exactly, we can expect that the interpretation of the MT data is greatly facilitated through using additional information about the structure derived from geology and other geophysical methods, including the borehole geophysics.

11.6.3. The Main Steps of Interpretation

This discussion allows us to distinguish the following steps of an interpretation.

1. Making use of a priori information about the resistivity structure parameters, we formulate the first guess. As was mentioned above, this is the very important step, which is necessary and may greatly facilitate the interpretation.
2. The second step is a solution of the forward problem; that is, calculation of the impedance using parameters of the first guess.
3. As a result of this calculation, we obtain a set of values of the impedance which can be graphically represented as a curve $Z_U(T)$ (Figure 11.8). Suppose that this curve is situated beyond the interval. Then after

changing model parameters, we solve the forward problem again and obtain a curve of the impedance or apparent resistivity, which is closer to the interval of useful signal. This process is continued until the calculated impedance is located inside the interval (bar). So far, the steps of interpretation are identical to those when we would consider in the case when the calculated and measured impedances are known exactly. Now, we will observe the fundamental difference.

4. When values of calculated impedance are located inside the observation interval (bar), further improvement of matching between the useful and calculated impedances does not make sense because we do not know where inside the interval the useful signal is located. Therefore, we stop the process of fitting of impedances and start a new procedure, which also requires a solution of the forward problem. At the last iteration of matching, we obtained a set of parameters,

$$p_1, p_2, p_3, \ldots, p_n,$$

that placed the calculated response inside the bar region. Our goal is to determine the range of change of each parameter, so that the useful signal remains inside the interval. This procedure is usually repeated several times and is accompanied by a solution of the forward problem. Of course, every step causes a movement of the curve of the useful signal and, so long as its position remains inside the interval, it is equivalent to the previous one. As a result of these steps for each parameter, we obtain its range as follows:

$$p_1^{\min} < p_1 < p_1^{\max}, \ldots, p_n^{\min} < p_n < p_n^{\max}$$

As was pointed out earlier, any particular values of parameters within these ranges result in the same fit goodness for the measured and useful signal. Certainly, knowledge of variations of these parameters is the most important step in solving the inverse problem because this allows us to separate the stable from unstable ones, and correspondingly performs a transition from the ill-posed problem to a well-posed one. Thus, interpretation gives us a set of stable parameters of the resistivity structure and the range of its change (model errors).

From a review of the solution of the inverse problem, it is clear that with a decrease of the interval width, the range of each possible parameter of the model decreases too and the number of unstable parameters may become smaller. At the same time, stable parameters can be determined with higher confidence. For this reason, reduction of different types of noise (first of all the geological noise) is very important subject. We outlined the main features of interpretation for the class of models where uniqueness takes place. In regard to interpretation within the class of models with no uniqueness, the situation is completely different. We can say that a solution of the inverse problem in such cases is hardly possible (speaking strictly, it is senseless). In fact, even in the absence of noise, we always have an infinite number of models, which display exactly the same measured field. At the same time, it may be possible to determine some generalized features of the structure. Thus, uniqueness theorems, which appear to be mainly of academic interest, are actually very important in solving practical inverse problems.

REFERENCES AND FURTHER READING

[1] M.N. Berdichevskiy, Electric Prospecting by the Method of Magnetotelluric Profiling, Nedra, Moscow, 1968.
[2] M.N. Berdichevskiy, V.I. Dimitriev, Models and Methods of Magnetotellurics, Springer-Verlag, 2008.
[3] L. Cagniard, Basic theory of the magnetotelluric method of geophysical prospecting, Geophysics 18 (1953).
[4] A. Jones, A. Chave, The Magnetotelluric Method: Theory and Practice, Cambridge University Press, 2012.
[5] A.A. Kaufman, G.V. Keller, The Magnetotelluric Sounding Method, Elsevier, 1981.
[6] J.R. Wait, Geoelectromagnetism, Academic Press, New York, NY, 1982.

CHAPTER TWELVE

Electromagnetic Soundings

12.1. DEVELOPMENT OF THE FREQUENCY AND TRANSIENT SOUNDINGS

The concept of the frequency and transient soundings emerged among the geophysicists almost by 1910. It was realized that the subsurface resistivity structure could be studied with an array, which consists of the generator, transmitting either the sinusoidal oscillations or impulses to the grounded line or loop, and a receiver measuring the electric or magnetic field at some distance from the current source. It was natural to expect that the frequency and transient responses of the electromagnetic field at the receiver contain information about geoelectric section. In other words, the current source and received signal are considered the input and output of the linear system (earth), respectively. Before the Second World War, there were attempts to apply the frequency and transient soundings but without noticeable success. First of all, it happened because of absence of the theoretical foundation of these methods. For instance, one of such techniques was Electrical transient (Eltran) method, suggested in the USA, which allowed one to measure the nonstationary field using either two- or four-electrode array. In order to have an idea about the theory level behind the Eltran, it is sufficient to say that earth was assumed to be a simple circuit characterized by a time constant, which was the only parameter to be determined.

The interest to the frequency and transient soundings was gained in the middle of 1950s and especially great achievements were made in the USSR, where the state industry was strongly involved in development of this method as well as other geophysical methods. There were several factors, providing favorable conditions for making progress in this area, such as the following: (1) After the Second World War, it was a great need for mineral resources and first of all for oil and gas deposits. (2) Vast areas in the USSR, especially in West and East Siberia, were promising for petroleum resources but poorly studied in terms of geology. (3) Practical

Methods in Geochemistry and Geophysics, Volume 45
ISSN 0076-6895,
http://dx.doi.org/10.1016/B978-0-444-53829-1.00012-5

interest to the basement structures of a relatively large size. (4) In many regions, a highly resistive layer was found to be present in sedimentary strata, affecting the Schlumberger soundings (screening effect) and it did not allow one to study the structure beneath this layer. (5) Seismic surveys were relatively expensive at that time and level of interpretation of reflection wave data was not so high, which meant that the frequency sounding might deliver the similar information but with lower expenses. (6) It was an impression that the electric and electromagnetic methods are able to yield physical properties of the layers, while seismic methods give only geometrical structure. All these factors as well as some others together played an important role in development of the theory, equipment, and implementation of the frequency sounding, which began almost simultaneously with practical application of magnetotellurics. At the beginning, following the classical traditions of the Russian geophysics, the main attention was paid to the theory of the method. As a result, the boundary value problems for a horizontally layered medium were solved and asymptotic formulas for the low- and high-frequency spectrum were obtained. Also the algorithms for calculation of the field caused by the horizontal electric and vertical magnetic dipoles were developed and all this information became available through numerous publications. This was mainly done by the group of researchers working at the Institute of Physics of the Earth under guidance of A. Tikhonov. Earlier, in 1947, a distinguished Russian geophysicist S. Sheinman had published the paper *On Transient Electromagnetic Fields in the Earth*, which was very useful for understanding the transient field on the earth's surface. In this paper, the author described in detail the behavior of the transient field on the surface of a uniform half space, and special attention was paid to relation between the electromagnetic field at different time ranges and resistivity of a medium. Certainly, this paper caused the initial interest to the electromagnetic methods and their development in Russia. It is interesting to notice that application of Schlumberger soundings with large separations between current electrodes has shown that it takes some noticeable time for the voltage to reach its maximum value, and there were attempts to explain this phenomenon by induced polarization effect in sediments. Sheinman clearly demonstrated that such behavior of the measured voltage was of purely electromagnetic nature. In parallel, the well-known American theoretical geophysicist J. Wait and many others also made very strong contribution to the theory of the electromagnetic fields in a horizontally layered medium.

12.2. FREQUENCY SOUNDINGS IN THE FAR ZONE

In the middle of 1950s, the use of computers was very limited and even the calculation of the Schlumberger apparent resistivity curves was not a simple task. At that time in the USSR, an arithmometer was the main tool for geophysical calculations. Of course, calculation of the complex valued electromagnetic responses for various resistivity models, which required numerical integration, was even much more complicated problem. For this reason, the discovery done by Tikhonov and his colleagues, namely, that in the wave zone, if the separation between the transmitter and receiver is sufficiently large, the electromagnetic field on the earth's surface is described through elementary functions, was extremely important fact defining the main features of the frequency sounding at that time. Certainly, calculation of these functions was even simpler than numerical integration required in case of constant electric field. Besides, unlike Schlumberger soundings, the measurements were performed with one separation between the current source and a receiver that was undoubtedly important advantage of the frequency sounding from the practical point of view. Note that at any distance it is possible to satisfy the condition of the wave zone, but with the decrease of the distance, it is necessary to use higher frequencies which lead to smaller depth of investigation. For this reason only, sufficiently low frequencies providing the required depth of investigation are of practical interest at large separations, in which case the wave zone approximation remains valid for the most part of the frequency response. Correspondingly, the main area of the method was the far zone when

$$\frac{r}{\lambda} > 1 \quad \text{and} \quad \frac{r}{H} \gg 1$$

Here, H is the total thickness of sediments above the basement, and λ is a quantity, which has a dimension of the wavelength, depending on the frequency and resistivity structure parameters. At the beginning, there were attempts to design the instruments for measuring both amplitude and phase of the field at the observation point, because at that time, amplitude and phase responses were considered containing different (independent) information on resistivity structure. However, it required transmitting the reference signal from the current source to the receiver located at dozens kilometers apart, often in the presence of rough topography, which was not a simple task to realize at that time. At the end of 1950s, Russian geophysicists, involved in work with electromagnetic methods, learned that Hilbert transform allows calculating the phase from the amplitude spectrum. Note that after they understood this idea, it has not been realized for quite a long time that such approach is possible if only the amplitude spectrum does not have zeroes in the upper part of the complex plane of ω. Fortunately, as follows from the numerical modeling, such singularities are absent in the case of a horizontally layered model. Correspondingly, the equipment for frequency soundings was greatly simplified and was based only on measurements of the field amplitude at different frequencies.

Thus at the beginning, the focus was paid to the frequency soundings at the far zone when separation between transmitter and receiver r was much greater than the total thickness H of sediments above the basement $(r \gg H)$, so that the main part of the spectrum of the field would correspond to the wave zone. The realization of the method was performed under the guidance of L. Vanyan, who developed the method's theory and approach to data interpretation, and was also deeply and successfully involved in its practical application.

Next, let us briefly describe the main features of the equipment which was used in this method. The transmitter provided the current of about 70 A and voltage of about 400 V. At the same time, the resistance of the current line normally did not exceed 6.0 Ω. The time-constant voltage, produced by the motor generator, was transformed into a sequence of the alternating periodical impulses. In order to measure the magnetic field, the square loops with side around 100–200 m were used, while the electric field was measured with a help of the receiver dipoles. Receiver system used filters to pass only the first harmonic of the impulse-periodic signal. The range of frequencies typical for petroleum applications was 10^{-2} Hz $< f <$ 10^2 Hz. The higher frequencies (10 Hz $< f <$ 10^4 Hz) were recorded in those cases when the method was used to study shallow-depth subsurface resistivity (up to several hundred meters). The most typical array for soundings consisted of the electric source dipole and magnetic receiver measuring the vertical magnetic field in the equatorial plane, while for the shallow exploration, the current loop was often used as a source instead of the electric dipole. Sometimes for the deeper studies, in the absence of the screening layers in the strata, instead of measuring magnetic field, the electric field was observed, often using nonpolarizable electrodes.

As follows from the theory of the fields in the far zone, the complex amplitude of the electromotive force induced in the horizontal receiver coil $\left(\varphi = \pi/2\right)$ is

$$\Xi^* = -\frac{3M_0\rho_1}{2\pi r^4} R_N^2(0) I_0 AB,$$

provided that an influence of the wave traveling from the source to a basement and then upward from the relatively resistive basement is negligible. Here, I_0 is the current in the electric dipole and AB is its length. $M_0 = nS$ is the product of the area and number of turns of the receiver coil. Note that in the case of the insulating basement, the expression for the electromotive force is slightly different (Chapter 9).

As usual, the apparent resistivity is introduced by the ratio

$$\frac{\rho_a}{\rho_1} = \left|\frac{\Xi^*}{\Xi^*(\rho_1)}\right|,$$

where $\Xi^*(\rho_1)$ is the complex amplitude of the electromotive force in the far zone in the case of a uniform half space when $R_N = 1$. Thus,

$$\rho_a = k\frac{\Xi}{I_0} = \rho_1 \left|R_N^2(0)\right|,$$

that is, the array coefficient is equal to

$$k = \frac{2\pi r^4}{3M_0 AB}$$

It is clear that the curves of the apparent resistivity of the frequency soundings in the far zone and magnetotelluric soundings coincide with each other. In this sense, both methods are equivalent to each other but there are some differences. For instance, the impedance and the vertical component of the magnetic field may have different sensitivity to resistivity structure parameters. For instance, the impedance is directly proportional to the square root of resistivity of a uniform half space and inversely proportional to the conductance S in the zone S, while the magnetic field of the electric dipole in the far zone is directly proportional to the resistivity and inversely proportional to the square of the same conductance. In other words, given the same accuracy of measurements, the frequency sounding is able to detect smaller changes in some parameters. Besides the use of a controlled source greatly simplifies determination of a signal at given frequency, but unlike the magnetotelluric soundings, it requires a man-made generator that makes the system more complicated. There is one more difference. Magnetotellurics normally operates in the wave zone regardless how low the frequency is and correspondingly the apparent resistivity curves reflect main features of the resistivity structure within the whole range of recorded frequencies. In contrast, in the case of the frequency soundings in the far zone, there are usually frequencies which result in wavelengths greater than the separation; that is, the measured the field does not meet the wave zone approximation. Respectively, there was necessity to perform full calculation involving numerical integration to determine the field instead of using simple expressions for the wave zone when the ratio $\frac{r}{\lambda} \to \infty$. It is natural that with an increase of the separation, this behavior extends to larger wavelengths.

In conclusion, it may be proper to emphasize again that although the period of application of this method was relatively short, the development of the frequency soundings in the far zone in 1950s is very important contribution to geophysical methods and in many cases allowed replacing the Schlumberger soundings, which required one to use different separations and had strong limitations for mapping the basement in the presence of highly resistive screens in a sedimentary strata. Besides, and what is the most important, it helped to investigate vast areas which were of great interest for petroleum geology.

12.3. TRANSIENT SOUNDING IN THE FAR ZONE

In the second half of 1950s, it was realized that the same information about subsurface resistivity can be derived from transient responses of either the electric or magnetic field instead of the frequency responses. With this purpose, the appropriate instruments were designed and manufactured, and during several years, the method was very popular for petroleum applications, especially in Siberia. The leading role in development of this approach again belonged to L. Vanyan. Later this modification of the transient method was applied for different purposes, such as geothermal studies in the USA (G. Keller) and is still in use (quite rarely) under different names, for instance, LOTEM and others. From the beginning, the measurements were performed only during time off when the current in the transmitter was absent and the field was caused by the currents induced inside the medium. To obtain the apparent resistivity as a function of time, the Fourier transform was used:

$$\rho_a(t) = \frac{1}{2\pi} \int_{-\infty}^{\infty} \frac{\rho_a(\omega)}{-i\omega} \exp(-i\omega t)\, d\omega,$$

Calculation of the apparent resistivity in wave zone approximation was not difficult problem even at that time. Moreover, as was demonstrated in Chapter 9, in case of two-layer model, the integral is expressed through elementary functions. In reality, the field is measured within a time range which partially corresponds to intermediate and near zones as well as the wave zone itself. For this reason, in order to calculate the transient responses of the apparent resistivity, it was necessary to perform double integration: first of all the calculation of the spectrum, using integrals with Bessel functions and then applying Fourier transform. Bearing in mind that at that time, an interpretation was based on a comparison of the master (theoretical) curves of the apparent resistivity being functions of time with experimental ones, there was a need to perform numerous calculations for different parameters of resistivity models and store them in the form of master curves sets. Inasmuch as the method of filters for numerical integration was not known among geophysicists and the use of the first generation of computers was rather limited, this problem was very difficult, and it was resolved by application of the Fillon's method for numerical integration of both integrals containing rapidly oscillating functions. Examples of three-layer curves of the apparent resistivity are given in Figure 12.1(a) and (b).

Here, ρ_1 and H_1 are resistivity and thickness of the upper layer, respectively, and

$$\tau_1 = \left(2\pi\rho_1 t 10^7\right)^{1/2} \mathrm{m}$$

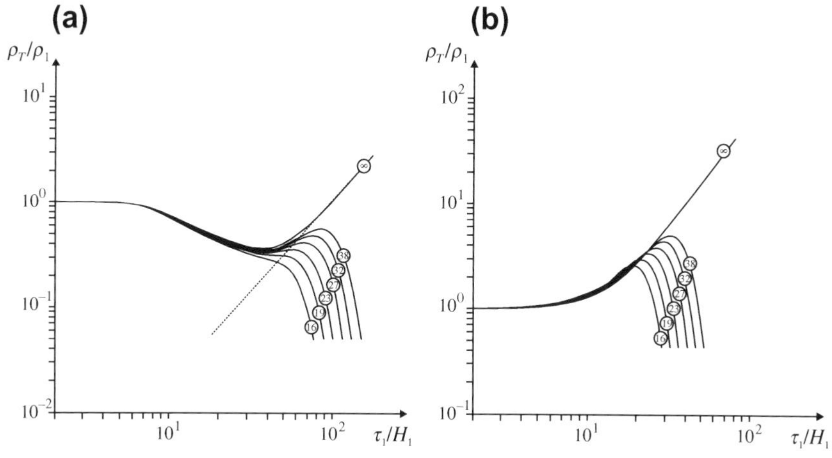

Figure 12.1 Three-layer apparent resistivity curves in wave zone (early stage) approximation. Curve index is r/H_1. Model parameters are (a) $\rho_2/\rho_1 = 1/4$, $H_2/H_1 = 4$, $\rho_3/\rho_1 = \infty$; (b) $\rho_2/\rho_1 = 4$, $H_2/H_1 = 4$, $\rho_3/\rho_1 = \infty$.

These curves adequately resemble general features of the resistivity structure within time range corresponding to the early stage. As in the case of the frequency soundings, the right part of the curves still contains information about the medium but does not image the resistivity-depth distribution since in this case, apparent resistivity relates the observed field with that for a uniform half space at the early stage. As we already know, for example, for the electric field E_φ, we have

$$\frac{\rho_a}{\rho_1} = \frac{E_\varphi}{E^{\text{un}}_{\varphi e.s}}, \tag{12.1}$$

where $E^{\text{un}}_{\varphi e.s}$ is the electric field caused by the magnetic dipole source on the surface of a uniform half space with the resistivity ρ_1, provided that measurements are performed at the early stage. Bearing in mind that

$$E^{\text{un}}_{\varphi e.s} = \frac{3M_0\rho_1}{2\pi r^4},$$

we have

$$\rho_a = \frac{2\pi r^4}{3M_0}E_\varphi \quad \text{or} \quad \rho_a = k_E E_\varphi, \tag{12.2}$$

where

$$k_E = \frac{2\pi r^4}{3M_0} \tag{12.3}$$

It is important to recognize that the coefficient k_E is independent of the resistivity ρ_1 and is defined only by the dipole moment and the separation, r. As time approaches zero, the apparent resistivity computed from Eq. (12.2) tends to the value ρ_1:

$$\rho_a \to \rho_1$$

However in the opposite case, for relatively large times, the apparent resistivity goes to zero no matter what the real resistivity–depth distribution is. In essence, the apparent resistivity function, defined in this way, differs from the field only by some multiplicative factor k_E. When the time range being considered corresponds to the early stage of the apparent resistivity response, Eq. (12.2) clearly reflects the actual changes in the resistivity with depth, while beyond this time range, over the intermediate and late stages of the transient responses, the shape of the apparent resistivity curve bears little resemblance to the resistivity–depth profile. As was mentioned earlier, this manner of defining the apparent resistivity is useful for the transient soundings in the far zone when the separation between transmitter and receiver is many times greater than the depth of investigation. We can also define apparent resistivity for components of the magnetic field or their time derivatives in a manner similar to that for Eq. (12.2). Namely, for the time derivatives of the magnetic field, we have

$$\frac{\rho_a}{\rho_1} = \frac{\dot{B}_z}{\dot{B}^{\mathrm{un}}_{z,e.s}}, \quad \text{or} \quad \frac{\rho_a}{\rho_1} = \frac{\dot{B}_r}{\dot{B}^{\mathrm{un}}_{r,e.s}} \tag{12.4}$$

As was pointed out, the development of the transient sounding in the far zone was very important step in the theory and practice of electromagnetic soundings. At the same time, it may be proper to notice the following. At the end of 1950s and beginning of 1960s, specialists in electromagnetic methods did not have a sufficient knowledge about the behavior of induced currents in the layered medium as a function of time despite the fact that a solution of the forward problem for the field in such models had been known for a long time. Otherwise, it would be possible to realize that the transient soundings could be carried out with much smaller separations that became clear later. Perhaps this opportunity was missed also due to the strong belief that the source–receiver separation is the main factor controlling the depth of investigation. Certainly, this illusion had formed due to a strong influence of the main concept of the Sclumberger soundings.

12.4. TRANSIENT SOUNDING

In the middle of 1960s, Russian geophysicists recognized that transient sounding, unlike the frequency sounding based on measuring an amplitude or phase, is capable of imaging resistivity structure regardless of a transmitter–receiver separation. In particular, this separation can be smaller than the total thickness of sediments. Such strong reduction of separation compared to the far-zone transient method ($r \gg H$) was a natural step in method's development, since it had been already in use for mining applications, where it was demonstrated that one can detect a confined body with the array having the size comparable to or even less than the depth to this body. Besides, it was shown that depth of investigation in the induction logging based on measuring the in-phase component of the secondary field can be much greater than the probe length. As soon as it was understood that the depth of investigation is controlled by the process of induced currents diffusion rather than separation, this approach was accepted and its theory, interpretation schemes, and instrumentation were developed relatively quickly. According to its name, transient sounding in the far zone emphasizes the fact that measurements are performed at distances greatly exceeding the total thickness of sediments $(r \gg H)$, while the near-zone transient soundings imply that measurements can be carried out at relatively small distances $r \approx H$. The passage to the transient soundings with small source–receiver separations occurred when the physical principles of the method became clear. As we already know, when the current is suddenly changed either in a loop or grounded line located on the earth's surface, the induced currents instantly appear at the earth's surface in the vicinity of the source (quasistationary approximation). In the case of the current in the circular loop, induced currents have only azimuthal component at the initial moment and later. In other words, in the horizontally layered medium, charges do not move in the vertical direction. At the same time, the magnitude of these currents obey the diffusion equation and only in this sense we can speak about diffusion of currents which have very simple geometry, namely, they form rings coaxial with the current loop. Inasmuch as influence of displacement currents is neglected, induced currents appear everywhere at the initial instant but their amplitude is equal to zero except for the vicinity of the source. Diffusion of the currents takes place with time, and at each point of a medium (except points on the axis of symmetry), the current's magnitude increases, reaches its maximum and after it becomes smaller and tends to zero. As depth increases, the maximum becomes smaller and wider. At the early stage, the field is mainly defined by currents in the upper layer and correspondingly it depends on the resistivity ρ_1, $(r \le H_1)$. With time, we observe two important features of a current distribution, namely, a decrease of currents in the upper part of a medium and a relative increase of current density at larger depth. As was demonstrated earlier, with an increase of time the depth of investigation also increases. For this reason, performing measurements within the intermediate stage, we obtain information about deeper layers. Moreover, as the theory shows, the influence of the upper layers becomes negligible regardless of the resistivity. This is a very important feature of the field which of course manifests itself at any separation r. Finally, at the late stage of the transient response of the field, the currents are mainly located in the lowermost layer, and the field is defined by its resistivity. Similar behavior is observed in the case when the primary source is grounded line but geometry of currents is more complicated. Thus, understanding of behavior of the induced currents clearly demonstrated that the time of measurements is the leading factor which controls the depth of investigation. In order to emphasize the difference with the transient soundings in the far zone, this modification of the method is usually called the *transient soundings* in the near zone.

Now, we describe the first steps in development of this method. In 1965, Soviet geophysicist P. Frolov published the paper in the journal *Physics of the Earth* describing the asymptotic behavior of the transient magnetic field generated by the electric dipole on the earth's surface at large separations (far zone), when a basement of a two-layer medium being either an insulator or ideal conductor. Perhaps, since he mainly paid attention to some details of the field behavior in the far zone, which was already well developed, this paper did not attract at the beginning attention of geophysicists. Considering the case of the nonconducting basement, he derived the asymptotic formula for the magnetic field at the late stage caused by the electric dipole:

$$B_z = \frac{3I\,\mathrm{d}xy\,\mu_0^4 S^3}{32\pi t^4}$$

Here, S is the longitudinal conductance of the upper layer, r is the separation, I and dl is the current and length of the dipole and measurements are performed in an equatorial plane. This asymptotic equation was derived earlier by L. Vanyan for the far zone when separation is much greater than the layer thickness. However, P. Frolov applying a different procedure of derivation came to an unexpected conclusion that this equation is valid regardless of the separation between the transmitter and receiver. In other words, the transient method can be applied for even very small separations, and this fact was strongly emphasized in the paper. Thus, P. Frolov was the first who shown that the transient method can be used with small separations but not only in the far zone, and his paper has to be considered as the beginning of the transient sounding in the near zone. Note that much later in fact, there was a natural merge of transient method proposed earlier for mining applications with that used for sedimentary structure exploration. Thus, eventually, the transient method has found a broad application beyond petroleum geology.

Further steps in development of this method were started almost the following year at three different places of the USSR. In 1968, V. Sidorov (Saratov) performed some experimental studies of the method in the field conditions and, having an experience in the mining prospecting, he suggested an approximate method for interpretation based on the model of a thin conducting layer with lateral conductance *S*. In the same year, G. Obukhov published a paper in the journal *Physics of the Earth*, where he considered the same limiting cases of two-layer model as P. Frolov and, applying different mathematical approach, came to the same conclusion: the transient sounding can be used with small separations. Both P. Frolov and G. Obukhov (Moscow) obtained only the leading term of the series describing the vertical component of the magnetic field of the electric dipole at the late stage when the basement is either insulator or ideal conductor. Starting from 1966, A. Kaufman and his colleagues (Novosibirsk) had been developing the theory, interpretation schemes, and equipment for the transient sounding in the near zone, being also involved in its practical application. First (in 1967), the field at the late stage on the surface of a uniform medium was studied and special attention was paid to the formulas at the late stage. Later, these expressions were used to define the apparent resistivity. Then in the same year 1967, asymptotic formulas at the late stage for the two- and three-layer models were obtained and published (in 1968) in the journal *Geology and Geophysics*. Unlike the previous papers, the models with conducting and nonconducting basement were considered and the asymptotic series were not restricted to only the leading terms but also included subsequent ones. These formulas have shown that there is always time when induced currents are mainly located in the underlying layer and what was very important an influence of sediments above is negligible. At that time, this result was not obvious. One of the reasons of such doubts was the fact that in the frequency domain, it is impossible to perform soundings at small distances, measuring the amplitude or phase of the field, regardless how low the frequency is. Inasmuch as the procedure of deriving the asymptotic formulas was based on the knowledge of the low-frequency part of the spectrum, it helped to understand the following. At this part of the spectrum, the secondary field is mainly defined by the quadrature component which at relatively small separations depends mainly on parameters of the upper layer, and frequency does not affect the depth of investigation. This is not surprising because this term is defined entirely by the primary electric field and does not have any relation to diffusion. In other words, the late stage of transient field is independent on this term but it is defined by the next term of the low-frequency spectrum. As concerns the in-phase component of the secondary field at the low-frequency spectrum, which contains information about subsurface resistivities regardless of separation, it is in many orders smaller than the primary field. This analysis allowed one to better understand relations between the frequency and transient soundings in the near zone. Besides, the study of the asymptotic formulas permitted one to evaluate times when the late stage takes place and develop the proper algorithm for calculation of the transient field proceeding from a solution of the boundary problem and Fourier integral. As a result at the end of 1960s, several albums of two- and three-layer apparent resistivity curves were published (Novosibirsk), and they became available to Soviet geophysical organizations and research institutes, and much later, to the western companies. Each of the albums contained apparent resistivity curves and approximate formulas for determining layers' parameters. Later, they were replaced by software tools, as a part of the conventional equipment. This work was started by A. Kaufman and his colleagues (G. Morozova, G. Isaev, B. Rabinovich and others), working at Institute of Geology and Geophysics and Siberian Institute of Mineral Resources (Novosibirsk), and one can say that at the end of 1960s, the theory and interpretation of the transient soundings in a horizontally layered medium was accomplished. As a result of their efforts, there were many publications, including the first book describing this method. At that time in Novosibirsk, Saratov, and Moscow, the equipment was also designed which allowed one to start the surveys in different areas across the country.

The first remarkable result of application of the method took place at the very beginning of 1970s, when the method was applied by Novosibirsk geophysicists for detecting oil–water contact at the known deposit at the Yaracta block of the East Siberia, characterized by pretty favorable conditions for its application. Namely, the sediments overlying a thin monocline reservoir, located at 2 km depth, had relatively low conductance while the conductance of the water-saturated part of the reservoir was twice higher. It allowed one, performing soundings along profiles between productive and nonproductive wells, to determine water–oil contact position with higher accuracy. The success of this survey demonstrated that under certain conditions, the method can be useful in petroleum geophysics, although seismics, having much higher resolution, was in absolute dominance. At the same time later this example inspired application of the method in other areas, for instance, in hydrogeology for detecting a contact between fresh and salt water. In essence, now the transient sounding with small transmitter–receiver separations is used in the surface and airborne prospecting for solving numerous problems, including its original area—petroleum exploration. Unlike the far-zone techniques, this method remains a successful tool of applied geophysics for more than 40 years.

12.5. APPARENT RESISTIVITY CURVES

Now we describe in some detail the relation between measured fields and parameters of a horizontally layered medium. With this purpose, it is natural to take advantage of experiences that has been gained in interpretation of the Schlumberger and other electromagnetic soundings and to introduce the apparent resistivity. As we already know, such modified representation of the transient response does not mean that recalculated data contain more information than the original field responses. However, as in other types of sounding, the shape of the apparent resistivity curve might express more clearly some of the important characteristics of the resistivity structure and, therefore, their use can significantly facilitate the process of interpretation. In principle, the manner of defining apparent resistivity for the transient sounding is the same as in other conventional methods. In fact, the procedure consists of normalizing the observed voltage by the voltage, which would be recorded on the surface of a uniform half space having the resistivity of the uppermost layer. The apparent resistivity curve then characterizes the differences in the response caused by deviation of real structure from a uniform half space. These deviations can be caused by the existence of horizontal stratification with the resistivities and thicknesses of the layers being the targets of interpretation, or by the lateral changes in resistivity. In this chapter, it is assumed that no lateral inhomogeneities are present. There are several ways to define the apparent resistivity because the various expressions for the field in a uniform half space with resistivity ρ_1 can be used for this purpose. Let us first consider as an example when the electric field generated by vertical magnetic dipole source is measured. As was pointed out, this is equivalent to the case when the vertical component of the magnetic field is caused by the horizontal electric dipole. Thus, we mainly focus at two cases when the derivative of the vertical component of the magnetic field with time is measured, and the source of the primary field is either the vertical magnetic dipole or the electric dipole. At the same time, definition of the apparent resistivity is based on comparison of the measured field with the field at the late stage on the surface of a uniform half space characterized by the resistivity ρ_1. Then this function is given by the expressions:

$$\frac{\rho_a}{\rho_1}(E_\varphi) = \left(\frac{E^{\text{un}}_{\varphi,l.s}}{E_\varphi}\right)^{2/3}, \quad \frac{\rho_a}{\rho_1}(B_z) = \left(\frac{B^{\text{un}}_{z,l.s}}{B_z}\right)^{2/3} \tag{12.5}$$

$$\frac{\rho_a}{\rho_1}(\dot{B}_z) = \left(\frac{\dot{B}^{\text{un}}_{z,l.s}}{\dot{B}_z}\right)^{2/3}, \quad \frac{\rho_a}{\rho_1}(B_r) = \left(\frac{B^{\text{un}}_{r,l.s}}{B_r}\right)^2$$

As was shown in Chapter 9, at the late stage,

$$E^{\text{un}}_{\varphi,l.s} = \frac{\mu_0^{5/2}\gamma^{3/2} r M_0}{40\pi^{3/2} t^{5/2}}, \quad B^{\text{un}}_{z,l.s} = \frac{\mu_0 M_0}{30\pi^{1/2} t^{3/2}}(\gamma\mu_0)^{3/2} \tag{12.6}$$

$$B^{\text{un}}_{r,l.s} = \frac{\mu_0 M_0}{4\pi}\frac{r}{32t^2}(\gamma\mu_0)^2$$

and

$$\frac{\partial B^{un}_{z,l.s}}{\partial t} = -\frac{\mu_0 M_0}{20\pi^{3/2}}\frac{(\gamma\mu_0)^{3/2}}{t^{5/2}}, \quad \frac{\partial B^{un}_{r,l.s}}{\partial t} = -\frac{\mu_0 M_0}{64\pi} r \frac{(\gamma\mu_0)^2}{t^3} \tag{12.7}$$

For example, if the electric field and $\partial B_z/\partial t$ are measured, we obtain the following expressions for the apparent resistivity:

$$\rho_a(E_\varphi) = \frac{\mu_0}{4\pi t}\left(\frac{\mu_0 r M_0}{5tE_\varphi}\right)^{2/3}, \quad \rho_a(\dot{B}_z) = \frac{\mu_0}{4\pi t}\left(\frac{2\mu_0 M_0}{5t\dot{B}_z}\right)^{2/3} \tag{12.8}$$

Similar expressions can be written for apparent resistivity when the components of the magnetic field, B_z and B_r are measured, as well as when the electric dipole is used as the source. In particular, if the quantity $\partial B_z/\partial t$, generated by the electric dipole located on the earth's surface is measured, we have

$$\rho_a = \frac{\mu_0}{4\pi t}\left(\frac{\mu_0 r I_0 ABnS\sin\varphi}{5t\Xi}\right)^{2/3} \tag{12.9}$$

Here, I_0 and AB are the current and length of the electric dipole, respectively, n and S are the number of turns and the area of the horizontal loop serving as a receiver, and Ξ is the electromotive force induced in the receiver, while φ is the angle between the orientation of the dipole and the direction to an observation point. Before we start considering curves of the apparent resistivity, it may be proper to notice the following: (1) As is seen from Eq. (12.8), the array coefficient is independent on resistivity of the upper layer. (2) Within the most important range of time, the behavior of the field in the upper layer corresponds to the late stage that justifies such method of introduction of the apparent resistivity. (3) Of course, the most logical way is to use the exact expression of the field in a uniform half space but it requires the knowledge of resistivity ρ_1. Now let us discuss the apparent resistivity curves and start from the simplest case of two-layer medium.

12.5.1. Apparent Resistivity Curves for Two-Layer Model

Apparent resistivity curves for a two-layer sequence can naturally be separated into two subgroups depending on the ratio of resistivity values. These two subgroups are curves, in which the ratio $\mu = \rho_2/\rho_1$ is greater than unity and those with μ less than unity. Examples of the apparent resistivity curves for a two-layer sequence are shown in Figure 12.2(a) and (b). Taking into account the fact that regardless of the method of interpretation it is always useful to know how different parts of the apparent resistivity curves are related to resistivity model parameters, let us discuss this question in some detail. First, as can be seen from these curves when the parameter τ_1/H_1 takes values less than 4, where $\tau_1 = (2\pi\rho_1 t 10^7)^{1/2}$ m, all the apparent resistivity curves merge with the curve corresponding to a uniform half space for the resistivity ρ_1. This is a consequence of the fact that during the initial part of the transient response, induction currents concentrate near the source and are localized in the uppermost layer. If an observation site is located close enough to the

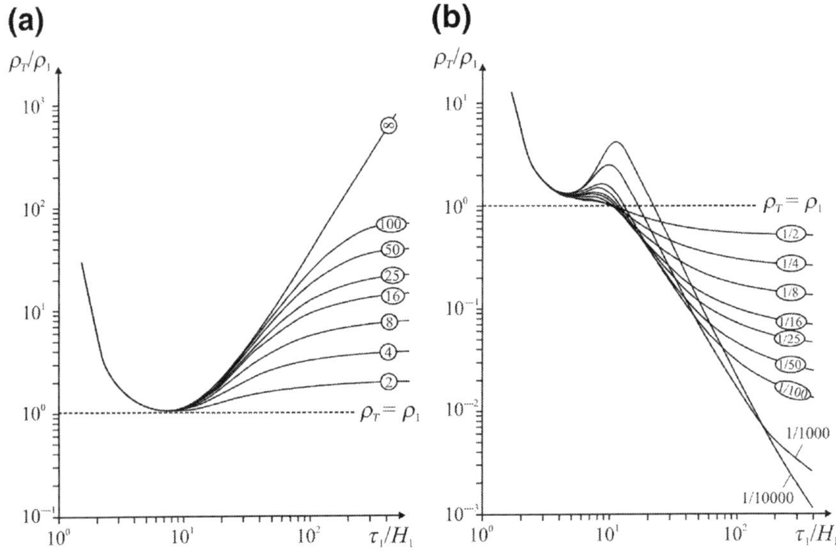

Figure 12.2 Two-layer apparent resistivity curves for (a) $\mu = \rho_2/\rho_1 > 1$ and (b) $\mu < 1$. $r/H_1 = 1/2$.

dipole source, there is nearly horizontal curve segment where $\rho_a = \rho_1$. This characteristic behavior of the apparent resistivity curves means that under certain conditions, the field component coincides with that for the late stage on the surface of a uniform half space. With further decrease of time, all curves have ascending branch, which only depends on the resistivity ρ_1.

In contrast, at the late times, the apparent resistivity curves approach an asymptotic value which is the resistivity of the lowermost medium. This happens because at such late times, the induction currents are almost completely confined to the basement and the change in the magnetic field caused by these currents does not induce appreciable currents in the upper layer. It is a simple matter to see from curves that in order for the apparent resistivity curve to reach an asymptotic value of ρ_a, it is often necessary to use very large ratios of τ_1/H_1, and in many practical cases, it is extremely challenging or even impossible to get field readings for such long times. For this reason, it is useful to find a resistivity of the lowermost layer at much smaller times. If the basement is virtually an insulator, the ascending branch of the apparent resistivity curve for $\tau_1/H_1 > 20 - 30$, presented as usual in log–log scale, approaches an asymptote inclined at an angle $63°25'$ to the abscissa. This line (the S asymptote) is described by equation:

$$\frac{\rho_a}{\rho_1} = \left[0.06\left(\tau_1/\rho_1 S\right)\right]^2 \quad \text{or} \quad \rho_a^{1/2} = \frac{189.3 \cdot (2\pi t)^{1/2}}{S}, \tag{12.10}$$

where S is the longitudinal conductance of the upper layer. The intersection of this line with the $\rho_a = \rho_1$ axis occurs at

$$\frac{\tau_1}{H_1} \approx 17 \tag{12.11}$$

It should be noted that the S-zone behavior is observed as well as for the Schlumberger apparent resistivity curves when the basement has high but finite resistivity with $\mu > 40$. At this time range, the induction currents in the upper layer are distributed uniformly in the vertical direction, and the magnetic field of currents induced in the basement is relatively small. Therefore, if the time interval over which the transient response can be measured is not great enough for the resistivity model with $\mu > 40$, it is virtually impossible to determine the resistivity of the basement.

In case of apparent resistivity curves with $\mu \ll 1$, there is also a time interval over which the inclination of the apparent resistivity curve is the same as that for the curve for which basement is an ideal conductor, $\rho_2 = 0$.

However, in contrast to the cases with $\mu \gg 1$, these curves do not merge into a single curve over this interval, but they are parallel to each other, having a slope of approximately -2 (forming the slope of $-68°30'$ with horizontal line). When relatively short separations are used, the apparent resistivity curves for cases with $\mu < 1$ exhibit both the minimum and the maximum. The appearance of this maximum on these curves over a range of small values of τ_1/H_1 can be explained as follows. During the very early stage of the transient process, induction currents concentrate near the surface, and the field coincides with that for a uniform half space having resistivity ρ_1. The change of the magnetic field of these currents with time in turn induces secondary currents whose intensity is a function of the coordinates of a point, the conductivity, and time. If the basement has a greater conductivity, one might expect that at some moment close to the beginning of the transient response, the effect of the lower medium with high conductivity will be relatively large. Since these currents tend to reduce the field strength, the effect manifests itself in the appearance of maximum on the apparent resistivity curve.

12.5.2. Apparent Resistivity Curves for Three-Layer Model

Next we consider apparent resistivity curves for three-layer media starting with case of an H-type of curve. The H-type sequence represents a set of layers in which the intermediate layer has the lowest resistivity. Examples of the apparent resistivity for three-layer medium in which the basement resistivity ρ_3 is either infinite or equal to ρ_1 are shown in Figures 12.3 and 12.4. Comparison of these curves with two-layer curves with corresponding values of μ shows that when $\frac{\tau_1}{H} < 7$ (H is defined as the combined thickness of the first two layers, $H_1 + H_2$), they almost coincide. As was demonstrated earlier, apparent resistivity curves over this range of $\frac{\tau_1}{H}$ are characterized by both a minimum and a maximum for relatively small values of $\frac{r}{H}$. For a small array separation, the value for $\rho_{\min}$ of two-layer curve with $\mu > 1$ is greater than the resistivity of the first layer while for three-layer H-type curves, we have $\rho_{\tau\min} < \rho_1$.

A study shows that for curves with $\nu = H_2/H_1 \geq 1$ and $\frac{r}{H} \leq 0.6$, the ordinate of the minimum may be defined from the following equation with an error no >5%:

$$\rho_{\tau\min} = \rho_l \left(\rho_1/\rho_2\right)^{1/2}$$

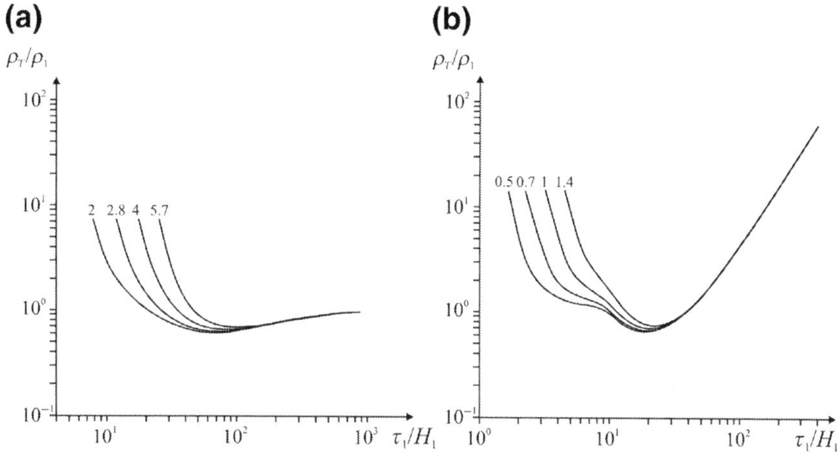

Figure 12.3 Three-layer *H*-type apparent resistivity curves. Curve index is r/H_1. Model parameters are (a) $\rho_2/\rho_1 = 1/2$, $H_2/H_1 = 4$, $\rho_3/\rho_1 = 1$; (b) $\rho_2/\rho_1 = 1/4$, $H_2/H_1 = 1/2$, $\rho_3/\rho_1 = \infty$.

where

$$\rho_l = \frac{H}{S}$$

is the longitudinal resistivity of the two upper layers. Note that there are several approximate ways of determining the total conductance *S* if the apparent resistivity curve lies within the *S* range, and one of them was used for two-layer curve:

$$S = 474\left(t/\rho_\tau\right)^{1/2}$$

Examples of three-layer curves of types *A*, *K*, and *Q* are given in Figure 12.5. We briefly outlined some relations between apparent resistivity curves and resistivity model parameters, which, as in the case of magnetotellurics, may be useful for interpretation of transient responses, especially to determine the initial guess parameters. In conclusion, we note that at the beginning, the main area of the transient sounding application was petroleum geophysics but with time, it found a broad application in the groundwater and engineering, as well as environmental geophysics. In this light, let us point out again that before the use of this method for studying stratified sediments, it was applied in mineral exploration for detecting massive ore bodies, but it was not called *sounding* in spite of the fact that

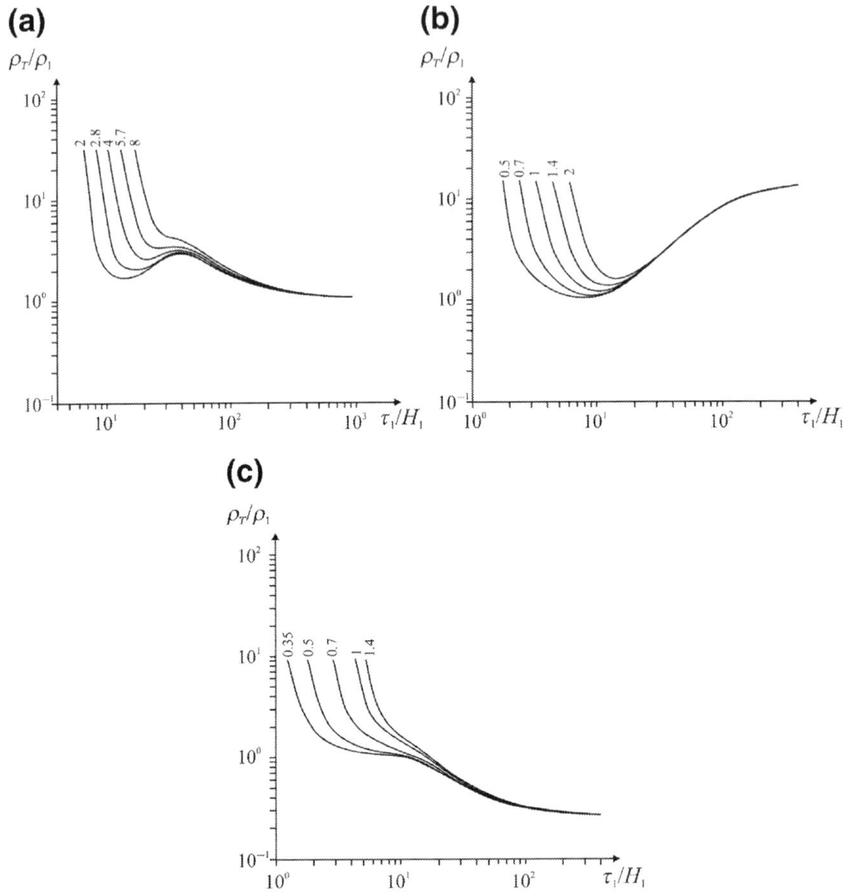

Figure 12.4 Three-layer apparent resistivity curves. Curve index is r/H_1. Model parameters are (a) $\rho_2/\rho_1 = 8$, $H_2/H_1 = 6$, $\rho_3/\rho_1 = 1$ (Type K); (b) $\rho_2/\rho_1 = 4$, $H_2/H_1 = 1$, $\rho_3/\rho_1 = 16$ (Type A); (c) $\rho_2/\rho_1 = 1/2$, $H_2/H_1 = 1/2$, $\rho_3/\rho_1 = 1/4$ (Type Q).

measurements were performed at different times. At the beginning of 1980s, the same method found application for a solution of completely different problems of the near-surface geophysics, and it was natural to design new equipment and its main features will be described later.

12.6. FREQUENCY SOUNDING

Earlier we described the frequency and transient sounding in the far zone, as well as demonstrated that the latter can yield resistivity structure

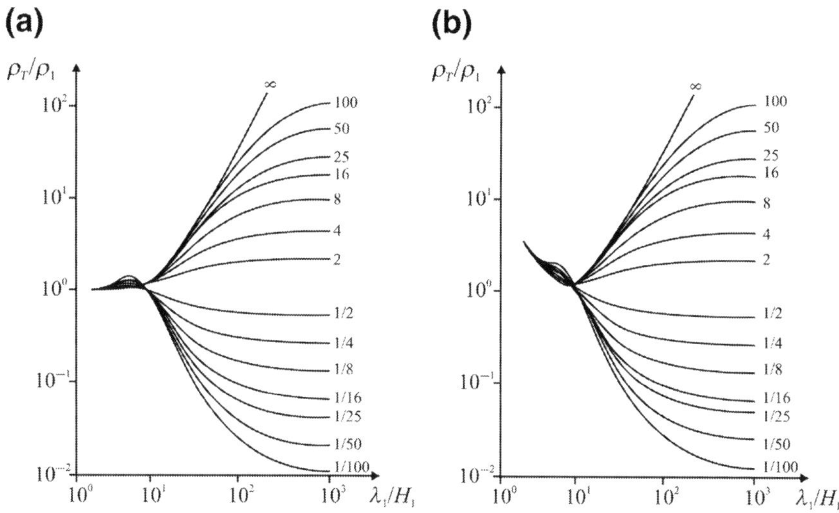

Figure 12.5 Two-layer apparent resistivity curves for two-frequency method. The curves are calculated from Eqs 12.14 and 12.15 for two values of r/H_1 parameter: (a) 0.25 and (b) 0.7.

when used at any distance from the source. Bearing in mind the relation between the frequency and transient responses (Fourier integral), it is natural to raise the following question. How in principle it is possible to perform the frequency soundings not only in the far zone but also near the source? Earlier we mentioned that measuring the amplitude and phase of the field, such soundings can be performed only at the far zone. Let us discuss this subject in some detail and for illustration, we consider two-layer medium. As was shown in studying the low-frequency parts of the spectrum for the vertical magnetic dipole situated on the earth's surface, the asymptotic expressions for QB_z and InE_φ can be written as

$$QB_z = \frac{\mu_0 M_0}{4\pi}\left\{\frac{\omega\mu_0\gamma_1}{4r} + \frac{\omega\mu_0(\gamma_2-\gamma_1)}{4}\frac{1}{\left(4H_1^2+r^2\right)^{1/2}} - \frac{2^{3/2}}{15}\left(\omega\mu_0\gamma_2\right)^{3/2} + \cdots\right\}$$

and

$$InE_\varphi = \frac{\mu_0 M_0}{4\pi}\,\omega\left\{\frac{\omega\mu_0\gamma_1}{4} + \frac{\omega\mu_0(\gamma_2-\gamma_1)}{4}\,\frac{\left(4H_1^2+r^2\right)^{1/2}-2H_1}{r} - \frac{r2^{1/2}}{15}(\omega\mu_0\gamma_2)^{3/2} - \cdots\right\} \tag{12.12}$$

and for a nonconducting basement,

$$QB_z = \frac{\mu_0 M_0}{4\pi}\left\{\frac{\omega\mu_0\gamma_1}{4r} - \frac{\omega\mu_0\gamma_1}{4}\,\frac{1}{\left(4H_1^2+r^2\right)^{1/2}} + \frac{H_1^3}{8}(\omega\mu_0\gamma_1)^3\ln\omega\mu_0\gamma_1 r^2 + \ldots\right\}$$

and

$$InE_\varphi = \frac{\mu_0\omega M_0}{4\pi}\left\{\frac{\omega\mu_0\gamma_1}{4} - \frac{\omega\mu_0\gamma_1}{4}\,\frac{\left(4H_1^2+r^2\right)^{1/2}-2H_1}{r} + \frac{rH_1^3}{16}(\omega\mu_0\gamma_1)^3\ln\omega\mu_0\gamma_1 r^2\right\} \tag{12.13}$$

Before it was demonstrated that if the separation r is relatively small, the first term in brackets of each expression, which is directly proportional to the frequency, describes the field when it is possible to neglect by interaction of currents and it is mainly controlled by the conductivity and thickness of the upper layer. In other words, the depth of investigation is small regardless how low the frequency is. However, the second term in brackets depends on the conductivity of the basement only (if $\gamma_2 \neq 0$). For this reason, having somehow eliminated the term proportional to ω, one can obtain, in principle, any depth of investigation, measuring components of the field, shifted by phase in 90° with respect to the primary field. It is clear that the functions

$$\Delta E_\varphi = InE_\varphi(\omega_1) - \left(\frac{\omega_1}{\omega_2}\right)^2 InE_\varphi(\omega_2), \quad \Delta B_z = QB_z(\omega_1) - \frac{\omega_1}{\omega_2}QB_z(\omega_2) \tag{12.14}$$

at low frequencies do not depend on the conductivity and the thickness of the upper layer, regardless of the separation, but they are controlled only by the resistivity of the basement. In the limiting case of the nonconducting basement by removing the first term proportional to ω, one can increase the sensitivity of the field to the longitudinal conductance of layers located above the basement. In fact at the low frequencies, the functions, given by Eq. (12.14), turn out to be proportional to S^3; that is in the same manner, as in the transient sounding when the field is measured in the late stage. Then let us introduce an apparent resistivity as

$$\frac{\rho_a^B}{\rho_1} = \left| \frac{\Delta \dot{B}_z^{un}(\rho_1)}{\Delta \dot{B}_z} \right|^{2/3} \quad \text{and} \quad \frac{\rho_a^E}{\rho_1} = \left| \frac{\Delta E_\varphi^{un}}{\Delta E_\varphi} \right|^{2/3}, \tag{12.15}$$

where numerator describes functions corresponding to uniform half space at small induction numbers. Examples of the apparent resistivity for two-layer medium are shown in Figure 12.5(a) and (b). Comparing these curves to those for the transient soundings in the near zone, we see that they almost coincide. The similar result can be obtained measuring the in-phase component of the secondary magnetic field, as well as that of the electric field.

It may be proper to mention that at the beginning of the development of the transient soundings with relatively small separations, the question about equivalence of the frequency and transient soundings was very important.

REFERENCES AND FURTHER READING

[1] A.A. Kaufman, G.M. Morozova, Theoretical Principles of Transient Soundings in the Near Zone, Nauka, Novosibirsk, 1970.
[2] A.A. Kaufman, G.V. Keller, Frequency and Transient Sounding, Elsevier, Amsterdam, 1983.
[3] L.L. Vanyan, Principles of Electromagnetic Soundings, Nedra, Moscow, 1965.
[4] L.L. Vanyan, Transient Sounding and Its Application in the Petroleum Geology, Nauka, Novosibirsk, 1966.

CHAPTER THIRTEEN

Quasi-Stationary Field of Electric Dipole in a Horizontally Layered Medium

INTRODUCTION

In previous chapters, we investigated the electromagnetic field caused by the vertical magnetic dipole located on the surface of a horizontally layered medium. Remarkable feature of this field in this case is the absence of charges and as a result the relative simplicity of the field geometry. Regardless of the number of layers, the electric field has a single component, that being of tangential direction. In accord with this, the induction currents flowing in the Earth are situated in horizontal planes and individual current flow lines do not intersect the surfaces between layers that are characterized by different values of electric conductivity. In other words, the electric charges do not arise at these boundaries. The similar situation takes place if the primary field is generated by a horizontal magnetic dipole. In this last case, charges appear only on the Earth's surface but they do not contain information about the conductivity of a medium. Such a field behavior is the most favorable for detecting relatively conducting layers even those which are relatively thin. However, as has been shown earlier, the use of a loop lying in the horizontal plane as a source for the electromagnetic field is improper method of excitation when it is desirable to detect the presence of relatively resistive layers. These are of some practical interest in cases, such as in an exploration for oil- or gas-saturated zones. Note that in spite of a relatively high level of a geological noise, the grounded line as the current source is sometimes used when the vertical component of the transient magnetic field is measured. In this chapter, we consider the behavior of the electric field caused by the electric dipole (Figure 13.1(a)) with the moment

$$\mathbf{p} = p_0 \mathbf{z}_0, \tag{13.1}$$

where the magnitude of the moment is equal to

$$p_0 = \varepsilon_0 \rho I dl \tag{13.2}$$

Methods in Geochemistry and Geophysics, Volume 45
ISSN 0076-6895,
http://dx.doi.org/10.1016/B978-0-444-53829-1.00013-7

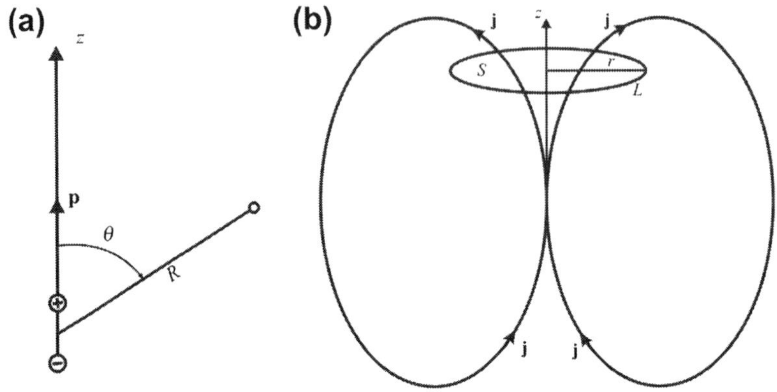

Figure 13.1 (a) Electric dipole. (b) Illustration used in derivation of field expressions.

Here, ρ is the resistivity of a medium surrounding the electrodes, I_0 is the current amplitude, dl is the dipole length which is much smaller than the distance to an observation point, and $\mathbf{z}_0$ is the unit vector showing the dipole direction.

As usual, we will proceed from Maxwell's equation for the quasi-stationary field:

$$\begin{aligned} \text{curl}\,\mathbf{E} &= -\frac{\partial \mathbf{B}}{\partial t}, \quad \text{div}\,\mathbf{E} = 0 \\ \text{curl}\,\mathbf{B} &= \gamma\mathbf{E}, \quad \text{div}\,\mathbf{B} = 0 \end{aligned} \tag{13.3}$$

In our study, we start from the simplest case of a uniform medium and at the beginning, consider time-invariant field.

13.1. THE CONSTANT ELECTRIC AND MAGNETIC FIELDS ($\omega = 0$) IN A UNIFORM MEDIUM

The symmetry of the problem makes it natural to introduce a spherical set of coordinates R,θ,φ as well as a cylindrical system of coordinates r,φ,z, and both of them have a common origin. The dipole is located at the origin with its moment directed along the z-axis ($\sin\theta = 0$). As current flows from the wire to the surrounding medium and back, electric charges arise at the surface of the electrodes (terminal points of the dipole):

$$e_0 = \pm\varepsilon_0\rho I_0 \tag{13.4}$$

The positive and negative charges appear at electrodes where the current density is directed either into a medium or in the wire, respectively. In a uniform conducting space, these charges give rise to an electric field which is equivalent to that of an electrical dipole with the moment p_0. Referred to the spherical coordinates, the electric field of these charges has two components E_R and E_θ:

$$E_R = \frac{2p_0}{4\pi\varepsilon_0 R^3}\cos\theta \quad E_\theta = \frac{p_0}{4\pi\varepsilon_0 R^3}\sin\theta \quad E_\varphi = 0 \tag{13.5}$$

As follows from Ohm's law and Eq. (13.5), the currents arise and its lines are situated only on the meridian planes since the component of the current density j_φ is zero. In turn, the electric current is the source of magnetic field, such that each element of the current flow $\mathbf{j}dl$ gives rise to an elementary magnetic field described by Biot–Savart law:

$$\mathrm{d}\mathbf{B}(p) = \frac{\mu_0}{4\pi}\frac{\left(\mathbf{j}(q)\times\mathbf{L}_{qp}\right)}{L_{qp}^3} \tag{13.6}$$

Making use of the principle of superposition, the total magnetic field is given by

$$\mathbf{B}(p) = \frac{\mu_0}{4\pi}\int_V \frac{\mathbf{j}(q)\times\mathbf{L}_{qp}}{L_{qp}^3}\,\mathrm{d}V, \tag{13.7}$$

where $\mathbf{j}(q)$ is the current density and L_{qp} is the distance between the points q and p. From Eqs (13.6) and (13.7) follows that the magnetic field from the electric dipole in a uniform conducting medium has but a single component B_φ. Because the current density is known at every point in the medium,

$$j_R = \frac{Idl}{2\pi R^3}\cos\theta, \quad j_\theta = \frac{Idl}{4\pi R^3}\sin\theta, \tag{13.8}$$

Equation (13.7) permits us to find the magnetic field. However, by taking into account the cylindrical symmetry of the electric and magnetic field components, that is,

$$\frac{\partial E}{\partial\varphi} = \frac{\partial B}{\partial\varphi} = 0,$$

one markedly simplifies the procedure for determining the magnetic field by making use of the second equation of the set (Eq. (13.3)) in the integral form:

$$\oint_L \mathbf{B}\cdot d\mathbf{l} = \mu_0 \int_S \mathbf{j}\cdot d\mathbf{S} = \mu_0 I_S, \tag{13.9}$$

where L is an arbitrary path of integration, dS is an elementary area, I_S is the flux of the current piercing the area S as shown in Figure 13.1(b). Considering the path of integration in a horizontal plane centered on the z-axis, we can rewrite Eq. (13.9) as

$$2\pi r B_\varphi = \int_S j_z dS = 2\pi \int_0^r j_z r dr$$

Thus,

$$B_\varphi = \frac{1}{r}\int_0^r j_z r dr \tag{13.10}$$

Here, we have made use of the fact that the magnetic field and the current density do not depend on the coordinate φ, and that the element of the area can be written as $dS = 2\pi r dr$. It is clear that the current I in the dipole return through the conducting medium, and therefore, there are positive and negative fluxes of current density through any path of integration as is shown in Figure 13.1(b). In accord with Eq. (13.8), the vertical component of the current density is

$$\begin{aligned} j_z &= j_R \cos\theta - j_\theta \sin\theta = \frac{Idl}{4\pi R^3}\left(2\cos^2\theta - \sin^2\theta\right) \\ &= \frac{Idl}{4\pi R^3}\left(3\cos^2\theta - 1\right) = \frac{Idl}{4\pi}\left(\frac{3z^2}{R^5} - \frac{1}{R^3}\right) \\ &= \frac{Idl}{4\pi}\left[\frac{3z^2}{(r^2+z^2)^{5/2}} - \frac{1}{(r^2+z^2)^{3/2}}\right] \end{aligned} \tag{13.11}$$

Substituting Eq. (13.11) into Eq. (13.10),

$$B_\varphi = \frac{\mu_0 Idl}{4\pi r}\left[3z^2\int_0^r \frac{rdr}{(r^2+z^2)^{5/2}} - \int_0^r \frac{rdr}{(r^2+z^2)^{3/2}}\right]$$

Both integrals are tabular such that

$$B_\varphi = \frac{\mu_0 Idl}{4\pi r}\left\{\left[\frac{1}{z} - \frac{z^2}{(r^2+z^2)^{3/2}}\right] - \left[\frac{1}{z} - \frac{1}{(r^2+z^2)^{1/2}}\right]\right\}$$

or

$$B_\varphi = \frac{\mu_0 Idl}{4\pi R^2}\sin\theta \tag{13.12}$$

so that the magnetic field is absent along the axis of the dipole source and is maximal in the equatorial plane. Thus, the electromagnetic field at zero frequency is

$$E_R = \frac{\rho Idl}{2\pi R^3}\cos\theta, \quad E_\theta = \frac{\rho Idl}{4\pi R^3}\sin\theta \quad B_\varphi = \frac{\mu_0 Idl}{4\pi R^2}\sin\theta \tag{13.13}$$

13.2. QUASI-STATIONARY FIELD OF THE ELECTRIC DIPOLE IN A UNIFORM MEDIUM

13.2.1. Derivation of Equations for the Field

Next we consider the general case of a quasi-stationary field proceeding from Eq. (13.3). As in the case of the magnetic dipole, in order to simplify the solution of the boundary value problem, we introduce a vector potential of the electric field, based on the fact that div $\mathbf{B} = 0$, in the following way:

$$\mathbf{B} = \text{curl}\,\mathbf{A} \tag{13.14}$$

Of course, the function $\mathbf{A}$ is not uniquely defined by Eq. (13.14). Substituting the last expression in the first Maxwell's equation,

$$\text{curl}\,\mathbf{E} = -\frac{\partial \mathbf{B}}{\partial t}$$

we have

$$\text{curl}\,\mathbf{E} = -\text{curl}\frac{\partial \mathbf{A}}{\partial t} \quad \text{or} \quad \text{curl}\left(\mathbf{E} + \frac{\partial \mathbf{A}}{\partial t}\right) = 0$$

Whence,

$$\mathbf{E} = -\frac{\partial \mathbf{A}}{\partial t} - \text{grad}\,U \tag{13.15}$$

Here, U is the scalar potential. Replacement of **E** and **B** in the second Maxwell's equation by the functions **A** and U gives

$$\text{curl curl}\,\mathbf{A} = -\mu_0\gamma\frac{\partial \mathbf{A}}{\partial t} - \gamma\mu_0\,\text{grad}\,U \tag{13.16}$$

Assuming that the electromagnetic field is sinusoidal,

$$\mathbf{E} = \text{Re}[\mathbf{E}^*\exp(-i\omega t)], \quad \mathbf{B} = \text{Re}[\mathbf{B}^*\exp(-i\omega t)]$$

and

$$\mathbf{A} = \text{Re}[\mathbf{A}^*\exp(-i\omega t)], \quad U = \text{Re}[U^*\exp(-i\omega t)]$$

and making use of the vector identity

$$\text{curl curl}\,\mathbf{A} = \text{grad div}\,\mathbf{A} - \Delta\mathbf{A}$$

we obtain from Eq. (13.16) the following equation for the complex amplitudes:

$$\text{grad div}\,\mathbf{A}^* - \Delta\mathbf{A}^* = i\gamma\mu_0\omega\mathbf{A}^* - \gamma\mu_0\,\text{grad}\,U^* \tag{13.17}$$

As in the case of potentials describing the field of the magnetic dipole, the functions $\mathbf{A}^*$ and U^* are not uniquely defined from Eqs (13.14) and (13.15), and therefore, there is freedom to choose a pair of functions in such a way as to simplify Eq. (13.17) and eliminate from further consideration the complex amplitude of the scalar potential. Choosing a pair of functions U^* and A^* that satisfy the gauge condition,

$$\text{div}\,\mathbf{A}^* = -\gamma\mu_0 U^*, \tag{13.18}$$

we obtain the Helmholtz equation for the complex amplitude of the vector potential $\mathbf{A}^*$

$$\nabla^2\mathbf{A}^* + k^2\mathbf{A}^* = 0, \tag{13.19}$$

where $k^2 = i\gamma\mu_0\omega$ is the square of the wave number. In accord with Eqs (13.14), (13.15), and (13.18), the electromagnetic field has been expressed solely in terms of the vector potential:

$$\mathbf{B}^* = \text{curl}\,\mathbf{A}^* \quad \mathbf{E}^* = i\omega\mathbf{A}^* + \frac{1}{\gamma\mu_0}\text{grad div}\,\mathbf{A}^* \tag{13.20}$$

The physical intuition suggests that the quasi-stationary field has as the constant field the φ−component of the magnetic field only. Then bearing in mind Eq. (13.14), it is reasonable to find an expression for all components of electromagnetic field with help of only one component of the vector potential A_z. Besides we suppose that this component is a function of the coordinate R; that is, $A_z^*(k, R)$. Then in the spherical system of coordinates, Eq. (13.19) has the form

$$\frac{1}{R^2}\frac{\mathrm{d}}{\mathrm{d}R}R^2\frac{\mathrm{d}A_z^*}{\mathrm{d}R} + k^2 A_z^* = 0, \tag{13.21}$$

since it is assumed that

$$\frac{\partial A_z}{\partial \theta} = \frac{\partial A_z}{\partial \varphi} = 0.$$

As was shown earlier, the solution of Eq. (13.21) which decreases as a function of R is

$$A_z^* = C \exp(ikR)/R. \tag{13.22}$$

It is clear that this expression is the same as that for the complex amplitude of the vector potential for the magnetic dipole but the constants may be different. From Eq. (13.22), we have

$$\mathrm{div}\, \mathbf{A}^* = \frac{\partial A_z^*}{\partial z} = C\frac{\exp(ikR)}{R^2}(ikR - 1)\cos\theta. \tag{13.23}$$

In the spherical system of coordinates, the vector potential is characterized by two components A_R and A_θ:

$$A_R = A_z \cos\theta, \quad A_\theta = -A_z \sin\theta,$$

and in accordance with Eq. (13.14), the magnetic field can be written as

$$\mathbf{B}^* = \frac{1}{R^2 \sin\theta}\begin{vmatrix} \mathbf{1}_R & R\mathbf{1}_\theta & R\sin\theta\mathbf{1}_\varphi \\ \dfrac{\partial}{\partial R} & \dfrac{\partial}{\partial \theta} & \dfrac{\partial}{\partial \varphi} \\ A_z^*\cos\theta & -A_z^* R\sin\theta & 0 \end{vmatrix}.$$

Whence,

$$B_R^* = B_\theta^* = 0$$

and

$$B^*_\varphi = \frac{C}{R^2}(1 - ikR)\exp(ikR)\sin\theta. \tag{13.24}$$

As the frequency goes to zero, Eq. (13.24) becomes equivalent to Eq. (13.12), so that we are able to find value for the constant C:

$$C = \frac{\mu_0 Idl}{4\pi}. \tag{13.25}$$

Thus, we have the following expression for the magnetic field:

$$B^*_\varphi = \frac{\mu_0 Idl}{4\pi R^2}(1 - ikR)\exp(ikR)\sin\theta \tag{13.26}$$

Making use of Eq. (13.20), expressions for the complex amplitudes of the electric field are

$$E^*_R = \frac{2p_0}{4\pi\varepsilon_0 R^3}\exp(ikR)(1 - ikR)\cos\theta$$

and

$$E^*_\theta = \frac{p_0}{4\pi\varepsilon_0 R^3}\exp(ikR)\left(1 - ikR - k^2R^2\right)\sin\theta \tag{13.27}$$

Inasmuch as the electromagnetic field, described by Eqs (13.26) and (13.27), satisfies the Helmholtz equation for the complex amplitude as well as boundary conditions near the source and at great distances, one can say that a unique solution has been found. In other words, our assumptions about the behavior of the vector potential were correct. In accord with Eq. (13.26), the geometry of the magnetic field is remarkably simple: its vector lines form circles situated in horizontal planes centered on the z-axis. As in the case of constant field currents, which are the sources of the magnetic field, are located in the meridian planes and in accord with Eq. (13.27),

$$\begin{aligned} j^*_R &= \frac{Idl}{2\pi R^3}\exp(ikR)(1 - ikR)\sin\theta, \\ j^*_\theta &= \frac{Idl}{4\pi R^3}\exp(ikR)\left(1 - ikR - k^2R^2\right)\sin\theta. \end{aligned} \tag{13.28}$$

In contrast to the case of the magnetic dipole, there are two sources for the electric field from the electric dipole, namely, the electric charges on the surface of the dipole electrodes and time variations of the magnetic field B_φ. In order to study the field of the magnetic dipole, it is convenient to normalize it by the primary field caused by the current in a loop. Here it is natural to consider the ratio between the total field and the stationary (direct current) field; that is, let us represent Eqs (13.26) and (13.27) as

$$B_\varphi^* = \frac{\mu_0 Idl}{4\pi R^2} b_\varphi^* \sin\theta \quad E_R^* = \frac{2p_0}{4\pi\varepsilon_0 R^3} e_R^* \cos\theta \quad E_\theta^* = \frac{p_0}{4\pi\varepsilon_0 R^3} e_\theta^* \sin\theta, \tag{13.29}$$

where

$$b_\varphi^* = e_R^* = \exp(ikR)(1 - ikR) \quad e_\theta^* = \exp(ikR)\left(1 - ikR - k^2R^2\right) \tag{13.30}$$

and

$$kR = \frac{R}{\delta}(1+i) = p(1+i)$$

It is proper to notice that expressions for the electric and magnetic fields caused by the electric dipole with the accuracy of a constant are the same as those for the magnetic and electric fields caused by the magnetic dipole. As usual, we distinguish three zones where the behavior of the sinusoidal waves in a conducting medium is different: the near, intermediate, and wave zones. Let us first consider in detail the range of small values for induction number p.

13.2.2. The Near Zone $p < 1$

Representing exp (ikR) as a series and substituting this into the first equation of the set (Eq. (13.30)), we obtain

$$e_R^* = b_\varphi^* = 1 + \sum_{n=2}^{\infty} \frac{1-n}{n!}(ikR) \tag{13.31}$$

Thus for the quadrature and in-phase components e_R^*, we have

$$Qe_R^* \approx p^2 - \frac{2}{3}p^3, \quad Ine_\varphi^* \approx 1 - \frac{2}{3}p^3, \tag{13.32}$$

and correspondingly,

$$QE_R^* \approx \frac{2p_0}{4\pi\varepsilon_0 R^3}\cos\theta\left(p^2 - \frac{2}{3}p^3\right)$$

or

$$QE_R^* \approx \frac{Idl}{2\pi}\cos\theta\left[\frac{\mu_0\omega}{2R} - \frac{1}{3\cdot 2^{1/2}}(\mu_0\omega)^{3/2}\gamma^{1/2}\right] \tag{13.33}$$

and

$$InE_R^* \approx \frac{2p_0}{4\pi\varepsilon_0 R^3}\cos\theta\left(1 - \frac{2}{3}p^3\right)$$

or

$$InE_R^* \approx \frac{Idl}{2\pi}\cos\theta\left[\frac{\rho}{R^3} - \frac{1}{3\cdot 2^{1/2}}(\mu_0\omega)^{3/2}\gamma^{1/2}\right] \tag{13.34}$$

Applying the same approach to the expression for the component E_θ^*, we have

$$e_\theta^* = 1 + \sum_{n=2}^{\infty}\frac{(n-1)^2}{n!}(ikR)^n \tag{13.35}$$

Hence,

$$Qe_\theta^* \approx -p^2 + \frac{4}{3}p^3, \quad Ine_\theta^* \approx 1 + \frac{4}{3}p^3 \tag{13.36}$$

and therefore,

$$QE_\theta^* \approx \frac{Idl}{4\pi}\sin\theta\left[-\frac{\mu_0\omega}{2R} + \frac{2^{1/2}}{3}(\mu_0\omega)^{3/2}\gamma^{1/2}\right]$$

and

$$InE_\theta^* \approx \frac{Idl}{4\pi}\sin\theta\left[\frac{\rho}{R^3} + \frac{2^{1/2}}{3}(\mu_0\omega)^{3/2}\gamma^{1/2}\right] \tag{13.37}$$

By analogy,

$$QB_\varphi^* \approx \frac{\mu_0 Idl}{4\pi}\sin\theta\left[\frac{\gamma\mu_0\omega}{2} - \frac{1}{3\cdot 2^{1/2}}(\gamma\mu_0\omega)^{3/2}R\right]$$

and

$$InB_\varphi^* \frac{\mu_0 Idl}{4\pi}\sin\theta\left[\frac{1}{R^2} - \frac{1}{3\cdot 2^{1/2}}(\gamma\mu_0\omega)^{3/2}R\right] \tag{13.38}$$

Equations (13.33)–(13.38) allow one to understand better the range of small parameters (near zone) or the low-frequency part of the spectrum. The in-phase component of the electric field can be thought of as a sum of the galvanic and vortex ones. The first one is directly proportional to the resistivity of the medium and coincides with the stationary field when the frequency is equal to zero. The second term is purely inductive; it does not depend on distance and is proportional to $\omega^{3/2}$ and $\gamma^{1/2}$. The first term of the quadrature component of the electric field is directly proportional to the frequency but is independent of conductivity; that is, it has pure inductive character. It arises as a consequence of the change with time of the leading term of the expression for the in-phase component of the magnetic field (Eq. (13.38)), which coincides with the stationary magnetic field. The second term in the expression for the quadrature component of E_θ is the same as that for the in-phase component which is proportional to $\gamma^{1/2}$. Thus, even the second term in the expansion for the electric field for small values of the parameter p is less sensitive to changes in conductivity than are terms in the approximation series for the magnetic field when the source is the magnetic dipole. Now let us consider briefly the series in Eq. (13.38). As was mentioned earlier, the first term in the series for the in-phase component of B_φ represents the stationary field which is independent on conductivity. As a consequence of variation of this field with time, a vortex electric field is generated (the first term of the quadrature component of the expression for the electric field). According to Ohm's law, this electric field gives rise to current, which is directly proportional to frequency as well as conductivity. These currents are the sources of a magnetic field described by the leading term in the expression for the quadrature component (Eq. (13.38)), and as one might expect, it is directly proportional to both conductivity and frequency. The second term in these series results from interaction between currents and it possesses the same, relatively high sensitivity to changes in conductivity as do the second terms of the series for the magnetic and electric fields excited by the magnetic dipole. Behavior of the electric and magnetic fields as a function of the parameter p is shown in Figure 13.2(a) and (b).

Until now, we assumed that the electric dipole is located in a conducting medium and influence of displacement currents is negligible. Now let us briefly consider the case when the dipole is surrounded by a nonconducting medium. If the voltage V applied to the dipole is constant, then the electric charges of the opposite sign appear at both pieces of the wire and their distribution is such that the electric field inside wires is zero. At the same

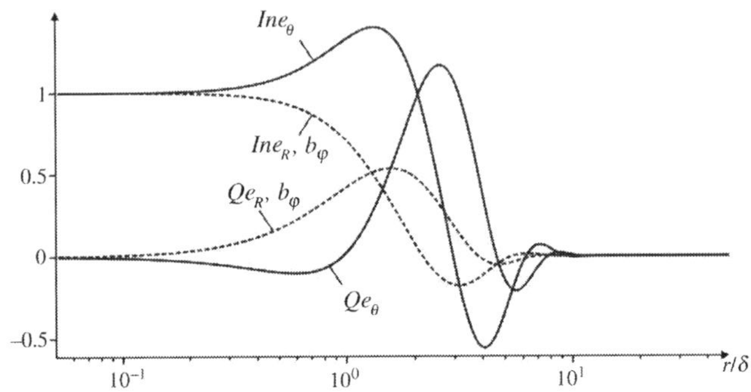

Figure 13.2 Frequency responses of the electric and magnetic fields of a horizontal electric dipole in a uniform conductive space.

time, these surface charges create the electric field outside the dipole. For instance, if the dipole is located above a conducting medium, then at the Earth's surface charges are induced, and in accordance with the electrostatic induction phenomenon, the total electric field in the conducting medium is zero. In the quasi-stationary approximation, neglecting displacement currents in air, the field of dipole charges obeys Coulomb's law, and correspondingly, they do not cause a field in a conducting medium.

13.3. THE HARMONIC FIELD OF THE HORIZONTAL ELECTRIC DIPOLE ON THE SURFACE OF A UNIFORM HALF SPACE

13.3.1. Formulation of Boundary Value Problem

Next suppose that a horizontal electric dipole is located in the lower half space as shown in Figure 13.3 at the distance h beneath the Earth's surface.

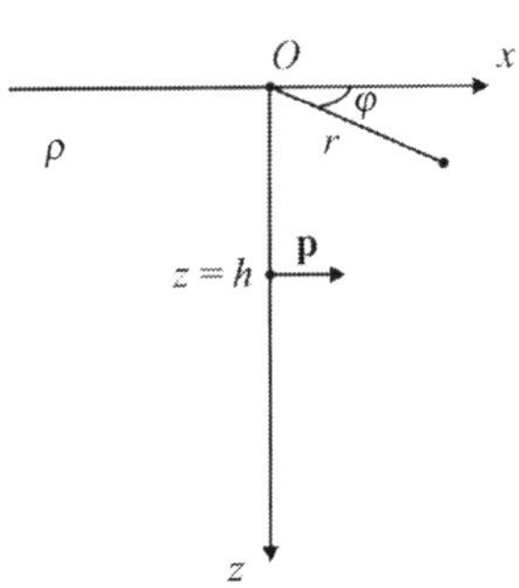

Figure 13.3 The horizontal electric dipole in the presence of the horizontal boundary.

Cartesian and cylindrical coordinate systems are chosen with a common origin at the boundary, $z=0$, and the z-axis is directed downward. The dipole moment is oriented along the x-axis. As follows from Eqs (13.19) and (13.20), the complex amplitude of the vector potential obeys the Helmholtz equation at regular points:

$$\nabla^2 \mathbf{A}^* + k^2 \mathbf{A}^* = 0,$$

and it is related to the field as

$$\mathbf{B}^* = \text{curl}\,\mathbf{A}^* \quad \text{and} \quad \mathbf{E}^* = i\omega \mathbf{A}^* + \frac{1}{\gamma \mu_0} \text{grad div}\,\mathbf{A}^*$$

Using the similar arguments as in the case when the field is caused by the horizontal magnetic dipole, we assume that the vector potential has two components: one of them is directed as the dipole moment, while the other is perpendicular to the interface:

$$\mathbf{A}^* = \left(A_x^*, 0, A_z^*\right) \tag{13.39}$$

In the Cartesian coordinate system, the boundary conditions can be written as

$$E_{1x}^* = E_{2x}^*, \quad E_{1y}^* = E_{2y}^*, \quad B_{1x}^* = B_{2x}^*, \quad B_{1y}^* = B_{2y}^* \quad \text{if } z = 0 \tag{13.40}$$

In order to satisfy these conditions in terms of the vector potential, the following equalities hold at the Earth's surface:

$$i\omega A_{1x}^* + \frac{1}{\gamma_1 \mu_0} \frac{\partial}{\partial x} \text{div}\,\mathbf{A}_1^* = i\omega A_{2x}^* + \frac{1}{\gamma_2 \mu_0} \frac{\partial}{\partial x} \text{div}\,\mathbf{A}_2^*$$

$$\frac{1}{\gamma_1 \mu_0} \frac{\partial}{\partial y} \text{div}\,\mathbf{A}_1^* = \frac{1}{\gamma_2 \mu_0} \frac{\partial}{\partial y} \text{div}\,\mathbf{A}_2^*$$

and

$$\frac{\partial A_{1z}^*}{\partial y} = \frac{\partial A_{2z}^*}{\partial y}, \quad \frac{\partial A_{1x}^*}{\partial z} - \frac{\partial A_{1z}^*}{\partial x} = \frac{\partial A_{2x}^*}{\partial z} - \frac{\partial A_{2z}^*}{\partial x}$$

By analogy with the case with the horizontal magnetic dipole, we see that these conditions are equivalent to two much simple groups of conditions:

$$A_{1x}^* = A_{2x}^*, \quad \frac{\partial A_{1x}^*}{\partial z} = \frac{\partial A_{2x}^*}{\partial z} \tag{13.41}$$

and

$$A_{1z}^* = A_{2z}^*, \quad \frac{1}{\gamma_1}\operatorname{div}\mathbf{A}_1^* = \frac{1}{\gamma_2}\operatorname{div}\mathbf{A}_2^* \tag{13.42}$$

To conclude the formulation of the boundary value problem, we have to add the condition at infinity where the field vanishes, as well as the condition near the source: approaching the electric dipole, the field tends to that in a uniform medium. As follows from Eq. (13.41), the latter do not contain any terms with the vertical component A_z^*. For this reason, it is convenient to solve first the boundary value problem for the horizontal component of the vector potential A_x^* and then, making use of Eq. (13.42), to find the vertical component of the vector potential A_z^*. Inasmuch as the denominator in the second term of the expression for the electric field (Eq. (13.42)) is conductivity, we first assumed that the upper space is also conductive, but then in the derived formulas for the quasi-stationary field, consider the limiting case when $\gamma \to 0$.

13.3.2. Integral Representation for A_x^*

By definition, the horizontal component A_x must satisfy the equations:

$$\Delta A_{1x}^* + k_1^2 A_{1x}^* = 0 \quad \text{if } z < 0, \quad \Delta A_{2x}^* + k_2^2 A_{2x}^* = 0 \quad \text{if } z > 0$$

and

$$A_{1x}^* = A_{2x}^*, \quad \frac{\partial A_{1x}^*}{\partial z} = \frac{\partial A_{2x}^*}{\partial z} \quad \text{if } z = 0 \tag{13.43}$$

In the lower half space, the component A_{2x}^* can be written as a sum as follows:

$$A_{2x}^* = A_{2x}^{(0)^*} + A_{2x}^{(s)^*}, \tag{13.44}$$

where $A_{2x}^{(0)^*}$ is the component of the vector potential for the electric dipole in a uniform full space, and $A_{2x}^{(s)^*}$ is the vector potential term representing the secondary field. Earlier, we demonstrated that

$$A_{2x}^{(0)^*} = \frac{\mu_0 Idl}{4\pi}\frac{\exp(ik_2R)}{R} = \frac{\mu_0 Idl}{4\pi}\int_0^\infty \frac{m}{m_2}\exp(-m_2|z-h|)J_0(mr)dm \tag{13.45}$$

Here,

$$R=\left[\left(r^2+(z-h)^2\right]^{1/2},\quad m_2=\left(m^2-k_2^2\right)^{1/2}$$

Inasmuch as the vector potential $A_{2x}^{(0)^*}$ is independent of the coordinate φ, we can represent the vector potential of the secondary field also as a function of coordinates r and z only. By analogy with the case of the magnetic dipole and taking into account that the field should decrease with a distance from the dipole, expressions for the component A_x can be written as

$$A_{1x}^*=\frac{\mu_0 Idl}{4\pi}\int_0^\infty C_1\exp\left(m_1 z\right)J_0(mr)\mathrm{d}m$$

and

$$A_{2x}^*=\frac{\mu_0 Idl}{4\pi}\left[\int_0^\infty \frac{m}{m_2}\exp\left(-\,m_2|z-h|\right)+D_1\exp\left(-m_2 z\right)\right]J_0(mr)\mathrm{d}m \tag{13.46}$$

Substituting this into Eq. (13.43), we arrive at a system of equations for determination of C_1 and D_1:

$$\begin{aligned}C_1&=\frac{m}{m_2}\exp(-m_2 h)+D_1\\ m_1 C_1&=m\exp(-m_2 h)-m_2 D_1\end{aligned} \tag{13.47}$$

Whence,

$$C_1=\frac{2m}{m_1+m_2}\exp\left(-m_2 h\right),\quad D_1=\frac{m}{m_2}\frac{m_2-m_1}{m_2+m_1}\exp\left(-m_2 h\right) \tag{13.48}$$

Here, $m_1=\left(m^2-k_1^2\right)^{1/2}$. Therefore, we obtain the following expressions for the horizontal component A_x:

$$A_{1x}^*=\frac{\mu_0 Idl}{4\pi}\int_0^\infty \frac{2m}{m_1+m_2}\exp\left(-m_2 h\right)\exp\left(m_1 z\right)J_0(mr)\mathrm{d}m$$

and

$$A_{2x}^* = \frac{\mu_0 Idl}{4\pi} \int_0^\infty \frac{m}{m_2} \left\{ \exp\left(- m_2 |z - h)| \right) + \frac{m_2 - m_1}{m_2 + m_1} \exp(- m_2 (z + h)) \right\} J_0(mr) \mathrm{d}m \tag{13.49}$$

13.3.3. Integral Representation for A_z

Next, we determine an expression for the vertical component of vector potential A_z^*, which also satisfies the Helmholtz equations

$$\Delta A_{1z}^* + k_1^2 A_{1z}^* = 0 \quad \text{if} \quad z < 0, \qquad \Delta A_{2z}^* + k_2^2 A_{2z}^* = 0 \quad \text{if} \quad z > 0 \tag{13.50}$$

and the boundary conditions

$$A_{1z}^* = A_{2z}^*$$

and

$$\frac{1}{\gamma_1} \left(\frac{\partial A_{1x}^*}{\partial x} + \frac{\partial A_{1z}^*}{\partial z} \right) = \frac{1}{\gamma_2} \left(\frac{\partial A_{2x}^*}{\partial x} + \frac{\partial A_{2z}^*}{\partial z} \right) \tag{13.51}$$

As in the case of the horizontal magnetic dipole, the second boundary condition permits us to find an integral representation for the component A_z^*. This condition for divergence includes the term $\partial A_x^* / \partial x$, which in accordance with Eq. (13.49) has the following form:

$$\frac{\partial A_x^*}{\partial x} = \frac{\mu_0 Idl}{4\pi} \frac{\partial}{\partial x} \int_0^\infty F(m) \exp\left(\pm m_i z \right) J_0(mr) \mathrm{d}m$$

$$= \frac{\mu_0 Idl}{4\pi} \frac{\partial r}{\partial x} \frac{\partial}{\partial r} \int_0^\infty F(m) \exp\left(\pm m_i z \right) J_0(mr) \mathrm{d}m \tag{13.52}$$

$$= -\frac{\mu_0 Idl}{4\pi} \cos\varphi \int_0^\infty mF(m) \exp\left(\pm m_i z \right) J_1(mr) \mathrm{d}m,$$

since

$$\frac{\partial}{\partial r} J_0(mr) = -mJ_1(mr)$$

and

$$\frac{\partial}{\partial r} = \frac{\partial}{\partial x}\left(x^2 + y^2\right)^{1/2} = \frac{x}{r} = \cos\varphi,$$

where φ is the angle between the x-axis along which the dipole is directed, and the radius r, as shown in Figure 13.3. The second equality of the set (Eq. (13.51)) can be satisfied for any point of the surface if both terms

$$\frac{\partial A_x^*}{\partial x} \quad \text{and} \quad \frac{\partial A_z^*}{\partial z}$$

have the same dependence on coordinates r and φ. Correspondingly, the integral representation for A_z^* should have the following form:

$$A_{1z}^* = \frac{\mu_0 Idl}{4\pi}\cos\varphi \int\limits_0^\infty C_2 \exp\left(m_1 z\right) J_1(mr)\mathrm{d}m \quad \text{if} \quad z \le 0$$

and

$$A_{2z}^* = \frac{\mu_0 Idl}{4\pi}\cos\varphi \int\limits_0^\infty D_2 \exp\left(-\,m_2 z\right) J_1(mr)\mathrm{d}m \quad \text{if} \quad z \ge 0 \tag{13.53}$$

Unlike the horizontal component of the vector potential, this component does depend on the azimuthal angle φ and it also satisfies the Helmholtz equation, which contains the term with this angle:

$$\frac{\partial^2 A_z^*}{\partial r^2} + \frac{1}{r}\frac{\partial A_z^*}{\partial r} + \frac{\partial^2 A_z^*}{\partial z^2} + \frac{1}{r^2}\frac{\partial^2 A_z^*}{\partial \varphi^2} + k^2 A_z^* = 0$$

Now substituting Eq. (13.53) into the set (Eq. (13.51)), we obtain a system of equations from which the coefficients C_2 and D_2 are determined:

$$C_2 = D_2$$

$$\frac{1}{\gamma_1}(-\,mC_1 + mC_2) = \frac{1}{\gamma_2}\left\{-\,m\left[\frac{m}{m_2}\exp\left(-m_2 h\right) + D_1\right] - m_2 D_2\right\} \tag{13.54}$$

Taking into account Eq. (13.48), we obtain

$$(\rho_1 - \rho_2)\frac{2m^2}{m_1 + m_2}\exp(-2m_2 h) = (\rho_1 m_1 C_2 + \rho_2 m_2 D_2)$$

Thus,

$$C_2 = D_2 = \frac{(\rho_1 - \rho_2)2m^2 \exp(-m_2 h)}{(m_1 + m_2)(\rho_1 m_1 + \rho_2 m_2)} \tag{13.55}$$

Therefore, we obtain the following expressions for the complex amplitude of the vertical component of the vector potential:

$$A^*_{1z} = \frac{\mu_0 Idl}{4\pi}(\rho_1 - \rho_2)\cos\varphi \int\limits_0^\infty \frac{2m^2 \exp(-m_2 h)\exp(m_1 z)}{(m_1 + m_2)(\rho_1 m_1 + \rho_2 m_2)} J_1(mr)dm$$

and

$$A^*_{2z} = \frac{\mu_0 Idl}{4\pi}(\rho_1 - \rho_2)\cos\varphi \int\limits_0^\infty \frac{2m^2 \exp(-m_2 h)\exp(-m_2 z)}{(m_1 + m_2)(\rho_1 m_1 + \rho_2 m_2)} J_1(mr)dm \tag{13.56}$$

Now we assume that the upper half space is an insulator and that the dipole is actually located on the Earth's surface at $h=0$. Then, letting $m_1 = m$ and $m_2 = m_1$, we have

$$A^*_{1x} = \frac{\mu_0 Idl}{4\pi}\int\limits_0^\infty \frac{2m}{m + m_1}\exp(mz)J_0(mr)dm$$

$$A^*_{1z} = \frac{\mu_0 Idl}{4\pi}\cos\varphi\int\limits_0^\infty \frac{2m}{m + m_1}\exp(mz)J_1(mr)dm \quad \text{if} \quad z < 0 \tag{13.57}$$

and

$$A^*_{2x} = \frac{\mu_0 Idl}{4\pi}\int\limits_0^\infty \frac{2m}{m + m_1}\exp(-m_1 z)J_0(mr)dm$$

$$A^*_{2z} = \frac{\mu_0 Idl}{4\pi}\cos\varphi\int\limits_0^\infty \frac{2m}{m + m_1}\exp(-m_1 z)J_1(mr)dm \quad \text{if} \quad z > 0 \tag{13.58}$$

13.3.4. Expression for div A*

To determine the electric field on the Earth's surface, we have to find expression of div $\mathbf{A}^*$. As follows from Eqs (13.57) and (13.58),

$$\text{div } \mathbf{A}_1^* = 0,$$

$$\text{div } \mathbf{A}_2^* = -2\frac{\mu_0 Idl}{4\pi}\cos\varphi \int_0^\infty mJ_1(mr)\mathrm{d}m \quad \text{if } z = 0$$

The integral can be written as

$$\int_0^\infty mJ_1(mr)\mathrm{d}m = -\frac{\partial}{\partial r}\int_0^\infty J_0(mr)\mathrm{d}m = -\frac{\partial}{\partial r}\frac{1}{r} = \frac{1}{r^2}$$

and

$$\text{div } \mathbf{A}_2^* = -\frac{2\mu_0 Idl}{4\pi r^2}\cos\varphi \tag{13.59}$$

From the condition,

$$\text{div } \mathbf{A}^* = -\gamma\mu_0 U^*$$

we can see that on the Earth's surface, the potential U coincides with the potential of the stationary field ($\omega = 0$) at every frequency. In fact from Eq. (13.59), it follows

$$U^* = \frac{2\rho Idl}{4\pi r^2}\cos\theta \tag{13.60}$$

On the surface of the Earth, the galvanic part of the field caused by electric charges is the same as that which arises with a stationary field. This does not hold beneath the surface where div $\mathbf{A}_2$ is

$$\text{div } \mathbf{A}_2 = -\frac{2\mu_0 Idl}{4\pi}\int_0^\infty m\exp(-m_1 z)J_1(mr)\mathrm{d}m$$

13.3.5. Equations for the Electric Field on the Earth's Surface

Making use of Eqs (13.20) and (13.58), we find expressions for the electromagnetic field at the Earth's surface. In a cylindrical system of coordinates,

$$E_r^* = i\omega A_r^* + \frac{1}{\gamma\mu_0}\frac{\partial}{\partial r}\text{div}\,\mathbf{A}^*, \quad E_\varphi^* = i\omega A_\varphi^* + \frac{1}{\gamma\mu_0 r}\frac{\partial}{\partial\varphi}\text{div}\,\mathbf{A}^*$$

$$E_z^* = i\omega A_z^* + \frac{1}{\gamma\mu_0}\frac{\partial}{\partial z}\text{div}\,\mathbf{A}^* \tag{13.61}$$

and

$$B_r^* = \frac{1}{r}\left(\frac{\partial A_z^*}{\partial\varphi} - r\frac{\partial A_\varphi^*}{\partial z}\right), \quad B_\varphi^* = \frac{\partial A_r^*}{\partial z} - \frac{\partial A_z^*}{\partial r}$$

$$B_z^* = \frac{1}{r}\left(\frac{\partial r A_\varphi^*}{\partial r} - \frac{\partial A_r^*}{\partial\varphi}\right) \tag{13.62}$$

Inasmuch as

$$A_r^* = A_x^* \cos\varphi, \quad A_\varphi^* = -A_x^* \sin\varphi,$$

we obtain the following expressions for the horizontal components of the electric field:

$$E_r^* = \frac{Idl}{2\pi}\cos\varphi\left[i\omega\mu_0\int_0^\infty \frac{m}{m+m_1}J_0(mr)\mathrm{d}m - \frac{1}{\gamma}\frac{\partial}{\partial r}\frac{1}{r^2}\right]$$

and

$$E_\varphi^* = \frac{Idl}{2\pi}\sin\varphi\left[-i\omega\mu_0\int_0^\infty \frac{m}{m+m_1}J_0(mr)\mathrm{d}m + \frac{1}{\gamma r^3}\right]$$

In Chapter 9, we found out that

$$\int_0^\infty \frac{m}{m+m_1}J_0(mr)\mathrm{d}m = -\frac{1}{k^2 r^3}\left[1 - \exp(ikr)(1-ikr)\right]$$

Therefore,

$$E_r^* = \frac{2\rho Idl}{4\pi r^3}\cos\varphi\left[1 + \exp(ikr)(1-ikr)\right]$$

and

$$E_\varphi^* = \frac{2\rho Idl}{4\pi r^3}\sin\varphi[2-\exp(ikr)(1-ikr)] \tag{13.63}$$

For the vertical component of the electric field, we have

$$E_z^* = \frac{2\rho Idl\mu_0}{4\pi}\cos\varphi\left[i\omega\int_0^\infty \frac{m}{m+m_1}J_1(mr)dm + \frac{1}{\mu_0\gamma}\int_0^\infty mm_1J_1(mr)dm\right]$$

$$= \frac{2Idl\cos\varphi}{4\pi}\int_0^\infty m^2J_1(mr)dm = -\frac{2Idl}{4\pi}\frac{\partial}{\partial x}\int_0^\infty mJ_0(mr)dm = 0$$

Thus, the vertical component of the electric field in a conducting medium tends to zero as the observation point approaches the surface of the Earth. This is obvious conclusion, inasmuch as currents cannot move from the conducting lower half space into the insulating medium (quasi-stationary approximation).

13.3.6. Equations for the Magnetic Field on the Earth's Surface

As follows from Eq. (13.62) for the radial component of the magnetic field, we have

$$B_r^* = -\frac{\mu_0 Idl}{2\pi r}\sin\varphi\left[\int_0^\infty \frac{m}{m+m_1}J_1(mr)dm + r\int_0^\infty \frac{mm_1}{m+m_1}J_0(mr)dm\right] \tag{13.64}$$

The integrals in the last expression can be transformed using the identity given earlier:

$$L_\nu = \int_0^\infty \frac{1}{m_1}\left(\frac{m_1-m}{m_1+m}\right)^\nu J_0(mr)dm = I_\nu\left(\frac{ikr}{2}\right)K_\nu\left(\frac{ikr}{2}\right) \tag{13.65}$$

Let us consider the first integral

$$\int_0^\infty \frac{m}{m+m_1} J_1(mr)\mathrm{d}m = -\frac{\partial}{\partial r}\int_0^\infty \frac{1}{m+m_1} J_0(mr)\mathrm{d}m$$

and introduce a new function F_1 defined as

$$\frac{2}{m+m_1} = \frac{1}{m_1} + F_1$$

Whence,

$$F_1 = \frac{1}{m_1}\frac{m_1-m}{m_1+m}$$

Therefore,

$$\int_0^\infty \frac{2m}{m+m_1} J_1(mr)\mathrm{d}m = -\frac{\partial}{\partial r}\left[I_0\left(\frac{ikr}{2}\right)K_0\left(\frac{ikr}{2}\right) + I_1\left(\frac{ikr}{2}\right)K_1\left(\frac{ikr}{2}\right)\right] \tag{13.66}$$

The second integrand is

$$\frac{2mm_1}{m+m_1} = m + F_2$$

and so

$$F_2 = \frac{m(m_1-m)}{m_1+m} = -\frac{k^2 m}{(m_1+m)^2}$$

or

$$F_2 = -\frac{k^2}{4m_1}\left[1 - \left(\frac{m_1-m}{m_1+m}\right)^2\right]$$

Thus, we have

$$2\int_0^\infty \frac{mm_1}{(m+m_1)} J_0(mr)\mathrm{d}m = -\frac{k^2}{4}\left[I_0\left(\frac{ikr}{2}\right)K_0\left(\frac{ikr}{2}\right) - I_2\left(\frac{ikr}{2}\right)K_2\left(\frac{ikr}{2}\right)\right] \tag{13.67}$$

Whence,

$$B_r^* = \frac{\mu_0 Idl}{4\pi r}\sin\varphi\left[\frac{\partial}{\partial r}(I_0K_0 + I_1K_1) + \frac{k^2 r}{4}(I_0K_0 - I_2K_2)\right] \tag{13.68}$$

Here, the argument of Bessel functions is $ikr/2$. Making use of series of identities,

$$\begin{aligned} &I_0'(x) = I_1(x), \quad K_0'(x) = -K_1(x) \\ &I_1'(x) = I_0(x) - \frac{1}{x}I_1(x), \quad K_1'(x) = -K_0(x) - \frac{1}{x}K_1(x) \\ &I_2(x) = I_0(x) - \frac{2}{x}I_1(x), \quad K_2(x) = K_0(x) + \frac{2}{x}K_1(x) \end{aligned} \tag{13.69}$$

and after substituting these identities into Eq. (13.68), we have

$$\begin{aligned} B_r^* = -\frac{\mu_0 Idl}{4\pi r^2}\sin\varphi\Bigg\{6I_1\left(\frac{ikr}{2}\right)K_1\left(\frac{ikr}{2}\right) + ikr\Bigg[I_1\left(\frac{ikr}{2}\right)K_0\left(\frac{ikr}{2}\right) \\ - I_0\left(\frac{ikr}{2}\right)K_1\left(\frac{ikr}{2}\right)\Bigg]\Bigg\}, \end{aligned} \tag{13.70}$$

From Eq. (13.62), we have the following expression for the azimuthal component of the magnetic field

$$B_\varphi^* = \frac{\mu_0 Idl}{4\pi}\cos\varphi\left[-\int_0^\infty \frac{2mm_1}{m_1+m}J_0(mr)dm - \int_0^\infty \frac{2m^2}{m_1+m}J_1'(mr)dm\right]$$

Inasmuch as

$$J_1'(mr) = -\frac{1}{mr}J_1 + J_0,$$

we obtain

$$\begin{aligned} B_\varphi^* &= -\frac{\mu_0 Idl}{4\pi}\cos\varphi\left[2\int_0^\infty mJ_0(mr)dm - \frac{1}{r}\int_0^\infty \frac{2m}{m_1+m}J_1(mr)dm\right] \\ &= \frac{\mu_0 Idl}{4\pi r}\cos\varphi\int_0^\infty \frac{2m}{m+m_1}J_1(mr)dm \end{aligned} \tag{13.71}$$

As has been shown earlier, this integral is

$$\int_0^\infty \frac{2m}{m+m_1} J_1(mr)\mathrm{d}m = \frac{2}{r} I_1\left(\frac{ikr}{2}\right) K_1\left(\frac{ikr}{2}\right)\cos\varphi$$

Therefore,

$$B_\varphi^* = \frac{\mu_0 Idl}{2\pi r^2} I_1\left(\frac{ikr}{2}\right) K_1\left(\frac{ikr}{2}\right)\cos\varphi \qquad (13.72)$$

Finally, the expression for the vertical component of the magnetic field is

$$\begin{aligned} B_z^* &= \frac{\partial A_\varphi}{\partial r} - \frac{\partial A_r}{r\partial\varphi} + \frac{1}{r}A_\varphi \\ &= \frac{\mu_0 Idl}{4\pi}\left[\int_0^\infty \frac{2m^2}{m+m_1} J_1(mr)\mathrm{d}m + \frac{1}{r}\int_0^\infty \frac{2m}{m+m_1} J_0(mr)\mathrm{d}m \right. \\ &\qquad \left. - \frac{1}{r}\int_0^\infty \frac{2m}{m+m_1} J_0(mr)\mathrm{d}m\right]\sin\varphi \\ &= -\frac{\mu_0 Idl}{2\pi}\sin\varphi \frac{\partial}{\partial r}\int_0^\infty \frac{2m}{m+m_1} J_0(mr)\mathrm{d}m \end{aligned}$$

or

$$B_z^* = -\frac{\mu_0 Idl}{2\pi k^2}\sin\varphi \frac{\partial}{\partial r}\left[\int_0^\infty m^2 J_0(mr)\mathrm{d}m - \int_0^\infty m m_1 J_0(mr)\mathrm{d}m\right]$$

Making use of the known expressions for these integrals (Chapter 9), we have

$$B_z^* = \frac{\mu_0 Idl}{2\pi k^2}\frac{\partial}{\partial r}\left[\frac{1}{r^3} - \frac{1}{r^3}(1-ikr)\exp(ikr)\right]$$

or

$$B_z^* = -\frac{3\mu_0 Idl}{2\pi k^2 r^4}\sin\varphi\left[1 - \left(1 - ikr - \frac{1}{3}k^2 r^2\right)\exp(ikr)\right] \qquad (13.73)$$

Before we continue, let us notice that with an accuracy of constant, the latter coincides with the expression for the electric field of the vertical magnetic dipole located at the Earth's surface, and as was pointed out, it is an example illustrating the principle of reciprocity. This fact allowed us to consider the frequency and transient soundings, proceeding from the expression for the electric field caused by the vertical magnetic dipole. In order to study a field behavior, it is convenient to represent the electric and magnetic field in the form

$$\begin{aligned} E_r^* &= E_r^0 e_r^*, \quad E_\varphi^* = E_\varphi^0 e_\varphi^*, \quad E_z^* = 0, \\ B_r^* &= B_r^0 b_r^*, \quad B_\varphi^* = B_\varphi^0 b_\varphi^*, \quad B_z^* = B_z^0 b_z^*, \end{aligned} \tag{13.74}$$

where $E_r^0, E_\varphi^0, B_r^0, B_\varphi^0$ and B_z^0 are the components of the stationary electric field ($\omega = 0$), namely,

$$\begin{aligned} E_r^0 &= \frac{4p_0}{4\pi\varepsilon_0 r^3}\cos\varphi, \quad E_\varphi^0 = \frac{2p_0}{4\pi\varepsilon_0 r^3}\sin\varphi, \quad p_0 = \varepsilon_0 \rho Idl \\ B_r^{(0)} &= -\frac{\mu_0 Idl}{4\pi r^2}\sin\varphi, \quad B_\varphi^0 = \frac{\mu_0 Idl}{4\pi r^2}\cos\varphi, \quad B_z^0 = \frac{\mu_0 Idl}{4\pi r^2}\sin\varphi \end{aligned} \tag{13.75}$$

and

$$\begin{aligned} e_r^* &= \frac{1}{2}[1 + (1 - ikr)\exp(ikr)], \quad e_\varphi^* = 2 - (1 - ikr)\exp(ikr) \\ b_r^* &= \left\{6I_1\left(\frac{ikr}{2}\right)K_1\left(\frac{ikr}{2}\right) + ikr\left[I_1\left(\frac{ikr}{2}\right)K_0\left(\frac{ikr}{2}\right) - I_0\left(\frac{ikr}{2}\right)K_1\left(\frac{ikr}{2}\right)\right]\right\} \\ b_\varphi^* &= 2I_1\left(\frac{ikr}{2}\right)K_1\left(\frac{ikr}{2}\right) \\ b_z^* &= -\frac{6}{k^2r^2}\left[1 - \left(1 - ikr - \frac{1}{3}k^2r^2\right)\exp(ikr)\right] \end{aligned} \tag{13.76}$$

The functions e and b represent the influence of skin effect and the departure between the actual and stationary fields. The electromagnetic field of an electric dipole situated on the Earth's surface depends on the azimuthal angle φ in the same way as does the stationary field regardless of the value of induction number r/δ, and all functions e and b depend only on this parameter.

13.3.7. The Range of Small Parameter $p = r/\delta$, (the Low-Frequency Spectrum)

First of all, consider the limiting case when the frequency is equal to zero. Inasmuch as

$$\exp x \approx 1 + x + \frac{x^2}{2}, \quad I_0(x) \to 1, \quad I_1(x) \to \frac{x}{2},$$

$$K_0(x) \to -\ln x, \quad K_1(x) \to \frac{1}{x} \quad \text{if } x \to 0$$

we see that the electromagnetic field tends to the stationary one $\mathbf{E}^0$, $\mathbf{B}^0$ (Eq. (13.75)), that of course is obvious. In such case, the electric field is caused by two opposite charges on the surface of the dipole electrodes, located on the Earth's surface:

$$e_A = 2\varepsilon_0 \rho I \quad \text{and} \quad e_B = -2\varepsilon_0 \rho I$$

The vertical component of the magnetic field on the Earth's surface is caused by the current along wire, connecting the electrodes, while the horizontal components are generated by volume currents beneath this surface. Note that in the quasi-stationary approximation, these charges and currents cause at each instant the same field as the stationary one, which obeys Coulomb's and Biot–Savart's laws but in addition, there is an effect of the electromagnetic induction. Here, we considered the limiting case when the frequency is equal to zero; next let us discuss the low-frequency part of the spectrum in some detail and start from the electric field. Expanding exp (*ikr*) in a series and substituting the latter in the expression for the quadrature and in-phase components, we obtain

$$e_r^* = 1 - \frac{1}{2}\sum_{n=2}^{\infty} \frac{n-1}{n!}(ikr)^n, \quad e_\varphi^* = 1 + \sum_{n=2}^{\infty} \frac{n-1}{n!}(ikr)^n \tag{13.77}$$

Whence, we have the following expressions for the in-phase and quadrature components:

$$\begin{aligned} Ine_r^* &\approx 1 - \frac{1}{3}p^3, \quad Qe_r^* \approx \frac{1}{2}p^2 - \frac{1}{3}p^3 \\ Ine_\varphi^* &\approx 1 + \frac{2}{3}p^3, \quad Qe_\varphi^* \approx -p^2 + \frac{2}{3}p^3 \end{aligned} \tag{13.78}$$

or

$$
\begin{aligned}
InE_r^* &\approx \frac{\rho Idl}{\pi r^3}\left[1-\frac{1}{6\cdot 2^{1/2}}(\gamma\mu_0\omega)^{3/2}r^3\right]\cos\varphi \\
QE_r^* &\approx \frac{\rho Idl}{\pi r^3}\left[\frac{\gamma\mu_0\omega r^2}{4}-\frac{1}{6\cdot 2^{1/2}}(\gamma\mu_0\omega)^{3/2}r^3\right]\cos\varphi \\
InE_\varphi^* &\approx \frac{\rho Idl}{2\pi r^3}\left[1+\frac{1}{3\cdot 2^{1/2}}(\gamma\mu_0\omega)^{3/2}r^3\right]\sin\varphi \\
QE_\varphi^* &\approx \frac{\rho Idl}{2\pi r^3}\left[-\frac{\gamma\mu_0\omega r^2}{2}+\frac{1}{3\cdot 2^{1/2}}(\gamma\mu_0\omega)^{3/2}r^3\right]\sin\varphi \quad p\ll 1
\end{aligned}
\tag{13.79}
$$

From Eq. (13.79), it can be readily seen that at the low-frequency part of the spectrum, the electric field can be approximately represented as a sum of two fields of the galvanic and vortex origin. More precisely, the in-phase component is mainly caused by charges of the dipole, while the quadrature component is directly proportional to frequency and it is independent on conductivity. It is a simple matter to see that this component arises due to a change of the primary magnetic field $\mathbf{B}^0$ with time. It may be proper to note that the study of this part of the spectrum of the electric field may be useful for different purposes, in particular, for decreasing the inductive effect in the induced polarization method and understanding the transient field at the late stage. The following term in the expression representing the quadrature component is proportional to $\omega^{3/2}$ but does not depend on the separation between the dipole source and an observation point. As in the case of the magnetic dipole, it reflects the current diffusion but is less sensitive to conductivity. Taking into account that the leading term in the expression for the quadrature component of the electric field is directly proportional to frequency for small values of the induction number, it seems reasonable to examine the function

$$
\frac{QE^*(\omega_1)}{\omega_1}-\frac{QE^*(\omega_2)}{\omega_2}
$$

which allows us to eliminate this term, so that the function turns out to be proportional to $\gamma^{1/2}$. Now consider an asymptotic representation for the magnetic field valid for small values of the induction number. Again expanding exp (ikr) in a series, we have the following expression for the vertical component b_z^*:

$$b_z^* = 1 + 2\sum_{n=4}^{\infty}\frac{(n-1)(3-n)}{n!}(ikr)^{n-2}$$

Correspondingly, the expressions for the quadrature and in-phase components are

$$\begin{aligned} Qb_z^* &\approx \frac{1}{2}p^2 - \frac{4}{15}p^3 + \frac{4}{105}p^5 \\ &= \frac{1}{4}\gamma\mu_0\omega r^2 - \frac{2^{1/2}}{15}\left(\gamma\mu_0\omega r^2\right)^{3/2} + \frac{1}{105\cdot 2^{1/2}}\left(\gamma\mu_0\omega r^2\right)^{5/2} + \cdots \end{aligned}$$

and

$$\begin{aligned} Inb_z^* &\approx 1 - \frac{4}{15}p^3 + \frac{1}{6}p^4 - \frac{4}{105}p^5 \\ &= 1 - \frac{2^{1/2}}{15}\left(\gamma\mu_0\omega r^2\right)^{3/2} + \frac{1}{24}\left(\gamma\mu_0\omega r^2\right)^2 - \frac{1}{105\cdot 2^{1/2}}\left(\gamma\mu_0\omega r^2\right)^{5/2} - \cdots \end{aligned} \tag{13.80}$$

Making use of identities for the Bessel functions given earlier, we have the following expressions for the horizontal components of the magnetic field:

$$b_\varphi^* \approx 1 - \frac{(ikr)^2}{32} + \frac{(ikr)^2}{8}\ln ikr\frac{\alpha}{2}$$

or

$$Qb_\varphi^* \approx -\frac{p^2}{4}\ln\alpha\frac{p}{2^{1/2}} + \frac{p^2}{16} = -\frac{1}{8}(\gamma\mu_0\omega)r^2\ln\frac{(\gamma\mu_0\omega)^{1/2}}{2}r\alpha + \frac{(\gamma\mu_0\omega)r^2}{32} + \cdots$$

$$Inb_\varphi^* \approx 1 + \frac{3\pi}{16}p^2 = 1 + \frac{3\pi}{32}\gamma\mu_0\omega r^2 \quad \text{and} \quad \ln\alpha = -0.11593 \tag{13.81}$$

By analogy,

$$b_r^* \approx 1 - \frac{1}{2}\left(\frac{ikr}{2}\right)^2\ln\frac{ikr}{2}\alpha - \frac{3}{32}(ikr)^2$$

or

$$Qb_r^* \approx \frac{1}{4}p^2 \ln p\frac{1}{2^{1/2}} + \frac{3}{16}p^2 = \frac{1}{8}\left(\gamma\mu_0\omega r^2\right)\ln \frac{\left(\gamma\mu_0\omega\right)^{1/2}}{2} r\alpha + \frac{3}{32}\gamma\mu_0\omega r^2$$

$$Inb_r^* \approx 1 - \frac{3\pi}{32}p^2 = 1 - \frac{3\pi}{64}\gamma\mu_0\omega r^2$$

(13.82)

From these last three equations, we see that the vertical and horizontal components of the magnetic field are functions of frequency and conductivity to different degrees. While the first term in the expression for the in-phase component corresponds to the stationary magnetic field, the second term in each of expressions reflects the skin effect which manifests itself differently for the vertical and horizontal components. Inasmuch as there is proportionality to $\omega^{3/2}$ for B_z and only to ω for the horizontal component, we can say that with decreasing frequency, the influence of the inductive effect on the in-phase component of the field B_z becomes less significant than the effect on the corresponding terms for the in-phase component of B_r or B_φ. The fact that the second term in the expression for the in-phase component for the horizontal magnetic field is directly proportional to frequency and conductivity does not mean that these terms describe a field in which the interaction between currents is negligible, as was in the case for the magnetic dipole source. It is significant that the leading term in the expression for the quadrature component B_z^* and the second term in the expression for the in-phase component are more sensitive to a change of conductivity than are corresponding terms for the electric and horizontal magnetic fields.

13.3.8. The Range of Large Parameters (Wave Zone)

Now we investigate the field for large induction number p when the separation between the dipole source and the observation point is significantly larger than the wave length. Assuming that the parameter r/λ is greater than the unity and making use of the asymptotic expansions for the Bessel functions, we obtain

$$e_r^* = \frac{1}{2}, \quad e_\varphi^* = 2, \quad b_r^* = \frac{4}{ikr}, \quad b_z^* = -\frac{6}{k^2r^2}, \quad b_\varphi^* = \frac{2}{ikr} \tag{13.83}$$

Thus, we have the following expressions for the field components:

$$E_r^* = \frac{\rho Idl}{2\pi r^3}\cos\varphi, \quad E_\varphi^* = \frac{\rho Idl}{\pi r^3}\sin\varphi$$
$$B_r^* = \frac{\mu_0 Idl}{\pi ikr^3}\sin\varphi, \quad B_\varphi^* = \frac{\mu_0 Idl}{2\pi ikr^3}\cos\varphi \quad B_z^* = -\frac{\mu_0 Idl}{2\pi ik^2r^4}\sin\varphi \tag{13.84}$$

In accordance with these equations, the electric field is in phase with the stationary field, has the same behavior, and differs only by a multiplicative factor:

$$E_r^* = \frac{1}{2}E_r^0, \quad E_\varphi^* = 2E_\varphi^0 \tag{13.85}$$

It should be noted that the electric field for large induction numbers is independent of frequency. Unlike the electric field, the magnetic field is a function of frequency and is significantly smaller than the stationary field because $|1/kr| \ll 1$. Also, the horizontal components of the magnetic field dominate, emphasizing the fact that the Poynting vector is mainly directed downward. Correspondingly, the ratio of mutually perpendicular components of the electric and magnetic field is equal to the impedance of a plane wave. In fact according to Eq. (13.84), we have

$$Z_{r\varphi}^* = \frac{E_r^*}{B_\varphi^*} = \frac{ik\rho}{\mu_0}, \quad Z_{\varphi r}^* = \frac{E_\varphi^*}{B_r^*} = -\frac{ik\rho}{\mu_0} \tag{13.86}$$

13.3.9. Behavior of the Field on the Earth's Surface

We have investigated the asymptotic behavior of the electric and magnetic fields of the electric dipole situated on the Earth's surface. It would be appropriate to make some comments concerning the behavior of the field independently of the induction number r/δ. As follows from Eq. (13.63),

$$E_r^* \sin\varphi + E_\varphi^* \cos\varphi = \frac{6\rho Idl}{4\pi r^3}\sin\varphi\cos\varphi = \frac{3\rho Idl}{4\pi r^3}\sin 2\varphi \tag{13.87}$$

It is a simple matter to see that this sum defines the component of the electric field perpendicular to the dipole axis; that is, E_y^*. Therefore, at any point on the surface of a uniform half space, the electric field component E_y is independent on the frequency and it is the same as the stationary field. For the component of the electric field which is parallel to the dipole axis, we have

$$E_x^* = E_r^* \cos\varphi - E_\varphi^* \sin\varphi = \frac{\rho I dl}{2\pi r^3}\left[\left(1 - 3\sin^2\varphi\right) + (1 - ikr)\exp(ikr)\right] \tag{13.88}$$

This means that there are no points where the component E_x^* is zero, unlike the stationary field component E_x^0. Also, Eq. (13.88) gives

$$E_x^* = \frac{\rho I dl}{2\pi r^3}[1 + (1 - ikr)\exp(ikr)] \quad \text{if} \quad \varphi = 0$$

and

$$E_x^* = \frac{2\rho I dl}{2\pi r^3}[-2 + (1 - ikr)\exp(ikr)] \quad \text{if} \quad \varphi = \frac{\pi}{2} \tag{13.89}$$

In comparing the electric field caused by the vertical magnetic dipole with the vertical component of the magnetic field due to the horizontal electric dipole, we early discovered very interesting fact, namely,

$$e_\varphi^* = b_z^*$$

As was pointed out, this reflects the principle of reciprocity and it will be demonstrated to hold as well as for horizontally layered medium. In conclusion, the functions e^* and b^* are shown in Figures 13.4 and 13.5.

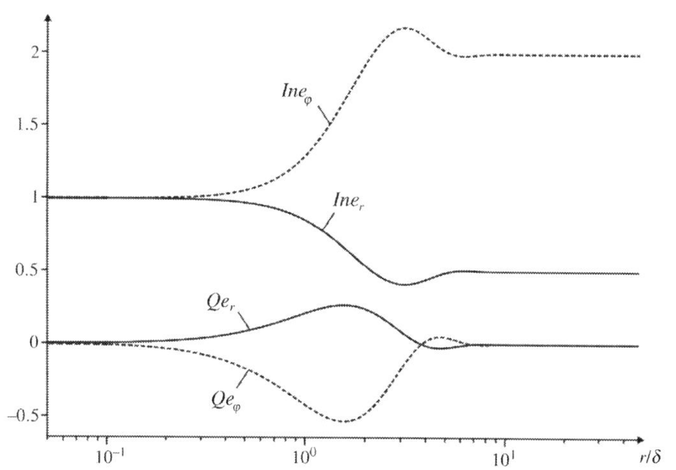

Figure 13.4 Frequency responses of the electric field of a horizontal electric dipole at the surface of a uniform half space.

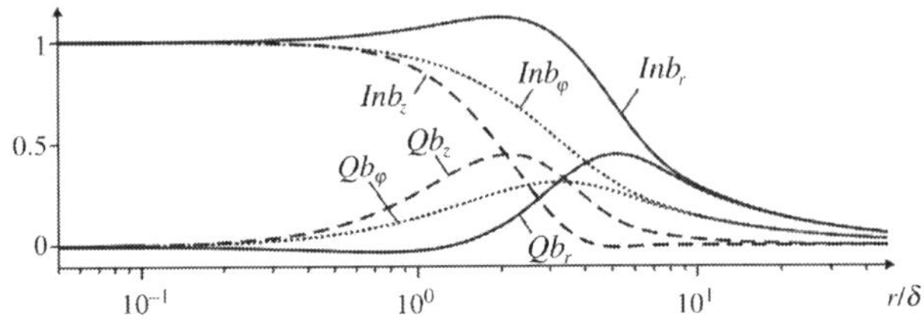

Figure 13.5 Frequency responses of the magnetic field of a horizontal electric dipole at the surface of a uniform half space.

13.4. THE HORIZONTAL ELECTRIC DIPOLE ON THE SURFACE OF A HORIZONTALLY LAYERED MEDIUM

13.4.1. Integral Representation for Components of the Vector Potential

As before, we will precede from relations Eqs (13.19) and (13.20)

$$\mathbf{B}^* = \text{curl } \mathbf{A}^*, \mathbf{E}^* = i\omega\mathbf{A}^* + \frac{1}{\gamma\mu_0}\text{grad div } \mathbf{A}^*, \nabla^2\mathbf{A}^* + k^2\mathbf{A}^* = 0,$$

and assumption that the field is described by only two components of the vector potential A_x^* and A_z^*, which obey at regular points the Helmholtz equations. At each uniform layer, the complex amplitude of the horizontal component A_x can be written as

$$A_x^* = \frac{\mu_0 Idx}{4\pi}\int_0^\infty X(m,z,k)J_0(mr)dm \tag{13.90}$$

The complex amplitude of the vertical component of the vector potential is represented in a slightly different form:

$$A_z^* = \frac{\partial W^*}{\partial x}, \tag{13.91}$$

where

$$W^* = \frac{\mu_0 I\mathrm{d}x}{4\pi}\int_0^\infty Z(m,k,z)J_0(mr)\mathrm{d}m \tag{13.92}$$

The reason for introduction of this function will be seen later. Here, functions X and Z satisfy the same ordinary differential equations of the second order

$$\frac{d^2X}{dz^2} - n^2X = 0, \qquad \frac{d^2Z}{dz^2} - n^2Z = 0, \tag{13.93}$$

and $n = \left(m^2 - k^2\right)^{1/2}$. Note that these equations appeared when the Helmholtz equation was solved using the method of separation of variables. Letters X and Z indicate that these functions describe the complex amplitudes of the corresponding components of the vector potential.

13.4.2. Boundary Conditions for the Functions X and Z

Earlier we demonstrated that from continuity of tangential components of the electric and magnetic fields at boundaries follows that

$$A_x^*, \qquad \frac{\partial A_x^*}{\partial z} \tag{13.94}$$

are continuous functions, as well as

$$A_z^*, \qquad \frac{1}{\gamma}\left(\frac{\partial A_x^*}{\partial x} + \frac{\partial A_z^*}{\partial z}\right) \tag{13.95}$$

As in the case of a uniform half space, the boundary value problem for the horizontal component A_x^* can be solved without knowledge of the vertical component. At the same time in accordance with the second equation of the set (Eq. (13.95)), in order to determine the vertical component A_z^*, it is necessary to know the component A_x^*. Correspondingly, method of solving of the boundary value problem for the electromagnetic field consists of two main steps. The first one is a solution of the value boundary problem for the horizontal component of the vector potential, and then knowing A_x^*, the boundary value problem is solved but respect to A_z^*; namely, this approach was used in the case of a uniform half space. However, for a general case of layered medium, this method is rather cumbersome, and for this reason, we describe a different way which allows one simultaneously to obtain formulas for both components of the vector potential. First of all, taking into account orthogonality of Bessel functions from Eq. (13.94) follows continuity of functions

$$X, \quad X' \tag{13.96}$$

where $X' = \frac{dX}{dz}$.

Making use of Eq. (13.91), we can represent Eq. (13.95) as

$$\frac{\partial W^*}{\partial x}, \quad \frac{1}{\gamma}\frac{\partial}{\partial x}\left(A_x^* + \frac{\partial W^*}{\partial z}\right) \tag{13.97}$$

Bearing in mind that derivative with respect to x is tangential to boundaries, the latter is equivalent to continuity of functions:

$$Z, \quad \frac{1}{k^2}(X + Z') \tag{13.98}$$

Here, $Z' = \frac{dZ}{dz}$.

Thus, in order to provide continuity of tangential components of the electromagnetic field, the following functions have to be continuous at the boundaries beneath the Earth's surface:

$$X, \quad X' \text{ and } Z, \quad \frac{1}{k^2}(X + Z') \tag{13.99}$$

Now let us transform the last expression in this set in such a way that boundary conditions for both components became similar. With this purpose in mind, we introduce new function V in the following way:

$$Z = V - \frac{1}{m^2}X', \tag{13.100}$$

and describe its behavior. First of all, it is important to show that at regular points, V obeys the same equation as function Z. In fact, substitution of Eq. (13.100) into Eq. (13.93) gives

$$\frac{d^2V}{dz^2} - \frac{1}{m^2}\frac{d^3X}{dz^3} - n^2\left(V - \frac{1}{m^2}X'\right) = \frac{d^2V}{dz^2} - n^2V - \frac{1}{m^2}\frac{\partial}{\partial z}\left(X'' - n^2X\right)$$

or

$$\frac{d^2V}{dz^2} - n^2V = 0 \tag{13.101}$$

Next consider the behavior of function V at a boundary. Inasmuch as Z and X' are continuous functions at the boundary, the function V (Eq. (13.100)) is also continuous. In other words, functions X, Z, and V behave in the same manner. Now we find the second condition for the function V. Substituting Eq. (13.100) into the last expression of the set (Eq. (13.99)), we have

$$\frac{1}{k^2}(X+Z')=\frac{1}{k^2}\left(X+V'-\frac{1}{m^2}X''\right)=\frac{1}{k^2}\left(X+V'-\frac{n^2}{m^2}X\right)$$
$$=\frac{1}{k^2}\left(X+V'-X+\frac{k^2}{m^2}X\right)=\frac{1}{k^2}V'+\frac{1}{m^2}X \tag{13.102}$$

Taking into account that the left-hand side is continuous, as well as X/m^2, we see that $1/k^2(V')$ is also continuous function. Thus in place of function Z, we have introduced a function V which satisfies the same differential equations as X, and at interfaces between layers beneath the Earth's surface

$$V \quad \text{and} \quad \frac{1}{k^2}V'$$

are continous. It is obvious that the equations and the boundary conditions for X and V are similar. This parallelism permits us to find both these functions simultaneously, and then from Eq. (13.100), the function Z becomes determined.

13.4.3. Derivation of Recursive Relations for Functions *X* and *V*

In order to determine the components of the vector potential, using the same approach as in the case of the vertical magnetic dipole, we seek a solution for the function Q satisfying the equation

$$\frac{d^2Q}{dz^2}-a_p^2Q=0 \tag{13.103}$$

with the following boundary conditions between layers

$$Q_p=Q_{p+1}, \quad b_pQ'_p=b_{p+1}Q'_{p+1} \tag{13.104}$$

It is obvious that function Q corresponds to the functions X and V if $b_p=1$ and $1/k_p^2$, respectively. A solution of Eq. (13.103) in a layer identified with the index p is

$$Q_p=D_p\exp(-a_pz)+C_p\exp(a_pz) \tag{13.105}$$

Here,

$$a_p = \left(m^2 - k_p^2\right)^{1/2}$$

Let us make use of a tactic, developed by Lipskaya, in which the ratio of the function Q to its vertical derivative is introduced as follows:

$$\frac{Q_p}{Q_p'} = -\frac{R_p}{a_p} \tag{13.106}$$

In the uppermost layer,

$$R_1 = \frac{D_1 \exp(-a_1 z) + C_1 \exp(a_1 z)}{D_1 \exp(-a_1 z) - C_1 \exp(a_1 z)}$$

and if $z = 0$,

$$R_1(0) = \frac{D_1 + C_1}{D_1 - C_1},$$

while at $z = H_1$:

$$R_1(H_1) = \frac{D_1 \exp(-a_1 H_1) + C_1 \exp(a_1 H_1)}{D_1 \exp(-a_1 H_1) - C_1 \exp(a_1 H_1)}$$

In the second layer,

$$R_2 = \frac{D_2 \exp(-a_2 z) + C_2 \exp(a_2 z)}{D_2 \exp(-a_2 z) - C_2 \exp(a_2 z)}$$

In particular, if $z = H_1$,

$$R_2(H_1) = \frac{D_2 \exp(-a_2 H_1) + C_2 \exp(a_2 H_1)}{D_2 \exp(-a_2 H_1) - C_2 \exp(a_2 H_1)}$$

From the condition for continuity of Q and bQ as z approaches H_1, we have

$$D_1 \exp(-a_1 H_1) + C_1 \exp(a_1 H_1) = D_2 \exp(-a_2 H_1) + C_2 \exp(a_2 H_1)$$
$$a_1 b_1 [D_1 \exp(-a_1 H_1) - C_1 \exp(a_1 H_1)] = a_2 b_2 [D_2 \exp(-a_2 H_1) - C_2 \exp(a_2 H_1)]$$

Dividing the first of these equalities by the second, we have:

$$\frac{D_1 \exp(-a_1 H_1) + C_1 \exp(a_1 H_1)}{D_1 \exp(-a_1 H_1) - C_1 \exp(a_1 H_1)} = \frac{a_1 b_1}{a_2 b_2} \frac{D_2 \exp(-a_2 H_1) + C_2 \exp(a_2 H_1)}{D_2 \exp(-a_2 H_1) - C_1 \exp(a_2 H_1)} \tag{13.107}$$

Then we can express the function R_1 at the Earth's surface in terms of its value on the upper surface of the second layer:

$$R_1(0) = \coth\left[a_1 H_1 + \coth^{-1}\left(\frac{a_1 b_1}{a_2 b_2} R_1(H_1)\right)\right] \tag{13.108}$$

By mathematical induction, Eq. (13.108) can be generalized for any number of layers:

$$R_1(0) = \coth\left\{a_1 H_1 + \coth^{-1}\frac{a_1 b_1}{a_2 b_2}\coth\left[a_2 H_2 + \cdots + \coth^{-1}\left(\frac{a_{N-1} b_{N-1}}{a_N b_N} R_N(H_{N-1})\right)\right]\right\}$$

The condition at infinity requires the choice of the Q_N function in the form:

$$Q_N = D_N \exp(-a_N z)$$

and it gives $R_N = 1$. Therefore,

$$R_1 = \coth\left\{a_1 H_1 + \coth^{-1}\frac{a_1 b_1}{a_2 b_2}\coth\left[a_2 H_2 + \cdots \coth^{-1}\left(\frac{a_{N-1} b_{N-1}}{a_N b_N}\right)\right]\right\} \tag{13.109}$$

In this last expression, taking $a = n$ and $b = 1$, we obtain

$$\frac{X_1}{X_1'} = -\frac{R_x}{n_1} = -\frac{1}{n_1}\coth\left\{n_1 H_1 + \coth^{-1}\frac{n_1}{n_2}\coth\left[n_2 H_2 + \cdots \coth^{-1}\left(\frac{n_{N-1}}{n_N}\right)\right]\right\} \tag{13.110}$$

By analogy, if $a = n$ and $b = 1/k^2$, we have

$$\frac{V_1}{V_1'} = -\frac{R_z}{n_1}\coth\left\{n_1H_1 + \coth^{-1}\frac{n_1\rho_1}{n_2\rho_2}\coth\left[n_2H_2 + \cdots + \coth^{-1}\left(\frac{n_{N-1}\rho_{N-1}}{n_N\rho_N}\right)\right]\right\} \tag{13.111}$$

Now we are prepared to find formulas for both components of the vector potential.

13.4.4. Expression for the Component A_x

Determination of the functions X and V requires the application of appropriate boundary conditions at the Earth's surface. Unlike the case of a uniform half space, let us assume that the electric dipole is located in the upper half space. Then the electromagnetic field in the upper half space is usually represented as a sum of two fields; one being the electromagnetic field of the dipole in a uniform full space, and the other representing the secondary field because medium is not uniform. Thus, the complex amplitude of the component A_x can be written as

$$A_{x0}^* = \frac{\mu_0 Idx}{4\pi}\int_0^\infty\left[\frac{m}{n_0}\exp\left(-n_0|z+h_0|\right) + C_0\exp\left(n_0z\right)\right]J_0(mr)\mathrm{d}m \tag{13.112}$$

In the first layer, this component has the following form:

$$A_{x1}^* = \frac{\mu_0 I\mathrm{d}x}{4\pi}\int_0^\infty\left[C_1\exp\left(n_1z\right) + D_1\exp\left(-n_1z\right)\right]J_0(mr)\mathrm{d}m \tag{13.113}$$

From continuity of A_x and $\partial A_x/\partial z$ at the Earth's surface ($z = 0$), we have

$$\frac{m}{n_0}\exp\left(-n_0h_0\right) + C_0 = C_1 + D_1 = X_1$$

and

$$-m\exp\left(-n_0h_0\right) + n_0C_0 = n_1(C_1 - D_1) = X_1' \tag{13.114}$$

Solution of this system with respect to C_0 gives

$$C_0 = \frac{\frac{m}{n_0}\frac{X_1'}{X_1} + m}{n_0 - \frac{X_1'}{X_1}} \exp(-n_0 h_0) \tag{13.115}$$

Thus at the Earth's surface,

$$\exp(-n_0 h_0)\frac{m}{n_0} + C_0 = \frac{2m \exp(-n_0 h_0)}{n_0 - X_1'/X_1}$$

In accordance with Eq. (13.110), the latter can be written as

$$X_1 = \frac{2m \exp(-n_0 h_0)}{n_0 + n_1/R_x}$$

and

$$X_1' = -\frac{n_1}{R_x} X_1 = -\frac{2m \exp(-n_0 h_0)}{n_0 + n_1/R_x} \cdot \frac{n_1}{R_x} \tag{13.116}$$

Thus, expression for the complex amplitude of the function A_x^* is

$$A_{x0}^* = \frac{\mu_0 I \mathrm{d}x}{2\pi} \int_0 \frac{m \exp(-n_0 h_0)}{n_0 + n_1/R_x} J_0(mr)\mathrm{d}m \tag{13.117}$$

As follows from the second equation of the set (Eq. (13.114)), we also have

$$\frac{\partial A_x^*}{\partial z} = -\frac{\mu_0 I \mathrm{d}x}{2\pi} \int_0^\infty \frac{n_1}{R_x} \frac{m \exp(n_0 h_0)}{n_0 + n_1/R_x} J_0(mr)\mathrm{d}m \tag{13.118}$$

13.4.5. Expressions for Functions V_1 and V_1'

First consider the behavior of the function V at the Earth's surface. As was shown above,

$$X_0(0) = C_0 + \frac{m}{n_0} \exp(-n_0 h_0), \quad X_0'(0) = n_0 C_0 - m \exp(-n_0 h_0)$$

Hence,

$$X_0'(0) = n_0 X_0(0) - 2m \exp(-n_0 h_0) \tag{13.119}$$

By definition,

$$V = Z + \frac{X'}{m^2}$$

Thus,

$$V_0 = Z_0 + \frac{n_0 X_0 - 2m \exp(-n_0 h_0)}{m^2} \qquad V_0' = Z_0' + \frac{n_0^2 X_0}{m^2}$$

Note that in deriving the last equality, Eq. (13.93) was used. Inasmuch as

$$Z_0(z) = \alpha_0 \exp(n_0 z)$$

at the Earth's surface, we have

$$Z_0' = n_0 Z_0$$

Taking into account continuity of $\frac{V'}{k^2}$, we can write

$$V_0' = \frac{V_1' k_0^2}{k_1^2}$$

These relationships permit us to obtain next formula:

$$\frac{k_0^2}{k_1^2 n_0} V_1' - V_1 = \frac{2}{m} \exp(-n_0 h_0) \tag{13.120}$$

Using Eq. (13.111), we obtain a set of equations for determining V_1' and V_1:

$$\frac{k_0^2}{k_1^2 n_0} V_1' - V_1 = \frac{2}{m} \exp(-n_0 h_0)$$
$$\frac{V_1}{V_1'} = -\frac{R_z}{n_1} \tag{13.121}$$

Therefore,

$$V_1 = -\frac{R_x}{n_1} \frac{2 \exp(-n_0 h_0)}{m\left(k_0^2/k_1^2 n_0 + R_z/n_1\right)} \tag{13.122}$$

and

$$V_1' = \frac{2 \exp(-n_0 h_0)}{m\left(k_0^2/k_1^2 n_0 + R_z/n_1\right)} \tag{13.123}$$

13.4.6. Expressions for A_z^* and U^*

Using these expressions for X_1, X_1', V_1, and V_1' along with Eq. (13.120), we obtain

$$A_z^* = \frac{\mu_0 I \mathrm{d}x}{2\pi}\frac{\partial}{\partial x}\int_0^\infty \left[\frac{\exp(-n_0 h_0)}{m}\cdot\frac{n_1/R_x}{n_0 + n_1/R_x} - \frac{R_z}{n_1}\frac{\exp(-n_0 h_0)}{m\left(k_0^2/k_1^2 n_0 + R_z/n_1\right)}\right] J_0(mr)\mathrm{d}m \tag{13.124}$$

Also, by definition,

$$U^* = -\frac{1}{\gamma_1}\left(\frac{\partial A_x^*}{\partial x} + \frac{\partial A_z^*}{\partial z}\right) = -\frac{1}{\gamma_1}\frac{\partial}{\partial x}\left(A_x^* + W'\right)$$

$$= \frac{iI\omega\mu_0 \mathrm{d}x}{4\pi k_1^2}\frac{\partial}{\partial x}\int_0^\infty \left(X_1 + Z_1'\right) J_0(mr)\mathrm{d}m$$

$$= -\frac{iI\omega\mu_0 \mathrm{d}x}{4\pi k_1^2}\frac{\partial}{\partial x}\int_0^\infty \left(X_1 + V_1' - \frac{X_1''}{m^2}\right) J_0(mr)\mathrm{d}m$$

$$= -\frac{iI\omega\mu_0 \mathrm{d}x}{4\pi k_1^2}\frac{\partial}{\partial x}\int_0^\infty \left(X_1 + V_1' - \frac{n_1^2}{m^2}X_1\right) J_0(mr)\mathrm{d}m$$

$$= -\frac{iI\omega\mu_0 \mathrm{d}x}{4\pi k_1^2}\frac{\partial}{\partial x}\int_0^\infty \left(V_1' + \frac{k_1^2}{m^2}X_1\right) J_0(mr)\mathrm{d}m$$

$$= \frac{iI\omega\mu_0 \mathrm{d}x}{2\pi k_1^2}\cos\varphi\int_0^\infty \left[\frac{1}{\left(k_0^2/k_1^2 n_0 + R_z/n_1\right)} + \frac{k_1^2}{n_0 + n_1/R_x}\right]\exp(-n_0 h_0) J_1(mr)\mathrm{d}m \tag{13.125}$$

13.4.7. Potentials on the Earth's Surface when $k_0 = 0$

Letting $k_0 = 0$, $n_0 = m$, and $h_0 = 0$, we obtain

$$A_x^* = \frac{\mu_0 I \mathrm{d}x}{2\pi} \int_0^\infty \frac{m}{m + n_1/R_x} J_0(mr)\mathrm{d}m \tag{13.126}$$

$$\frac{\partial A_x^*}{\partial z} = -\frac{\mu_0 I \mathrm{d}x}{2\pi} \int_0^\infty \frac{n_1}{R_x} \frac{m}{m + n_1/R_x} J_0(mr)\mathrm{d}m \tag{13.127}$$

$$A_z^* = \frac{\mu_0 I \mathrm{d}x}{2\pi} \frac{\partial}{\partial x} \int_0^\infty \frac{1}{m} \left[\frac{n_1}{R_x} \frac{1}{m + n_1/R_x} - 1 \right] J_0(mr)\mathrm{d}m$$

$$= -\frac{\mu_0 I \mathrm{d}x}{2\pi} \frac{\partial}{\partial x} \int_0^\infty \frac{1}{m + n_1/R_x} J_0(mr)\mathrm{d}m$$

Thus,

$$A_z^* = \frac{\mu_0 I \mathrm{d}x}{2\pi} \cos\varphi \int_0^\infty \frac{1}{m + n_1/R_x} J_1(mr)\mathrm{d}m \tag{13.128}$$

and

$$U^* = -\frac{\operatorname{div} \mathbf{A}^*}{\gamma_1}$$

$$= \frac{\rho_1 I}{2\pi} \mathrm{d}x \cos\varphi \int_0^\infty \frac{n_1}{R_z} J_1(mr)\mathrm{d}m + \frac{i\omega\mu_0 I \mathrm{d}x}{2\pi} \cos\varphi \int_0^\infty \frac{J_1(mr)\mathrm{d}m}{m + n_1/R_x} \tag{13.129}$$

where

$$R_x = \coth\left[n_1 H_1 + \coth^{-1}\left\{ \frac{n_1}{n_2} \coth\left(n_2 H_2 + \cdots + \coth^{-1} \frac{n_{N-1}}{n_N} \right) \right\} \right] \tag{13.130}$$

and

$$R_z = \coth\left[n_1 H_1 + \coth^{-1}\left\{\frac{n_1}{n_2}\frac{\rho_1}{\rho_2}\coth(n_2 H_2 + \cdots + \coth^{-1}\frac{n_{N-1}}{n_N}\frac{\rho_{N-1}}{\rho_N}\right)\right\}\right] \tag{13.131}$$

13.4.8. Expressions for the Electric and Magnetic Fields

Taking into account that in the cylindrical system of coordinates,

$$A_r^* = A_x^* \cos\varphi \quad \text{and} \quad A_\varphi^* = -A_x^* \sin\varphi$$

we have

$$\begin{aligned} B_r^* &= \frac{\mu_0 I dx}{2\pi r}\left[-\sin\varphi \int_0^\infty \frac{m}{m+n_1/R_x} J_1(mr)dm \right. \\ &\quad \left. - r\sin\varphi \int_0^\infty \frac{n_1}{R_x}\frac{m}{m+n_1/R_x} J_0(mr)dm\right] \\ &= -\frac{\mu_0 I dx}{2\pi r}\sin\varphi\left[\int_0^\infty \frac{m}{m+n_1/R_x} J_1(mr)dm \right. \\ &\quad \left. + r\int_0^\infty \frac{n_1}{R_x}\frac{m}{m+n_1/R_x} J_0(mr)dm\right] \end{aligned} \tag{13.132}$$

$$B_\varphi^* = -\frac{\mu_0 I dx}{2\pi}\cos\varphi \int_0^\infty \frac{n_1}{R_x}\frac{m}{m+n_1/R_x} J_0(mr)dm - \frac{\mu_0 I dx}{2\pi}\cos\varphi$$

$$\cdot\left[\int_0^\infty \frac{n_1}{R_x}\frac{m}{m+n_1/R_x} J_0(mr)dm + \int_0^\infty \frac{m^2}{m+n_1/R_x} J_1'(mr)dm\right]$$

Inasmuch as

$$J_1'(mr) = J_0(mr) - \frac{1}{mr} J_1(mr)$$

we have

$$\int_0^\infty \frac{n_1}{R_x}\frac{m}{m+n_1/R_x}J_0(mr)\mathrm{d}m+\int_0^\infty \frac{m^2}{m+n_1/R_x}J_0(mr)\mathrm{d}m$$

$$-\frac{1}{r}\int_0^\infty \frac{m}{m+n_1/R_x}J_1(mr)\mathrm{d}m$$

$$=\int_0^\infty \frac{m}{m+n_1/R_x}\left(m+n_1/R_x\right)J_0(mr)\mathrm{d}m-\frac{1}{r}\int_0^\infty \frac{m}{m+n_1/R_x}J_1(mr)\mathrm{d}m$$

$$=-\frac{1}{r}\int_0^\infty \frac{m}{m+n_1/R_x}J_1(mr)\mathrm{d}m \tag{13.133}$$

For the vertical component B_z, we have

$$B_z^*=\frac{\mu_0 I\mathrm{d}x}{2\pi r}\sin\varphi\int_0^\infty \frac{m^2}{m+n_1/R_x}J_0(mr)\mathrm{d}m \tag{13.134}$$

From Eqs (13.18) and (13.20), we have

$$E_r^*=i\omega A_r^*-\frac{1}{\mu_0}\frac{\partial U^*}{\partial r}$$

or

$$E_r^*=\frac{I\mathrm{d}x}{2\pi}\cos\varphi\left[i\omega\mu_0\int_0^\infty \frac{m}{m+n_1/R_x}J_0(mr)\mathrm{d}m-\rho_1\int_0^\infty \frac{n_1 m}{R_z}J_1'(mr)\mathrm{d}m\right.$$
$$\left.-\,i\omega\mu_0\int_0^\infty \frac{mJ_1'(mr)}{m+n_1/R_x}\mathrm{d}m\right]$$

or

$$E_r^* = \frac{I\mathrm{d}x}{2\pi}\cos\varphi\left[\frac{i\omega\mu_0}{r}\int_0^\infty \frac{J_1(mr)}{m+n_1/R_x}\mathrm{d}m\right.$$

$$\left.-\rho_1\int_0^\infty \frac{n_1 m}{R_z}J_0(mr)\mathrm{d}m + \frac{\rho_1}{r}\int_0^\infty \frac{n_1}{R_z}J_1(mr)\mathrm{d}m\right] \tag{13.135}$$

and

$$E_\varphi^* = i\omega A_\varphi^* - \frac{1}{r\mu_0}\frac{\partial U^*}{\partial\varphi}$$

or

$$E_\varphi^* = \frac{I\mathrm{d}x}{2\pi}\left[-i\omega\mu_0\sin\varphi\int_0^\infty \frac{m}{m+n_1/R_x}J_0(mr)\mathrm{d}m\right.$$

$$\left.+\frac{\rho_1}{r}\sin\varphi\int_0^\infty \frac{n_1}{R_z}J_1(mr)\mathrm{d}m + \frac{i\omega\mu_0}{r}\sin\varphi\int_0^\infty \frac{J_1(mr)}{m+n_1/R_x}\mathrm{d}m\right]$$

$$= \frac{I\mathrm{d}x}{2\pi}\sin\varphi\left[\frac{\rho_1}{r}\int_0^\infty \frac{n_1}{R_z}J_1(mr)\mathrm{d}m - i\omega\mu_0\int_0^\infty \frac{mJ_0(mr)}{m+n_1/R_x}\mathrm{d}m\right.$$

$$\left.+\frac{i\omega\mu_0}{r}\int_0^\infty \frac{J_1(mr)}{m+n_1/R_x}\mathrm{d}m\right] \tag{13.136}$$

Thus, expressions characterizing the various components of the electromagnetic field from a horizontal electric dipole situated on the surface of a horizontally stratified earth are written in the form:

$$B_r^* = -\frac{\mu_0 I\mathrm{d}x}{2\pi r}\sin\varphi\left[\int_0^\infty \frac{m}{m+n_1/R_x}J_1(mr)\mathrm{d}m\right.$$

$$\left.+r\int_0^\infty \frac{n_1}{R_x}\,\frac{m}{m+n_1/R_x}J_0(mr)\mathrm{d}m\right]$$

$$B_\varphi^* = \frac{\mu_0 I \mathrm{d}x}{2\pi r} \cos\varphi \int_0^\infty \frac{m}{m + n_1/R_x} J_1(mr)\mathrm{d}m \tag{13.137}$$

$$B_z^* = \frac{\mu_0 I \mathrm{d}x}{2\pi} \sin\varphi \int_0^\infty \frac{m^2}{m + n_1/R_x} J_1(mr)\mathrm{d}m$$

and

$$E_r^* = \frac{I\mathrm{d}x}{2\pi} \cos\varphi \left[\frac{i\omega\mu_0}{r} \int_0^\infty \frac{J_1(mr)}{m + n_1/R_x} \mathrm{d}m - \rho_1 \int_0^\infty \frac{n_1 m}{R_z} J_0(mr)\mathrm{d}m + \frac{\rho_1}{r} \int_0^\infty \frac{n_1}{R_z} J_1(mr)\mathrm{d}m \right]$$

$$E_\varphi^* = \frac{I\mathrm{d}x}{2\pi} \sin\varphi \left[\frac{\rho_1}{r} \int_0^\infty \frac{n_1}{R_z} J_1(mr)\mathrm{d}m - i\omega\mu_0 \int_0^\infty \frac{m}{m + n_1/R_x} J_0(mr)\mathrm{d}m + \frac{i\omega\mu_0}{r} \int_0^\infty \frac{J_1(mr)}{m + n_1/R_x} \mathrm{d}m \right] \tag{13.138}$$

Using the appropriate transform relationships between cylindrical and Cartesian coordinates, we have

$$E_x = E_r \cos\varphi - E_\varphi \sin\varphi \quad B_x = B_r \cos\varphi - B_\varphi \sin\varphi$$

and

$$E_y = E_r \sin\varphi + E_\varphi \cos\varphi \quad B_y = B_r \sin\varphi + B_\varphi \cos\varphi$$

Now let us examine these expressions for the electromagnetic field components. First of all, the functions R_x and R_z differ from each other in important ways. The function R_x was already derived in describing the electromagnetic field components caused by a vertical magnetic dipole source situated over a horizontally stratified medium, and it depends on the thickness of layers normalized by the appropriate skin depth, as well as a variable of integration. For example, for a uniform half space,

$$R_{1x} = 1$$

for a two-layer sequence,

$$R_{2x} = \coth\left[n_1 H_1 + \coth^{-1}\left(\frac{n_1}{n_2}\right)\right] \tag{13.139}$$

and for three-layer medium,

$$R_{3x} = \coth\left\{n_1 H_1 + \coth^{-1}\left(\frac{n_1}{n_2}\right)\coth\left[n_2 H_2 + \coth^{-1}\frac{n_2}{n_3}\right]\right\}$$

In contrast, in the case of the electric dipole, both functions R_z and R_x must be used in describing the electromagnetic field components. It is essential that function R_z depends on the ratio of resistivity, as well as the ratio of H/δ and m. For example, for a uniform half space,

$$R_z = 1,$$

for two-layer medium,

$$R_z = \coth\left[n_1 H_1 + \coth^{-1}\frac{\rho_1}{\rho_2}\frac{n_1}{n_2}\right] \tag{13.140}$$

and a three-layer sequence,

$$R_3 = \coth\left\{n_1 H_1 + \coth^{-1}\frac{\rho_1}{\rho_2}\frac{n_1}{n_2}\coth\left[n_2 H_2 + \coth^{-1}\frac{\rho_2}{\rho_3}\frac{n_2}{n_3}\right]\right\}$$

Considering Eq. (13.137), we see that expressions for the magnetic field on the Earth's surface do not contain the function R_z. For example, this means that a thin resistive layer within the medium has a minimal response when the vertical or horizontal components of the magnetic field are measured. The same was shown earlier for the case of a vertical magnetic dipole source. In contrast, the electric field components contain terms with both R functions, and as a result, they are more sensitive to the presence of a resistive layer within the sequence. In comparing the expressions for the electric field about the vertical magnetic dipole source E_φ, and the expression for the vertical component of the magnetic field about a horizontal electric dipole source, we see that they are the same except for a multiplying constant. Thus, the behavior of the vortex electric field E_φ, generated by a vertical magnetic dipole source, provides information on the behavior of the vertical magnetic field for the case in which the source is a grounded electric dipole. In other words, we again demonstrated the principle of reciprocity. Also it is proper to mention that dependence of the electromagnetic field components on the Earth's surface to the azimuthal angle φ is the same as that for the stationary field caused by the direct current.

13.5. TRANSITION TO THE STATIONARY FIELD

Now we consider some general features of the behavior of the electromagnetic field for small and large values of the induction number. Let us start with the case of the stationary field when the frequency is zero. In this case, the function R_x, regardless of the number of layers, has the value unity. In fact, the last term on the right-hand side of the expression for R_x is

$$\coth^{-1}\left(\frac{n_{N-1}}{n_N}\right)$$

and in the limit as frequency goes to zero, we have

$$\coth^{-1} 1 \to \infty$$

As an example,

$$R_{2x} = \coth\left[n_1 H_1 + \coth^{-1}\left(\frac{n_1}{n_2}\right)\right] = \coth\left(mH_1 + \coth^{-1} 1\right) = 1$$

By analogy,

$$R_{2n} \to 1 \quad \text{if} \quad \omega \to 0$$

Therefore, from Eq. (13.137),

$$B_r = -\frac{\mu_0 I dx}{4\pi r}\sin\varphi\left[\int_0^\infty J_1(mr)dm + r\int_0^\infty mJ_0(mr)dm\right] = -\frac{\mu_0 I dx}{4\pi r^2}\sin\varphi$$

$$B_\varphi = \frac{\mu_0 I dx}{4\pi r}\cos\varphi\int_0^\infty J_1(mr)dm = \frac{\mu_0 I dx}{4\pi r^2}\cos\varphi$$

$$B_z = \frac{\mu_0 I dx}{4\pi}\sin\varphi\int_0^\infty mJ_1(mr)dm = \frac{\mu_0 I dx}{4\pi r^2}\sin\varphi \tag{13.141}$$

These are precisely the same as the expressions for stationary magnetic field on the surface of a uniform half space. Thus, as has been noted previously, the magnetic field above the Earth's surface does not reflect changes in resistivity within a horizontally stratified earth when the frequency is zero. In accord with Eq. (13.138), the asymptotic expressions for the electric field are

$$E_r = \frac{\rho_1 I dx}{2\pi}\cos\varphi\left[-\int\limits_0^\infty \frac{m^2}{R_z}J_0(mr)dm + \frac{1}{r}\int\limits_0^\infty \frac{m}{R_z}J_1(mr)dm\right] \quad (13.142)$$

where

$$R_z = \coth\left\{mH_1 + \coth^{-1}\left(\frac{\rho_1}{\rho_2}\right)\coth\left[mH_2 + \cdots + \coth^{-1}\left(\frac{\rho_{N-1}}{\rho_N}\right)\right]\right\} \quad (13.143)$$

In the case of a uniform half space,

$$R_z = 1$$

For two-layer sequence,

$$R_z = \coth\left[mH_1 + \coth^{-1}\left(\frac{\rho_1}{\rho_2}\right)\right],$$

and for three-layer sequence,

$$R_z = \coth\left\{mH_1 + \coth^{-1}\left(\frac{\rho_1}{\rho_2}\right)\coth\left[mH_2 + \coth^{-1}\left(\frac{\rho_2}{\rho_3}\right)\right]\right\} \quad (13.144)$$

Equations (13.142)–(13.144) are basic to the theory of geometric soundings in a horizontally stratified medium for the dipole axial array. In this case, as is widely known, the depth of investigation is determined solely by the distance between the dipole source and the observation point. To show the effect of the separation r, consider a two-layer sequence with a conducting basement $r \to \infty$. Transforming to a new variable $x = mr$, we have the following expression for the radial component E_r:

$$E_r = \frac{\rho_1 I dx}{2\pi r^3}\cos\varphi\left[-\int\limits_0^\infty \frac{x^2}{R_z}J_0(x)dx + \int\limits_0^\infty \frac{x}{R_z}J_1(x)dx\right] \quad (13.145)$$

$$\text{Here}\quad R_z = \coth\left[x\left(\frac{H_1}{r}\right) + \coth^{-1}\left(\frac{\rho_1}{\rho_2}\right)\right] \tag{13.146}$$

As the separation r decreases, the function R_z tends to unity, and we obtain

$$E_r = \frac{\rho_1 I \mathrm{d}x}{2\pi r^3}\left[-\int_0^\infty x^2 J_0(x)\mathrm{d}x + \int_0^\infty xJ_1(x)\mathrm{d}x\right]\cos\varphi = \frac{\rho_1 I\mathrm{d}x}{\pi r^3}\cos\varphi$$

By analogy,

$$E_\varphi = \frac{\rho_1 I\mathrm{d}x}{2\pi r^3}\sin\varphi$$

We see that when the separation is significantly smaller than the layer thickness, the field components are defined by the charges at the electrode surfaces and do not depend on the resistivity of the second layer. From Eq. (13.146), it follows that when the separation r is increased, the limit for the function R_z is

$$R_z = \frac{\rho_1}{\rho_2}$$

and therefore,

$$E_r = \frac{\rho_2 I\mathrm{d}x}{\pi r^3}\cos\varphi, \quad E_\varphi = \frac{\rho_2 I\mathrm{d}x}{2\pi r^3}\sin\varphi \quad \text{if} \ \ r \to \infty$$

The field is defined by the resistivity of the second layer only. Of course, in general, in n-layered medium, ρ_2 is replaced by ρ_N. This behavior is the remarkable feature of the stationary field and basic of geometric soundings. Now suppose that in the case of two-layer medium, $\rho_2 = \infty$; that is, the basement is insulating. Writing R_z in the form

$$R_z = \frac{1 + \rho_1/\rho_2 \coth mH_1}{\rho_1/\rho_2 + \coth mH_1}$$

we see that

$$R_z = \tan mH_1 \quad \text{or} \quad R_z = \tan m\frac{x}{r}$$

Substituting this last result into Eq. (13.145) and making the substitution xH_1/r in place of $\tan(xH_1/r)$ as r tends to infinity, we have

$$E_r = \frac{\rho_1 I\mathrm{d}x}{2\pi r^2 H_1}\cos\varphi\left[-\int_0^\infty xJ_0(x)\mathrm{d}x + \int_0^\infty J_1(x)\mathrm{d}x\right] = \frac{I\mathrm{d}x}{2\pi r^2 S_1}\cos\varphi \tag{13.147}$$

By analogy,

$$E_\varphi = \frac{I\mathrm{d}x}{2\pi r^2 S_1}\sin\varphi$$

Thus in the far zone, the electric field is inversely proportional to the longitudinal conductance S_1. This behavior of the electromagnetic field for a two-layer sequence is readily generalized to a case of N layers. It should be obvious that the components of the stationary electric and magnetic fields describe the respective leading terms of the series expansion for the range of small induction number, when the skin depth is much greater than the separation r, that is, over the low-frequency part of the spectrum. As in the case of the vertical magnetic dipole, the following term of the series describes the quadrature component of the electric and magnetic fields, and it is directly proportional to ω. In fact, a change of the stationary magnetic field with time causes the vortex electric field which is shifted by 90° with respect to the leading term of this field and proportional to frequency. Unlike the corresponding term of the field generated by the magnetic dipole, this part of the field cannot be represented through geometrical factors because of the presence of the surface charges. Inasmuch as this vortex part of the field gives rise to currents, the second term of the series describing the magnetic field also characterizes the quadrature component and it is proportional to ω. To obtain higher order terms of this series, one can use the methods described in Chapter 9. In particular, the series will contain integer and fractional powers of ω along with logarithmic terms. Using the methods developed in that chapter, it is possible to show that only the initial portion of the range of integration is responsible for the fractional powers of ω and logarithmic terms, and corresponding leading term of this part of the spectrum is a function of the conductivity of the lower medium only.

13.6. THE RANGE OF LARGE INDUCTION NUMBER (WAVE ZONE)

Next consider the field for large values of the induction number when the separation becomes greater than the wave length. We will start from expressions for the components of the vector potential and its derivatives. Assume that the parameter r/λ is greater than unity, so that the integral is defined by very small values of m. This allows us, as in the case of the vertical magnetic dipole, to replace in Eqs (13.126)–(13.129) n_1 by ik, which gives

$$A_x^* = \frac{\mu_0 I \mathrm{d}x}{2\pi}\left(\frac{R_x(0)}{-ik_1}\right)\int_0^\infty \frac{m}{1+\left(R_x(0)/-ik_1\right)m} J_0(mr)\mathrm{d}m,$$

since $\mathrm{Re}\,m_i > 0$. Thus,

$$A_x^* = -\frac{\mu_0 I \mathrm{d}x}{2\pi}\frac{R_x(0)}{ik_1}\left[\int_0^\infty mJ_0(mr)\mathrm{d}m + \frac{R_x(0)}{ik_1}\int_0^\infty m^2 J_0(mr)\mathrm{d}m\right]$$
$$= -\frac{\mu_0 I \mathrm{d}x}{2\pi k_1^2}\frac{R_x^2(0)}{r^3}, \tag{13.148}$$

where $R_x(0) = \lim R_x \quad \text{if} \quad m \to 0$.

Correspondingly, for the vertical component,

$$A_z^* = -\frac{\mu_0 I \mathrm{d}x}{2\pi}\frac{R_x(0)}{ik_1}\cos\varphi \int_0^\infty mJ_1(mr)\mathrm{d}m = -\frac{\mu_0 I \mathrm{d}x}{2\pi}\frac{R_x(0)}{ik_1 r^2}\cos\varphi \tag{13.149}$$

Thus in the wave zone, the vertical component of the vector potential is dominant:

$$A_z^* \gg A_x^*$$

By definition,

$$\frac{\partial A_x^*}{\partial z} = -\frac{\mu_0 I \mathrm{d}x}{2\pi}\int_0^\infty \frac{m}{1-\left(R_x(0)/ik_1\right)m} J_0(mr)\mathrm{d}m$$
$$= -\frac{\mu_0 I \mathrm{d}x R_x(0)}{2\pi i k_1}\int_0^\infty m^2 J_0(mr)\mathrm{d}m$$

or

$$\frac{\partial A_x^*}{\partial z} = \frac{\mu_0 I \mathrm{d}x}{2\pi i k_1 r^3} R_x(0) \tag{13.150}$$

It is readily seen that

$$R_x(0)=\frac{1}{R_z(0)} \tag{13.151}$$

For example, for a two-layer sequence,

$$R_x(0)=\coth\left[-ik_1H_1+\coth^{-1}\left(\frac{\rho_2}{\rho_1}\right)^{1/2}\right]$$

and

$$R_z(0)=\coth\left[-ik_1H_1+\coth\left(\frac{\rho_1}{\rho_2}\right)^{1/2}\right]$$

Inasmuch as

$$\tan\mathrm{h}\left(a+\tan\mathrm{h}^{-1}\,b\right)=\coth\left(a+\coth^{-1}\,b\right)$$

we have the equality expressed in Eq. (13.151). Taking into account expression for U and Eq. (13.151), we have

$$U=\frac{\rho_1 I\mathrm{d}x}{2\pi}\cos\varphi\left[\frac{1}{R_z(0)}\int\limits_0^\infty n_1J_1(mr)\mathrm{d}m-\frac{k_1^2R_x(0)}{ik_1}\int\limits_0^\infty\frac{J_1(mr)}{1-R_x(0)m/_{ik_1}}\mathrm{d}m\right]$$

$$=\frac{\rho_1 I\mathrm{d}x}{2\pi}\cos\varphi\left[-i_1k_1R_x(0)\int\limits_0^\infty J_1(mr)\mathrm{d}m-ik_1R_x(0)\int\limits_0^\infty J_1(mr)\mathrm{d}m\right.$$

$$\left.+R_x^2(0)\int\limits_0^\infty mJ_1(mr)\mathrm{d}m\right]$$

Therefore,

$$U=\frac{\rho_1 I\mathrm{d}x}{2\pi}R_x^2(0)\frac{\cos\varphi}{r^2} \tag{13.152}$$

Now we can write expressions for the components of the electromagnetic field in the wave zone:

$$B_r^* = \frac{\mu_0 I dx}{ik_1 \pi} \frac{R_x(0)}{r^3} \sin \varphi, \quad B_\varphi^* = \frac{\mu_0 I dx R_x(0)}{2\pi i k_1} \frac{\cos \varphi}{r^3},$$

$$B_z^* = -\frac{3\mu_0 I dx}{2\pi k_1^2 r^4} R_x^2(0) \sin \varphi$$

and

$$E_r^* = \frac{\rho_1 I dx}{2\pi r^3} R_x^2(0) \cos \varphi, \quad E_\varphi^* = \frac{\rho_1 I dx}{\pi r^3} R_x^2(0) \sin \varphi \tag{13.153}$$

Thus, as we should expect, the horizontal component of the magnetic field is dominant over the vertical component while the former is more sensitive to a change in the electrical properties in the medium. Also the ratio of the horizontal components of the electric and magnetic fields is numerically equal to the impedance of the plane wave, but in reality, we observe again as in the case of the magnetic dipole source the quasi-plane wave propagating downward.

13.7. THE TRANSIENT FIELD FROM THE ELECTRIC DIPOLE SOURCE ON THE SURFACE OF A UNIFORM HALF SPACE

Earlier, the behavior of time-domain electromagnetic field caused by a vertical magnetic dipole source was considered in some detail. Now we make use of the results of the analysis of the fields of a current dipole source at the frequency domain and, applying the Fourier transform, investigate the time-domain responses for the electric dipole over a uniform half space and a two-layer medium.

13.7.1. Equations of the Field

Proceeding from expressions for the field in the frequency domain and using the technique of deformation for the path of integration, it is possible to obtain the following formulas of the transient electric field of the electric dipole when the current in the dipole is turned off:

$$E_x^-(t) = \frac{2p_0}{4\pi\varepsilon_0 r^3} e_x \cos\varphi$$

$$B_x^-(t) = \frac{\mu_0 I\mathrm{d}x}{4\pi r^2} b_x \sin 2\varphi, \quad B_y^-(t) = -\frac{\mu_0 I\mathrm{d}x}{4\pi r^2} b_y \quad B_z^-(t) = \frac{\mu_0 I\mathrm{d}x}{4\pi r^2} b_z \sin\varphi \tag{13.154}$$

Here,

$$p_0 = \varepsilon_0 \rho I\mathrm{d}x,$$

$$e_x = \Phi(u) - \frac{2}{\pi^{1/2}} u \exp(-u^2) \tag{13.155}$$

$$u = \frac{\pi 2^{1/2} r}{\tau}, \quad \tau = 2\pi\left(\frac{2t}{\gamma\mu_0}\right)^{1/2} \tag{13.156}$$

and

$$\Phi(u) = \frac{2}{\pi^{1/2}} \int_0^u \exp(-x^2)\,\mathrm{d}x \tag{13.157}$$

Also,

$$b_x = \exp\left(-\frac{u^2}{2}\right)\left[I_0\left(\frac{u^2}{2}\right) + 2I_1\left(\frac{u^2}{2}\right)\right] - 1 \tag{13.158}$$

$$b_y = \left\{\left[I_0\left(\frac{u^2}{2}\right) + 2I_1\left(\frac{u^2}{2}\right)\right]\cos 2\varphi - I_1\left(\frac{u^2}{2}\right)\right\}\exp\left(-\frac{u^2}{2}\right) - \cos 2\varphi \tag{13.159}$$

and

$$b_z = \frac{3}{\pi^{1/2}} \frac{\exp(-u^2)}{u} + \left(1 - \frac{3}{2u^2}\right)\Phi(u), \tag{13.160}$$

where $I_0(x)$ and $I_1(x)$ are modified Bessel functions of the order 0 and 1.

13.7.2. The Early Stage

First suppose that the ratio τ/r goes to zero. Then making use of asymptotic expansions for the functions $\Phi(u)$, $I_0(u)$, and $I_1(u)$, we have

$$
\begin{aligned}
E_x^- &= \frac{p}{2\pi\varepsilon_0 r^3}, \quad E_y^- = 0 \\
B_x^- &= -\frac{\mu_0 I dx}{4\pi r^2}\left[1 - \frac{6t^{1/2}}{r(\pi\gamma\mu_0)^{1/2}}\right]\sin 2\varphi \\
B_y^- &= \frac{\mu_0 I dx}{4\pi r^2}\left[\cos 2\varphi - \frac{2(3\cos 2\varphi - 1)}{r(\pi\gamma\mu_0)^{1/2}}t^{1/2}\right] \\
B_z^- &= \frac{\mu_0 I dx}{4\pi r^2}\left[1 - \frac{6t}{r^2\mu_0\gamma}\right]\sin\varphi \quad \text{if} \quad \frac{\tau}{r} < 2
\end{aligned}
\qquad (13.161)
$$

Thus at the first moment when the current is turned off, the charges on the electrode surface and currents in a conducting medium remain the same as in the case of the time-invariant field.

13.7.3. The Late Stage

Next we consider the late stage of the transient field when the parameter τ/r is significantly greater than unity. Expanding the error function and the Bessel function in terms of small values for the parameter u,

$$
\Phi(u) \approx \frac{2}{\pi^{1/2}}\left(u - \frac{u^3}{3} + \frac{u^5}{10}\right),
$$

$$
I_0(u) \approx 1 + \frac{u^2}{4} + \frac{u^4}{10}, \quad I_1(u) \approx \frac{u^2}{2} + \frac{u^3}{16}
$$

we obtain

$$
E_x^- = \frac{I dx}{12\pi^{3/2}}\frac{\gamma^{1/2}\mu_0^{3/2}}{t^{3/2}}, \quad E_y^- = 0
$$

and

$$
B_x^- = -\frac{\mu_0 I dx \gamma^2\mu_0^2 r^2}{1024\pi t^2}\sin 2\varphi, \qquad (13.162)
$$

$$
B_y = \frac{\mu_0 I dx \gamma\mu_0}{64\pi t}, \quad B_z = \frac{\mu_0 I dx \gamma^{3/2}\mu_0^{3/2} r}{60\pi^{3/2}t^{3/2}}\sin\varphi \quad \text{if} \quad \frac{\tau}{r} > 20
$$

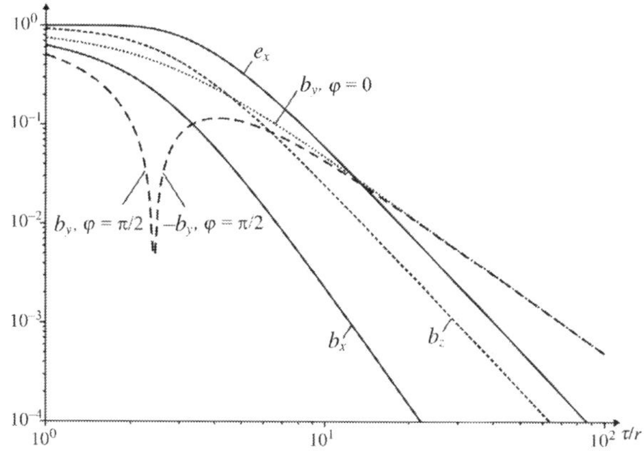

Figure 13.6 Behavior of transient responses of the electric and magnetic fields.

It is obvious that as in the case of the magnetic dipole source, we observe the main features of the diffusion of the field magnitude. At the same time, the electric field at the late stage is less sensitive to a change in the conductivity of the medium than the electric field generated by a magnetic dipole source. Behavior of the transient responses of the electric and magnetic fields are shown in Figure 13.6.

13.8. TRANSIENT FIELD ON THE SURFACE OF TWO-LAYER MEDIUM

13.8.1. Relationship between Fields Caused by Current when it is Turned off and Turned on

Before we describe some features of a transient field on the surface of a two-layer medium, it is reasonable to discuss the relationship between fields observed when the dipole current is switched on and when it is turned off. As was shown earlier,

$$E^{-}(t) = \frac{2}{\pi} \int_0^\infty \frac{QE(\omega)}{\omega} \cos \omega t d\omega$$

and

$$E^{-}(t) = E_0 - \frac{2}{\pi} \int_0^\infty \frac{InE(\omega)}{\omega} \sin \omega t d\omega \tag{13.163}$$

Also,

$$E^{+}(t) = E_0 - \frac{2}{\pi}\int\limits_0^{\infty} \frac{QE(\omega)}{\omega}\cos \omega t d\omega$$

and

$$E^{+}(t) = \frac{2}{\pi}\int\limits_0^{\infty} \frac{InE(\omega)}{\omega}\sin \omega t d\omega, \tag{13.164}$$

where E_0 is the stationary electric field caused by the direct current. In accord with these equations at any instant t, the components of the nonstationary field corresponding to either turn on or turn off conditions differ from each other by amount which is the magnitude of the direct current field; that is,

$$E^{+}(t) + E^{-}(t) = E_0 \tag{13.165}$$

13.8.2. The Early Stage of the Transient Field

As was demonstrated above at the initial moment after switching on, we have

$$E_x^{+} = \frac{p}{4\pi\varepsilon_0 r^3}\left(3\cos^2\varphi - 2\right)$$

For instance, we have for the electric fields along the dipole axis and on the equatorial plane, respectively,

$$E_x^{+} = \frac{p}{4\pi\varepsilon_0 r^3}, \quad E_x^{+} = -\frac{2p}{4\pi\varepsilon_0 r^3}, \tag{13.166}$$

It is instructive to write the stationary electric field ($\omega = 0$) on the surface of a horizontally layered medium as the sum of two terms:

$$E_{0x} = E_{0x}^{(1)} + E_{0x}^{(2)}$$

Here, the first term is the field of the electric dipole on the surface of a uniform half space with resistivity ρ_1, created by charges located on the electrode surface. It is equal to

$$E_{0x}^{(1)} = \frac{p}{4\pi\varepsilon_0}\left(3\cos^2\varphi - 1\right) \tag{13.167}$$

The origin of the field $E_{0x}^{(2)}$ is the electric charges arising on interfaces of media with different resistivity. This part of the field contains the information on the geoelectric properties of the medium, but when measurements are made near the dipole source, the primary field dominates. Thus, in accord with Eqs (13.165) and (13.167), when the current is turned off, the early stage of the electric field is

$$\begin{aligned} E_x^- &= E_{0x} - E_x^+ = E_{0x}^{(1)} + E_{0x}^{(2)} - E_x^+ \\ &= \frac{p}{4\pi\varepsilon_0 r^3}\left(3\cos^2\varphi - 1\right) + E_{0x}^{(2)} - \frac{p}{4\pi\varepsilon_0 r^3}\left(3\cos^2\varphi - 2\right) \\ &= E_{0x}^{(2)} + \frac{p}{4\pi\varepsilon_0 r^3} \quad \text{as} \quad t \to 0 \end{aligned} \tag{13.168}$$

This last equation is valid regardless of the distance from the source. As was shown, the magnetic field caused by direct current flow in a horizontally layered medium is the same as that for a uniform half space, and in accord with that during the early stage, we have

$$B_x^- = \frac{\mu_0 I dx}{4\pi r^2}\sin 2\varphi, \quad B_y^- = -\frac{\mu_0 I dx}{4\pi r^2}\cos 2\varphi,$$

$$B_z^- = \frac{\mu_0 I dx}{4\pi r^2}\sin\varphi, \quad \text{if } \frac{\tau}{r} < 3$$

13.8.3. The Late Stage of the Field on the Surface of Two-Layer Medium (Horizontal Components)

Next, applying the approach described in Chapter 9, we can derive the following asymptotic expressions for horizontal components of the transient field during the late stage. As an example, consider two-layer medium:

$$E_x(t) = \frac{p}{4\pi\varepsilon_0}\left[\frac{s^{1/2}}{6\pi^{1/2}}\left(\frac{\gamma_1\mu_0}{t}\right)^{3/2} - \frac{1}{16}(s-1)\left(\frac{\gamma_1\mu_0}{t}\right)^2 H_1 + \cdots\right],$$

and

$$B_y(t) = \frac{\mu_0 I dx}{4\pi}\left[\frac{\gamma_2\mu_0}{16t} - \frac{2}{15}\left(\frac{s}{\pi}\right)^{1/2}(s-1)\left(\frac{\gamma_1\mu_0}{t}\right)^{3/2} H_1 + \cdots\right], \tag{13.169}$$

where

$$s = \gamma_2/\gamma_1 \quad \text{and} \quad \rho_2 \neq \infty .$$

Thus, the transient field of the electric dipole source, as in the case of the magnetic dipole, tends to that corresponding to a uniform half space having resistivity of the basement during the late stage. Of course, it is not surprising because in both cases, the field magnitude displays main features of diffusion. The same conclusion is valid for any number of layers of the medium. Earlier it was pointed out that the vertical component of the magnetic field coincides with the electric field of the magnetic dipole. If the basement can be treated as an insulator, then in the case of two-layer medium, the late stage of the horizontal components of field is described as

$$\begin{aligned} E_x(t) &= \frac{\mu_0 I dx}{4\pi}\left(\frac{\mu_0 S_1}{4t^2} - \frac{\mu_0^2 S_1^2 H_1}{3t^3} + \cdots\right) \\ B_y(t) &= \frac{\mu_0 I dx}{4\pi}\left(\frac{\mu_0^2 S_1^2}{8t^2} - \frac{5}{24}\frac{\mu_0^3 S_1^3}{t^3} H_1 + \cdots\right), \end{aligned} \tag{13.170}$$

where S_1 is the longitudinal conductance of the upper layer. It is possible to show that in the case of many layered medium, we can use the leading terms of these equations if S_1 is replaced by the total conductance S. Note that at the late stage, the horizontal component of the magnetic field is dominant, and this component as well as the electric field depend on the longitudinal conductance to a lesser degree than the same components of the magnetic dipole source.

13.8.4. Apparent Resistivity Curves

As usual, we define the apparent resistivity as

$$\frac{\rho_\tau}{\rho_1} = \left(\frac{E_x^{\text{un}}}{E_x}\right)^2 \quad \text{and} \quad \frac{\rho_\tau}{\rho_1} = \frac{B_y^{\text{un}}}{B_y}, \tag{13.171}$$

where E_x^{un} and B_y^{un} are the electromagnetic field components on the surface of a uniform half space with the resistivity ρ_1 at the late stage. The apparent resistivity curves for two-layer medium are shown in Figure 13.7(a) and (b). The index of each curve is the value for the parameter ρ_2/ρ_1. During the late stage of the transient, these curves approach a right-hand asymptotic value of ρ_2/ρ_1. The left-hand asymptote of these curves is defined by the stationary field contributed by only electric charges at interfaces. As in the

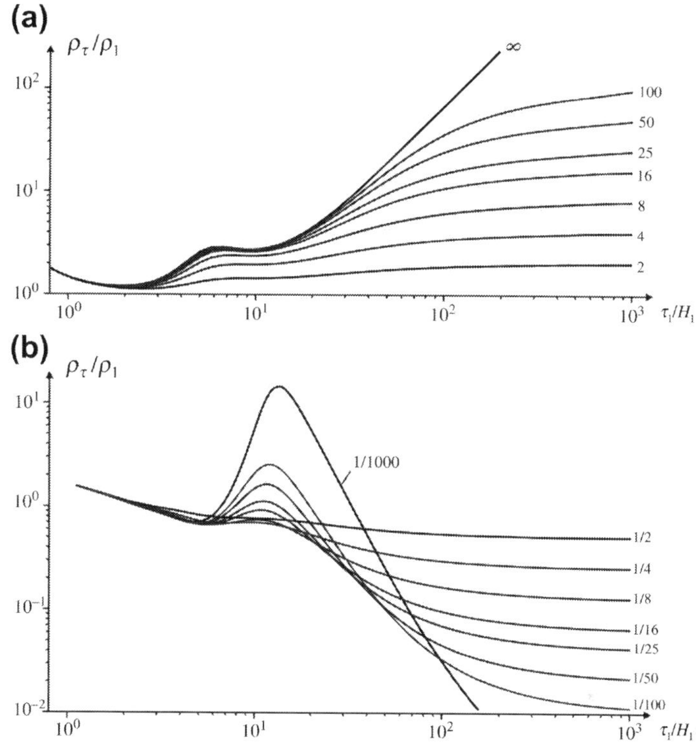

Figure 13.7 Two-layer apparent resistivity curves simulated for models with (a) Resistive and (b) Conductive basement. Curve index is ρ_2/ρ_1 ratio.

case of the magnetic dipole, the apparent resistivity curves corresponding to media with relatively resistive basement have an intermediate zone S when currents are distributed almost uniformly along the vertical in the layer but do not yet penetrate significantly into the basement.

13.8.5. Screening Effect

Now let us briefly consider the influence of an intermediate resistive layer on the transient electric field. Assume a three-layer model with a high resistive second layer. In the late stage, we have

$$E_x(t) = \frac{p}{8\pi\varepsilon_0}\left[\left(\frac{s_3}{\pi}\right)^{1/2}\frac{1}{3}\left(\frac{\gamma_1\mu_0}{t}\right)^{3/2} - A_1\left(\frac{\gamma_1\mu_0}{t}\right)^2\frac{1}{8}\right] \qquad (13.172)$$

Here,

$$A_1 = H_1(s_3 - 1) + H_2(s_3 - s_2),$$

and H_1 and H_2 are the thicknesses of the first and second layers, respectively; $s_2 = \gamma_2/\gamma_1, s_3 = \gamma_3/\gamma_1$, but p is the dipole moment.

Thus, the late stage of the electric field, observed on the surface of the earth containing an intermediate layer with either finite or infinite large resistivity ρ_2 depends on only the basement resistivity. As is generally known, with transient soundings made during "time on" and with frequency soundings when amplitude is measured, a layer characterized by infinite resistivity completely screens layers at greater depth. It is related to the fact that the late stage of electric field or the low-frequency part of the spectrum is practically the same as the stationary field, and the influence of the vortex part of the field is small. If the transient soundings is performed during "time off", the component E_x differs slightly from that for the stationary field ($\omega \to 0$) only at the beginning of the response, and with increasing time, the vortex character of the field becomes dominant. At sufficiently large times, induced currents in the medium become practically horizontal and they do not create electric charges on the surface of a resistive layer. Thus, if transient soundings based on the observation of the electric field are done during "time off", even resistive layer with relatively large thickness do not screen the section located at greater depth, since one can always find some moment when the field on the Earth's surface is controlled by the geoelectric parameters of this part of the section. In conclusion, let us note again that the transient soundings based on measuring of the electric field are used much less than arrays with measuring the magnetic field and it is mainly related to the fact that the electric field is more sensitive to the presence of inhomogeneities located relatively close to the surface.

REFERENCES AND FURTHER READING

[1] A.A. Kaufman, G.M. Morozova, Theoretical Principles of Transient Soundings in the Near Zone, Nauka, Novosibirsk, 1970.
[2] A.A. Kaufman, G.V. Keller, Frequency and Transient Sounding, Elsevier, Amsterdam, 1983.
[3] L.L. Vanyan, Principles of Electromagnetic Soundings, Nedra, Moscow, 1965.
[4] L.L. Vanyan, Transient Soundings, Nauka, Novosibirsk, 1966.
[5] J.R. Wait, Geoelectromagnetism, Academic Press, New York, NY, 1982.

PART FIVE

Principles of Inductive Mining Prospecting

CHAPTER FOURTEEN

Behavior of the Fields Caused by Currents in Confined Conductors

INTRODUCTION

Inductive methods of mining prospecting, based on the use of the quasi-stationary electromagnetic fields, appeared almost 100 years ago, and at the beginning they have been applied mainly to discover rather large and conductive bodies situated in almost nonconductive structure like Scandinavian and Canadian shields. Such favorable conditions allowed one to apply successfully these methods in spite of the fact that geophysicists had rather very modest understanding of the electromagnetic field behavior caused by induced currents in a conductive body. Only after the Second War, there was started a systematical development of the theory of these methods. At the beginning, the frequency and transient responses of the magnetic field caused by conductors of a simple form (sphere, cylinder) were investigated. These studies were based on a solution of the boundary-value problems that were well known from the beginning of the twentieth century. In such cases, the fields were expressed in terms of elementary or well-known special functions. The main focus of these investigations was directed to understand a relationship between the field and parameters of a conductor at different frequencies and times. With development of computers, it became possible to enlarge a class of models and consider more complicated bodies, when study of fields requires application of numerical methods, such as integral equations and finite difference methods. Next stage of development of the theory was related with investigation of influence of the surrounding medium, which represents a geological noise. The importance of these studies is difficult to overestimate, because they demonstrated that there is a difference between the conventional frequency and transient methods and often the latter allowed one to reduce the influence of geological noise in a greater extent. In other words, they usually have the greater depth of investigation. Certainly, such studies had strong influence on the recent development of this branch of the surface and airborne mining

Methods in Geochemistry and Geophysics, Volume 45
ISSN 0076-6895,
http://dx.doi.org/10.1016/B978-0-444-53829-1.00014-9

geophysics. These investigations also helped to realize that in general the magnetic field is caused by both the induction currents inside and outside of a conductor and currents that intersect the boundary between this body and surrounding medium. Correspondingly, it became clear that usually electric charges appear at this boundary and often it is possible to distinguish the vortex and galvanic parts of the field, which are differently related with parameters of the conductor. This study was very helpful to understand limitations of inductive methods from point of resolving capabilities methods when it is necessary to separate anomalies caused by good and poor conductors. Besides, it allowed one to understand that influence of geological noise changes with frequency and time, as well as it is different in measuring the electric and magnetic field and depends on the method of field excitation. Certainly, these studies had direct effect on understanding of the role of the lateral changes of resistivity on the measured response in the magnetotelluric and magnetovariational soundings. Now we consider in detail the main aspects of the theory of the inductive mining prospecting and start from the simplest cases when a surrounding medium is an insulator.

14.1. CONDUCTIVE SPHERE IN A UNIFORM MAGNETIC FIELD (FREQUENCY DOMAIN)

14.1.1. Formulation of Boundary-Value Problem

Consider a sphere with radius a, conductivity γ, and magnetic permeability μ, placed in a uniform harmonic magnetic field $\mathbf{B}_0^*$ directed along the z-axis, as is shown in Figure 14.1(a). Note that in this case, it is assumed that a sphere can be magnetic. We use a spherical coordinate system R,θ,φ with its origin at the center of the sphere. It is obvious that in general a nearly uniformly magnetic field can be created over a limited range in space using various types of current sources. However, here it is assumed that the source of the primary field has only the φ component of the current density, which is independent of the angle φ, and that the center of the sphere is located on the axis of symmetry. In practice, a current ring of large diameter can be used as the source for an almost uniform magnetic field in the area occupied by such a sphere.

Correspondingly, one can expect that the vortex electric field $\mathbf{E}_0^*$ caused by the time rate of change of the primary field $\mathbf{B}_0^*$ is not uniform, and it is equal to zero on the axis of symmetry. The field $\mathbf{E}_0^*$ has a single component

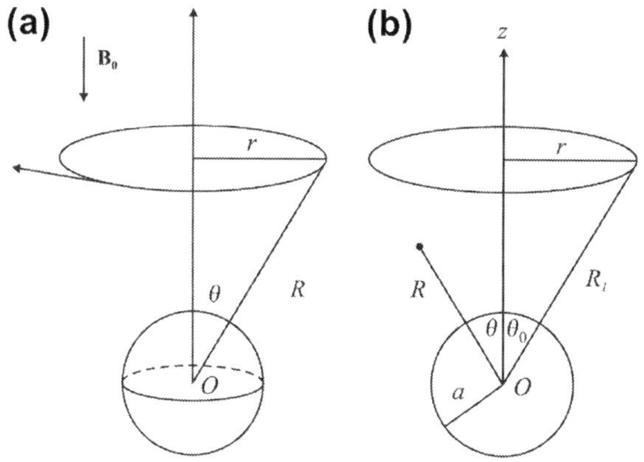

Figure 14.1 (a) Conductive sphere in a uniform magnetic field; (b) Conductive sphere in presence of magnetic field generated by circular current loop.

in the spherical system, this being $E^*_{0\varphi}$, which can be readily found from the first of Maxwell's equations expressed in integral form as:

$$\oint_L \mathbf{E}_0 \cdot d\mathbf{l} = -\frac{\partial}{\partial t}\int_S \mathbf{B}_0 \cdot d\mathbf{S} \tag{14.1}$$

Applying this expression along the circular path L situated in a horizontal plane as shown in Figure 14.1(a) and taking into account the axial symmetry and uniformity of the primary field, we obtain

$$E^*_{0\varphi}2\pi r = i\omega B^*_{0z}\pi r^2$$

or

$$E^*_{0\varphi} = i\omega B^*_{0z}\frac{r}{2}\sin\theta \tag{14.2}$$

Thus, the electric field $E^*_{0\varphi}$ increases linearly with distance from the z-axis. Inasmuch as the radial component of the primary field $\mathbf{E}^*_0$ is zero, that is it does not cross the surface of the sphere, no electric charge is formed, and therefore the only sources for the secondary magnetic field are currents arising due to the vortex electric field. It is obvious that current filaments within the sphere are located in horizontal planes and have the form of circles with their centers situated on the z-axis. Because of the interaction between induced currents the current density at any point is a relatively complicated function of frequency, conductivity, and the radius of the sphere, as we observed in a layered medium. The secondary electromagnetic field can be represented as being the sum of fields created by the individual elementary current rings, and for this reason in the spherical system the field is described by only one electric component, $E^*_{1\varphi}$ and two magnetic components B^*_{1R} and $B^*_{1\theta}$. Since there is only one single component of the electric field, it is proper at the beginning to define this component and then using Maxwell's equations to determine the magnetic field. Before we continue, let us point out again that unlike other parts of the book, we here consider the case when a conducting body is also magnetic. From the physical point of view, this means that along with the conduction currents, there are currents of magnetization and, correspondingly, the second Maxwell's equation has the form:

$$\text{curl}\,\mathbf{B} = \mu_0(\mathbf{j}_c + \mathbf{j}_m)$$

Here $\mathbf{j}_c$ and $\mathbf{j}_m$ are conduction and magnetization currents, respectively. As is known, the latter is related with the vector of magnetization $\mathbf{P}$ in the following way

$$\mathbf{j}_m = \text{curl}\,\mathbf{P},$$

where $\mathbf{P}$ is the vector of induced magnetization; that is the dipole moment of the unit volume. Correspondingly, the Maxwell's equation can be written as

$$\text{curl}\,(\mathbf{B} - \mu_0\mathbf{P}) = \mu_0\mathbf{j}_c, \quad \text{or} \quad \text{curl}\,\mathbf{H} = \mathbf{j}_c$$

Here

$$\mathbf{B} = \mu_0(\mathbf{H} + \mathbf{P})$$

Letting $\mathbf{P} = \chi\mathbf{H}$, we have

$$\mathbf{B} = \mu_0(1 + \chi)\mathbf{H} = \mu_i\mathbf{H},$$

where χ and μ_i are susceptibility and magnetic permeability of a sphere, respectively. Thus, the system of equations describing the field at regular points is

$$\begin{aligned} &\text{curl}\,\mathbf{E}^* = i\omega\mathbf{B}^*, \quad \text{div}\,\mathbf{E}^* = 0 \\ &\text{curl}\,\frac{\mathbf{B}^*}{\mu} = \mathbf{j}^*_c, \quad \text{div}\,\mathbf{B}^* = 0, \end{aligned} \tag{14.3}$$

where $\mathbf{E}^*$ and $\mathbf{B}^*$ are the complex amplitudes of the field, and μ is the magnetic permeability of either the sphere or a surrounding medium. From the first and second equations of the system and Ohm's law, we have

$$\text{curl curl}\,\mathbf{E}^* = i\omega\mu\,\text{curl}\,\frac{\mathbf{B}^*}{\mu} = i\mu\omega\gamma\mathbf{E}^*,$$

that is the electric field both outside and inside the sphere satisfies the following conditions respectively:

$$\text{curl curl}\,\mathbf{E}^* = 0, \quad \text{if } R > a$$

and

$$\text{curl curl}\,\mathbf{E}^* - k^2\mathbf{E}^* = 0, \quad \text{if } R < a \tag{14.4}$$

where $k^2 = i\gamma\mu\omega$.

For the tangential components of the electromagnetic field at the interface $R = a$, we have

$$E_\varphi^{e^*} = E_\varphi^{i^*}, \quad \frac{B_\theta^{e^*}}{\mu_e} = \frac{B_\theta^{i^*}}{\mu_i} \quad \text{if } r = a \tag{14.5}$$

Using the first Maxwell's equation:

$$B_\theta^* = -\frac{1}{i\omega}\frac{1}{R}\frac{\partial}{\partial R}RE_\varphi^*,$$

the boundary conditions can be written as

$$E_\varphi^{e^*} = E_\varphi^{i^*}, \quad \frac{1}{\mu_e}\frac{\partial}{\partial R}RE_\varphi^{e^*} = \frac{1}{\mu_i}\frac{\partial}{\partial R}RE_\varphi^{i^*} \quad \text{if } R = a \tag{14.6}$$

As usual, we represent the electric field outside the sphere as being the sum of primary and secondary parts:

$$E_\varphi^{e^*} = E_{o\varphi}^* + E_{1\varphi}^* \tag{14.7}$$

From the physical point of view, it should be clear that the field inside the sphere remains finite, while outside the sphere the secondary field tends to zero for large distances, R. Before proceeding with a solution of the boundary-value problem, let us make necessary changes in algebraic form of the equations in the set 14.4. In spherical coordinates

$$\text{curl}_R\mathbf{E}^* = \frac{1}{R\sin\theta}\frac{\partial}{\partial\theta}\sin\theta E_\varphi^* = a_R$$

$$\mathrm{curl}_\theta \mathbf{E}^* = -\frac{1}{R}\frac{\partial}{\partial R} R E^*_\varphi = a_\theta \quad \text{and} \quad \mathrm{curl}_\varphi \mathbf{E} = 0$$

since $E_R = E_\theta = 0$. Therefore,

$$\begin{aligned}\mathrm{curl}_\varphi \mathrm{curl}\,\mathbf{E}^* &= \mathrm{curl}_\varphi(a_R\mathbf{R}_0 + a_\theta\boldsymbol{\theta}_0)\\ &= -\frac{1}{R}\frac{\partial^2}{\partial R^2}\left(RE^*_\varphi\right) - \frac{1}{R^2}\frac{\partial}{\partial\theta}\left(\frac{1}{\sin\theta}\frac{\partial}{\partial\theta}\sin\theta E^*_\varphi\right),\end{aligned} \tag{14.8}$$

where $\mathbf{R}_0$ and $\boldsymbol{\theta}_0$ are unit vectors directed along the coordinates R and θ.

14.1.2. Solution of Helmholtz Equation

Substituting Eq. (14.8) into Eq. (14.4), we obtain equations in partial derivatives of the second order for the electric field both outside and inside the sphere, respectively:

$$\frac{1}{R}\frac{\partial^2}{\partial R^2}\left(RE^{e^*}_\varphi\right) + \frac{1}{R^2}\frac{\partial}{\partial\theta}\left(\frac{1}{\sin\theta}\frac{\partial}{\partial\theta}\sin\theta E^{e^*}_\varphi\right) = 0 \tag{14.9}$$

$$\frac{1}{R}\frac{\partial^2}{\partial R^2}\left(RE^{i^*}_\varphi\right) + \frac{1}{R^2}\frac{\partial}{\partial\theta}\left(\frac{1}{\sin\theta}\frac{\partial}{\partial\theta}\sin\theta E^{i^*}_\varphi\right) + k^2E^{i^*}_\varphi = 0 \tag{14.10}$$

These are Helmholtz equations in the spherical system of coordinates, providing that the field is independent on the $\varphi-$ coordinate. Making use of the method of separation of variables, we solve Eq. (14.10) with the axial symmetry $\partial E_\varphi/\partial\varphi = 0$ being taken into account. The form of solution will be

$$E^{i^*}_\varphi = T(R)\Phi(\theta) \tag{14.11}$$

Substituting this expression into Eq. (14.10) and multiplying both sides by $R^2/T\Phi$, we have

$$\frac{R}{T}\frac{\partial^2}{\partial R^2}RT + \frac{1}{\Phi}\frac{\partial}{\partial\theta}\left(\frac{1}{\sin\theta}\frac{\partial}{\partial\theta}\sin\theta\Phi\right) + k^2R^2 = 0 \tag{14.12}$$

As is always the case when a method of separation of variables is used, this equation separates into two ordinary differential equations, which are

$$\frac{R}{T}\frac{\mathrm{d}^2RT}{\mathrm{d}R^2} + k^2R^2 = m \tag{14.13}$$

$$\frac{1}{\Phi}\frac{\partial}{\partial\theta}\left(\frac{1}{\sin\theta}\frac{\partial}{\partial\theta}\sin\theta\Phi\right) = -m, \tag{14.14}$$

where m is the separation constant having dimension of the inverse distance. After differentiation of the product RT Eq. (14.13) takes the form

$$\frac{\mathrm{d}^2T}{\mathrm{d}R^2} + \frac{2}{R}\frac{\mathrm{d}T}{\mathrm{d}R} + \left(k^2 - \frac{m}{R^2}\right)T = 0 \tag{14.15}$$

This is Bessel equation for which the solution can be written as:

$$T(k, R) = R^{-1/2}Z_{\left(\frac{1}{4}+m\right)^{\frac{1}{2}}}(ikr), \tag{14.16}$$

where

$$Z_{\left(\frac{1}{4}+m\right)^{\frac{1}{2}}}(ikr)$$

are modified Bessel's functions of the first and second kind. Carrying out differentiation on the left-hand side of Eq. (14.14), we have

$$\left(1-u^2\right)\frac{\mathrm{d}^2\Phi}{\mathrm{d}u^2} - 2u\frac{\mathrm{d}\Phi}{\mathrm{d}u} - \frac{\Phi}{1-u^2} + m\Phi = 0, \tag{14.17}$$

where

$$u = \cos\theta.$$

This equation is Legendre's equation for associated Legendre's functions with the index $l = 1$. From the behavior of these functions, we recognize that a nontrivial solution of Eq. (14.17) is possible only if the parameter m has values:

$$m = n(n+1). \tag{14.18}$$

Thus, partial solutions of Eqs (14.15) and (14.17) are

$$T_n(k, R) = \frac{1}{R^{1/2}}\left\{A_n^1 I_{\left[\frac{1}{4}+n(n+1)\right]^{1/2}}(ikR) + B_n^1 K_{\left[\frac{1}{4}+n(n+1)\right]^{1/2}}(ikR)\right\} \tag{14.19}$$

and

$$\Phi(\theta) = C_n^1 P_n^{(1)}(u) + D_n^{(1)} Q_n^{(1)}(u), \tag{14.20}$$

where

$$I_{\left[\frac{1}{4}+n(n+1)\right]^{1/2}}(ikR) \quad \text{and} \quad K_{\left[\frac{1}{4}+n(n+1)\right]^{1/2}}(ikR)$$

are modified Bessel functions of the first and second kind, $P_n^{(1)}$ and $Q_n^{(1)}$ are associated Legendre's functions, $u = \cos\theta$ and A_n^1, B_n^1, C_n^1, and D_n^1 are constants representing the amplitudes of various radial and angular harmonics in the solutions. The coefficient for the function $Q_n^{(1)}$ must be identically zero because at all points along the z-axis ($\theta = 0$) inside and outside the sphere this function becomes infinitely large. At the center of the sphere the Bessel function,

$$K_{\left[\frac{1}{4}+n(n+1)\right]^{1/2}}(ikR)$$

also tends to infinity, and for this reasons its coefficient must be zero in the general solution for the electric field inside the sphere:

$$E_\varphi^{i^*} = R^{-1/2} \sum_{n=0}^{\infty} B_n I_{\left[\frac{1}{4}+n(n+1)\right]^{1/2}}(ikR) P_n^{(1)}(\cos\theta) \tag{14.21}$$

Equation 14.15 is considerably simpler in the nonconducting medium outside the sphere because the wave number is zero. Therefore,

$$\frac{\mathrm{d}}{\mathrm{d}R}\left(R^2 \frac{\mathrm{d}T}{\mathrm{d}R}\right) - n(n+1)T = 0. \tag{14.22}$$

The solution of this equation is well known

$$T_n(R) = C_n^1 R^n + F_n^1 R^{-n-1}. \tag{14.23}$$

Inasmuch as the field created by induced currents tends to zero with increasing R, the expression for the secondary electric field outside the sphere can be written as

$$E_{1\varphi}^{e^*} = \sum_{n=0}^{\infty} F_n^1 R^{-n-1} P_n^{(1)}(\cos\theta). \tag{14.24}$$

Considering Eqs (14.2), (14.7), (14.21) and (14.24) the expression for the electric field outside and inside the sphere can be written as:

$$E_\varphi^{e^*} = \frac{i\omega B_0}{2} R P_1^{(1)}(u) + \frac{i\omega B_0}{2} \sum_{n=0}^{\infty} D_n R^{-n-1} P_n^{(1)}(u) \tag{14.25}$$

$$E_\varphi^{i^*} = \frac{i\omega B_0}{2} R^{-1/2} \sum_{n=0}^{\infty} C_n I_{\left[\frac{1}{4}+n(n+1)\right]^{1/2}}(ikR) P_n^{(1)}(u), \tag{14.26}$$

inasmuch as

$$P_1^{(1)}(\cos\theta) = \sin\theta.$$

14.1.3. Expressions for the Field

To determine the unknown coefficients D_n and C_n, we apply the boundary conditions, Eq. (14.6). Taking into account orthogonality of the associated Legendre's functions instead of a system of an infinite number of equations with an infinite number of unknowns, we obtain two equations and two unknowns, D_n and C_n for each harmonic with the index n. Considering that the primary electric field $E_{0\varphi}$ is described by only the first harmonic ($n = 1$), we can readily see that all of the coefficients C_n and D_n, except for C_1 and D_1, are zero and that for their determination we can write the following system of equations:

$$\begin{aligned} a^3 + D_1 &= a^{3/2} C_1 I_{3/2}(ika) \\ \mu_i\left(2a^3 - D_1\right) &= \mu_e a^{3/2} C_1 \left[I_{3/2}(ika) + ikaI'_{3/2}(ika)\right] \end{aligned} \tag{14.27}$$

where

$$I'_{3/2}(x) = \frac{\partial}{\partial x} I_{3/2}(x).$$

Bessel functions are related through recurrence relationships:

$$I_{3/2}(x) = I_{-1/2}(x) - \frac{1}{x} I_{1/2}(x) \quad xI'_{3/2}(x) = xI_{1/2}(x) - \frac{3}{2} I_{3/2}(x)$$

Substituting these expressions into Eq. (14.27), after some straightforward algebraic operations, we have

$$D_1 = \frac{(2\mu_i + \mu_e)xI_{-1/2}(x) - \left[\mu_e(1 + x^2) + 2\mu_i\right] I_{1/2}(x)}{(\mu_i - \mu_e)xI_{-1/2}(x) + \left[\mu_e(1 + x^2) - \mu_i\right] I_{1/2}(x)} a^3 \tag{14.28}$$

and

$$C_1 = \frac{3\mu_i x a^{3/2}}{(\mu_i - \mu_e)xI_{-1/2}(x) + \left[\mu_e(1 + x^2) - \mu_i\right] I_{1/2}(x)}, \tag{14.29}$$

where $x = ika, k = (i\gamma\mu_i\omega)^{1/2}$; μ_i and μ_e are the magnetic permeability values for the sphere and for a surrounding medium, respectively. Inasmuch as

$$C_n = D_n = 0, \quad \text{if} \;\; n \neq 1$$

we obtain a relatively simple set of expressions for the electric field outside and inside the sphere

$$E_\varphi^{e^*} = E_{o\varphi}^* + \frac{i\omega B_{0z}}{2} D_1 R^{-2} \sin\theta$$
$$E_\varphi^{i^*} = \frac{i\omega B_{0z}}{2} C_1 R^{-1/2} I_{3/2}(ikR) \sin\theta. \tag{14.30}$$

From the first Maxwell's equation

$$\text{curl}\, \mathbf{E}^* = i\omega\mathbf{B}^*,$$

we have

$$B_R^* = \frac{1}{i\omega}\frac{1}{R\sin\theta}\frac{\partial}{\partial\theta}\left(\sin\theta E_\varphi^*\right), \quad B_\theta^* = -\frac{1}{i\omega}\frac{1}{R}\frac{\partial}{\partial R}\left(RE_\varphi^*\right). \tag{14.31}$$

Thus, we obtain the following expressions for the secondary magnetic field as observed outside the sphere

$$B_{sR}^{e^*} = \frac{D_1}{R^3} B_{0z} \cos\theta \quad B_{s\theta}^{e^*} = \frac{D_1}{2R^3} B_{0z} \sin\theta \quad \text{if} \;\; R \geq a. \tag{14.32}$$

Certainly, understanding the behavior of Legendre's and Bessel functions allowed us to perform transition from rather complicated expressions to amazingly simple ones. Before we continue, let us make several comments: (1) The simplicity of the field expression is related to the fact that a sphere is placed into a uniform primary field. For instance, if the latter is caused by current loop, located relatively close to the sphere, then in general the secondary field is described by a sum of harmonics and with approaching the current to the sphere the harmonics with larger number of n begin to play more important role. As we saw, both the primary and secondary electric fields are described by the same spherical harmonic $n = 1$. In other words, the spherical harmonic of the primary field gives rise to a secondary field that has the same dependence on the angle θ. Due to this fact the geometry of currents induced in the sphere, as well as that of the field, is the same for all frequencies. In the general case, when the field $\mathbf{B}_0$ is not uniform, the primary electric field is described by a sum of spherical harmonics. For instance, if the current source is a horizontal circle, Figure 14.1(b), the field $E_{0\varphi}^*$ can be represented as

$$E^*_{0\varphi} = \frac{i\omega\mu_0 I}{2}\sum_{n=1}^{\infty}\frac{\sin\theta_0}{n(n+1)}\left(\frac{R}{R_1}\right)^n P_n^{(1)}(\cos\theta_0)P_n^{(1)}(\cos\theta).$$

Inasmuch as each harmonic of the primary field generates the corresponding harmonic of the secondary field, it is also represented by a sum of spherical harmonics with different complex amplitudes. (1) The closer the current loop to the sphere the more harmonics contribute to the field. (2) If the primary electric field intersects the sphere surface then electric charges appear, and we observe more complicated dependence of the currents on coordinates. (3) Applying the method of separation of variables, we found out that the function $\Phi(\theta)$ is the same inside and outside of the sphere; in other words, it is independent on the wave number. Because of this and using the orthogonality of these functions, we were able to obtain a very simple system of equations for determination of unknown coefficients. As will be shown later, the same simplification takes place in the case of the circular cylinder placed in an alternating field. However, in general case, for example of a conductor having a shape of a spheroid or an elliptical cylinder the situation is completely different, even if the primary filed is uniform. As before, it is still possible to apply the method of separation of variables, but a solution of each ordinary differential equation depends on the wave number. For instance, functions describing dependence of the field on points of the conductor surface outside and inside contain different wave numbers. This means that they are not orthogonal and correspondingly the infinite system of equations doesn't transform into a simple system of two equations with two unknowns. This is a reason why in the past the analysis of the electromagnetic fields, applied in the mining prospecting, was mainly based on a study of only two cases of a confined conductor, namely the sphere and cylinder. Of course, with the use of computers and application of integral equations as well as finite difference and finite elements methods, the class of models was greatly extended.

14.2. BEHAVIOR OF THE FIELD CAUSED BY CURRENTS IN A NONMAGNETIC SPHERE (THE FREQUENCY DOMAIN)

14.2.1. Equivalence to the Magnetic Dipole

First of all, it is clear that the electromagnetic field from induced currents is equivalent outside the sphere to the field that would be observed from a magnetic dipole situated at the center of the sphere, and the complex amplitude of its moment is

$$M^* = 2\pi D_1 B_{0z}, \tag{14.33}$$

directed along the z-axis. Because D_1 is in general complex, the secondary field differs from the primary field both in magnitude and phase, and for this reason it can be written as the sum of the in-phase and quadrature components. The in-phase component of the secondary field is either actually in-phase or reversed in-phase by 180° with respect to the primary field, while the quadrature component is shifted by $\pm 90^\circ$ with respect to B_{0z}. At the beginning, we investigate the secondary field when the sphere and surrounding medium are not magnetic; that is $\mu_i = \mu_e = 4\pi 10^{-7}$ H/m. In this case, the expression for the function D_1 is simplified markedly:

$$D_1 = Da^3, \tag{14.34}$$

and we have:

$$D = \frac{3x\cosh x - (3 + x^2)\sinh x}{x^2 \sinh x} = \frac{3\coth x}{x} - \frac{3}{x^2} - 1, \tag{14.35}$$

because

$$I_{1/2}(x) = \left(\frac{2}{\pi x}\right)^{1/2} \sinh x, \quad I_{-1/2}(x) = \left(\frac{2}{\pi x}\right)^{1/2} \cosh x,$$

where

$$x = ika = i(i\gamma\omega\mu)^{1/2} a = p(i-1) \tag{14.36}$$

14.2.2. Presentation of the Function *D* as a Series

The parameter p is the ratio between the radius of the sphere a and the thickness of the skin layer δ

$$p = \frac{a}{\delta} = \frac{a(\gamma\mu\omega)^{1/2}}{2^{1/2}} \quad \text{and} \quad \delta = \frac{\lambda}{2\pi} = \frac{10^3}{2\pi}\left(\frac{10\rho}{f}\right)^{1/2} m.$$

In order to investigate the low-frequency part of the spectrum, assume that the skin layer thickness δ is much greater than the radius of the sphere, a. Then the function coth x can be written as a series:

$$\coth x = \frac{1}{x} + \frac{x}{3} - \frac{x^3}{45} + \frac{2x^3}{45} - = \frac{1}{x} + \sum_{k=1}^{\infty} \frac{2^{2k} B_{2k}}{(2k)!} x^{2k-1} \quad \text{if } |x^2| < \pi^2$$

where B_{2k} are Bernoulli numbers. Substituting this series into Eq. (14.35), we have

$$D = 3 \sum_{k=2}^{\infty} \frac{2^{2k} B_{2k}}{(2k)!} x^{2k-2} \quad \text{if } |x^2| < \pi^2$$

or

$$D = 3 \sum_{k=2}^{\infty} \frac{(-i)^{k-1} 2^{2k} B_{2k}}{(2k)!} \left(\gamma \mu_0 \omega a^2\right)^{k-1}. \tag{14.38}$$

In accordance with this last equation, the low-frequency part of the spectrum of the field contains only integer powers in ω (it is a MacLauren series), and as will be shown later, this feature of the spectrum is diagnostic for fields created by currents induced in confined conductors situated in insulating medium. Shortening Eq. (14.38) to the first few terms:

$$D = i \frac{\gamma \mu_0 a^2}{15} \omega - \frac{2}{315} \left(\gamma \mu_0 a^2\right)^2 \omega^2 - i \frac{1}{1575} \left(\gamma \mu_0 a^2\right)^3 \omega^3, \tag{14.39}$$

it is easily seen that the ratio of coefficients for this series tends to a constant value:

$$-i \frac{\pi^2}{\gamma \mu_0 a^2}.$$

The latter defines the radius of convergence for a MacLauren series and also, as will be proved later, the character of the decay of a transient field at the late stage.

14.2.3. The Low-Frequency Part of the Spectrum

For very small values of the parameter (low frequencies, large resistivity or small values of radius), the quadrature component of the magnetic field is dominant and increases in direct proportion to the frequency and conductivity, while the in-phase component is much more sensitive to a change of these both parameters, but it is much smaller:

$$QB^{e^*}_{sR} \approx \frac{\gamma\mu_0\omega a^2}{15} B_{0z}\frac{a^3}{R^3}\cos\theta, \quad InB^{e^*}_{sR} \approx -\frac{2}{315}\left(\gamma\mu_0\omega a^2\right)^2 B_{0z}\frac{a^3}{R^3}\cos\theta$$

and

$$QB^{e^*}_{s\theta} \approx \frac{\gamma\mu_0\omega a^2}{30} B_{0z}\frac{a^3}{R^3}\sin\theta, \quad InB^{e^*}_{s\theta} \approx -\frac{1}{315}\left(\gamma\mu_0\omega a^2\right)^2\frac{a^3}{R^3}\sin\theta. \tag{14.40}$$

Now consider the behavior of the currents when $p < 1$. From Eq. (14.29), we have

$$j^*_\varphi = \gamma E^*_\varphi = \frac{i\gamma\omega}{2}B_{0z}C_1 R^{-1/2} I_{3/2}(ikR)\sin\theta, \tag{14.41}$$

where

$$C_1 = \frac{3a^{3/2}}{xI_{1/2}(x)}, \quad I_{1/2}(ikR) = \left(\frac{2}{i\pi kR}\right)^{1/2}\sinh ika$$

and

$$I_{3/2}(ikR) = \left(\frac{2}{i\pi kR}\right)^{1/2}\left(\cosh ikR - \frac{1}{ikR}\sinh ikR\right).$$

Then the expression for the current density can be written as

$$j^*_\varphi = -\frac{3}{2R\mu_0}\frac{x}{\sinh x}B_{0z}\left(\cosh ikR - \frac{1}{ikR}\sinh ikR\right)\sin\theta. \tag{14.42}$$

Expanding the hyperbolic functions in power series and retaining only the first two terms in each series, we obtain

$$j^*_\varphi \approx i\frac{B_{0z}}{2\mu_0}\gamma\mu_0\omega R\left[1 - k^2\left(\frac{R^2}{10} - \frac{a^2}{6}\right)\right]\sin\theta. \tag{14.43}$$

The first term in Eq. (14.43) is

$$j^*_{0\varphi} = i\frac{B_{0z}\gamma\omega}{2}R\sin\theta = \gamma E^*_{0\varphi}. \tag{14.44}$$

Thus, for small values of p, the quadrature component of the current density is defined by the primary electric field $E^*_{0\varphi}$. In other words, the interaction between current filaments is neglected. In accordance with Eq. (14.44), the quadrature component of the current density increases linearly toward the surface of the sphere. However, the in-phase component of j^*_φ caused by interaction of currents increases almost linearly near the center of the sphere ($R \ll a$), reaches a maximum value and then decreases as it approaches its appropriate value on the surface of the sphere. The in-phase component of j^*_φ for small values of p, in contrast to the behavior of the quadrature component, is proportional to k^4. It should be clear that by neglecting the interaction of current filaments, we have markedly simplified the procedure for determining the quadrature component of the magnetic field, so that it consists merely of the calculation of current density: $j^*_\varphi = \gamma E^*_{0\varphi}$ and the use of the Biot–Savart law.

14.2.4. The High-Frequency Part of the Spectrum ($p \gg 1$)

First, let us note that the high- or low-frequency spectrums don't mean that frequency should be either low or high, since the field behavior is defined by the value of the parameter p, which also depends on conductivity and a sphere radius. Now we consider the high-frequency portion of the spectrum for which the parameter $p \gg 1$. Inasmuch as $\coth x \to 1$ when x tends to infinity, the function D approaches the value -1, and the appropriate formulas for the field are those for ideal conductor, that is

$$B^{e^*}_R = B_{0R} - \frac{a^3}{R^3} B_{0z} \cos\theta, \quad B^{e^*}_\theta = B_{0\theta} - \frac{a^3}{2R^3} B_{0z} \sin\theta$$

and

$$E^{e^*}_\varphi = E^*_{0\varphi} - \frac{i\omega B_{0z}}{2} \frac{a^3}{R^2} \sin\theta \quad \text{if } R \geq a. \tag{14.45}$$

As a particular case, on the surface of the sphere, we have

$$B^{e^*}_R = 0, \quad B^{e^*}_\theta = -\frac{3}{2} B_{0z} \sin\theta, \quad E^e_\varphi = 0. \tag{14.46}$$

Letting $\coth x = 1$ and keeping the rest of the terms in Eq. (14.35), we obtain an asymptotic representation for the function D that describes the high frequency portion of the spectrum

$$D \approx \left(-1 + \frac{3}{2p}\right) + \frac{3i}{2p}\left(1 - \frac{1}{p}\right) \quad \text{if } p \gg 1. \tag{14.47}$$

As may be seen from this last equation, both the in-phase and quadrature components approach their asymptotic values relatively slowly. It should be noticed that the same result can be derived assuming that in the vicinity of the sphere surface the relation between tangential components of the electric and magnetic field is equal to the impedance of the plane wave. In fact, from Eq. (14.47), we have the following expressions for the total field:

$$E_\varphi^{e^*} = 3B_{0z}\frac{i\omega}{2ki}\sin\theta, \quad B_\theta^{e^*} = \frac{3}{2\mu_0}B_{0z}\sin\theta$$

and therefore

$$Z = \frac{E_\varphi^{e^*}}{B_\theta^{e^*}} = \frac{\omega\mu_0}{k} = \left(\frac{\omega\mu_0}{k}\right)^{1/2} = \left(-\frac{i\omega\mu}{\gamma}\right)^{1/2} = (\omega\mu_0\rho)^{1/2}\exp\left(-\frac{\pi}{4}\right). \tag{14.48}$$

The latter allows one to avoid a study of the field inside a body applying the well-known boundary condition, which is often used to evaluate an influence of conductors of an arbitrary shape provided that the skin depth is much smaller than the radius of curvature of the conductor surface. At this point, we consider some features of the distribution of the induced currents for this portion of the spectrum. For large values of ka, the expression for current density in Eq. (14.42) can be replaced by the following:

$$j_\varphi^* \approx -\frac{3i}{2}\frac{ka}{R\mu_0}B_{0z}\exp\left[ik(R-a)\right]\sin\theta, \tag{14.49}$$

because

$$\sinh x = \cosh x = \frac{\exp(x)}{2} \quad \text{as } |x| \gg 1.$$

At all points inside the sphere, the current density tends to zero with increasing wave number, and if the skin depth is significantly less than the radius of the sphere, the major part of the current flows near the surface of the conductor. In accordance with Eq. (14.49) at $R=a$, the volume current density is

$$j_\varphi^* \approx -\frac{3}{2\mu_0}ikB_{0z}\sin\theta, \quad \text{if } |ka| \gg 1$$

and increases without limit as frequency increases. However, the cross-sectional area through which most of the current passes becomes smaller, and from the physical point of view, we can expect that the magnitude of the surface current in the limit is finite. In fact, the current passing through elementary surface dS is

$$\mathrm{d}I = j_\varphi \mathrm{d}S = -\frac{3i}{2\mu_0} B_{0z} ka \exp\left[ik(R-a)\right]\mathrm{d}R\mathrm{d}\theta \sin\theta.$$

For the total current through a sector subtending an angle dθ, we have:

$$\mathrm{d}I = -\frac{3i}{2\mu_0} B_{0z} ka \sin\theta\mathrm{d}\theta \int_0^a \exp\left[ik(R-a)\right]\mathrm{d}R$$

$$= -\frac{3}{2\mu_0} aB_{0z}\left[1-\exp\left(-ika\right)\right]\sin\theta\mathrm{d}\theta.$$

From this, we have

$$\mathrm{d}I = -\frac{3a}{2\mu_0} B_{0z} \sin\theta\mathrm{d}\theta \quad \text{as } k \to \infty.$$

Integrating this last expression with respect to θ over the limits from π to zero, we find that in the limiting case as frequency becomes infinite the total surface current is

$$I = -\frac{3a}{\mu_0} B_{0z}. \tag{14.50}$$

Curves showing the distribution of current density along the radius of the sphere for various values of the p are shown in Figure 14.2(a) and (b), where the ratio of the current density at an arbitrary point R to that on the surface of the sphere is plotted along the ordinate.

14.2.5. The Second Form of Function *D* and Time Constant

As will be seen later, the main features of the field caused by currents of the sphere are inherent to the quasi-stationary fields that are created by induced currents flowing in confined conductors of arbitrary shape, embedded in an insulating host medium. In this respect, it is useful to represent the function D characterizing the spectrum of the field in another form. Making use of the identity:

$$\coth x = \frac{1}{x} + 2x\sum_{n=1}^{\infty} \frac{1}{\pi^2 n^2 + x^2}$$

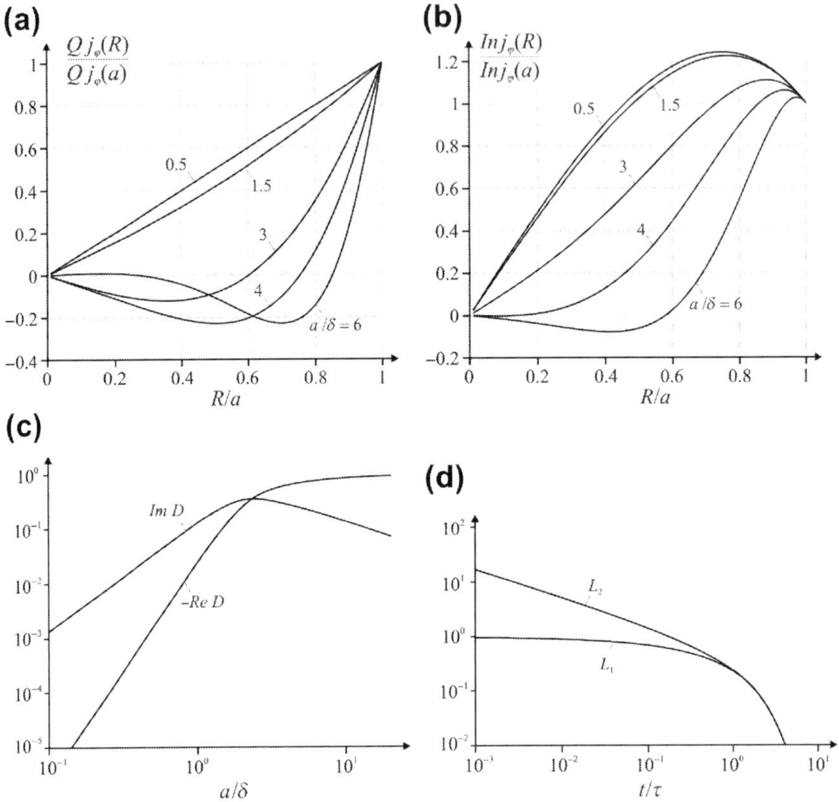

Figure 14.2 Distribution of the quadrature (a) and in-phase (b) components of the current density inside the conductive sphere; real and imaginary parts of the function D (c); functions L_1 and L_2 (d).

along with Eq. (14.35), we have

$$D = 6\sum_{n=1}^{\infty}\frac{1}{\pi^2 n^2 - i\gamma\mu_0 a^2\omega} - 1 \tag{14.51}$$

Inasmuch as

$$1 = 6\sum_{n=1}^{\infty}\frac{1}{\pi^2 n^2}$$

the function D can be represented as

$$D = \frac{6i\omega\tau}{\pi^2}\sum_{n=1}^{\infty}\frac{1}{n^2}\frac{1}{(n^2 - i\omega\tau)}, \tag{14.52}$$

where

$$\tau = \frac{\gamma \mu_0 a^2}{\pi^2} \tag{14.53}$$

is the parameter, which is called the time constant of a nonmagnetic sphere. Later, we will explain the origin of such name. Thus, the spectrum of the field, generated by currents in the sphere, can be described as an infinite sum of simple fractions. It is useful to consider the behavior of the function D in the complex plane of ω. The denominator of every term in the right-hand side of Eq. (14.53) becomes equal to zero if

$$\omega_n = -i\,\frac{n^2}{\tau}. \tag{14.54}$$

Therefore, the spectrum has singularities or poles, which are located beneath the real axis of ω along the imaginary axis Im ω. In accordance with Eq. (14.54) the distance between the pole and the origin ($\omega = 0$) increases rapidly with increase of n. For instance,

$$\omega_1 = -\frac{i}{\tau}, \quad \omega_2 = -\frac{4i}{\tau}, \quad \omega_3 = -\frac{9i}{\tau}$$

and so on. Correspondingly, for the function D we have

$$D \approx i.0.66\omega\tau - 0.62(\omega\tau)^2 - i0.61(\omega\tau)^3 + 0.61(\omega\tau)^4 - \cdots \quad \text{if } \omega\tau < 1 \tag{14.55}$$

and

$$D \approx -1 + \frac{3}{\pi(2\omega\tau)^{1/2}} + \frac{3i}{\pi(2\omega\tau)^{1/2}}\left[1 - \frac{1}{\pi}\left(\frac{2}{\omega\tau}\right)^{1/2}\right] \quad \text{if } \omega\tau \gg 1. \tag{14.56}$$

14.2.6. Frequency Responses of the Secondary Magnetic Field

Until now, we discussed only the asymptotic behavior of the field; next consider the spectrum of the quadrature and in-phase components of the field and for illustration functions Re D and Im D are shown in

Figure 14.2(c). In considering these curves, we can easily recognize three ranges of the frequency response, namely:

1. At the low frequencies, $\omega\tau < 1$, the quadrature component of the magnetic field increases almost directly proportional to the frequency and conductivity, while the in-phase component is considerably smaller and grows approximately as a square of ω.
2. Within the intermediate range of frequencies, the quadrature component of the magnetic field reaches a maximum and then with further increase of the frequency it decreases. This maximum takes place when ($\omega\tau \approx 1$). At the same time, the in-phase component gradually increases.
3. At high frequencies due to the skin-effect, induced currents are mainly concentrated near the surface of the sphere. The in-phase component of the magnetic field tends to a constant, which corresponds to the case of an ideal conductor. At the same time, the quadrature component QB tends to zero at a rate inversely proportional to $\omega^{1/2}$.

14.2.7. Sensitivity of the Field to the Parameter τ

As is well known, one of the main goals of the inductive methods of mining prospecting is detection of good conductors. In this light, let us discuss influence of the frequency on the sensitivity of the field to the parameters characterizing a conductor. In accordance with Eq. (14.32), the expressions for the magnetic field can be represented as a product of two terms, namely:

1. The function D, which depends only on the parameter τ.
2. The functions

$$\frac{a^3}{R^3} B_0 \cos\theta \quad \text{or} \quad \frac{a^3}{R^3} B_0 \sin\theta,$$

which are defined by the primary field, the sphere radius a, and the coordinates of an observation point, R and θ, but does not contain information about the conductivity. Correspondingly, we will mainly pay attention to the relationship between the field and the parameter τ at different frequency ranges. Our study of the low-frequency spectrum shows that the quadrature and in-phase components of the magnetic field have different sensitivities to the parameter τ. For instance, for the radial component B_{1R}, we obtain

$$QB^*_{sR} \approx 0.66\omega\tau \frac{a^3}{R^3} B_0 \cos\theta, \quad InB^*_{sR} \approx -0.62(\omega\tau)^2 \frac{a^3}{R^3} B_0 \cos\theta \quad \omega\tau \ll 1. \tag{14.57}$$

Thus, the in-phase component of the secondary field is much more sensitive to a change of conductivity and a sphere radius than the quadrature component, but it is essentially smaller. As follows from Eq. (14.57), the parameter τ can be approximately defined from the ratio of components measured at one frequency. In fact, we have

$$\tau \approx -\frac{1}{\omega}\frac{InB^*_{sR}}{QB^*_{sR}}, \quad \omega\tau \ll 1, \tag{14.58}$$

and this means that we do not need to measure the field at different frequencies to determine the parameter τ. In principle, we could markedly increase the sensitivity of the field to a change of parameter τ, if it would be possible to measure separately the terms of the series Eq. (14.38). As is seen from Figure 14.2(c), with an increase of the frequency, the relationship between the field and parameter τ generally becomes weaker. For instance, at high frequencies, (where the in-phase component of the field prevails), a change of τ practically does not make influence on the field. Thus, we can conclude that the low-frequency part of the spectrum is the most useful range, since the field can have a higher sensitivity to parameter τ. This means that within this range, it is often possible to a greater extent to separate fields caused by currents in conductors having different values of τ and therefore to reduce the influence of "geological noise".

14.2.8. The Role of This Example for Development of Inductive Methods

Inductive mining prospecting appeared almost 90 years ago but until the middle of the 1950s the geophysical community didn't know practically at all about the behavior of frequencies responses caused by induced currents in confined conductors. The courses of electroprospecting, given students, didn't contain any useful information about this subject, except some semi-quantitative facts derived from the experiments. And it happened in spite of the fact that a solution of the boundary-value problem for sphere, as well as for cylinder, were known almost from the beginning of the last century and perhaps earlier and long ago found place in the most classical books of electrodynamics. For instance, W. Smythe in his book "Static and Dynamic Electricity" described a solution of this problem in the frequency- and time domains and emphasized that the field can be expressed through elementary functions. This last factor was very attractive at the time of calculators. At the middle of 1950s, J. Wait described again the solution of this problem, but now it was published in the journal "Geophysics" that made this subject much more available for geophysical community, educated mainly in the geology-geophysical departments where courses of electrodynamics didn't cover such topics. It is natural that after this publication, the treatment of these solutions from geophysical point of view was undertaken and became possible to understand the main features of the mining inductive methods based on the use of the frequency and transient fields. It also helped to understand the difference of these methods from point of resolution and an influence of geological noise, and of course this allowed one to choose more optimal parameters of arrays and equipment.

14.3. THE CONDUCTING SPHERE IN A UNIFORM MAGNETIC FIELD (TIME DOMAIN)

14.3.1. Expressions for the Field Outside the Sphere

Next, we study a field caused by induced currents in the sphere, when the primary uniform magnetic field changes with time as the step function:

$$B_0 = \begin{cases} B_0 & t < 0 \\ 0 & t > 0 \end{cases}. \tag{14.59}$$

In order to determine the transient field, we make use of a Fourier transform. For instance, for the radial component B_{1R}, we have

$$B_{1R}(t) = -\frac{1}{2\pi}\int_{-\infty}^{\infty}\frac{B^*_{1R}(\omega)}{\omega}\exp\,(-\,i\omega t)\mathrm{d}\omega.$$

Then, taking into account Eq. (14.52), we obtain

$$B_{1R} = -\frac{B_0 a^3}{2\pi R^3}\cos\theta\frac{6\tau}{\pi^2}\sum_{n=1}^{\infty}\frac{1}{n^2}\int_{-\infty}^{\infty}\frac{\exp\,(-\,i\omega t)\mathrm{d}\omega}{n^2 - i\omega\tau}. \tag{14.60}$$

As we know, all integrands on the right-hand side of Eq. (14.60) have poles that are defined from Eq. (14.54). Applying the residue theorem, we have

$$\int_{-\infty}^{\infty}\frac{\exp\,(-\,i\omega t)}{n^2 - i\omega\tau}\mathrm{d}\omega = \frac{2\pi i\exp\left(-\frac{n^2 t}{\tau}\right)}{-i\tau} = -\frac{2\pi}{\tau}\exp\left(-\frac{n^2 t}{\tau}\right).$$

Therefore,

$$B_{sR}(t) = \frac{B_0 a^3}{R^3}\frac{6}{\pi^2}\cos\theta\sum_{n=1}^{\infty}\frac{1}{n^2}\exp\left(-\frac{n^2 t}{\tau}\right). \tag{14.61}$$

By analogy

$$B_{s\theta}(t) = \frac{B_0 a^3}{R^3}\frac{6}{\pi^2}\sin\theta\sum_{n=1}^{\infty}\frac{1}{n^2}\exp\left(-\frac{n^2 t}{\tau}\right)$$

and

$$E_{s\varphi} = \frac{3B_0 a^3}{R^2}\alpha\sin\theta\sum_{n=1}^{\infty}\exp\left(-\frac{n^2 t}{\tau}\right), \tag{14.62}$$

where

$$\alpha = \frac{1}{\pi^2\tau} = \frac{1}{\gamma\mu_0 a^2}. \tag{14.63}$$

Thus, the electromagnetic field outside the sphere is equivalent to that of the magnetic dipole, just as in the case of sinusoidal oscillations. The moment of the dipole is directly proportional to the primary field B_0 and decreases with time. As usual in studying the transient responses, we distinguish three stages, namely (1) the early stage ($t < \tau$), (2) an intermediate one, and (3) the late stage when the time of observation exceeds the parameter τ,($t > \tau$).

14.3.2. The Early and Late Stage

Let us begin with the early stage. In the limiting case, when $t \to 0$ from Eqs (14.61) and (14.62) follows

$$B_{1R} = \frac{6B_0 a^3}{\pi^2 R^3}\cos\theta\sum_{n=1}^{\infty}\frac{1}{n^2}, \quad B_{1\theta} = \frac{3B_0 a^3}{\pi^2 R^3}\sin\theta\sum_{n=1}^{\infty}\frac{1}{n^2},$$

Taking into account that

$$6\sum_{n=1}^{\infty}\frac{1}{\pi^2 n^2} = 1,$$

we have

$$B_{1R} = B_0\frac{a^3}{R^3}\cos\theta, \quad B_{1\theta} = B_0\frac{a^3}{2R^3}\sin\theta. \tag{14.64}$$

These expressions are exactly the same as those that correspond to the high-frequency spectrum. This coincidence is not accidental, since at the initial moment induced currents also flow on the surface of the sphere. They create a magnetic field equal in magnitude to the field at high frequencies ($\omega \to \infty$). At the same time, the magnetic field inside the sphere is equal to B_0 and induced currents are absent, except those on the surface of the sphere. Consider the early stage in some detail. Inasmuch as the early stage of the transient field is defined almost entirely by the high-frequency part of the spectrum, we make use of the limit theorem for Fourier transform. For instance, proceeding from the quadrature component of B_{1R}, we have

$$\frac{\partial B_{1R}}{\partial t} = -\frac{2}{\pi} \int_0^\infty QB_{1R} \sin \omega t \mathrm{d}\omega. \tag{14.65}$$

Bearing in mind Eq. (14.47), we have

$$\frac{\partial B_{1R}}{\partial t} = -\frac{2}{\pi}\frac{3}{2}(2\alpha)^{1/2}\frac{a^3}{R^3} B_0 \cos \theta \int_0^\infty \frac{\sin \omega t}{\omega^{1/2}} \mathrm{d}\omega \quad \text{if} \quad t \ll \tau.$$

Since

$$\int_0^\infty \frac{\sin \omega t}{\omega^{1/2}} \mathrm{d}\omega = \frac{1}{t^{1/2}} \int_0^\infty \frac{\sin x}{x^{1/2}} \mathrm{d}x = \left(\frac{\pi}{2t}\right)^{1/2},$$

we have

$$\frac{\partial B_{1R}}{\partial t} = -\frac{3B_0}{\pi^{3/2}(t\tau)^{1/2}} \frac{a^3}{R^3} \cos \theta, \tag{14.66}$$

By analogy

$$\frac{\partial B_{1\theta}}{\partial t} = -\frac{3B_0}{2\pi^{3/2}(t\tau)^{1/2}} \frac{a^3}{R^3} \sin \theta. \tag{14.67}$$

Thus, during the early stage, the derivative of the magnetic field with respect to time is inversely proportional to $\tau^{1/2}$ and decreases with time as $t^{-1/2}$. Now we derive an expression for the magnetic field. Since

$$B_{1R} = B_{1R}(0) + \int_0^t \frac{\partial B_{1R}}{\partial t} \mathrm{d}t$$

and taking into account Eq. (14.66), we obtain

$$B_{1R}(t) = B_0 \frac{a^3}{R^3} \cos\theta \left[1 - \frac{6}{\pi^{3/2}}\left(\frac{t}{\tau}\right)^{1/2}\right]$$

and

$$B_{1\theta}(t) = B_0 \frac{a^3}{2R^3} \sin\theta \left[1 - \frac{6}{\pi^{3/2}}\left(\frac{t}{\tau}\right)^{1/2}\right] \qquad t \ll \tau. \tag{14.68}$$

It is obvious that with an increase of the conductivity or the radius of the sphere, the early stage behavior persists over the longer times. In particular, in the case of perfectly conducting sphere, induced currents are present only on the surface of a conductor at all times and they do not decay. For this reason, the magnetic field of these currents is constant if $\tau = \infty$, and its value is equal to that for $t = 0$. As follows from Eq. (14.68), the magnetic field is only weakly related to conductivity and mainly depends on the radius of the sphere and its location. At the same time, electric field caused by a change of the magnetic field with time is inversely proportional to $\tau^{1/2}$ and increases without limit when time t tends to zero. This peculiarity is consequence of the step function form of excitation of the primary field. In practice, there is always some range of time ("a ramp time") over which the current source decreases and therefore the electric field remains finite at the initial instant. In accordance with Eqs (14.66) and (14.67), with an increase of parameter τ; that is with increase in conductivity the electric field becomes smaller at the early stage. In other words, the more resistive the conductor the stronger the electric field; and this occurs due to an increase in the rate of a change of the magnetic field with time. Calculations show that the early stage of the transient field practically takes place if $t/\tau < 0.3$. In deriving formulas for the field at the early stage, we have used the high-frequency spectrum, when at the boundary the impedance relation for horizontal components of the electric and magnetic field is valid. Therefore, it is natural to expect that all features of the early stage of the field caused by currents in a sphere are also typical in a case of an arbitrary conductor of finite dimensions.

Now, we consider the field behavior at the late stage when $t > \tau$. As follows from Eqs (14.61)–(14.64), no matter what the conductivity of this sphere at sufficiently large times, the field is almost entirely determined by the first exponent of these series

$$B_{1R} = \frac{6B_0}{\pi^2} \frac{a^3}{R^3} \exp\left(-\frac{t}{\tau}\right) \cos\theta, \quad B_{1\theta} = \frac{3B_0}{\pi^2} \frac{a^3}{R^3} \exp\left(-\frac{t}{\tau}\right) \sin\theta$$

and

$$E_\varphi = \frac{3B_0\, a^3}{\pi^2 \tau\, R^2}\exp\left(-\frac{t}{\tau}\right)\sin\theta \quad \text{if } t > \tau. \tag{14.69}$$

The parameter τ, given by Eq. (14.53), characterizes the rate of the change of the field with time at the late stage, and correspondingly it is called the time constant. Later, we demonstrate that the remarkable simplicity of Eq. (14.69) occurs at the late stage of the transient field caused by currents in any confined conductor. At the same time, it turns out that the time constant τ is a function of conductivity and the shape and dimensions of a conductor but is independent of the type of the primary field and the position of the observation point.

14.3.3. About Sensitivity of the Field at the Late Stage

The transient responses of functions L_1 and L_2, which describe the field **B** and its derivative with respect to time, that include the early, intermediate, and the late stages are shown in Figure 14.2(d):

$$L_1 = \frac{6}{\pi^2}\sum_{n=1}^{\infty}\frac{1}{n^2}\exp\left(-\frac{n^2 t}{\tau}\right), \quad L_2 = \frac{6}{\pi^2}\sum_{n=1}^{\infty}\exp\left(-\frac{n^2 t}{\tau}\right).$$

As is seen from these curves and a study of the asymptotic behavior of the field, the highest sensitivity to a change of parameter τ occurs at the late stage and it increases with an increase of time. Due to this fact, the transient method became the most efficient approach, allowing us to recognize fields caused by currents in conductors having different time constants. In this light, it may be proper to compare briefly the low-frequency part of the spectrum and the late stage. As we know, with a decrease of frequency ($\omega\tau < 1$) the sensitivity of the field to parameter τ does not change. In fact, the quadrature and in-phase components of the magnetic field are directly proportional to $\omega\tau$ and $(\omega\tau)^2$, respectively. However, at the late stage, this sensitivity changes and it grows with an increase of time. Of course, by separately measuring terms of the series Eq. (14.39), we can also increase the sensitivity of the field in the frequency domain to the parameter τ, but it is hardly possible to realize. Let us notice that the time constant τ can be easily determined from the late stage. For instance, in accordance with Eq. (14.69), we have

$$\ln B_{1R} = \ln\frac{6B_0 a^3}{R^3}\cos\theta - \frac{t}{\tau}, \tag{14.70}$$

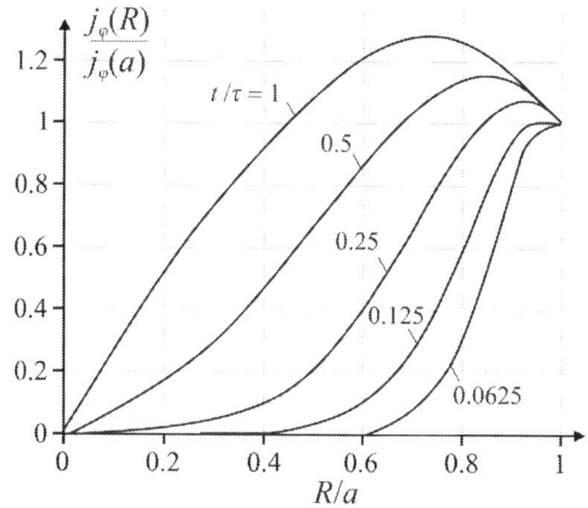

Figure 14.3 Behavior of induced currents inside the conductive sphere. Curve index is t/τ.

and τ defines the slope of the corresponding line, which represents Eq. (14.70).

14.3.4. Induced Currents in the Sphere

Consider also the transient responses of induced currents, Figure 14.3. They are obtained by performing a Fourier's transform of the function $j^*_\varphi(\omega)$. The similarity of these curves and those for the in-phase component is obvious since both of them reflect diffusion. In particular, at the late stage near the sphere center the current density increases almost linearly with increase of R. Then it reaches maximum and afterward it approaches the proper values at the surface. During the late stage, all induced currents decay in the same manner with time, $\exp(-t/\tau)$. Therefore, the relationship between current density at various points within the sphere is independent on time. This interesting feature of the current behavior at the late stage takes place in any confined conductor.

14.4. INFLUENCE OF MAGNETIZATION ON THE FIELD BEHAVIOR

14.4.1. Frequency Domain

Now consider the effect of the magnetic permeability, μ_i, on the behavior of the field and start from the frequency domain. In accordance with Eq. (14.28), we have

$$D = \frac{(2K+1)x - \left[(1+2K) + x^2\right]\tanh x}{(K-1)x + \left[(1-K) + x^2\right]\tanh x}, \tag{14.71}$$

where $K = \mu_i/\mu_e$. Letting $|x|\langle 1$ and replacing tan h x by the first few terms of its power series expansion, we obtain

$$D \approx \frac{2(K-1)}{2+K} + i\frac{3K\gamma\mu_i a^2}{5(2+K)^2}\omega - \cdots \tag{14.72}$$

Thus, in contrast to the behavior in the case of a nonmagnetic sphere, at low frequencies, the in-phase component of the magnetic field is usually dominant and with a decrease of the frequency tends to the limit corresponding to a constant magnetic field given by

$$D = \frac{2(K-1)}{2+K} = 2\frac{\mu_i - \mu_e}{\mu_i + 2\mu_e} \quad \text{if } \omega \to 0. \tag{14.73}$$

Correspondingly, we have the following expressions for the magnetic field in this approximation:

$$InB_{sR} = 2\frac{\mu_i - \mu_e}{\mu_i + 2\mu_e}\frac{a^3}{R^3}B_{0z}\sin\omega t\cos\theta$$

and

$$InB_{s\theta} = \frac{\mu_i - \mu_e}{\mu_i + 2\mu_e}\frac{a^3}{R^3}B_{0z}\sin\omega t\sin\theta. \tag{14.74}$$

At any instant, this magnetic field coincides with that when the primary field is constant and is equal to $B_{0z}\sin\omega t$. Equation (14.74) describe the field caused by the magnetization currents inside the sphere. As a result of a change of this field with time, a vortex electric field is generated. In accordance with Eq. (14.30), we have

$$E^e_{s\varphi} = \omega B_{0z}\frac{\mu_i - \mu_e}{\mu_i + 2\mu_e}\frac{a^3}{R^2}\sin\theta\cos\omega t \quad \text{if } \omega \to 0. \tag{14.75}$$

In order to derive corresponding expressions for the field inside the sphere, we can make use of asymptotic expressions for the Bessel's functions

$$I_{1/2}(x) \approx \left(\frac{2}{\pi x}\right)^{1/2}\left(x + \frac{x^3}{6}\right), \quad I_{-1/2}(x) \approx \left(\frac{2}{\pi x}\right)^{1/2}\left(1 + \frac{x^2}{2}\right).$$

Substituting these into Eq. (14.29), we have

$$C_1 \approx \left(\frac{\pi x}{2}\right)^{1/2} \frac{3Ka^{3/2}}{x^2\left[1 + \frac{(K-1)}{3}\right]}$$

Because

$$I_{3/2}(\alpha x) \approx \left(\frac{2}{\pi x \alpha}\right)^{1/2} \frac{\alpha x}{3}, \quad \alpha = \frac{R}{2}$$

we have

$$E_{\varphi}^{i} \approx i\omega B_{0z} \frac{3K}{2(2+K)} R \sin\theta. \tag{14.76}$$

Making use of Maxwell's equation

$$\text{curl}\, \mathbf{E}^* = i\omega \mathbf{B}^*,$$

it can be readily shown that inside the sphere the magnetic field is uniform and equal to:

$$\mathbf{B}_i = \frac{3K}{2+K} \mathbf{B}_0. \tag{14.77}$$

For this reason, the induced current density at small values of p increases with increase of μ_i, and for large values of K it is almost three times greater than that for a nonmagnetic sphere. From expression of the function D, it follows that

$$\frac{QD(K,p)}{QD(1,p)} = \frac{9K^2}{(2+K)^2}, \quad \text{if } p \ll 1 \tag{14.78}$$

that is the quadrature component of the magnetic field outside the sphere increases to a greater extent than does the current density as a consequence of the presence of the magnetic medium.

It may be proper to note that with an increase in magnetic permeability μ_i, the skin-effect expresses itself at lower frequencies than would otherwise be the case. At the high frequencies, both the magnetic field and induced currents vanish inside the sphere and the sole source for the field is a surface current. Correspondingly, the influence of the magnetic permeability disappears, and in the limit the function D tends to the value -1. Curves for the real and imaginary parts of this function are shown in Figure 14.4.

It should be noted that in the case of the magnetic sphere the in-phase component of the secondary magnetic field changes sign.

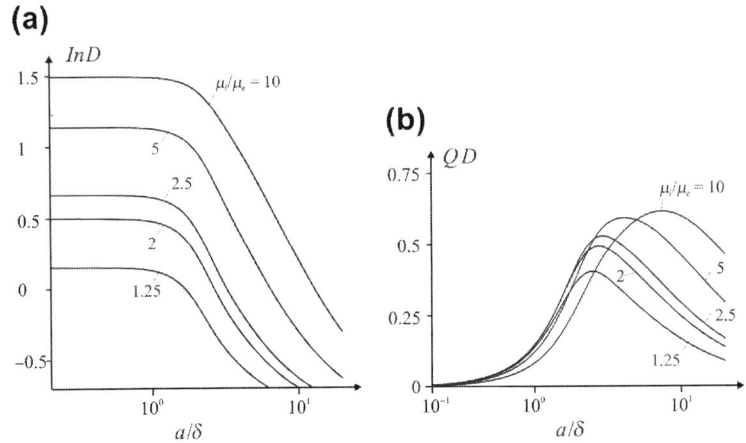

Figure 14.4 Real (a) and imaginary (b) parts of the function *D*.

14.4.2. Transient Responses of the Field

By analogy with a nonmagnetic sphere, the function D can be represented as a sum of fractions. Then, performing the Fourier transform, we obtain

$$B^e_{sR} = B_{0z}\frac{a^3}{R^3}\cos\theta\cdot K\pi^2 L_1(K,t)$$

$$B^e_{s\theta} = B_{0z}\frac{a^3}{2R^3}\sin\theta\cdot K\pi^2 L_1(K,t) \tag{14.79}$$

$$E^e_{s\varphi} = B_{0z}\frac{a^3}{2R^2}\sin\theta\cdot \alpha K\pi^2 L_2(K,t),$$

and

$$L_1 = \frac{6}{\pi^2}\sum_{s=1}^{\infty}\frac{\exp(-q_s\alpha t)}{q_s+(K-1)(K+2)},$$

$$L_2 = \frac{6}{\pi^2}\sum_{s=1}^{\infty}\frac{q_s\exp(-q_s\alpha t)}{q_s+(K-1)(K+2)}, \tag{14.80}$$

where q_s is defined from Eq. (14.28) when the denominator is equal to zero.

Behavior of these two functions on the parameter t/τ, where

$$\tau = \gamma\mu_i a^2/\pi^2$$

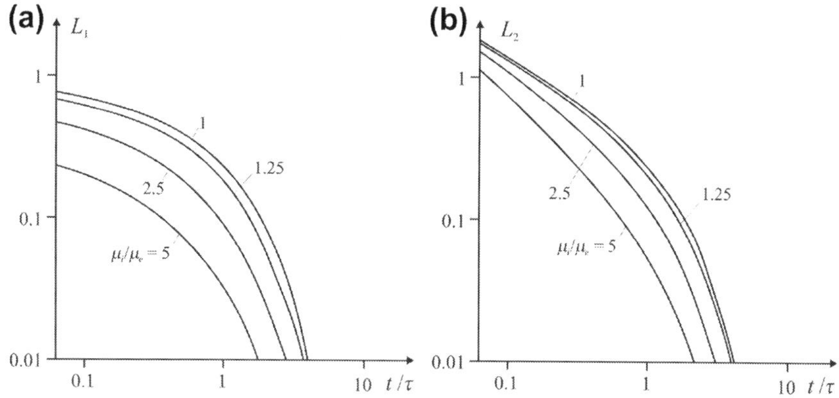

Figure 14.5 Functions L_1 (a) and L_2 (b). Curve index is μ_i/μ_e.

is shown in Figure 14.5. As is readily seen from the curves during the early stage, just as at high frequencies the effect of magnetic permeability vanishes and the function $KL_1(K,t)$ tends to unity. With increase in time, the influence of μ_i becomes stronger, especially at the late stage because the time constant depends strongly on magnetic permeability.

14.5. CONDUCTIVE SPHERE IN THE FIELD CAUSED BY A CURRENT LOOP WITH AXIAL SYMMETRY

Introduction

In deriving formulas for the field caused by induced currents in the sphere, it was assumed that the latter is placed in a uniform time-varying electromagnetic field. As was mentioned earlier, such behavior of the primary (normal) field in the volume occupied by a conductor can be observed with a high accuracy, if the source of the field is located at sufficient distance.

14.5.1. Expressions for the Field

Now consider the influence of nonuniformity in the primary magnetic field on the frequency responses of the field caused by induced currents in the sphere. We assume that the source of the primary field is a horizontal loop with radius r and its current is $I \cos \omega t$. As before, the primary electric field has only a single component E_φ; that is it doesn't intersect the sphere, and therefore the surface charges are absent. This means that the secondary electric field is caused by only a change of the magnetic field of induced currents with time, and correspondingly has also only the $\varphi-$ component.

In fact, as was shown in Chapter 1, the vector potential caused by a constant current in the circle loop has only the component A_φ

$$A_\varphi = \frac{\mu_0 I}{2\pi} \int_0^\pi \frac{r \cos \varphi d\varphi}{R}, \tag{14.81}$$

and the latter is valid for the quasi-stationary field too. By definition

$$\mathbf{B} = \text{curl } \mathbf{A}.$$

Then from the Faraday's law, we have

$$\text{curl } \mathbf{E} = -\text{curl } \frac{\partial \mathbf{A}}{\partial t} \quad \text{or} \quad \mathbf{E} = -\frac{\partial \mathbf{A}}{\partial t} - \text{grad } U.$$

In principle, there are electric charges located on the surface of the current loop, but their influence is negligible and correspondingly

$$\mathbf{E} = -\frac{\partial \mathbf{A}}{\partial t}. \tag{14.82}$$

Taking into account Eq. (14.81), we have for the normal field

$$E^*_{0\varphi} = \frac{i\omega\mu_0 r I}{2\pi} \int_0^\pi \frac{\cos \varphi d\varphi}{R}.$$

Then replacing the function $1/R$ by a sum containing spherical functions and performing integration, we arrive at the well-known expression for the primary electric field as a sum of spherical harmonics in the form:

$$E^*_{0\varphi} = \frac{i\omega\mu_e I}{2} \sum_{n=1}^{\infty} \frac{\sin \theta_0}{n(n+1)} \left(\frac{R}{R_1}\right)^n P^1_n(\cos \theta_0) P^1_n(\cos \theta), \tag{14.83}$$

which was given earlier. Here, R,θ are spherical coordinates of an observation point with the origin being at the center of the sphere. The parameters θ_0 and R_1 are shown in Figure 14.1(b). P^1_n is the associated Legendre function. For example,

$$P^1_1(x) = \left(1 - x^2\right)^{1/2}, \quad P^1_2(x) = 3x\left(1 - x^2\right)^{1/2}$$

$$P^1_3(x) = \frac{3}{2}\left(1 - x^2\right)^{1/2}\left(5x^2 - 1\right), \quad P^1_4(x) = \frac{5}{2}\left(1 - x^2\right)^{1/2}\left(7x^2 - 3x\right).$$

There are recurrence relationships for these functions:

$$nP_{n+1}^1(x) - (2n+1)xP_n^1 + (n+1)P_{n-1}^1(x) = 0.$$

By analogy with the case of a uniform field, described by the function $P_n^1(x)$, the electric field outside and inside the sphere can be written in the form:

$$\overset{*}{E}{}_\varphi^{e} = \frac{i\omega\mu_e I}{2}\sum_{n=1}^{\infty}\frac{\sin\theta_0}{n(n+1)}\left(\frac{R}{R_1}\right)^n P_n^1(\cos\theta_0)P_n^1(\cos\theta)$$

$$+\frac{i\omega\mu_e I}{2}\sum_{n=1}^{\infty}\frac{\sin\theta_0}{n(n+1)}\frac{D_{1n}}{R^{n+1}}\frac{1}{R_1^n}P_n^1(\cos\theta_0)P_n^1(\cos\theta)\quad \text{if } R\geq r$$

$$\overset{*}{E}{}_\varphi^{i} = \frac{i\omega\mu_e I}{2}R^{-1/2}\sum_{n=1}^{\infty}\frac{\sin\theta_0}{n(n+1)}$$

$$\times\frac{1}{R_1^n}C_n I_{\left[\frac{1}{4}+n(n+1)\right]}(ikR)P_n^1(\cos\theta_0)P_n^1(\cos\theta)\quad \text{if } R\leq r, \tag{14.84}$$

where D_{1n} and C_n are coefficients whose values are yet to be determined.

14.5.2. Formulas for Coefficients and Their Asymptotic Behavior

In order to find them, we can use the boundary conditions provided that $\mu_i = \mu_e = \mu_0$. In view of orthogonality of associated Legendre's functions, it follows from the equality of sums of the type in Eq. (14 84) that the spherical harmonics with the same index n can be equated. Thus, for each harmonic, we have two equations with two unknowns

$$\begin{aligned} a^n + \frac{D_{1n}}{a^{n+1}} &= a^{-1/2}C_n I_{n+\frac{1}{2}}(ika) \\ (n+1)a^n - \frac{nD_{1n}}{a^{n+1}} &= \left\{\frac{1}{2}a^{-1/2}I_{n+\frac{1}{2}}(ika) + ika^{1/2}I'_{n+\frac{1}{2}}(ika)\right\}C_n. \end{aligned} \tag{14.85}$$

Hence

$$C_n = \frac{\left(2n+1\right)a^{n+\frac{1}{2}}}{\left[\left(n+\frac{1}{2}\right)I_{n+\frac{1}{2}}(x) + xI'_{n+\frac{1}{2}}(x)\right]}, \qquad x = ika$$

Inasmuch as

$$I'_m(x) = I_{m-1}(x) - \frac{m}{x} I_m(x),$$

we obtain

$$C_n = \frac{(2n+1)a^{n+\frac{1}{2}}}{xI_{n-\frac{1}{2}}(x)}. \tag{14.86}$$

Substituting this last equation into the first equation of set 14.85, we have

$$D_{1n} = a^{2n+1} \frac{(2n+1)I_{n+\frac{1}{2}}(x) - xI_{n-\frac{1}{2}}(x)}{xI_{n-\frac{1}{2}}(x)}.$$

Since

$$\frac{2m}{x} I_m(x) = I_{m-1}(x) - I_{m+1}(x),$$

we have

$$D_{1n} = -a^{2n+1} \frac{I_{n+\frac{3}{2}}(x)}{I_{n-\frac{1}{2}}(x)} = -a^{2n+1} D_n,$$

where

$$D_n = \frac{I_{n+\frac{3}{2}}(x)}{I_{n-\frac{1}{2}}(x)}. \tag{14.87}$$

and $I_\nu(x)$ is a modified Bessel function. Correspondingly, the expression for the secondary electric field has the form:

$$E^{e^*}_{s\varphi} = -\frac{i\omega\mu_e I}{2} \sum_{n=1}^{\infty} \frac{\sin\theta_0}{(n+1)n} \left(\frac{a}{R_1}\right)^{n+1} D_n P_n^1(\cos\theta_0) P_n^1(\cos\theta). \tag{14.88}$$

Similar expressions can be written for both components of the magnetic field, $B^{e^*}_R$ an $B^{e^*}_\theta$. As follows from Eq. (14.88), each spherical harmonic of the primary field gives rise to a spherical harmonic of the secondary field for the same index n. The complex function D_n describes the spectrum for every harmonic of the field. In particular, at the low frequencies when $|ka|\langle 1$ we can obtain the following expression replacing Bessel function by the first few terms in its series expansion

$$D_n \approx -i\frac{\gamma\mu_0\omega a^2}{(2n+3)(2n+1)}\left[1+i\gamma\mu_0 a^2\frac{2\omega}{(2n+5)(2n+1)}\right] \quad \text{if } |ka|<1. \tag{14.89}$$

Therefore, the ratio of the in-phase component to the quadrature one tends to the following limit for each of the harmonics

$$\frac{\operatorname{Re} D_n}{\operatorname{Im} D_n}\to -\gamma\mu_0 a^2\frac{2\omega}{(2n+5)(2n+1)} \quad \text{as} \quad \omega\to 0. \tag{14.90}$$

This means that at low frequencies for spherical harmonics with larger number n, the role of the quadrature component becomes more significant; that is at this part of the spectrum, the greater the nonuniformity of the primary field, the less will be extent of the skin effect. In order to consider the high-frequency part of the spectrum, one should make use of asymptotic expressions of Bessel functions. For instance

$$I_{n-\frac{1}{2}}(x)\approx\frac{\exp(x)}{(2\pi x)^{1/2}}\left[1-\frac{n(n-1)}{2x}\right],$$

$$I_{n+\frac{3}{2}}(x)\approx\frac{\exp(x)}{(2\pi x)^{1/2}}\left[1-\frac{(n+2)(n+1)}{2x}\right].$$

Then, in accordance with Eq. (14.87), we have

$$D_n\approx 1-\frac{2n+1}{ika}, \tag{14.91}$$

Whence

$$\operatorname{Re} D_n\approx 1-\frac{2n+1}{2p}, \quad \operatorname{Im} D_n\approx -\frac{2n+1}{2p}, \tag{14.92}$$

where

$$p=\frac{a}{\delta}.$$

The real and imaginary parts of functions are shown in Figure 14.6. Regardless of the number n of the spherical harmonic, at high frequencies

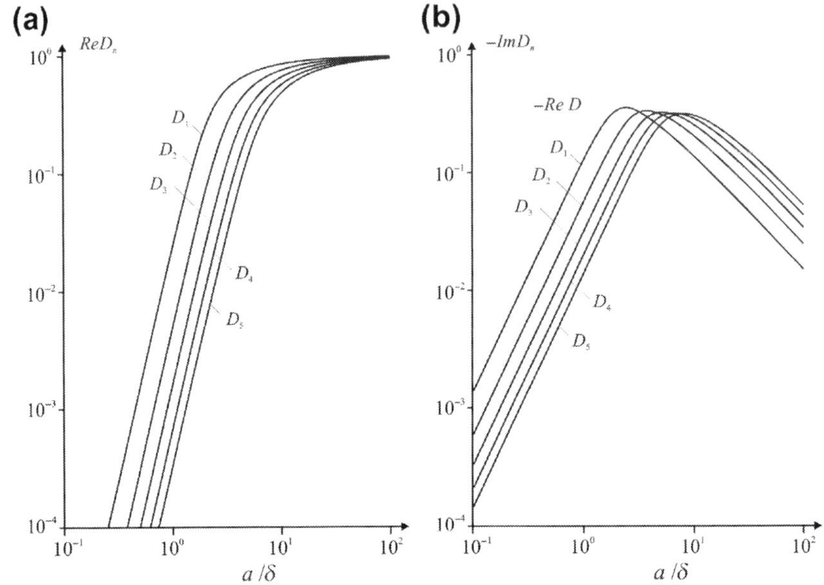

Figure 14.6 Behavior of the real and imaginary parts of the harmonics D_n.

the spectrum of the function D_n tends toward unity, as a consequence of the concentration of induced currents on the surface of the conductive sphere. At the same time, these responses clearly demonstrate that with an increase of n, the skin effect expresses itself at progressively higher frequencies; in particular, coincidence of the impedances for the plane and spherical surfaces of the conductor occurs also at higher frequencies.

14.5.3. Transition to a Uniform Field

Also let us note the following: the relative contribution of the first harmonic of the secondary field, which corresponds to the uniform primary field, increases in three cases, namely: (1) The distance from the current loop to the center of the sphere increases. This occurs because the primary field becomes more uniform in the vicinity of the sphere. (2) When the distance from the observation point to the center of the sphere increases. This can be explained in the following way. In accordance with Eq. (14.84), a real distribution of currents is represented as being the sum of various systems of currents, each of which corresponds to a specific spherical harmonic. Every such system except the first one for the first harmonic consists of a group of currents having different directions. Due to the change of the sign of these currents, with an increase in the harmonic number n the corresponding harmonic of the field decreases more rapidly with an increase of the distance. Therefore, the effect of the first harmonic becomes dominant. (3) In this third case, the first harmonic plays the most important role regardless of the distance from the source or an observation point from the sphere's center, but implies that measurements are performed at low frequencies. Using the expressions for Bessel functions, one can show that for each spherical harmonic the function D_n can be expressed in a series containing only integer powers of ω,

$$D_n = a_{1n}\omega + a_{2n}\omega^2 + a_{3n}\omega^3 + a_{4n}\omega^4 + \cdots \tag{14.93}$$

Now suppose that in measuring a secondary field, that only the term proportional to ω is defined; following which only the term proportional to ω^2 is measured, and so on. It is obvious that the effect of the spherical harmonics with various numbers n is mainly defined by the rate of decrease of the corresponding coefficients a_n. In accordance with Eq. (14.89), the ratio of the coefficients a_{1n} and a_{11}, characterizing the contribution of a spherical harmonic with index n in measuring the part of the field proportional to ω is

$$\frac{a_{1n}}{a_{11}} = \frac{15}{(2n+3)(2n+1)}. \tag{14.94}$$

This shows that when the quadrature component of the magnetic field is observed at low frequencies, contribution of harmonics decreases relatively slowly with increasing n. Now for a moment, suppose that the secondary field proportional to ω^2 (the in-phase component of the magnetic field at low frequencies) is measured. Then, from Eq. (14.89), we have

$$\frac{a_{2n}}{a_{21}} = \frac{315}{(2n+5)(2n+3)(2n+1)^2}, \tag{14.95}$$

and from comparison with Eq. (14.94), it follows that the effect of the first harmonic becomes greater. One can show that in measuring the terms in the spectrum containing higher powers of ω, the effect of the first harmonic gradually increases. Thus, no matter what the geometric factor, when the latter terms in a series describing the field at low frequencies are measured, the effect of the nonuniformity of the primary field disappears.

We have considered the frequency domain and shown three cases when the influence of uniformity of the primary field becomes insignificant. In the time domain, we have the same two first cases, but the third case corresponds to the late stage. In other words, with an increase of index n the spherical harmonics decay more rapidly at the late stage and therefore again the first harmonic plays the dominant role.

14.6. THE CIRCULAR CYLINDER IN A UNIFORM MAGNETIC FIELD (FREQUENCY DOMAIN)

Introduction

Next consider the second example of a conductor when it is possible applying the method of separation of variables to avoid a solution of an infinite system of equations with infinite number of unknowns and obtain the expression of the field in the explicit form. First, assume that a markedly elongated conductor is oriented along the primary electric field and surrounded by a nonconducting medium. Electric charges arise on the edges and lateral surface near the edges, and the normal component of the electric field caused by these charges is equal to zero at points of the internal side of this surface. Correspondingly, induced currents form closed loops within the conductor and are almost parallel to the conductor surface. From a physical point of view, it is clear that with an increase of the length of the conductor in the direction of the primary field, $\mathbf{E}_0$, in comparison to the cross section and the distance from the body to an observation point, the influence from the edge effect will decrease progressively. The contribution to the total magnetic field from currents that are not parallel to the lateral surface of the conductor becomes smaller and smaller. Therefore, under certain conditions, the electromagnetic field from the currents in the confined conductive body is almost exactly the same as that from currents in an infinitely long cylinder. A similar conclusion can apply when the primary electromagnetic field vanishes near the edges of the elongated conductive body. We will investigate this problem in some detail later, but now consider the model of an infinitely long cylinder.

14.6.1. Solution of the Boundary-Value Problem

Suppose that a right circular conducting cylinder is situated in a uniform magnetic field

$$\mathbf{B} = \mathrm{Re}\, \mathbf{B}_{0z} \exp(-i\omega t),$$

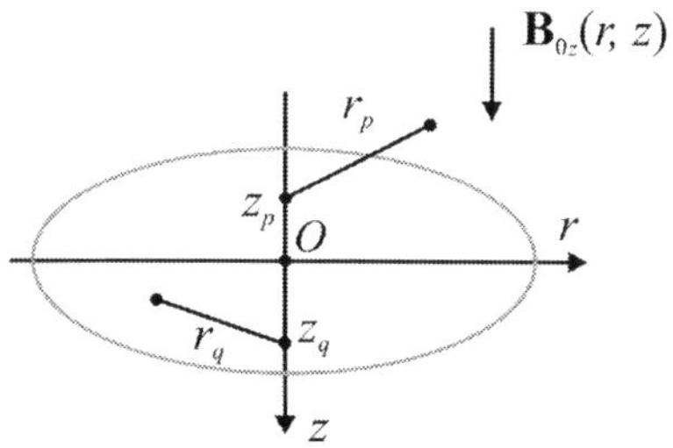

Figure 14.7 The picture illustrating derivation of integral equation.

that is directed perpendicularly to the axis of the cylinder. Also assume that the primary field is caused by two systems of linear current filaments, which are oriented parallel to the axis of the conductor and with the currents of the same magnitude flowing in the opposite directions. In this case, the primary electric field has only a single component E_{0x}, and making use of the first Maxwell's equation in its integral form:

$$\oint \mathbf{E}\cdot d\mathbf{l} = -\frac{\partial \Phi}{\partial t},$$

we find the following relation for an arbitrary path L, laying in the plane xy, Figure 14.7,

$$E_{0x}^{*} = -i\omega B_{0z} r \sin\varphi, \qquad (14.96)$$

where

$$\sin\varphi = \frac{y}{r}, \quad r = \left(y^2 + z^2\right)^{1/2}.$$

We see from this equation that the primary electric field is directed oppositely on either sides of the plane $z0x$. This field causes vortex currents to flow in the conductor with the current filaments being oriented parallel to the x-axis and closing themselves at infinite. The secondary field can be represented as being the sum of fields due to the system of linear currents, and in particular, in the cylindrical system of coordinates, this field can be described by three components: B_r, B_φ, E_x. First, we find the electric field, which at regular points obey the Helmholtz equation. Taking into account that the field is independent on the coordinate x this equation becomes:

$$\frac{\partial^2 E_x^{i^*}}{\partial r^2} + \frac{1}{r}\frac{\partial E_x^{i^*}}{\partial r} + \frac{1}{r^2}\frac{\partial^2 E_x^{i^*}}{\partial \varphi^2} + k^2 E_x^{i^*} = 0, \quad \text{if } r < a \qquad (14.97)$$

and

$$\frac{\partial^2 E_x^{e^*}}{\partial r^2} + \frac{1}{r}\frac{\partial E_x^{e^*}}{\partial r} + \frac{1}{r^2}\frac{\partial^2 E_x^{e^*}}{\partial \varphi^2} = 0, \quad \text{if } r > a \tag{14.98}$$

At the cylinder surface, we have:

$$E_x^{i^*} = E_x^{e^*} \quad \text{and} \quad \frac{1}{\mu_i} B_\varphi^{i^*} = \frac{1}{\mu_e} B_\varphi^{e^*} \quad \text{if } r = a. \tag{14.99}$$

Making use of Maxwell's equation,

$$\text{curl}\,\mathbf{E} = -\partial \mathbf{B}/_{\partial t},$$

these conditions can be written as

$$E_x^{i^*} = E_x^{e^*}, \quad \frac{1}{\mu_1}\frac{\partial E_x^{e^*}}{\partial r} = \frac{1}{\mu_i}\frac{\partial E_x^{i^*}}{\partial r}. \tag{14.99}$$

We will seek for a solution of Eqs (14.97) and (14.98) in the form of the product of two functions $R(r)\Phi(\varphi)$. Substituting this assumed solution into Eq. (14.97) and multiplying both sides by $r^2/_{R\Phi}$, we obtain

$$\frac{r^2}{R}R'' + \frac{r}{R}R' + \frac{\Phi''}{\Phi} + k^2 r^2 = 0.$$

This equation can be easily separated into two ordinary differential equations

$$\frac{\mathrm{d}^2 R}{\mathrm{d}r^2} + \frac{1}{r}\frac{\mathrm{d}R}{\mathrm{d}r} - \left[(ik)^2 + \frac{n^2}{r^2}\right]R = 0 \tag{14.100}$$

and

$$\frac{\mathrm{d}^2 \Phi}{\mathrm{d}\varphi^2} + n^2 \Phi = 0. \tag{14.101}$$

The second of these equations is very common one, being the equation for a simple harmonic oscillator that has the well-known solution:

$$\Phi_n(\varphi) = A_n^1 \cos n\varphi + B_n^1 \sin n\varphi. \tag{14.102}$$

Equation (14.100) is a form of the Bessel equation, for which the solutions are modified Bessel functions:

$$R_n = C_n^1 I_n(ikr) + D_n^1 K_n(ikr). \tag{14.103}$$

Outside the cylinder, Eq. (14.100) is greatly simplified

$$\frac{d^2R}{dr^2}+\frac{1}{r}\frac{dR}{dr}-\frac{n^2}{r^2}R=0. \tag{14.104}$$

Particular solutions to this equation are:

$$R_n = C_n^2 r^n + D_n^2 r^{-n} \quad \text{if} \quad n \neq 0 \tag{14.105}$$

and

$$R_n = C_0^2 \ln r + D_0^2 \quad \text{if} \quad n=0. \tag{14.106}$$

Inasmuch as the magnetic field from induced currents in a cylinder observed at distances that are considerably greater than the radius of the cylinder is equivalent to the field from a simple linear dipole; that is it decreases as $1/_{r}2$ if $r \to \infty$, the secondary electric field will tend to zero as $\frac{1}{r}$. In other words, the zero harmonic, Eq. (14.106), has to be discarded, and the field can be represented as

$$E_x^{e^*} = -i\omega B_0\left[r\sin\varphi+\sum_{n=1}^{\infty} r^{-n}(B_n\cos n\varphi+T_n\sin n\varphi)\right]$$

and

$$E_x^{i^*} = -i\omega B_0\sum_{n=1}^{\infty} I_n(ikr)(D_n\cos n\varphi+C_n\sin n\varphi). \tag{14.107}$$

Note that Bessel function $K_n(ikr)$ cannot be used to describe the field inside the cylinder because

$$K_n(ikr)\to\infty \quad \text{as} \quad r\to 0.$$

Applying the boundary conditions, Eq. (14.99), and making use of orthogonality of trigonometric functions, we obtain the system of equations for unknown coefficients for each harmonic:

$$a+a^{-1}T_1 = I_1(ika)C_1$$

and

$$\frac{1}{\mu_e}\left(1-a^{-2}T_1\right)=\frac{1}{\mu_i}ikI_1'(ika)C_1 \quad \text{if} \quad n=1 \tag{14.108}$$

and

$$a^{-n}T_n - I_n(ika)C_n = 0$$
$$\frac{n}{\mu_e}a^{-n-1}T_n + \frac{1}{\mu_i}ikI_n'(ika)C_n = 0 \quad \text{if } n \neq 1. \tag{14.109}$$

The determinant for the system of Eq. (14.109) is nonzero, and therefore

$$T_n = C_n = 0 \quad \text{if} \quad n \neq 1.$$

In the same way, we can demonstrate that all the coefficients B_n and D_n are zero. Thus, the field is completely described by a single cylindrical harmonic, $n = 1$. Solving the system of Eq. (14.108) and making use of the recurrence relationships for modified Bessel functions, which are as follows:

$$I_n'(x) = I_{n-1}(x) - \frac{n}{x}I_n(x)$$
$$\frac{2n}{x}I_n(x) = I_{n-1}(x) - I_{n+1}(x),$$

after some algebraic operations we arrive at the result:

$$T_1 = a^2\frac{(\mu_i - \mu_e)I_0(ika) - (\mu_i + \mu_e)I_2(ika)}{(\mu_i + \mu_e)I_0(ika) - (\mu_i - \mu_e)I_2(ika)} \tag{14.110}$$

and

$$C_1 = \frac{4\mu_i}{ik[(\mu_i + \mu_e)I_0(ika) - (\mu_i - \mu_e)I_2(ika)]}. \tag{14.111}$$

From Eq. (14.107), we see that the electromagnetic field from currents induced in a right circular cylinder is the same as that generated by a linear dipole:

$$E_{1x}^{e^*} = -i\omega B_0 T_1\frac{\sin\varphi}{r}, \quad B_{ir}^{e^*} = -B_0 T_1\frac{\cos\varphi}{r^2}, \quad B_{1\varphi}^{e^*} = -B_0 T_1\frac{\sin\varphi}{r^2}. \tag{14.112}$$

These equations describe a secondary field, which is characterized by unusually simple geometry. In particular, the current density in the half-planes $0 < \varphi < \pi$ and $\pi < \varphi < 2\pi$ has opposite directions. The function T_1 is particularly simple when the magnetic permeability in both media has the same value. Then

$$T_1 = Ta^2 = -\frac{I_2(ika)}{I_0(ika)}a^2 \quad \text{if} \quad \mu_i = \mu_e. \tag{14.113}$$

The function $I_0(ika)$ has simple roots; that is the singularities of the spectrum for the complex amplitude of T are simple poles situated on the imaginary ω axis, and the first of them defines the radius of convergence of the low-frequency spectrum. In this part of the spectrum $|ka|\langle 1$ replacing the Bessel functions by the first few terms in their series expansion we have, if $\mu_i = \mu_e = \mu$

$$T \approx i\frac{\gamma\mu\omega a^2}{8} - \frac{1}{48}\left(\gamma\mu\omega a^2\right)^2 + \cdots \tag{14.114}$$

In contrast, at the high-frequency spectrum, applying asymptotic expressions for the Bessel functions

$$I_n(x) \approx \frac{\exp x}{(2\pi x)^{1/2}}\left[1 - \frac{4n^2-1}{1!8x} + \frac{\left(4n^2-1\right)\left(4n^2-3\right)}{2!(8x)^2} - \cdots\right]$$

and keeping only terms up to the second order, we have

$$T \approx -1 + \frac{2}{ika}, \quad \text{if } |ka| \gg 1. \tag{14.115}$$

As in the case of spherical conductor, the quadrature and in-phase approach their limit in inverse proportional to $\omega^{1/2}$. This coincidence is not occasional and it follows from the approximate boundary condition valid regardless of a conductor shape, provided that the wave length is much smaller than the radius of curvature. Similarity with the function D is obvious.

14.7. TRANSIENT RESPONSES OF THE FIELD CAUSED CURRENTS IN A CIRCULAR CYLINDER

In principle, we can obtain expression for the transient field in the same way as in the case of the spherical conductor, namely applying Fourier's transform. However, it is instructive to use a different approach based on a solution of diffusion equation for the component of the vector potential A_x:

$$\nabla^2 A_x^i = \gamma\mu\frac{\partial A_x^i}{\partial t} \quad r < a$$

and

$$\nabla^2 A_x^e = 0 \quad r > a \tag{14.116}$$

Applying the method of separation of variables as has been done previously and taking into account that the primary field is uniform, the component A_x can be written as

$$A_x^i = \sin\varphi \sum_{s=1}^{\infty} C_s J_1(k_s r)\exp(-q_s \alpha t)$$

and

$$A_x^e = \frac{\sin\varphi}{r} \sum_{s=1}^{\infty} T_s \exp(-q_s \alpha t), \tag{14.117}$$

where

$$k_s^2 a^2 = q_s, \quad \alpha = \frac{1}{\gamma\mu a^2}. \tag{14.118}$$

Using the equation

$$\mathbf{B} = \text{curl}\,\mathbf{A},$$

we obtain the following expression for the vector potential characterizing the uniform primary field

$$A_{0x} = -B_0 r \sin\varphi. \tag{14.119}$$

Assuming a step function form of excitation the initial condition at $t=0$ is

$$B_0 r = -\sum_{s=1}^{\infty} C_s J_1(k_s r), \quad \text{if } r < a \tag{14.120}$$

that is we derived the first equation for determination of unknown coefficients C_s. Next, making use of the boundary conditions at the surface of the cylinder, we obtain a uniform system of equations:

$$aC_s J_1(k_s a) = T_s \quad \text{and} \quad a^2 k_s C_s J_1'(k_s a) = -T_s, \tag{14.121}$$

which has nonzero solution for C_s and T_s only if the determinant for the system is zero; that is

$$J_1(k_s a) + k_s a J_1'(k_s a) = 0.$$

Taking into account the recurrence relationship for Bessel functions, the latter is greatly simplified, and it gives

$$J_0(k_s a) = 0. \tag{14.122}$$

Table 14.1 Roots of equation (14.122)

$s = 1$	2	3	4
$k_s a = 2.405$	5.521	8.654	11.792
$q_s = 5.783$	30.472	74.887	139.04

Therefore, continuity of the tangential components of the electric and magnetic field in the time domain on the surface of a circular cylinder is satisfied providing that the values for $k_s a$ are roots of the Bessel function of the zero order. The first five values of these roots are given in Table 14.1

From the initial condition, Eq. (14.120), and making use of formulas for determination of the expansion of the function $B_0 r$ in a series in terms of the function $J_1(k_s r)$, we find

$$C_s = -\frac{2B_0}{k_s}\frac{J_2(k_s a)}{[J_1(k_s a)]^2}, \qquad T_s = -\frac{4B_0 a^2}{(k_s a)^2} \tag{14.123}$$

Whence for the transient field outside the cylinder, we obtain the following expressions:

$$E_x^e = -B_0 \frac{4a^2}{r} \alpha \sin\varphi \sum_{s=1}^{\infty} \exp(-q_s \alpha t)$$

$$B_r^e = -\frac{B_0 a^2}{r^2} 4 \cos\varphi \sum_{s=1}^{\infty} \frac{1}{q_s} \exp(-q_s \alpha t) \tag{14.124}$$

$$B_\varphi^e = -\frac{B_0 a^2}{r^2} 4 \sin\varphi \sum_{s=1}^{\infty} \frac{1}{q_s} \exp(-q_s \alpha t).$$

From these expressions, it is clear that during a very early stage of the time response, the magnetic field is independent on the parameter α and we have

$$B_r^e = -4B_0 \frac{a^2}{r^2} \cos\varphi \sum_{s=1}^{\infty} \frac{1}{q_s} \quad \text{and} \quad B_\varphi^e = -4B_0 \frac{a^2}{r^2} \sin\varphi \sum_{s=1}^{\infty} \frac{1}{q_s}. \tag{14.125}$$

At the same time, the electromotive force tends to infinity. Applying Fourier's transform to the high-frequency spectrum of the function T, we can see again that with a decrease of time the electromotive force increases in inverse proportion to the square root of time, while the magnetic field approaches the asymptote, Eq. (14.125), as a function that is proportional to $t^{1/2}$. In contrast, at the late stage, the field is described by a single exponential term, just as was in the case of a spherical conductor. In fact, Eq. 14.124 give

$$E_x^e = -4B_0 \frac{a^2}{r} \alpha \exp(-q_1 \alpha t) \sin \varphi, \quad B_r^e = -4B_0 \frac{a^2}{r^2} \frac{1}{q_1} \exp(-q_1 \alpha t) \cos \varphi$$

and

$$B_\varphi^e = -4B_0 \frac{a^2}{r^2} \frac{1}{q_1} \exp(-q_1 \alpha t) \sin \varphi \quad \text{if} \quad \frac{t}{\tau} \gg 1. \qquad (14.126)$$

Here $\tau = \gamma\mu a^2 / q_1$ and q_1 is approximately 5.78. In comparison with the late stage of the field by currents in a spherical body, we can see that if $\alpha_{\text{sph}} = \alpha_c$ the field generated by linear current filaments decays more slowly than that caused by circular filaments. In conclusion, note that as in the case of the sphere, the ratio of the in-phase and quadrature components, Eq. (14.114), defines the position of the first pole in the spectrum within reasonable accuracy as well as the corresponding time constant. We have

$$\left| \frac{\text{Re } T}{\text{Im } T} \right| \approx \frac{\gamma\mu a^2}{6} \approx \omega\tau.$$

14.8. EQUATIONS FOR THE FIELD CAUSED BY CURRENTS IN A CONFINED CONDUCTOR

Introduction

Until now, we have considered the frequency and transient responses for two very special cases of a conductor: sphere and circular cylinder. Next, we demonstrate that the most important features discovered in these examples remain valid for any primary field and an arbitrary confined conductor surrounded by an insulator. Let us begin with the model when the primary field and conductors are characterized by the axial symmetry with the common axis. In all these cases, the primary magnetic field is caused by a current ring or a system of current rings, and the primary electric field has only a single component $E_{0\varphi}$. In particular, the primary magnetic field can be practically uniform in the neighborhood of the conductor. It is obvious that due to geometry the surface electric charges are absent and the current density at every point of a conductor is

$$j_\varphi = \gamma\left(E_{0\varphi} + E_{1\varphi}\right), \tag{14.127}$$

where $E_{0\varphi}$ is the primary electric field, and $E_{1\varphi}$ is a secondary vortex electric field caused by the magnetic field of induced currents. Also, γ is the conductivity and in the cylindrical system of coordinates (r,φ,z); it can be an arbitrary function of r and z, but not φ. It is clear that the term $E_{1\varphi}$ is unknown before the field is calculated. It is almost obvious that the interaction between currents does not change the direction of the current flow, and it remains the same regardless of either frequency or the time, but later this question will be discussed in some detail. As was mentioned earlier, in contrast to the model of the spherical conductor determination of induced currents and the field for these more general cases is much more complicated problem in spite of the axial symmetry, and it requires application of numerical methods. Even when the conductor has the relatively simple shape of a spheroid, the solution using the method of separation of variables is still cumbersome regardless of the simplicity of the primary field. In this case, the method of separation of variables permits one to replace the Helmholtz equation in spheroidal coordinates by two ordinary differential equations, for which the solutions are the associated Legendre's functions. The field inside and outside the spheroid can be written as an infinite sum, where the amplitudes of the spheroidal harmonics are unknown. In principle, these amplitudes can be determined from an infinite system of equations that describes the required continuity of tangential components of the electric and magnetic fields at the surface of the spheroid. However, in contrast to the same problem for a sphere, this system cannot be reduced to sets of two equations with two unknowns for each spheroidal harmonic. It is a consequence of the fact that both spheroidal functions describing the field inside and outside of the conductor depend on the wave number. As was pointed out, a computational algorithm for the solution of an infinite system of equations is in general not available, and therefore the application of the method of separation of variables leads to extremely cumbersome numerical problems. For this reason, the field is usually determined even in this case using the method of finite difference or integral equations.

14.8.1. The Integral Equation for the Current Density

In order to derive equations for the frequency domain and then obtain transient fields caused by currents in confined conductors, we make use of the integral equation, and with this purpose in mind proceed from equation for the current density, Eq. (14.127). Applying the principle of superposition, the current density at any point of a conductor can be written in the form:

$$j_\varphi(p) = j_{0\varphi}(p) + i\omega\mu\gamma(p)\int\limits_S G(p,q) j_\varphi(q)\mathrm{d}S, \tag{14.128}$$

where

$$j_{0\varphi}(p) = \gamma(p)E_{0\varphi}(p)$$

is the current density caused by the primary vortex electric field, and

$$i\omega\mu G(p,q) j_\varphi(q)\mathrm{d}S$$

is the vortex electric field caused by a circular current filament passing through a cross section of an elementary current tube at the point q, as is shown in Figure 14.7. There are several known expressions for the function G. For instance

$$G(p,q) = \frac{1}{\pi u}\left(\frac{r_q}{r_p}\right)^{1/2}\left[\left(1-\frac{1}{2}u^2\right)K(u) - E(u)\right], \tag{14.129}$$

where

$$u^2 = \frac{4r_p r_q}{\left(r_p + r_q\right)^2 + \left(z_p - z_q\right)^2}$$

and r_q, z_q, r_p, z_p are coordinates of the points q and p, respectively. Also, $K(u)$ and $E(u)$ are the complete elliptical integrals of the first and second kinds, and S is the half cross-section of a conductor. Thus, we have obtained an integral equation with respect to the current density j_φ. The first term on the right-hand side, $j_{0\varphi}(p)$ is easily calculated.

14.8.2. Transition to the System of Linear Equations

In general, the integral equation can be reduced to a system of linear equations with constant coefficients, and it is useful to examine this equivalence from the physical point of view. Let us represent a current distribution within a conductor as a system of currents flowing in elementary tubes; due to symmetry their centers are situated on the z-axis. We suppose that the cross-sectional area of these elementary tubes is small, so that the change of current density or the field across this section can be neglected. Correspondingly, Eq. (14.127) can be rewritten as

$$j_\varphi(p) = j_{0\varphi}(p) + i\omega\mu\gamma(p)\sum_{q\neq p} G(p,q)I(q) + i\omega\mu\gamma(p)\int\limits_{S(p)} G(p,q)j_\varphi(q)\mathrm{d}S. \tag{14.130}$$

The summation on the right-hand side of this equation is carried out over all elementary tubes except the tube at the point p. Over the cross section of this tube, the integration must still be performed, and it is described by the last term in Eq. (14.130). Also here $I(q)$ is the complex amplitude of the current flowing in the elementary tube in which point q is located at the center of the cross-sectional area. Multiplying both sides of Eq. (14.130) by the value $2\pi r_p/\gamma(p)$ and considering that

$$j_\varphi(p)\frac{2\pi r_p}{\gamma(p)} = \frac{I(p)}{\mathrm{d}S(p)}\frac{2\pi r_p}{\gamma(p)} = I(p)R(p),$$

we obtain

$$I(p)R(p) = \Xi_0(p) + i\omega\mu\sum_{q\neq p} G^*(p,q)I(p) + i\omega\mu\int\limits_{S(p)} G^*(p,q)j_\varphi\mathrm{d}S, \tag{14.131}$$

where $R(p)$ is the resistance of the elementary tube passing through the point p and

$$\Xi_0(p) = \frac{2\pi r_p}{\gamma(p)}j_{0\varphi}(p) = 2\pi r_p E_{0\varphi}(p)$$

is the electromotive force around the ring p caused by the primary electric field.

$$G^*(q,p) = 2\pi r_p G(p,q),$$

and finally

$$i\omega\mu G^*(p,q)I(q)$$

is the electromotive force in the ring p, caused by the current flowing in the ring q, which is $I(q)$. The last term in Eq. (14.131) is then electromotive force in the ring p due to the current $I(p)$ flowing in it, and this can be written as

$$i\omega\mu\int\limits_{S(p)} j_\varphi G^*\mathrm{d}S = \frac{i\omega\mu I(p)}{S(p)}\int\limits_{S(p)} G^*\mathrm{d}S.$$

For convenience, we make use of the following notations:

$$M_{pq} = -\mu G^*(p, q)$$

and

$$L_p = -\frac{\mu}{S(p)} \int\limits_{S(p)} G^*(p, q)\mathrm{d}S, \tag{14.132}$$

where M_{pq} and L_p are the mutual inductance between coaxial rings passing through points q and p, and the self-inductance of the ring p, respectively. Substituting these notations into Eq. (14.131), we have

$$I(p)R(p) = \Xi_0(p) - i\omega \sum_{q \neq p} M_{pq} I(q) - i\omega L_p I(p)$$

or

$$\left(R_p + i\omega L_p\right) I(p) + i\omega \sum_{q \neq p} M_{pq} I(q) = \Xi_0(p). \tag{14.133}$$

Having written Eq. (14.133) for every ring, in place of the integral Eq. (14.128), we arrive at the system of linear equations with respect to currents in all elementary tubes:

$$\begin{aligned}
&Z_{11} I_1 + i\omega M_{12} I_2 + \cdots + i\omega M_{1n} I_n + \cdots + i\omega M_{1N} I_N = \Xi_{01}\\
&i\omega M_{21} I_1 + Z_{22} I_2 + \cdots + i\omega M_{2n} I_n + \cdots + i\omega M_{2N} I_N = \Xi_{02}\\
&i\omega M_{n1} I_1 + Z_{n2} I_2 + \cdots + i\omega M_{nn} I_n + \cdots + i\omega M_{nN} I_N = \Xi_{0n}\\
&i\omega M_{N1} I_1 + Z_{N2} I_2 + \cdots + i\omega M_{Nn} I_n + \cdots + i\omega M_{NN} I_N = \Xi_{0N},
\end{aligned} \tag{14.134}$$

where

$$Z_{nn} = R_n + i\omega L_n$$

is the impedance of the nth ring. The right-hand side of Eq. (14.134) is the electromotive force caused by the primary electric field $E_{0\varphi}$, and in every ring, it is directly proportional to ω. It is obvious that the conversion from the integral equation to the system of equations of the last set (Eq. (14.134)) is based on the equivalence of fields caused by a volume distribution and a system of currents forming linear rings with resistance as well as self- and mutual inductance. The coefficients in Eq. (14.134) can be evaluated using the well-established expressions for R, L, and M for linear rings with a circular cross section. As is well known, for a thin circular ring, we have

$$R = \frac{2\pi r}{\gamma S}, \quad L = r\mu\left(\ln\frac{8r}{r_0} - 1.75\right), \quad M_{pq} = -2\pi r_p G_{pq},$$

where r_0 is the radius of the cross section of the ring. Of course, with an increase in number of tubes with which the current distribution is approximated, the errors involved in replacing a volume tube by a linear current become smaller. It can readily be seen that the inductive resistance of an elementary tube must be significantly less than the resistance R; that is $R > \omega L$. Otherwise, the radius of the cross section of the tube r_0 becomes greater than the skin depth, and when this happens, there is no uniformity of the current density and the field over the cross section. Therefore, in order to carry out calculations at relatively high frequencies, it is necessary to increase the number of equations. This is one of the reasons why other types of integral equations are usually used to solve the boundary-value problem.

14.8.3. Representation of the Currents and the Field as a Sum of Simple Fractions

However, as has already been mentioned, this system is extremely useful, since it allows one to establish some very fundamental equations that describe the field caused by currents in conductors surrounded by an insulator. In fact, since coefficients of the system (Eq. (14.134)) contain integer powers of ω, the expression for the current in a ring through any point q can be written as

$$I(q, i\omega) = i\omega\frac{P_{N-1}(q, i\omega)}{P_N(i\omega)}, \tag{14.135}$$

where $P_N(i\omega)$ is a polynomial of order N with respect to ω and is the determinant of the system, $P_{N-1}(q, i\omega)$ is a polynomial of order N-1 with respect to $i\omega$. The multiplier $i\omega$ appears because the primary electromotive force in each ring is directly proportional to $i\omega$. I is the complex amplitude of the current. As is known from the theory of polynomials, the right-hand side of Eq. (14.135) can be represented as a sum of simple fractions:

$$I(q) = i\omega\sum_{n=1}^{N}\frac{u_n}{\omega_n - i\omega}, \tag{14.136}$$

where $-i\omega_n$ are roots of the polynomial P_N, and ω_n are real numbers such that

$$\omega_1 < \omega_2 < \omega_3 < \omega_4 < \cdots$$

With an increase of number of elementary tubes, the accuracy of determination of currents increases, and in the limit, we can represent Eq. (14.136) as

$$I(q) = \sum_{n=1}^{\infty} u_n^* \frac{i\omega\tau_n}{1 - i\omega\tau_n}. \tag{14.137}$$

In accordance with the Biot–Savart law, every current ring generates a magnetic field. Then, applying the principle of superposition, we have for the complex amplitude of an arbitrary component B_l:

$$B_l^*(p) = B_{0l}(p) \sum_{n=1}^{\infty} d_{nl} \frac{i\omega\tau_n}{1 - i\omega\tau_n} \tag{14.138}$$

or

$$QB_l^*(p) = B_{0l}(p) \sum_{n=1}^{\infty} d_{nl} \frac{\omega\tau_n}{1 + (\omega\tau_n)^2}$$

and

$$InB_l^*(p) = B_{0l}(p) \sum_{n=1}^{\infty} d_{nl} \frac{(\omega\tau_n)^2}{1 + (\omega\tau_n)^2}, \tag{14.139}$$

where p is an observation point, B_{0l} is the l-component of the primary field, while $B_l^*(i\omega, p)$ is the complex amplitude of the l-component of the secondary field. It is essential that d_{nl} are coefficients, which depend on geometrical factors such as the position of an observation point, the shape and size of a conductor, its location, as well as geometry of the primary field, but they are independent of the conductivity. In general, every component of the field B_l is characterized by its own set of coefficients d_{nl}. From the physical point of view, it is obvious that the series:

$$\sum_{n=1}^{\infty} d_{nl}$$

converges, and therefore starting from some term of this series, we have

$$d_{nl} > d_{n+1,l} > d_{n+2,l} > \cdots \tag{14.140}$$

The parameters τ_n are functions of the conductivity γ as well as the dimensions and a shape of the conductor, but they are independent of the position of the observation point and magnitude of the primary field. In general, the set of these parameters does not change when the geometry of the primary field changes. However, there are exceptions. For instance, the eigenfunctions describing the uniform and nonuniform primary field are different and, correspondingly, the set of parameters τ_n differs too. Let us notice that in Eqs (14.138) and (14.139) τ_n obey the inequality

$$\tau_1 > \tau_2 > \tau_3 > \cdots \tag{14.141}$$

and the largest parameter τ_1 is called the time constant of the conductor:

$$\tau_1 = \tau. \tag{14.142}$$

Thus, we see that the spectrum of the field and currents induced in a confined conductor can be represented as a sum of simple fractions. This means that at the low part of the complex plane of ω the spectrum has singularities that are simple poles:

$$\omega_n = -\frac{i}{\tau_n}, \tag{14.143}$$

situated on the imaginary axis of ω. It is important to emphasize that the distribution of these poles controls the principal features of the spectrum for real values of ω. As we already know, the behavior of the spectrum of a field in a horizontally layered medium is completely different.

Now, assuming a step function excitation of the primary field and applying Fourier's transform,

$$B_l(t) = \frac{1}{2\pi} \int\limits_{-\infty}^{\infty} \frac{B_l(i\omega)}{i\omega} \exp\,(-\,i\omega t)\mathrm{d}\omega$$

to each term of the series describing the current density and field, we obtain the following expressions for the transient responses of the current density and the field:

$$j_\varphi(t,q) = \sum_{n=1}^{\infty} a_n^*(q) \exp\left(-{}^{t}\!/_{\tau_n}\right) \tag{14.144}$$

and

$$B_l(t,p) = B_{0l}(p) \sum_{n=1}^{\infty} d_{nl} \exp\left(-{}^{t}\!/_{\tau_n}\right). \tag{14.145}$$

Thus, as in the case of a sphere and a circular cylinder, the transient responses are represented as the sum of exponents. Comparing Eqs (14.137) and (14.138) with Eqs (14.144) and (14.145), we see that the currents and fields in the frequency- and time domains are described by the same set of parameters that of course is obvious. Note, at the beginning of development of the theory of the mining inductive methods, this fact was hardly known. In deriving equations for the field, we have assumed that both the conductor and the primary field have the same axis of symmetry. Because of this currents induced in a conductor are described by a single component j_φ, and the currents filaments are circles that do not change its shape with either frequency or time change. This means that coefficients M_{pq}, L_q, and R_q are independent of the frequency in Eq. (14.134), while the right-hand side of the system Ξ_{on} is directly proportional to frequency. Now we demonstrate that in spite of our assumptions Eqs (14.137), (14.138), (14.144) and (14.145) correctly describe the currents and a field in a general case, when a conductor has an arbitrary shape and size, and it is surrounded by an insulator. In general, the primary electric field of the vortex origin $\mathbf{E}_0$ intersects the surface of the conductor and surface charges arise, and they certainly have a strong influence on electric field. Correspondingly, the geometry of the current filaments becomes sufficiently more complicated and it changes inside the conductor. Moreover, when a frequency varies the relative contribution of the vortex and galvanic parts of the electric field also changes. This means that the shape of the current filaments is usually a function of frequency. For this reason, the system of equations with respect to the currents $I\,(q)$ is characterized by coefficients that depend on the frequency, except in some special cases, and therefore Eq. (14.135) becomes invalid. In order to avoid this problem, we mentally represent the primary field as a sum of eigenfunctions, which correspond to a given conductor. For instance, in the case of sphere, every term of this sum includes a combination of cylindrical and Legendre functions. As was pointed out in a previous section, every term of this sum generates a field that is described by the corresponding eigenfunction. This means that the geometry of this field as well as the currents induced in a conductor are independent of the frequency. Therefore, if instead of the total primary field, we consider only one term then the coefficients of the system Eq. (14.134) still remain constant with respect to frequency. Respectively, the spectrum of currents and the field caused by an eigenfunction of the primary field as before is described by a sum of simple fractions, while the transient responses can be represented as a sum of exponents. We can say that every eigenfunction describing the secondary field is characterized by two sets of parameters, d_{nl} and τ_{nl}. Thus, taking into account the influence of all eigenfunctions composing the primary field, we again conclude that the frequency and transient responses of the field and currents induced in an arbitrary confined conductor are described by Eqs (14.137), (14.138), (14.144) and (14.145) regardless of the type of the primary field. From the geometry of current system, corresponding to different eigenfunctions, and as was shown earlier, it is not difficult to notice that with an increase of their index their influence becomes relatively stronger at higher frequencies. Of course, this means that with an increase of time, they decay more rapidly; that is their time constant becomes smaller. Also, with an increase of the distance from the conductor the relative contribution of the first eigenfunction becomes bigger. In conclusion, it may be proper to point out that in some cases of two-dimensional conductors (not confined bodies) Eqs (14.137), (14.138), (14.144) and (14.145) still remain valid. For instance, such behavior is observed if the primary field does not change along an infinitely long cylindrical conductor with an arbitrary cross section.

14.9. BEHAVIOR OF THE FIELD DUE TO CURRENTS IN A CONFINED CONDUCTOR

Assume that an arbitrary confined conductor is surrounded by highly resistive medium, which can be treated as an insulator, and that the primary electromagnetic field is caused by some system of currents. In studying the secondary magnetic field, we will proceed from Eqs (14.138) and (14.145).

14.9.1. The Low-Frequency Part of the Spectrum

First, consider the case when

$$\omega\tau < 1. \tag{14.146}$$

This condition usually defines the so-called the low-frequency part of the spectrum, and it is natural that with a decrease of the time constant τ, this range is observed at higher frequencies. Taking into account inequality Eq. (14.141) any term of the series in Eq. (14.138):

$$d_{nl}\frac{i\omega\tau_n}{1 - i\omega\tau_n}$$

can be expanded in a series in $\omega\tau_n$, if $\omega\tau < 1$. Then we obtain

$$d_{nl}\frac{i\omega\tau_n}{1-i\omega\tau_n} = d_{nl}\sum_{k=1}^{\infty}(i\omega\tau_n)^k. \tag{14.147}$$

Substituting Eq. (14.147) into Eq. (14.138) and changing the order of summation and then collecting all terms having the same power of $i\omega$, we have

$$B_l^*(p) = B_{0l}(p)\sum_{k=1}^{\infty}(i\omega)^k\sum_{n=1}^{\infty}d_{nl}\tau_n^k \quad \text{or} \quad B_l^*(p) = B_{0l}(p)\sum_{k=1}^{\infty}C_{kl}(i\omega\tau)^k, \tag{14.148}$$

where

$$C_{kl} = \sum_{n=1} d_{nl}\left(\frac{\tau_n}{\tau}\right)^k. \tag{14.149}$$

For instance

$$\begin{aligned} C_{1l} &= d_{1l} + \frac{\tau_2}{\tau}d_{2l} + \frac{\tau_3}{\tau}d_{3l} + \cdots \\ C_{2l} &= d_{1l} + \left(\frac{\tau_2}{\tau}\right)^2 d_{2l} + \left(\frac{\tau_3}{\tau}\right)^2 d_{3l} + \cdots \\ C_{kl} &= d_{1l} + \left(\frac{\tau_2}{\tau}\right)^k d_{2l} + \left(\frac{\tau_3}{\tau}\right)^k d_{3l} + \cdots \end{aligned} \tag{14.150}$$

Inasmuch as the radius of convergence of the power series corresponding to the first fraction in Eq. (14.138) is the smallest, the series Eq. (14.148) converges if

$$\omega\tau < 1. \tag{14.151}$$

Thus, we demonstrated that in a general case of an arbitrary confined conductor, the low-frequency spectrum is described by a series in ω containing only its integer powers, and the radius of convergence of this series is defined by the time constant. As follows from Eq. (14.151), the maximal frequency that still corresponds to the low-frequency part of the spectrum is defined from the equality

$$\omega = \frac{1}{\tau}. \tag{14.152}$$

The series Eq. (14.148) has one very remarkable feature, namely, that with an increase in the order k of the term, the coefficients C_{kl} tend to the same limit. In fact, taking into account Eq. (14.150), we see that

$$C_{kl} \rightarrow d_{1l} \quad \text{if} \quad k \rightarrow \infty \tag{14.153}$$

This means that the high terms of the series describing the field depend on the same parameters d_{1l} and τ. Correspondingly, we can write

$$\begin{aligned} B_l^*(p) = B_{0l}\Big\{ & C_{1l}(i\omega\tau) + C_{2l}(i\omega\tau)^2 + C_{3l}(i\omega\tau)^3 + \cdots \\ & + d_{1l}\Big[(i\omega\tau)^k + (i\omega\tau)^{k+1} + (i\omega\tau)^{k+2} + \cdots\Big]\Big\} \end{aligned} \tag{14.154}$$

Thus, terms of the series with large values of k for the quadrature and in-phase components can be written as

$$(-1)^{k+1} d_{1l}(\omega\tau)^{2k-1} \quad \text{and} \quad (-1)^k d_{1l}(\omega\tau)^{2k}. \tag{14.155}$$

It is appropriate to notice that very often such a behavior is practically observed when order of a series term exceeds three; that is instead of Eq. (14.153), one can write

$$C_{kl} \approx d_{1l} \quad \text{if} \quad k > 3. \tag{14.156}$$

In accordance with Eq. (14.148) for the quadrature and in-phase components of the field, we have

$$QB_l^*(p) = \sum_{k=1}^{\infty} C_{2k-1,l}(-1)^{k+1}(\omega\tau)^{2k-1}$$

and

$$InB_l^*(p) = \sum_{k=1}^{\infty} C_{2k,l}(-1)^k(\omega\tau)^{2k} \tag{14.157}$$

and, correspondingly, at the low-frequency part of the spectrum, the leading terms of the quadrature and in-phase components of the magnetic field are proportional to $\omega\tau$ and $(\omega\tau)^2$, respectively. Note, that a solution of the integral equation, for example, with respect to the current density by the method of subsequent approximations allows one to obtain terms of the series Eq. (14.157) and evaluate the time constant of a conductor.

14.9.2. Approximate Representation of the Spectrum

Strictly speaking, the spectrum depends on both sets of parameters d_{nl} and τ_n for all values of $\omega\tau$. However, as follows from inequalities Eqs (14.140) and (14.141) and numerical modeling, in many cases that are of the practical interest in geophysical applications the field is mainly defined by the two parameters only: d_{1l} and τ. Consequently, the quadrature and in-phase components can be approximately represented as

$$QB_l^*(p) \approx B_{0l}(p)d_{1l}\frac{\omega\tau}{1+(\omega\tau)^2}$$

and

$$InB_l^*(p) \approx B_{0l}(p)d_{1l}\frac{(\omega\tau)^2}{1+(\omega\tau)^2}, \tag{14.158}$$

provided that $\omega\tau < 1$. The study of the fields shows that the more uniform field $\mathbf{B}_0$ in the vicinity of the conductor and further an observation point from it, the more accurately Eq. (14.158) describe this part of the spectrum. In accordance with the Biot–Savart law, Eq. (14.158) implies that within this range of frequencies, we can treat the distribution of induced currents as a system of elementary current filaments with the same time constant. It is proper to notice that Eq. (14.158) will be used later in deriving some equations for the field when the medium surrounding a conductor has a finite value of resistivity.

14.9.3. The High-Frequency Part of the Spectrum

Next consider the opposite case; that is the high-frequency part of the spectrum and at the beginning assume that the frequency is infinitely high. Then, due to the skin effect, induced currents concentrate at the surface of a conductor and they are distributed in such a way that the electromagnetic field inside the conductor $\mathbf{E}^i$ and $\mathbf{B}^i$ vanish:

$$\mathbf{E}_0 + \mathbf{E}_s^i = 0, \quad \mathbf{B}_0 + \mathbf{B}_s^i = 0. \tag{14.159}$$

In particular, on its surface, the tangential component of the electric field and the normal component of the magnetic field are equal to zero:

$$E_t = 0 \quad \text{and} \quad B_n = 0 \quad \text{on } S \tag{14.160}$$

From Eq. (14.159), it directly follows that the surface currents and currents creating the primary magnetic field are shifted in phase by 180°.

Therefore, in the limiting case, the secondary magnetic field has only an in-phase component that is independent of frequency, while the quadrature component is absent. Of course, we can obtain the same result from Eqs (14.138) and (14.139) when ω increases without limit. In fact, we have

$$InB_l^*(p) \to -\sum_{n=1}^{\infty} d_{nl} \quad \text{and} \quad QB_l^*(p) \to 0 \quad \text{if} \quad \omega \to \infty. \tag{14.161}$$

To obtain a more accurate representation of the high-frequency spectrum suppose that the skin-depth is smaller than the minimum radius of the curvature of the conducting surface. Then, as was pointed earlier, in the vicinity of any point inside the conductor but located near the surface, the distribution of the field is close to that of the plane wave. Correspondingly, the relationship between the tangential components of the field $\mathbf{E}^i$ and $\mathbf{B}^i$ at such points can be approximately written as

$$\frac{E_{t_1}^{i*}}{B_{t_2}^{i*}} = \left(\frac{i\omega}{\gamma\mu}\right)^{1/2}, \tag{14.162}$$

where t_1,t_2 characterize two mutually perpendicular directions, which are tangential to the conductor surface. Taking into account the fact that the tangential components of the electromagnetic field are continuous on the surface, we also have

$$\frac{E_{t_1}^{e*}}{B_{t_2}^{e*}} = \left(\frac{i\omega}{\gamma\mu}\right)^{1/2}. \tag{14.163}$$

Here $E_{t_1}^e$, $B_{t_2}^e$ are the tangential components of the field on the external side of the surface. This equation greatly simplifies determination of the field outside the conductor, since it doesn't require its knowledge inside the conducting medium. One can show that from Eq. (14.163) it follows that

$$InB_l^*(p) \approx B_{0l}(p)\left(M_l - \frac{N_l}{(\omega\tau)^{1/2}}\right)$$

and

$$QB_l^*(p) \approx B_{0l}(p)\frac{N_l}{(\omega\tau)^{1/2}}, \quad \text{if} \quad \omega\tau \gg 1 \tag{14.164}$$

where M_l and N_l are functions depending on geometric factors only. Thus, in this part of the spectrum both components of the magnetic field approach

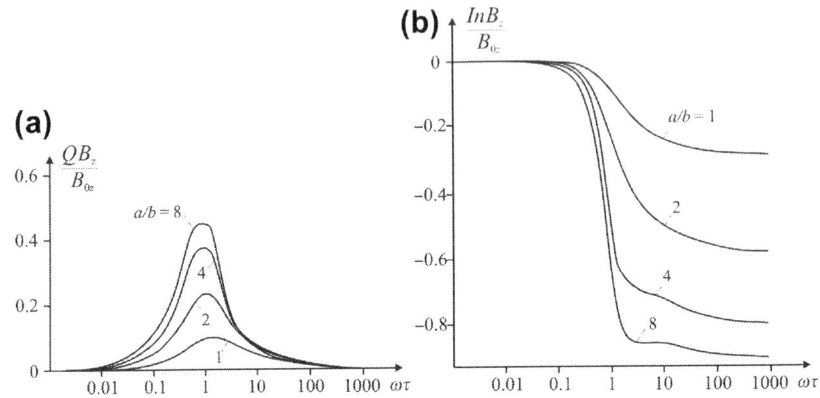

Figure 14.8 Frequency responses of the quadrature (a) and in-phase (b) components of the magnetic field, caused by currents in spheroids; $z/b = 1.5$. Curve index is a/b ratio.

their asymptotes relatively slowly. As one more example, consider the frequency responses of the magnetic field caused by currents in a spheroid when the uniform magnetic field $\mathbf{B}_0$ is directed along the minor axis, Figure 14.8(a) and (b). The observation point is located on the axis of symmetry z. Here, a and b are the major and minor semi-axes, respectively and z is the distance between the spheroid center and an observation point. The argument $\omega\tau$ is plotted in such a way, that for each value of the ratio of axes a/b an appropriate value of the time constant is chosen. As is seen from these curves, the maximal value of the quadrature component is observed when the argument $\omega\tau$ is close to unity, and it happens regardless of the ratio a/b. Also, it is clear that the low-frequency part constitutes the essential part of the spectrum. In particular, for the quadrature component its upper limit corresponds almost exactly to the position of the maximum, while the in-phase component magnitude reaches practically 50% of that for the high-frequency limit. Certainly, this comparison illustrates one aspect of why the study of the low-frequency spectrum is important.

14.9.4. Early Stage of the Transient Field

Next consider the transient field and start with the early stage when the time of observation t is much less than the time constant τ:

$$t \ll \tau. \tag{14.165}$$

Taking into account Eq. (14.164) and applying Fourier transform, we obtain

$$B_l(p,t) \approx B_{0l}(p)\left\{M_l - 2N_l\left(\frac{2}{\pi}\right)^{1/2}\left(\frac{t}{\tau}\right)^{1/2}\right\}$$

and

$$\frac{\partial B_l}{\partial t} \approx -B_{0l}\left(\frac{2}{\pi}\right)^{1/2}\frac{N_l}{(t\tau)^{1/2}}. \tag{14.166}$$

Consequently, the electromotive force that is proportional to $\frac{\partial B_l}{\partial t}$ increases without limit as t decreases. It is also clear that at the early stage the influence of parameter τ, as in the case of the high-frequency spectrum, is relatively small.

14.9.5. The Late Stage of the Transient Field

Now let us discuss the late stage of the transient response. As follows from Eq. (14.145) with an increase of time, the first term of the sum becomes dominant and in the limit we obtain

$$B_l(p,t) = B_{0l}(p)d_{1l}\exp\left(-\frac{t}{\tau}\right) \quad \text{if} \quad t > \tau. \tag{14.167}$$

Thus, regardless of the shape and size of a confined conductor, as well as its resistivity, the field is defined at the late stage by only two parameters d_{1l} and τ, and it always decays exponentially. It is essential that the time constant is the same for any component of the field. It is proper to note that such remarkable simplicity of the field behavior greatly facilitates interpretation when the transient method is used for detecting relatively good conductors. Moreover, as follows from Eq. (14.167), the field at the late stage is very sensitive to a change of the parameter τ. In principle, this means that measurements of the field at the late stage allow us to distinguish the fields caused by currents in conductors with different time constants. In this light, it is interesting to compare the late stage and the low-frequency spectrum. In general, as was shown earlier, this part of the spectrum, as well as others, is defined by both sets of parameters d_{nl} and τ_n, while the late stage depends only on d_{1l} and τ. However, if we consider the high terms of the series Eq. (14.157), then it turns out that they are also functions d_{1l} and τ. This fact allows us to conclude that the high terms of the series describing the low-frequency spectrum contain the same information about conductor as the late stage of the transient field. Taking into account the fact that the time constant τ is the important parameter defining the frequency and transient responses, we illustrate its relationship with the size and a shape of a confined conductor, Table 14.2. Here it is appropriate to make several comments:

1. Time constants were determined by solving the boundary-value problems using either the method of separation of variables or numerical methods, such as integral equations.
2. In determining τ, it was assumed that the primary field $\mathbf{B}_0$ is uniform in the vicinity of the conductor and that it is directed in such a way as to provide a maximal length of the current filaments. These conditions guarantee a relatively slow decay of the field at the late stage.
3. At the same time, we can imagine cases when the time constant is relatively small. For instance, in the case of the elliptical cylinder this happens when the primary uniform field is directed along its major axis.

Table 14.2 Time Constants of Conductors

Conductor	Time Constant	Notes
Sphere	$\tau = \dfrac{\gamma\mu_0 a^2}{\pi^2}$	Sphere radius is a
Thin disk with constant conductance S; γ and h are conductivity and thickness of the disk, respectively	$\tau = \dfrac{\mu_0 Sa}{5.51}$	Disk radius is a; $S = \gamma h$
Thin disk with varying conductance $S^* = S_0(1-\rho^2)$; S_0 is the conductance at the disk center; r is the distance from the center	$\tau = \dfrac{\mu_0 S_0 a}{7.71}$	$\rho = \frac{r}{a}$
Cylinder with the circular cross section; a is its radius	$\tau = \dfrac{\gamma\mu_0 a^2}{5.78}$	
Infinitely long plate with constant conductance S; γ and h are conductivity and thickness of the plate, respectively; width of the plate is $2a$	$\tau = \dfrac{\mu_0 Sa}{4}$	$h \ll a$
Infinitely long plate with elliptical cross section; a and b are major and minor axis, respectively	$\tau = \dfrac{\mu_0 Sa}{5.5}$	$S = 2\gamma b$; $b \ll a$
Infinitely long plate with elliptical cross section and primary field directed along the major axis	$\tau = \dfrac{\gamma\mu_0 b^2}{3}$	$S = 2\gamma b$; $b \ll a$

4. When the primary field is not uniform, then it can be mentally represented as a combination of eigenfunctions, and the function that corresponds to a uniform field provides the slowest decay with time at the late stage. In other words, the transient field at the late stage coincides with that for which the primary field is uniform.
5. This parameter is always directly proportional to conductivity and the cross-section of an area, perpendicular to the current flow.

Relation between the Time Constant and Conductor's Shape

Conductor Shape							Notes
			Spheroid				
a/b	1	2	4	8	16	32	$\tau_d = \frac{\mu_0 S_0 a}{7.71}$
τ/τ_d	0.39	0.56	0.71	0.84	0.92	0.99	$S = 2\gamma b$
			Elliptical cylinder				Primary field is not directed along the major axis
a/b	1	2	4	8	16		
τ/τ_p^e	0.48	0.64	0.77	0.87	0.93		$\tau_p^e = \frac{\mu_0 S a}{5.5}$
			Elliptical cylinder				Primary field along the major axis
a/b	1	2	4	8	16		$\tau_d = \frac{\gamma\mu_0 b^2}{3}$
τ/τ_d	0.48	0.64	0.84	0.95	0.99		
			Thin conducting plate				l_1 and l_2 are sides of the plate; $l_1 \leq l_2$
l_2/l_1	1	2	4				$\tau_p = \frac{\mu_0 S l_1}{8.02}$
τ/τ_p	0.77	0.94	1				

In conclusion, let us remind that regardless of the shape and size of a confined conductor their frequency and transient responses of the field have the similar behavior and it is always possible to distinguish the low-frequency, intermediate- and high-frequency parts of the spectrum, as well as the early, intermediate, and late stages, where the field has completely different behavior.

14.10. INFLUENCE OF GEOLOGICAL NOISE REPRESENTED BY CONFINED CONDUCTORS

Introduction

As is well known, the main goal of the inductive methods of mining prospecting is a search for massive ore bodies made up of highly conductive minerals, so that ore body has a conductivity that is much greater than that of the surrounding medium. At the same time, anomalies in the electric and magnetic fields can also be caused by accumulations of conductive but noneconomic bodies. In reality, the sources for the quasi-stationary field consist both of currents in the conductive medium and electric charges on the surfaces between media with different resistivity. For this reason, the electromagnetic field depends also on electric properties of the surrounding medium, which is host for the ore bodies. The ratio between the fields caused by the presence of economic conducting ore bodies and other conductors, commonly called geological noise, determines the resolution of prospecting methods and depth of investigation. Clearly, the effectiveness of the induction methods in the frequency- and time domains also depend on a number of other factors, including cultural noise, magnetotelluric noise, errors of the geometrical parameters of the array, the stability with which the primary field is cancelled or taken into account, and so on. By decreasing the effect of these various factors to the level of the geological noise, we can increase the effectiveness of the induction method to the maximal amount possible. For this reason, the analysis of geological noise is highly important, because it permits us to choose the optimal method for a given geoelectric section.

14.10.1. Direct Current Method and Geological Noise

From the beginning of application of geophysical methods for mining exploration, the use of direct current became the conventional approach for detecting good conductors located relatively close to the earth's surface, and they have been widely and successfully used over the world. However, this method has very low resolution; that is relatively good conductors with respect to surrounding medium create practically the same anomalies in spite of the fact that their resistivity may differ by orders. In other words, the influence of the geological noise is rather high, that of course makes an influence on the depth of investigation. The low sensitivity of the direct currents methods to a conductivity of a body is related to the fact that sources of the measured electric fields are electric charges that develop on the surfaces, which are only slightly dependent on the resistivity of the conductive body and in the surrounding medium. Understanding the physical principles of the inductive methods suggested that their application can essentially improve the resolution by conductivity, and this was one of the main reasons for the use of the frequency and then time-domain methods for looking for good conductors. Now we consider in some detail the resolution of these methods.

14.10.2. Frequency-Domain Methods and Geological Noise (Confined Inhomogeneity)

First, consider the high-frequency part of the spectrum where the in-phase component of the secondary field dominates. The quadrature component decreases slowly with increasing of the parameter $\omega\tau$; that is with an increase of frequency, conductivity, dimensions of the body, and in the limit it approaches zero. As was shown earlier at this range, we have

$$InB \approx B_0\left[A - \frac{C}{(\omega\tau)^{1/2}}\right] \quad \text{and} \quad QB \approx B_0\frac{C}{(\omega\tau)^{1/2}} \tag{14.168}$$

In accordance with Eq. (14.168), the influence of the time constant (conductivity) on the in-phase component is small. This is readily explainable inasmuch as at the high-frequency limit currents concentrate near the surface of the conductive body and the total current approaches to that of the ideal conductor. Therefore, if the magnetic field due to the currents in a relatively resistive body corresponds to the high-frequency part of the spectrum, the resolution of the method based on use of either direct or alternating current will be practically the same. Certainly, the relationship between the quadrature as well as in-phase components of the field and resistivity of the conductive body for this part of the spectrum does not facilitate discrimination of anomalies caused by conductors. Next, consider the contrary case of the low-frequency spectrum, where the quadrature component dominates over the in-phase component and both of them increase with time constant:

$$QB \approx B_0\left[C_1(\omega\tau) - C_3(\omega\tau)^3 + C_5(\omega\tau)^5 - \cdots\right]$$

and

$$InB \approx B_0\left[-C_2(\omega\tau)^2 + C_4(\omega\tau)^4 - C_6(\omega\tau)^4 + \cdots\right], \quad \omega < \frac{1}{\tau}, \tag{14.169}$$

where QB and InB represent components of the secondary field in any direction. At the very low frequencies, the quadrature component is defined almost entirely by the first term in the series:

$$QB \approx B_0 C\omega\tau, \tag{14.170}$$

and it is directly proportional to conductivity. Thus, when induced currents induced in the ore body and in an inhomogeneity of higher resistivity generate fields corresponding to the low-frequency part of the spectrum, the ratio of quadrature components of the magnetic field is directly proportional to the ratio of time constants:

$$\frac{B_N}{B_R} \approx \frac{C_{1N}}{C_{1R}} \frac{\tau_N}{\tau_R}. \tag{14.171}$$

Here B_N and B_R are respectively the quadrature components generated by currents in an inhomogeneity of higher resistivity (N) and in the target ore body (R). This relationship indicates that under certain conditions the inductive methods are characterized by better resolution that would be a method based on the use of direct current fields. However, to a considerable extent, the level of geological noise depends on the ratio between time constants that characterize the sources of the geological noise and an ore body. In particular, when $\tau_N = \tau_R$, the relationship between these two fields is practically defined in the same way as in the case of the direct current methods with only geometrical factors being involved. For example, when a high-resistivity inhomogeneity that has dimensions greater than those of a target ore body is located closer to the observation point than the ore body, the anomaly due to the noise can be significantly greater. As follows from Eq. (14.171) at low frequencies, the ratio B_N/B_R is independent of the frequency and with further decrease in frequency there is no reduction of a geological noise. On the other hand, with increasing frequency, as the response contributed by currents in the target ore body no longer reflects the behavior at low frequencies, the level of geological noise observed in measuring quadrature component becomes greater. Note, that inasmuch as the quadrature component is shifted in phase by 90° with respect to the primary field, the measurement of quite small signals is accomplished with relative ease. Now we consider the in-phase component of the secondary field at the low frequencies, more precisely the leading term of the series in Eq. (14.169). Then we have

$$InB \approx -B_0 C_2 (\omega\tau)^2, \tag{14.172}$$

which shows that the in-phase component is directly proportional to square of conductivity. For this reason, when the in-phase component is measured the inductive methods are characterized by higher resolution, and the ratio between geological noise and useful signal is given by

$$\frac{B_N}{B_R} \approx \frac{C_{2N}}{C_{2R}} \frac{\tau_N^2}{\tau_R^2}. \tag{14.173}$$

Comparing Eqs (14.171) and (14.173), we can see that if $\tau_R > \tau_N$ one can markedly reduce an influence of geological noise by measuring the in-phase component instead of the quadrature component. However, the in-phase component is more difficult to measure at low frequencies because it is much smaller than the quadrature component, and it is also in phase with the primary field. It should be pointed out that in measuring the phase and amplitude of the total field at low frequencies, one can in principle achieve the same resolution as in the case of the individual quadrature and in-phase components. In fact, from Eq. (14.169), we have the following expressions for the amplitude and phase of the total field:

$$A \approx B_0\left\{\left[1 - C_2(\omega\tau)^2\right]^2 + C_1^2(\omega\tau)^2\right\}^{1/2} \approx \left[1 - \left(C_2 - \frac{1}{2}C_1^2\right)(\omega\tau)^2\right] B_0 \tag{14.174}$$

and

$$\varphi \approx \tan^{-1} \frac{C_1}{1 - C_2(\omega\tau)^2}\omega\tau \approx C_1\omega\tau. \tag{14.175}$$

The second term in Eq. (14.174) is directly proportional to the square of time constant as in the case of the in-phase component. Since the contribution of the secondary field is usually small, any method of measurement that reduces the contribution of the primary field is of a great practical interest. Suppression of this field can be done by different ways, including measurements simultaneously at two frequencies.

As follows from Eq. (14.169), the use of the higher order terms of the series can markedly increase the resolution obtaining in the induction methods. For example, the difference between measurements of the quadrature component made at two frequencies can be used to eliminate the term proportional to time constant in the first power:

$$\Delta B = QB(\omega_1) - \frac{\omega_1}{\omega_2} QB(\omega_2) \quad \text{or} \quad \Delta\Xi = Q\Xi(\omega_1) - \left(\frac{\omega_1}{\omega_2}\right)^2 Q\Xi(\omega_2). \tag{14.176}$$

It is clear that this difference is directly proportional to cube of time constant and therefore they allow one to reduce an influence of geological noise even in a greater extent.

Now let us pose the following question: Under what conditions it is possible to remove an influence of geological noise of this type (confined conductor) by measuring only the higher order terms in a series Eq. (14.169)? The ratio of fields corresponding to the i-th term in these series is

$$\frac{B_{iN}}{B_{iR}} = \frac{C_{iN}}{C_{iR}}\left(\frac{\tau_N}{\tau_R}\right)^i, \tag{14.177}$$

where time constants characterize the geological noise and the target ore body. From Eq. (14.177), we can recognize three different cases as follows:

1. The time constant of the ore body is greater than for the geological noise:

$$\tau_R > \tau_N. \tag{14.178}$$

In this case, in principle, one can decrease the contribution of geological noise to any degree desired by making measurements only of later terms in the series, Eq. (14.169). In fact, when the inequality Eq. (14.178) is satisfied, as the number of terms is increased the ratio $\left(\tau_N/\tau_R\right)^i$ decreases. It is proper to notice, that the condition Eq. (14.178) does not depend on the strength of the primary field in the vicinity of the source of geological noise or the vicinity of the target ore body; that is on the depth of burial of the ore body. It may occur that in measuring leading terms in the series Eq. (14.169), describing the quadrature and in-phase components of the field at low frequencies, the influence of geological noise may prevail. Then, in measuring the successive terms in these series, the contribution of geological noise becomes less and less, provided of course that the condition Eq. (14.178) is met. It is clear, if the frequency at which measurements are being made actually corresponds to the low-frequency part of the spectrum, $\omega < 1/\tau_R$, a further decrease of the frequency will not permit us to reduce the relative contribution of geological noise. Instead of a change of frequency, it is necessary to measure the latter terms in a series Eq. (14.169). In order to illustrate this approach, let us imagine two models, shown in Figure 14.9(a) and (b). First, suppose that the body contributing the geological noise and the target ore body are situated at practically the same depth from the surface of the earth, Figure 14.9(a). If the quadrature component (leading term) is being measured, it may happen that the signal contributed by the geological noise is greater than that due to the target ore body. In measuring the in-phase component, that is its leading term, the signal observed over the body, representing geological noise, can still be greater than that over the ore body, even though the ratio B_N/B_R is smaller than in the previous case. In both cases, the signal over the target ore body

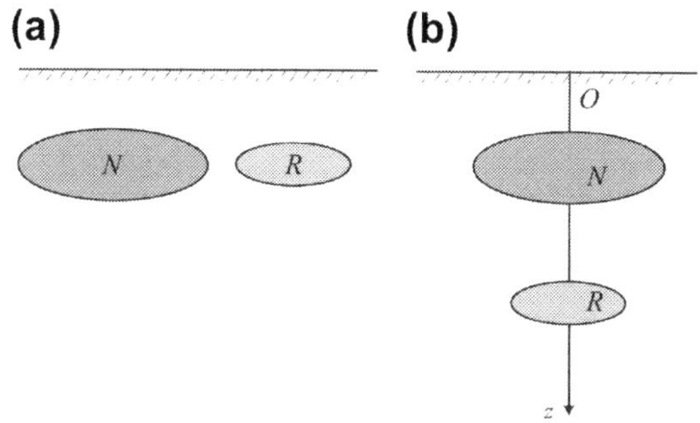

Figure 14.9 A confined inhomogeneity treated as geological noise. (a) Bodies at the same depth; (b) Bodies at different depth.

may be less than over the source of geological noise, and therefore it is hard to recognize the presence of an ore body. However, if it would be possible to perform measurements of the latter terms in the series Eq. (14.169), the relation between these anomalies changes and the signals over the ore body become dominant. This means that the approach based on measuring the latter terms of the series describing the low-frequency spectrum allows one in principle markedly increase the resolution of the inductive methods when $\tau_R > \tau_N$. Let us now assume that the ore body is situated beneath some also conductive body, which is considered to be a source of geological noise, Figure 14.9(b) and that $\tau_R > \tau_N$. In this case, the measured field will be a sum of fields caused by currents in the geological noise and the ore body:

$$B = B_R + B_N, \tag{14.179}$$

where B is any component of the field being observed. It may occur that in measuring the quadrature component of the field, the principal part of the signal is obtained from the source of geological noise, that is, $QB_N > QB_R$. In other words, the inhomogeneity lying above the ore body serves as an electromagnetic screen, and therefore, the ore body located at greater depth cannot be detected. In measuring the in-phase component (leading part), the ratio InB_R/InB_N markedly increases, and one can consider two situations. The first one is the case in which the in-phase component of the magnetic field contributed by currents within the target ore body is dominant: $InB_R \gg InB_N$. Then, it can be said that the body contributing geological noise becomes "transparent", and the ore body itself can be recognized. This means that the depth of investigation achieved by making measurements of the in-phase component is greater than that achieved with the quadrature component. If it is impossible to remove the screening effect of the geological noise even by measuring the in-phase component ($InB_N \geq InB_R$), then following terms of the series Eq. (14.169) have to be measured. In accordance with Eq. (14.177), the relative contribution of geological noise can be reduced to any degree desired, regardless of a position of the ore body relative to the source of geological noise, provided that $\tau_R > \tau_N$. In principle at least, by measuring the latter terms in the series representing the low-frequency part of the spectrum, one might achieve any depth of investigation for this type of geological noise (confined conductor) whenever condition Eq. (14.178) is met. It is reasonable to expect that the near the observation point to the source of the noise and the deeper the ore body is, the more difficult it will be to separate the useful signal. In such cases, it is necessary to measure higher order terms of the series. On the other hand, if the ore body lies at relatively shallow depths, and there is a large contrast in time constants between the ore body and the screening inhomogeneity, a significant reduction of this noise can be obtained by measuring the in-phase component or the amplitude of the total field with compensation of the primary field or, for example, the difference between the quadrature components at two frequencies, Eq. (14.176).

1. So far, we have considered only the case in which the time constant for the ore body is greater than the time constant for the source of geological noise. Now let us assume the contrary case when

$$\tau_N > \tau_R. \tag{14.180}$$

Then, by measuring the latter terms in the series Eq. (14.169), the effect of geological noise is increased, causing a reduction in resolution and depth of investigation. For example, when measurements are being made of the quadrature component, and the signal contributed by currents induced in a body representing geological noise is dominant, the effect of this body will be even greater when the in-phase component is measured, and it becomes greater and greater as higher order terms in the series Eq. (14.169) are measured. In this case, one can say that the use of induction methods will not permit one to distinguish between anomalies caused by "ore" and "non-ore" conductors.

2. Finally, if these time constants are equal

$$\tau_N = \tau_R \tag{14.181}$$

the ratio of the signals contributed by the currents in these two types of conductors is determined by geometrical factors. It is obvious that by making measurements of the latter terms of the series Eq. (14.169), there is no change in the relative contribution caused by currents in either of conductor. It can be said that the resolution and depth of investigation achievable with an induction method in this case are essentially the same as those obtainable with the direct current method.

14.10.3. Time-Domain Method and Geological Noise

Next, consider the capability of a time-domain or transient method for increasing the depth of investigation and for separating magnetic field anomalies caused by bodies with different time constants. During the early stage of the transient process, currents within a conductive body are concentrated almost entirely in a thin surface layer and, as consequence, the magnetic field only weakly depends on conductivity. Therefore, we focus on the late stage during which the field contributed by currents in the confined conductive ore body decays exponentially. As has been shown

$$B \approx D \exp\left(-\frac{t}{\tau}\right) \quad \text{and} \quad \frac{\partial B}{\partial t} \approx -\frac{D}{\tau}\exp\left(-\frac{t}{\tau}\right). \tag{14.182}$$

Assuming that measurements are performed during time off (the primary field is absent) and level of ambient noise is reduced to the level of a geological noise, consider again the ratio between the geological noise and the useful signal caused by an ore body:

$$\frac{B_N}{B_R} \approx \frac{D_N}{D_R}\exp\left[-\left(\frac{1}{\tau_N}-\frac{1}{\tau_R}\right)t\right]$$

and

$$\frac{\Xi_N}{\Xi_R} \approx \frac{\tau_R}{\tau_N}\frac{D_N}{D_R}\exp\left[-\left(\frac{1}{\tau_N}-\frac{1}{\tau_R}\right)\right]t. \tag{14.183}$$

Here Ξ indicates the electromotive force induced in the receiver coil. Just in the case of the harmonic fields, we can recognize three cases, namely

$$1.\ \tau_R > \tau_N \quad 2.\ \tau_R < \tau_N \quad 3.\ \tau_R = \tau_N$$

In the first case when the time constant in the ore body is greater than that in the noise source, as time increases the effect of the geological noise becomes smaller, and there will always be some time after which the field is predominantly determined by currents induced in the ore body. In the second case, with increasing time, the effect of geological noise becomes greater, and therefore the depth of investigation decreases. Finally, when both constants are the same, the ratio between the useful signal and geological noise is independent of observation time, and is a function only of the geometry of the body and the coordinates of an observation point.

As follows from the Fourier's integral, the resolution and depth of investigation for the frequency- and time-domain methods are the same. This statement implies that measurements in the frequency domain, as well as accuracy of calculation of Fourier's integral, have to be performed with a reasonable high accuracy, that it isn't simple to achieve. However, realization of this equivalence can be done differently. In the frequency-domain methods it requires a measurements of new quantities such as subsequent terms of the series Eq. (14.169), which are not usually measured in practice. Without any doubts their measurements, unlike the transient response, is hardly simple task. Perhaps this analysis already allows us to understand one of the important advantages of the transient method and its broad application for detecting different conductive bodies in the mining and environmental geophysics.

14.11. INFLUENCE OF A SURROUNDING MEDIUM ON THE FIELD DUE TO A CONFINED CONDUCTOR (CHARGES ARE ABSENT)

Until now, it was assumed that a surrounding medium is an insulator, and only a confined inhomogeneity with a relatively small resistivity can be source of a geological noise. Such model is proper for conditions similar to the Scandinavian and Canadian shields, but in general a massive ore body is located beneath an overburden and surrounded by a conducting medium influence of which cannot be neglected. In order to study their contribution to the field, we restrict ourselves to the case when the electric field is tangential to the conductor surface. In other words, charges are absent and the induced currents are the sole generators of the electromagnetic field. For instance, if the current source of the primary field and the conducting medium possesses axial symmetry with a common axis, the normal component of the electric field on the surface of a confined conductor is equal to zero, and therefore charges do not arise. Correspondingly, we can mentally represent the current field at two systems of current filaments so that one of them is located inside a confined conductor, but the other in the surrounding medium. Inasmuch as we consider the quasi-stationary fields, each current filament generates a magnetic field that can be calculated by the

Biot–Savart's law. Of course, in general, there is an interaction between induced currents, and the current density at any point is a function of the conductivity of both the surrounding medium and the confined conductor. It is convenient to represent the total electromagnetic field as a sum

$$\mathbf{B}(p) = \mathbf{B}_0(p) + \mathbf{B}_1(p)$$

and

$$\mathbf{E}(p) = \mathbf{E}_0(p) + \mathbf{E}_1(p), \tag{14.184}$$

where $\mathbf{B}_0,\mathbf{E}_0$ and $\mathbf{B}_1,\mathbf{E}_1$ are usually called the normal and secondary fields, respectively. The normal field is caused by the current source of the primary field and currents induced in the surrounding medium, provided that a confined conductor is absent. Thus, the normal field can be written in the following form

$$\mathbf{B}_0 = \mathbf{B}_0^{(0)} + \mathbf{B}_0^{(1)} \quad \text{and} \quad \mathbf{E}_0 = \mathbf{E}_0^{(0)} + \mathbf{E}_0^{(1)}. \tag{14.185}$$

Here, $\mathbf{B}_0^{(0)}$ and $\mathbf{E}_0^{(0)}$ form the primary field at an observation point p, generatedby its current source, while$\mathbf{B}_0^{(1)}$ and $\mathbf{E}_0^{(1)}$ are caused by currents induced in the surrounding medium in the absence of the confined body. Let us notice that the normal fields were studied in some details in the previous chapters. At the beginning, we consider the field in the frequency domain and then, applying a Fourier's transform, the transient field will be studied. It is obvious that at the high frequencies, as well as at the early stage, due to the skin effect, currents are mainly concentrated near the current source, while the electromagnetic field in the vicinity of the confined conductor is very small. This means that within this range of frequencies or times, the influence of the conductor is negligible, and therefore from the geophysical point of view this part of the spectrum and the transient response at the early stage do not have any practical interest. For this reason, we mainly pay attention to the intermediate- and low-frequency part of the spectrum, where the skin depth in the surrounding medium δ_e is essentially greater than the distance from the current source to a confined conductor. In the same manner, we will consider the intermediate and late stages of the transient response. In most cases that are of practical interest in the exploration geophysics, we can assume that δ_e greatly exceeds a dimension of a body or its cross section, when it is treated as a two-dimensional one. By definition, the normal field ($\mathbf{E}_0,\mathbf{B}_0$) does not contain an information about a confined conductor and, correspondingly, it represents a "geological noise", when the electromagnetic methods are used to detect a confined conductor and determine its parameters. In contrast, the secondary field ($\mathbf{E}_1$,$\mathbf{B}_1$) arises due to the presence of the conductive target; that is currents that are located inside and outside the confined conductor. Respectively, the secondary field is a function of the resistivity of the confined conductor and that of the surrounding medium, and its appearance can be described in the following way. Under the action of the normal electric field, induced currents arise inside the conductor, and their magnetic field gives rise to the secondary currents in the surrounding medium. They in their own turn influence the currents inside the conductor. Thus, due to interaction, some current distribution occurs, and in general it is rather a complicated function of parameters of a medium, and usually in order to determine this field, it is necessary to apply either the integral equations or other numerical method.

14.11.1. Approximate Method of Field Calculation

Now, we describe an approach that allows us in many cases to avoid application these numerical methods and calculate the field with a sufficient accuracy. Of course, it can be used under certain conditions, namely

1. The conductivity of a confined conductor is significantly greater than that of the surrounding medium and
2. The skin depth in the surrounding medium δ_e is much greater than the linear dimensions L of the conductor. Thus, we assume that

$$\frac{\gamma_i}{\gamma_e} \gg 1 \quad \text{and} \quad \frac{L}{\delta_e} \ll 1. \tag{14.186}$$

It is appropriate to note that both of these conditions are very often observed in the practice of the mining prospecting. From the physical point of view, these conditions indicate that the interaction between currents induced in the conductor practically is independent on the conductivity of the surrounding medium. In other words, the current density in the confined conductor, surrounded by either by a conducting medium or by an insulator, is the same if in both cases the normal fields $\mathbf{E}_0$ coincide at each point of the conductor. In contrast, the magnitude and phase of these currents are functions of a conductivity of this medium, and this happens due to the normal field $\mathbf{E}_0$. Thus, determination of the secondary field consists of two steps, namely:

1. Calculation of the current density in the conductor located in a free space.
2. Determination of the field caused by these currents in the surrounding medium when the conductor is absent.

It is obvious that with an increase in the frequency, the secondary currents in a surrounding medium begin to influence the current density in the conductor. For this reason, the high-frequency portion of the spectrum, as well as the early stage of the transient field, cannot be calculated using this approximate method. Therefore, in determining the current density j_φ, we make use of the fact that over a relatively wide range of frequencies, the strength of currents inside the conductor depends mainly on the normal electric field $E_{0\varphi}$ and the first pole of the spectrum of these currents ω_1, where

$$\omega_1 = -\frac{1}{\tau}. \tag{14.187}$$

As was shown in the previous chapter, the equation for the complex amplitude of the induced currents in the conductor placed in an insulator can be written as

$$j_\varphi(q) = i\omega B_0(q)\tau \sum_{n=1}^{\infty} \frac{\beta_n(q)}{1 - i\omega\tau_n}, \tag{14.188}$$

where $\tau_1 = \tau$ is the time constant and B_0 is the magnitude of the normal magnetic field at some point q of the conductor. Moreover, we demonstrated earlier that for a sufficiently wide range of frequencies, including the low-frequency part of the spectrum, the first term of the sum in Eq. (14.188) is the most important one. For this reason, we assume that the current density at any point in the conductor is given by

$$j_\varphi(q) = \frac{i\omega\tau\beta_1(q)}{1 - i\omega\tau} B_0(q). \tag{14.189}$$

Thus, we have a system of current rings inside of the conductor and in order to find the field at an observation point we have to perform a summation of the fields caused by each current filament situated in a surrounding medium, as if the conductor were absent. Later, we describe the method of integral equations that allowed one to determine this field at all parts of the spectrum, and comparison with this method has shown that an error in determining the field by an approximate approach in the case of a spheroidal conductor is not greater than 10% when

$$\omega\tau < 1 \quad \text{and} \quad \frac{a}{\delta_e} < 0.4$$

Here, a and b are the major and minor semi-axes of the spheroid, respectively, and the normal field $\mathbf{B}_0$ is directed along the minor axis. The procedure of an approximate calculation of the field is greatly simplified when the normal magnetic field in the vicinity of this conductor is uniform. For instance, as we already know, the field of currents in the sphere is equivalent to that of the magnetic dipole with the moment

$$\mathbf{M} = 2\pi D_1 a^3 \mathbf{B}_0, \tag{14.190}$$

where

$$D_1 = \frac{3}{x}\coth x - \frac{3}{x^2} - 1 \quad \text{and} \quad x = ik_i a.$$

Correspondingly, for the vertical component of the magnetic field in a uniform medium and along the z-axis, we have

$$B_{1z} = \frac{\mu_0 M}{2\pi z^3}\exp(ik_e z)(1 - ik_e z). \tag{14.191}$$

With a proper generalization, this equation can be used to describe the secondary field of currents in the sphere or in a spheroid, located in a horizontally layered medium.

14.11.2. Influence of Surrounding Medium on the Depth of Investigation

Introduction

Now, we illustrate an application of this method, as well as an exact solution, to study an influence of a surrounding medium on the depth of investigation in the frequency- and time domains. At the beginning for simplicity, suppose that a confined conductor is located in a uniform medium with conductivity γ_e, and the normal field is caused by a current in a loop of radius r_l and consider the total field at the center of the loop:

$$B_z = B_{0z} + B_{1z}. \tag{14.192}$$

As was pointed out in the previous chapter, when the surrounding medium is an insulator, the sole conductor can be determined by measuring either the frequency or the transient responses, inasmuch as the generators of the field are induced currents in a single conductor, and the interpretation of observed data is relatively simple. Geometric characteristics like shape, dimensions, orientation, and depth of location of the conductor can usually be determined from the geological information and the behavior of the field at different observation points. Then, making use of the frequency or transient responses, the time constant τ can be calculated. Thus, the frequency and transient methods provide the same information about the conductor, clearly demonstrating their equivalence. However, as in the case of two conductors, if the medium surrounding a conductor has a finite resistivity, we often do not observe this equivalence between these two conventional electromagnetic methods. It is clear that this is related to the fact that the total magnetic field measured at an observation site consists of two parts, namely.

(1) The field $\mathbf{B}_0$ caused by currents induced in the surrounding medium, which does not contain any information about the conductor. As in the case of a confined conductor, it is called the geological noise. (2) The secondary field $\mathbf{B}_1$, which depends on the conductivity and geometrical parameters of the confined conductor, as well as, in general, on the conductivity of the surrounding medium, and naturally it represents the useful signal. It is clear that the ratio of the useful signal and geological noise is what unlimitedly characterizes the efficiency of the electromagnetic methods in discovering a conductor and determining its time constant and geometric parameters. For this reason, we treat the ability to reduce the influence of the geological noise as the most important feature of the frequency and transient methods. The importance of this approach is emphasized by the fact that the parameters of the surrounding medium are usually unknown, and, correspondingly, the geological noise can hardly be reduced by using analytical means. To illustrate how the ratio of the useful signal and geological noise changes, we assume that the magnetic field is measured at the loop center, and correspondingly the normal field is defined as

$$B^*_{0z}(\omega) = \frac{\mu_0 I}{2r_l}\exp(ik_e r_l)(1 - ik_e r_l). \tag{14.193}$$

In deriving the expression for the secondary field, we make two assumptions, which drastically simplify calculations of the field:

1. The normal field $\mathbf{B}_0$ caused by currents in the medium and by the current source is uniform in the vicinity of the confined conductor.
2. The magnitude of the current density in the conductor is directly proportional to the normal field B_{0z} and it depends on the product $\omega\tau$ as

$$j_\varphi(q) \sim \frac{i\omega\tau}{1 - i\omega\tau} B_{0z}(q).$$

As was shown earlier (Chapter 9), the vertical component of the field of a circular loop on the z-axis is given by

$$B_{0z} = \frac{\mu_0 I r_l^2}{2R_0^3} \exp(ik_e R_0)(1 - ik_e R_0),$$

where

$$k_e = (i\gamma_e \mu_0 \omega)^{1/2}, \quad R_0 = \left(r_l^2 + z^2\right)^{1/2}.$$

Proceeding from the latter and our assumptions, we have for the useful signal

$$B_{1z}(\omega) = \frac{\mu_0 I}{2r_l} d \frac{i\omega\tau}{1 - i\omega\tau} \exp(ik_e R_0)(1 - ik_e R_0)\exp(ik_e z)(1 - ik_e z), \tag{14.194}$$

where d is the function depending on geometric parameters only. Eqs (14.193) and (14.194) clearly demonstrate that the behavior of the field caused by induced currents in the surrounding medium differs significantly from that of the field generated by currents in a confined conductor. As was pointed out earlier, due to the skin effect, the high-frequency part of the spectrum as well as the early stage are not of great practical interest, and for this reason attention will be mainly concentrated on the low-frequency spectrum and the late stage of transient field. At the low-frequency spectrum, the quadrature and in-phase components of the normal field B_{0z} can be represented as

$$QB_{0z} \approx \frac{\mu_0 I}{2r_l} \left[\frac{\omega\tau_0}{2} - \frac{(\omega\tau_0)^{3/2}}{3 \cdot 2^{1/2}} + \frac{(\omega\tau_0)^{5/2}}{30 \cdot 2^{1/2}} - \cdots \right]$$

and

$$InB_{0z} = \frac{\mu_0 I}{2r_l} \left[1 - \frac{(\omega\tau_0)^{3/2}}{3 \cdot 2^{1/2}} + \frac{(\omega\tau_0)^2}{8} - \cdots \right]. \tag{14.195}$$

Here, τ_0 is the time constant of the surrounding medium, provided that the field is measured at the loop center, and it is equal to

$$\tau_0 = \gamma_e \mu_0 r_l^2. \tag{14.196}$$

Correspondingly, with a decrease of the radius r_l or, in general, the distance between the current source and the observation point, this time constant decreases rather rapidly. First, we investigate the ratio between the useful signal and geological noise in the case when the surrounding medium is relatively resistive, and the skin depth in this medium δ_e is much greater than such parameters as R_0, r_l and L, where the latter characterizes the dimension of the conductor. Then, the secondary field B_{1z} is not practically subjected to the influence of the surrounding medium and, in accordance with Eq. (14.194), we have

$$B_{1z}(\omega) = \frac{\mu_0 I}{2r_l} d \frac{i\omega\tau}{1 - i\omega\tau}. \tag{14.197}$$

As follows from Eqs (14.195) and (14.196) in this limiting case, the quadrature components are

$$QB_{1z}(\omega) \approx \frac{\mu_0 I}{2r_l} d \frac{\omega\tau}{1 + (\omega\tau)^2}, \quad QB_{0z}(\omega) \approx \frac{\mu_0 I}{2r_l} \frac{\omega\tau_0}{2}. \tag{14.198}$$

Thus, the ratio between the useful signal and the geological noise is

$$\frac{QB_{1z}(\omega)}{QB_{0z}(\omega)} = 2.0 \cdot d \frac{\tau}{\tau_0} \frac{1}{1 + (\omega\tau)^2}.$$

In particular, if $\omega\tau$ is small, we obtain

$$\frac{QB_{1z}(\omega)}{QB_{0z}(\omega)} \approx 2.0 d \frac{\tau}{\tau_0}, \tag{14.199}$$

that is at the low-frequency spectrum ($\omega\tau < 1$), this ratio reaches its maximal value, and it depends on geometrical factors and the ratio of time constants. Next, consider the in-phase components of the field, assuming that the primary field is cancelled. Again, proceeding from Eqs (14.195) and (14.197), we have:

$$InB_{1z}(\omega) \approx -\frac{\mu_0 I}{2r_l} d \frac{(\omega\tau)^2}{1 + (\omega\tau)^2}, \quad InB_{0z}(\omega) \approx -\frac{\mu_0 I}{2r_l} \frac{\tau_0^{3/2}}{3 \cdot 2^{1/2}} \omega^{3/2}$$

Whence

$$\frac{InB_{1z}(\omega)}{InB_{0z}(\omega)} \approx 3 \cdot 2^{1/2} \left(\frac{\tau}{\tau_0}\right)^{3/2} \frac{(\omega\tau)^{1/2}}{1 + (\omega\tau)^2}. \tag{14.200}$$

Thus, the behavior of this ratio differs essentially from that for the quadrature components. In fact, with an increase of frequency, the ratio increases, reaches a maximum, and then approaches zero. The maximum occurs when

$$\omega\tau \approx 0.6, \tag{14.201}$$

and its value is

$$\left[\frac{InB_{1z}(\omega)}{InB_{0z}(\omega)}\right]_{\max} \approx 2.4d\left(\frac{\tau}{\tau_0}\right)^{3/2}. \tag{14.202}$$

Therefore, by measuring the in-phase component one can significantly reduce the influence of the geological noise if $\tau > \tau_0$.

Next, we study the relationship between the secondary and normal transient fields measured at the loop center, assuming again that the magnetic field from the secondary currents in the surrounding medium does not have any effect on the interaction of currents within a confined conductor. In this approximation, the development of the transient field can be described as follows. When the current source changes as a step function, normal fields $\mathbf{B}_0$ and $\mathbf{E}_0$ arise instantly at each point in the medium, provided that displacement currents are neglected. The character of the field behavior with time depends on the distance from the current source and conductivity of a medium. In particular, for relatively high resistivity of the surrounding medium, the field $\mathbf{B}_0$ changes more rapidly, while a maximum of the electric field manifests itself at earlier times. In describing the secondary field, we distinguish as usual the early, intermediate, and late stages. At the early stage, due to the skin effect, the normal electric field $\mathbf{E}_0$ inside the conducting medium is very small, and therefore the secondary field is negligible. For this reason, we pay attention only in the intermediate and late stages.

Proceeding from our assumptions, we obtain at the late stage

$$B_{1z}(t) \approx \frac{\mu_0 I}{2r_l} d \exp\left(-\frac{t}{\tau}\right), \quad B_{0z}(t) \approx \frac{\mu_0 I}{12r_l}\frac{\tau_0^{3/2}}{\pi^{1/2}t^{3/2}}. \tag{14.203}$$

Consequently, the ratio of the transient responses of the magnetic field caused by currents in the confined conductor and in the surrounding medium is

$$\frac{B_{1z}(t)}{B_{0z}(t)} \approx 6\pi^{1/2}d\left(\frac{t}{\tau_0}\right)^{3/2} \exp\left(-\frac{t}{\tau}\right). \tag{14.204}$$

Thus, with an increase of time this ratio increases, reaches a maximum and then tends to zero, since the field caused by induced currents in the surrounding medium decreases more slowly than due to currents in the confined conductor. It is simple matter to see that the ratio reaches the maximum for

$$t = 1,5\tau, \tag{14.205}$$

and its value is

$$\left(\frac{B_{1z}}{B_{0z}}\right)_{\max} \approx 4.4d\left(\frac{\tau}{\tau_0}\right)^{3/2}. \tag{14.206}$$

Now, let us consider the ratio of the useful signal and geological noise when the electromotive forces are measured. In accordance with Eq. (14.203), we have

$$\dot{B}_{1z}(t) \approx -\frac{\mu_0 I}{2r_l}\frac{d}{\tau}\exp\left(-\frac{t}{\tau}\right), \quad \dot{B}_{0z}(t) \approx -\frac{\mu_0 I}{2r_l}\frac{\tau_0^{3/2}}{4.0\pi^{1/2}}, \tag{14.207}$$

where $\dot{B}(t) = \dfrac{\partial B}{\partial t}$.

For the ratio of these functions, we have

$$\dot{B}_{1z}(t)/\dot{B}_{0z}(t) = 4.0\pi^{1/2}d\left(\frac{t}{\tau_0}\right)^{3/2}\frac{t}{\tau}\exp\left(-\frac{t}{\tau}\right) \tag{14.208}$$

and it reaches a maximum when

$$t = 2.5\tau. \tag{14.209}$$

Thus,

$$\dot{B}_{1z}(t)/\dot{B}_{0z}(t_{\max}) \approx 6.d\left(\frac{\tau}{\tau_0}\right)^{3/2} \tag{14.210}$$

Comparison of Eqs (14.199), (14.206) and (14.210) clearly shows that by measuring the transient magnetic field or electromotive force, it is possible to reduce the influence of the geological noise to a greater extent than in the case when the quadrature component of the magnetic field is measured. Using a very simple and approximate method, we were able to understand the essential features of the frequency and transient methods. Taking into account the fact that they are of a great practical interest, let us also discuss the results of calculations based on the exact solution of the boundary-value problem, using the integral equations. With this purpose in mind, first suppose that a conducting sphere with radius b is surrounded by a uniform conducting medium, and the field is measured at the center of the current loop. At the beginning, consider the frequency responses of the function

$$Qb_z = \frac{QB_{0z} + QB_{1z}}{QB_{0z}} = 1 + \frac{QB_{1z}}{QB_{0z}}.$$

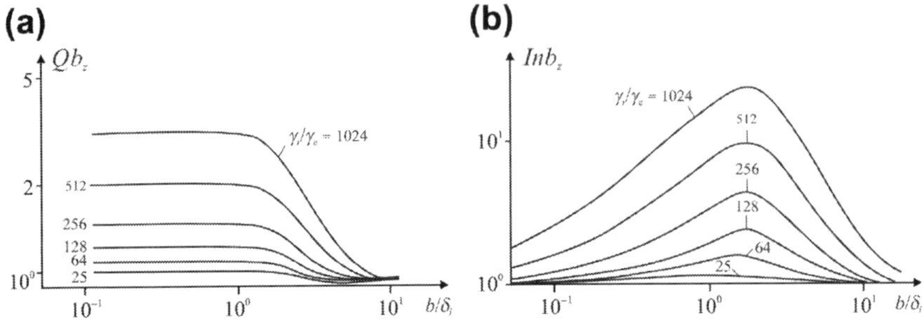

Figure 14.10 Frequency responses of the functions Qb_z (a) and Inb_z (b); $r_l/b = 1$, $z/b = 2$.

The latter characterizes the influence of the surrounding medium and therefore the depth of investigation when the quadrature component is measured. The frequency responses of the function Qb_z are shown in Figure 14.10(a). The index of curves is the ratio of conductivities γ_i/γ_e. The left-hand asymptote of the curves corresponds to the low-frequency part of the spectrum over which the normal and secondary fields are directly proportional to the frequency. It is clear that a decrease of the frequency within this range does not reduce the influence of the geological noise. With an increase in the value of the parameter b/δ_i, the secondary field begins to increase more slowly than does the normal field, and therefore the relative anomaly QB_{1z}/QB_{0z} becomes smaller, and correspondingly the function Qb_z approaches unity. Thus, in measuring the quadrature component of the field $\boldsymbol{B}$, the influence of the geological noise is minimal within the low-frequency spectrum, where the ratio Qb_z is independent of the frequency. In particular, this means that a decrease of the frequency in this end of the spectrum does not result in an increase of the depth of investigation, even though the depth of penetration of the electromagnetic field of course becomes greater.

Next, we study the behavior of the function Inb_z, Figure 14.10(b), where

$$Inb_z = \frac{InB_{0z}^{(1)} + InB_{1z}}{InB_{0z}^{(1)}} = 1 + \frac{InB_{1z}}{InB_{0z}^{(1)}},$$

since we assume that the primary field $B_{0z}^{(0)}$ is canceled instrumentally. As is seen from the curves for Inb_z at low frequencies, the ratio Inb_z tends toward unity in contrast to the behavior of the function Qb_z. This happens because over this part of the spectrum the secondary field, that is the useful signal

InB_{1z} decreases more rapidly with a decrease in frequency (ω^2) than does the geological noise $InB_0^{(1)}$, which varies as $\omega^{3/2}$. Thus, in this part of the spectrum with a decrease of the frequency, the depth of investigation becomes smaller (when in-phase component is measured), in spite of the fact that the skin depth δ_e increases unlimitedly. With an increase of the frequency, the function Inb_z also increases, (influence of the geological noise becomes smaller) passes through a maximum and then approaches unity. This behavior of Inb_z, as well as that of the function Qb_z in the high and intermediate parts of the spectrum, is mainly governed by the skin effect in the surrounding medium. Note that the position of the maximum and its value are well described by Eqs (14.201)–(14.209), respectively if $\gamma_i/\gamma_e \gg 1$. Thus, we have found that there is an intermediate range of frequencies, a "window", where the influence of the geological noise on the in-phase component can be reduced to a greater extent than at the low- and high-frequency parts of the spectrum. In other words, within this intermediate range of frequencies, it is possible to achieve the maximal depth of investigation, which exceeds that when the quadrature component QB_z is measured. In this light, it is natural, as in the case of confined conductors, to consider the difference of the quadrature components measured at two frequencies

$$\Delta QB_{1z} = QB_z(\omega_1) - \frac{\omega_1}{\omega_2} QB_z(\omega_2).$$

Correspondingly, the behavior of the function

$$\Delta Qb_z = \frac{\Delta QB_{0z} + \Delta QB_{1z}}{\Delta QB_{0z}}$$

are shown in Figure 14.11(a). Again as before, there is a range of frequencies where it is possible to reach a maximal reduction of a geological noise in a degree that is even greater than in the case of the in-phase component.

Next, we consider the influence of the geological noise on the transient field. With this purpose in mind, we study the function

$$e(t) = \frac{\Xi_0(t) + \Xi_1(t)}{\Xi_0(t)} = 1 + \frac{\Xi_1(t)}{\Xi_0(t)},$$

where $\Xi_0(t)$ and $\Xi_1(t)$ are the electromotive forces caused by the normal and secondary fields, respectively. Examples of the behavior of the function $e(t)$ are given in Figure 14.11(c).

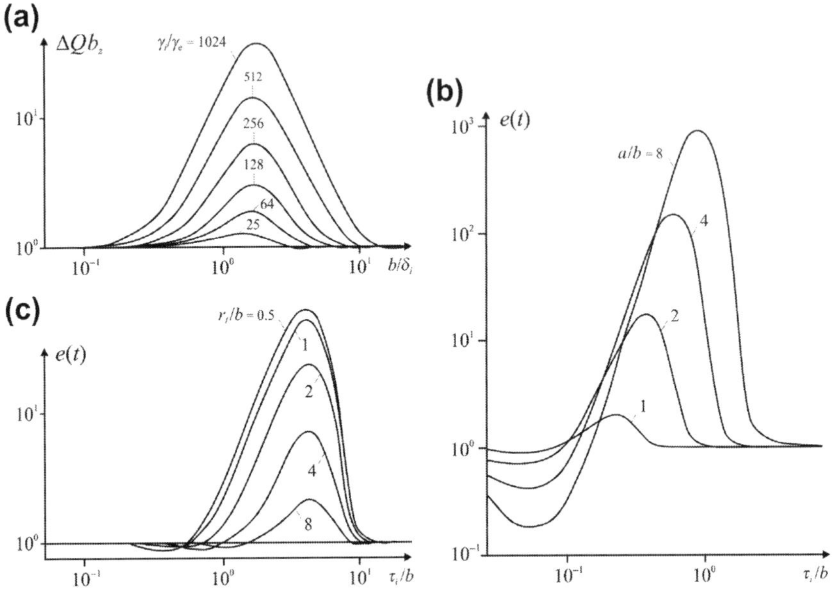

Figure 14.11 (a) Frequency responses of the function ΔQb_z, $r_l/b = 1$, $z/b = 2$, curve index is γ_i/γ_e; (b) Transient responses of the function $e(t)$, $\gamma_i/\gamma_e = 1024$, $r_l/b = 8$, $z/b = 2$, curve index is a/b; (c) Transient responses of the function $e(t)$, $\gamma_i/\gamma_e = 1024$, $z/b = 2$, curve index is r_l/b.

Unlike the previous cases, the index of curves is the ratio r_l/b, while the parameter τ_i is

$$\tau_i = \left(2\pi\rho_i t 10^7\right)^{1/2}.$$

Let us note that the transient responses of the function $e(t)$ were calculated by using a Fourier's transform. As is seen from these curves, at the early stage the influence of the useful signal is negligible and the ratio $e(t)$ tends to unity. As was pointed out earlier, this behavior is obvious, since in the early stage of the transient response of the normal electric field $\mathbf{E}_0$ is very small in the vicinity of the conductor. With an increase of time, the function $e(t)$ also increases, reaches a maximum and then with further increase in time again approaches unity. Therefore, as in the case of the in-phase component InB_z and difference of the quadrature components ∇QB_z, there is a "window" that is, an intermediate range of times where it is possible to achieve the maximal reduction of geological noise, which is slightly more than in measuring ∇QB_z. If the surrounding medium is sufficiently resistive ($\gamma_e < \gamma_i$), the secondary field $B_{1z}(t)$ is practically independent of the parameters of this medium within this intermediate interval of times. It is easy to see that the maximum of $e(t)$ takes place when time t exceeds the time constant τ and, correspondingly the secondary field exponentially decays. As was mentioned above, at the late stage the normal field changes relatively slowly:

$$E_0 \sim \frac{1}{t^{5/2}}$$

and therefore currents inside the confined conductor vary in the same manner. For this reason, the secondary magnetic field changes with time as

$$B_{1z} \sim \frac{1}{t^{5/2}} \quad \text{if} \quad t \gg \tau, \quad \text{and} \quad t > \tau_0.$$

At the same time, the normal electric field B_{0z}, except some special cases, decreases in the late stage more slowly:

$$B_{0z} \sim \frac{1}{t^{3/2}}$$

Consequently, the electromotive forces $\Xi_0(t)$ and $\Xi_1(t)$ behave in the late stage in the following way:

$$\Xi_1(t) \sim \frac{1}{t^{7/2}} \quad \text{and} \quad \Xi_0(t) \sim \frac{1}{t^{5/2}},$$

and with an increase of time, the influence of the useful signal vanishes. This analysis suggests that the first equation of the set Eq. (14.203), describing the secondary field at the late stage, can be improved and represented in the form:

$$B_{1z}(t) \sim \frac{\mu_0 I}{2r_l}\left[d \exp\left(-\frac{t}{\tau}\right) + c/t^{5/2}\right],$$

where c is a function, which depends on conductivities and geometrical parameters.

Proceeding from the physical point of view, we can expect that all conclusions concerning the influence of the useful signal and geological noise at different parts of the frequency and transient responses remain valid for any surrounding medium and a confined conductor provided that charges are absent. For illustration results of calculations of the function $e(t)$ for different spheroid, located beneath the earth surface, are given in Figure 14.11(b). Let also briefly consider an influence of the radius of the current loop, Figure 14.11(c). The curves of the function $e(t)$ vividly demonstrate how the radius of the current loop r_l makes influence on the depth of investigation. First, with a decrease of r_l, the distance between the current loop and the confined conductor also becomes smaller. Correspondingly, the normal electric field, which generates currents on the conductor surface at the initial moment, becomes bigger. In addition with a decrease of the loop radius, the late stage of the normal field is observed at the earlier times. Because of these factors, the influence of the geological noise decreases with a decrease of r_l, as soon as it greater than the distance from the conductor to an observation point. Further decrease of the loop radius, when $r_l < z$ doesn't practically reduce the influence of the geological noise, since the behavior of the normal field is mainly defined in this case by the distance z. This is the reason why in practice, except the cases of a very resistive surrounding medium, the radius of the current loop is comparable with the distance from an observation point to a conducting body.

14.11.3. Equivalence and Difference between the Frequency and Transient Methods

Modern methods of processing data, measured in the frequency- and time domains, allow one to reach practically the same level of reduction of the cultural and natural types of noise. For this reason, efficiency of the frequency and transient methods, in particular, the depth of investigation, is determined by their ability to reduce the geological noise. In favorable conditions when a surrounding medium is very resistive measurements of the transient response or the quadrature component of the field in the frequency domain may give the same result; that is in such conditions both methods are equivalent. However, with an increase of a conductivity of an overburden or a host rocks or both of them the situation changes, and in order to achieve equivalence it may require to measure instead of the quadrature component either the in-phase component of the secondary field or the difference of quadrature components at two frequencies. As was shown above, the latter allows one to reach practically the same degree of reduction of the geological noise as in the time domain and there is no need to measure the terms of the low-frequency spectrum of the higher order. Thus, as follows from the theory of Fourier's integral, the low-frequency part of the spectrum defines the late stage of the transient response. In other words, both the frequency and transient methods are always equivalent to each other in any geoelectric conditions. However, measurements of the in-phase component of the secondary field or difference of the quadrature component at two frequencies are usually much more complicated problem than observation of the transient field in the absence of the primary field. This is one of the two main reasons why the transient method for detecting good conductors became the conventional method of the exploration geophysics. The other reason is related to the fact that it is possible to measure the secondary field in the absence of the primary one.

14.12. ELLIPTICAL POLARIZATION OF THE ELECTRIC AND MAGNETIC FIELD

Until now, we have studied the frequency and transient responses of the electric and magnetic fields, considering their components mainly along Cartesian axes and didn't pay attention to a behavior of the total vector of the field in the space and its change with time. It turns out that this subject has not only theoretical interest but also found a practical application because allowed one to develop the electromagnetic methods with relatively low sensitivity to a change of the primary field. We assume that the field is a sinusoidal function of time and start from the electric field measured at the earth's surface. Here, it may be proper to make one comment. In the mining prospecting, except some cases, the electric field isn't usually measured, but it is an important element in magnetotellurics, as well as in the method of the "infinitely" long wire (CSMT). Introducing the Cartesian system x,y,z, where axes x and y are located on the earth's surface and the z-axis is directed downward, we represent the electric field in the form:

$$E_x = E_{x0}\cos(\omega t + \varphi_x), \quad E_y = E_{y0}\cos\left(\omega t + \varphi_y\right),$$

and

$$E_z = 0. \tag{14.211}$$

Also

$$B_x = B_{x0}\cos(\omega t + \psi_x), \quad B_y = B_{x0}\cos\left(\omega t + \psi_y\right),$$

and

$$B_z = B_{z0}\cos(\omega t + \psi_z). \tag{14.212}$$

Note that we consider the electric field in the conducting medium in the vicinity of the earth's surface, where this field is tangential to the boundary with a free space.

14.12.1. The Elliptical Polarization of the Electric Field

As follows from Eq. (14.211), the magnitude and direction of the total electric field, located in the plane of the earth's surface, change with time and this behavior has a periodical character. In other words, during each period the vector **E** returns to its initial position and the higher frequency is, the more quickly this vector rotates. To visualize this movement, let us find the curve along which the end of the vector **E** moves during each period. We can obtain the equation of this path by eliminating the time from Eq. (14.211). First, let us rewrite two equations of this system as

$$E_x(t) = E_{x0}\cos\omega t\cos\varphi_x - E_{x0}\sin\omega t\sin\varphi_x$$

$$E_y(t) = E_{y0}\cos\omega t\cos\varphi_y - E_{y0}\sin\omega t\sin\varphi_y$$

or

$$\frac{E_x(t)}{E_{x0}} = \cos\omega t\cos\varphi_x - \sin\omega t\sin\varphi_x$$

$$\frac{E_y(t)}{E_{y0}} = \cos\omega t\cos\varphi_y - \sin\omega t\sin\varphi_y.$$

Multiplying the first of these last equalities by $\sin\varphi_y$, but the second one by $\sin\varphi_x$ and taking the difference, we obtain

$$\begin{aligned}\frac{E_x}{E_{x0}}\sin\varphi_y - \frac{E_y}{E_{y0}}\sin\varphi_x &= \cos\omega t\left(\sin\varphi_y\cos\varphi_x - \cos\varphi_y\sin\varphi_x\right)\\ &= \cos\omega t\sin\left(\varphi_y - \varphi_x\right).\end{aligned}$$

By analogy, multiplying the same first equality by $\cos\varphi_y$, but the second one by $\cos\varphi_x$ and also taking the difference, we obtain

$$\begin{aligned}\frac{E_x}{E_{x0}}\cos\varphi_y - \frac{E_y}{E_{y0}}\cos\varphi_x &= \sin\omega t\left(\sin\varphi_y\cos\varphi_x - \cos\varphi_y\sin\varphi_x\right)\\ &= \sin\omega t\sin\left(\varphi_y - \varphi_x\right).\end{aligned}$$

Taking square of each of the last equalities and adding them, we have

$$\left(\frac{E_x}{E_{x0}}\sin\varphi_y-\frac{E_y}{E_{y0}}\sin\varphi_x\right)^2+\left(\frac{E_x}{E_{x0}}\cos\varphi_y-\frac{E_y}{E_{y0}}\cos\varphi_x\right)^2$$
$$=\sin^2(\varphi_y-\varphi_x)$$

or

$$\frac{E_x^2}{E_{x0}^2}+\frac{E_y^2}{E_{y0}^2}-2\frac{E_x}{E_{x0}}\frac{E_y}{E_{y0}}\cos(\varphi_y-\varphi_x)=\sin^2(\varphi_y-\varphi_x). \qquad (14.213)$$

This equation describes the ellipse, because

$$4\left[\frac{\cos^2(\varphi_y-\varphi_x)}{E_{x0}^2E_{y0}^2}-\frac{1}{E_{x0}^2E_{y0}^2}\right]<0$$

and the right-hand side of Eq. (14.213) is positive. To simplify Eq. (14.213), let us introduce new system of coordinates x', y', so that its axes form with the original ones the angle α:

$$\tan 2\alpha=\left(2\frac{\cos\left(\varphi_y-\varphi_x\right)}{E_{x,0}E_{y0}}\right):\left(\frac{1}{E_{x0}^2}-\frac{1}{E_{y0}^2}\right)=\frac{2E_{x0}E_{y0}\cos(\varphi_y-\varphi_x)}{E_{y0}^2-E_{x0}^2}$$

Then, expressing E_x and E_y in terms of $E_x^{'}$, $E_y^{'}$, we obtain the equation of the ellipse that doesn't contain the product of components of the electric field in the new system of coordinates, while coefficients in front of $E_x'^2$ and $E_y'^2$ are

$$\frac{1}{2}\left\{\frac{1}{E_{x0}^2}+\frac{1}{E_{y0}^2}+\left[\left(\frac{1}{E_{x0}^2}-\frac{1}{E_{y0}^2}\right)^2+\frac{4}{E_{x0}^2E_{y0}^2}\cos^2(\varphi_y-\varphi_x)\right]^{1/2}\right\}$$

and

$$\frac{1}{2}\left\{\frac{1}{E_{x0}^2}+\frac{1}{E_{y0}^2}-\left[\left(\frac{1}{E_{x0}^2}-\frac{1}{E_{y0}^2}\right)^2+\frac{4}{E_{x0}^2E_{y0}^2}\cos^2(\varphi_y-\varphi_x)\right]^{1/2}\right\}.$$

The latter can be represented as

$$\frac{1}{2E_{x0}^2E_{y0}^2}\left\{E_{x0}^2+E_{y0}^2+\left[E_{x0}^4+E_{y0}^4+2E_{x0}^2E_{y0}^2\cos^2\left(\varphi_y-\varphi_x\right)\right]^{1/2}\right\}$$

and

$$\frac{1}{2E_{x0}^2 E_{y0}^2}\left\{E_{x0}^2 + E_{y0}^2 - \left[E_{x0}^4 + E_{y0}^4 + 2E_{x0}^2 E_{y0}^2 \cos^2\left(\varphi_y - \varphi_x\right)\right]^{1/2}\right\}.$$

Thus, equation of the ellipse in the canonical form is

$$\frac{E_x'^2}{a^2} + \frac{E_y'^2}{b^2} = 1.$$

This shows that the major and minor axes are oriented along the axes of the new system of coordinates. Here

$$a^2 = \frac{2E_{x0}^2 E_{y0}^2 \sin^2\left(\varphi_y - \varphi_x\right)}{E_{x0}^2 + E_{y0}^2 + \left[E_{x0}^4 + E_{y0}^4 + 2E_{x0}^2 E_{y0}^2 \cos^2\left(\varphi_y - \varphi_x\right)\right]^{1/2}}$$

and

$$b^2 = \frac{2E_{x0}^2 E_{y0}^2 \sin^2\left(\varphi_y - \varphi_x\right)}{E_{x0}^2 + E_{y0}^2 - \left[E_{x0}^4 + E_{y0}^4 + 2E_{x0}^2 E_{y0}^2 \cos^2\left(\varphi_y - \varphi_x\right)\right]^{1/2}}.$$

Thus, we see that at points of the earth surface the end of the vector of the electric field describes during each period the ellipse, and the ratio of its axes is

$$\frac{E_{x0}^2 + E_{y0}^2 + \left[E_{x0}^4 + E_{y0}^4 + 2E_{x0}^2 E_{y0}^2 \cos^2\left(\varphi_y - \varphi_x\right)\right]^{1/2}}{E_{x0}^2 + E_{y0}^2 - \left[E_{x0}^4 + E_{y0}^4 + 2E_{x0}^2 E_{y0}^2 \cos^2\left(\varphi_y - \varphi_x\right)\right]^{1/2}}.$$

It is clear that if we know the magnitude and phase of each component of the electric field, $E_{x0}, E_{y0}, \varphi_x, \varphi_y$ all elements of the ellipse can be determined—namely its minor and major axes and orientation with respect to the original coordinate axes. Such field in which the vector changes its orientation, tracing the ellipse by its end during one period, is called the elliptically polarized field. Now we illustrate this phenomenon, and with this purpose briefly describe one of the oldest methods of mining prospecting.

14.12.2. Method of Charged Body with Alternating Current

One of modifications of the method of charged body, designed for imaging the geometry of the known ore body is based on the use of alternating current. As is well known, in this case one current electrode is connected with the ore body, while the other is located far away in order to neglect its influence. Suppose that we have a receiver line that consists of two electrodes and the indicator (earphone). Let us fix a position of one electrode at some point of the earth surface and move the other along the circle around the first one. Since there is an alternating field in the earth, we observe the electromotive force with the frequency ω, which has the amplitude a_s and phase φ_s. The latter characterizes the phase shift with respect to the source current. Now, we find how the amplitude and phase depend on the direction s, and for simplicity assume that at all points of the area where the receiver line rotates around the point the field changes in the same manner. In other words, the elliptical polarization is the same at this part of the area. Consider the ellipse, described by the electric field, Figure 14.12(a), and s- is a chosen direction, which forms with the major axis of the ellipse the angle α. It is obvious that the amplitude of the electromotive force along this direction s is defined by the maximal projection of the radius vector of the ellipse on this direction. In order to find this amplitude a_s, it is sufficient to draw the perpendicular to the line s and tangential to the ellipse. This simple procedure gives two points K and L, and half-distance between them is equal to a_s. This maximal value of the electromotive force occurs when the field is directed to the point where perpendicular are tangential to the ellipse. Correspondingly, the phase is the same as for vectors PK' and PL'. Repeating this procedure for different directions s, we obtain the set of points K and L, which define the amplitude of the electromotive force in the receiver line. This curve has more complicated shape than the curves of the second order, and it

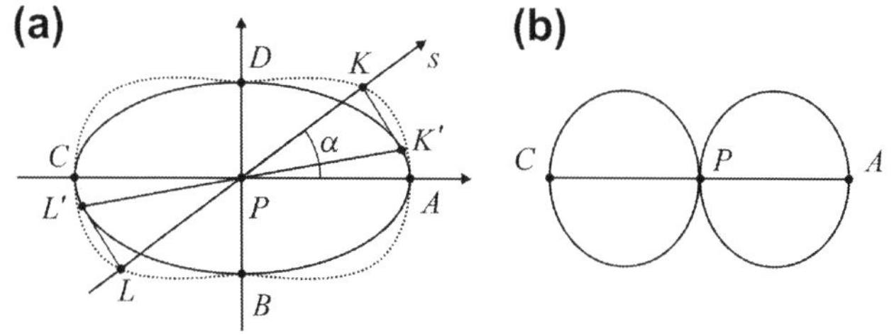

Figure 14.12 Measuring electric field with elliptical (a) and linear (b) polarization.

touches ellipse at points A, B, C, and D. The difference between them increases with elongation of the ellipse. In particular, when instead of the ellipse, we have a straight line AC, Figure 14.12(b), this curve is transformed into two circles tangent to each other at the point P. In contrast, when semi-axes are equal, the curve of the electromotive force amplitude in the receiver line coincides with this circle. The shape of this curve allows one to understand how the electromotive force changes with direction. This change becomes stronger with the ellipse elongation. If the receiver line is directed along the major axis, the sound in earphone is the strongest. In contrast, along the minor axis this signal is minimal. For instance, when instead of the ellipse we have a straight line (linear polarization), in the direction perpendicular to this line the signal vanishes. Thus, we see that a change of a direction of the field vector during one period is the reason why the signal doesn't completely disappear. In this light, it is proper to notice the following. In the case of the constant field, this method allows to trace the equipotential lines: it happens when the voltage between two points located at the same equipotential line is equal to zero. In the case of the alternating current, we draw lines where the electromotive force is minimal; that is we draw lines showing a direction of the minimal axis of the ellipse of polarization. It is obvious that with a decrease of the frequency, the inductive effect becomes smaller and the linear polarization takes place. As a result, both these curves approach each other.

14.12.3. Elliptical Polarization of the Magnetic Field

Now, we consider a behavior of the vector of the magnetic field in the space as a function of time, and this will be done in a slightly different manner than in the case of the electric field. As follows from Eq. (14.212) we have

$$B_x = B_{0x}\cos(\omega t + \psi_x) = \mathrm{Re}B^*_{0x}\exp(i\omega t)$$

$$B_y = B_{0y}\cos\left(\omega t + \psi_y\right) = \mathrm{Re}B^*_{0y}\exp(i\omega t) \qquad (14.214)$$

$$B_z = B_{0z}\cos(\omega t + \psi_z) = \mathrm{Re}B^*_{0z}\exp(i\omega t),$$

where B^*_{0x}, B^*_{0y}, and B^*_{0z} are complex amplitudes of the field components, which can be written as

$$B^*_{0x} = \alpha_x + i\beta_x, \quad B^*_{0y} = \alpha_y + i\beta_y, \quad B^*_{0z} = \alpha_z + i\beta_z.$$

Thus, in place of Eq. (14.214), we have

$$B_x = \alpha_x \cos \omega t - \beta_x \sin \omega t$$
$$B_y = \alpha_y \cos \omega t - \beta_y \sin \omega t \qquad (14.215)$$
$$B_z = \alpha_z \cos \omega t - \beta_z \sin \omega t.$$

It is obvious that amplitudes and phases of each component are

$$a_x = \left|\left(\alpha_x^2 + \beta_x^2\right)^{1/2}\right|, \quad a_y = \left|\left(\alpha_y^2 + \beta_y^2\right)^{1/2}\right|, \quad a_z = \left|\left(\alpha_z^2 + \beta_z^2\right)^{1/2}\right|,$$

and

$$\cos \delta_x = \frac{a_x}{\left|\left(\alpha_x^2 + \beta_x^2\right)^{1/2}\right|}, \quad \sin \delta_x = \frac{\beta_x}{\left|\left(\alpha_y^2 + \beta_y^2\right)^{1/2}\right|} \quad \text{and so on.}$$

The latter also gives

$$\tan \delta_x = \frac{\beta_x}{\alpha_x},$$

but in order to determine the quadrant where the angle δ_x or other angles are located it is necessary to take into account the sign of sinus and cosines functions. It is convenient to treat $\alpha_x,\alpha_y,\alpha_z$ as components of the real vector $\boldsymbol{\alpha}$ and β_x,β_y,β_z as components of the vector $\boldsymbol{\beta}$. Then, as follows from Eq. (14.215), we can consider the total vector of the magnetic field as a sum of two vectors

$$\mathbf{B}_1 = \boldsymbol{\alpha} \cos \omega t \quad \text{and} \quad \mathbf{B}_2 = -\boldsymbol{\beta} \sin \omega t. \qquad (14.216)$$

Each of them describes the linear oscillation with the same frequency ω, and the phase shift between them is $-\frac{\pi}{2}$. As is well known from the theory of elementary oscillations, two linear oscillations with different amplitudes and directions, but with the same frequency, produce the elliptical oscillation. This means that the end of the vector of the magnetic field moves along the ellipse, located in the plane defined by vectors $\boldsymbol{\alpha}$ and $\boldsymbol{\beta}$. In other words, these vectors are located in the same plane as the ellipse of polarization. The unit vector normal to this plane is defined from the equation

$$\mathbf{n} = \frac{\boldsymbol{\beta} \times \boldsymbol{\alpha}}{|\boldsymbol{\beta} \times \boldsymbol{\alpha}|}. \qquad (14.217)$$

This formula gives uncertainty, if $\boldsymbol{\alpha}=0$ or $\boldsymbol{\alpha}\|\boldsymbol{\beta}$ or $\boldsymbol{\beta}=0$. In all cases, there is a linear polarization. From Eq. (14.215), we have

$$\mathbf{B}^2=\alpha^2\cos^2\omega t+\beta^2\sin^2\omega t-(\boldsymbol{\alpha}\cdot\boldsymbol{\beta})\sin 2\omega t.$$

Substituting

$$\cos^2\omega t=\frac{1+\cos 2\omega t}{2},\quad \sin^2\omega t=\frac{1-\cos 2\omega t}{2},$$

we have

$$\mathbf{B}^2=\frac{1}{2}\left[\left(\alpha^2+\beta^2\right)+\left(\alpha^2-\beta^2\right)\cos 2\omega t-2(\boldsymbol{\alpha}\cdot\boldsymbol{\beta})\sin 2\omega t\right]. \quad (14.218)$$

It is clear that minimums and maximums of $\mathbf{B}^2$ correspond to the minor and major axes of the ellipse. Making the first derivative with respect to time equals zero, we obtain

$$\frac{\mathrm{d}}{\mathrm{d}t}\mathbf{B}^2=-\omega\left[\left(\alpha^2-\beta^2\right)\sin 2\omega t+2(\boldsymbol{\alpha}\cdot\boldsymbol{\beta})\cos 2\omega t\right]=0. \quad (14.219)$$

If $\alpha^2=\beta^2$ and $\boldsymbol{\alpha}\cdot\boldsymbol{\beta}=0$, then the circle polarization takes place and it happens as result of summation of two perpendicular vectors with the phase shift 90°. In all other cases, the solution of Eq. (14.219) is

$$\tan\omega t_0=\frac{2(\boldsymbol{\alpha}\cdot\boldsymbol{\beta})}{\beta^2-\alpha^2}. \quad (14.220)$$

Here, t_0 is the instant when the vector B coincides with the ellipse axis. In order to have maximum, it is necessary that

$$\left(\frac{\mathrm{d}^2\mathbf{B}^2}{\mathrm{d}t^2}\right)_{t=t_0}=-2\omega^2\left[\left(\alpha^2-\beta^2\right)\cos 2\omega t_0-2(\boldsymbol{\alpha}\cdot\boldsymbol{\beta})\sin 2\omega t_0\right]<0. \quad (14.221)$$

To satisfy this inequality, we take into account Eq. (14.220) and define $2\omega t_0$ from conditions

$$\cos 2\omega t=\frac{\beta^2-\alpha^2}{\left[\left(\beta^2-\alpha^2\right)^2+4(\boldsymbol{\beta}\cdot\boldsymbol{\alpha})^2\right]^{1/2}}$$

and

$$\sin 2\omega t_0 = -\frac{2(\boldsymbol{\beta}\cdot\boldsymbol{\alpha})}{\left[\left(\beta^2-\alpha^2\right)^2+4(\boldsymbol{\beta}\cdot\boldsymbol{\alpha})^2\right]^{1/2}}.$$

Then in accordance with Eq. (14.218)

$$B_{\max}^2 = \frac{1}{2}\left\{\left(\alpha^2+\beta^2\right)+\left[\left(\beta^2-\alpha^2\right)^2+4(\alpha\cdot\beta)^2\right]^{1/2}\right\},$$

and the major axis is

$$c = \left[\frac{1}{2}\left\{\left(\alpha^2+\beta^2\right)+\left[\left(\beta^2-\alpha^2\right)^2+4(\alpha\cdot\beta)^2\right]^{1/2}\right\}\right]^{1/2} \quad (14.222)$$

Taking the different moment:

$$2\omega t_0' = 2\omega t_0 + \pi,$$

we obtain the expression for the minor axes

$$d = \left[\frac{1}{2}\left\{\left(\alpha^2+\beta^2\right)-\left[\left(\beta^2-\alpha^2\right)^2+4(\alpha\cdot\beta)^2\right]^{1/2}\right\}\right]^{1/2}.$$

Sometimes, it is also useful to know projections of the ellipse of polarization on the coordinate plane. Assuming that α_x,β_x and α_y,β_y are given on the plane xy, we have for components of the field in this plane:

$$B_x = \alpha_x \cos\omega t - \beta_x \sin\omega t \quad \text{and} \quad B_y = \alpha_y \cos\omega t - \beta_y \sin\omega t$$

Then for its amplitude, we have

$$\begin{aligned} B^2 &= B_x^2 + B_y^2 \\ &= \left(\alpha_x^2+\alpha_y^2\right)\cos^2\omega t + \left(\beta_x^2+\beta_y^2\right)\sin^2\omega t - \left(\alpha_x\beta_x+\alpha_y\beta_y\right)\sin 2\omega t, \end{aligned}$$

and the latter can be written as

$$B^2 = \frac{1}{2}(T - P\cos 2\omega t - Q\sin 2\omega t),$$

Here

$$T = \beta_x^2+\beta_y^2+\alpha_x^2+\alpha_y^2; \quad P = \beta_x^2+\beta_y^2-\alpha_x^2-\alpha_y^2,$$

$$Q = 2\left(\alpha_x\beta_x+\alpha_y\beta_y\right)$$

The conditions of the maximum are

$$\frac{dB^2}{dt} = \omega(P \sin 2\omega t - Q \cos 2\omega t) = 0$$

and

$$\frac{d^2B^2}{dt^2} = 2\omega^2(P \cos 2\omega t + Q \sin 2\omega t) < 0.$$

The first condition gives

$$\tan 2\omega t_0 = \frac{Q}{P}, \tag{14.223}$$

and the second one shows that the maximum takes place if

$$\cos 2\omega t_0 = -\frac{P}{(P^2 + Q^2)^{1/2}}, \quad \text{and} \quad \sin 2\omega t_0 = -\frac{Q}{(P^2 + Q^2)^{1/2}}.$$

Substituting these values into expression for B^2, we obtain

$$B^2_{max} = \frac{1}{2}\left[T + \left(P^2 + Q^2\right)^{1/2}\right]. \tag{14.224}$$

Having taken $\sin 2\omega t_0$ and $\cos 2\omega_0 t$ with opposite signs, we obtain

$$B^2_{min} = \frac{1}{2}\left[T - \left(P^2 + Q^2\right)^{1/2}\right]. \tag{14.225}$$

We have described some features of the elliptical polarization of the electric and magnetic fields, and it will help us to understand some of the electromagnetic methods, which were widely used in the past. In practice, the plane of the ellipse of polarization is defined from the fact that if the receiver coil is located in this plane the flux of the magnetic field through coil is absent, and the voltage is equal to zero. In those cases when the receiver is solenoid the electromotive force is zero, provided that the receiver axis is perpendicular to the ellipse plane.

14.13. DEVELOPMENT OF THE INDUCTIVE METHODS OF MINING PROSPECTING

14.13.1. Equipotential Lines Method

As usually introduction of an alternating current into the electroprospecting took place gradually. In 1900, Daft and Williams applied the sinusoidal current for the method of equipotential lines that allowed one to trace these lines with help of airphone and very light electrodes. In 1906, Peterson applied this method in Sweden, where it became popular, and already in 1914, Bergstrom using the method discovered the first mining deposit of commercial value. In 1913, in order to discover ore bodies Shillowsky suggested using inductive coils to measure magnetic fields, and in 1918, Konklin introduced the inductive method of the field excitation, while Hankel carried out both amplitude and phase measurements.

14.13.2. "Infinitely" Long Cable Method (CSMT)

Now we will describe the main steps in development of the inductive methods of the surface and airborne geophysics, which were found at the beginning application in the mining prospecting and then also became useful for solution of different problems in the engineering, environmental, and groundwater geophysics. The inductive method for looking for good conductors appears in Sweden at the first part of the last century, namely the method of the long cable, and its modifications, that have different names such as the Sunberg method, Turam, the method of infinitely long cable, and much later CSMT. The most essential feature of this method is the long current line grounded in two points with the earth or a large current loop, which creates the electromagnetic field. The magnetic field is usually measured along profiles perpendicular to this current line at one or several frequencies. As a rule, orientation of this current line is placed in parallel to a strike of a conductive body. This arrangement allows rather quick measurements at relatively large area, and it is not accidental that much later the first transient method of the surface geophysics in USA was based on the use of this kind of array. At distances from the current line that are smaller than separation between electrodes the electric field in the earth is almost parallel to the cable, and in general, it consists of two parts. One of them has the vortex character and arises due to a change of the magnetic field with time; the other is the galvanic part and it is initially caused by charges at the electrodes. If observations are made far away from electrodes and relatively close to the current line, it is often possible to neglect their influence, as it was assumed at the beginning. It is natural that this pioneering work was started in Sweden, where the mining industry was very active at that time, and the high resistive environment of the Scandinavian shield allowed in many cases to neglect by influence of induced currents in host rocks.

14.13.2.1. Sunberg Method of Measurement

At the beginning, K. Sunberg suggested to measure the in-phase and quadrature components of the magnetic field, for instance, B_z and for this purpose, the following approach was used. Suppose that the source current is the sinusoidal function of time. Then any component of the magnetic field at the point of observation can be represented as

$$B(t) = B_0 \sin \omega t + B_s \sin(\omega t + \varphi), \tag{14.226}$$

where B_0 and B_s are the amplitudes of the primary and secondary magnetic fields, respectively, and φ is the phase shift between these fields. In most practical cases, the secondary field is much smaller than the primary field:

$$B_s \ll B_0, \tag{14.227}$$

For this reason, measurements of the useful signal caused by induced currents in a conductor in the presence of much stronger primary field, as well as different types of noises that may change with time, became the main problem for the development of the inductive systems in the frequency domain. In this light, it is also useful to represent Eq. (14.226) in a different form

$$B(t) = a \sin \omega t + b \cos \omega t, \tag{14.228}$$

where a and b are amplitudes of the in-phase and quadrature components, respectively. It is clear, that the amplitude of the in-phase component is the sum of the primary a_0 and secondary fields, and that the latter is usually much smaller: $a_s \ll a_0$.

For simplicity, we assume that surrounding medium is very resistive, and it is possible to neglect its influence. Usually, especially at the low-frequency spectrum, we have

$$a_s < b \ll a_0. \tag{14.229}$$

Since the quadrature component is shifted by phase in 90° with respect to the primary field, K. Sunberg suggested using the Geiger compensator to measure this part of the secondary field. First, consider for illustration the measuring set up that includes a receiver coil, voltmeter, and a resistor, which can be changed. Under the action of the alternating magnetic field, the electromotive force induced in the coil is

$$\Xi = k\omega(a \cos \omega t - b \sin \omega t), \tag{14.230}$$

where k is the coefficient proportionality depending on the area and number of the coil turns. Thus, there are two electromotive forces in the receiver circuit shifted in phase by 90°. Now, we imagine the compensating circuit, which introduces through the resistance R the compensating voltage:

$$V_k = d \cos \omega t. \tag{14.231}$$

Thus, the total voltage acting in the receiver circuit is

$$V = k\omega(a \cos \omega t - b \sin \omega t) + d \cos \omega t$$

or

$$V = (d + k\omega a)\cos \omega t - k\omega b \sin \omega t. \tag{14.232}$$

Changing the amplitude of the compensating voltage and properly choosing its sign, it is possible to eliminate the first term in Eq. (14.232) and it happens when

$$a = -\frac{d}{k\omega}. \tag{14.233}$$

Then the measured voltage is minimal, and it changes as

$$V = V_0 \sin \omega t.$$

Therefore, we have

$$b = -\frac{V_0}{k\omega}, \tag{14.234}$$

that is applying the compensating signal and reaching the minimal voltage, we determine the quadrature component of the magnetic field. Note that only in the absence of this component, it is possible to compensate the electromotive force in the receiver and obtain zero signal. Of course, if the compensating signal is known, Eq. (14.231) permits us also to find the in-phase component of the field, but it is hardly interesting information since the dominant role plays by the primary field. On the other hand, the idea of compensation found later the very important application in the conventional systems of the electromagnetic profiling and replaced practically all other means, which take into account the effect of variation of the primary field. Perhaps, the first method of measuring the both components of the field was done using the Geiger compensator. In essence, it is a transformer that consists of two circuits inductively connected with each other. In the primary circuit, the current and voltage between its two points is in in-phase with the current source:

$$V_1 = V_{01} \sin \omega t. \tag{14.235}$$

Inasmuch as the resistance and inductance of the secondary circuit are chosen in such way that the latter is negligible, the voltage between two points of this circuit changes as

$$V_2 = V_{02} \cos \omega t. \tag{14.236}$$

Both compensating voltages are known, and they change during measurements. The receiver line includes the coil where the unknown electromotive force is

$$\Xi = In\Xi \sin \omega t + Q\Xi \cos \omega t \tag{14.237}$$

and two resistors, which are elements of the first and second circuits, respectively.

Also, there is an inductive connection of the measuring line with the indicator, for example, geophone, which shows either the presence or absence of the signal in this line. Thus, in the measuring line the total voltage is

$$V_t = (V_{01} + In\Xi)\sin \omega t + (V_{02} + Q\Xi)\cos \omega t. \tag{14.238}$$

Changing voltage of these resistors, it is possible to find the moment when the sound in the geophone is absent and this means that

$$In\Xi = -V_{01} \quad \text{and} \quad Q\Xi = -V_{02}, \tag{14.239}$$

that is we can determine the in-phase and quadrature components of the electromotive force, induced in the receiver coil. Of course, the in-phase component is caused by both the primary and secondary fields. Since the measurements required the reference signal, it was necessary to bring this signal from the current line that was not convenient, especially when observation points were far away from the current line. Besides, this first variant of the method was sensitive to a change of the current in the cable.

14.13.2.2. Turam Modification

For these reasons Hedstrom later introduced the second version, called Turam, where the receiver line was different. Instead of one receiver coil, there were two coils measuring the vertical component of the magnetic field and located at the same profile separated by the distance of 10–50 m. The voltages in each coil were connected with the device (bridge) and two quantities were measured: the ratio of amplitudes of the fields and difference of phase between them:

$$\frac{A_2}{A_1} \quad \text{and} \quad \varphi_2 - \varphi_1.$$

The measuring circuit consists of two parts. Each of them contains one measuring coil and has impedance Z. These parts have one common side with the indicator of the current. Changing these impedances, it is possible to provide the same current in the common line but with the opposite direction. Then the indicator shows zero, and this happens when

$$\frac{\Xi_1}{Z_1} = \frac{\Xi_2}{Z_2}.$$

Here

$$\Xi_1 = A_1 \exp(i\varphi_1), \quad \text{and} \quad \Xi_2 = A_2 \exp(i\varphi_2).$$

From the last equality, the ratio of amplitudes and difference of phase are determined. It is clear that both calculated quantities are independent on variation of the source current. This fact strongly increased accuracy of measurement, and during rather a long time, Turam method was successfully applied around the world. During the first part of last century, the development of seismic methods for petroleum exploration was only at the beginning and Turam found some application in Sweden for looking for relatively shallow oil reservoirs. In Canada and Russia as well as in other countries, for instance Australia, the method was used in the mining prospecting during rather long time and was accompanied by the theoretical studies, performed by many scientists and first of all by V. Bursian, A. Price, and J. Wait. Their work helped to understand an influence of a surrounding medium, especially an overburden. In this light, it may be proper to notice the following: during long time a solution of the forward problems in either a horizontally layered medium or with coaxially cylindrical interfaces, as well as in the presence of a spherical or cylindrical inhomogeneities attracted many outstanding scientists who worked mainly in areas of physics and mathematics, and their contribution to geophysics is difficult to overestimate. At the same time, there was the second group of scientists who were usually geophysicists by education that used the results of the first group and focused on developing the theory of specific methods that helped to understand their efficiency in different geoelectric conditions. For instance, their work helped to evaluate the effect of the frequency or time or separation between the current source and a receiver on the useful signal, as well as influence of the geological noise, determine optimal parameters of equipment, and also to develop the main tools for interpretation. Now we make several comments: (1) The use of this method to look directly for good conductors started to decline with an appearance of the dipole electromagnetic profiling. (2) At the same time one modification of this method (CSMT) is still used for mapping and study lateral changes of resistivity, in particular, for detecting conducting and resistive zones. Unlike Turam, this system performs measurements of the horizontal components of the electric and magnetic fields at different frequencies. Then, as in the magnetotelluric method, the impedance is calculated at each observation point. (3) Note that there were attempts to use the method of a long cable for airborne prospecting when receivers were located on an airplane, but long current line is grounded on the surface. (4) Measurements of the ratio of amplitudes and difference of phases are the first example of reduction of influence of the primary field, or more precisely, the change of the current in a source. In a fact, a long history of the equipment of the inductive methods of mining prospecting is a history of improvement of a compensation of the primary field and therefore an increase of an accuracy of measurements of the magnetic field.

14.14. DIPOLE ELECTROMAGNETIC PROFILING

Introduction

As the method of an "infinitely" long cable, the electromagnetic profiling was invented in Sweden and called Slingram. Its appearance was related to the following. There are two reasons why the method of an infinitely long cable isn't convenient to create a strong secondary field when the target is a relatively thin and vertical conductor, surrounded by very resistive medium. One of them is related to the fact that the primary magnetic field is mainly vertical, especially near the earth's surface. Of course, currents arise in the horizontal planes of the body, but their effect is usually small. The secondary field is also caused by the normal electric field, which is in parallel to the strike of the body, and it may generate a strong current following in a horizontal direction. But in the case of a very resistive surrounding medium charges, arising on the surface of a body, generate the electric field that practically compensate this normal field, and correspondingly, the current along the strike is rather small, and this is the second reason why anomaly of the magnetic field due to such bodies is small.

The Slingram array consists of the transmitter and receiver coils moving either along the line of measurements or along two parallel lines, so that the separation between coils remains the same. In this sense, there is some analogy with the electric dipole profiling, based on the use of the constant current. In both cases the transmitter and receiver coils or lines can be treated as the magnetic or electric dipoles, respectively. Bearing in mind that the primary field intersects the lateral surface of the body, induced currents are located in the vertical planes, and the simplicity of the geometry of expected anomaly is almost obvious. Inasmuch as currents are located in the vertical plane, the vertical component of the secondary magnetic field has characteristic zero above the body and maximum and minimum from sides, which are equal in the case of the vertical body but differs in the presence of target inclination. Distance between maximum and minimum is related to the depth of the target. The simplicity of anomalies and small influence of a surrounding medium created very favorable conditions for the successful application of the method, which practically didn't require a theoretical foundation.

One can imagine several orientations of coils, but usually they both are either horizontal or vertical, but in the past there were important exceptions. Note, that in the case when the goal is to detect a thin vertical body the vertical coils are usually used. As in the case of the method of long cable, it became clear that instability of the transmitter current doesn't allow one to measure the small secondary magnetic field with a high accuracy and somewhere at the end of 1920s, there was developed a new approach, especially in Australia, which didn't require the measurements of the components of the magnetic field. As we already know, the field is in general elliptically polarized and an orientation of the ellipse of polarization in a space is independent of a change of the transmitter current. Of course, it is also true for a projection of this ellipse on some plane. Thus, measurements of any parameter, characterizing the position of the ellipse of polarization, are not subjected to the fluctuations of the current, and this is essence of the dip angle method. One can imagine different modifications of this approach. For instance, rotating the receiver loop until the electromotive force becomes zero, we determine the plane where the ellipse of polarization is located. Measuring the angles characterizing a position of this plane, we obtain quantities that are independent on the transmitter current. Also, it is possible to have a receiver, which consists of three mutually perpendicular coils arbitrary oriented in space, where the amplitude and phase are measured. This information is sufficient to calculate angles defining position of the ellipse. Moreover, such approach allows one to greatly remove an effect caused by disorientation of coils, when we deal with a two-coil system. However, both these approaches are rather cumbersome and didn't find a broad application. Instead of it, the different and much simpler approach was suggested almost 90 years ago, which allows one to determine one parameter, characterizing a position of a projection of the polarization ellipse at one of the planes. This method is illustrated by two examples.

Example One

Consider a system of profiles, perpendicular to a strike of a target, and two vertical coils moving simultaneously along profiles, so that they are located at the same distance from the conducting body and their axis are horizontal, directed along profiles. If an ore body is absent and an influence of a surrounding medium is insignificant, there is only the primary field, which is directed along the profile. In other words, the field is linearly polarized and located in a horizontal plane. Then, rotating the receiver coil around the horizontal axis, perpendicular to the transmitter we obtain the zero signal. At this instant, this coil is located in the horizontal plane and, correspondingly, its axis coincides with the vertical; that is the angle between them is zero. In the presence of a conductor, the field is elliptically polarized, because the secondary field is shifted by phase and has different amplitude. In general, this ellipse is arbitrary oriented in a space and its position changes with an observation point. Consider the projection of the polarization ellipse on the vertical plane, perpendicular to the line *T-R*. Now in order to obtain the minimum signal, we have to turn the horizontal receiver coil around this line as the axis, until the minor axis of the projection of the ellipse on the vertical plane will be along the coil axis. The angle between the vertical and this axis is either called the tilt or dip angle. The behavior of the dip angle has one important feature, the so-called "cross-over", which is useful for detecting of a conducting body. The same feature takes place when the combined profiling method with the constant current is applied. It is obvious that zero angle over the body is observed because both the primary and secondary fields have the same direction. It is essential that measurements of the angle can be performed with a relatively high accuracy and its value is independent on variations of the transmitter current. Of course, measurements of the dip angle are not free from an influence of the topography.

Example Two (Shoot-Back Technique)

This variant of the dip method was developed in Canada by Crone geophysical company and successfully used during long time in areas with sufficiently complicated topography. The array consists of two coils located one behind the other. The axes of these coils are in the vertical plane containing the line connecting both coils and these axes are perpendicular to each other, if both coils are located at the same horizontal plane. Suppose the transmitter coil is behind and its axis is along the vertical. Then the axis of the receiver coil has horizontal direction. In the presence of topography and absence of a conductor, the orientation of the receiver when signal is equal to zero, "null" position, is characterized by the angle α. This is the first measurement. Then the role of coils is changed and the horizontal coil in front becomes transmitter and the "null" position of the receiver coil behind is measured. As follows from the geometry of the primary field, these two angles have the same magnitude but opposite sign, and correspondingly, the sum of these angles is equal to zero. Of course, appearance of these angles does not have any relation to the elliptical polarization and it is a result of geometry of the primary field only. Thus, performing measurements in the presence of a conducting body and taking an average from two measurements of angles, it became possible to greatly reduce an influence of topography, as well as errors related with disorientation of coils. Note that application of the dip angle method was accompanied as in the case of other electromagnetic methods by development of the theory and interpretation. Before we describe the next generation of the dipole profiling method, it may be proper to mention again that there were also different attempts to use the phenomenon of elliptical polarization in order to decrease an influence of the current change in the transmitter. In one method, the receiver included three coils perpendicular to each other. As was mentioned earlier, the knowledge of amplitudes and phase in each of them allows one to find all parameters of the polarization vector, including its orientation, which is independent on the current in the transmitter coil, as well as the ratio of the major and minor axes. This method was called the method of elliptical polarization. The other array consists of two pairs of the transmitter and receiver coils, and in each pair coils have the same parameters, and they are perpendicular to each other. Current in each transmitter coil has the same amplitude but is shifted by 90° with respect to each other. As a result, the primary magnetic field is characterized by the circular polarization. This means that the electromotive force Ξ_1 and Ξ_2 induced in each coil due to this field has the same magnitude, but with the phase shift equal 90°. Correspondingly, the following quantities, being independent on the current, are measured

$$\frac{\Xi_2 - \Xi_1}{\Xi_1}, \quad \text{and} \quad \Delta\varphi = \varphi_2 - \varphi_1 - 90^\circ.$$

Note that due to the rotation of the field, the electromotive forces, caused by the primary field, coincide with each other. At the same time in the presence of a conducting body, the circular polarization is transformed into the elliptical one, and the amplitude and phase at each receiver coil became different. This approach is called the method of rotating field and perhaps as the previous one was suggested in Russia, mainly for air borne prospecting. One can say that this group of methods, related to the elliptical polarization of the field, can be treated as the second approach to reduce an influence of the primary field variations on an accuracy of measurements of the useful signal.

14.15. MODERN SYSTEMS OF ELECTROMAGNETIC PROFILING

In the mid 1950s, electronic engineer Ronka (Canada) formed the geophysical company and developed the profiling method with the horizontal loops, where the electromotive force in the receiver due to the primary field was in a great extend compensated. It was a beginning of the third approach for reducing the effect of the primary field, which allowed one measuring again the magnetic field itself instead of some quantities, which are only indirectly related to the field, like dip angle, and what is the most important to increase an accuracy of measurements of the secondary field, even when it is only extremely small parts of the primary field. Compensation of the primary field, which is often in many orders of magnitude greater the secondary one, can be done at least by two methods. The first approach is mainly used when the distance between the transmitter and receiver coils is relatively small and both coils are placed on a rigid structure. The array consists of three coils: two of them are receivers: R_1 and R, and the third is transmitter T. The receiver coils are connected in such way that the primary electromotive forces have opposite signs. Choosing properly number of turns, it is possible to obtain zero in the receiver channel. In other words, the electromotive forces arising in both receiver coils due to the current in the transmitter coil have the same amplitudes but opposite signs; that is

$$\frac{M_{R_1}}{L_c^3} = \frac{M_R}{L^3}.$$

Here M_{R_1} and M_R are moments of receiver coils, and L_c and L are distances from the transmitter to corresponding receiver coil. It is essential that $L_c \ll L$. When the last equality takes place, the primary electromotive force is completely absent in the receiver channel, and any change of the transmitter current doesn't affect this result. However, in real conditions, it is impossible to reach this equality and there is always some very small signal that varies with a change of the transmitter current, temperature, or system geometry. But if the degree of compensation of the primary electromotive force is high, this system allows one to measure both the quadrature and in-phase components of the magnetic field, even when they are relatively small. It is proper to point out that this array can be treated as a sum of two pairs of coil system: T-R_c with separation L_c and the T-R with length L. Since

$$M \gg M_c, \quad \text{and} \quad L \gg L_c$$

contribution of the secondary field into the array T-R_c can be neglected, and the secondary field is practically measured by the main coil pair T-R. Perhaps, the compensation of the primary electromotive force with help of the additional coil in the receiver line was first used in the induction logging at the end the 1940s by the Sclumberger company. In such a method, the secondary field is also very small with respect to the primary one. The second approach, which was implemented by the Ronka geophysical company, was only slightly different: the compensating receiver coil was constructed together with the transmitter coil and the compensating voltage is brought to the receiver coil with help of a compensating cable. These approaches were later implemented in the systems developed by Geonics and Apex companies, which have continued the work started by Ronka and greatly improved the original electromagnetic systems based on the compensation of the primary electromotive force. It may seem that relative simplicity of electric circuits means that it isn't difficult to compensate the main part of the primary field. However, in order to measure very small secondary field, it was necessary to overcome influence of many factors, including contribution of the capacitive currents between wires, change of system parameters with time, temperature and most importantly effects of external and internal noise. It was a great engineering achievement. During the last 50 years, this principle of compensation of the primary magnetic field became the basis for the modern electromagnetic profiling arrays in the frequency domain. For illustration, there are several examples of the instruments (listed below), which became popular in the West. Of course, now there are other profiling systems developed in Canada and other countries, in particular, in Sweden and in Russia.

Example One: MaxMin System

This system was developed in the 1970s and it consists of two horizontal coils, and allows one to measure the vertical component of the magnetic field at broad range of frequencies (110–1777 Hz). Compensating cable has two functions: for compensating the main part of the primary field and as well serves as a reference for measuring the in-phase and the quadrature components of the secondary field with different separations between coils (30 m $< L <$ 400 m). Possibility of measurements at different frequencies, as well as different separations, allows one not only to discover good conductor but also obtain some information about its conductivity, size and shape, as well as some information about surrounding medium. Besides horizontal loop measurements, MaxMin permits one to use different orientation of coils and, in particular, measure the dip angle.

The next three examples briefly describe systems that are mainly designed to solve problems of the mining, engineering, and environmental geophysics.

Example Two: EM-31

This system, as well as two others described later, was developed by Geonics (Canada) and its main area of application is a mapping of near surface of a geoelectric section, for instance, groundwater contaminants. The depth of investigation of this array is around 6 m, but of course, it depends on a distribution of resistivity with depth. The system consists of two horizontal coils rigidly connected with each other, and the distance between them of 3.66 m and the frequency 9.8 kHz. Both coils are mounted on the same boom. Cancellation of the primary electromotive force is realized with help of the second receiver coil, located very close to the transmitter coil. The system allows one to measure the quadrature component of the magnetic field within the range from 1–1000 Ω that is directly proportional to conductivity, if the layer with the thickness around 6 m is uniform beneath the earth's surface. As was shown in Chapter 9, this means that the field is defined by induced currents that are caused only by the primary electric field and, therefore it is possible to use the concept of the geometric factor.

Example Three: EM-34

This system allows one to solve different problems of mining and engineering geophysics, and it is also useful for mapping relatively deep contaminant plumes, as well as for groundwater exploration. As in the case of the EM-31 frequency and separations between the transmitter and receiver coils are chosen in such a way that it is possible to apply the concept of geometrical factor that greatly simplifies interpretation of measured quadrature component of the field. Measurements can be performed at three fixed distances that allows in many cases to reach depth of around 60 m. Intercool spacing and corresponding frequencies are shown below:

$$10\text{ m} - 6.4\text{ kHz};\quad 20\text{ m} - 1.6\text{ kHz};\quad 40\text{ m} - 0.4\text{ kHz}.$$

As we see in the case of a uniform half-space, the ratio between the separation and skin depth in a uniform half-space remains the same. The system is designed to measure resistivity of a medium within the range of 3–300 Ω. Since measurement can be performed with three separations and two modes (vertical and horizontal coils orientation), it is also possible to some extent to carry out the geometrical soundings. The system also permits measurements of the in-phase component.

Example Four: EM-38

It is a very small array with separation between coils of 1 m, mounted on the same frame, which allows one to measure the quadrature and in-phase component of the secondary field at the frequency of 14.6 kHz. The working range of measuring resistivity is from 1–1000 Ω, while the in-phase component of the secondary field, normalized by the primary field may be measured from $0.3 \cdot 10^{-4}$ to $30 \cdot 10^{-3}$. This device is useful for agricultural surveys measuring soil salinity, as well as in some geotechnical applications such as archeology where measurements of the in-phase component allow one to determine the soil magnetic susceptibility.

Starting from 1950s, the inductive profiling was actively used in Russia, where different systems were developed, allowing one to measure different quantities, such as the quadrature and in-phase components, amplitude and phase, elements of the ellipse of polarization, its orientation, including the difference of the quadrature components at two frequencies:

$$\Delta QB_z = QB_z(f_2) - \frac{f_2}{f_1} QB_z(f_1),$$

where $QB_z(f_1)$, $QB_z(f_2)$ are quadrature components at frequency f_1 and f_2. The latter was introduced to reduce an influence of an overburden and host rocks, and as the theory shows it is practically equivalent in this sense to the time-domain system. Important role in development of the dipole electromagnetic system was played by B. Svetov. Along with the profiling arrays the method of intensity found some application in Russia. In this system, the quadrature and in-phase components of the field were measured at the center of the loop and its radius was comparable with depth of investigation.

We briefly described a development of the main surface electromagnetic methods for the mining prospecting and some other areas, based on the use of a sinusoidal current. We will consider the time-domain systems that appear at the second part of the 1950s in Canada and Russia almost at the same time.

14.16. THE TRANSIENT METHOD OF THE MINING PROSPECTING

14.16.1. The Beginning

In 1951, J. Wait published in Geophysics the paper that described the transient field caused by induced currents in the sphere, when the primary field changes abruptly with time. Solution of this problem was known much earlier and it can be found in some earlier text books. However, three facts made this paper very important: first it was published in the reputable geophysical journal, second the author pointed out in the paper that measuring the transient response it is possible to determine the parameters of a sphere; and finally the author worked in Newmont mining company, which obtained the patent and at the end of the 1950s, started to develop the transient system (McLaughlin and Dolan). The system consisted of a relatively large transmitter loop where the current changes almost as the step function, and measurements of the magnetic field as a function of time were performed along profiles. It is essential, that the field was observed during the current off-time, when the primary field was absent. As it was pointed out earlier, the main goal in development of different arrays in the frequency domain was to reduce influence of the primary field. In this sense, the transient method provided the best solution for this problem. As we recall, the first electromagnetic system in Sweden was exactly the same but instead of measuring the magnetic field in the frequency domain, the Newmont system allowed one to measure the field at different times without the primary field. This was beginning of the surface transient method in the West. Almost at the same time and perhaps independently, the outstanding geophysicist T. Barringer suggested using the transient field for airborne geophysics. With time, other companies started to produce time-domain equipment and performed surveys. For instance, in the 1960s, Crone Company (Canada) made equipment for the transient method with small transmitter and receiver coils, as well as a system for measuring the transient field in boreholes. Later at the end of the 1970s and at the beginning of the 1980s, Geonics (Canada) started to develop transient systems designed to solve different problems of in mining, engineering, groundwater, and petroleum geophysics. One can continue the list of companies involved in development of this method, but for us it is proper to emphasize the fact that this very successful technique was initiated by the J. Wait paper. Now we will briefly describe the beginning of development of the transient method in Russia, where at the middle of the 1950s, there were several groups of geophysicists who were involved in development of the inductive electromagnetic methods. We will describe only two of them because they had direct relation to our topic. At that time, private companies were absent and all organization were supported and controlled by government. One such group was in the geophysical department of the Moscow Institute of Geological Prospecting, where two geophysicists A. Kaufman and F. Kamenetskiy worked under the guidance of an outstanding geophysicist and professor Y. Yakubovskiy who realized that the direct current methods have very poor resolution; that is anomalies caused by good and poor conductors are practically the same, if they differ only by resistivity. For this reason, he suggested somewhere in 1955 to study the frequency responses caused by conductors, when the source is a current loop and measurements are performed at its center. In other words, the array of the so-called "phase" or intensity method was used, which was rather popular in Russia and described in the several text books. At that time, this group didn't know anything about the first Wait paper, published also 1951, and their work was based on the physical modeling and the well-known solution of the boundary problem for the spherical conductor. As was described in detail earlier, the study of frequency responses caused by currents in the sphere clearly show that at the low-frequency part of the spectrum the electromagnetic method has much higher resolution than the direct-current method. This result was very useful, and improvement of the frequency method was continued by Y. Yakubovskiy and F. Kamenetskiy and their group in order to prepare it for production of the equipment by state companies. It may be proper to notice that the question of a reduction of the influence of the primary field didn't play an important role in this research. Somewhere in 1957, professor Yakubovsky was able to find a copy of the journal "Geophysics", where the paper of J. Wait about the transient field was published. Note that at that time, it wasn't simple task to discover such publication. This paper was translated and study of resolution of this method was performed from only one point of view, namely classification of anomalies by conductivity. It was immediately discovered that it has much higher resolution than the measurements of the quadrature and in-phase components at the low-frequency part of the spectrum. That was the reason why Prof. Yakubovsky changed direction of his research, and development of the transient method became his main goal. At the beginning, theoretical and experimental studies were performed by Y. Yakubovskiy, A. Kaufman, and V. Kovalenko applying the center loop array, which was later called MPP, including the first testing of equipment in the field in 1959. Later, when the last two members left this group work was continued by Y. Yakubovskiy, F. Kamenetskiy, and several talent engineers. At almost the same time, the outstanding geophysicist A. Velikin with his colleagues started to develop the transient method with the array, where the transmitter and receiver coils were placed together. This took place in Institute of Applied geophysics in San Petersburg. The purpose of their work was to create the relatively simple system, which was convenient to use in the field. Then, in accordance with agreement between Russia and Australia, this technology was transferred into Australia and became popular in this and other countries. Perhaps, this is a rare case when the equipment of the electromagnetic method, developed in Russia, was used in West without essential modifications as it happened in the 1960s.

In both groups, development of the equipment went along with the understanding of the favorable conditions for application of the method, as well as its limitations. Practically from the start, due to the physical and numerical modeling, it was understood an influence of surrounding medium, in particular an overburden on the transient responses, and it helped to choose proper parameters of the transmitter and receiver components of the system. For instance, it was known that with an increase of a radius of the transmitter loop an influence of the geological noise greatly increases, and the most optimal radius of this loop has to be compared with the distance to the probable position of ore body. Taking into account that, unlike in Sweden and Canada, in many mining areas of Russia there is a conductive overburden as well as a sufficiently conductive host rocks, the transient method with large loop was rarely used. With the time, interaction between developments of different system of the transient method in different countries took place. In conclusion of this review, it may be proper to emphasize the following: first, the transient method allows without additional efforts to perform simultaneously profiling and soundings, and secondary, this method proved that it is possible to reach the depth of investigation with separation between the source and receiver of less than target depth. In other words, application of this method prepared background for introduction of the transient soundings in the near zone. It is not surprising that the latter was first introduced mainly by those who were earlier involved in application of time domain for the mining prospecting.

14.16.2. The Principle of Measurement of Time Domain

Before we describe some modern variants of the method, let us briefly characterize the main parts of the transient system. The transmitter current is a system of pulses of either the same or (and) opposite polarity. As a function of time, it is characterized by three time intervals, namely time-on, where the current gradually increases and becomes constant, then the ramp time, where it relatively quickly decreases to zero and finally the time-off where the current is absent and measurements are performed. Of course, stability of the current during time-on is important, but within certain range of its changes it is not as essential unlike the frequency domain, because the secondary field is measured only during the time-off, except of airborne systems. The ramp time is very important range because only within this time the primary electric field appears in a medium and causes the secondary currents. Inasmuch the theory and correspondingly an interpretation is usually based on assumption that the current is a step function, it is necessary to take into account the behavior of the current during the ramp time. Certainly, it should be the same as much is possible after each impulse and with the simplest form as the linear function. The width of ramp time should be also sufficiently narrow and much smaller than the smallest time of measurement. The measurement is performed at certain instant after the ramp time. Of course, it is impossible to measure simultaneously at all times of this range, and correspondingly certain intervals are chosen for these measurements. They have finite width, since it is impossible to measure the electromotive force instantly. The width of these intervals becomes larger with increase of time of measurements. On one hand, a decrease of intervals allows more accurate recording of the transient response, on the other hand the measured signal becomes smaller. The distance between these intervals and therefore number of them (gates) as well as their width are chosen in such way that they permit one to obtain the transient response with a sufficient accuracy. There are several different techniques to record transient response, but we will describe one of the most effective in terms of noise reduction and dynamic range. During the first time interval, the receiver is connected with some type of capacitor and the electromotive force causes the current, which accumulate a charge on this capacitor, while at other gates the similar circuits are open and accumulation of charges doesn't take place. Inasmuch as the time interval is very small, the charge is also extremely small, and it is practically impossible to measure the electromotive force after the single impulse. After it the capacitor (storage) is disconnected. Then at the next time interval, the corresponding capacitor is connected with the coil and this procedure is repeated. Thus, after each impulse capacitors at all gates are open and they have some small charges that are directly proportional to the electromotive forces at different times. The same takes after next impulse and so on. If the current impulses have the same sign, then after summation of charges at each capacitor (stacking) and averaging we obtain the charge that can be measured. Often the system of alternating impulses is used and it is done to reduce an influence of relatively low-frequency noise. Subtracting signals from two alternating neighbor impulses the influence of such noise becomes smaller, since it only slightly varies during this time interval. Of course, different types of filters are also used to improve the ratio between the useful signal and a nongeological noise.

14.16.3. General Features of the Transient Systems

With time, the transient method became useful in solving many different problems beyond the mining prospecting, including general geological mapping, groundwater and engineering geology, and environmental geophysics, and also sometimes it is used for petroleum exploration. For this reason, the transient system is usually designed in such way that it is able to provide a survey for such different purposes. For instance, Geonics Time-Domain System usually consists of the five parts:

(1) Transmitter, (2) transmitter coil or loop, (3) receiver, (4) receiver coil and (5) software.

The system has the same digital receiver for different types of transmitters. The latter are designed in such way that they maintain the high degree of linearity during ramp time and remaining current after turn off as small and stable as possible. As was pointed out earlier, the linear behavior of the current is important for more accurate measurements, particularly at early times to deconvolve data for the case of the step function. Also, the maximal reduction of the current after the ramp time is also essential at early times especially when measurements are performed near the transmitter loop. It is clear that the remaining current will introduce a measurement error. Now consider the main features of three types of transmitters of this system.

Example One: TEM-47

This transmitter is designed for relatively shallow studies from 2 to 150 m. For these depths the proper current amplitude can be easily achieved. For instance, a few amperes in the 5×5 m loop with a single turn can provide depth of investigation around 40 m. The important feature of the transmitter is extremely fast current, turn-off time, less than 1 μs for the shallowest depth, while the current after the ramp time is practically absent. Because of this measurements can be started at few μs after current turn-off, and it is possible to measure resistivity of the uppermost layer. Also measurements can be performed very close to transmitter loop. The variety of transmitter loop is used with this transmitter. For shallow depth of exploration (few tens of meters), a rigid multiturn loop of dimensions 1×1 m is used. For deeper applications such as groundwater exploration, a single turn loop of dimensions from 5×5 m up to 100×100 m is employed. The larger loop gives exploration depths of up to 150 m. The TEM-47 transmitter can be used in the Slingram mode, for example, for fracture location, in which case a flexible 5 × 5 m multiturn loop is used.

Example Two: TEM57-MK2

It is designed for intermediate depths of exploration: from 10–20 m to 400 m. For carrying out the transient soundings for the midrange depth the loop 250 × 250 m with current around 20 A is usually used. At the same time, for shallow exploration the smaller loop 40 × 40 m is often applied. A system with this transmitter is deployed for several configurations other than the central loop soundings. For example, a transmitter loop of dimensions 200 × 300 m can be used in mineral exploration when the position of the transmitter loop is fixed, but receiver moves along profiles. Also the TEM-57 can be used in conjunction with the borehole probe for looking for good conductors around a borehole. Finally, it can be applied as the TEM47 in the Slingram mode survey; that is normally referred as a moving transmitter and receiver mode of operation (profiling).

Example Three: TEM-67

This transmitter is designed to perform transient soundings for depth from 20 m to 1 km and maximum output current of 28 A. It can be used with loops up to 1 × 1.5 km in size.

The receiver of the system allows one to perform measurements in a broad range of time depending on the purpose of the geophysical survey. Correspondingly different measuring coils are used.

14.17. INFLUENCE OF CHARGES ON RESOLUTION OF ELECTROMAGNETIC METHODS OF MINING PROSPECTING

Earlier, we have described resolution of the inductive methods of mining prospecting assuming that the secondary electromagnetic field is caused by vortex currents arising inside a conductor and in a surrounding medium. In other words, it was assumed that the electric field doesn't intersect the boundary between a confined conductor and a surrounding medium that is hardly realistical. At the same time, as it was shown in Chapter 2, each time when there is a component of the electric field normal to the boundary electric charges arise, and their density is related to the quasi-stationary field in the same manner as in the case of the time-invariant field, Eq. (2.54):

$$\sigma = 2\varepsilon_0 K_{12} E_n^{av}.$$

These charges give rise to the electric field, and correspondingly to currents that flow through a confined conductor and a surrounding medium, and it is proper to expect that both the electric and magnetic fields, caused by these charges, are related to conductivity, or more precisely, to time constant of a conductor in a different manner than fields generated by vortex currents. It is also natural to assume that the relation between these two parts of the secondary field depends on the type of field excitation, the zone where field is measured, electric or magnetic field are measured, frequency or time of observation, and of course a shape of a conductor and its orientation with respect to the normal field. An influence of the last two factors will be not considered in this chapter, and correspondingly in order to simplify our analysis we choose the simplest model of a conductor, namely, sphere, located in a uniform conducting medium, and restrict ourself by the frequency domain. Before we will continue, let us notice the following. As we know, the inductive methods were first developed for mining prospecting on the Scandinavian shield, which is usually very resistive and it is difficult to make galvanic contact. This was the main reason for introducing systems like Turam and Slingram. Then with time, it became clear that the inductive methods can be very useful in areas with much lower resistivity of a surrounding medium, and moreover under certain conditions they have much higher resolution and depth of investigation than methods based on the use of the direct current. Such conclusion was based on an assumption that the sources of the secondary field are vortex currents arising in a conductor, and only at the end of 1950 of the last century it became clear that the presence of the surface electric charges can be also important. Certainly, from the point of view of resolution by conductivity this was hardly pleasant discovery.

14.17.1. Expressions for the Field in the Frequency Domain

The boundary-value problem, describing a diffraction of electromagnetic waves in the presence of a sphere located in a uniform medium, was solved long ago. At the beginning, a solution was found when the normal field is a plane wave, and it happened at the very beginning of the last century. Later, this problem was solved for arbitrary oriented source. The essence of the solution was the fact that the field was presented in terms of two scalar functions, which describe two parts of the field. One of them is called oscillations of the electric type, while the other is oscillations of the magnetic type, and they are independent from each other. It turns out that the first and second ones are related to surface charges and induced currents inside the sphere, respectively. Note that boundary-value problem for each potential was formulated independently. Both potentials satisfy Helmholtz equation, and their solution is the product of Bessel and Legendre's functions. Inasmuch as the plane wave can be represented as an infinite sum of terms, containing the product of these functions, the secondary field is also described as an infinite sum. Bearing in mind that this boundary problem found very broad application in different areas of physics and it described in detail in numerous books and papers, especially in problems related to diffraction of electromagnetic field, we omit derivations but focus our attention to one special case, which is the most important for practical applications of inductive methods. Suppose that the skin depth in a surrounding medium is in several times greater than the sphere radius and observations are performed relatively far away from this conductor:

$$\frac{a}{\delta_e} \ll 1 \quad \text{and} \quad \frac{r}{a} > 2. \tag{14.240}$$

If these conditions are met, then the first term in each series, describing the field, plays the dominant role and with sufficient accuracy represents the field. Before we write down these approximate expressions for the field, let us notice the following. (1) It is assumed that at the center of the sphere the normal electric and magnetic fields are directed along the x- and y-axes, respectively. (2) These expressions remain valid for any type of the field excitation. (3) In order to create a relatively strong anomaly, the normal field in the vicinity of a conductors shouldn't be small, that is the distance between the source and a sphere has to be smaller than the parameter δ_e. Correspondingly, the first inequality of the set Eq. (14.240) is justified, since this distance is usually exceeds the dimension of a conductor.

Taking into account the last conditions expressions for the secondary field approximately can be represented as

$$E_r^* = 2\frac{\gamma_i - \gamma_e}{\gamma_i + 2\gamma_e}\frac{a^3}{r^3}E_0^* \exp(-ik_e r)(1 + ik_e r)\cos\varphi \sin\theta$$

$$E_\theta^* = \frac{\gamma_i - \gamma_e}{\gamma_i + 2\gamma_e}\frac{a^3}{r^3}E_0^* \exp(-ik_e r)\left(k_e^2 r^2 - ik_e r - 1\right)\cos\varphi \cos\theta$$

$$-\frac{i\omega}{2}D\frac{a^3}{r^3}B_0^* \exp(-ik_e r)(1 + ik_e r)\cos\varphi$$

$$E_\phi^* = -\frac{\gamma_i - \gamma_e}{\gamma_i + 2\gamma_e}\frac{a^3}{r^3}E_0^* \exp(-ik_e r)\left(k_e^2 r^2 - ik_e r - 1\right)\sin\varphi$$

$$+\frac{i\omega}{2}D\frac{a^3}{r^3}B_0^* \exp(-ik_e r)(1 + ik_e r)\sin\varphi \sin\theta \tag{14.241}$$

$$B_r^* = DB_0^*\frac{a^3}{r^2}\exp(-ik_e r)(1 + ik_e r)$$

$$B_\theta^* = -\gamma_e E_0^*\frac{\gamma_i - \gamma_e}{\gamma_i + 2\gamma_e}\frac{a^3}{r^2}\exp(-ik_e r)(1 + ik_e r)\sin\varphi$$

$$+\frac{D}{2}B_0^*\frac{a^3}{r^3}\exp(-ik_e r)\left(k_e^2 r^2 - ik_e r - 1\right)\sin\varphi \cos\theta$$

$$B_\varphi^* = -\gamma_e E_0^*\frac{\gamma_i - \gamma_e}{\gamma_i + 2\gamma_e}\frac{a^3}{r^2}\exp(-ik_e r)(1 + ik_e r)\cos\varphi \cos\theta$$

$$+\frac{D}{2}B_0^*\frac{a^3}{r^3}\exp(-ik_e r)\left(k_e^2 r^2 - ik_e r - 1\right)\cos\varphi.$$

Here $k_e^2 = -i\gamma_e\mu_0\omega$ and r,θ,φ are the spherical coordinates of an observation point, provided that the origin of coordinates coincides with the sphere center. The set of these equations permit us to arrive at some useful conclusions concerning the behavior of the electromagnetic field. First of all, the secondary field in such approximation is equivalent to the field of two dipoles situated in a uniform conducting medium at the origin; that is at the center of the sphere. The first of these dipoles is the electric dipole with the moment P_e directed along the x-axis (direction of the normal electric field $\mathbf{E}_0$) and its strength is:

$$P_e^* = 4\pi \frac{\gamma_i - \gamma_e}{\gamma_i + 2\gamma_e} a^3 E_{0x}^*. \tag{14.242}$$

The second dipole is the magnetic one with the moment given by

$$P_m^* = 2\pi D a^3 B_{0y}^* \tag{14.243}$$

and directed along the y-axis (direction of the normal magnetic field $\mathbf{B}_0$). As was shown earlier, the function D can be written in the form:

$$D = \frac{3 \coth ik_i a}{ik_i a} + \frac{3}{k_i^2 a^2} - 1. \tag{14.244}$$

This function was earlier investigated in detail, but it is important to note that this function depends only on the parameter a/δ_i, defined by the conductivity and radius of the sphere. Next, let us make one more assumption that will not affect our conclusions, but greater simplifies analysis, namely suppose that the skin depth in a surrounding medium is greater than the distance between an observation point and the sphere center:

$$\frac{r}{\delta_e} < 1. \tag{14.245}$$

Then the set Eq. (14.241) has much simpler form

$$E_r^* = 2E_{0x}^* \frac{\gamma_i - \gamma_e}{\gamma_i + 2\gamma_e} \frac{a^3}{r^3} \cos\theta \sin\varphi \tag{14.246}$$

$$E_\theta^* = -E_{0x}^* \frac{\gamma_i - \gamma_e}{\gamma_i + 2\gamma_e} \frac{a^3}{r^3} \cos\theta \cos\varphi - \frac{i\omega}{2} D B_{0y}^* \frac{a^3}{r^2} \cos\varphi \tag{14.247}$$

$$E_\varphi^* = E_{0x}^* \frac{\gamma_i - \gamma_e}{\gamma_i + 2\gamma_e} \frac{a^3}{r^3} \sin\varphi + \frac{i\omega}{2} D B_{0y}^* \frac{a^3}{r^2} \cos\theta \sin\varphi \tag{14.248}$$

$$B_r^* = D B_{0y}^* \frac{a^3}{r^3} \sin\theta \sin\varphi \tag{14.249}$$

$$B_\theta^* = -\gamma_e E_{0x}^* \frac{\gamma_i - \gamma_e}{\gamma_i + 2\gamma_e} \frac{a^3}{r^2} \sin\varphi - \frac{D}{2} B_{0y}^* \frac{a^3}{r^3} \sin\varphi \cos\theta \tag{14.250}$$

$$B_\varphi^* = -\gamma_e E_{0x}^* \frac{\gamma_i - \gamma_e}{\gamma_i + 2\gamma_e} \frac{a^3}{r^2} \cos\theta \cos\varphi - \frac{D}{2} B_{0y}^* \frac{a^3}{r^3} \cos\varphi \tag{14.251}$$

It is clear that simplicity of the last set of equations is based on the assumption that measurements are performed in the near zone of the secondary field; that is $r/\delta_e < 1$. The set of Eqs (14.246)–(14.251) permit us to interpret the behavior of the secondary field in the following way. Because of the change of the normal magnetic field with time-induced currents appear in the sphere. In our approximation, the electromagnetic field from these currents is equivalent to that from the magnetic dipole located in a uniform conducting medium and these currents are tangential to the surface of the sphere. Inasmuch as the expression for the moment of the magnetic dipole contains the function D, this part of the secondary field is naturally called the vortex or inductive one, and it emphasizes the fact that the electromagnetic induction plays the dominant role. In the last approximation $\left(r/\delta_e < 1\right)$, this field depends on only conductivity of the sphere. For example, when resistivity of the sphere is an infinitely large (insulator), the electromagnetic field of the magnetic dipole is zero. In contrast, when the sphere has a very high conductivity, the function D approaches its asymptote of -1, and the magnitude of the secondary field is maximum. Next, let us examine the secondary part of the field, usually called the galvanic part. It is clear that under action of the normal field $\mathbf{E}_0$, the electric charges of both signs appear on the surface of the sphere. The electromagnetic field of these alternating charges is equivalent to that of the electric dipole in a uniform conducting medium, and the charges cause the electric field in accordance of Coulomb's law. The electric field in correspondence with Ohm's law causes the secondary currents to arise, and they intersect the surface of the sphere. In turn, these currents create the magnetic field. It is obvious that if the conductivity of the surrounding medium is zero, the galvanic part of the magnetic field disappears. In fact, as follows from Eqs (14.250) and (14.251) this part of the field is directly proportional to $\gamma_e\left(\gamma_i/\gamma_e \gg 1 \quad \text{or} \quad \gamma_i/\gamma_e \ll 1\right)$ while the galvanic part of the electric field does not vanish when γ_e is zero, because the surface charges still remain. In the case of an insulating sphere, the galvanic part of the magnetic and electric fields still exist because the surface charges appear, and there are secondary currents in the surrounding medium flowing around the insulator. It is useful to emphasize that the galvanic part of the field is only weakly relate to the conductivity of the sphere as follows from Eqs (14.246)–(14.251). In fact, the coefficient:

$$\frac{\gamma_i - \gamma_e}{\gamma_i + 2\gamma_e}$$

can be written as

$$\frac{\frac{\gamma_i}{\gamma_e}-1}{\frac{\gamma_i}{\gamma_2}+2}$$

and it is obvious that it varies from values $-\frac{1}{2}$ to $+1$ as the ratio of conductivities varies from zero to infinity. We should note that even when the ratio γ_i/γ_e amounts to a factor only as large as 5, the coefficient

$$\frac{\frac{\gamma_i}{\gamma_e}-1}{\frac{\gamma_i}{\gamma_2}+2},$$

is close to unity, and the influence of the conductivity of the sphere practically vanishes. In other words, a change of its conductivity doesn't practically make influence on the galvanic part of the field as soon as $\gamma_i/\gamma_e \gg 1$. It is obvious that the galvanic part of the field coincides with the direct current field behavior, provided that observations are taken place in the near zone of the secondary field. In other words, the influence of the skin effect is negligible. In contrast, the vortex part of the field is subjected to the skin effect; that is interaction of currents inside the sphere, where its behavior is controlled by the function $D\left(a/\delta_i\right)$. Until now, we have made use equations that describe the secondary field in the near zone when $r/\delta_e < 1$. If an observation point is located at an intermediate and wave zones, but still meets condition $r/a > 2$, one can use of the equations of the set Eq. (14.241). In this case, the skin effect in the host medium can be significant. Thus, both parts of the secondary field turn out to be functions of the conductivity of a surrounding medium. However, the effect of the conductivity γ_e is exerted on the vortex and galvanic parts in different ways. First of all, in the near zone $\left(r/\delta_e < 1\right)$, the vortex part of the secondary field is nearly totally insensitive to an influence from the induced currents in the host medium. With an increase in values for the parameter r/δ_e, this influence can become significant. In contrast, the galvanic part is subjected to the influence of the conductivity γ_e as a consequence of two factors, namely:

1. The magnetic field associated with the galvanic part, at least in the near zone of the secondary field, turns out to be directly proportional to the conductivity γ_e, when the contrast coefficient $\gamma_i - \gamma_e/\gamma_i + 2\gamma_e$ approaches to its limited value. However, the charge density and the

corresponding electric field E^g are almost independent of the conductivities when $\gamma_i/\gamma_e \gg 1$ or $\gamma_i/\gamma_e \ll 1$ in the near zone.

2. The second factor is the same as that for the vortex part; that is the appearance of induced currents due to the secondary field in the host medium.

It may be appropriate at this point to emphasize again that in accordance with Eq. (14.241), in contrast to the vortex part of the field, the galvanic part is practically independent of the sphere conductivity. This means that separation of anomalies by conductivity based on measuring the galvanic part of the field is not likely to be feasible, just in the case for the direct current methods. Therefore, it is of a considerable practical interest to investigate the relationship between the galvanic and vortex parts of the electric and magnetic fields. With this purpose in mind, we consider three cases.

Case One. Wave Zone of the Normal Field

As was shown in the previous chapters in the wave zone, the ratio between the normal magnetic and electric field is equal to

$$B_0 = \left(\frac{-i\gamma_e\mu_0}{\omega}\right)^{1/2} E_0 = \frac{k_e}{\omega} E_0. \tag{14.252}$$

Here, E_0 and B_0 are components of the normal field, which are perpendicular to each other. It is essential that Eq. (14.252) is valid only in the wave zone; that is when the distance from the source to an observation point exceeds the wave length in the surrounding medium regardless of the type of the field excitation. It can be either the magnetic or electric dipole or the grounded line of finite dimensions or a quasi-plane wave. This means that our study is useful for understanding a resolution of such methods as the frequency soundings with relative large separations or magnetotelluric soundings or CSMT. First, we examine the ratio between the galvanic and vortex parts of the electric field. As an example, in accordance with Eq. (14.248), we have

$$\left|\frac{E_\varphi^v}{E_\varphi^q}\right| = \frac{|i\omega r||D||B_0|}{2|E_0|}|\cos\theta| \tag{14.253}$$

or taking into account Eq. (14.252)

$$\eta_E = \left|\frac{E_\varphi^v}{E^g}\right| \le \frac{|k_e r|}{2}|D| \tag{14.254}$$

Considering that the parameter $k_e r$ is usually quite small, the electric field of the galvanic part is significantly greater than the electric field of the vortex origin. This means that in measuring electric field in the wave zone significant anomalies can be caused equally by both conductive and relatively resistive inhomogeneities. In other words, the resolution with these electromagnetic techniques is practically impossible.

Next, let consider the relationship between the magnetic fields caused by both parts of the magnetic field. As an example, the magnitude of the ratio φ-components of the magnetic field is

$$\eta_B = \left|\frac{B_\varphi^v}{B_\varphi^g}\right| = \frac{|D|B_0}{2\mu_0 r E_0 \gamma_e \cos\theta} = \frac{|D|}{2|k_e r||\cos\theta|} \tag{14.255}$$

Let $2\cos\theta = 1$ and with definitions: $\gamma_i/\gamma_e = s$, $r = na$ we have

$$\left|\frac{B_\varphi^v}{B_\varphi^g}\right| = \frac{s^{1/2}|D|}{n|k_i a|} \tag{14.256}$$

As follows from Eq. (14.255) and (14.256)

$$\eta_E \ll \eta_B. \tag{14.257}$$

This means that in measuring the electric field, the influence of the galvanic part is much stronger that in the case of the magnetic field. Now we consider how the function η_B varies with the frequency. At the high frequencies, the function D takes on the value unity, and we have

$$\left|\frac{B_\varphi^v}{B_\varphi^g}\right| = \frac{s^{1/2}}{n|k_i a|}, \tag{14.258}$$

that is with an increase of frequency the magnetic field of the galvanic part becomes greater. In contrast, at the sufficiently low frequencies the function D is nearly equal to $|k_i^2 a^2|/15$. Substituting this approximation into Eq. (14.256), we have

$$\left|\frac{B_\varphi^v}{B_\varphi^g}\right| = \frac{s^{1/2}|k_i a|}{15n}. \tag{14.259}$$

Therefore, within this range with a decrease of frequency, the relative importance of the galvanic part of the magnetic field grows; that is the magnetic field from currents that close through the surrounding medium begins to dominate over that of the induced currents within the sphere. This behavior can be easily explained. In fact, with a decrease in frequency the magnetic field from the eddy currents flowing in the sphere decreases in direct proportion to ω, while the magnetic field of the galvanic part is proportional to the square root of ω, because the primary electric field is related to the normal magnetic field as

$$E_0 = \left(\frac{\omega}{-i\gamma_e\mu_0}\right)^{1/2} B_0.$$

It is obvious from physical point of view and Eq. (14.241) that with an increase in the resistivity of the host medium, the relative contribution of the vortex of the magnetic field increases. Whence it is possible to say that only for very large ratios of the resistivity between the host medium and a confined conductor, the galvanic part of the secondary field can be neglected. In other words, we arrive at the conclusion that electromagnetic methods can be applied in the wave zone to classify anomalies on the basis of conductivity only in the rare cases when the surrounding medium is very resistive and only the magnetic field is measured. Note that from equations given above, it follows that in principle the possibility exists of making measurements in a way in which the galvanic part of the magnetic field would be largely suppressed. For instance, the difference

$$\frac{B_\varphi(\omega_1)}{\omega_1^{1/2}} - \frac{B_\varphi(\omega_2)}{\omega_2^{1/2}}$$

is defined almost entirely by the vortex part of the field.

Case Two. Near-Zone for the Normal Field

First of all, suppose that the normal field is caused by inductively; that is the source of the primary field is a current loop, in particular it can be magnetic dipole. By definition, within this zone, the primary field is dominant, and in the case of the magnetic dipole, we have for the ratio of the magnetic and electric field

$$\frac{B_0}{E_0} = \frac{1}{i\omega l}, \tag{14.260}$$

where l is the distance between the dipole and an observation point. Substituting the latter into Eq. (14.248), we have

$$\eta_E = \frac{r}{l}|D|. \tag{14.261}$$

This indicates that even in the case of the inductive excitation of the field, the galvanic part usually prevails when the electric field is measured. As concerns the magnetic field Eq. (14.255), we have

$$\eta_B = \frac{|D|}{k_e^2 rl}. \tag{14.262}$$

Bearing in mind that the denominator is less than one, we see that in measuring the magnetic field the vortex part of the field is dominant and therefore the inductive methods have relatively high resolution by conductivity. Next, suppose that measurements are performed in the near zone when the field is caused by a grounded line. As we know, in the case of the electric dipole

$$\frac{B_0}{E_0} = \gamma_e \mu_0 l. \tag{14.263}$$

Then we obtain

$$\eta_E = \frac{|i\gamma_e \mu_0 \omega rlD|}{2} \tag{14.264}$$

and it means that the galvanic part of the field prevails. For the magnetic field the ratio η_B is

$$\eta_B = \frac{l}{r}|D|, \tag{14.265}$$

and therefore the influence on the galvanic part of the field is rather strong, regardless which component of the electromagnetic field is measured. This fact is well known, and it isn't occasional that arrays, measuring electrical field and caused by the electrical dipole, are successfully used for mapping, but they are not applied for separating anomalies by conductivity.

We demonstrated that practically independently on the method of field excitation the electric field is mainly defined by the galvanic part, especially in the wave zone, as well as in the case of the grounded line as the source, and this is one of the reasons why the magnetic field is measured, when the primary source is a nongrounded loop. Let us look at these results slightly different, from point of view of a geological noise. Inasmuch as the galvanic part of the field only slightly depends on conductivity, inhomogeneities having different resistivities may cause almost the same anomalies of the electric field, as it takes place in methods of the direct current. Such situation often takes place in magnetotelluric soundings and CSMT. As it will be shown in the next chapter, influence of near surface inhomogeneities can strongly affect the behavior of the magnetotelluric apparent resistivity curves, and in order to understand this phenomenon and take into account the contribution caused by lateral changes of resistivity, it became necessary to greatly modify the theory of magnetotelluric soundings, developed for a horizontally layered medium. From the historical point of view, it may be proper to notice that the subject of relation between the vortex and galvanic parts of the field was first studied in the frame of the inductive mining prospecting and later was applied for magnetotelluric soundings. It is one of several examples of a useful interaction between different branches of electromagnetic methods.

REFERENCES AND FURTHER READING

[1] F.S. Grant, G.F. West, Interpretation Theory in Applied Geophysics, McGraw-Hill, New York, NY, 1965.
[2] A.A. Kaufman, Basic Theory of Inductive Mining Prospecting, Nauka, Novosibirsk, 1974.
[3] A.A. Kaufman, G.V. Keller, Inductive Mining Prospecting: Theory, Elsevier, 1985.
[4] M.N. Nabighian (Ed.), Electromagnetic Methods in Applied Geophysics, vol. 1, SEG, 1991.
[5] J.R. Wait, Geoelectromagnetism, Academic Press, New York, NY, 1982.

CHAPTER FIFTEEN

Magnetotelluric Soundings in a Laterally Inhomogeneous Medium

INTRODUCTION

As we already know (Chapter 11), the Cagniard's theory implies that a medium is horizontally layered one and the electromagnetic field is a linearly polarized plane wave traveling downward. Under such assumption, the impedance and correspondingly the apparent resistivity curves are independent of orientation of measured components of the electromagnetic field. However, very quickly, almost at the end of 1950s, it became clear that the behavior of the apparent resistivity curves very often changes with this orientation, and their shape strongly differs from Cagniard's curves. Moreover, it was found that repeated measurements at the same observation point could yield different apparent resistivity values at the same frequency that resulted in the scattering of the curve. Such situation became a formidable obstacle for application of the classical theory of magnetotelluric soundings. Understanding of such distortions and attempts to still carry out the quantitative interpretation became the main purpose of the development of much more sophisticated theory of this method, in which the Cagniard's theory still plays a certain role. Of course, assumption of the Cagniard's theory that electromagnetic field is a plane wave from the beginning caused suspicious, since in reality such wave does not exist. However, as was shown earlier, the primary field of any source at the wave zone behaves as a quasi-plane wave, and its impedance at the earth surface coincides with that of the plane wave. For this reason, one can think that usually this assumption could not be the reason for either distortion of the apparent resistivity curves or its change with an orientation of sensors, as soon as a horizontally layered medium is considered. At the beginning, when we introduced the impedance (Chapter 11), it was assumed that a field is caused by a single source of a given orientation. In other words, the electromagnetic field is linearly polarized and its direction does not change. Now let us make several obvious generalizations. First, suppose that direction of the source of the primary field changes. Of course, projections of

Methods in Geochemistry and Geophysics, Volume 45
ISSN 0076-6895,
http://dx.doi.org/10.1016/B978-0-444-53829-1.00015-0

the fields along the x- and y-coordinates also change but the impedance remains the same. Next, assume that there are simultaneously two or more sources of the primary field, and these fields differ by the direction, amplitude, and phase. In this case, an elliptical polarization takes place, but still the ratio of the total electric and magnetic fields defines the same impedance as for the single source. In fact, elliptically polarized field can be represented as a sum of two linearly polarized fields along two perpendicular directions, and measuring the electric and magnetic fields along two perpendicular directions. We deal with the linearly polarized fields, and their ratio gives the same impedance as before. Thus, the elliptical polarization of the field and its change do not affect the impedance in a horizontally layered medium.

However, completely different situation takes place in case of lateral variations of resistivity. Suppose that beneath the earth surface, there is an arbitrary inhomogeneity with resistivity, which is different from that of a surrounding medium. Then, the total measured field can be represented as a sum of the normal and secondary fields:

$$\mathbf{E} = \mathbf{E}_n + \mathbf{E}_s \quad \text{and} \quad \mathbf{B} = \mathbf{B}_n + \mathbf{B}_s$$

Unlike the normal field, the behavior of the secondary field measured on the earth surface does not correspond to the wave zone, and therefore, the ratio of the electric and magnetic components of the total field depends on an orientation of the measuring lines. This means that instead of the impedance of the plane wave, we obtain some function, which is called the *apparent impedance*, Z_a, since it varies with a change of an orientation of measured components. Therefore, the Cagniard's theory becomes invalid when there is an influence of lateral inhomogeneities. Besides, from the physical point of view, it is obvious that the secondary field, caused by the presence of an inhomogeneity beneath the Earth's surface, depends strongly on the normal field direction. Indeed, a change of its direction causes a change in distribution of charges on the surface of an inhomogeneity as well as induced currents inside and outside of it. Respectively, the effect of lateral changes of resistivity is a function of the location of current sources in the ionosphere which may vary. This usually causes scattering of the apparent resistivity curve, obtained from measuring with the same sensors. Note, that since the normal and secondary fields practically always have different magnitudes and phases, the total field is elliptically polarized and orientation of the ellipse on the Earth's surface may vary with time. Thus, the apparent impedances are usually functions of (1) the resistivity of the inhomogeneous medium, (2) the orientation of coordinate axis (orientation of sensors), and

(3) the direction of the normal field which often changes with time. Because of the influence of these factors, the quantitative interpretation based on the Cagniard's theory and the use of the apparent impedance became impossible, except those cases when a medium could be treated as a horizontally layered one (1D model). It forced researches to look for a new approach, based on the use of quantities which are not affected by elliptical polarization of the field, and at the middle of 1960s such a method, namely, the impedance tensor method was proposed. Certainly, that was very important step in development of the magnetotelluric soundings. Before we describe the role of the impedance tensor for interpretation, let us make several comments. (1) Transition from the scalar impedance to the impedance tensor and development of interpretation based on the use of this new concept was not simple process, and it has required efforts of an unusually large group of very talented geophysicists, mathematicians, and the electrical engineers working in many different countries of the world, mainly in USA, Russia, and Europe. (2) Fifty years ago, the interaction and cooperation between scientists of the West and East were very poor, and often they obtained the same or similar results almost at the same time independently, without knowledge of achievements of colleagues abroad. At that time two large groups were formed: one of them was around Prof. T. Madden (Boston); the other one was around Prof. M. Berdichevsky (Moscow) and his colleagues. Among those who contributed at the beginning to a development of a new approach following Cagniard were T. Madden, T. Cantwell, C. Swift, and M. Berdichevsky. (3) As was pointed out, in 1951, Cagniard had introduced the concept of admittance, which is closely related to the impedance. (4) Perhaps, one say that even at this first stage an application of magnetotelluric soundings allowed mainly to map a highly resistive basement surface, and results of interpretation were represented in the form of a distribution of the total longitudinal conductance S that had a great practical interest. Relative simplicity of the field measurements with respect to Schlumberger soundings made this method very popular and it was widely used. For instance, the magnetotelluric soundings found a broad application for mapping vast areas in West and East Siberia, where it was conducted by several state companies. The results of the surveys allowed one to locate zones of decreased conductance, which were usually associated with anticlines on the basement surface. Several companies, for instance Phoenix, were formed in North America and abroad, which created the equipment and provided magnetotelluric services over the world. Application of this method was much simpler than the other electromagnetic

soundings since it did not require the current source. As is well known, in the case when sediments contain a highly resistive layer (shield), the direct current soundings are not able to image the structure beneath this layer because the electric field of charges located above the shield induces the charges on its surface, which makes the field beneath the shield practically equal to zero (electrostatic induction). The situation is slightly better in magnetotellurics, since the normal field (the field in case of laterally homogenous medium) penetrates through this layer, while the secondary field, emerging in presence of inhomogeneity, again causes the charges to appear on the both surfaces of a resistive layer. As a result, the secondary field is either shielded or becomes much weaker, and it usually produces anomaly which is essentially wider than the subshield structure itself. This phenomenon sometimes is called the *screening effect*. (5) Measurements of the scalar impedance allow one to determine its magnitude and phase. At that time, it was shown that in the case of a horizontally layered medium, the phase could be calculated if the frequency response of the impedance magnitude was known. In other words, it was demonstrated that the frequency responses of the amplitude and phase are related with each other. This follows from the fact that in the case of 1D model, the amplitude is an analytical function on the complex frequency plane and it differs from zero at all points of the upper half plane. Certainly it was useful fact because earlier it was an impression that unlike Schlumberger soundings, magnetotelluric data, measuring two independent quantities, contain more information. We briefly described the first very promising stage of application of the method when the scalar impedance was used, and it was assumed that one could perform the quantitative interpretation as in the case of the Schlumberger soundings.

Let us also add the following. From the beginning of it was clear that development of acquisition instruments and algorithms for processing observation data was very important, especially taking into account the fact that the incident waves essentially differ from sinusoidal waves, for which the theory was developed. Another problem was cultural electromagnetic noise, affecting and severely distorting the data in some areas. At the beginning of practical application of magnetotellurics, the typical accuracy of measurements was about 10%, which often resulted in extremely large equivalence in the inverse problem solution. After more than 20 years of work, an amazing success in this area was achieved. Application of complicated processing procedures including the robust methods based on coherence weighting and remote reference technique resulted in the possibility to measure the magnetotelluric response with an accuracy of a few percentage for the magnitude and about 1–2° for the phase.

Before we describe the impedance tensor method, it may be useful to summarize our discussion and with this purpose in mind, consider again three cases.

Case 1

Suppose that a medium is a horizontally layered one and the normal field with components and is linearly polarized. Then, changing an orientation of the field sensors causes the measured values of the electric and magnetic fields to change in the same manner (they are simply multiplied by the same scalar), and therefore, the impedance remains the same. This means that the impedance does not depend on direction of the electric and magnetic fields of the incident wave, as well as sensors orientation. Of course, it is assumed that the incident wave is a quasi-plane one; otherwise, the ratio of components would not be equal to the impedance of the plane wave even in a horizontally layered medium. It is clear that the non-quasi-plane behavior of the incident wave can be a reason for scattering of measured values of the impedance. In this light, let us remind that the wave zone practically takes place when the horizontal distance from the source of the magnetotelluric field, situated above the Earth's surface, exceeds the wavelength in a conducting medium.

Case 2

Now assume again that that medium is a horizontally layered one, but the incident quasi-plane wave is elliptically polarized in the horizontal plane. The latter means that two linearly polarized waves with different directions and the same period of oscillations arrive at an observation point or, alternatively, the source of the incident wave is elliptically polarized. Correspondingly, the measured field is also elliptically polarized, that is, vector tips of the electric and magnetic fields move along ellipses, and this motion repeats every period. As was demonstrated earlier, the amplitude of the elliptically polarized sinusoidal oscillations along some particular direction is defined by such position of the vector, which yields maximal length of the projection onto this direction, while the phase depends on its orientation. Mutual position of both ellipses changes with the period of oscillations and parameters of layers. In principle, considering the impedance, it is possible to proceed from both ellipses, but much simpler to take into account the principle of superposition and the fact that for each linearly polarized field, regardless of its orientation, the ratio of the fields is equal to impedance of the plane wave. Then we can say that the same is true for elliptically polarized field. Of course, the same conclusion is correct when two linearly polarized incident waves of the same period but with different directions arrive at observation point at different times.

Case 3

Next suppose that the incident wave is linearly polarized field and within a horizontally layered medium, there is an inhomogeneity. It is obvious that the total field is a sum of the normal and secondary fields, where the latter is caused by the presence of an inhomogeneity. Also it is natural to expect that, in general, the total field is elliptically polarized. From the physical point of view, it follows that a change of an orientation of the vectors **E** and **B** of the normal waves makes an influence on the secondary field, and therefore, the measured impedance becomes a function of this orientation. In other words, measurements of the impedance at the same frequency performed at different times can give different values. In particular, it may happen if the incident wave is elliptically polarized, that is, it is formed by two linearly polarized waves having different orientations of field vectors and nonzero phase shift between them. Prior to describing a method, which allows one to eliminate an influence of a direction of field vectors of the quasi-plane incident wave or its elliptical polarization, it is proper to indicate factors which can still cause scattering of measured values of the tensor after application of this method. First, it may happen that a source of some incident wave is located relatively close to an observation point and, correspondingly, this wave is not quasi-plane one. As was pointed out earlier, even in a horizontally layered medium, the impedance of such wave does not coincide with that of the plane wave. The second reason for scattering is the following. Suppose that two incident quasi-plane waves with the same orientation of field components have different spatial spectrum in a volume, occupied by an inhomogeneity. In other words, they change differently from point to point, and therefore, generate different secondary fields. It may be proper to note that in the exploration geophysics, the sources of the incident wave are usually located sufficiently far away from an area of a survey, and for this reason, this field does not practically change within this area, and in this sense, such wave is called the *plane one.*

15.1. THE IMPEDANCE TENSOR

15.1.1. Relation between the Normal and Total Field

Suppose that a medium is nonuniform; that is, its resistivity varies arbitrarily as a function of a point, and the quasi-plane wave propagates along the

z-axis into a conducting half space. Let us also assume that the normal field $\mathbf{E}^0$, $\mathbf{B}^0$ is elliptically polarized in the horizontal x–y plane. By definition, such a field can be represented as a sum of two linearly polarized quasi-plane waves with the complex amplitudes E_x^0, B_y^0 and E_y^0, B_x^0, where x and y characterize the directions along which the field is measured. Thus, the complex amplitudes of the horizontal component of the normal field are

$$\mathbf{E}^0 = E_x^0\mathbf{1}_x + E_y^0\mathbf{1}_y \quad \text{and} \quad \mathbf{B}^0 = B_x^0\mathbf{1}_x + B_y^0\mathbf{1}_y \tag{15.1}$$

In general, the corresponding components of linearly polarized fields have different amplitudes and phases, so that elliptical polarization of the normal field takes place. Since the normal field is considered as a field in a horizontally layered medium with no lateral inhomogeneities, we have

$$E_x^0 = ZB_y^0, \quad E_y^0 = -ZB_x^0,$$

where Z is the impedance at the surface of a horizontally layered medium described earlier. As the first step, we consider the field caused by each of these linearly polarized fields (E_x^0, B_y^0 and E_y^0, B_x^0). Then, making use of the principle of superposition, the total field will be obtained. Let us start with the field E_x^0, B_y^0. In the presence of lateral inhomogeneities, this field generates secondary currents and electrical surface charges, and as a result, a field with the complex amplitudes $\mathbf{E}_1$, $\mathbf{B}_1$ arises. In general, this field can have all components, three electrical and three magnetic, and the electromagnetic field caused by the normal field E_x^0, B_y^0 is usually elliptically polarized. Thus,

$$E_{1x} = a_{1x}B_y^0, \quad E_{1y} = a_{1y}B_y^0, \quad E_{1z} = a_{1z}B_y^0 \tag{15.2}$$

and

$$B_{1x} = b_{1x}B_y^0, \quad B_{1y} = b_{1y}B_y^0, \quad B_{1z} = b_{1z}B_y^0, \tag{15.3}$$

where coefficients a_{1i} and b_{1i} are complex. They depend on several factors, namely, the electrical properties of the medium, the period of oscillations, the orientation of the axes x and y; that is, an orientation of the normal field, and of course, the location of the observation site. They do not depend on the strength of this normal field E_x^0, B_y^0. These coefficients characterize the electromagnetic field at the surface of a conducting medium when the amplitude of the magnetic component of a linearly polarized normal field is of a unit strength. The second part of the normal field E_y^0, B_x^0 also gives rise to a field in the medium, which may also have all six components, that is

$$E_{2x} = a_{2x}B_x^0, \quad E_{2y} = a_{2y}B_x^0, \quad E_{2z} = a_{2z}B_x^0 \tag{15.4}$$

and

$$B_{2x} = b_{2x}B_x^0, \quad B_{2y} = b_{2y}B_x^0, \quad B_{2z} = b_{2z}B_x^0 \tag{15.5}$$

Here a_{2i}, b_{2i} are the coefficients that depend on the same parameters as a_{1i}, b_{1i} and represent the electromagnetic field when the normal field E_y^0, B_x^0 is linearly polarized and $\left|B_x^0\right| = 1$. Thus, the total field caused by normal elliptically polarized field can be written in the form

$$\mathbf{E} = \mathbf{E}_1 + \mathbf{E}_2 \quad \mathbf{B} = \mathbf{B}_1 + \mathbf{B}_2, \tag{15.6}$$

and therefore,

$$E_x = E_{1x} + E_{2x} = a_{1x}B_y^0 + a_{2x}B_x^0 \tag{15.7}$$

$$E_y = E_{1y} + E_{2y} = a_{1y}B_y^0 + a_{2y}B_x^0 \tag{15.8}$$

$$E_z = E_{1z} + E_{2z} = a_{1z}B_y^0 + a_{2z}B_x^0 \tag{15.9}$$

$$B_x = B_{1x} + B_{2x} = b_{1x}B_y^0 + b_{2x}B_x^0 \tag{15.10}$$

$$B_y = B_{1y} + B_{2y} = b_{1y}B_y^0 + b_{2y}B_x^0 \tag{15.11}$$

$$B_z = B_{1z} + B_{2z} = b_{1z}B_y^0 + b_{2z}B_x^0 \tag{15.12}$$

In particular, in the case of a horizontally layered medium, we have

$$b_{1x} = b_{2y} = b_{1z} = b_{2z} = a_{2x} = a_{1y} = 0, \quad a_{1x} = Z, \ a_{2y} = -Z,$$

$$b_{2x} = b_{1y} = 1$$

It is obvious that at the surface of the conducting earth, the vertical component of the electric field is zero ($E_z = 0$), and therefore coefficients a_{1z} and a_{2z} must be zero:

$$a_{1z} = a_{2z} = 0 \tag{15.13}$$

Because the normal field is unknown, it is natural to eliminate its components and find relationship between the measured components of the total field making use of Eqs (15.7)–(15.12). Assuming that

$$b_{1x}b_{2y} - b_{2x}b_{1y} \neq 0 \tag{15.14}$$

Then having expressed B_x^0 and B_y^0 in terms of the total field B_x and B_y in Eqs (15.10) and (15.11), we obtain

$$B_x^0 = \frac{b_{1x}B_y - b_{1y}B_x}{b_{1x}b_{2y} - b_{2x}b_{1y}} \quad \text{and} \quad B_y^0 = \frac{b_{2y}B_x - b_{2x}B_y}{b_{1x}b_{2y} - b_{2x}b_{1y}} \tag{15.15}$$

Here, it is essential that an orientation of the normal fields is different and, correspondingly, they cause the total fields with different elliptical polarization. Later we consider cases for which Eq. (15.14) is invalid.

15.1.2. Elements of the Impedance Tensor

Now, substituting Eq. (15.15) into Eqs (15.7) and (15.8), we obtain the linear relationships between the horizontal components of the total magnetotelluric field, measured at the earth surface:

$$E_x = Z_{xx}B_x + Z_{xy}B_y \quad \text{and} \quad E_y = Z_{yx}B_x + Z_{yy}B_y \tag{15.16}$$

where

$$Z_{xx} = \frac{a_{1x}b_{2y} - a_{2x}b_1}{b_{1x}b_{2y} - b_{2x}b_{1y}}, \qquad Z_{xy} = \frac{a_{2x}b_{1x} - a_{1x}b_{2x}}{b_{1x}b_{2y} - b_{2x}b_{1y}}$$

and

$$Z_{yx} = \frac{a_{1y}b_{2y} - a_{2y}b_{1y}}{b_{1x}b_{2y} - b_{2x}b_{1y}} \qquad Z_{yy} = \frac{a_{2y}b_{1x} - a_{1y}b_{2x}}{b_{1x}b_{2y} - b_{2x}b_{1y}} \tag{15.17}$$

By analogy, from Eq. (15.12), we have

$$B_z = X_{zx}B_x + X_{zy}B_y \tag{15.18}$$

where

$$X_{zx} = \frac{b_{1z}b_{2y} - b_{2z}b_{1y}}{b_{1x}b_{2y} - b_{2x}b_{1y}}, \qquad X_{zy} = \frac{b_{2z}b_{1x} - b_{1z}b_{2x}}{b_{1x}b_{2y} - b_{2x}b_{1y}} \tag{15.19}$$

The complex quantities Z_{xx}, Z_{xy}, Z_{yx}, and Z_{yy} along with X_{zx} and X_{zy} are combinations of coefficients a_{1i}, b_{1i}, a_{2i}, b_{2i} ($i = x, y, z$), and therefore, they are functions of frequency, electrical properties of the medium, orientation of sensor axes, and the location of the observation site. Since these complex quantities Z_{ij} are derived for any elliptical polarization of the normal field, they do not depend on the direction of the arrival of the normal field. Thus, the problem posed at the beginning was in principle solved. In a fact, in place of apparent impedances $Z_{xy}^{a} = E_x/B_y$ and

$Z^{\rm a}_{yx} = E_y/B_x$, usually varying with a change of an orientation of the normal field, we have arrived at four new functions Z_{xx}, Z_{xy}, Z_{yx}, and Z_{yy}, which are insensitive to changes in the elliptical polarization of the normal electromagnetic field. Now let us write Eq. (15.16) in the form

$$\begin{pmatrix} E_x \\ E_y \end{pmatrix} = \begin{pmatrix} Z_{xx} & Z_{xy} \\ Z_{yx} & Z_{yy} \end{pmatrix} \cdot \begin{pmatrix} B_x \\ B_y \end{pmatrix} \tag{15.20}$$

or in the compact form

$$\mathbf{E} = Z \cdot \mathbf{B}, \tag{15.21}$$

where

$$Z = \begin{pmatrix} Z_{xx} & Z_{xy} \\ Z_{yx} & Z_{yy} \end{pmatrix} \tag{15.22}$$

According to Eq. (15.22), the function Z can be treated as an operator which transforms the horizontal component of the magnetic field $\mathbf{B}$ to the horizontal component of the electric field $\mathbf{E}$. In the vector analysis, such an operator is called a *tensor* and the ordered combination of four functions Z_{xx}, Z_{xy}, Z_{yx}, and Z_{yy} forms the tensor of the second order, which is called the *impedance tensor.* The theory of the tensor or the tensor analysis was developed in detail at the beginning of the twentieth century, and it represents the well-known part of mathematics. Off-diagonal elements Z_{xy} and Z_{yx} are termed the *principal impedances*, while Z_{xx} and Z_{yy} are the *auxiliary impedances*. Thus, in order to remove an influence of an orientation of the electromagnetic field, we have to deal with a tensor of the second order instead of the scalar quantities $Z^{\rm a}_{xy}$ and $Z^{\rm a}_{yx}$. Taking into account importance of the impedance Z, let us emphasize again that even in the presence of lateral inhomogeneities, the change of an orientation of the normal field does not make influence on all elements of this tensor. Before we continue, it is proper to notice the following. Of course, introduction of the impedance tensor was very important step in developing the magnetotellurics, in spite of the fact that its elements are subjected to an influence of lateral changes of a resistivity of a medium. Correspondingly, it has been still necessary to solve the forward and inverse problems in order to carry out a quantitative interpretation. For this reason, as soon as it was realized that in many cases Cagniard's theory cannot be applied, the main attention has been paid to the forward problem solution, allowing one to understand a field behavior in an inhomogeneous medium, as well as development of the algorithms for a solution of the magnetotelluric inverse problem.

15.1.3. Main Features of the Impedance Tensor

Next we examine the main features of the tensor Z, which are useful in interpreting magnetotelluric data. First, let us demonstrate that if the diagonal and off-diagonal impedances are known for two mutually perpendicular directions x and y, then these elements can be evaluated for any pair of directions x' and y'; that is, the values Z_{xx}, Z_{xy}, Z_{yx}, and Z_{yy} uniquely define the impedance Z in a nonuniform conducting medium. To obtain the values for the impedance Z for any pair of directions x' and y', it is not necessary to make additional measurements of the electric and magnetic field components along these directions if a set of values for the impedance Z is known for any single pair of directions (x, y), that is of a great practical importance for interpretation. Now we demonstrate this fact and show that E_x, E_y, and B_x, B_y contain all information about the impedance tensor in the horizontal plane. Suppose that we have two pairs of Cartesian axes, x, y and x', y'; one coordinate system is rotated relative to the other through angle α (Figure 15.1(a)). As it can be seen, the components of the electric and magnetic fields along these four directions are related through the equations:

$$E_{x'} = E_x \cos\alpha - E_y \sin\alpha, \quad E_{y'} = E_x \sin\alpha + E_y \cos\alpha \tag{15.23}$$

$$B_{x'} = B_x \cos\alpha - B_y \sin\alpha, \quad B_{y'} = B_y \sin\alpha + B_x \cos\alpha \tag{15.24}$$

According to Eq. (15.16), we have

$$E_x = Z_{xx}B_x + Z_{xy}B_y, \quad E_y = Z_{yx}B_x + Z_{yy}B_y \tag{15.25}$$

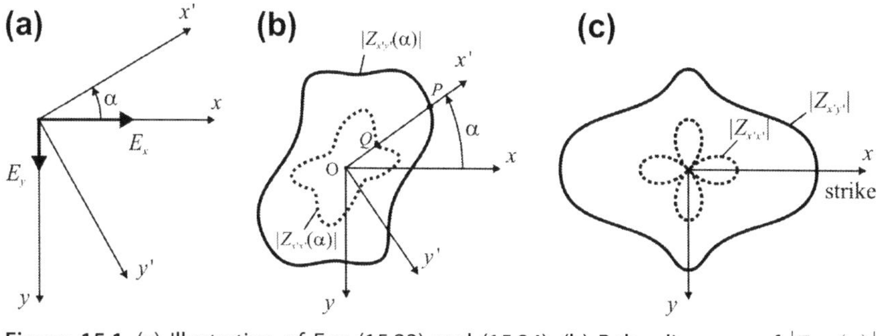

Figure 15.1 (a) Illustration of Eqs (15.23) and (15.24). (b) Polar diagrams of $|Z_{x'y'}(\alpha)|$ and $|Z_{x'x'}(\alpha)|$ for three-dimensional model (3D). (c) Polar diagrams for two-dimensional model (2D).

and

$$E_{x'} = Z_{x'x'}B_{x'} + Z_{x'y'}B_{y'}, \quad E_{y'} = Z_{y'x'}B_{x'} + Z_{y'y'}B_{y'} \tag{15.26}$$

where $Z_{x'x'}$, $Z_{x'y'}$, $Z_{y'x'}$ and $Z_{y'y'}$ are the values for the impedance tensor corresponding to the axes x', y'. Substituting Eqs (15.23) and (15.24) into Eq. (15.26), we obtain

$$\begin{aligned}
Z_{x'x'}(\alpha) &= Z_{xx}\cos^2\alpha + Z_{yy}\sin^2\alpha - \left(Z_{xy} + Z_{yx}\right)\sin\alpha\cos\alpha \\
Z_{x'y'}(\alpha) &= Z_{xy}\cos^2\alpha - Z_{yx}\sin^2\alpha + \left(Z_{xx} - Z_{yy}\right)\sin\alpha\cos\alpha \\
Z_{y'x'}(\alpha) &= Z_{yx}\cos^2\alpha - Z_{xy}\sin^2\alpha + \left(Z_{xx} - Z_{yy}\right)\sin\alpha\cos\alpha \\
Z_{y'y'}(\alpha) &= Z_{yy}\cos^2\alpha + Z_{xx}\sin^2\alpha + \left(Z_{xy} + Z_{yx}\right)\sin\alpha\cos\alpha
\end{aligned} \tag{15.27}$$

Equation (15.27) specifies the relationship between impedance tensors Z' and Z; that is, they show how the tensor elements change with rotation of the coordinate axes (measurement directions) x, y. Moreover, Eq. (15.27) permits us to calculate the diagonal and off-diagonal tensor elements along any direction if these values are known for one set of measuring directions, for example, x and y. Now we will show another interesting feature of the impedance tensor. As immediately follows from Eq. (15.27), the functions

$$\begin{aligned}
I_1 &= Z_{x'x'}(\alpha)Z_{y'y'}(\alpha) - Z_{x'y'}(\alpha)Z_{y'x'}(\alpha) = Z_{xx}Z_{yy} - Z_{xy}Z_{yx} \\
I_2 &= Z_{x'x'}(\alpha) + Z_{y'y'}(\alpha) = Z_{xx} + Z_{yy} \\
I_3 &= Z_{x'y'}(\alpha) - Z_{y'x'}(\alpha) = Z_{xy} - Z_{yx}
\end{aligned} \tag{15.28}$$

are independent of α, which means they have the same values for any orientation of the measurement array. For this reason, they are called *invariants* of the impedance tensor Z. The invariant I_1 is the difference of paired products of the diagonal and off-diagonal impedance elements, that is, it is determinant of matrix Z. The invariant I_2 is the sum of the two diagonal terms, and the invariant I_3 is the difference between the two off-diagonal terms. As well as the diagonal and off-diagonal impedance terms, the invariants I_1, I_2, and I_3 are the functions of the position of observation site, the frequency and resistivity of a medium. It is essential that unlike the impedance terms themselves, they are not functions of the orientation of the sensor axes. From comparison of these invariants with impedances Z^{a}_{xy} and Z^{a}_{yx}, we see that the latter quantities depend on two more factors, namely, the polarization of the field and the orientation of the sensor axes. Clearly, such behavior of functions I_1, I_2, and I_3 is of interest for interpretation. Also, the independence of invariants of orientation can be useful in calculating special parameters, which help to distinguish between the models of different dimensionality and will be considered later. From Eq. (14.27), one can expect periodicity of their values with rotation of coordinate axes. In fact, we have

$$Z_{x'y'}(\alpha) = Z_{x'y'}(\alpha \pm \pi), \quad Z_{x'x'}(\alpha) = Z_{x'x'}(\alpha \pm \pi)$$

as well as

$$Z_{x'y'}(\alpha) = -Z_{y'x'}\left(\alpha \pm \frac{\pi}{2}\right), \quad Z_{x'x'}(\alpha) = Z_{y'y'}\left(\alpha \pm \frac{\pi}{2}\right) \tag{15.29}$$

15.1.4. Relation of the Apparent Impedance with Elements of the Impedance Tensor

Let us return to the linear relationships in Eq. (15.16) and interpret the diagonal and off-diagonal impedance terms in the following way. Suppose that a horizontal magnetic field is linearly polarized at the Earth's surface and is directed along x' axis. Then, according to Eq. (15.16), we have

$$E_{x'} = Z_{x'x'}(\alpha)B_{x'} \quad \text{and} \quad E_{y'} = Z_{y'x'}(\alpha)B_{x'}$$

and

$$Z_{x'x'}(\alpha) = \frac{E_{x'}}{B_{x'}} \quad \text{and} \quad Z_{y'x'}(\alpha) = \frac{E_{y'}}{B_{x'}} \tag{15.30}$$

Thus, in this special case, the off-diagonal impedance $Z_{y'x'}$ coincides with apparent impedance $Z_{y'x'} = E_{x'}/B_{y'}$. By analogy, assuming that the magnetic field is linearly polarized and directed along the y' axis, we obtain

$$E_{x'} = Z_{x'y'}(\alpha)B_{y'}, \quad E_{y'} = Z_{y'y'}(\alpha)B_{y'}$$

Therefore, in the latter case, the off-diagonal impedance $Z_{x'y'}$ coincides with impedance $Z_{x'y'}(\alpha) = E_{x'}/B_{y'}$. By convention, the diagonal and off-diagonal impedances are associated with the direction along which an electrical field is measured. Then, $Z_{z'y'}$ is the off-diagonal impedance in the direction x', and $Z_{y'x'}$ is the off-diagonal impedance in the direction y', and so on.

15.1.5. Polar Diagrams of Impedance Tensor

It is natural to represent the dependence of these impedances on a direction with help of graphs. Let us plot segments OP and OQ which are proportional to the values $|Z_{x'y'}(\alpha)|$, $|Z_{x'x'}(\alpha)|$ along the axis x', respectively. As the angle α varies from 0 to 2π, the points P and Q trace the curves which are called *impedance polar diagrams* (Figure 15.1(b),(c)). The equations of these polar diagrams can readily be derived from Eq. (15.27):

$$|Z_{x'y'}(\alpha)| = \left\{|Z_{xy}|^2 \cos^4 \alpha + |Z_{yx}|^2 \sin^4 \alpha + 2U \sin^2 \alpha \cos^2 \alpha\right\}^{1/2} \tag{15.31}$$

$$|Z_{x'x'}(\alpha)| = \left\{|Z_{xx}|^2 \cos^4 \alpha + |Z_{yy}|^2 \sin^4 \alpha + 2V \sin^2 \alpha \cos^2 \alpha\right\}^{1/2} \tag{15.32}$$

Here,

$$\begin{aligned} U = {} & \frac{|Z_{xx} - Z_{yy}|^2}{2} + |Z_{xy}|\cdot|Z_{yx}|\cos(\arg Z_{xy} - \arg Z_{yx}) \\ & +|Z_{xy}|\cdot|Z_{xx} - Z_{yy}|\cot\alpha\cdot\cos[\arg Z_{xy} - \arg(Z_{xx} - Z_{yy})] \\ & -|Z_{yx}|\cdot|Z_{xx} - Z_{yy}|\cdot\tan\alpha\cdot\cos[\arg Z_{xy} - \arg(Z_{xx} - Z_{yx})] \end{aligned} \tag{15.33}$$

and

$$\begin{aligned} V = {} & \frac{|Z_{xy} + Z_{yx}|^2}{2} + |Z_{xx}|\cdot|Z_{yy}|\cos(\arg Z_{xx} - \arg Z_{yy}) \\ & -|Z_{xx}|\cdot|Z_{xy} + Z_{yx}|\cot\alpha\cdot\cos[\arg Z_{xx} - \arg(Z_{xy} + Z_{yx})] \\ & -|Z_{yy}|\cdot|Z_{xy} + Z_{yx}|\tan\alpha\cdot\cos[\arg Z_{yy} - \arg(Z_{xy} + Z_{yx})] \end{aligned} \tag{15.34}$$

As is seen from Eq. (15.29), the polar diagrams for $|Z_{x'y'}(\alpha)|$ and $|Z_{x'x'}(\alpha)|$ are symmetric about any axis passing through the origin. In general case, impedance polar diagrams have appearance of four asymmetric leaves, each of which repeats itself through an angle π. As follows from Eq. (15.30), the polar diagrams $|Z_{y'x'}(\alpha)|$ and $|Z_{y'y'}(\alpha)|$ are shifted by $\pi/2$ with respect to the polar diagrams for $|Z_{x'y'}(\alpha)|$ and $|Z_{y'y'}(\alpha)|$. In addition to the polar diagrams of the absolute value of the impedance $|Z_{x'y'}(\alpha)|$ or $|Z_{x'x'}(\alpha)|$, one can plot a polar diagram for the real and imaginary parts of the diagonal and off-diagonal terms. As is seen from Eq. (15.27), the expressions for the real and imaginary parts are the same, and are much simpler than the equations for $|Z_{x'y'}(\alpha)|$ and $|Z_{x'x'}(\alpha)|$. For example, for the real part, we have

$$\begin{aligned}\text{Re } Z_{x'x'}(\alpha) &= \text{Re } Z_{xx}\cos^2\alpha + \text{Re } Z_{yy}\sin^2\alpha \\ &\quad - \text{Re}(Z_{xy} + Z_{yx})\sin\alpha\cos\alpha\end{aligned}$$

and

$$\begin{aligned}\text{Re } Z_{x'y'}(\alpha) &= \text{Re } Z_{xy}\cos^2\alpha - \text{Re } Z_{yx}\sin^2\alpha \\ &\quad + \text{Re}(Z_{xx} - Z_{yy})\sin\alpha\cos\alpha\end{aligned}$$

15.2. BEHAVIOR OF THE IMPEDANCE TENSOR

15.2.1. Horizontally Layered (1D) Medium

As was shown earlier, in a horizontally layered medium, we have

$$\frac{E_x}{B_y} = Z_{xy} = Z_0, \quad \text{and} \quad \frac{E_y}{B_x} = Z_{yx} = -Z_0,$$

where Z_0 is the impedance of a plane wave at the Earth's surface, and **E** and **B** are the complex amplitudes of these fields. Comparing these relationships with Eq. (15.16), we see that the principles of impedances are

$$Z_{xy} = Z_0 \quad \text{and} \quad Z_{yx} = -Z_0 \tag{15.35}$$

and the auxiliary impedances are

$$Z_{xx} = Z_{yy} = 0,$$

and only if the medium is nonuniform in a horizontal direction, the latter are not zeros. Thus, the impedance tensor is written as

$$Z = \begin{pmatrix} 0 & Z_{xy} \\ -Z_{xy} & 0 \end{pmatrix} \tag{15.36}$$

From Eq. (15.28), it follows that

$$I_1 = Z_0^2, \quad I_2 = 0, \quad \text{and} \quad I_3 = 2Z_0 \tag{15.37}$$

that is, the invariant I_2 characterizes the effect of lateral inhomogeneities while the invariants I_1 and I_2 are related to the impedance Z_0:

$$Z_0 = I_1^{1/2} = \frac{I_3}{2}$$

It can readily be seen that the polar diagram for $|Z_{x'y'}(\alpha)|$ is a circle with the radius $|Z_0|$, which usually changes as a function of frequency except in the range of S-zone. As was demonstrated previously, the polarization ellipses for the electric and magnetic field components are mutually perpendicular.

15.2.2. Two-Dimensional Model

Suppose that resistivity of the medium is invariant in one direction, for example, in the x direction. As noted earlier, such models are termed two-dimensional (2D) ones. Studying the impedance tensor for such cases is of a great practical importance since the real resistivity structure can often be approximately considered as two-dimensional. The axis along which the resistivity does not vary is called the *axis of uniformity*, or the *strike direction*. For example, if there is a horst or trough-like structure, its strike is the axis of uniformity of the model. Assume that the linearly polarized normal field is arbitrary oriented with respect to the strike (the x-axis) of the structure, and it can be divided into two parts:

$$E_x^{(0)}, B_y^{(0)} \quad \text{(A)} \quad \text{and} \quad E_y^{(0)}, B_x^{(0)} \quad \text{(B)}$$

The first part of the field corresponds to E polarization, in which the electric field has only the component E_x, and magnetic field has components B_y and B_z with $B_x = 0$. Then, from Eqs (15.2) and (15.3), we have

$$a_{1y} = a_{1z} = b_{1x} = 0 \tag{15.38}$$

The second part of the field (B) corresponds to H-polarization, which is represented at the Earth's surface solely by components E_y and B_x. Applying Eqs (15.4) and (15.5), we obtain

$$a_{2x} = b_{2x} = a_{2z} = 0 \tag{15.39}$$

Substituting Eqs (15.38) and (15.39) into Eq. (15.17), we have

$$Z_{xx} = Z_{yy} = 0 \tag{15.40}$$

Thus, in the case of a two-dimensional model, the auxiliary impedances are zero in the structure's strike and dip directions. This conclusion is not valid for any other direction with respect to the strike axis. If the y-axis is directed along the uniformity axis (strike), the field components E_y, B_x, and B_z correspond to E-polarization, and from Eqs (15.4) and (15.5), we have

$$a_{2x} = b_{2x} = a_{2z} = 0 \tag{15.41}$$

The field components E_x and B_y correspond to H-polarization, and from Eqs (15.2) and (15.3), we have

$$a_{1y} = a_{1z} = b_{1x} = b_{1z} = 0 \tag{15.42}$$

Again, we see

$$Z_{xx} = Z_{yy} = 0$$

Moreover, in accordance with Eq. (15.17), the off-diagonal impedances are

$$Z_{xy} = \frac{a_{1x}}{b_{1y}} \quad \text{and} \quad Z_{yx} = \frac{a_{2y}}{b_{2x}},$$

which correspond to E- and H-polarizations, respectively. From Eq. (15.40), it follows that the invariants of the impedance tensor are

$$I_1 = -Z_{xy}Z_{yx}, \quad I_2 = 0, \quad \text{and} \quad I_3 = Z_{xy} - Z_{yx} \tag{15.43}$$

Thus, in the case of a two-dimensional structure, just as for the case of a horizontally layered medium, the invariant $I_2 = 0$. For this reason, the ratios of invariants I_2/I_1 and I_2/I_3 can be useful in distinguishing between two-dimensional (2D) and three-dimensional (3D) models. These ratios are defined as L and P:

$$L = \frac{I_2}{I_1} = \frac{Z_{xx} + Z_{yy}}{Z_{xx}Z_{yy} - Z_{yx}Z_{xy}}, \quad P = \frac{I_2}{I_3} = \frac{Z_{xx} + Z_{yy}}{Z_{xy} - Z_{yx}}$$

The parameters L and P are zero in 1D and 2D structures when the normal field is a quasi-plane wave. If the normal field does not behave as such wave, and it changes relatively rapidly along the strike direction, the diagonal impedances can be nonzero. For this reason, these parameters may be useful in separating electromagnetic fields caused by sources close to the observation sites from those, which behave as quasi-plane wave.

15.2.3. Polar Diagrams of Impedance for Two-Dimensional Models

Assume that x- and y-axes coincide with the strike and dip of the structure, respectively, so that

$$Z_{xx} = Z_{yy} = 0$$

From Eq. (15.27), we have

$$Z_{x'x'}(\alpha) = -\left(Z_{xy} + Z_{yx}\right)\sin\alpha\cos\alpha \tag{15.44}$$

$$Z_{x'y'}(\alpha) = Z_{xy}\cos^2\alpha - Z_{yx}\sin^2\alpha \tag{15.45}$$

Thus,

$$\left|Z_{x'x'}(\alpha)\right| = \left|Z_{xy} + Z_{yx}\right| \cdot \frac{\sin 2\alpha}{2} \tag{15.46}$$

and

$$\left|Z_{x'y'}(\alpha)\right| = \left[\left|Z_{xy}\right|^2\cos^4\alpha + \left|Z_{yx}\right|^2\sin^4\alpha - 2\left|Z_{xy}\right| \cdot \left|Z_{yx}\right|\sin^2\alpha\cos^2\alpha \cdot \cos\left(\arg Z_{xy} - \arg Z_{yx}\right)\right]^{1/2} \tag{15.47}$$

Equation (15.46) represents a curve which is symmetric with respect to the x- and y-axes (the strike and dip of the structure) and it has four leaves with a common origin (Figure 15.1(c)). In the principal directions $(\alpha = 0, \pi/2, \pi$ and $3\pi/2)$, leaves meet each other at the origin. Thus, polar diagram for diagonal impedances can be used for classifying the dimensionality of the resistivity structure, as well as for determining the principal directions in case of 2D model. Equation (15.47) describes the curve with shape resembling a four-leaf clover which is symmetric with respect to x- and y-axes. For the polar plot representing $\left|Z_{x'y'}(\alpha)\right|$, these leaves are expressed more or less clearly depending on the relationship between the amplitudes and phases of the electromagnetic fields, which correspond to E- and H-polarizations (Figure 15.1(c)). Although the impedance polar diagrams for $\left|Z_{x'x'}(\alpha)\right|$ and $\left|Z_{x'y'}(\alpha)\right|$ permit us to identify the axial directions for the models, determination of strike and dip is possible only if additional information is available. For instance, it can be done by comparing magnetotelluric responses measured at different sites. In contrast to the impedance polar diagrams for two-dimensional models, the orientation of ellipses of polarization for the electric and magnetic fields can be in general arbitrary with respect to strike and dip and to each other.

Note that the results obtained for two-dimensional model can also be applied for models possessing axial symmetry with respect to the z-axis. If the x-axis along which the electric field is measured coincides with a radial direction, then because of symmetry, Eqs (15.38) and (15.39) are valid, and therefore, the auxiliary impedances along the radial direction are zero:

$$Z_{xx} = Z_{yy} = 0$$

15.2.4. Relationship between Impedance Tensor and Apparent Impedances Z^a_{xy} and Z^a_{yx}

As follows from Eq. (15.16), we have for the apparent impedance

$$Z^a_{xy} = \frac{E_x}{B_y} = Z_{xy} + Z_{xx}\frac{B_x}{B_y}$$

and

$$Z^a_{yx} = \frac{E_y}{B_x} = Z_{yx} + Z_{yy}\frac{B_y}{B_x}$$

or

$$Z^a_{xy} = Z_{xy} + Z_{xx}h_{xy} \quad \text{and} \quad Z^a_{yx} = Z_{yx} + h^{-1}_{xy}Z_{yy}, \tag{15.48}$$

where h_{xy} is the complex parameter characterizing the ratio of amplitudes and phase difference of components of the magnetic field along the x- and y-axes. As was demonstrated above, the diagonal and off-diagonal impedances do not depend on the field polarization. Thus, a change in the apparent impedance is caused by a change in elliptical polarization of the field, being the main reason for scattering of apparent resistivity curves obtained from the apparent impedance. However, there are situations when the apparent impedance values coincide with off-diagonal values. First, if a medium is horizontally layered (1D),

$$Z^a_{xy} = Z_{xy} = Z_0 \quad \text{and} \quad Z^a_{yx} = Z_{yx} = -Z_0, \tag{15.49}$$

and Z_0 is the impedance of a plane wave (or quasi-plane wave) at the Earth's surface. Next, suppose that a medium is two-dimensional one and the x-axis coincides with the strike. Then

$$Z^a_{xy} = Z_{xy}, \quad Z^a_{yx} = Z_{yx},$$

that is in this case, the ratio of the orthogonal components of the electric and magnetic fields is equal to the antidiagonal impedances. It is obvious that the same results can be obtained with a model which has an axial symmetry and x is in a radial direction.

15.3. The Wiese–Parkinson Vector (Tipper)

From Eq. (15.18), we have the following relationship between the horizontal and vertical components of the magnetic field:

$$B_z = X_{zx}B_x + X_{zy}B_y,$$

where X_{zx} and X_{zy} are the complex-valued functions, associated with perpendicular directions x and y, respectively, and they form so-called Wiese–Parkinson vector, or tipper. It is clear that in general in order to determine its components, at least two different polarizations of the normal field have to be involved. In such case, we obtain two equations with two unknowns:

$$B_z^{(1)} = X_{zx}B_x^{(1)} + X_{zy}B_y^{(1)} \quad \text{and} \quad B_z^{(2)} = X_{zx}B_x^{(2)} + X_{zy}B_y^{(2)}, \qquad (15.50)$$

Its solution is

$$X_{zx} = \frac{B_z^{(1)}B_y^{(2)} - B_z^{(2)}B_y^{(1)}}{B_x^{(1)}B_y^{(2)} - B_x^{(2)}B_y^{(1)}} \quad \text{and} \quad X_{zy} = \frac{B_x^{(1)}B_z^{(2)} - B_x^{(2)}B_z^{(1)}}{B_x^{(1)}B_y^{(2)} - B_x^{(2)}B_y^{(1)}} \qquad (15.51)$$

If the magnetic field is linearly polarized, for example, along the x' axis, we obtain

$$B_z = X_{zx'}B_{x'} \qquad (15.52)$$

For linear polarization of the magnetic field along the axis y', we have

$$B_z = X_{zy'}B_{y'} \qquad (15.53)$$

In practice, the vertical component of the magnetic field is the sum of a vertical component B_{z1}, corresponding to a quasi-plane wave primary field (this term is very small except for polar regions), the vertical component B_{z2} caused by external sources located relatively close to the observation point, and the third term B_{z3} which arises as the result of the Earth's inhomogeneous structure:

$$B_z = B_{z1} + B_{z2} + B_{z3}$$

If the vertical component B_{z3} is dominant, it can be used for the interpretation of magnetotelluric data. With this purpose in mind, let us study how the magnetic parameters change as the sensor orientation changes. Suppose the coordinate system x, y, and z is rotated about the z-axis through the angle α, as shown in Figure 15.2. Then considering

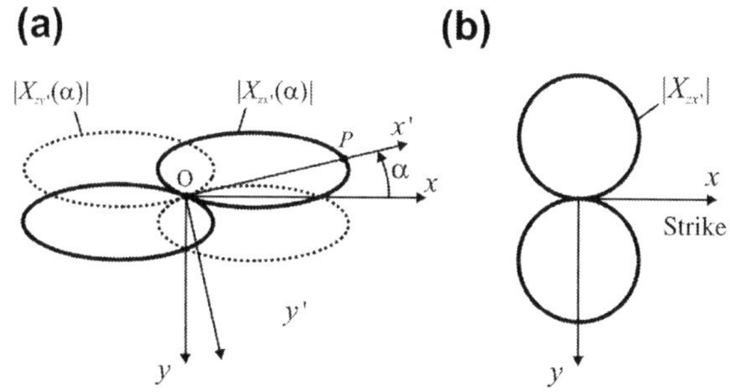

Figure 15.2 (a) Tipper polar diagram for 3D structure. (b) Tipper polar diagram for 2D model.

Eq. (15.18) and taking into account the change in the components B_x and B_y, we have

$$X_{zx'}(\alpha) = X_{zx}\cos\alpha - X_{zy}\sin\alpha$$

and

$$X_{zy'}(\alpha) = X_{zx}\sin\alpha + X_{zy}\cos\alpha \qquad (15.54)$$

This demonstrates that the relationship of the tipper parameters for various orientations of the sensor axes are in principle the same as those for the diagonal and off-diagonal impedances but their behavior is much simpler. It is obvious that

$$X_{zx'}(\alpha) = -X_{zx'}(\alpha \pm \pi), \quad \text{and} \quad X_{zx'}(\alpha) = \pm X_{zy'}\left(\alpha \pm \frac{\pi}{2}\right) \qquad (15.55)$$

Following the approach used for an impedance polar diagram, we will now draw a tipper polar diagram. Plotting the segment OP with length equal to $|X_{zx'}(\alpha)|$ along the x'-axis and changing the angle α from 0 to 2π, we obtain the amplitude polar diagram for $X_{zx'}$ characterized by the following equation:

$$|X_{zx'}| = \Big[|X_{zx}|^2\cos^2\alpha + |X_{zy}|^2\sin^2\alpha - 2|X_{zx}|$$

$$\cdot |X_{zy}|\sin\alpha\cos\alpha\cdot\cos(\arg X_{zx} - \arg X_{zy})\Big]^{1/2} \qquad (15.56)$$

From Eq. (15.55), it is clear that the magnetic polar diagram for $|X_{zy'}|$ can be obtained from the diagram for $|X_{zx'}|$ by rotation through $\pm\pi/2$. In the general case of a three-dimensional medium, the polar magnetic diagram $|X_{zx'}|$ has a character of a symmetrical two leaved curve with mutually perpendicular axes. If the x-axis coincides with the axis of uniformity for a two-dimensional medium, we have

$$b_{1z} = b_{1y} = b_{2x} = 0 \tag{15.57}$$

and therefore,

$$X_{zx} = 0 \quad \text{and} \quad |X_{zx'}(\alpha)| = |X_{zx}||\sin \alpha| \tag{15.58}$$

Curves for the function $|X_{zx}(\alpha)|$, defined by Eq. (15.58), are shown in Figure 15.2(b). They have two equal leaves intersecting at the origin. The leaves are elongated in the direction perpendicular to the axis of uniformity and, therefore, provide information about the structure. If the medium is characterized by axial symmetry, the polar diagrams have a similar form and the leaves are elongated in the radial direction.

15.4. BEHAVIOR OF THE PLANE WAVE IN A NONHORIZONTAL LAYERED MEDIUM

15.4.1. Introduction

As was pointed out, earlier transition from the scalar impedance to the tensor impedance allowed one to remove in the most cases an influence of the direction of the arrival of the normal field in the presence of inhomogeneities beneath the Earth's surface. Certainly, it was very important step in developing of the magnetotellurics. However, in general, every element of this tensor is subjected to an influence of the lateral changes of resistivity, and correspondingly, it became clear that the quantitative interpretation in such conditions is extremely difficult task and is often hardly possible. When this formidable problem was realized, that happened at the end of 1950s, geophysicists started to work in two directions related with each other. One of them is a solution of the forward problems, which allowed one to understand a behavior of the electric and magnetic fields, as well as impedance, caused by a plane wave in the presence of inhomogeneities. At the beginning, several boundary value problems were solved analytically which allowed one to obtain an expression for the field in the explicit form. Then with the development of computers, numerical methods were used to solve more complicated 2D and 3D problems. Of course, the information on specific features of the fields obtained from simulation for some particular resistivity models was not sufficient to solve an inverse problem but certainly facilitated this procedure. In parallel, there was development of algorithms of the inverse problem solution, which, of course, was to a great extent based on the solution of the forward problem. In this light, it may be proper to note the following. Long experience of application of different types of electromagnetic sounding shows that these methods give valuable information mainly in those cases when at each observation site, the medium can be treated as a horizontally layered one. The magnetotellurics is not an exception and in general, the lateral changes of resistivity became, as it was correctly foresaw by Cagniard, its formidable obstacle for performing a quantitative interpretation of an observed data. At the same time, it is proper to point out that there are cases of inhomogeneous medium when it is possible to obtain some very useful from a practical point of view information about the resistivity. It mainly occurs when it is known beforehand a source of the useful signal and a level of a geological noise is relatively small. For instance, sometimes in geothermal studies, the main goal is to locate a relatively resistive intrusion of a volcanic origin surrounded by a sufficiently uniform and conducting medium. The other example is mineral exploration geophysics when the magnetotellurics is applied to detect sufficiently wide zones of relatively low resistivity hosted in basement. In both cases, interpretation is carried out within a rather narrow class of models. Of course, some other model classes exist which also allow a quantitative interpretation. Inasmuch as a solution of inverse problem is always based on a comparison of the observed and simulated data, an understanding of a field behavior in laterally inhomogeneous medium is very useful. At the beginning, this information was vital for interpretation and it is still important when the modern algorithms are used for data inversion. Bearing this fact in mind, we analyze here the behavior of the electric and magnetic fields as well as impedance and apparent resistivity responses for different resistivity models.

15.4.2. Galvanic and Vortex Parts of the Field due to an Inhomogeneity

As was pointed out in the previous chapters, the sources of the electromagnetic field are charges and currents and, correspondingly, this field represents a sum of the galvanic and vortex parts. In general, they are related to each other and their contribution to the total field essentially depends on resistivity distribution and frequency. At the same time, from the physical consideration, it is almost clear that at the low-frequency range, the galvanic and vortex parts of the field are practically independent. This means that behavior of the galvanic part of the field is mainly governed by Coulomb, Ampere, and Ohm's laws, while the interaction of currents and electromagnetic induction play the leading role in the vortex part of the field. In this light, let us notice the following. Except for some cases, we will consider an inhomogeneity as a source of a geological noise located relatively close to the Earth's surface, while the useful signal is supposed to be caused by currents and charges, situated at the greater depth. For this reason, the low-frequency part of the spectrum, when the skin depth in the medium surrounding an inhomogeneity is greater than its dimensions is of a special interest. This means that the plane wave is almost uniform within a volume, occupied by an inhomogeneity where the normal electric and magnetic fields are practically constant. After these comments, let us briefly review both parts of the field.

15.4.3. The Main Features of Galvanic Part of the Field

As we already know, at each instant, the time-varying normal electric field $\mathbf{E}_0$, intersecting the inhomogeneity surface, causes the same effect as if it would be time-constant field. First of all, the electric charges appear at the inhomogeneity surface and their density is equal to (Chapter 1)

$$\sigma(q) = 2\varepsilon_0 K_{12} E_n^{av}(q), \tag{15.59}$$

where

$$K_{12} = \frac{\rho_2 - \rho_1}{\rho_2 + \rho_1}$$

and $E_n^{av}(q)$ is the normal component of the field, caused by all charges except that located in the vicinity of the point q. Inasmuch as the density of charges is weakly related to the resistivity of both media, the secondary field depends only slightly on the ratio of ρ_2/ρ_1. The latter is an essential feature of the galvanic part of the field. Let us distinguish two cases:

$$1.\ \rho_2 > \rho_1 \quad \text{and} \quad 2.\ \rho_2 < \rho_1,$$

where ρ_2 and ρ_1 are the resistivity of an inhomogeneity and host medium, respectively. Suppose that at some instant of time, the normal electric field E_x^0 is directed toward a body as is shown in Figure 15.3(a),(b). Then distribution of charges and the total field is shown in both cases.

If an inhomogeneity is more resistive, then in accordance with Eq. (15.59) at the back and front sides of the body, the positive and negative

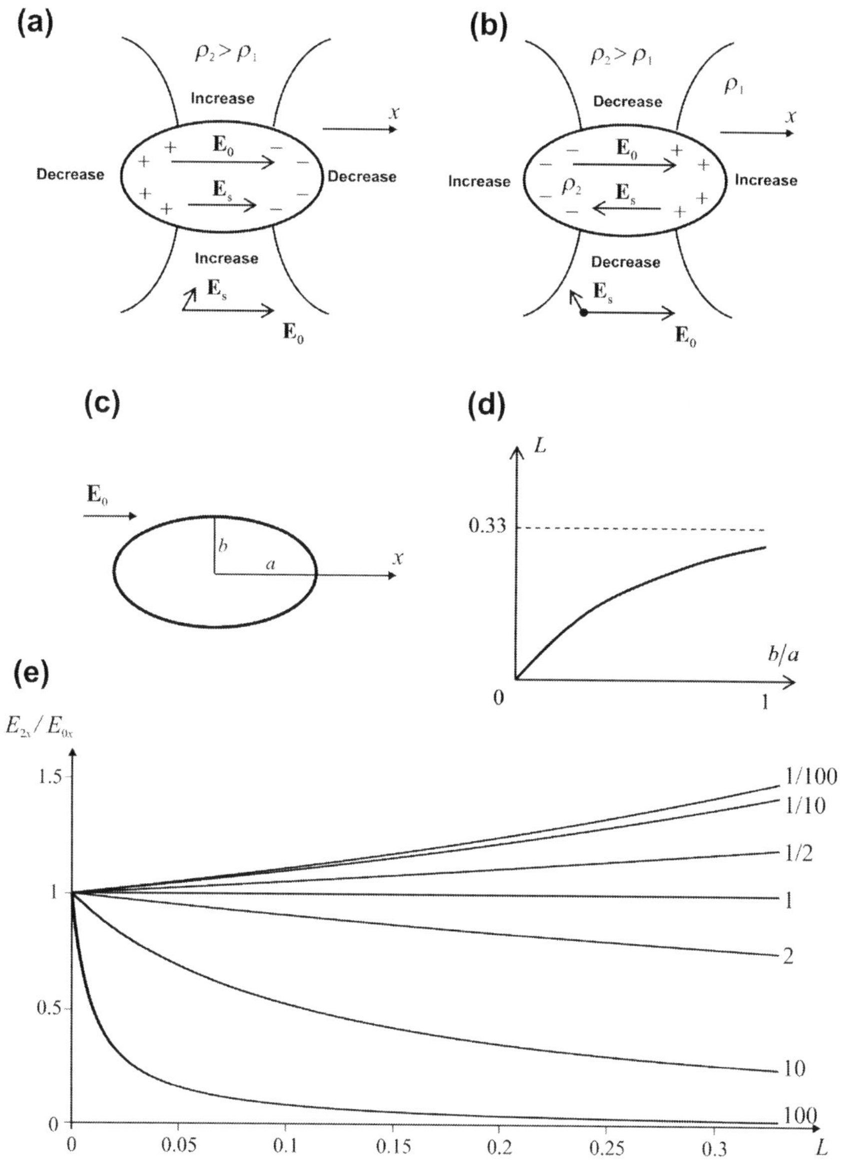

Figure 15.3 (a) Distribution of charges when $\rho_2 > \rho_1$. (b) Distribution of charges when $\rho_2 < \rho_1$. (c) Elongated ellipsoid in a uniform electric field. (d) Behavior of the function L. (e) Behavior of the function E_{2x}/E_{0x} (curve index is ρ_1/ρ_2).

charges arise, respectively. As follows from Coulomb's law, the total electric field E_x inside the body increases but outside, in front, and behind the body, it becomes smaller while at equatorial plane and around, it increases. In the case of more conductive body, we observe opposite pattern: inside inhomogeneity and outside at the equatorial plane, the field becomes smaller but in front and behind, this component of the total field increases. Of course, such behavior of the field is a result of a distribution of the surface charges which is characterized by the magnitude and sign. Until now, we gave only qualitative description of the field distribution. Now it is also instructive to consider how the field inside the body depends on resistivity of both media and its geometrical parameters. With this purpose in mind, assume that an inhomogeneity has a shape of an elongated spheroid (Figure 15.3(c)) with semi axes a and b, $a > b$. Then, if the normal field E_{0x} is uniform within the body and directed along the x-axis, the total field inside is defined by a very simple expression:

$$E_{2x} = \frac{E_{0x}}{1 + \left(\frac{\gamma_2}{\gamma_1} - 1\right)L} \tag{15.60}$$

Here, γ_1 and γ_2 are the conductivities, and

$$L = \frac{1 - e^2}{2e^3}\left(\ln\frac{1+e}{1-e} - 2e\right), \tag{15.61}$$

where

$$e = \left(1 - \frac{b^2}{a^2}\right)^{1/2} \tag{15.62}$$

By definition, the function L depends only on geometrical factors and its behavior is shown in Figure 15.3(d). As is seen, a change of a shape of the spheroid from the circular cylinder to a sphere causes a relatively small change of the function L:

$$0 < L < 0.33, \tag{15.63}$$

and this explains why in the case of a relatively resistive inhomogeneity, there is a rather small increase of the field inside. In fact, if $\gamma_2/\gamma_1 \ll 1$, in place of Eq. (15.60), we have

$$E_{2x} \approx \frac{E_{0x}}{1-L} \tag{15.64}$$

Thus, the field inside a body is practically independent on conductivity and its maximum increase is only around 50% of the normal field when the spheroid is transformed into a sphere. Such behavior is another feature of the galvanic part of the field, and because of it, the current density inside the body becomes much smaller than that in a uniform medium with resistivity ρ_1:

$$\gamma_2 E_{2x} \ll \gamma_1 E_{0x}$$

Because of the behavior of the electric field inside and outside, the current tends to flow around a highly resistive inhomogeneity that of course seems obvious. In this light, it may be proper to point out that such behavior of the current density is hardly reasonable calling as the "flow around" effect because it is directly defined by the electric field.

Next consider a change of the electric field when an ellipsoid is transformed into a cylinder. As follows from Figure 15.3(d) with a decrease of the ratio b/a, the function L rapidly decreases and the field inside tends to be the normal one. In other words, the density of charges becomes very small, and correspondingly, the electric field inside as well as outside approaches the normal field:

$$E_{1x} \to E_{0x}, \quad E_{2x} \to E_{0x} \tag{15.65}$$

Such behavior of the galvanic part of the field is called E-polarization, and it can be easily explained in the following way. With an extension of the spheroid along the x-axis, the normal component of the field E_{0x} at the spheroid surface becomes smaller, and therefore, the charge density also decreases. As concerns the ends of the body, it is clear that due to the large distance from them, their influence becomes negligible. From Eq. (15.90) and Figure 15.3(d), it follows that transition to E-polarization inside a body is practically independent on a resistivity of the inhomogeneity and a surrounding medium if $\rho_2 > \rho_1$, and it takes place even when an extension is relatively small: $a > 5b$. The same behavior takes place outside if observations are performed somewhere near equatorial plane.

Now we study a case of a more conductive spheroid, when $\gamma_2 > \gamma_1$. As was mentioned earlier, at the back and front sides, the negative and positive charges arise, respectively, and therefore, the secondary field inside the body is directed opposite to the normal one:

$$E_{2x} < E_{0x} \tag{15.66}$$

The behavior of the function

$$\frac{E_{2x}}{E_{0x}} = \frac{1}{1 + \left(\frac{\gamma_2}{\gamma_1} - 1\right)L} \tag{15.67}$$

is shown in Figure 15.3(e). With an increase of the ratio a/b, the function L decreases and the electric field E_{2x} approaches the normal one. In other words, the influence of surface charges becomes negligible and again the E-polarization is observed. However, as the conductivity contrast γ_2/γ_1 increases, such behavior takes place only for greater values of the ratio a/b. Making use of Eq. (15.61) and expanding the right-hand side in the series, the function L can be approximately represented in the form

$$L \approx \frac{b^2}{a^2} \ln\frac{a}{b} \quad \text{if } a/b \gg 1$$

Correspondingly, when

$$\frac{\gamma_2}{\gamma_1} \frac{b^2}{a^2} \ln\frac{a}{b} \ll 1$$

the field inside approaches the normal field:

$$E_{2x} \rightarrow E_{0x}$$

The same behavior is observed outside the body except for places near its edges. It happens because with an extension of the body, the surface charges, creating the secondary field, become smaller. Of course, when the surrounding medium is insulating, the secondary field is equal to and opposite the primary field inside a body independently of its linear dimensions; that is, the phenomenon of the electrostatic induction takes place. This means that in such limiting case, the transition to the E-polarization is impossible. Next consider the current density inside the spheroid:

$$j_{2x} \approx \frac{\gamma_2 E_{0x}}{1 + \left(\frac{\gamma_2}{\gamma_1} - 1\right)L} \tag{15.68}$$

Here, it is instructive to distinguish two limiting cases, namely,

$$1.\ \frac{\gamma_2}{\gamma_1}L \gg 1 \quad \text{and} \quad 2.\ \frac{\gamma_2}{\gamma_1}L \ll 1$$

In the first case, the influence of charges is great and the electric field inside very small and currents density turns out to be proportional to the conductivity of the surrounding medium:

$$j_{2x} \approx \frac{\gamma_1 E_{0x}}{L}, \tag{15.69}$$

With an elongation of the body, it becomes much stronger than the current density corresponding to the normal field. In the second case, the current density becomes proportional to the conductivity of a body that corresponds to E-polarization. In conclusion, note the following: (1) We considered the field in the presence of the spheroid. From the physical point of view, it is obvious that the similar dependence on conductivity and linear dimensions takes place for various shape of an inhomogeneity. (2) If the skin depth inside and outside the body is larger than its cross-section, it is natural to expect that an influence of the inductive effect is small. Correspondingly, in the second case, the electric field, unlike the magnetic field, is not practically affected by the body. (3) As was pointed out in the case of E-polarization for the more conductive body, there is a strong increase of the current density inside since the influence of surface charges is negligible. Correspondingly, in order to explain this behavior, it is hardly proper to introduce such concept as channeling effect. In fact, we simply multiply an almost normal field inside a body by a large value of conductivity (Ohm's law). (4) It is obvious that if the normal electric field is directed along the minor axis of the ellipsoid, a change of the major axis does not make a strong influence on the galvanic part of the field ($a \gg b$). In particular, when this axis becomes so large that its change does not make any influence on the field, we arrive at so-called H-polarization.

15.4.4. Main Features of the Inductive Part of the Field

The study of the electromagnetic field generated by the induced currents was the subject in several previous chapters and, for this reason, we only briefly describe its main features. First of all, if a conducting body is surrounded by an insulator, where the source of the primary field is located, the field magnitude inside tends to increase toward the surface. This behavior occurs regardless of the frequency, but with its increase, such tendency becomes more noticeable, and the skin effect is clearly observed. It means that inside the body, both the electric and magnetic fields tend to disappear. Of course, it is also true for the current density which is mainly concentrated near the surface at sufficiently high frequencies. It may be interesting to notice that with its increase, the current density also increases near the surface, but the volume occupied by currents becomes smaller in such a way that the total current tends to a finite value which is independent on the frequency. Some of these features remain valid even when the conductor is surrounded by a medium with much higher resistivity. Now we consider several examples of the field behavior where either the galvanic or vortex or both parts of the field play the dominant role.

15.5. EXAMPLES OF THE FIELD BEHAVIOR

15.5.1. Vertical Contact

Suppose that the normal field is a plane wave propagating downward along the z-axis and there is a vertical contact between media with different resistivities. We choose the Cartesian system of coordinates with the axes x and y directed perpendicular and parallel to the contact, respectively, while the z-axis is perpendicular to the earth surface (Figure 15.4(a)).

Case 1: H-polarization

At the beginning, consider the case when the magnetic field is directed along the contact. As follows from the geometry of the field, the total current passing through any cross-section, parallel to the plane Y0Z and shown in Figure 15.4(a), remains the same. Then from Ampere's law, we conclude that the magnetic field at the Earth's surface is not subjected to the

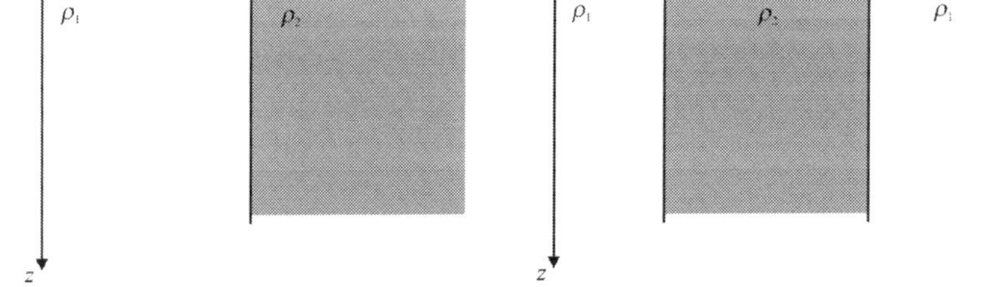

Figure 15.4 (a) Vertical contact model, $\rho_2/\rho_1 = 100$. (b) Vertical dyke model.

influence of the contact, and there are only two components of the electromagnetic field:

$$E_x, B_y^0 \tag{15.90}$$

Correspondingly, we will pay attention to the electric field which behavior coincides with that of the impedance. It is clear that under an action of the normal electric field, the charges arise at the contact as well as at the Earth's surface and inside a medium, they give rise to both components of the electric field:

$$E_x \quad \text{and} \quad E_z,$$

while the vertical component vanishes at the Earth's surface. It is clear that in the presence of the contact, the secondary field differs from the plane wave. Because of an attenuation of the field with depth and an interaction of charges at both surfaces, the density of charges depends on a point and, in particular, it decreases with a depth, and its rate of change varies with a frequency. It is natural to expect that with an increase of the distance from the Earth's surface, the charge density almost exponentially decreases and periodically changes a sign. From geometry of the field and a model of a medium follows that current lines are almost horizontal and straight far away from the contact but in approaching, they change shape and remain in the plane perpendicular to the y-axis.

Vector lines of the electric field behave almost in the same manner, except the fact that they are discontinuous at the points where charges at the contact are present. Bearing in mind that these charges are main sources of the secondary field, let us consider the behavior of this field at the different distances from the contact. When the distance between the contact and observation point is smaller than the skin depth, the charges create practically the same field as if it were time independent. For instance, suppose that the point is located in more conductive medium and at some instant the normal field is directed toward to more a resistive medium. As we know in such case, the density of charges at the contact and near the Earth's surface is positive and, therefore, the normal and secondary fields have opposite directions and the total field becomes smaller than the normal one. Correspondingly, in approaching the contact, the impedance as well as the apparent resistivity decreases. From the theory of the constant field, it follows that an influence of charges induced on the surface of the contact is equivalent to that of the point charge. Then it is a simple matter to show that in the case of a uniform primary field near the contact, the total field is equal to

$$E_x(\rho_1) = (1 - K_{12})E_x^0(\rho_1) = \frac{2\rho_1}{\rho_1 + \rho_2}E_x^0(\rho_1)$$

or

$$Z_{xy}(\rho_1) = \frac{2\rho_1}{\rho_1 + \rho_2} Z_{xy}^0(\rho_1) \quad \text{if } x = 0 \tag{15.91}$$

Here,

$$K_{12} = \frac{\rho_2 - \rho_1}{\rho_2 + \rho_1}$$

For sufficiently strong contrast of resistivity, $\rho_1 \ll \rho_2$, the electric field becomes very small:

$$E_x \approx \frac{2\rho_1}{\rho_2}E_x^0 \quad \text{if } x = 0$$

With an increase of the distance from the contact, its influence decreases, and it happens due to smaller effect of charges and attenuation of electromagnetic field with the distance from the sources (charges). As a result, the impedance gradually increases and approaches to its limit, $Z_{xy}^0(\rho_1)$. Next suppose that an observation point is located at the opposite side of the contact where medium is more resistive. Then, as we know, the normal component of the electric field is discontinuous function at the points of the contact and it is equal to

$$E_x^{(2)}(\rho_2) = (1 + K_{12})E_x^0(\rho_1) = \frac{2\rho_2}{\rho_1 + \rho_2} E_x^0(\rho_1)$$

or

$$Z_{xy}^{(2)}(\rho_2) = \frac{2\rho_2}{\rho_1 + \rho_2} Z_{xy}^{(0)}(\rho_1) = \frac{2\rho_2}{\rho_1 + \rho_2}\left(\frac{\rho_1}{\rho_2}\right)^{1/2} Z_{xy}^{(0)}(\rho_2) \quad \text{if } x = 0 \tag{15.92}$$

Of course, the latter follows from the continuity of the normal component of the current density. Then with an increase of the distance from the contact, the electric field approaches to that of a uniform half space with resistivity ρ_2. We described a behavior near the contact properly assuming that charges play the dominant role. Now let us discuss the influence of the frequency; that is, interaction of currents. As is well known, the main feature of the skin effect with an increase of frequency is a tendency of currents to concentrate near sources which create field. The sources of the primary field are located above the Earth's surface and first of all, induced currents tend to concentrate near the Earth's surface. Besides the sources of the secondary field are surface charges which appear on the vertical contact, and therefore, there is also a tendency of currents with an increase of frequency be mainly located near contact. The skin effect displays at its both parts of a medium, but mainly in more conductive medium. As is seen from numerical analysis, an influence of the contact practically vanishes at the distance $x \approx \lambda$. Note that sometimes such behavior of the field is called a *lateral effect* that is hardly proper, since it simply illustrates the skin effect which displays regardless of an orientation of a surface. Apparent resistivity profiles ρ_a for H-polarization are given in Figure 15.5(a). The graphs of the component E_x, normalized by the field at the surface of a uniform half space with resistivity ρ_1, are shown in Figure 15.6(a).

Case 2: E-polarization

Now assume that the electric field of the plane wave is directed along the y-axis; that is, it is parallel to the contact. In such case, the galvanic part of the field is absent since the normal electric field E_y^0 does not intersect the contact surface. Correspondingly, both the normal and secondary fields have the inductive origin and their behavior follows from the law of electromagnetic induction as well as Ampere's and Ohm's laws. In fact, the change of the magnetic field with time gives rise to the inductive electric field that causes the currents and the latter generates the magnetic field. As a result, the

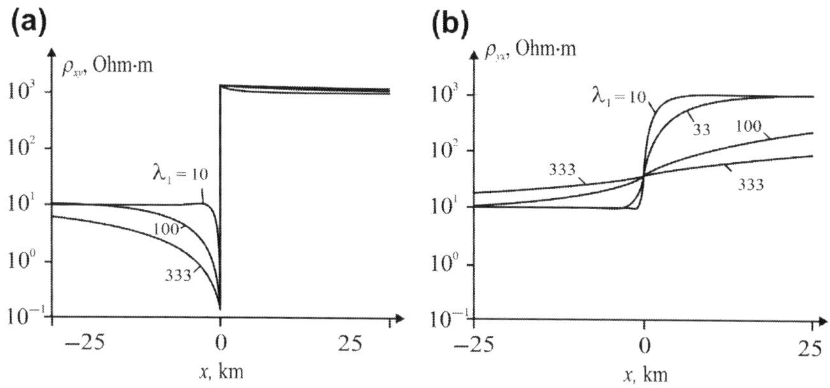

Figure 15.5 Apparent resistivity profiles for the vertical contact: (a) H-polarization; (b) E-polarization.

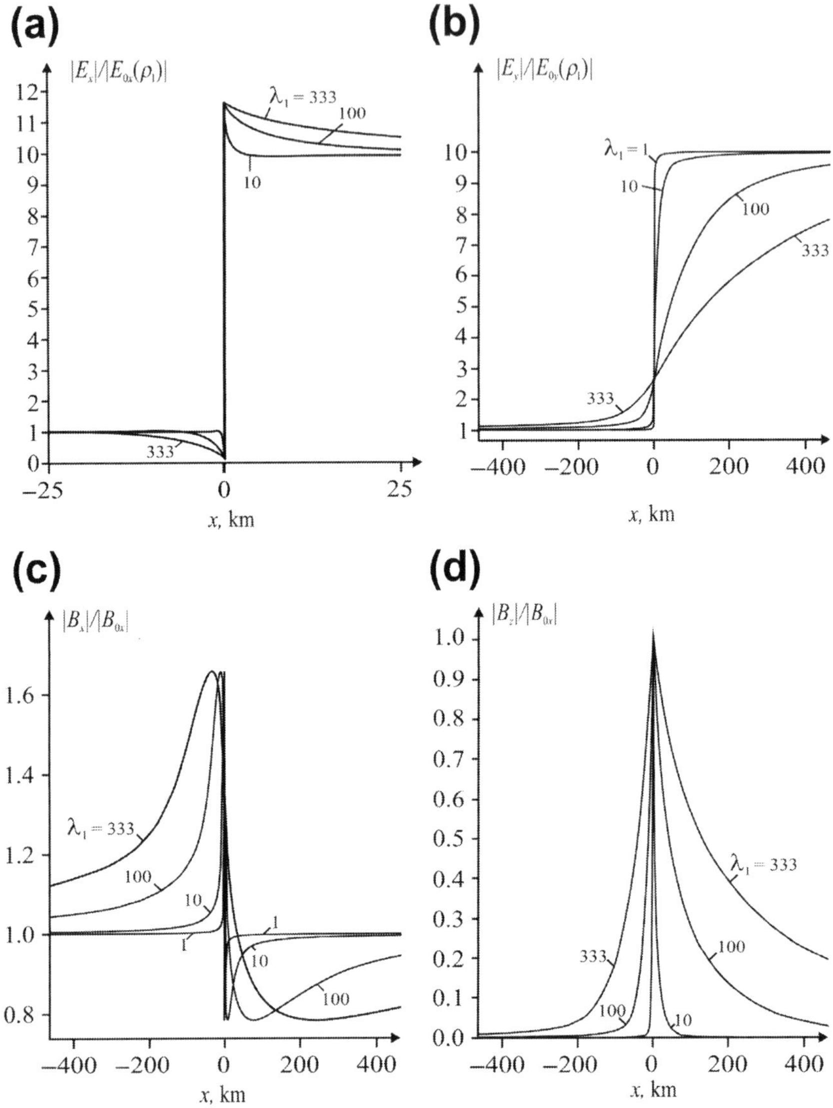

Figure 15.6 Normalized (a) E_x, (b) E_y, (c) B_x, and (d) B_z magnitudes across the vertical contact. Curve index is λ_1.

electromagnetic field is established. By analogy with the previous case, we consider the ratio between the total field and that in a uniform half space with the resistivity ρ_1, $(\rho_2 > \rho_1)$. It is simple manner to imagine that the normal magnetic field is caused by a system of uniformly distributed currents within thin conducting layers, and therefore, they create only the horizontal component, B_{0y}. In the presence of the contact, the density of currents is a function of the coordinate x, and for this reason, the vertical component of the field appears. Correspondingly, the total field is not plane wave and it is described by the following components:

$$E_y, B_x, B_z \tag{15.93}$$

To illustrate an influence of the contact, the functions $\left|E_y/E_{0y}(x, \rho_1)\right|$, $|B_x(x)/B_{0x}|$, and $|B_z(x)/B_{0x}|$ are shown in Figure 15.6(b)–(d). Let us note that the normal magnetic field at the Earth's surface is independent on a medium resistivity. As in previous case, the index of curves is the wavelength λ_1. First, consider the behavior of the electric field above the more conducting part of a medium surrounded by the Earth's surface and a contact. As was mentioned earlier due to interaction of currents, the electric field increases toward both boundaries and it happens regardless how small the frequency is. With its increase, such behavior displays more vividly near the Earth's surface and the vertical contact (skin effect). Thus, the electric field approaches to the normal one far away of the contact but rapidly increases at its vicinity. At the boundary, the tangential component of the electric field is a continuous function and gradually increases approaching to the normal field over more resistive medium. This part of the curve shows that an interaction of currents results in a decrease of the electric field near contact and, with increase of the frequency, this transition zone becomes narrower. From Ohm's law follows the same behavior of the current density. Of course, the horizontal component of the magnetic field is defined by all currents in a medium but the main contribution comes from the currents located somewhere beneath the point of observation. For this reason, it is natural to expect an increase of this component over a more conductive medium near the contact and its decrease from the other side (Figure 15.6(c)).

Behavior of the vertical component normalized by the horizontal component is shown in Figure 15.6(d). As we may expect, the field almost coincides with that in a uniform medium when $x > \lambda$.

15.5.2. The Vertical Dyke

Now consider a slightly more complicated model of the dyke, as shown in Figure 15.4(b). If the dyke width exceeds its wavelength, the field almost coincides with that in the presence of one vertical contact. A different picture takes place when $\lambda > l$. As before, in the case of H-polarization, the galvanic part of the field prevails and with a decrease of the frequency, the

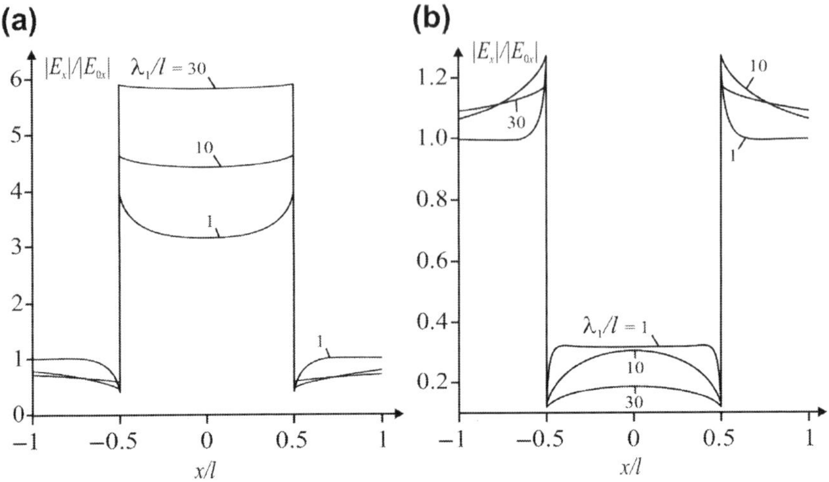

Figure 15.7 Normalized E_x magnitude across the (a) resistive dyke, $\rho_2/\rho_1 = 10$, and (b) conductive dyke, $\rho_2/\rho_1 = 1/10$. Curve index is λ_1/l.

positive and negative charges practically define the electric field. Because of them, the field inside essentially increases if $\rho_2 > \rho_1$, but outside rather quickly decays with the distance (Figure 15.7(a)). The curves of the electric field, when the dyke is more conductive, are shown in Figure 15.7(b). If the electric field is directed along the dyke (E-polarization, Figure 15.8) the field is defined by interaction of currents only and it has the same features as in the case of a single contact. In particular, when the dyke is more resistive, the electric field displays a small maximum, which increases with an increase of frequency (Figure 15.8(a)), while the horizontal magnetic component has a minimum over the dyke (Figure 15.8(c)). Behavior of the vertical component of the magnetic field is illustrated by Figures 15.8(e) and (f).

15.5.3. The Horst on the Basement Surface

This two-dimensional model is shown in Figure 15.9(a) and is characterized by the following parameters: $\rho_1/\rho_2 = 100$, $h_2/h_1 = 5$, $l/h_1 = 20$. Here, ρ_1 and ρ_2 are the resistivities of the two horizontally layered medium, h_1 is the distance from the top of the horst to the Earth's surface, h_2 is the thickness of the upper layer, and l is the structure width. The x-axis is directed across the structure, but the y-axis coincides with the strike.

Figure 15.10 illustrates the behavior of field components, normalized by the normal field, across the structure as functions of ratio.

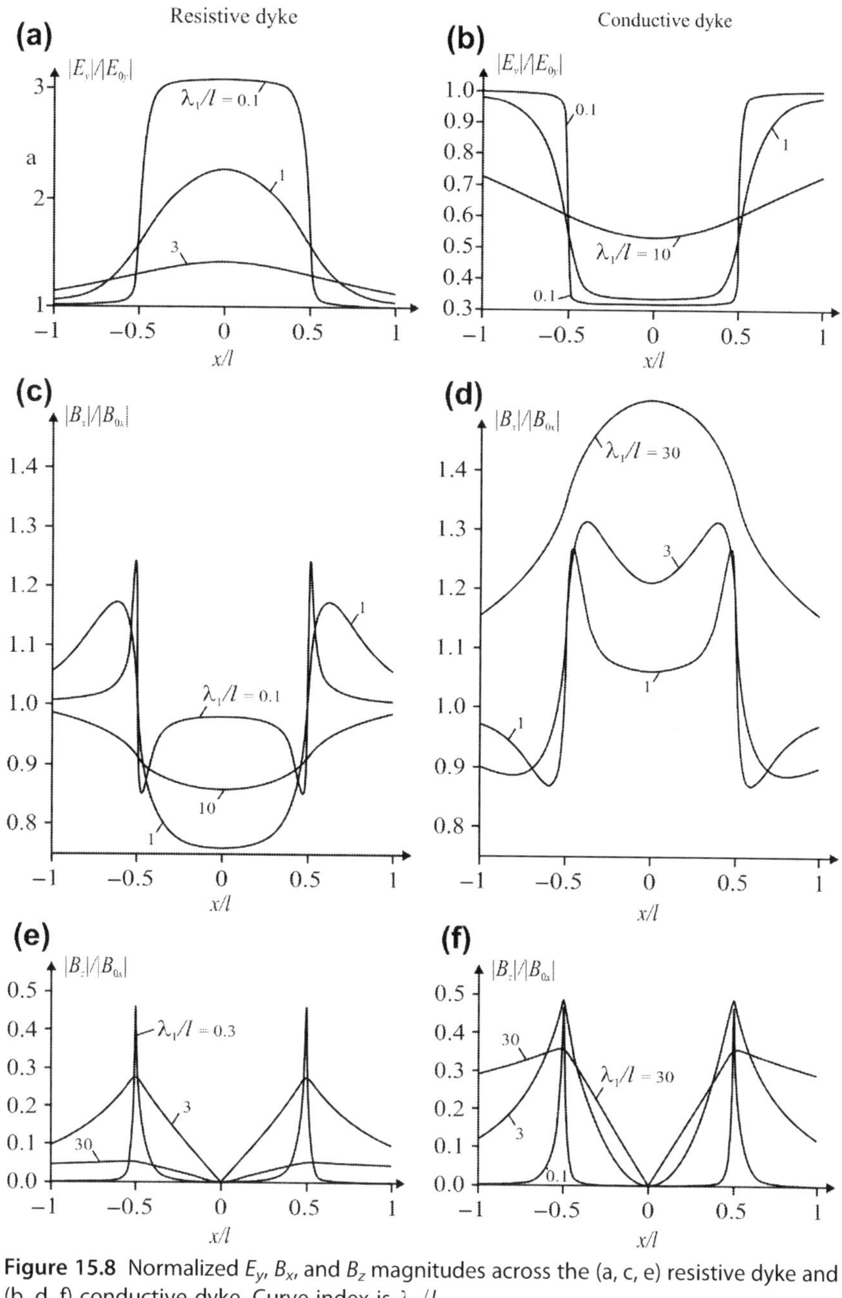

Figure 15.8 Normalized E_y, B_x, and B_z magnitudes across the (a, c, e) resistive dyke and (b, d, f) conductive dyke. Curve index is λ_1/l.

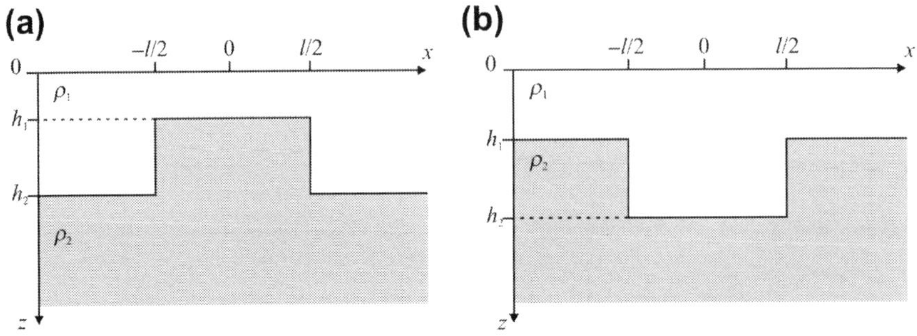

Figure 15.9 (a) 2D horst model and (b) 2D trough model.

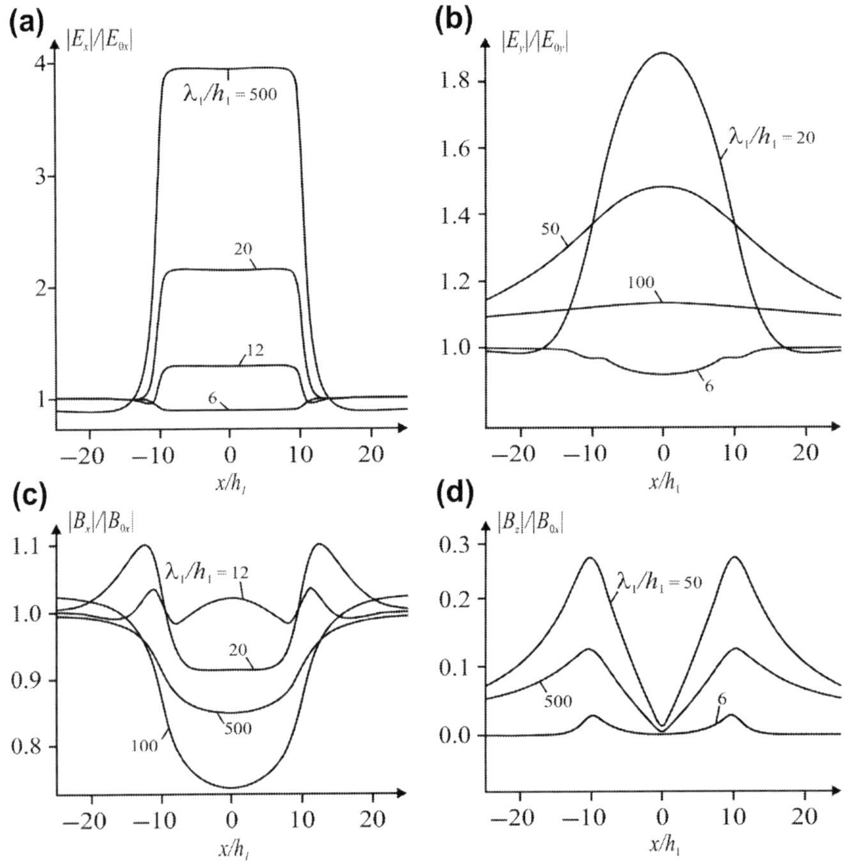

Figure 15.10 Normalized E_x, E_y, B_x, and B_z magnitude across the horst. Curve index is λ_1/h_1.

15.5.4. *H*-polarization

Unlike *E*-polarization, the electric field intersects boundaries of the horst and surface electric charges arise. For instance, if the normal field is directed along the x-axis, the positive and negative charges appear at the left and right sides, respectively, while at the top, there are charges of both signs. Correspondingly, at the low-frequency part of the spectrum, the galvanic part of the field is dominant, and relative anomaly is large and it is independent of frequency. Note that a positive anomaly of the field above horst directly follows from Coulomb's law. With a decrease of period, anomaly becomes smaller due to skin effect (Figure 15.10(a)).

15.5.5. *E*-polarization

First consider the behavior of the electric field. Because of the skin effect, curves with relatively small values of λ_1/h_1 display only slight negative anomalies above the uplift structure. At the same time, with larger values of λ_1/h_1, the intense positive anomaly is observed with electric field reaching a double value of the normal field.

With further increase of relative anomaly becomes wider and decreases. In fact, as was pointed out earlier, we can expect that at sufficiently low frequencies, the currents arise mainly due to the normal field:

$$j_y \approx (\gamma_2 - \gamma_1)E_{0y}$$

Therefore, the secondary magnetic fields is proportional to

$$A(\gamma_2 - \gamma_1)E_{0y}$$

where A is the geometric factor. Consequently, the secondary field turns out to be proportional to

$$\omega A(\gamma_2 - \gamma_1)E_{0y} \quad \text{or} \quad \frac{E_{sy}}{E_{0y}} \approx A(\gamma_2 - \gamma_1)\omega$$

and with a decrease of frequency, an influence of horst vanishes.

Next consider the behavior of the horizontal component of the magnetic field B_x, as shown in Figure 15.10(c). It is natural that in general the function $B_x(x)$ resembles the field caused by a linear current and located at some depth. Inasmuch as above the horst amount of currents is smaller than in the case of the two-layer medium, the relative anomaly is negative. It is interesting to notice that the presence of the resistive inhomogeneity may cause much stronger anomalies of the electric field than the magnetic one. As is seen from Figure 15.10(c), a relatively small maximum is observed above each horst edge, and it takes place at sufficiently high frequencies when due to the skin effect, there is some concentration of currents in the vicinity of this inhomogeneity.

Earlier we demonstrated that with a decrease of frequency, the relative anomaly of tends to either zero or very small values if the normal field meets the S-zone approximation ($S \gg 1$) and such behavior is clearly seen in Figure 15.10(b). In fact, as was pointed out, the secondary magnetic field is proportional to

$$A(\gamma_2 - \gamma_1)E_{0y} \approx A(\gamma_2 - \gamma_2)ZB_{0y}$$

Here, Z is the impedance of two-layer medium. Therefore, at the low-frequency spectrum,

$$B_x \approx B_{0x}[1 + A(\gamma_2 - \gamma_1)Z]$$

As in the case of the electric field, there is a range of periods where anomalies of the magnetic field are maximal. Behavior of the vertical component of the magnetic field is shown in Figure 15.10(d). Of course, we again see that the field is caused by a system of linear currents, and it is natural that above the middle of the horst, this component is equal to zero.

15.5.6. The Trough on the Basement Surface

This model is shown in Figure 15.9(b) and it is characterized by the following parameters: $\rho_1/\rho_2 = 100$, $h_2/h_1 = 3$, $l/h_2 = 10/3$. Inasmuch as behavior of the fields can be easily explained in the same manner as in the previous case, we restrict ourselves by giving only examples shown below (Figure 15.11).

15.5.7. Three-Dimensional Model

Next consider the behavior of the field in the presence of three-dimensional structure as shown in Figure 15.12. It is important to notice that in case of 3D structure, the simulation impedance tensor can be found only if one has measured (or simulated) for two independent directions of the primary field.

Figure 15.13 shows the behavior of the normalized field components along the profile parallel to x-axis and crossing the model through the center. As we can expect with a decrease of frequency, the secondary electric field does not vanish and becomes practically independent on the period. The latter happens due to the dominant role of the surface electric charges (galvanic effect), and of course, this behavior resembles the case of

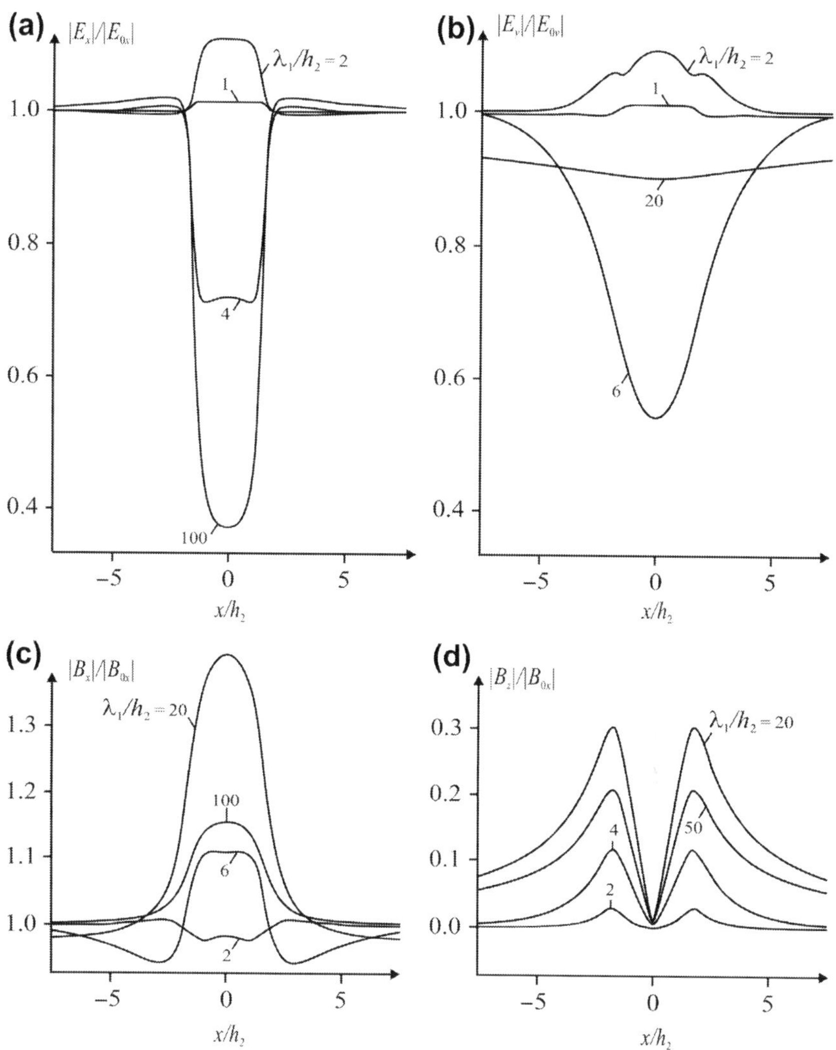

Figure 15.11 Normalized E_x, E_y, B_x, and B_z (a, b, c, d) magnitudes across the trough. Curve index is λ_1/h_1.

H-polarization. It is also obvious that due to the skin effect at sufficiently high frequencies, the secondary electric and magnetic fields become very small. It may be proper to note that an anomaly of the magnetic field becomes small at the low-frequency spectrum where the field of charges plays the main role. Thus, the maximum of the secondary magnetic field may display within the intermediate range of frequencies.

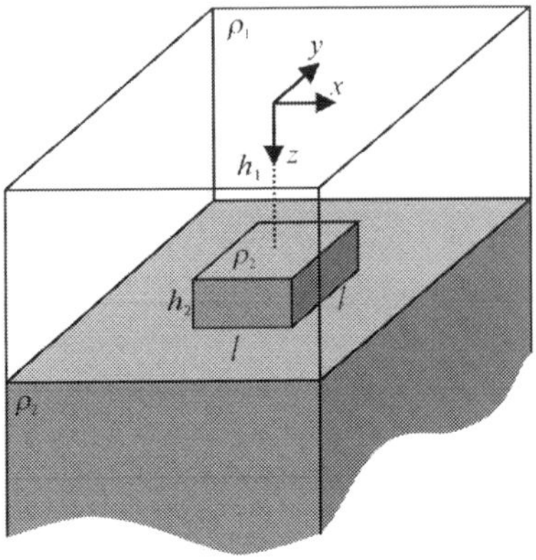

Figure 15.12 3D horst model, Characterized by following parameters: $\rho_2/\rho_1 = 100$, $h_2/h_1 = 5$, and $l/h_1 = 20$.

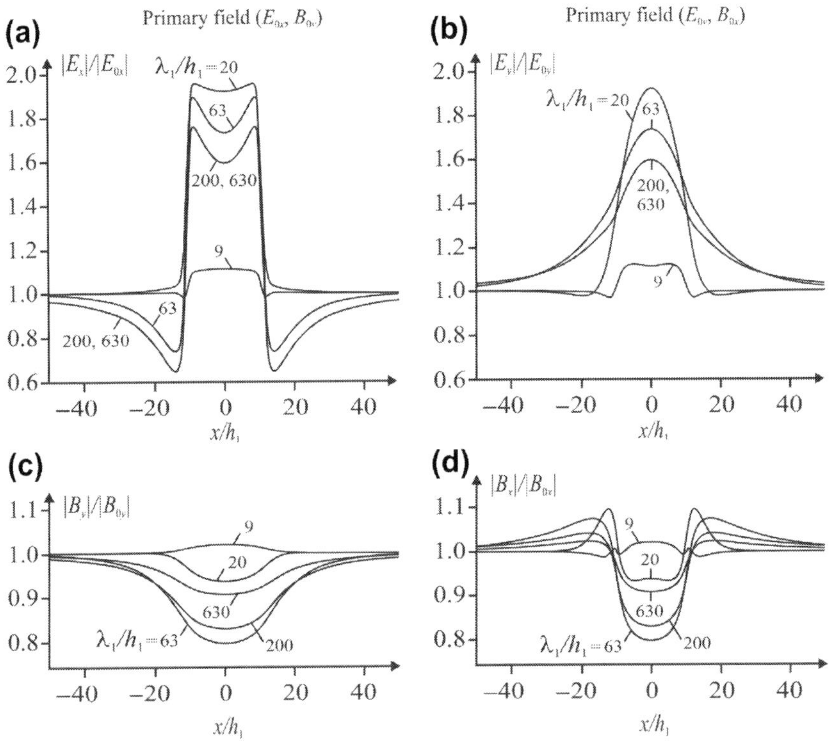

Figure 15.13 Normalized EM field magnitudes across the center of the model for two polarizations of the primary field. (a) E_x component and (c) B_y component calculated for source polarization (E_{0x}, B_{0y}); (b) E_y component and (d) B_x component calculated for source polarization (E_{0y}, B_{0x}).

It may be proper to mention that our discussion of the field behavior for different models of a medium is also applied to the case when there is a lateral change of resistivity, located relatively close to the Earth's surface. As is well known, this is one of the main sources of a geological noise and, correspondingly, it is useful to describe its influence on the electric and magnetic fields. First, we focus on 2D models at the low-frequency part of the spectrum. At the beginning, suppose that the electric field is directed along the strike (*E*-polarization) and represents the total field inside an inhomogeneity as a sum:

$$E_x = E_{x0} + E_{xs}$$

where E_{x0} and E_{xs} are the normal field in a horizontally layered medium and the secondary field, caused by the presence of an inhomogeneity. Taking into account the inductive origin of the field, it is natural to expect that with a decrease of frequency, the second term becomes negligible and the total field inside, as well as outside, practically coincides with the normal one:

$$E_x \to E_{x0} \tag{15.94}$$

Thus, at the low-frequency part of the spectrum, the electric field is not subjected to an influence of an inhomogeneity. As a result, the secondary current density within an inhomogeneity can be written as

$$j_x \approx (\gamma_2 - \gamma_1)E_{x0},$$

where γ_2 and γ_1 are the conductivities of an inhomogeneity and surrounding medium, respectively. In accordance with Ohm's law for the secondary magnetic field, we have

$$B_{ys} \approx F_1 E_{x0} \tag{15.95}$$

Here, F_1 is independent of the frequency. Therefore, unlike the electric field, the magnetic field is subjected to an influence of frequency regardless how it is small. In particular, for impedance, we have

$$Z_{zy} \approx \frac{E_{x0}}{B_{0y} + F_1 E_{0y}} = Z_0 \frac{1}{1 + F_1 Z_0} \tag{15.96}$$

It may be instructive to distinguish three cases. In the first case, when the basement is very conductive, the impedance Z_0 at the low-frequency spectrum is directly proportional to ω; that is, $Z_{xy} \to Z_0$ and influence of a surface inhomogeneity vanishes rather quickly. The same is observed when the basement resistivity relatively slightly differs from that of the upper medium. In fact at the low frequencies, the impedance Z_0 becomes proportional to the square root of ω and, therefore, a contribution of an inhomogeneity decreases with a decrease of frequency. Finally, when the range of frequencies corresponds to S-zone, we have

$$Z_{xy} \approx Z_0 \frac{1}{1 + \dfrac{F_1}{S}} \tag{15.97}$$

and it is independent of frequency. Here, S is the total conductance of sediments above a basement. Usually, it is relatively a large number and, for this reason, an influence of an inhomogeneity is small.

Next consider a more general case, including H-polarization, when the electric field intersects a surface of an inhomogeneity and a frequency is sufficiently low. Then, it is obvious that as before, an influence of the galvanic part of the secondary field prevails and it can be strong, especially when an inhomogeneity is close to the Earth's surface. In this approximation, the density of surface charges is directly proportional to the normal electric field,

$$\sigma = 2\varepsilon_0 K_{12} E_{n0},$$

and in accordance with Coulomb's law, for instance, x-component of the secondary field can be represented in the form,

$$E_{xs} = F_2 E_{0x},$$

where F_2 is independent of a frequency. Respectively, for the total field, we have

$$E_x \approx (1 + F_2) E_{0x} \tag{15.98}$$

Since the current density is proportional to the normal field for the y-component of the total magnetic field, we obtain

$$B_y \approx B_{0y} + F_3 E_{0x} = B_{0y}(1 + F_3 Z_0) \tag{15.99}$$

As follows from Biot–Savart's law, the function F_3 is also independent of a frequency. Earlier, we pointed out that at low frequencies, the term $F_3 Z_0$ is usually small, and it allows us to write

$$Z_{xy} \approx (1 + F_2) Z_0 \tag{15.100}$$

The latter shows that an influence of an inhomogeneity produces a shift of the apparent resistivity curves either up or down, as soon as the frequency is so low that permits to neglect by an inductive part of the field. Often this displacement of a curve of the apparent resistivity is almost the same within a wide range of frequencies where the field is measured, and correspondingly, such behavior is called the *static effect*. Note that in some cases applying, for example, a transient soundings, it is possible to remove this kind of distortion, and such procedure is called the *static shift correction*.

In conclusion, let us notice that we considered the presence of an inhomogeneity as source of a geological noise, causing a distortion of magnetotelluric response calculated for a horizontally layered medium, and our goal was study of its influence and determination of conditions when it is possible to neglect by their contribution. In principle, it is possible to use a different approach and perform a solution of the inverse problem in a nonhorizontally layered medium. However, in this case, an accuracy of determination of parameters of a geoelectric section becomes smaller. In other words, a range of a change of these parameters, corresponding to given accuracy of measurements, may essentially increase. As we already know even in the simplest case of a horizontally layered medium, an error of determination of some of its parameters is very large, but for 2D and 3D models, situation gets even much worse.

REFERENCES AND FURTHER READING

[1] M.N. Berdichevsky, V.I. Dimitriev, Basic Principles of Interpretation of Magnetotellurics Soundings Curves, in: Geoelectric and Geothermal Studies, Akademiai Kiado, Budapest, 1976.

[2] M.N. Berdichevsky, V.I. Dimitriev, Models and Methods of Magnetotellurics, Springer-Verlag, 2008.

[3] A.A. Kaufman, G.V. Keller, The Magnetotelluric Sounding Method, Elsevier, 1981.

[4] A.A. Kaufman, Basic Theory of Inductive Mining Prospecting, Nauka, Novosibirsk, 1974.

[5] A.A. Kaufman, Three methods of field excitation for low frequency exploration, Izv. Vyssh. Ucgebn.Zaved., Ser. Geol. Geofiz 5 (1961).

[6] A. Jones, A. Chave, The Magnetotelluric Method: Theory and Practice, Cambridge University Press, 2012.

APPENDIX ONE

Airborne Electromagnetic Prospecting Systems

Richard S. Smith[a], Andrey K. Volkovitsky[b]

[a]Laurentian University, Sudbury, Canada
[b]Trapeznikov Institute of control sciences of Russian academy of sciences, Moscow, Russia; GeoTechnologies, Moscow, Russia (see www.gtcomp.ru)

On February 13, 1948, Stanley Davidson, filed an application for a US patent. The invention was an airborne electromagnetic (AEM) prospecting system and was subsequently awarded US patent 2652530A, Canadian patent 519,679, and Australian patent 149,124. The system was developed for him by McPhar Engineering Limited of Toronto, Canada. Davidson and McPhar had previously developed a mobile ground electromagnetic (EM) system and the airborne system was a natural evolution. The patent filing marked the beginning of a new era in exploration technology development; it happened during a period of intense development of the postwar economy, with a corresponding high demand for metals. This demand was stimulating the exploration for new ore deposits. There were large terrains recognized to be poorly explored with modern technology and airborne methods were one way of doing this exploration quickly and cost effectively.

Previously, during World War II, many servicemen and civilians had been exposed to aircraft being flown for military purposes. New reliable airplanes along with special tools for technical surveillance were developed. At the same time, the availability of highly skilled personnel such as pilots and navigators, aviation engineers, and technicians facilitated the expansion of aviation. The beginning of airborne geophysical methods came out of successful experience in conducting aeromagnetic surveys that incorporated the use of magnetometers. These magnetometers were initially developed during the war to search for submarines. As well, during the war, large EM transmitters had been installed on some aircraft and the magnetic field from these transmitters was demonstrated to be capable of triggering the safe explosion of mines laid by enemy combatants. This EM minesweeping technology was used to make the harbors and shipping lanes safe for military and merchant traffic. In addition, experience prior to the war had identified ground EM

Methods in Geochemistry and Geophysics, Volume 45
ISSN 0076-6895,
http://dx.doi.org/10.1016/B978-0-444-53829-1.19001-X

systems as being effective in the search for conductive ore deposits. These ground EM systems were only capable of surveying about five-line-km in a day. The development of AEM prospecting systems therefore seemed opportune; combining the need for an exploration tool and the skill and interests of former military aviators. The ability to collect hundreds of line-kms of data in a single day, made the techniques very attractive.

A1.1. FREQUENCY-DOMAIN AEM SYSTEMS

Reviews by [1–4], and [5] summarize in some detail the development of AEM exploration methods in the early years. The first flight experiments were carried out by engineers from the McPhar Company in partnership with Stanmac (a small exploration company lead by Stan Davidson) using equipment installed on an Anson aircraft. The success of the experiment was such that in 1949, the International Nickel Company of Canada (Inco) ordered the development of a production system (Figure A1.1).

The initial Stanmac equipment utilized a transmitter that was a vertical coil wound around the fuselage, with a wire running along the top from back to front, round the nose, and then back along the bottom of the fuselage to the back. The fuselage frame was made of wood and the cover of cloth, so this material did not substantially distort the field from the transmitter. The receiver was a horizontal induction loop (vertical dipole) that was towed below and behind the aircraft in a bird (bottom right of Figure A1.1). Typically flights were undertaken using an aircraft altitude of 100–130 m, with the towed bird usually 60 m below and 140 m behind the

Figure A1.1 Inco system, 1950. *After Ref. [5].*

plane, at the altitude of 40–70 m above the ground surface. In theory, the vertical transmitter and horizontal receiver coils are null or minimum coupled to each other, so the only time a field would be measured was if there was a conductor in the ground. The minimum coupling created problems as small variations in the receiver position or aircraft orientation resulted in nonzero coupling and a false anomaly.

The 1950 Inco system consisted of two independent transmitter–receiver pairs, each operating at different frequencies. The first pair had the transmitter dipole pointing towards the receiver bird and the corresponding receiver in the bird was a parallel dipole. These dipoles were on a common axis and so this configuration is known as a coaxial pair. The second pair has the transmitter loop orthogonal to the first with the plane of the loop sloping down and pointing towards the bird. The corresponding receiver loop is in the same plane (coplanar coils). This coil pair is in an inclined coplanar configuration. Both these transmitter–receiver coil pairs were maximum coupled, so small variations in geometry did not give false anomalies as large as the null-coupled systems. However, geometric variations were still evident in the amplitude data. The frequencies of operations of the two coil pairs were 1800 and 2000 Hz [6] and the system also measures the phase in both systems, which was used for estimating the conductance [7]. A subsequent Inco system developed in the late 1960s had transmitter coils mounted externally on a Twin Otter aircraft, as the fuselage was metal and conducting. In this case the coplanar coil pair was a set of vertical coils (Figure A1.2). The coplanar pairs could transmit at two frequencies and the coaxial one worked at one frequency. Usually these frequencies were somewhere between 100 and 2500 Hz [8].

Also an additional third receiver loop was placed in the towed bird perpendicular to the two basic ones. This coil is null coupled to both transmitters, so the signal in this coil should be equal to zero. If it is not equal

Figure A1.2 Inco system on a twin otter aircraft from the 1960s.

to zero, then it is an indication of changes in the geometry of the coils from their expected values or the presence of a conductive body in the ground.

The rationale for having a coplanar and coaxial set of coil pairs is that the amplitude of the response increases and decreases when the receiver gets closer or further from the transmitter; however, the rate of increase or decrease is identical for both pairs, so that the effect of changes in system geometry can be removed. For example, the coaxial field from the transmitter should always be twice as big as the coplanar field. If the coplanar is multiplied by two and subtracted from the coaxial, the result is a special combination of measured parameters that will always be zero—unless there is an additional field coming from a conductor present in the subsurface. The system also measured the phase and calculated other special combinations of measured parameters called *indicator parameters.*

The system was successful at detecting conductors in the subsurface; the system being used extensively in Canada from 1950 to 1984; as well as being flown on surveys in Australia and South Africa. During this time, more than 1.5 million km of surveys were flown, leading to the discovery of large sulfide deposits: Thompson Manitoba (150 million ton 3% of Ni), Heath Steele (20,7 million ton 4.79% of Zn, 1.75% of Pb, 1% of Cu, 65 g/ton of Ag), and Casa Berardi (10.5 million ton 8.2 g/ton of Au).

Issues dealt with by the engineers/designers of AEM prospecting systems are discussed below.

1. Calibration of Equipment-Specific Drift.

Electronic equipment responds differently when the temperature of the circuits change and this will adversely impact the measurements. Hence specific reference signals can be passed through the circuits and changes in the electronics monitored. This calibration process is usually carried out by the operator before and after surveying each line.

2. Suppression of Influence of Unwanted Signal

The signals associated with changes in flux of the natural magnetic noise field of the earth and time-varying magnetic fields from power lines and other artificial sources are also measured at the receiver. These can overpower the signal from the transmitter and conductors in the ground. In order to reduce this unwanted signal two methods are traditionally used. The first is band-pass filtering—suppression of signals at all frequencies, except the frequency of interest; normally this was done using electronic filters. The second method is to increase the amplitude of the energizing field. Each method has its limitations: band-pass filters are difficult to implement in practice and the transmitter cannot use more power than the aircraft electrical system can supply.

3. Suppression of Influence of Sound and Vibration of the Induction Coil (Microphonic Noise)

When the bird and tow cable move through the atmosphere, turbulence induces vibrations and these travel mechanically to the induction coils. Movement of the induction coil can change the flux of the magnetic field through the coil and this will induce an unwanted voltage in the coil (Faraday's law). As these vibrations occur at audio frequencies and can often be heard, they are called microphonic noise. The way to suppress these vibrations is to mechanically isolate the coils from the tow cable and/or the bird housing (shell). This can be done with foam or elastic bands. One example of a suspension system to isolate vibrational noise is suggested by Kuzmin and Morrison in US patent 8,030,933 B2, assigned to Geotech Airborne Inc. Unfortunately, such materials have different behavior at different temperatures, so this can sometimes be problematic when moving from hot to cold climates and vice versa.

4. Suppression of the influence of uncontrolled changes in the geometry of the equipment (geometric noise).

In most active AEM exploration systems, the primary field at the receiver is much larger than the field from the subsurface body of interest. This is because the fields from the transmitter and the ground fall off as a function of distance from the source and the transmitter is much closer than the subsurface body. Changes in the distance between the transmitter and receiver and the orientation of either will result in large changes in the primary field. This influence is called geometric noise. There are three ways to suppress this type of noise:

a. Instead of measuring the amplitude, some systems are designed to measure the phase or delay of a signal at a specific frequency. The delay is measured with respect to the signal in the transmitter (the primary field). In air, the phase delay does not depend on the position of the receiver with respect to the transmitter; however, the signal will be delayed if there are eddy currents induced in the ground. The phase delay is expressed as an angle where 360° represents a whole period of the sinusoidal waveform. Another way electrical engineers like to describe phase delays is by decomposing a signal into a component that is in phase (no delay, so identical to the primary) and a component that is shifted by 90° or one quarter of a cycle. The latter component is called the quadrature component or phase. Signals with no phase delay have no quadrature component. Systems that measure only the quadrature component are called quadrature systems. These systems are not impacted by system geometry changes, which are by definition in phase.

b. Controlling the geometric alignment of the system. The most obvious way to do this is to make the system rigid. Despite the seeming simplicity of this method, in practice this is difficult to achieve. On the one hand, to ensure the receiver can measure small signals from the ground and the signals from the transmitter, it should be distanced as far as possible from the transmitter, and on the other hand, it is straightforward to make a structure bigger and rigid at the same time. These types of systems are called rigid-boom or towed-tube systems.

Another option is to measure the distance between the transmitter and the receiver and the orientation of both sensors in order to be able to predict the primary field and account for it. Because the field changes very rapidly as a function of distance from the transmitter (inverse distance cubed) to predict the primary accurately requires very precise measurements of distance and orientation [9,10] and hence no commercial AEM systems use this as their primary way of removing geometric noise. An experimental system [11] had the transmitter towed below one helicopter and a receiver towed below a second helicopter, with the two being separated by approximately 400 m. The system was able to detect a highly conductive body in the Sudbury area and was projected to be able to detect this body to depths of 200 m below surface.

c. Finding some special combination of the measured parameters, which is not dependent on the distance between the transmitter and receiver and/or the geometric alignment of the transmitter–receiver. In the Inco system discussed above the special combination of measured parameters are the indicator parameters. The Inco system relies on the receiver coils being on the axis of the coaxial transmitter dipole. This is difficult to achieve on a windy day. Hence the Inco system would only fly on relatively calm days. More recently, a patent application has proposed another set of indicator parameters for a system with a three-component transmitter and a three-component receiver (Smith, WO 2012/129,654 A1). There is a possibility that this proposed system will not be dependent on calm flight conditions.

5. Effect of the conductivity of the aircraft.

As a consequence of the time-varying field from the transmitter, eddy currents arise not only in conductive zones in the subsurface, but also in metal parts of the aircraft. Measuring and removing the impact of these eddy currents is known as the compensation process. This is normally done by measuring the impact on a flight when the system is about 1000 m above the ground and well away from the impact of the ground. This compensation signal, or a signal proportional to it is removed while the aircraft is closer to the ground [10]. Systems towed by helicopters usually assume that the impact of aircraft eddy currents is small. The much older systems that used wood and cloth fuselages had minimal aircraft eddy currents, but this approach is no longer possible as all modern airplanes utilize a metal fuselage.

6. Effect of surface objects in the probed environment (geological noise). As the fields from conductive zones falls off rapidly as a function of distance from a conductor, the response from conductors (e.g., overburden, clays, bogs) close to the surface can dominate those from the deeper conductors of interest at depth. Generally, the near-surface conductors have larger responses at high frequencies and responses that are smaller at low frequencies, while the response from more conductive bodies at depth are relatively large at low frequencies. Hence, if the targets of interest at depth are more conductive than the near-surface features, then their response can be made relatively large by making the frequency low. For frequency-domain systems, frequencies close to 1000 Hz or less are considered low and designed to see the deeper conductors. For example, the Inco system used frequencies of 285 and 1215 Hz.

A1.1.1. Quadrature Systems

The bird towed behind the aircraft moves relative to the transmitter, generating geometric noise. The most common way of dealing with this issue it to design and build a system that measures only the quadrature component (see 4(a) above). Between 1950 and 1960, a number of quadrature systems were utilized for mineral exploration. In 1953 the two-frequency quadrature system was developed in Finland by Puaranen, Kama, and Ronka and successfully deployed. Later, in 1954 a similar system was used in Canada by the Hunting Survey Company. The system mounted on a Canso aircraft had a horizontal triangular transmitter loop (vertical dipole) attached to the ends of the aircraft's wings and tail. The receiver was installed in a towed bird on a connecting cable—120 m behind and 60 m below the aircraft. Quadrature measurements were made at frequencies of 400 and 2300 Hz [3]. The system measured the phase angle and the anomaly associated with the Mattagami Lake Mine discovery was fractions of a degree and barely above the noise level; the magnetic anomaly was however clearly anomalous. This system was used extensively in the period 1955–1959, as part of an intense airborne EM campaign where 800,000 line-km of airborne EM were flown in Canada using a wide variety of systems and 16 potential ore bodies were discovered [12].

Another system with both a fixed-wing and helicopter towed-bird configuration was popular in the 1970s: the two-frequency quadrature F-400 system developed by McPhar Geophysics Limited in 1968. Parameters measured by this system were phase values at two operating frequencies—340 and 1070 Hz [6]. This system was initially developed as a "button on system" to allow ease of installation near the survey area and thus reduce high mobilization costs. Because of this design factor both the fixed-wing and helicopter systems were utilized throughout North America, Australia, and Africa. In addition systems were sold to mining companies (South Africa) and government geological survey groups (Turkey and USSR). The transmitter of this system consisted of two symmetrically mounted iron-core transmitting coils connected in series to form a single transmitting dipole. The splitting of the coils introduced a symmetry that reduces any system noise due to airfoil or airframe flexure. The two operating frequencies were transmitted on a time-share basis. These transmitting coils were mounted under the wings of high-wing aircraft; within special landing skids for helicopters; and above the fuselage of low-wing aircraft. The transmitter coils were all oriented along the flight direction [5]. The receiving system is a vertical dipole receiver in a bird towed behind the aircraft [13]. The length of a connecting cable was traditionally about 122 m [6]. Subsequent developments of the system included adding additional transmitters to provide higher frequency data.

Quadrature systems have a common problem: the response of weakly conductive and highly conductive features can be below the noise level of the system depending upon the range of transmitter frequencies employed. Therefore quadrature systems with limited range of transmitter frequencies may possess low sensitivity to highly conductive bodies, which is frequently the target of interest in mineral exploration.

A towed-bird system developed by GeoTechnologies in Russia is the EM-4H (Figure A1.3) is a quadrature and an in phase system.

The main feature of this system is measuring and accounting for the geometrical configuration of the system. The quantities measured are the three components of the offset of the receiver from the transmitter (in the x,

Figure A1.3 EM4H (Geotechnologies, Russia).

y, and z directions) and the orientation of the receiver. The geometric measurements are achieved using two additional vertical loop transmitter coils operating at 670 and 830 Hz. The horizontal loop transmitter operates at four frequencies: 130, 520, 2080, and 8320 Hz with dipole moments of approximately 20,000, 10,000, 4000, and 2000 Am^2 respectively. The geometric measurements allow these dipole moments to be estimated more precisely and then the data are scaled so that in effect the dipole moments are the same. Then if the measurements at different frequencies at the receiver are subtracted then in phase component of the primary field can be removed and the in phase secondary from the ground remains [14]. Hence, both the quadrature and in phase components are measured and the EM-4H is no longer a quadrature only system. This enables the system to see conductive bodies. The system will not see an in phase response when the conductor in the ground is extremely conductive, as in this case the in phase response from the ground is the same at all frequencies and will also sum to zero, making the conductor invisible.

A1.1.2. Rigid Frequency-Domain Systems

The geometrically stable (rigid) frequency-domain systems are capable of detecting an in phase response from all bodies, including extremely conductive bodies. This allows the interpreter to distinguish between a quadrature response that might be associated with a weakly conductive body and a highly conductive body. Measuring the secondary in phase requires that the primary in phase can be calculated or removed. Rigid systems achieve this removal by placing the transmitter and receiver a constant distance apart. The field from the transmitter is nulled at the receiver by using a smaller bucking coil that has a field that is equal in magnitude, but opposite in direction. The bucking coil is closer to the receiver than the transmitter, so the dipole moment is much smaller. Hence the field from the bucking coil can be ignored in the subsurface. One of the first rigid systems was developed in 1955 by Nucom (a company associated with McPhar). The transmitting and receiving coils were a coaxial pair with the axial direction being the flight direction and the coil separation being about 6 m [5]. The entire construction was towed by a helicopter 30 m above the ground. The system worked at only one frequency—1000 Hz. After calibration at high altitude the device measured both the in phase and quadrature components of the response; however, the noise levels were rather high [6]. Nonetheless, other rigid systems followed. As a rule these early rigid systems were single frequency with different coil separations, coil orientations, and operating frequencies. The rigid structure was generally a rigid boom or a cylindrical tube; another semirigid system used the wing as the boom. The Mullard system, introduced by the Rio Tinto company in 1956, utilized two coplanar vertical coils on the ends of the wings, termed the wing-tip configuration [5]. The vertical coils will couple well to a vertical conductive body with a strike perpendicular to the strike direction. A modern system similar to this, but operating at two frequencies has been developed in Finland and used there and in the UK [15,16]. In 1998, the system had four vertical coplanar coils with the dipole axis pointing in the flight direction. The two transmitters are on the end of one wing, and the two receivers on the end of the other wing 21.4 m distant. One transmitter–receiver pair operates at 3125 Hz, the other at 14.4 kHz. Poikonen et al. [16] argue that these frequencies are equivalent to higher frequencies (8 and 38 kHz) on a helicopter system with a coil separation of 8 m. More recently, the system was upgraded to four coil pairs operating at 912, 3005, 11,962, and 24,510 Hz [17] and in 2009 the system was sold, so it is now owned and operated by Sander Geophysics (a company which first collected frequency-domain EM data in 1958).

Other systems employing a coaxial configuration were the fixed-wing Tridem system (Scintrex, Canada), and the helicopter towed-tube systems operated by Dighem and Aerodat [5]. The latter two systems added a second set of coplanar coils. Initially these were for helping to interpret the geometry of conductive bodies, but were later used for mapping horizontal layers—resistivity mapping [18]. The DIGHEM system is still offered by CGG. As the resistivity mapping became more useful, further coplanar coils were added at multiple frequencies to map layers at multiple depths [19]. The RESOLVE system (currently offered by CGG) has five coplanar coils for this purpose operating at 400, 1800, 8200 Hz, 40 kHz, and 140 kHz, and one additional coaxial coil operating at approximately 3300 Hz for vertical conductor mapping.

The rigid systems are very sensitive to mechanical rigidity when measuring the in phase response. A minute increase in the distance between the receiver and the transmitter (e.g., due to temperature change) will mean that the bucking does not work correctly, so the in phase response will be incorrect. This drift is sometimes monitored by taking calibration measurements before or after or even during a flight, but this is complicated by the fact that the calibration measurements are taken at higher altitudes away from the influence of the ground where the in phase response of the ground can be ignored. However, the temperature at altitude is often different from the temperature at survey altitude, sometimes making these measurements problematic. If the offset distance between the transmitter and receiver could be greater, then the effect of small changes in distance would be less; however, systems with greater offsets are much less rigid, nullifying potential gains.

In the 65-year history of AEMs, various rigid systems have been created. They differed in coil configuration, geometrical parameters, and operating frequencies. One of the most financially successful is the Texas Gulf Sulphur system (Figure A1.4), which is attributed with the discovery of the huge sulfide ore deposit at Kidd Creek, near Timmins, Ontario [5]. The system used coaxial coils displaced by 15 m, and operated at 400 Hz [20].

Figure A1.4 Rigid system of the texas gulf sulphur company. *After Ref. [5].*

A1.1.3. Depth of Exploration of Frequency-Domain Systems

AEM systems have a wide range of applications: geophysical mapping of quaternary deposits, exploration for kimberlites, bathymetry in fresh and salty water areas, sea ice thickness, and many engineering and hydrogeological problems. However, an important goal is still the search for conductive ore bodies in various geological conditions. In order to choose the most appropriate system for geophysical surveys, it is important to accurately estimate their sensitivity to detect the target of interest. Generally this requires modeling the expected response with the system and comparing this with the noise level.

Generally, the noise is determined experimentally, as it is difficult to predict theoretically. Noise levels depend on mechanical rigidity, coil motions in the earth's field, vibrations, temperature variability, electronic noise, etc. In addition there is natural noise associated with changes in the EM fields of the earth; in the bandwidth of AEM systems, this is primarily atmospheric noise associated with atmospheric discharges (lightning) that is difficult to identify and remove [21]. This noise is highly dependent on the geographical position, time of day, and other additional factors. Furthermore, even in the quiet conditions the noise can be 0.1–0.2 $nT/Hz^{1/2}$. This value defines the minimum possible level of measurement noise.

The amplitude of the signal is determined by the energizing field of the transmitter. In any AEM prospecting system this field is limited by the power and voltage that can be supplied by the aircraft power system. The field at the ground is proportional to the dipole moment of the transmitter loop $M = I \cdot \frac{\pi d^2}{4} \cdot N$. (Where I = current, d = diameter of a loop, and N = the number of turns of the loop). For the majority of the rigid-boom systems, which have coil diameters not more than 1 m, the current and turns are such that the dipole moment does not exceed 200–400 Am^2. The fixed-wing systems can have larger loops wound around the aircraft, so d is larger, also, the larger aircraft can carry the weight associated with more turns (N larger) and can supply a larger voltage, so the current I is also greater. The very large four-engine Dash-7 aircraft used by the MEGATEM system can supply a dipole moment of 2,000,000 Am^2.

The primary field to noise ratio can be calculated for AEM systems by calculating the primary field for the known dipole moment and assuming a typical atmospheric noise level. For a coplanar rigid system with distance between coils of 8 m, the relative sensitivity is about 1.5 ppm, and for the EM-4H equipment with a towed cable length of 75 m it is about 100 ppm.

Let us give examples of the depth sensitivity for two contemporary frequency-domain AEM systems: the separated—EM-4H system (Geo-Technologies, Russia) and the rigid—RESOLVE system (CGG, Canada).

The depth sensitivity is calculated by estimating the maximum distance of detection of an ideal conductive half-space. We can use the image dipole method to calculate the secondary field. For a rigid system with coplanar coils we equate a calculated value of a signal at height h over the ideal conductor with the minimum distinguishable signal at relative sensitivity of s (ppm) and a coil separation of p, i.e., $\frac{2M}{(2h)^3} = s\frac{M}{p^3}$, where the left hand side is the field from a coaxial image dipole $2h$ below the system and the right hand side is the smallest detectable field from a coplanar dipole. Solving for the largest value of h we get a height of detection $h = p\left(\frac{1}{4s}\right)^{1/3}$. For regular parameters of rigid equipment ($p = 8$ m, $s = 1.5 \cdot 10^{-6}$) we get $h = 440$ m. Carrying out similar calculations for the separated unit (vertical dipolar transmitter and vertical receiver), we get $h = \frac{p}{2}(s^{-2/3} - 1)^{1/2}$, where we have assumed the vertical offset of the receiver from the transmitter is small compared with the depth of the ideal conductor. For this case $s = 100$ ppm and $p = 75$ m, so h is 807 m. It is important to notice that the height of detection of the ideal conductor does not depend on the working frequency of the system.

For an assessment of the depth penetration of the fields into a half-space with a low conductivity or high resistivity the value equal to the skin depth for the lowest operating frequency is often used as an estimate of depth of exploration. However, this value does not always adequately convey the penetration because it does not consider the noise level of the equipment. Hence the depth of investigation for resistive ground calculated using half the skin depth could be greater than the depth of detection of the ideal conductor, which is illogical. In this case the lesser of the depth of detection of an ideal conductor and the skin depth should be used.

One of the essential advantages of rigid towed installations compared with separated systems (which have a larger flying height) is that rigid systems are able to resolve small objects at shallow depths. The best way to determine the depth of exploration of systems is to compare the response from the target and ensure that this is greater than the noise level [22].

In general, rigid-boom frequency-domain systems are recommended for surveys exploring for conducting objects down to 80 or 100 m depth when embedded in resistive terrain, or for resolving the near-surface resistivity structure of more conductive terrain (typically down to about 60 m). Generally, the fixed-wing systems can use a wider line spacing [22].

A1.1.4. Systems Without an Associated Transmitter

There are two types of airborne methods that do not require a transmitter to be mounted on or towed behind an aircraft, the AFMAG method and the VLF method.

The airborne AFMAG method (US patent 2,931,974), developed by McPhar/Nucom in the late 1950s [23], tows only a receiver and uses natural variations in the vertical component of the earth's field to investigate the subsurface [8]. The magnetic fields are generated from distant lightning strikes. If the earth is layered, then the secondary currents induced in the ground will only have horizontal components. If the *tipper* (ratio of the measured vertical component to the magnitude of the horizontal components at a distant base station) is calculated, then the amount of current induced in an anomalous (nonlayered) structure can be estimated. The current systems are large and the size of the vertical field can be proportional to the conductivity of the structure but will also depend on the conductivity of the surrounding structure. Hence estimating the ground conductivity at a specific location requires an understanding of the background conductivity. AFMAG systems tend to be good for recognizing large structures (like porphyries) and can be poor for identifying small structures like massive sulfides [24]. Due to limited recording and data processing technology in the 1950s and 1960s, the results from early AFMAG systems were complex and sometimes difficult to interpret. Airborne AFMAG was not used very extensively for direct target detection, however it had proven success in structural geology mapping, especially when flown in conjunction with magnetics and spectrometry. A more recent system introduced by Geotech called Z component Tipper EM (ZTEM) has coils for measuring the tipper and base station that are significantly larger than the older systems [25,26]. Fields are measured in the frequency range between 25 and 720 Hz. The system has been used successfully for mapping large structures and porphyries [27–30].

The vertical tipper field is normally small compared with the horizontal fields; hence a slight rotation of the vertical dipole receiver into the horizontal plane will measure a component of the horizontal field and be interpreted erroneously as a vertical field. One way of correcting for this is to measure the orientation of the tipper sensor. An alternate approach was introduced by Geotech and is called the AirMt system [31]. This system measures three components at the base station and three components at the aircraft. A tensor relating these two vector quantities is calculated and from this a complex scaler called the amplification parameter is derived. This parameter is independent of rotation of the survey sensor. The comparison presented by [31] shows promise, being comparable to the ZTEM results.

The VLF method [32] is conceptually similar to the AFMAG method, except the sources are not naturally occurring lightning strikes, but radio transmitters established for communication purposes (primarily intended for communication with submarines). The frequency of these transmitters are around 20,000 Hz, so the signals are strongly attenuated in conductive ground. VLF anomalies are also generated by strong lateral changes in resistivity associated with topographic features like hills and cliffs. The system also works well when there are many VLF transmitters in a variety of different directions from the survey area. The VLF technique is used extensively in Sweden [33], as the ground is relatively flat, resistive, and there are a number of VLF transmitters surrounding Sweden.

A1.2. AIRBORNE TRANSIENT EM SURVEYS

A1.2.1. Fixed-Wing Transient Systems

The innovation of using transient EM system for AEM prospecting was proposed in the late 1950s by Anthony (Tony) R. Barringer [4–6]. The principle behind this method was that an energizing field was created by a powerful transmitter not continuously, but in short impulses with a pause ("silence") between them. The time when the transmitter was on is known as the *on time* and the time when it is off is known as the *off time*. During the off time there is no primary field from the transmitter so that bird movements do not create geometric noise in the off time. Hence overcoming issue 4 discussed in the system design section above. Also, the receiver gain can be turned up to measure the small secondary field, as there is no primary field to saturate the electronics. It was observed that poor conductors would decay quickly and good conductors slowly. By recording the response in three time windows [4], it was possible to identify the poor conductors (overburden) with the fast decays and the good conductors with the slow decays. However, the very good conductors with very slow decays have very low-amplitude responses so the conductors of greatest interest cannot

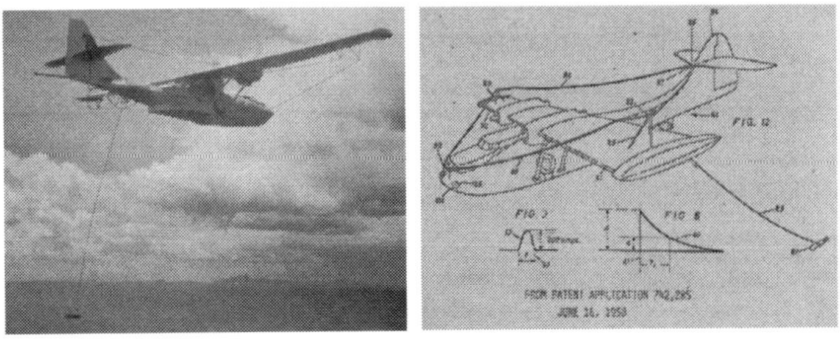

Figure A1.5 The INPUT time-domain system in fight (top left). *The diagram bottom right is from the original patent documents filed by Tony Barringer.*

be seen [34]. This is similar to the quadrature systems that also cannot see very conductive bodies. The transmitter configuration of the induced pulse transient (INPUT) system was very similar to the Hunting system, with a powerful transmitter and a large loop wound around a Canso aircraft. The Hunting loop was triangular with vertices at the tail and wing tips, but the INPUT loop, was a diamond, with an additional vertex at the nose, making it larger. The system development was funded by The Selection Trust Company (Selco) and the system was flown on their projects and those of their partners. The system was given the name INPUT (Figure A1.5). By 1961 a separate company (Barringer Research) had been formed to operate the system for Selco and another company called MacIntyre [35].

A key innovation of the INPUT system was the transmitter waveform. The half-sine shape was output by a resonant circuit, with the electrical energy oscillating between being stored in the loop inductance and in a bank of very large capacitors. This meant that a very large transmitter current could be obtained with moderate power being drawn from the aircraft power supply. This key aspect of the Barringer patent meant that the technical development of time-domain fixed-wing airborne system was controlled by the Barringer Research company for nearly 30 years.

Selco and McIntyre were unable to support the cost of development and operations alone so in 1967 they commercialized the system. Selco created a wholly owned subsidiary (Questor Surveys) and granted them a license to use the system worldwide. A second license was granted to CGG for all countries, except Canada. CGG transferred the license to a Canadian subsidiary, Geoterrex Ltd., that CGG held a controlling interest in. As Questor had a monopoly in Canada they were kept busy serving this market, leaving Geoterrex to develop strong markets in Australia, Africa, and the USA.

The INPUT system was very successful in Canada. Selco used it to discover the South Bay deposit (1967) and the Selbaie deposit (1974) in the Mattagami camp. Other companies also used the system to discover deposits. Of 26 discoveries in Quebec attributed to all types of geophysics—eight of these are attributed to INPUT [36]. The system was less successful in Australia where the conductive overburden masked the response of deep conductors. In Africa it was used successfully to explore for kimberlites [37].

The initial system used a 1.5 ms pulse. A second version, the INPUT Mark II system was being talked about in 1962; by the end of INPUT's lifetime, the system was up to Mark VI. Over this time, the system underwent many upgrades. By the late 1980s there were 12 measurement windows and the power of the transmitter was significantly greater with the dipole moment being more than 300,000 Am^2.

The INPUT receiver sensor was initially a horizontal dipole oriented in the flight direction that is housed in a bird towed behind the aircraft using a 137 m long cable. The sensor was sometimes about 70 m below the aircraft and 50 m above the ground. The signal came up the tow cable and passed through the receiver electronics and was recorded on an analogue chart. There was associated equipment for measuring the earth's magnetic field, the aircraft altitude and a camera was used to take images of the ground that the aircraft had flown over. These images were used to determine where the geophysical measurements were located. Hence it was important to synchronize the camera clock with the clock of the geophysical system. There is a story (possibly apocryphal) involving Tony Barringer and his ability to quickly improvise. One day the mechanism, which was used to show the time on the camera broke down in the middle of a flight. Tony did not want to abort the flight, so he lay down on the floor of the aircraft and looking at his watch, passed a pencil backward and forward every second in front of the camera to mark the passage of time.

The receiver electronics recorded the decay of the transient in the measurement windows. This was the average value between the start and the end of the window (at the time these windows were called channels). The average was obtained by electronically integrating the voltage from the sensor and dividing by the width of the window Figure A1.6 shows the window position for the INPUT Mark VI receiver [38].

In order to reduce the high-frequency noise, the measured signal in each window was averaged with a small portion of the previous measurement in that window. This was known as the time-constant filter and sometimes gave geophysical anomalies an asymmetric shape.

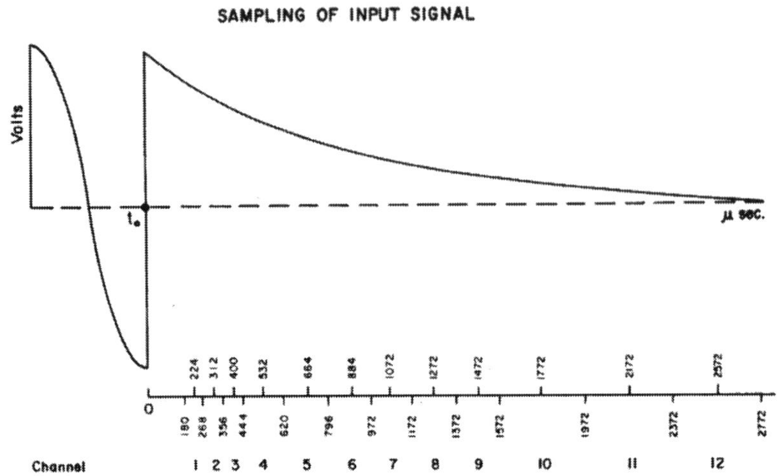

Figure A1.6 Distribution of windows or channels after the INPUT transmitter pulse.

In the late 1970s and the early 1980s, Barringer Research realized that the patent on the INPUT transmitter would soon be expiring and that a new upgraded transient EM system was required. A radical and ambitious new system, using correlation of transients (and named COTRAN), was designed that incorporated a number of radical new features—a new waveform, on-time and off-time measurements, a digital receiver, two-component measurements (x and z) and correlation processing of the decays so as to more easily distinguish the slow decays from the rapid decays [1]. This development work was undertaken in partnership with Geoterrex and Elf Acquitaine. The receiver was also capable of capturing full-bandwidth stream data, but the storage devices available at the time were only capable of handling and storing a few seconds worth of data.

At the same time Questor Surveys were also designing their own ambitious new system also with multicomponent measurements [39,40]. Questor was also experimenting with longer pulse widths and lower base frequencies [41].

Another ambitious systems being developed at this time included the SWEEPEM system, being developed by Kenting and funded by Royal Dutch/Shell [1]. The system was a hybrid time/frequency-domain system with the waveform switching on and off but the data being digitally recorded and processed in the frequency domain. The developers believed that the system could be used to identify extremely good conductors, but this was never demonstrated in practice.

In the mid-1980s, these exciting developments efforts all drew to a close when there was an economic downturn in the mineral exploration industry. Unfortunately, none of these systems flew commercially. In the space of a few years, Kenting went bankrupt; Geoterrex had severe financial difficulties, so CGG took 100% control of the company; and Questor was eventually taken over by Aerodata, later to become World Geoscience.

Some of the people working on the COTRAN system at Barringer Research later formed a new company A-Cubed and commenced development of an airborne EM systems called PROSPECT in association with Anglo American; the system was later renamed SPECTREM. Later, individuals from A-Cubed contributed to the subsequent development of less ambitious upgrades of the INPUT system: the Geoterrex system, named GEOTEM, was essentially INPUT with a digital receiver now mounted on a CASA C-212 aircraft and Questor/World Geoscience developed an INPUT upgrade named Questem. The Questem was mounted on a Short Brothers Skyvan aircraft as their INPUT system had been put on this aircraft some years before. These systems were generally operational in the late 1980s and seeking to regain the markets lost before the downturn.

A lot of the developments that were in the pipeline in the early 1980s were eventually realized in the 1990s. GEOTEM developed a multicomponent capability [42,43] and on-time capability [44], Geoterrex was also involved in a project to improve the system for detecting conductors below conductive overburden in Australia (sponsored by BHP and Aberfoyle). In order to achieve these goals, the GEOTEM system was initially modified to operate at 90 Hz with a 2 ms pulse and then 25 Hz with a 4 ms pulse. During this period, BHP were also testing a SQUID magnetometer for use with ground and airborne EM systems [45,46] which lead to the concept of integrating the digital voltage or dB/dt data to give a B-field, equivalent to what would be measured with a SQUID sensor [47]. Late in the 1990s, the GEOTEM system was scaled up for installation on a larger Dash-7 aircraft, becoming the MEGATEM system [48].

During the 1990s, World Geoscience took their developments in a different direction, modifying the QUESTEM system to the SALTMAP system [49] for measuring near-surface saline zones. World Geoscience then got involved in a large research program called Cerberus, undertaken in conjunction with the Australian government research organization, CSIRO, and also commenced development of the TEMPEST system for salinity mapping as part of a government cofunded research project [50]. The TEMPEST system was a fixed-wing towed bird time-domain system that had a small dipole moment (55,800 Am^2) and a 50% duty-cycle waveform that was deconvolved to a 100% duty-cycle waveform. These government-funded research projects required a lot of resources and placed a strain on World Geoscience. The projects were drawing to a close when there was another economic downturn at the end of the 1990s. A Dutch geotechnical engineering company, Fugro, saw this downturn as an opportunity to buy a large number of struggling airborne geophysical companies and Fugro consolidated the airborne EM industry, forming Fugro Airborne Surveys. This consolidation included Geoterrex and World Geoscience, and included three companies that offered rigid towed-tube frequency-domain systems: High Sense Geophysics (which had previously bought Aerodat); Dighem (previously bought by CGG/Geoterrex), and Sial Geophysics. All the towed-tube frequency-domain systems were discontinued except the DIGHEM and RESOLVE systems. The fixed-wing time-domain systems were all continued, GEOTEM, MEGATEM, and TEMPEST (which had already replaced SALTMAP). The consolidation subsequently bought some improvements through cross-fertilization, the TEMPEST system adopted the GEOTEM three-component bird and the TEMPEST coils were used to acquire high-frequency data on GEOTEM and MEGATEM. A new system, GENESIS, mounted on a Cessna Caravan was an interesting mixture of all the existing fixed-wing airborne EM systems that was intended for low-cost regional mapping. The GEOTEM was upgraded to a dipole moment of 1,000,000 Am^2 and the MEGATEM to 2,000,000 Am^2. In addition, considerable effort was spent on consolidating aspects of the systems, both acquisition hardware and processing software. Another downturn in 2008 further slowed EM developments, as did a shift in emphasis to acquiring gravity and gravity gradiometer data.

Today the fixed-wing transient EM systems that are commercially available are the two remaining successors of INPUT (the GEOTEM (Figure A1.7) and MEGATEM systems) and the TEMPEST and GENESIS systems (all now operated by CGG, who purchased Fugro Airborne Surveys in September 2013). The SPECTREM system (for many years proprietary to Anglo American) is now also commercially available. One of the disadvantages of the fixed-wing systems is that the asymmetry of the transmitter–receiver system results in subsurface bodies that are symmetric having a response that is asymmetric; hence the response is different when flown in alternating flight directions [51]. This asymmetry means that EM maps from fixed-wing systems can contain distracting artifacts.

Figure A1.7 Time-domain system GEOTEM (CGG, Canada). The TEMPEST system can also be mounted on the same CASA aircraft.

A1.2.2. Helicopter Transient Systems

Around the turn of the millennium, there was research and development going into the development of time-domain helicopter systems. This development was implicitly encouraged by the mineral exploration companies, to counter the lack of competition bought about by the Fugro consolidation. The developments initially took place in Canada [52]: the THEM system was initially configured as a system with a transmitter towed below the helicopter and the receiver was mounted in a helium balloon that was towed directly behind the transmitter. Other systems developed at this time include the AeroTEM system [53] and the Versatile TEM (VTEM) system [54]. All of these systems were symmetric, so the response of symmetric subsurface bodies was not asymmetric.

The early versions of the AeroTEM system (Aeroquest) were designed to be rigid and included a bucking coil so that the system could potentially see extremely conductive bodies by measuring an in phase response in the on time. However, this was never demonstrated convincingly, probably because the configuration was not rigid enough. The latter version (IV), shown on Figure A1.8, has no bucking coil evident. The AeroTEM system uses a triangular waveform.

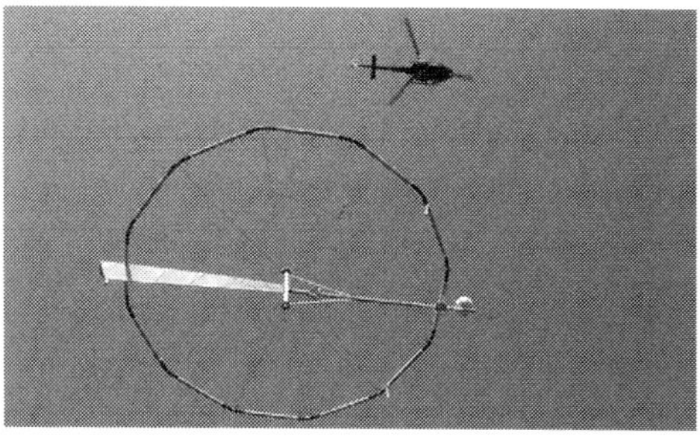

Figure A1.8 AEROTEM IV (Aeroquest, Canada). *From http://www.ngi.no/en/News-archive/News/NGI-into-airborne-EM-surveys-/, image courtesy of Aeroquest.*

Figure A1.9 The VTEM system. The large polygonal loop (partially shown) is the transmitter loop, the smaller loop is the bucking coil and the smallest loop is the receiver coil. The magnetometer is part way up the tow cable. *Image from Geotech web site www.geotech.ca/vtem.*

The VTEM system (Figure A1.9) had the receiver inside the transmitter loop. The term versatile was initially intended to describe the system's ability to change waveform and window position; however, the real advantage of the system was that it could be scaled up to increase the dipole moment so that the moment became comparable with the fixed-wing systems. A more recent addition to the system included a nonrigid bucking coil, which was intended to reduce the strength of the primary field in the on time, not completely, but to a sufficient extent that the secondary in the on time could be measured and integrated to give a B-field response.

The VTEM system achieved exploration success and hence considerable commercial success. To counter this success, Fugro bought the rights to the THEM transmitter and mounted a receiver part way up the tow cable to create the HeliGEOTEM system [55,56]. This system was weakly asymmetric. Subsequent modifications of this system have also created a scaled up version, judged sufficiently different to get a new name, HELITEM (Figure A1.10). For a number of years, the various companies were claiming greater and greater dipole moments for their transmitters. The currently available figures (2013) are AeroTEM 1,000,000 Am^2, HELITEM 2,000,000 Am^2, and VTEM 2,500,000 Am^2.

Although the initial developments took place in Canada, subsequent developments in other countries have resulted in a number of other systems (Vallee et al., 2011). In addition to the above, a partial list of systems previously and currently available (commercial or proprietary systems) contains HoistEM [57], REPTEM, SeaTEM [58], NEWTEM [59]. Information about other systems such as XTEM from GPX Surveys and EQUATOR from GeoTechnologies (Figure A1.11) can be found on the Internet.

Figure A1.10 The HELITEM system. *Image from www.cgg.com.*

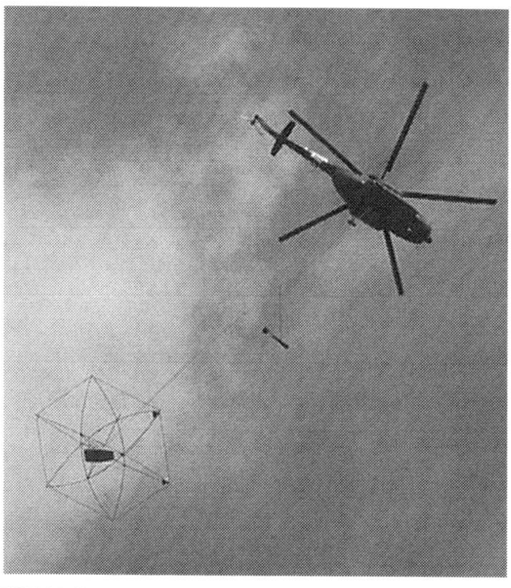

Figure A1.11 The EQUATOR system with a vertical dipole transmitter loop and a three-component receiver part way up the tow cable. (GeoTechnologies, Russia).

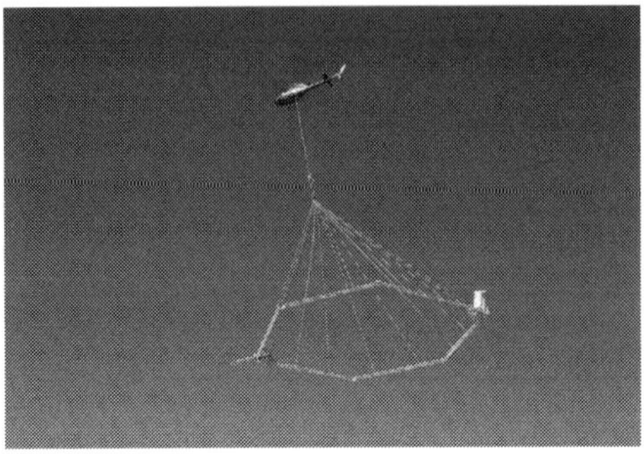

Figure A1.12 The SkyTEM system. *Image from http://danishresponsibility.dk/skytem-method-mapping-groundwater-resources.*

Worthy of particular mention is the SkyTEM system [60] shown in Figure A1.12. This system is a large horizontal loop, with the receiver coils mounted in null-coupled positions to reduce (but not remove) the size of the primary field and the effect of geometric noise. The system was designed for hydrogeological mapping, initially in Denmark. In order to resolve both shallow and deep features, the system has two transmitters; one is a low-power short-duration pulse, the other is a high-power longer duration pulse. The system operates these transmitters in sequence, with each transmitter generating a train of pulses. The current for the low-power pulse can be controlled very carefully at the switch off, so it is possible to get accurate information at early time for resolving the shallow conductivity structure. The high-power pulse provides good signal to noise at late time, so the deeper conductivity structure can be resolved. Significant effort has gone into developing procedures for processing the data and resolving the conductivity structure of the hydrogeological section [61,62].

A1.2.3. Signal to Noise in Time-Domain Systems

For the purposes of this discussion, assume that the magnetic field $B(t)$ induced in the ground after switch off of the magnetic field in the transmitter can be represented by four terms

$$B(t) = A_{\mathrm{T}}e^{-t/\tau_T} + A_{\mathrm{O}}e^{-t/\tau_{\mathrm{O}}} + n(t) + \alpha B(t)$$

Where A_T and A_{O} are the coupling coefficients of the system to the target (T) and other (O) geological features (e.g., overburden, power lines, fences, etc.), t is the time, and τ is the time constant of the decay; $n(t)$ represents noise terms like atmospheric noise, microphonic (vibrational) noise, thermal noise in the coil, etc.; and $\alpha B(t)$ is a term that depends on the magnitude of the primary field and hence the measured secondary (e.g., geometric noise, aircraft conductivity, etc.). The engineering design of the system will try to make $n(t)$ and α as small as possible. If $n(t)$ is the dominant term, then the standard way of increasing the signal A_T relative to $n(t)$ is to increase the dipole moment, either by increasing the loop size, the current in the loop or the number of turns in the loop. This approach requires that the system has been well engineered so that α is small.

If A_T is less than A_O then the target will not be evident. However, if $\tau_N > \tau_O$, then at late time (t large), the target term will eventually dominate. For this reason late times and low base frequencies are used to look for slowly decaying targets.

In the preceding, we have assumed that the EM system is measuring the magnetic field. Some sensors measure the voltage in a coil, which is proportional to the time derivative of the magnetic field ($\mathrm{d}B/\mathrm{d}t$). In the off time

$$\frac{\mathrm{d}B}{\mathrm{d}t}(t) = -\frac{1}{\tau_T}A_T e^{-t/\tau_T} + -\frac{1}{\tau_O}A_O e^{-t/\tau_O} + \frac{\mathrm{d}n}{\mathrm{d}t}(t) + \alpha\frac{\mathrm{d}B}{\mathrm{d}t}(t)$$

Except for the α term, the time derivative will change the relative sizes of the terms. If $n(t)$ is primarily low-frequency noise, this term will be suppressed, but if there is high-frequency noise, this term will be enhanced. The relative sizes of the exponential terms will change depending on the time constants. If τ is smaller, then the term will become larger. Hence, the voltage should be measured when the survey is looking for small time-constant targets like near-surface overburden or weakly conductive targets; if large time-constant targets are of interest, then it is better to measure the B-field, $B(t)$. The latter approach has the added advantage of suppressing the high-frequency noise in $n(t)$. It is for these reasons that integrating $\mathrm{d}B/\mathrm{d}t$ or measuring the B-field has become popular [47].

One difficulty in achieving low noise for the rapidly decaying responses such as near-surface overburden (τ small) is getting good signal at early time, just after the switch off. This requires a receiver sensor coil with a capability of measuring high frequencies and receiver electronics capable of sampling rapidly. The TEMPEST system was intended to measure frequencies as high as 50,000 Hz or as early as 13 μs [50]. In addition, the transmitter must have high-frequency information in the waveform. This requires that the current is switched off quickly, which is easier to achieve when the amount of current transmitted is smaller (see discussion of SkyTEM system above).

A1.2.4. Comparison of Fixed-Wing and Helicopter Time-Domain Systems

The dipole moments of the fixed-wing and helicopter systems are roughly comparable, so the depths of penetrations will be comparable as well. The fixed-wing moments are sometimes greater, but the lower terrain clearance of the helicopter-towed systems can compensate for this. Macnae [63] has compared the depth of exploration of airborne systems and found that the helicopter systems have larger responses for shallow bodies, but that the fixed-wing systems can have larger responses at greater depths. A study by [22] showed that the greater flying height and the larger transmitter–receiver separation meant that a fixed-wing system could see further to the side than a helicopter system; this means that the flight line spacing can be greater and that ground can be covered more quickly at less cost. The relative advantage of each system can only be determined by detailed modeling that takes into account the system geometry (flying height, transmitter–receiver separation, and offset), the waveform and window times and the conductivities of the target, overburden, and/or host. In many cases, the choice of system will depend on the ability to detect the target, the number of flights required to efficiently covering the survey area, the time the survey will take to be flown, the time of year it can be flown, and the ancillary data that will be acquired with the EM data. Virtually all EM surveys provide magnetic data and topographic data, but some systems can also provide other geophysical data: VLF data, gamma-ray spectrometry data, gravity data, or gravity gradiometry data. These additional data help to better understand the geological context.

A1.2.5. Comparative Analysis of Efficiency of Frequency and Time-Domain Systems

Time-domain systems are designed for detecting decays that are comparable to the off time of the EM system. If the decay is too short, the transmitter current and sensor will not detect the target and when the decay is too long, the effect of previous half cycles of the transmitter can reduce the size of the decay [64]. As noise associated with motion of the receiver coils generally prevents good quality data being collected when the base frequency is lower than 25 Hz, this means that the measurement time is about 20 ms or less. If the bandwidth of the receiver is such that decays as early at 20 μs can be measured, then this gives a time range over nearly three orders of magnitude, corresponding roughly to frequencies from 50 Hz to 50 kHz. From an engineering perspective, it appears easier to build a time-domain transmitter that can transmit over this range of frequencies in the time domain than multiple transmitters in the frequency domain. Having this range of frequencies allows the interpreter to distinguish between a wide range of conductive responses, for example, those associated with poorly conductive overburden and those associated with more slowly decaying conductive bodies. Historically, the time-domain systems worked very well in resistive environments free of thick conductive overburden like the Canadian and Scandinavian shields; however, recent successes have been documented in areas of thicker conductive overburden [65,66].

Time-domain methods are essentially quadrature methods and they work poorly in cases when the ground is very resistive or very conductive. The sensitivity at low conductivities is partially addressed by collecting on-time (quadrature) data (Annan et al., 1996). The frequency-domain systems (e.g., RESOLVE) can measure at frequencies as high as 140,000 Hz, so are sensitive to more resistive features. The problem of extremely high conductivity remains to be addressed by the time-domain systems; the rigid-boom frequency-domain systems address this very well with in phase data.

A1.2.6. Semiairborne Systems (Transmitters on the Ground and Receivers in the Air)

The semiairborne system Turair was introduced by [67]. This system used a large loop or grounded bipole transmitter laid out on the ground and a receiver was flown in the air. The system operated in the frequency-domain and did not require synchronization between the transmitter and receiver. A time-domain system was introduced in the 1990s [68]; in which the transmitter was a large loop and the receiver was towed by a helicopter. Synchronization between the transmitter and receiver system was via a radio link or crystal clocks. A little later, a system was introduced which did not require synchronization during the flight [69]; the data could be synchronized postflight. Modeling showed that these semiairborne systems are capable of seeing deeper than airborne systems, but not as deep as ground EM systems [69]. There are no semiairborne systems currently available for commercial use, but they could be quickly deployed if the terrain is not suitable for airborne or ground systems and penetration greater than the airborne systems is required.

REFERENCES

[1] L.S. Collett, Development of the airborne electromagnetic techniques, in: G.J. Palacky (Ed.), Airborne Resistivity Mapping, Geological Survey of Canada Paper, Vol. 86–22, 1986, pp. 9–18.

[2] G.J. Palacky, G.F. West, Airborne electromagnetic methods, in: M.N. Nabighian (Ed.), Electromagnetic Methods in Applied Geophysics – Applications Part B, Investigations in Geophysics, Vol. 3, Society of Exploration Geophysicists, 1991, pp. 811–879.

[3] N. Paterson, Experimental and field data for the dual-frequency phase-shift method of airborne electromagnetic prospecting, Geophysics 26 (5) (1961) 601–617, http://dx.doi.org/10.1190/1.1438921.

[4] R. Pemberton, Airborne electromagnetics in review, Geophysics 27 (5) (1962) 691–713, http://dx.doi.org/10.1190/1.1439081.

[5] D. Fountain, Airborne electromagnetic systems – 50 years of development, Exploration Geophysics 29 (1998) 1–11.

[6] S.H. Ward, Airborne electromagnetic methods, in: L.W. Morley (Ed.), Proceedings of the Canadian Centennial Conference on Mining and Groundwater Geophysics, Niagara Falls, Canada, in October 1967, Geological Survey of Canada, Economic Geology Report 26 (1967a) 81–108.

[7] J.S. Dowsett, Geophysical exploration for nickel, in: L.W. Morley (Ed.), Proceedings of the Canadian Centennial Conference on Mining and Groundwater Geophysics, Niagara Falls, Canada, in October 1967, Geological Survey of Canada, Economic Geology Report Vol. 26 (1967) 310–321.

[8] S.H. Ward, The electromagnetic method, in: S.H. Ward (Ed.), Mining Geophysics, Vol. 2, The Society of Exploration Geophysicists, 1967b, pp. 224–372.

[9] S.W. Hefford, R.S. Smith, C. Samson, Quantifying the effect that changes in transmitter-receiver geometry have on the capability of an airborne electromagnetic survey system to detect good conductors, Exploration and Mining Geology 15 (1–2) (2006) 43–52.

[10] R.S. Smith, On removing the primary field from fixed-wing time-domain airborne electromagnetic data: some consequences for quantitative modelling, estimating bird position and detecting perfect conductors, Geophysical Prospecting 49 (2001) 405–416.

[11] A. Smiarowski, J. Macnae, Detection of a perfect conductor with an airborne electromagnetic system: the Gemini Field test, Geophysics 78 (5) (2013) E249–E259, http://dx.doi.org/10.1190/geo2012-0469.1.

[12] N. Paterson, Mattagami lake mines – a discovery by geophysics, in: S.H. Ward (Ed.), Mining Geophysics, Vol. 1The Society of Exploration Geophysicists, 1966, pp. 185–196.

[13] McPhar Geophysics Ltd, 1969, Report on the combined airborne magnetic and electromagnetic survey 157, 145 and "U" claim groups for Mr. A. St. Denis. 41 pp. http://www.geologyontario.mndmf.gov.on.ca/mndmfiles/afri/data/imaging/41J10NE0020/41J10NE0020.pdf Accessed 20.09.13.

[14] Volkovitskiy, A., Karshakov, E., Trusov, A., 2008, Four frequency AEM system EM-4H, extended abstracts of AEM2008, 5th International Conference in Airborne Electromagnetics, 4 pp.

[15] D. Beamish, M. Young, Geophysics of Northern Ireland: the Tellus effect, First Break 27 (8) (2009) 43–49.

[16] A. Poikonen, K. Sulkanen, M. Oksama, I. Suppala, Novel dual frequency fixed wing airborne EM system of Geological Survey of Finland (GTK), Exploration Geophysics 29 (1998) 46–51.

[17] Leväniemi, H., Kurimo, M., Beamish, D., Suppala, I., Jokinen, T., Lohva, J., Tiainen, M., 2008, JAC 4-frequency AEM system with case example in gold exploration: extended abstracts of AEM2008, 5th International Conference in Airborne Electromagnetics, 4 pp.
[18] D. Fraser, Resistivity mapping with an airborne multicoil electromagnetic system, Geophysics 43 (1) (1978) 144–172, http://dx.doi.org/10.1190/1.1440817.
[19] H. Huang, D.C. Fraser, The differential parameter method for multifrequency airborne resistivity mapping, Geophysics 61 (1) (1996) 100–109.
[20] G. Podolsky, An evaluation of an airborne electromagnetic anomaly in northwestern Quebec, in: S.H. Ward (Ed.), Mining Geophysics, Vol. 1, The Society of Exploration Geophysicists, 1966, pp. 197–205.
[21] A. Bouchedda, M. Chouteau, P. Keating, R. Smith, Sferics noise reduction in time-domain electromagnetic systems: application to MEGATEMII signal enhancement, Exploration Geophysics 41 (2010) 225–239, http://dx.doi.org/10.1071/EG09007.
[22] R.S. Smith, R. Wasylechko, Sensitivity cross-sections in airborne electromagnetic methods using discrete conductors, Exploration Geophysics 43 (2012) 95–103.
[23] S.H. Ward, AFMAG - airborne and ground, Geophysics 24 (1959) 761–787, http://dx.doi.org/10.1190/1.1438657.
[24] R.S. Smith, Electromagnetic induction methods in mining geophysics from 2008 to 2012, Surveys in Geophysics (2013) 1–34, http://dx.doi.org/10.1007/s10712-013-9227-1.
[25] J. Legault, Ten Years of Passive Airborne AFMAG EM Development for Mineral Exploration, SEG Technical Program, 2012, http://dx.doi.org/10.1190/segam2012-1267.1. Expanded Abstracts, 2012: 1–6.
[26] B. Lo, M. Zang, Numerical modeling of Z-TEM (airborne AFMAG) responses to guide exploration strategies, SEG Expanded Abstracts 27 (2008) 1098–1102.
[27] J.M. Legault, H. Kumar, B. Milicevic, L. Hulbert, ZTEM airborne tipper AFMAG test survey over a magmatic copper-nickel target at Axis Lake in northern Saskatchewan, SEG Expanded Abstracts 28 (2009a) 1272–1276.
[28] J.M. Legault, H. Kumar, B. Milicevic, P. Wannamaker, ZTEM tipper AFMAG and 2D inversion results over an unconformity uranium target in northern Saskatchewan, SEG Expanded Abstracts 28 (2009b) 1277–1281.
[29] P.L. Kowalczyk, P.B.M. Van Kooten, Inverting ZTEM data in 3D: Process and Practice, in: R.J.L. Lane (Ed.), Natural Fields EM Forum 2012: Abstract from the Australian Society of Exploration Geophysicists Natural Fields EM Forum 2012, Published by Geoscience Australia, Geoscience Australia Record 2012/04, 2012, pp. 93–100.
[30] P. Paré, A.V. Gribenko, L.H. Cox, M. Čuma, G.A. Wilson, M.S. Zhdanov, J. Legault, J. Smit, L. Polomé, 3D inversion of SPECTREM and ZTEM airborne electromagnetic data from the Pebble Cu–Au–Mo porphyry deposit, Alaska, Exploration Geophysics 43 (2012) 104–115.
[31] J. Legault, G.A. Wilson, A.V. Gribenko, M.S. Zhdanov, S. Zhao, K. Fisk, An overview of the ZTEM and AirMt systems: a case study from the Nebo-Babel Ni–Cu–PGE deposit, West Musgrave, Western Australia, in: R.J.L. Lane (Ed.), Natural Fields EM Forum 2012: Abstract from the Australian Society of Exploration Geophysicists Natural Fields EM Forum 2012, Geoscience Australia Record, 2012/04, Geoscience Australia, 2012, pp. 101–121.
[32] J.D. McNeill, V.F. Labson, Geological mapping using VLF radio fields, in: M.N. Nabighian (Ed.), Electromagnetic Methods in Applied Geophysics – Applications Part B, Investigations in Geophysics, Vol. 3, Society of Exploration Geophysicists, 1991, pp. 521–640.

[33] L.B. Pedersen, L. Persson, M. Bastani, S. Byström, Airborne VLF measurements and mapping of ground conductivity in Sweden, Journal of Applied Geophysics 67 (2009) 250–258, http://dx.doi.org/10.1016/j.jappgeo.2007.11.002.
[34] F.S. Grant, G.F. West, Interpretation Theory in Applied Geophysics, McGraw-Hill, 1965.
[35] K. Witherly, The quest for the Holy Grail in mining geophysics: a review of the development and application of airborne EM systems over the last 50 years, First Break (2000 March) 270–274.
[36] P.M. Vallée, P. Keating, 400 years of exploration, The Leading Edge 27 (12) (2008) 1612–1613. http://dx.doi.org/10.1190/1.3036961.
[37] J.C. Macnae, Kimberlites and exploration geophysics, Geophysics 44 (8) (1979) 1395–1416.
[38] Questor Surveys, Interpretation Report INPUT Mark VI Electromagnetic/Magnetic Survey, project #88032, Cumberland Resources Ltd., February 1989, http://www.geologyontario.mndmf.gov.on.ca/mndmfiles/afri/data/imaging/52H06SW0002//52H06SW0002.Pdf, (accessed 20.09.13).
[39] P.G. Lazenby, A. Becker, Redefinition of the INPUT system: part A—instrumentation, in exploration technology program of the board of industrial leadership and development, summary of research 1981-1983, Ontario Geol. Surv. Misc. Paper 115 (1983) 1–9.
[40] P.G. Lazenby, A. Becker, Redefinition of the INPUT system: part B—signal processing and field results, in exploration technology program of the board of industrial leadership and development, summary of research 1983–1984, Ontario Geol. Surv. Misc. Paper 120 (1984) 5–16.
[41] A. Becker, R. DeCarle, P. Lazenby, Simplified prediction of transient electromagnetic response, Geophysics 49 (7) (1984) 913–917, http://dx.doi.org/10.1190/1.1441736.
[42] R.S. Smith, P.B. Keating, The usefulness of multicomponent, time-domain airborne electromagnetic measurements, Geophysics 61 (1) (1996) 74–81. http://dx.doi.org/10.1190/1.1443958.
[43] R.S. Smith, A.P. Annan, Advances in airborne time-domain EM technology, in: A.G. Gubins, (Ed.), Proceedings of Exploration 97: Fourth Decennial International Conference on Mineral Exploration, (1997) 497–504.
[44] A.P. Annan, R.S. Smith, J. Lemieux, M.D. O'Connell, R.N. Pedersen, Resistive-limit, time-domain AEM apparent conductivity, Geophysics 61 (1) (1996) 93–99. http://dx.doi.org/10.1190/1.1443960.
[45] J.B. Lee, D.L. Dart, R.J. Turner, M.A. Downey, A. Maddever, G. Panjkovic, C.P. Foley, K.E. Leslie, R. Binks, C. Lewis, W. Murray, Airborne TEM surveying with a SQUID magnetometer sensor, Geophysics 67 (2002) 468–477.
[46] J.B. Lee, R.J. Turner, M.A. Downey, A. Maddever, D.L. Dart, C.P. Foley, R. Binks, C. Lewis, W. Murray, G. Panjkovic, M. Asten, Experience with SQUID magnetometers in airborne TEM surveying, Exploration Geophysics 32 (2001) 9–13.
[47] R.S. Smith, A.P. Annan, Using an induction coil sensor to indirectly measure the B-field response in the bandwidth of the transient electromagnetic method, Geophysics 65 (2000) 1489–1494.
[48] R. Smith, D. Fountain, M. Allard, The MEGATEM fixed-wing transient EM system applied to mineral exploration: a discovery case history, First Break 21 (July 2003) 73–77.
[49] A.C. Duncan, G.P. Roberts, G. Buselli, J.P. Pik, D.R. Williamson, P.A. Roocke, R.G. Thorn, A. Anderson, SALTMAP—airborne EM for the environment, Exploration Geophysics 23 (1992) 123–126, http://dx.doi.org/10.1071/EG992123.
[50] R. Lane, A. Green, C. Golding, M. Owers, P. Pik, C. Plunkett, D. Sattel, B. Thorn, An example of 3D conductivity mapping using the TEMPEST airborne electromagnetic system, Exploration Geophysics 31 (2000) 162–172, http://dx.doi.org/10.1071/EG00162.

[51] R.S. Smith, M.C. Chouteau, Combining airborne electromagnetic data from alternating flight directions to form a virtual symmetric array, Geophysics 71 (2006) G35–G41, http://dx.doi.org/10.1190/1.2187745.
[52] Allard, M. (2007). On the origin of the HTEM species. In: Milkereit, B. (Ed.), 5th Decennial International Conference on Mineral Exploration, Proceedings, 355–374.
[53] S.J. Balch, W.P. Boyko, N.R. Paterson, The AeroTEM airborne electromagnetic system, The Leading Edge 22 (2003) 562–566.
[54] K. Witherly, R. Irvine, E.B. Morrison, The Geotech VTEM time domain helicopter EM system, SEG Expanded Abstracts 23 (2004) 1217–1220.
[55] D. Fountain, R. Smith, T. Payne, J. Lemieux, A helicopter time-domain EM system applied to mineral exploration, system and data, First Break 23, November (2005) 73–78.
[56] R.S. Smith, G. Hodges, J. Lemieux, Case histories illustrating the characteristics of the HeliGEOTEM system, Exploration Geophysics 40 (3) (2009) 246–256. http://dx.doi.org/10.1071/EG09006.
[57] G.W. Boyd, in: HoistEM – A new airborne electromagnetic system. PACRIM 2004 Congress, 19–22 September 2004, Adelaide, South Australia, Proceedings, Australian Institute of Mining and Metallurgy, 2004, pp. 211–218, 9781920806187.
[58] J. Vrbancich, G. Boyd, K. Mathews, SeaTEM – A New Airborne Electromagnetic System for Bathymetric Mapping and Seafloor Characterisation, EGM 2007 International Workshop, Capri, Italy, 15–18 April 2007, 4 pp.
[59] P. Eaton, B. Anderson, B. Nilsson, E. Lauritsen, S. Queen, C. Barnett, NEWTEM - a novel time-domain helicopter electromagnetic system for resistivity mapping, SEG Expanded Abstracts 21 (2002) 1–4.
[60] K.I. Sørensen, E. Auken, SkyTEM - a new high-resolution helicopter transient electromagnetic system, Exploration Geophysics 35 (2004) 194–202.
[61] E. Auken, J. Westergaard, A.V. Christiansen, K. Sørensen, Processing and Inversion of SkyTEM Data for High Resolution Hydrogeophysical Surveys, ASEG Extended Abstracts, 2007, 1–4.
[62] E. Auken, A.V. Christiansen, J.H. Westergaard, C. Kirkegaard, N. Foged, A. Viezzoli, An integrated processing scheme for high-resolution airborne electromagnetic surveys, the SkyTEM system, Exploration Geophysics 40 (2) (2009) 184–192. http://dx.doi.org/10.1071/EG08128.
[63] J. Macnae, Comparing airborne electromagnetic systems, Preview 133 (2008) 24–29.
[64] R.S. Smith, B. Neil, Precision requirements for specifying transmitter waveforms used for modelling the off-time electromagnetic response, Exploration Geophysics 44 (1) (2013) 1–5. http://dx.doi.org/10.1071/EG12040.
[65] M. Vallée, R. Smith, P. Keating, Metalliferous mining geophysics – State of the art after a decade in the new millennium, Geophysics 76 (4) (2011) W31–W50. http://dx.doi.org/10.1190/1.3587224.
[66] Combrinck, M., Mutton, P., Lo, B. (May 2008). Detecting basement conductors in a conductive environment: an example from the Nepean Mine VTEM survey: Paper presented at AEM2008 – 5th International Conference on Airborne Electromagnetics, Haikko Manor, Finland, 28–30.
[67] Bosschart, R.A., Seigel, H.O. (1972). Advances in deep penetration airborne electromagnetic methods: Conference Proceedings, 24th International Geological Congress, Section 9, 37–48.
[68] P. Elliott, The principles and practice of FLAIRTEM, Exploration Geophysics 29 (1998) 58–60.
[69] R.S. Smith, A.P. Annan, P.D. McGowan, A comparison of data from airborne, semi-airborne and ground electromagnetic systems, Geophysics 66 (5) (2001) 1379–1385, http://dx.doi.org/10.1190/1.1487084.

APPENDIX TWO

Estimation of the Impedance Tensor

INTRODUCTION

The impedance tensor by definition is a linear operator that transforms a measured horizontal magnetic field (input) into a measured horizontal electric field (output)—both of which are functions of time. Hence it can be treated by conventional techniques of linear system analysis that allow one to find the system response from known input and output. The usual techniques compute the (temporal) Fourier transform of the recorded time series and by further calculations estimate spectral characteristics of the signals that are statistically stable (robust). Once these spectral characteristics have been estimated, the response function (impedance tensor) can be calculated easily.

SINGLE-INPUT–SINGLE-OUTPUT SYSTEM

Let us describe briefly the key ideas of the linear systems approach. Consider first a simple case when the goal is to find a response of a single-input–single-output linear system, assuming that the input and output are noise free. We start with the (known) linear equation relating the system's input and output in the frequency domain:

$$Y(\omega) = K(\omega) \cdot X(\omega), \tag{A2.1}$$

where $X(\omega)$ and $Y(\omega)$ are input and output spectra, respectively, while $K(\omega)$ is a complex-valued system response. It is natural to suppose that the input and output time series $x(t)$ and $y(t)$, from which the spectra, are computed, are measured over a series of time intervals of finite length. In other words, we have N intervals, or time windows, and within each window, we use a finite Fourier Transform to compute the spectra, $X_i(\omega)$, $Y_i(\omega)$, where $i = 1,\ldots,N$ is the interval number.

The spectra computed in this way are not equal to the true spectra of the signals, because of the finite duration of the windows, the discrete-time

Methods in Geochemistry and Geophysics, Volume 45
ISSN 0076-6895,
http://dx.doi.org/10.1016/B978-0-444-53829-1.19002-1

sampling of the signals, and computation of the Fourier Transform by numerical integration. The equations expressing the system response

$$Y_i(\omega) \approx K(\omega)\cdot X_i(\omega), \quad i = 1, \ldots, N \tag{A2.2}$$

are therefore approximate within each window. Solution of the system (A2.2) for the system response function $K(\omega)$ should minimize the overall misfit between the left-hand and right-hand sides of Eq. (A2.2) over $i = 1,\ldots,N$. Inasmuch as this misfit

$$Y_i(\omega) - K(\omega)\cdot X_i(\omega)$$

is a complex number, it is natural to minimize the square of its absolute value, summed over all samples by a linear least-squares method. Thus, our goal is to determine a minimum of the following functional

$$\Phi(K) = \sum_{i=1}^{N}(Y_i - K\cdot X_i)(Y_i - K\cdot X_i)^*, \tag{A2.3}$$

where * denotes complex conjugate and the dependence on ω is omitted. The minimum occurs where the derivatives of $\Phi(K)$ with respect to real and imaginary parts of $K = A + iB$ are zero. Rearranging terms in Eq. (A2.3), we obtain

$$\Phi(K) = \sum_{i=1}^{N}\left\{Y_iY_i^* - Y_iK^*X_i^* - Y_i^*KX_i + KK^*X_iX_i^*\right\}.$$

Taking derivatives gives

$$\frac{\partial\Phi}{\partial A} = -\frac{\partial K^*}{\partial A}\cdot\sum_{i=1}^{N} Y_i\cdot X_i^* - \frac{\partial K}{\partial A}\cdot\sum_{i=1}^{N} Y_i^*\cdot X_i + \frac{\partial|K|^2}{\partial A}\cdot\sum_{i=1}^{N} X_i\cdot X_i^*$$

$$= 2\sum_{i=1}^{N}\left\{A\cdot X_i\cdot X_i^* - (\mathrm{Re}X_i\cdot\mathrm{Re}Y_i + \mathrm{Im}X_i\cdot\mathrm{Im}Y_i)\right\} = 0 \tag{A2.4}$$

$$\frac{\partial\Phi}{\partial B} = -\frac{\partial K^*}{\partial B}\cdot\sum_{i=1}^{N} Y_i\cdot X_i^* - \frac{\partial K}{\partial B}\cdot\sum_{i=1}^{N} Y_i^*\cdot X_i + \frac{\partial|K|^2}{\partial B}\cdot\sum_{i=1}^{N} X_i\cdot X_i^*$$

$$= 2\sum_{i=1}^{N}\left\{B\cdot X_i\cdot X_i^* - (\mathrm{Re}Y_i\cdot\mathrm{Im}X_i - \mathrm{Im}Y_i\cdot\mathrm{Re}X_i)\right\} = 0$$

Grouping these two real-valued equations into a single complex-valued equation, we obtain

$$\sum_{i=1}^{N}\{K\cdot X_i\cdot X_i^* - [(\mathrm{Re}X_i\cdot\mathrm{Re}Y_i + \mathrm{Im}X_i\cdot\mathrm{Im}Y_i) + i(\mathrm{Re}Y_i\cdot\mathrm{Im}X_i - \mathrm{Im}Y_i\cdot\mathrm{Re}X_i)]\} = 0$$

Thus, we arrive at the following system of linear equations:

$$\sum_{i=1}^{N}\{K\cdot X_i\cdot X_i^* - X_i\cdot Y_i^*\} = 0, \qquad \text{(A2.5)}$$

which can easily be solved for K, since it is independent of the summation index. The solution is

$$K(\omega) = \frac{S_{xy}(\omega)}{S_{xx}(\omega)}, \qquad \text{(A2.6)}$$

where $S_{xy} = \sum_{i=1}^{N} X_i Y_i^*$, $S_{xx} = \sum_{i=1}^{N} X_i X_i^*$ are called the cross-spectral density and autospectral density, respectively.

Note that in place of minimizing Eq. A2.3 one could solve the alternative problem of minimizing the functional that arises from viewing $X_i(\omega)$ and a function of $Y_i(\omega)$; that is,

$$\Omega(L) = \sum_{i=1}^{N}(X_i - L\cdot Y_i)(X_i - L\cdot Y_i)^*, \qquad \text{(A2.7)}$$

where $L = \frac{1}{K}$. In this case the least-squares method yields a new expression for K:

$$K(\omega) = \frac{1}{L(\omega)} = \frac{S_{yy}(\omega)}{S_{yx}(\omega)}, \qquad \text{(A2.8)}$$

where $S_{yx} = S_{xy}^* = \sum_{i=1}^{N} Y_i X_i^*$, $S_{yy} = \sum_{i=1}^{N} Y_i Y_i^*$.

If each of the equations of the set (A2.2) holds exactly, the solutions (A2.6) and (A2.8) would be identical:

$$K_2(\omega) = \frac{1}{L(\omega)} = \frac{S_{yy}(\omega)}{S_{yx}(\omega)} = \frac{\sum_{i=1}^{N} Y_i Y_i^*}{\sum_{i=1}^{N} Y_i X_i^*} = \frac{\sum_{i=1}^{N} K X_i Y_i^*}{\sum_{i=1}^{N} K X_i X_i^*} = \frac{K\sum_{i=1}^{N} X_i Y_i^*}{K\sum_{i=1}^{N} X_i X_i^*}$$
$$= \frac{S_{xy}(\omega)}{S_{xx}(\omega)} = K_1(\omega)$$

But if the relationships expressed by the set of Eq. (A2.2) are approximate, the two expressions for the system response may yield different values. For this reason, the values obtained are usually called least-squares estimates.

The latter are only approximations to the true response K. In order to understand the behavior of these estimates we consider a slightly more complicated system in which both input and output are affected by an additive noise:

$$x(t) = \widehat{x}(t) + n^x(t), \quad y(t) = \widehat{y}(t) + n^y(t),$$

where $\widehat{x}(t), \widehat{y}(t)$ are true signals, while $n^x(t), n^y(t)$ are noise components. Therefore, for each individual spectrum we have

$$X_i(\omega) = \widehat{X}_i(\omega) + N_i^x(\omega), \quad Y_i(\omega) = \widehat{Y}_i(\omega) + N_i^y(\omega),$$

where, for example, $N_i^x(\omega)$ is the spectral component of the noise in the signal $X_i(\omega)$ at frequency ω in window i (and should not be confused with the window number N in the summation). Correspondingly, from Eq. (A2.6) and (A2.8) we obtain the following equations for the least-squares estimates of the system function (subscripts d and u will be explained later):

$$K_d(\omega) = \frac{\sum_{i=1}^{N}\left(\widehat{X}_i\widehat{Y}_i^* + \widehat{X}_i N_i^{y^*} + \widehat{Y}_i^* N_i^x + N_i^x N_i^{y^*}\right)}{\sum_{i=1}^{N}\left(\widehat{X}_i\widehat{X}_i^* + \widehat{X}_i N_i^{x^*} + \widehat{X}_i^* N_i^x + N_i^x N_i^{x^*}\right)} = \frac{S_{\widehat{x}\widehat{y}} + S_{\widehat{x}n^y} + S_{n^x\widehat{y}} + S_{n^x n^y}}{S_{\widehat{x}\widehat{x}} + S_{\widehat{x}n^x} + S_{n^x\widehat{x}} + S_{n^x n^x}} \tag{A2.9}$$

$$K_u(\omega) = \frac{\sum_{i=1}^{N}\left(\widehat{Y}_i\widehat{Y}_i^* + \widehat{Y}_i N_i^{y^*} + \widehat{Y}_i^* N_i^y + N_i^y N_i^{y^*}\right)}{\sum_{i=1}^{N}\left(\widehat{X}_i\widehat{Y}_i^* + \widehat{X}_i N_i^{y^*} + \widehat{Y}_i^* N_i^x + N_i^x N_i^{y^*}\right)} = \frac{S_{\widehat{y}\widehat{y}} + S_{\widehat{y}n^y} + S_{n^y\widehat{y}} + S_{n^y n^y}}{S_{\widehat{x}\widehat{y}} + S_{\widehat{x}n^y} + S_{n^x\widehat{y}} + S_{n^x n^y}} \tag{A2.10}$$

In general, one may assume that noise is uncorrelated with the signal and also that is usually uncorrelated from channel to channel. These assumptions mean that

$$S_{\widehat{x}n^x} = S_{\widehat{x}n^y} = S_{\widehat{y}n^x} = S_{\widehat{y}n^y} = S_{n^x n^y} = 0$$

Therefore from Eqs (A2.9) and (A2.10) we have

$$K_d(\omega) \approx \frac{S_{x^0 y^0}}{S_{x^0 x^0} + S_{n^x n^x}}, \quad K_u(\omega) \approx \frac{S_{y^0 y^0} + S_{n^y n^y}}{S_{y^0 x^0}} \tag{A2.11}$$

From these expressions one can see that the presence of any noise biases the estimates of the response magnitudes. The bias is due to the presence of noise autospectral densities $S_{n^x n^x}$ and $S_{n^y n^y}$ in the denominator and numerator of appropriate expressions. Since these autospectral densities are positive real-valued quantities, they bias the estimates in such a way that the magnitude K_d is shifted downwards (underestimated), while the magnitude of K_u is shifted upwards (overestimated). These biases explain the choice of subscripts d and u.

Now we consider a more complex case, which has an exact analogy in magnetotelluric data processing. Suppose that a 3-channel linear system has two inputs and one output:

$$Y(\omega) = K^1(\omega)\cdot X^1(\omega) + K^2(\omega)\cdot X^2(\omega) \tag{A2.12}$$

First we find general expressions for response components K^1 and K^2. By performing the procedure similar to the sequence in Eqs (A2.3–A2.6), one obtains the following equations:

$$K^1(\omega) = \frac{S_{x^1x^2}S_{yx^2} - S_{x^2x^2}S_{yx^1}}{S_{x^1x^2}S_{x^2x^1} - S_{x^1x^1}S_{x^2x^2}}, \quad K^2(\omega) = \frac{S_{x^1x^2}S_{yx^1} - S_{x^1x^1}S_{yx^2}}{S_{x^1x^2}S_{x^2x^1} - S_{x^1x^1}S_{x^2x^2}} \tag{A2.13}$$

Next, assume that there is an additive noise in all three channels,

$$x^1(t) = \widehat{x^1(t)} + n^{x1}(t), \quad x^2(t) = \widehat{x^2(t)} + n^{x2}(t), \quad y(t) = \widehat{y(t)} + n^y(t) \tag{A2.14}$$

and that the noise is uncorrelated with signal,

$$S_{\widehat{x^1} n^{x1}} = S_{\widehat{x^2} n^{x1}} = S_{\widehat{y} n^{x1}} = S_{\widehat{x^1} n^{x2}} = S_{\widehat{x^2} n^{x2}} = S_{\widehat{y} n^{x2}} = S_{\widehat{x^1} n^y} = S_{\widehat{x^2} n^y} = S_{\widehat{y} n^y} = 0, \tag{A2.15}$$

and also that the noise is uncorrelated from channel to channel:

$$S_{n^{x1}n^{x2}} = S_{n^{x1}n^y} = S_{n^{x2}n^y} \tag{A2.16}$$

After substituting Eq. (A2.14) into Eq. (A2.13) and taking into account Eqs (A2.15, A2.16), we have

$$K^1(\omega) = \frac{S_{x^1x^2}S_{yx^2} - (S_{x^2x^2} + S_{n^{x2}n^{x2}})S_{yx^1}}{S_{x^1x^2}S_{x^2x^1} - (S_{x^1x^1} + S_{n^{x1}n^{x1}})(S_{x^2x^2} + S_{n^{x2}n^{x2}})}$$

$$K^2(\omega) = \frac{S_{x^1x^2}S_{yx^1} - \left(S_{x^1x^1} + S_{n^{x1}n^{x1}}\right)S_{yx^2}}{S_{x^1x^2}S_{x^2x^1} - \left(S_{x^1x^1} + S_{n^{x1}n^{x1}}\right)\left(S_{x^2x^2} + S_{n^{x2}n^{x2}}\right)} \tag{A2.17}$$

ESTIMATING IMPEDANCE TENSOR ELEMENTS WITH LEAST SQUARES

Now consider the problem of estimating the impedance and start with Eq. (15.16) (chapter 15):

$$E_x = Z_{xx}B_x + Z_{xy}B_y \quad \text{and} \quad E_y = Z_{yx}B_x + Z_{yy}B_y$$

Thus, we have a pair of input–single-output systems, where the inputs to each are B_x and B_y, whereas the output of the first system is E_x, and the output of the second system is E_y. Both inputs and outputs are affected by noise which is assumed to be additive. Suppose that for a given frequency we have M independent spectral estimates $E_x^i(\omega)$, $E_y^i(\omega)$, $B_x^i(\omega)$, $B_y^i(\omega)$ corresponding to different segments of the recorded data. Then applying least-squares method to the problem described by equations

$$E_x^i = Z_{xx}B_x^i + Z_{xy}B_y^i$$

$$E_y^i = Z_{yx}B_x^i + Z_{yy}B_y^i, \quad i = 1, \ldots, M$$

by analogy with Eq. (A2.3) we obtain the following minimization problem

$$\begin{cases} \sum\limits_{i=1}^{M}\left(E_x^i - Z_{xx}B_x^i - Z_{xy}B_y^i\right)\cdot\left(E_x^{i^*} - Z_{xx}^*B_x^{i^*} - Z_{xy}^*B_y^{i^*}\right) = \min \\ \sum\limits_{i=1}^{M}\left(E_y^i - Z_{yx}B_x^i - Z_{yy}B_y^i\right)\cdot\left(E_y^{i^*} - Z_{yx}^*B_x^{i^*} - Z_{yy}^*B_y^{i^*}\right) = \min \end{cases} \tag{A2.18}$$

Differentiating the first equation of this set with respect to real and imaginary parts of Z_{xx} and setting the result to zero gives:

$$\sum_{i=1}^{M} E_x^i B_x^{i^*} = Z_{xx}\sum_{i=1}^{M} B_x^i B_x^{i^*} + Z_{xy}\sum_{i=1}^{M} B_y^i B_x^{i^*}. \tag{A2.19}$$

A similar procedure for Z_{xy} gives:

$$\sum_{i=1}^{M} E_x^i B_y^{i^*} = Z_{xx}\sum_{i=1}^{M} B_x^i B_y^{i^*} + Z_{xy}\sum_{i=1}^{M} B_y^i B_y^{i^*}. \tag{A2.20}$$

We thus arrive at the following system of four linear equations:

$$\begin{cases} \langle E_x B_x^* \rangle = Z_{xx} \cdot \langle B_x B_x^* \rangle + Z_{xy} \cdot \langle B_y B_x^* \rangle \\ \left\langle E_x B_y^* \right\rangle = Z_{xx} \cdot \left\langle B_x B_y^* \right\rangle + Z_{xy} \cdot \left\langle B_y B_y^* \right\rangle \\ \langle E_y B_x^* \rangle = Z_{yx} \cdot \langle B_x B_x^* \rangle + Z_{yy} \cdot \langle B_y B_x^* \rangle \\ \left\langle E_y B_y^* \right\rangle = Z_{yx} \cdot \left\langle B_x B_y^* \right\rangle + Z_{yy} \cdot \left\langle B_y B_y^* \right\rangle \end{cases}, \tag{A2.21}$$

where $\langle\rangle$ denote an algebraic sum over i, that is, the quantities inside $\langle\rangle$ are the cross-spectral and auto-spectral densities.

An important feature of the cross-spectral density is that individual spectral estimates can be considered as random variables, whereas summed (or averaged) cross-spectra (spectral densities) are close to their expected values provided that the number of independent estimates is large enough. The system (A2.21) can be rewritten in matrix form as follows:

$$\begin{pmatrix} Z_{xx} & Z_{xy} \\ Z_{yx} & Z_{yy} \end{pmatrix} \cdot \begin{pmatrix} \langle B_x B_x^* \rangle & \left\langle B_x B_y^* \right\rangle \\ \langle B_y B_x^* \rangle & \left\langle B_y B_y^* \right\rangle \end{pmatrix} = \begin{pmatrix} \langle E_x B_x^* \rangle & \left\langle E_x B_y^* \right\rangle \\ \langle E_y B_x^* \rangle & \left\langle E_y B_y^* \right\rangle \end{pmatrix}$$

The so-called impedance estimate is obtained by solving this equation for the matrix on the left-hand side containing the impedance functions.

$$\begin{pmatrix} Z_{xx} & Z_{xy} \\ Z_{yx} & Z_{yy} \end{pmatrix} = \begin{pmatrix} \langle E_x B_x^* \rangle & \left\langle E_x B_y^* \right\rangle \\ \langle E_y B_x^* \rangle & \left\langle E_y B_y^* \right\rangle \end{pmatrix} \cdot \begin{pmatrix} \langle B_x B_x^* \rangle & \left\langle B_x B_y^* \right\rangle \\ \langle B_y B_x^* \rangle & \left\langle B_y B_y^* \right\rangle \end{pmatrix}^{-1}. \tag{A2.22}$$

Thus, the impedance estimate gives the for individual elements of impedance tensor:

$$\begin{aligned} Z_{xx} &= \frac{\langle B_y B_y^* \rangle \cdot \langle E_x B_x^* \rangle - \langle B_y B_x^* \rangle \cdot \langle E_x B_y^* \rangle}{\langle B_x B_x^* \rangle \cdot \langle B_y B_y^* \rangle - \langle B_x B_y^* \rangle \cdot \langle B_y B_x^* \rangle} \\ Z_{xy} &= \frac{\langle B_x B_x^* \rangle \cdot \langle E_x B_y^* \rangle - \langle B_x B_y^* \rangle \cdot \langle E_x B_x^* \rangle}{\langle B_x B_x^* \rangle \cdot \langle B_y B_y^* \rangle - \langle B_x B_y^* \rangle \cdot \langle B_y B_x^* \rangle} \\ Z_{yx} &= \frac{\langle B_y B_y^* \rangle \cdot \langle E_y B_x^* \rangle - \langle B_y B_x^* \rangle \cdot \langle E_y B_y^* \rangle}{\langle B_x B_x^* \rangle \cdot \langle B_y B_y^* \rangle - \langle B_x B_y^* \rangle \cdot \langle B_y B_x^* \rangle} \\ Z_{yy} &= \frac{\langle B_x B_x^* \rangle \cdot \langle E_y B_y^* \rangle - \langle B_x B_y^* \rangle \cdot \langle E_y B_x^* \rangle}{\langle B_x B_x^* \rangle \cdot \langle B_y B_y^* \rangle - \langle B_x B_y^* \rangle \cdot \langle B_y B_x^* \rangle} \end{aligned} \tag{A2.23}$$

The situation here is similar to the one described earlier for a single-input–single-output system, where one can derive two different estimates, described by Eqs (A2.6) and (A2.8), by interchanging the roles of the input and output. With two inputs and two outputs, one can construct a number of distinct alternative minimization problems, which in principle yield identical results for noise-free data, but lead to six different estimates in the presence of noise. It can be shown that two of these estimates are unstable, if the resistivity structure is close to 1D. The other four estimates, including Eq. (A2.23), can be used for efficient processing of magnetotelluric data. Each of the estimates is an approximation to the true response, because all estimates are biased by the noise: Even if the noise is uncorrelated with the signal and from channel to channel, it still affects the auto-spectra and causes either over- or underestimation of the response. The conventional impedance estimate Eq. (A2.23) is mostly used in practical applications, because it is less biased by noise in the electric-field channels, which is usually higher compared to noise in the magnetic-field channels.

Robust Estimation

So far we assumed that individual estimates are summed with equal weights to form averaged values denoted by brackets $\langle\rangle$ in Eqs (A2.21–A2.23). It is natural to expect, however, that different segments of the record are characterized by different signal-to-noise ratios and that there are different distributions of noise across channels measuring different field components. Each window and corresponding spectral estimate should ideally contribute to the calculated average with a weight proportional to its quality, that is, to its signal-to-noise ratio. For this reason, robust statistics are used to improve standard least-squares estimates, making the solution less sensitive to the distortions in individual spectral estimates.

In robust estimation, the quadratic minimization problem, Eq. (A2.18), is replaced by a more complicated nonlinear problem:

$$\begin{cases} \sum_{i=1}^{M} \gamma\left(E_x^i - Z_{xx}B_x^i - Z_{xy}B_y^i\right) = \sum_{i=1}^{M} \gamma\left(r_x^i\right) = \min \\ \sum_{i=1}^{M} \gamma\left(E_y^i - Z_{yx}B_x^i - Z_{yy}B_y^i\right) = \sum_{i=1}^{M} \gamma\left(r_y^i\right) = \min \end{cases} \tag{A2.24}$$

where $\gamma(r)$ is a function of the misfit r (the argument in parentheses above) and is called the loss function. Following the initial work by P. Huber on robust estimation, a number of different loss functions have been proposed for processing magnetotelluric data. It turns out that minimization of such functions can be reduced to problems of weighted least squares. As a result Eqs (A2.19, A2.20) are replaced with similar equations in which all sums are replaced by weighted sums:

$$\begin{aligned}\sum_{i=1}^{M} w_i E_x^i B_x^{i^*} &= Z_{xx}\sum_{i=1}^{M} w_i B_x^i B_x^{i^*} + Z_{xy}\sum_{i=1}^{M} w_i B_y^i B_x^{i^*}\\ \sum_{i=1}^{M} w_i E_x^i B_y^{i^*} &= Z_{xx}\sum_{i=1}^{M} w_i B_x^i B_y^{i^*} + Z_{xy}\sum_{i=1}^{M} w_i B_y^i B_y^{i^*},\end{aligned} \qquad \text{(A2.25)}$$

where w_i are weight coefficients. For example, for relatively small misfits the weights can be chosen equal to 1, while for larger misfits they are inversely proportional to misfit value. Notice that, since the misfit value is unknown in advance, the problem should be solved iteratively. At first the standard least squares with unity weights is applied to calculate the initial estimate, which allows one to obtain partial misfits. These partial misfits are used to calculate weights w_i and then the set (A2.25) is solved to produce a new least-squares estimate. This process is repeated until convergence.

Another robust approach is to reject some individual cross-spectra estimates (in specific time windows) based on predefined criteria—for example, to achieve higher correlation between appropriate channels. This procedure is called the jack-knife algorithm, and the measure of linear correlation is called coherence. The goal is to eliminate those estimates that decrease the total coherence value. The algorithm proceeds by trial and error; only cross-spectral estimates that maintain the reasonable coherence value are kept.

Let us discuss the concept of coherence in more detail by considering a single-input–single-output system. For this system one can examine the following relation between input $i(t)$ and output $o(t)$,

$$C_{io}^2 = \frac{|S_{io}|^2}{S_{ii}S_{oo}} \qquad \text{(A2.26)}$$

called squared coherence. Here S_{io} is the input–output cross-spectral density, S_{ii} and S_{oo} are input and output auto-spectral densities, respectively. Coherence C_{io} is a real-valued function of frequency with values in the interval [0, 1]. Zero coherence means there is no linear relation between input and output; whereas a coherence value of 1 implies that output is rigorously related to the input. A typical threshold used for rejection with jack-knife method is $C_{io} < 0.6$.

REMOTE-REFERENCE PROCESSING

Impedance estimates obtained from either Eq. (A2.23) or Eq. (A2.25) are unbiased only if there is no noise in the original magnetotelluric data. If the noise in different field components is uncorrelated, the cross-spectral densities will be unbiased, but the autospectral densities remain biased because of the noise autocorrelation. Thus, each one of the different possible impedance estimates is either over- or underestimated depending on where the autospectra appear in the estimator. In addition, it can happen that the noise in different components is correlated; this correlation causes the bias in the cross-spectral density and leads to inaccurate final estimates. More efficient noise suppression can be obtained using a remote-reference technique originally proposed by Gamble and his colleagues in late 1970s.

The remote-reference method requires the magnetotelluric field to be recorded synchronously at two stations. Suppose that the goal is to calculate impedance tensor at the first survey station, where all four horizontal components of magnetotelluric field E_x, E_y, B_x, B_y are measured. The reference station operates simultaneously with the first station and records, for example, two magnetic components B_{xr}, B_{yr}. Then, proceeding from the linear relations between horizontal magnetic fields at two stations

$$B_x = \lambda_{xx}B_{xr} + \lambda_{xy}B_{yr}, \quad B_y = \lambda_{yx}B_{xr} + \lambda_{yy}B_{yr}, \tag{A2.27}$$

which exist along with conventional impedance relations, Eqs (A2.19, A2.20) can be transformed as:

$$\sum_{i=1}^{M} E_x^i B_{xr}^{i*} = Z_{xx}\sum_{i=1}^{M} B_x^i B_{xr}^{i*} + Z_{xy}\sum_{i=1}^{M} B_y^i B_{xr}^{i*} \tag{A2.28}$$

$$\sum_{i=1}^{M} E_x^i B_{yr}^{i*} = Z_{xx}\sum_{i=1}^{M} B_x^i B_{yr}^{i*} + Z_{xy}\sum_{i=1}^{M} B_y^i B_{yr}^{i*} \tag{A2.29}$$

In this case, the impedance least-squares estimate can be obtained from cross-spectral densities only. Those cross-spectral densities are calculated across the pairs of different field components measured at two different stations. Thus, the system given by Eq. (A2.22) is replaced by

$$\begin{pmatrix} Z_{xx} & Z_{xy} \\ Z_{yx} & Z_{yy} \end{pmatrix} = \begin{pmatrix} \langle E_x B_{xr}^* \rangle & \langle E_x B_{yr}^* \rangle \\ \langle E_y B_{xr}^* \rangle & \langle E_y B_{yr}^* \rangle \end{pmatrix} \cdot \begin{pmatrix} \langle B_x B_{xr}^* \rangle & \langle B_x B_{yr}^* \rangle \\ \langle B_y B_{xr}^* \rangle & \langle B_y B_{yr}^* \rangle \end{pmatrix}^{-1} \tag{A2.30}$$

The replacement of autospectra by cross-spectra should reduce bias in the resulting estimates if the noise is uncorrelated between two stations while the signal remains correlated. It should be noticed that instead of two magnetic reference components two electric components can be used as well. The best results are obtained by measuring reference field components that are well correlated with the corresponding components at the survey station and that are little contaminated by noise. These conditions generally hold for the magnetic components. The distance between the survey station and remote reference station can vary significantly from tens of meters to hundreds of kilometers. For example, to suppress instrumentation noise, which is unique to each station, it is sufficient to use any neighboring survey station as a reference, even if it is close to the survey station under consideration. On the other hand, if intense cultural electromagnetic noise (for example, an electrified railway) affects a large area, distant reference stations (up to 200–300 km from the survey station) may be required to suppress the noise.

Of course, with this remote-reference approach, robust statistics can also be applied for more accurate calculation of averaged cross-spectra. Implementation of remote reference technique along with modern digital acquisition systems allows one to achieve the accuracy of about 1% for apparent resistivity and 1° for impedance phase in most surveys, even in noisy regions.

REFERENCES AND FURTHER READING

[1] T.D. Gamble, W.M. Goubau, J. Clarke, Magnetotellurics with a remote magnetic reference, Geophysics (44) (1979) 53–68.

[2] V.Yu Semenov, Magnetotelluric Data Processing, Nedra, Moscow, 1985, pp. 133.

[3] W.E. Sims, F.X. Bostick, H.W. Smith, The estimation of magnetotelluric impedance tensor elements from measured data, Geophysics 36 (1971) 938–942.

Relation between Amplitude and Phase for Magnetotelluric Impedance

In chapter 15 we have studied an influence of different inhomogeneities on the field amplitude but did not consider phases, assuming that the Hilbert transform allows one to calculate a phase knowing the amplitude (chapter seven). For instance, if the logarithm of the amplitude spectrum is the analytical function in the upper half-plane of the complex frequency we have for the apparent resistivity:

$$\varphi(\omega_o) = -\frac{\pi}{4} - \frac{\omega_o}{\pi}\,\text{p.v.}\int_0^\infty \ln \rho_a(\omega)\frac{d\omega}{\omega^2 - \omega_o^2}$$

$$\ln\frac{\rho_a(\omega_o)}{\rho_a(\infty)} = \frac{\pi}{4}\,\text{p.v.}\int_0^\infty \left[\frac{\pi}{4} + \varphi(\omega)\right]\frac{\omega\, d\omega}{\omega^2 - \omega_o^2}, \tag{A3.1}$$

where ρ_a is the apparent resistivity, φ is the impedance phase, ω is the frequency, and p.v. indicates the principal value of the integral. As was shown

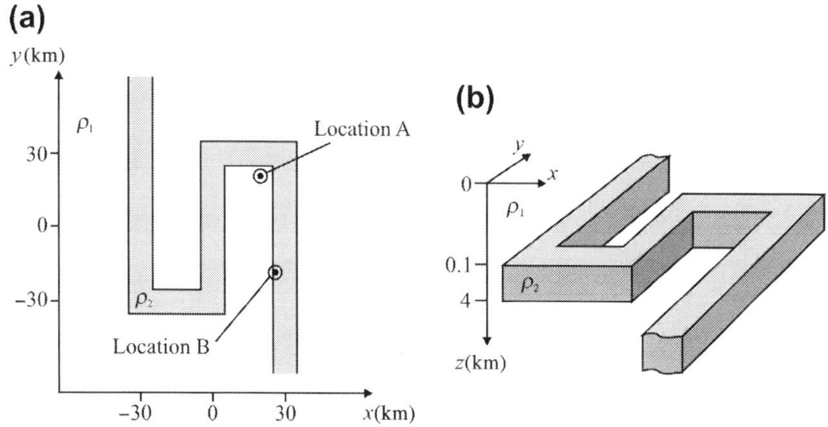

Figure A3.1 (a) Horizontal slice showing an S-shaped conductor with $\rho_1/\rho_2 = 3333$. (b) 3D-view of an S-shaped conductor.

Methods in Geochemistry and Geophysics, Volume 45
ISSN 0076-6895,
http://dx.doi.org/10.1016/B978-0-444-53829-1.19003-3

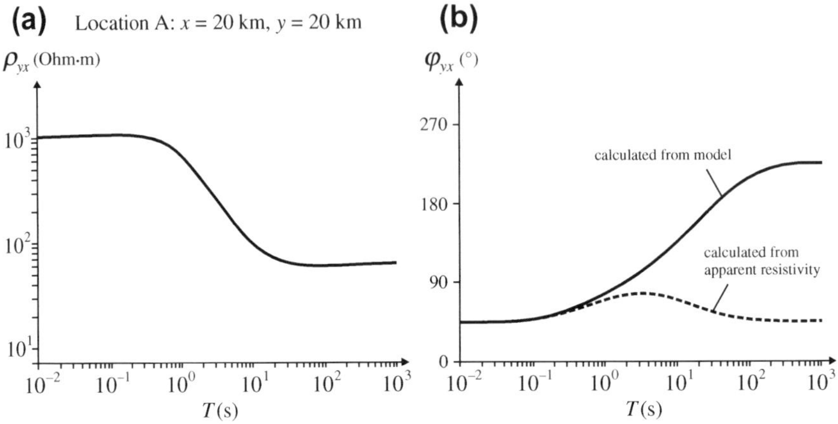

Figure A3.2 (a) Theoretical curve of the apparent resistivity ρ_{yx} at the point A in Figure A3.1. (b) Impedance phase. Solid line at the right panel shows phase directly calculated from the model, while dashed line corresponds to the phase, determined from apparent resistivity.

by P. Weidelt, Eqs (A3.1) can be always used for the magnetotelluric impedance in the case of one-dimensional model and H-polarization. At the same time for other models this relation may be sometimes invalid. Most of magnetotelluric data is consistent with this relation within a measurement error, and instead of measurements the phase can be usually calculated from Eqs (A3.1). However some amount of data mostly collected in mountain

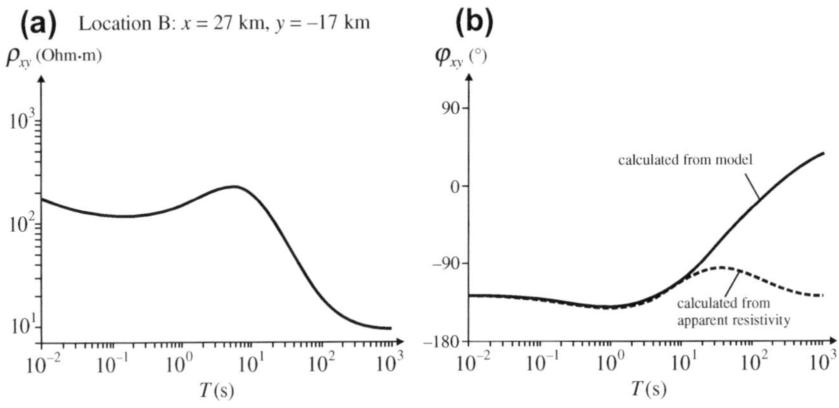

Figure A3.3 (a) Theoretical curve of the apparent resistivity ρ_{xy} at location B in Figure A3.1. (b) Impedance phase. Solid line shows phase directly calculated from modeling, while dashed line corresponds to the phase, determined from apparent resistivity.

regions or at the seafloor in the coastal areas clearly shows that this equation cannot be always applied. From the mathematical point of view, this indicates that there are complex frequencies where logarithm of the spectrum amplitude is not an analytical function (chapter seven).

The first example when Eqs (A3.1) becomes invalid is a three-dimensional model of the conductive inhomogeneity, which has a shape like letter S and surrounded much more resistive homogenous half-space (Figure A3.1). Its width is 10 km and has about 4 km in thickness with its top situated at 100 m beneath the earth surface. Also this conductor has an infinite extension along the y-axis. It may be useful to note that at the central

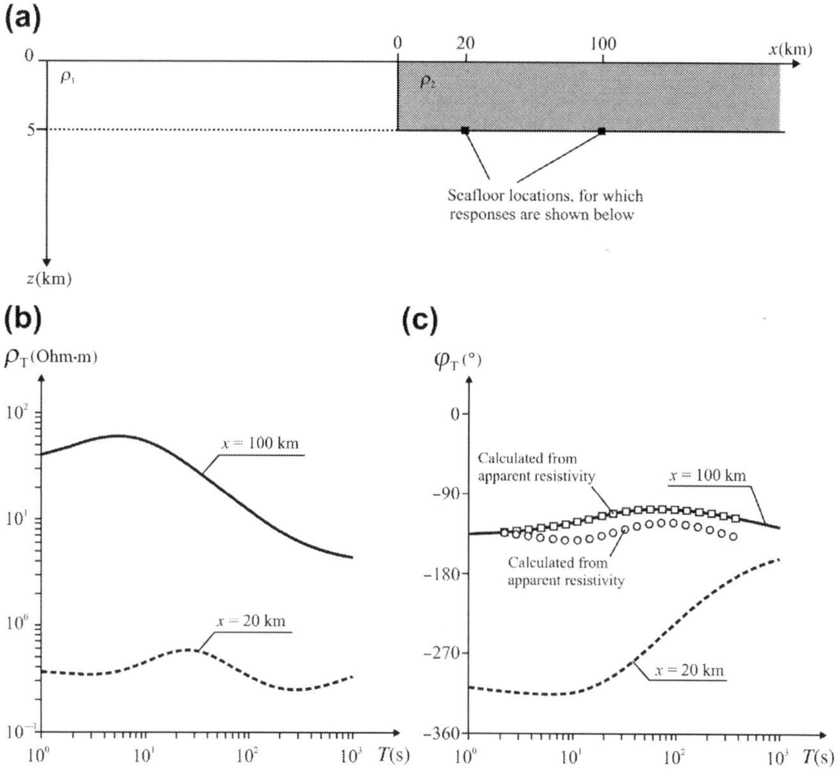

Figure A3.4 (a) Model of a medium, $\rho_1/\rho_2 = 300$. (b) Theoretical apparent resistivity curves at two locations: $x = 100$ km (solid line) and $x = 20$ km (dashed line). (c) Phase responses at the same points. The solid line describes the theoretical phase response but white squares show phase response calculated from Eqs (A3.1). White circles correspond to the phase, calculated from apparent resistivity for $x = 20$ km, while dashed line is the result of modeling.

part of the model the induced current changes its direction. Note that observation points are located at the earth's surface.

Results of calculation of the phase based on a solution of the forward problem and performed within range of 0.01–1000 s, clearly shows the dispersion Eqs (A3.1) gives incorrect values. This discrepancy is observed for both *xy* and *yx* components of the impedance tensor in some areas adjacent to the curved parts of the conductor (Figures A3.2 and A3.3).

Next consider a relation between the amplitude and phase at different observation points at the sea bottom in the case of E-polarization for a simplified 2D model, Figure A3.4(a). As is seen from Figure A3.4(c), at a sufficiently great distance from the coast ($x = 100$ km), calculation of the phase by Eqs (A3.1) gives correct values. However, relatively near the coast ($x = 20$ km) discrepancy between phases calculated from this equation and determined from modeling becomes very large and reaches almost 180°.

REFERENCES AND FURTHER READING

[1] D.A. Alekseev, N.A. Palshin, Iv.M. Varentsov, Magnetotelluric dispersion relations in a two-dimensional model of the coastal effect, Izvestiya, Physics of the Solid Earth 45 (2) (2009) 167–170.

[2] M.N. Berdichevsky, V.I. Dmitriev, Models and Methods of Magnetotellurics, Springer, 2008, p. 564.

APPENDIX FOUR

The Field of the Vertical Electric Dipole in the Layered Medium

INTRODUCTION

Here we consider the electromagnetic field generated in the Earth by a vertical electric dipole. The material complements Chapter 13, where we described the field generated by a horizontal electric dipole located at the Earth's surface. In electrical prospecting on land, a horizontal electric dipole, grounded at its ends, is the most natural electric-field source for use in exploration geophysics, except in the case where the field is being measured at relatively high frequencies (20 kHz or more), far away from the source. In marine electrical prospecting, however, it is easy to use vertical dipoles both as the source of the field and as receivers. This configuration has potential applications in low-frequency electromagnetic exploration for offshore oil fields. O. Nazarenko first suggested the vertical-dipole configuration for electromagnetic sounding in 1961 (Russia); the method was revived through work by Nigel Edwards and others starting around 1981 (Canada). A brief history of electrical methods in marine exploration and in exploration for oil follows.

The first application of electrical methods in a marine setting probably took place in Russia. Around 1930, large-scale electrical resistivity surveys for shallow oil fields were carried out both on land and offshore in the Caucasus region by teams of scientists working for a company owned by the Schlumberger brothers in a joint venture with the Soviet oil trusts. More than 70 years ago, profiling and sounding with electric dipoles were introduced for oil and mining exploration by L. Alpin, who developed the theory and interpretation of this method for direct (steady or time invariant) current flow. In contrast to the sounding methods introduced in the early 1900s by Conrad Schlumberger (later called "Schlumberger soundings"), dipole soundings do not require the use of long wires. For this reason, the dipole soundings were easily adapted to marine conditions, especially in shallow waters with depth less than 200 m. Marine dipole soundings were been used in the Azov Sea and Caspian Sea, as well as in the Sea of Okhotsk and other marine settings to study the basement surface, as well as the resistivity

Methods in Geochemistry and Geophysics, Volume 45
ISSN 0076-6895,
http://dx.doi.org/10.1016/B978-0-444-53829-1.19004-5

structure of sediments. Of course, dipole soundings using direct current (DC) can be used to probe deep inside the Earth only when very resistive layers are not present within sediments, since such layers will form a barrier to the penetration of the current, screening deeper layers.

Marine electric dipole measurements, like land measurements, were initially carried out with arrays arranged with different distances between the current dipole source and the dipole receivers. Later, at the beginning of the 1960s, transient soundings with measurements of the vertical component of the magnetic field were used in Baltic and Okhotsk Seas to provide sensitivity to deep layers even in the presence of very resistive layers. Dipole soundings were usually carried out with two ships—a source (current) ship and a receiver ship. The surveys generally used axial dipole arrays (recall that an axial dipole array is arranged so that the dipole direction for a given electrode pair lies along the same line as the profile direction). There were two main surveying methods: a two-sided method and a symmetrical (midpoint) method. In the first configuration, the ship towing the receiver array is held steady with the measurement line MN at a point along the profile where the natural noise level is low. The ship towing the current array then moves continuously toward the receiver array, until it passes by the receiver array and continues on a straight path away from it on the other side. In the second arrangement, the two ships start at a large distance apart and move toward each other at the same speed, so that their midpoint remains fixed.

Marine induced polarization surveys have been used for more than 30 years to detect ore bodies beneath the sea bottom. Induced polarization methods also had a brief history in oil exploration. During the 1960s it was discovered that there could sometimes be accumulations, above oil fields, of pyrites that generate an induced polarization effect. But attempts to use induced polarization to locate oil reservoirs did not show consistent correlations between shallow pyrite halos and deep oil fields, and the method has more or less been abandoned.

Standard electric and electromagnetic soundings have been used for many years as secondary techniques in oil exploration to delineate large structures beneath of the surface or sea bottom. But until recently these methods had rarely been deployed directly to detect the reservoir itself. Possibly the first successful applications of electromagnetic sounding for oil were transient soundings on land in the near zone that were carried out in the early 1970s to locate the depth of the oil–water contact in reservoirs. As shown in Chapter 9, when conditions are favorable, it is possible to find the boundary between the portion of a porous reservoir layer that is saturated with gas or oil—and is therefore highly resistive—and the conductive part that is saturated with saltwater. When the medium surrounding the reservoir is itself resistive, the presence of an oil–water contact can be detected by inducing currents in the zone saturated with saltwater, which becomes the source of the secondary electromagnetic field.

In 1981, Len Srnka (Exxon) suggested using marine electromagnetic surveys to study the distribution of resistivity in sedimentary basins that may contain petroleum systems. Sedimentary basins are generally composed of sandstone, limestone, and shale layers that, aside from the reservoir layer itself, are saturated with saltwater. The resistive oil-saturated reservoir layer cannot easily be detected by induction, because the induced currents will flow mainly in the surrounding conductive layers that can extend for large distances above and below the reservoir zone. For this reason, it was suggested that a horizontal grounded source near the sea or ocean bottom could be used to map the presence of an oil layer. A highly resistive layer in an otherwise conductive sedimentary column will impede the flow of vertical current from a grounded electrical dipole source and change the rate of decay of the electric field with distance from the source. This method, called marine controlled-source electromagnetic surveying, is now a standard technique in offshore oil exploration. Measurements of the electric and magnetic field are mainly performed with arrays having horizontal current and receiver dipoles, in configurations similar to the ones first used by Alpin.

In the future, arrays with the vertical electric dipoles will undoubtedly find more application in oil exploration, so it is interesting to study the field of such a current source in the presence of a resistive reservoir layer. But before we solve the problem for a layer of finite thickness, we examine the case where the thickness of the resistive part of reservoir is much smaller than the other dimensions, so that it can be replaced by a plane with transversal resistivity T. We derive boundary conditions for the field at such a surface, which enables full solution of the boundary value problem for a thin resistive layer without having to solve for the field in the layer itself.

A4.1. BOUNDARY CONDITIONS AT THE SURFACE OF A PLANE *T*

Suppose that a thin resistive layer is situated at a distance H from the origin and surrounded by a conducting medium as is shown in Figure A4.1(a).

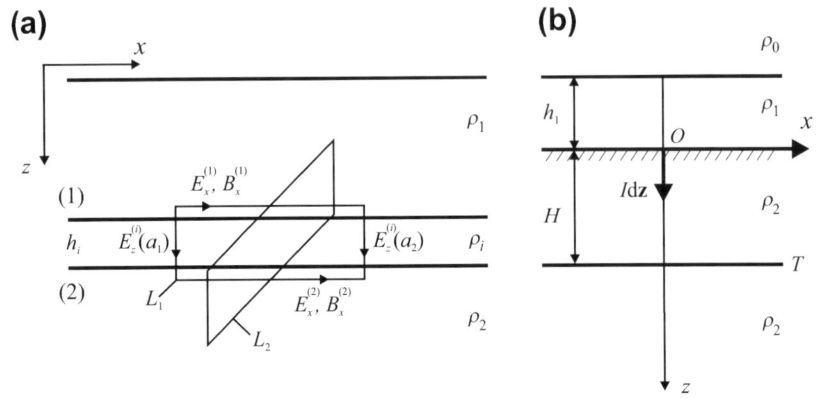

Figure A4.1 (a) Illustration used in deriving boundary conditions at plane *T*. (b) Two-layer structure with plane *T* embedded in a basement layer.

The transversal resistance of this layer T is defined as

$$T = h_i \rho_i, \tag{A4.1}$$

where h_i and ρ_i are the thickness and resistivity of this layer, respectively. If we assume that the layer is sufficiently thin, we can make use of the approximate boundary conditions at its surface. Applying the second Maxwell's equation in the integral form:

$$\oint_L \mathbf{B} \cdot \mathrm{d}\mathbf{l} = \mu_0 \int_S \mathbf{j} \cdot \mathrm{d}\mathbf{S}$$

to the path L_1 in Figure A4.1(a), we have

$$B_x^{(1)} \mathrm{d}x - B_x^{(2)} \mathrm{d}x = -\mu_0 j_y h_i \mathrm{d}x, \quad \text{or} \quad B_x^{(1)} - B_x^{(2)} = -\mu_0 j_y h_i = -\mu_0 \frac{h_i}{\rho_i} E_y. \tag{A4.2}$$

Similarly, along path L_2

$$B_y^{(1)} - B_y^{(2)} = \mu_0 j_x h_i = \mu_0 \frac{h_i}{\rho_i} E_x. \tag{A4.3}$$

Because we assume that the resistivity ρ_i is very high, it is proper to neglect any currents flowing along the thin layer, and so we have

$$B_x^{(1)} = B_x^{(2)}, \quad B_y^{(1)} = B_y^{(2)}. \tag{A4.4}$$

Thus, the horizontal components of the magnetic field have the same values at both surface of the thin layer. Using the first Maxwell equation

$$\oint\limits_{L} \mathbf{E}\cdot d\mathbf{l} = -\frac{\partial \Phi}{\partial t}$$

and applying it along the path L_1, we obtain

$$E_x^{(1)}dx - E_z^{(i)}(a_2)h_i - E_x^{(2)}dx + E_z^{(i)}(a_1)h_i = \frac{\partial B_y}{\partial t}h_i dx,$$

where $E_z^i(a_2)$ and $E_z^i(a_1)$ are values of the normal component of the electric field inside the layer at points a_2 and a_1, respectively. As the thickness h_i tends to zero, the flux caused by B_y also tends to zero, and we have for the complex amplitudes of the electric field

$$\left(E_x^{(1)^*} - E_x^{(2)^*}\right)dx + \left(E_z^{i^*}(a_2) - E_z^{i^*}(a_1)\right)h_i = 0.$$

It is obvious that

$$E_z^{i^*}(a_2) - E_z^{i^*}(a_1) = \frac{\partial E_z^{i^*}}{\partial x}dx$$

and hence

$$E_x^{(2)^*} - E_x^{(1)^*} = h_i\frac{\partial E_z^{i^*}}{\partial x}. \qquad (A4.5)$$

By analogy, applying the same equation to the path L_2, we have

$$E_y^{(2)^*} - E_y^{(1)^*} = h_i\frac{\partial E_z^{i^*}}{\partial y}. \qquad (A4.6)$$

The last two equations can be rewritten in the following form

$$E_x^{(2)^*} - E_x^{(1)^*} = h_i\rho_i\gamma_i\frac{\partial E_z^{(i)^*}}{\partial x} = T\frac{\partial \gamma_i E_z^{(i)^*}}{\partial x} = T\frac{\partial j_z^*}{\partial x}$$

and

$$E_y^{(2)^*} - E_y^{(1)^*} = h_i\rho_i\gamma_i\frac{\partial E_z^{(i)^*}}{\partial y} = T\frac{\partial \gamma_i E_z^{(i)^*}}{\partial y} = T\frac{\partial j_z^*}{\partial y}, \qquad (A4.7)$$

where $\gamma_i = 1/\rho_i$, j_z^* is the vertical component of current density inside the thin layer, and T is the transversal resistance. Inasmuch as the horizontal component of the current inside a thin resistive layer is negligible, its vertical component is a continuous function; that is it has the same values at both sides of the plate T:

$$\gamma_i E_z^{(i)^*} = \gamma_1 E_z^{(1)^*} = \gamma_2 E_z^{(2)^*}.$$

Here $\gamma_1, E_z^{(1)^*}$ and $\gamma_2, E_z^{(2)^*}$ are conductivity and the vertical component of the electric field outside the plate T, but at its vicinity. Correspondingly, the boundary conditions at the plane T assume the following form:

$$B_x^{(1)^*} = B_x^{(2)^*}, \quad B_y^{(1)^*} = B_y^{(2)^*},$$

$$E_x^{(2)^*} - E_x^{(1)^*} = T\gamma_1 \frac{\partial E_z^{(1)^*}}{\partial x} = T\gamma_2 \frac{\partial E_z^{(2)^*}}{\partial x}, \tag{A4.8}$$

$$E_y^{(2)^*} - E_y^{(1)^*} = T\gamma_1 \frac{\partial E_z^{(1)^*}}{\partial y} = T\gamma_2 \frac{\partial E_z^{(2)^*}}{\partial y}.$$

These boundary conditions do not require information about the behavior of the field components inside the layer, a feature that markedly simplifies a solution of the boundary value problem. Unlike the tangential components of the magnetic field, the horizontal components of the electric field at the surface of the thin resistive layer are discontinuous functions. From the theory for the stationary field it is known that the tangential components of the electric field can be discontinuous only across a surface of a double layer with the changing surface density of dipole moments:

$$E_t^{(2)} - E_t^{(1)} = -\frac{1}{\varepsilon_0}\frac{\partial \eta}{\partial t}, \tag{A4.9}$$

where ε_0 is the dielectric constant and "t" indicates a tangential direction. Comparing Eqs (A4.8) and (A4.9) we can say that there are electric charges with opposite signs on either side of the thin layer. For this reason, each surface element can be considered as being a dipole with a moment η given by

$$\eta = -T\gamma_1 \varepsilon_0 E_z^{(1)} = -T\gamma_2 \varepsilon_0 E_z^{(2)}. \tag{A4.10}$$

Since we usually assume that conductivities above and beneath plane T are equal, the normal components coincide too, that is the simple layer of charges is absent. Therefore in place of Eq. (A4.10) we have

$$\eta(p) = -T\gamma\varepsilon_0 E_z(p).$$

Note that the electric charges on the surface of the horizontal plane T create an electromagnetic field, which is equivalent to that caused by a system of vertical electric dipoles. Each elementary dipole is the source of a horizontal magnetic field, while the vertical component of this field is absent.

A4.2. MECHANISM OF APPEARANCE OF THE SECONDARY FIELD

In this light let us notice that the presence of the thin resistive layer may be detected only due to charges arising at both its sides. They have opposite signs and equal magnitudes if resistivity of a medium above and beneath is the same. In fact, as is well known from the theory of a quasi-stationary field the surface charge density on the boundary between two media (i) and (j) is defined from the same equation as in the case of the time invariant field:

$$\sigma(p) = 2\varepsilon_0 K_{ij} E_n^{\mathrm{av}}(p).$$

Here

$$\mathrm{K}_{ij} = \frac{\rho_j - \rho_i}{\rho_j + \rho_i},$$

where ρ_i and ρ_j are resistivity of a medium, but E_n^{av} is the normal component of the total field at the same point p caused by all charges, except charge at the point p. Also this field includes the vortex part generated by a change of the magnetic field with time. It is essential that the normal n is directed from medium (i) to medium (j) that gives charges of different sign at the upper and lower surfaces of the thin layer. By definition, we may assume that $K_{ij} = \pm 1$. As was pointed out it is natural to treat the field E_n^{av} as a sum of the normal and secondary fields, where the latter field is due to the presence of the plane T. It is obvious that if the normal electric field does not have component perpendicular to this layer the charges are absent, and the secondary field is equal to zero. The classical example of such field is the vertical magnetic dipole, since its vortex electric field is tangential to this horizontal layer, and correspondingly the secondary dipole moment is zero. Therefore, the necessary condition for detecting the thin resistive layer is the presence of the relatively strong component of the normal electric field perpendicular to this layer.

For simplicity we assume that the thin layer has an infinite extension in horizontal direction, which will allow us in a simple manner to solve boundary value problem and obtain an expression for the electric field in the explicit form. In spite of such approximation we will be able to obtain some useful information about an influence of frequency or time, as well as a separation between the current and receiver electrodes on the depth of investigation for different resistivity structure parameters.

In our case an array consists of two parallel vertical lines and length of which may be comparable with the distance between these current and receiver lines, again for simplicity we replace them by dipoles. At the same time transition to real system can be done by summation of the fields of dipoles along these lines. In such case the main attention will be paid to the vertical component of the electric field. At the beginning we consider the normal field in two-layer medium, shown in Figure A4.1(b).

A4.3. EXPRESSIONS FOR THE NORMAL FIELD AT THE SEA BOTTOM

Let us choose the Cartesian system coordinates with the origin 0, located on the sea bottom and suppose that the vertical electric dipole is placed beneath the sea surface at the point 0 and directed along the z-axis, Figure A4.1(b). As usual we seek a solution in the frequency domain in terms of the vector potential, $\mathbf{A}$. Its complex amplitude satisfies the Helmholtz equation in each layer:

$$\nabla^2 \mathbf{A}^* + k^2 \mathbf{A}^* = 0.$$

As is known,

$$\mathbf{B}^* = \text{curl}\,\mathbf{A}^*, \quad \mathbf{E}^* = i\omega \mathbf{A}^* + \frac{1}{\gamma \mu_0} \text{grad div}\,\mathbf{A}^*. \tag{A4.11}$$

A4.3.1. Boundary Conditions for the Vector Potential

Taking into account the axial symmetry of the field and the model of a medium, we will try to solve the boundary value problem with help of one component of the vector potential A_z. Then, as follows from Eq. (A4.11), we have

$$B_x^* = \frac{\partial A_z^*}{\partial y}, \quad B_y^* = -\frac{\partial A_z^*}{\partial x}, \quad B_z^* = 0$$

and

$$E_x^* = \frac{1}{\gamma\mu_0}\frac{\partial^2 A_z^*}{\partial x \partial z}, \quad E_y^* = \frac{1}{\gamma\mu_0}\frac{\partial^2 A_z^*}{\partial y \partial z}, \quad E_z^* = i\omega A_z^* + \frac{1}{\gamma\mu_0}\frac{\partial^2 A_z^*}{\partial z^2}. \quad \text{(A4.12)}$$

From continuity of the tangential components of the magnetic and electric fields at sea surface and sea bottom, we have

$$\begin{aligned} &A_z^{*(0)} = A_z^{*(1)} \quad \text{and} \quad \rho_0 \frac{\partial A_z^{*(0)}}{\partial z} = \rho_1 \frac{\partial A_z^{*(1)}}{\partial z}, \quad \text{if } z = -h_1. \\ &A_z^{*(1)} = A_z^{*(2)} \quad \text{and} \quad \rho_1 \frac{\partial A_z^{*(1)}}{\partial z} = \rho_2 \frac{\partial A_z^{*(2)}}{\partial z}, \quad \text{if } z = -h. \end{aligned} \quad \text{(A4.13)}$$

Here ρ_0, ρ_1, and ρ_2 are resistivity of the upper space (air), the seawater, and a medium beneath, respectively, and h_1 is the thickness of the seawater, Figure A4.1(b). At the surface T, we have

$$\begin{aligned} &A_z^{*(1)} = A_z^{*(2)}, \\ &\frac{\partial A_z^{*(2)}}{\partial z} - \frac{\partial A_z^{*(1)}}{\partial z} = T\gamma\left(k_2^2 A_z^{*(2)} + \frac{\partial^2 A_z^{*(2)}}{\partial z^2}\right), \quad \text{if } z = H. \end{aligned} \quad \text{(A4.14)}$$

Note that in deriving Eqs (A4.13) and (A4.14), we used the fact that from continuity of function's tangential derivatives at the boundary follows the continuity of the function itself.

A4.3.2. Expressions for the Vector Potential A_z^* of the Normal Field

As was shown in the Chapter 13 the vector potential in a uniform medium is written as

$$A_z^* = \frac{\mu_0 I \mathrm{d}z}{4\pi}\frac{\exp(ik_1 R)}{R} = \frac{\mu_0 I \mathrm{d}z}{4\pi}\int_0^\infty \frac{m}{m_1}\exp(-m_1|z|)J_0(mr)\mathrm{d}m. \quad \text{(A4.15)}$$

Correspondingly, we have the following expressions for the complex amplitudes of the vector potential

$$A_z^{*(0)} = \frac{\mu_0 I \mathrm{d}z}{4\pi}\int_0^\infty C_0 \exp(m_0 z)J_0(mr)\mathrm{d}m, \quad \text{if } z < -h_1,$$

$$A_z^{*(1)} = \frac{\mu_0 I dz}{4\pi} \int_0^\infty \left[\frac{m}{m_1} \exp(-m_1|z|) + C_1 \exp(m_1 z) + D_1 \exp(-m_1 z) \right]$$

$$\times J_0(mr)dm \quad \text{if} \quad -h_1 < z < 0, \tag{A4.16}$$

$$A_z^{*(2)} = \frac{\mu_0 I dz}{4\pi} \int_0^\infty D_2 \exp(-m_2 z) J_0(mr) dm, \quad \text{if} \ z > 0.$$

Here

$$m_0 = \left(m^2 - k_0^2\right)^{1/2}, \ m_1 = \left(m^2 - k_1^2\right)^{1/2} \ m_2 = \left(m^2 - k_2^2\right)^{1/2}.$$

Substituting the latter into boundary conditions, we obtain the system of equations for determination of unknowns

$$C_0 \exp(-m_0 h_1) = \frac{m}{m_1} \exp(-m_1 h_1) + C_1 \exp(-m_1 h_1) + D_1 \exp(m_1 h_1),$$

$$\rho_0 m_0 C_0 \exp(-m_0 h_1)$$

$$= \rho_1 [m \exp(-m_1 h_1) + m_1 C_1 \exp(-m_1 h_1) - m_1 D_1 \exp(m_1 h_1)], \tag{A4.17}$$

$$D_2 = \frac{m}{m_1} + C_1 + D_1,$$

$$-\rho_2 m_2 D_2 = \rho_1(-m + m_1 C_1 - m_1 D_1).$$

In deriving this system we assume at the beginning that the dipole was located slightly above the boundary $z = 0$. Eliminating C_0 and D_2, we have:

$$-\frac{m}{m_1} m_{01} \exp(-2m_1 h_1) = m_{01} \exp(-2m_1 h_1) C_1 + D_1,$$

$$\frac{m}{m_1} m_{12} = C_1 - m_{12} D_1,$$

where $m_{01} = \frac{\rho_o m_0 - \rho_1 m_1}{\rho_o m_0 + \rho_1 m_1}$, $m_{12} = \frac{\rho_1 m_1 - \rho_2 m_2}{\rho_1 m_1 + \rho_2 m_2}$.

This gives

$$C_1 = \frac{m}{m_1} m_{12} \frac{1 - m_{01} \exp(-2m_1 h_1)}{1 + m_{01} m_{12} \exp(-2m_1 h_1)}$$

and

$$D_1 = -\frac{m}{m_1} m_{01} \exp(-2m_1 h_1) \frac{1 + m_{12}}{1 + m_{01} m_{12} \exp(-2m_1 h_1)}. \qquad \text{(A4.18)}$$

Further we will be mainly interested by the case when the upper medium is an insulator, that is $m_{01} = 1$. Then we have

$$C_1 = \frac{m}{m_1} m_{12} \frac{1 - \exp(-2m_1 h_1)}{1 + m_{12} \exp(-2m_1 h_1)}$$

and

$$D_1 = -\frac{m}{m_1} \exp(-2m_1 h_1) \frac{1 + m_{12}}{1 + m_{12} \exp(-2m_1 h_1)}. \qquad \text{(A4.19)}$$

A4.3.3. Expressions for the Normal Field Beneath Sea Bottom

Taking into account that at the cylindrical system of coordinates with the origin 0, we have

$$\mathbf{B}^* = \frac{1}{r} \begin{vmatrix} \mathbf{1}_r & r\mathbf{1}_\varphi & \mathbf{1}_z \\ \frac{\partial}{\partial r} & \frac{\partial}{\partial \varphi} & \frac{\partial}{\partial z} \\ 0 & 0 & A_z^* \end{vmatrix}$$

Therefore

$$B_\varphi^* = -\frac{\partial A_z^*}{\partial r}.$$

As follows from Eq. (A4.16)

$$B_\varphi^* = B_\varphi^{*0}(k_1 R) + \frac{\mu_0 I dz}{4\pi} \int_0^\infty m[C_1 \exp(m_1 z) + D_1 \exp(-m_1 z)] J_1(mr) dm \qquad \text{(A4.20)}$$

Here $B_\varphi^{*0}(k_1 R)$ is the magnetic field of the electric dipole in a uniform medium with resistivity ρ_1. In particular, in the plane $z = 0$, we have

$$B_\varphi^* = B_\varphi^{*0}(k_1 r) + \frac{\mu_0 I dz}{4\pi} \int_0^\infty m(C_1 + D_1) J_1(mr) dm, \quad \text{if } z = 0. \qquad \text{(A4.21)}$$

For the radial and vertical components of the electric field, we have

$$E_{1r}^{*}=\frac{1}{\gamma_1\mu_0}\frac{\partial^2 A_{1z}^{*}}{\partial r\partial z}\quad\text{and}\quad E_{1z}^{*}=i\omega A_{1z}^{*}+\frac{1}{\gamma_1\mu_0}\frac{\partial^2 A_{1z}^{*}}{\partial z^2}.$$

Then Eq. (A4.16) gives

$$E_{1r}^{*}=E_{1r}^{*0}(k_1R)-\frac{\rho_1 I\mathrm{d}z}{4\pi}\int\limits_0^\infty mm_1[C_1\exp(m_1z)$$
$$-D_1\exp(-m_1z)]J_1(mr)\mathrm{d}m$$

and

$$E_{1z}^{*}=E_{1z}^{*0}(k_1R)+\frac{\rho_1 I\mathrm{d}z}{4\pi}\int\limits_0^\infty m^2[C_1\exp(m_1z)$$
$$+D_1\exp(-m_1z)]J_0(mr)\mathrm{d}m. \tag{A4.22}$$

At the points of the plane $z=0$ the latter gives

$$E_{1r}^{*}=E_{1r}^{*0}(k_1r)-\frac{\rho_1 I\mathrm{d}z}{4\pi}\int\limits_0^\infty mm_1(C_1-D_1)J_1(mr)\mathrm{d}m$$

and

$$E_{1z}^{*}=E_{1z}^{*0}(k_1r)+\frac{\rho_1 I\mathrm{d}z}{4\pi}\int\limits_0^\infty m^2(C_1+D_1)J_0(mr)\mathrm{d}m \tag{A4.23}$$

Note, that we will consider only the vertical component of this field. In general, determination of the field requires numerical integration, except one case when $\rho_1=\rho_2$, and it can be expressed by elementary functions.

A4.3.4. The Normal Field When $\rho_1=\rho_2$ (Uniform Half-Space)

As follows from Eq. (A4.19), we have

$$C_1=0\quad\text{and}\quad D_1=-\frac{m}{m_1}\exp(-2m_1h_1), \tag{A4.24}$$

since $m_{12}=0$. Therefore, taking into account Somerfield integral, we have

$$A_z^{*(1)}=\frac{\rho_1 I}{4\pi}\left[\frac{\exp(ik_1R)}{R}-\frac{\exp(ik_1R_1)}{R_1}\right],\quad\text{if}\ z<0. \tag{A4.25}$$

Here $R=\sqrt{x^2+y^2+z^2}, \quad R_1=\sqrt{x^2+y^2+(z+2h_1)^2}$.

This means that as in the case of the time invariant field the electromagnetic field is represented as the field of two electric dipoles, located symmetrically with respect to the boundary. Their moments have the same magnitude but opposite directions. It may be proper to note that this is the rare case when the electromagnetic field of the dipole located beneath the boundary is expressed in terms of elementary functions. It is obvious that the normal component of the electric field is equal to zero at the conducting side of the sea surface as well as the magnetic field.

$$B_\varphi^*(r,-h_1)=0. \tag{A4.26}$$

The latter follows from the fact that near the surface, currents have only tangential component and the field is independent on the angle φ. Correspondingly, in the quasi-stationary approximation the magnetic field is absent in the upper nonconducting half-space. Note that the same is true in the case when instead of a conducting uniform half-space there is a horizontally layered medium. Absence of the magnetic field indicates that electromagnetic energy does not propagate through the upper space, if displacement currents are neglected in the upper half-space.

From Eq. (A4.25) we see that the normal field beneath the sea surface consists of the direct and reflected wave, and the latter is equivalent to the field of the electric dipole in a uniform medium, as if it is located above the boundary at the point, which is mirror reflection of the origin where the source is placed. As follows from Eq. (A4.25) the amplitude of the reflected wave does not exceed that of the direct one and with an increase of the distance both waves begin to decay as exponential functions. In particular, at relatively large distances, $R \gg h_1$, these waves almost cancel each other. In a fact we have

$$A_z^{*(1)}=\frac{\rho_1 I}{4\pi}\frac{\exp(ik_1R)}{R}\left(1-\frac{\exp\left(ik_1\frac{2h_1^2}{R}\right)}{1+\frac{2h_1^2}{R^2}}\right)\approx$$

$$-i2k_1h_1^2\frac{\rho_1 I}{4\pi}\frac{\exp(ik_1R)}{R^2}, \quad \text{if } R\to\infty. \tag{A4.27}$$

In the Chapter 13 it was shown that in a uniform medium and in the spherical system of coordinates the complex amplitudes of the field components E_R and E_θ are

$$E_R^* = \frac{2\rho_1 I dz}{4\pi R^3}\exp(ik_1R)(1 - ik_1R)\cos\theta \text{ and } E_\theta^*$$
$$= \frac{\rho_1 I dz}{4\pi R^3}\exp(ik_1R)\left(1 - ik_1R - k_1^2R^2\right)\sin\theta. \tag{A4.28}$$

Also

$$B_\varphi^* = B_y^* = \frac{\mu_0 I dz}{4\pi r^2}\exp(ikR)(1 - ikR)\sin\theta$$

Then

$$E_z^* = E_R^* \cos\theta - E_\theta^* \sin\theta,$$

and therefore the field in the presence of the upper boundary is:

$$E_z^* = \frac{\rho_1 I dz}{4\pi R^3}\exp(ik_1R)\left[2(1 - ik_1R)\cos^2\theta\right.$$
$$\left. - \left(1 - ik_1R - k_1^2R^2\right)\left(1 - \cos^2\theta\right)\right] + \frac{\rho_1 I dz}{4\pi R_1^3}\exp(ik_1R_1)$$
$$\times \left[2(1 - ik_1R_1)\cos^2\theta_1 - \left(1 - ik_1R_1 - k_1^2R_1^2\right)\left(1 - \cos^2\theta_1\right)\right]$$

or

$$E_z^* = \frac{\rho_1 I}{4\pi}\frac{\exp(ik_1R)}{R^3}\left\{\left(-1 + ik_1R + k_1^2R^2\right) + [3 - ik_1R(3 - ik_1R)]\frac{z^2}{R^2}\right\}$$
$$- \frac{\rho_1 I}{4\pi}\frac{\exp(ik_1R_1)}{R_1^3}\left\{\left(-1 + ik_1R_1 + k_1^2R_1^2\right)\right.$$
$$\left. + [3 - ik_1R_1(3 - ik_1R_1)]\frac{(2h_1 + z)^2}{R_1^2}\right\}, \tag{A4.29}$$

since

$$\cos\theta = \frac{z}{R} \quad \text{and} \quad \cos\theta_1 = \frac{2h_1 + z}{R_1}$$

For the magnetic field, we have

$$B_\varphi^* = \frac{\mu_0 I dz}{4\pi}\left\{\frac{1}{R^2}\exp(ik_1R)(1 - ik_1R)\sin\theta - \frac{1}{R_1^2}\exp(ik_1R_1)\right.$$
$$\left. \times (1 - ik_1R_1)\sin\theta_1\right\} \tag{A4.30}$$

These equations, describing the field in a uniform half-space, can be used to introduce the apparent resistivity, and correspondingly, it is natural to study this field in some detail. First, consider the asymptotic behavior of the field in the frequency domain. Expanding Eqs (A4.29) and (A4.30) in the series by small parameters k_1R and k_1R_1, we have

$$E_z^* \approx \frac{\rho_1 I}{4\pi}\frac{1}{R^3}\left[\left(-1+\frac{1}{2}k_1^2R^2+\frac{2}{3}ik_1^3R^3\right)+\left(3+\frac{1}{2}k_1^2R^2\right)\frac{z^2}{R^2}\right]$$
$$-\frac{\rho_1 I}{4\pi}\frac{1}{R_1^3}\left[\left(-1+\frac{1}{2}k_1^2R_1^2+\frac{2}{3}ik_1^3R_1^3\right)+\left(3+\frac{1}{2}k_1^2R_1^2\right)\frac{(2h_1+z)^2}{R_1^2}\right] \tag{A4.31}$$

and

$$B_\varphi^* \approx \frac{\mu_0 I dz}{4\pi} r\left[\frac{1}{R^3}\left(1+\frac{k_1^2R^2}{2}+\frac{1}{3}ik_1^3R^3\right)-\frac{1}{R_1^3}\left(1+\frac{k_1^2R_1^2}{2}+\frac{1}{3}ik_1^3R_1^3\right)\right],$$
$$\text{if } k_1R \ll 1 \tag{A4.32}$$

First of all from these equations it follows that the series, describing the low-frequency part of the spectrum, does not contain the term in the power $\omega^{3/2}$. In other words, after the term with ω the next two terms of the series are proportional to ω^2 and $\omega^{5/2}$, correspondingly. And, as we know (Chapter 9) this means that at the late stage both fields decay with time as $t^{-5/2}$. Let us discuss Eqs (A4.31) and (A4.32), which are valid regardless of the frequency, as soon as the wavelength is much greater that the distance from the dipole and its mirror reflection, $(R/\lambda_1 \ll 1)$, that is they characterize the field at the near zone. The leading term coincides with that of the time invariant field:

$$E_{z1}^* = \frac{\rho_1 I}{4\pi}\left\{\left(\frac{1}{R_1^3}-\frac{1}{R^3}\right)+3\left[\frac{z^2}{R^5}-\frac{(2h_1+z)^2}{R_1^5}\right]\right\},$$

and

$$B_{\varphi 1}^* = \frac{\mu_0 I dz}{4\pi} r\left(\frac{1}{R^3}-\frac{1}{R_1^3}\right). \tag{A4.33}$$

It is clear that this field synchronously changes with the dipole current and by definition has a pure galvanic origin, since its electric field is caused by charges on the surface of the dipole electrodes and charges distributed on the boundary $z = -h_1$. The field of these charges gives rise to currents, which are the sources of the magnetic field $B^*_{1\varphi}$. As follows from Eqs (A4.31) and (A4.32), the next term for both fields is

$$E^*_{z2} = \frac{\rho_1 I dz}{8\pi} k_1^2 \left\{ \frac{1}{R}\left(1 + \frac{z^2}{R^2}\right) - \frac{1}{R_1}\left[1 + \frac{(2h_1 + z)^2}{R_1^2}\right] \right\}$$

and

$$B^*_{\varphi 2} = \frac{\mu_0 I dz}{8\pi} k_1^2 r \left(\frac{1}{R} - \frac{1}{R_1}\right). \tag{A4.34}$$

The latter arises in the following way. A change of the magnetic field $B_{\varphi 1}$ with time generates the vortex electric field E_{z2}, as well E_{r2}, and in accordance with Ohm's law, we have

$$\mathbf{j}_2 = \gamma_1 \mathbf{E}_2,$$

that causes the magnetic field $B_{\varphi 2}$. In the same manner the field $\mathbf{E}_3$ and $\mathbf{B}_3$, associated with the third term of the series and proportional to k_1^4, arises. It turns out that the first three terms of the series describe portion of the fields, which is hardly related to diffusion. In contrast, the next terms of the series, starting from the term with power of $k_1^5(\omega^{5/2})$, contains some information about a diffusion of the field. Note that considering separately the series for the in-phase and quadrature components of the field we can see that in the first case the term k_1^5 is only the second one, while for the in-phase component it is the third. Correspondingly, we can expect that the quantity $\frac{QE_z(\omega_1)}{\omega_1} - \frac{QE_z(\omega_2)}{\omega_2}$ depends on resistivity of layers practically in the same manner as the transient field does. Until now we focused on the low-frequency spectrum. As concerns the opposite case when the wavelength becomes comparable or smaller than the distance between the dipole and the boundary h, its influence vanishes, and happens due to attenuation of the dipole field. Now making use of Eqs (A4.31) and (A4.32), consider the dependence of the field, normalized by the field $(E_z^{\text{un}}, B_\varphi^{\text{un}})$, on distance from the origin, Figure A4.2(a)and (b).

Here

$$E_z^{\text{un}} = \frac{\rho_1 I dz}{4\pi x^3} \quad \text{and} \quad B_y^{\text{un}} = \frac{\mu_0 I dz}{4\pi x^2}$$

are the fields of the dipole in a uniform medium when frequency is equal to zero and an observation point is located at an equatorial plane, $z = 0$. Index of

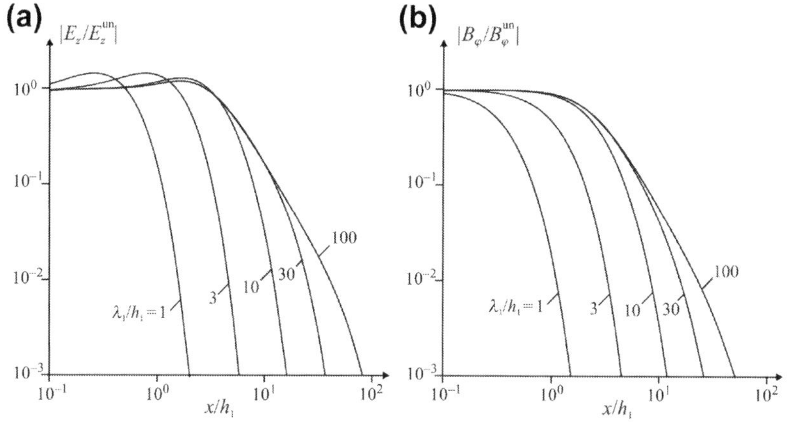

Figure A4.2 Dependence of normalized fields: $\left|E_z^*\right|/E_z^{un}$ and $\left|B_\varphi^*\right|/B_\varphi^{un}$ on the distance.

curves is the value of the ratio λ_1/h_1. Along the vertical axis we plot either the ratio $\left|E_z^*\right|/E^{un}$ or $\left|B_\varphi^*\right|/B^{un}$. First of all, it is clear that the ratio of the electric field tends to unity near the dipole. Then at the beginning with an increase of the distance x we first observe a maximum, and it happens due to opposite directions of the real and fictitious dipoles, but and after it the ratio starts to decrease. Behavior of normalized magnetic field, Figure A4.2(b), is sufficiently simpler and in the vicinity of the dipole, $x < \lambda_1$ is described by Eq. (A4.33).

Next consider the transient electric and magnetic fields when the dipole current is turned off. Unlike the frequency domain here we deal with the diffusion process only. First of it is clear that at the first instance when the current vanishes, the field cannot disappear instantly, and at each point of a medium it is equal to the field, caused by the constant current. Then with increasing of time the field starts to decay and at sufficiently large times (late stage) it decreases as $t^{-5/2}$. The latter follows from the fact that the leading term of the series with fractional power of ω, describing the low-frequency spectrum, has the power 5/2. Now we derive expressions for the transient field and with this purpose in mind, use again (Chapter 9) the equalities:

$$1-\frac{1}{2\pi}\int\limits_{-\infty}^{\infty}\frac{\exp(-i\omega t)}{-i\omega}d\omega=\begin{cases}0, & t<0\\ 1, & t>0\end{cases}$$

$$\frac{1}{2\pi}\int\limits_{-\infty}^{\infty}\exp(ikR)\frac{\exp(-i\omega t)}{-i\omega}d\omega=\begin{cases}0, & t<0\\ 1-\Phi(u), & t>0\end{cases},\tag{A4.35}$$

where $\Phi(u)$ is the probability integral:

$$\Phi(u) = \left(\frac{2}{\pi}\right)^{1/2} \int\limits_0^u \exp(-x^2/2)\,dx \tag{A4.36}$$

and

$$u = \frac{2\pi R}{\tau}, \; \tau = \left(2\pi\rho t 10^7\right)^{1/2}. \tag{A4.37}$$

Differentiating Eq. (A4.35) successively with respect to R, we obtain

$$\begin{aligned} &\frac{1}{2\pi}\int\limits_{-\infty}^{\infty} ikR\frac{\exp(-i\omega t)}{-i\omega}\exp(ikR)d\omega = -\left(\frac{2}{\pi}\right)^{1/2} u\exp\left(-\frac{u^2}{2}\right) \\ &\frac{1}{2\pi}\int\limits_{-\infty}^{\infty} (ikR)^2\frac{\exp(-i\omega t)}{-i\omega}\exp(ikR)d\omega = \left(\frac{2}{\pi}\right)^{1/2} u^3\exp\left(-\frac{u^2}{2}\right). \end{aligned} \tag{A4.38}$$

Taking Fourier's integral from expressions for the electric and magnetic fields, Eq. (A4.28), we obtain

$$\begin{aligned} E_R &= \frac{2\rho_1 I dz}{4\pi R^3}\left[\Phi(u) - \left(\frac{2}{\pi}\right)^{1/2} u\exp\left(-\frac{u^2}{2}\right)\right]\cos\theta, \\ E_\theta &= \frac{\rho_1 I dz}{4\pi R^3}\left[\Phi(u) - \left(\frac{2}{\pi}\right)^{1/2} u(1+u^2)\exp\left(-\frac{u^2}{2}\right)\right]\sin\theta \end{aligned} \tag{A4.39}$$

and

$$B_\varphi = -\frac{\mu_0 I dz}{4\pi R^2}\left[\Phi(u) - \left(\frac{2}{\pi}\right)^{1/2} u\exp\left(-\frac{u^2}{2}\right)\right]\sin\theta. \tag{A4.40}$$

Note that with accuracy of a constant, expressions for the electric field, Eq. (A4.39), coincide with those of the magnetic field, caused by the magnetic dipole situated in a uniform medium. Of course, as in the frequency domain, the electric field and the current are located in the vertical planes $Z0R$, but the vector lines of the magnetic field are horizontal circles with the centers on the z-axis. As usual the study of the transient responses of the field we start from its asymptotic behavior. Bearing in mind that

$$\Phi(u) \to 1, \quad \text{if } u \to \infty$$

at the initial stage, we have

$$E_R = \frac{2\rho_1 Idz}{4\pi R^3}\cos\theta,\ \ E_\theta = \frac{\rho_1 Idz}{4\pi R^3}\sin\theta,\ \ B_\varphi = \frac{\mu_0 Idz}{4\pi R^2}\sin\theta, \quad \text{if } t\to 0, \tag{A4.41}$$

that is due to inertia of the magnetic field, which cannot change instantly, at the moment when the dipole current is switched off, the electromagnetic field remains the same as before. Correspondingly, for the vertical component of the electric field, we have

$$E_z = \frac{Idz}{4\pi R^3}\left(3\frac{z^2}{R^2} - 1\right), \quad \text{if } t\to 0. \tag{A4.42}$$

To derive the formulas for the late stage the following expansion is used

$$\exp\left(-\frac{u^2}{2}\right) \approx 1 - \frac{u^2}{2} + \frac{u^4}{8} - \cdots, \qquad \Phi(u) \approx \left(\frac{2}{\pi}\right)^{1/2}\left(u - \frac{u^3}{6} + \frac{u^5}{40}\cdots\right)$$

Their substitution into Eqs (A4.39) and (A4.40) gives

$$\begin{aligned} E_R &\approx \frac{\rho_1 Idz}{4\pi R^3} 2\left(\frac{2}{\pi}\right)^{1/2}\left(\frac{u^3}{3} - \frac{u^5}{10}\right)\cos\theta, \\ E_\theta &\approx \frac{\rho_1 Idz}{4\pi R^3}\left(\frac{2}{\pi}\right)^{1/2}\left(-\frac{2}{3}u^3 - \frac{2}{5}u^5\right)\sin\theta \end{aligned} \tag{A4.43}$$

and

$$B_\varphi \approx -\frac{\mu_0 Idz}{4\pi R^2}\left(\frac{2}{\pi}\right)^{1/2}\left(\frac{u^3}{3} - \frac{u^5}{10}\right)\sin\theta, \quad \text{if } u < 1. \tag{A4.44}$$

Correspondingly, in the same manner as in the frequency domain we have for the vertical component of the electric field caused by the dipole

$$E_z \approx \frac{\rho_1 Idz}{4\pi R^3}\left[-\frac{3}{5}u^5\frac{z^2}{R^2} + \frac{2}{3}u^3 + \frac{2}{5}u^5\right]$$

and, therefore, the field in the presence of the boundary is

$$E_z \approx \frac{\rho_1 Idz}{4\pi}\left(\frac{2}{\pi}\right)^{1/2}\left[\frac{1}{R^3}\left(\frac{2}{3}u^3 + \frac{2}{5}u^5 - \frac{3}{5}\frac{z^2}{R^2}u^5\right)\right.$$

$$\left. - \frac{1}{R_1^3}\left(\frac{2}{3}u_1^3 + \frac{2}{5}u_1^5 - \frac{3}{5}\frac{(2h_1 + z)^2}{R_1^2}u_1^5\right)\right]$$

and

$$B_\varphi \approx -\frac{\mu_0 I dz}{4\pi}\left(\frac{2}{\pi}\right)^{1/2}\left[\frac{1}{R^2}\left(\frac{1}{3}u^3-\frac{1}{10}u^5\right)\frac{r}{R}-\frac{1}{R_1^2}\left(\frac{1}{3}u_1^3-\frac{1}{10}u_1^5\right)\frac{r}{R_1}\right] \tag{A4.45}$$

After simple algebra, we have

$$E_z \approx \frac{I dz}{20\pi(\pi)^{1/2}} h_1(h_1+z)\frac{\gamma_1^{3/2}\mu_0^{5/2}}{t^{5/2}}$$

and

$$B_\varphi \approx \frac{\mu_0 I dz}{40\pi(\pi)^{1/2}} h_1(h_1+z) r\frac{\gamma_1^{5/2}\mu_0^{5/2}}{t^{5/2}}. \tag{A4.46}$$

The latter is valid when

$$\left(\frac{\rho_1 t}{\mu_0}\right)^{1/2} \gg r. \tag{A4.47}$$

Applying the same approach we have also for the radial component of the electric field at the late stage, we have

$$E_r \approx \frac{I dz}{40\pi(\pi)^{1/2} t^{5/2}}\gamma_1^{3/2}\mu_0^{5/2} h_1 r. \tag{A4.48}$$

As we see the field at the late stage decays relatively quickly, as $t^{-5/2}$, while the components E_z and B_φ, caused by the single dipole decrease more slowly, as $t^{-3/2}$. This means that the field, associated with the surface charges, cancels at the late stage the term proportional to $t^{-3/2}$. At the same time the radial component E_r due to the single dipole also decays as $t^{-5/2}$. From Eq. (A4.46) it follows that between the dipole and a boundary, components E_z and B_y decrease toward the boundary, but beneath the dipole they linearly increase. Unlike them the radial component of the electric field is independent on z. In particular, at the plane $z=0$, we have

$$E_z \approx \frac{I dz}{20\pi(\pi)^{1/2}} h_1^2\frac{\gamma_1^{3/2}\mu_0^{5/2}}{t^{5/2}}, \quad E_r \approx \frac{I dz}{40\pi(\pi)^{1/2} t^{5/2}}\gamma_1^{3/2}\mu_0^{5/2} h_1 r$$

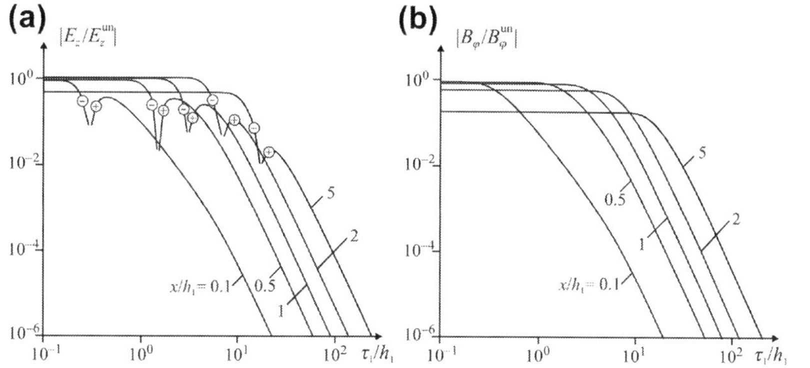

Figure A4.3 Normalized transient responses of the electric field $|E_z|/E_z^{un}$ (a) and magnetic field $|B_\varphi|/B_\varphi^{un}$ (b). Curve index is x/h_1.

and

$$B_\varphi \approx \frac{\mu_0 I dz}{40\pi(\pi)^{1/2}} h_1^2 r \frac{\gamma_1^{5/2} \mu_0^{5/2}}{t^{5/2}}. \tag{A4.49}$$

Correspondingly

$$\frac{E_r}{E_z} \approx \frac{1}{2} \frac{r}{h_1}$$

and with increasing the distance r the contribution of the horizontal component becomes stronger. Behavior of transient responses of the electric and magnetic fields is shown in Figure A4.3(a) and (b).

A4.3.5. The Normal Field at the Sea Bottom When $\rho_1 \neq \rho_2$

Unlike the previous case there are two boundaries, and at both of them electric charges arise, and naturally there is an interaction between them. Before we begin to discuss results of the field calculations let us briefly consider its behavior at the vicinity of the dipole and at the wave zone. As well known, the electric and magnetic fields change almost synchronously with the dipole current, when the distance from an observation point to the dipole is much smaller than the wavelength. In other words, it behaves as the time invariant field. For this reason, in order to obtain the asymptotic expressions at the near zone we can assume that $k \to 0$. Then Eq. (A4.19) gives

$$C_1 = K_{12} \frac{1 - \exp(-2mh_1)}{1 + K_{12}\exp(-2mh_1)},$$

$$D_1 = -\exp(-2mh_1) \frac{1 + K_{12}}{1 + K_{12}\exp(-2mh_1)}$$

and

$$K_{12} = \frac{\rho_1 - \rho_2}{\rho_1 + \rho_2}$$

Functions C_1 and D_1 contain information about an influence of both boundaries. To study their effect it is convenient to expand the fractions in the last equations in the series. Considering only the leading term, which does not contain h_1, we have

$$C_1 \approx K_{12} \text{ and } D_1 = 0$$

and therefore from Eq. (A4.21) it follows that at the point $z = 0$

$$B_\varphi^* \approx \frac{\mu_0 I dz}{4\pi}\left[\frac{1}{r^2} + K_{12}\int\limits_0^\infty mJ_1(mr)dm\right] = \frac{\mu_0 I dz}{2\pi r^2}\frac{\rho_1}{\rho_1 + \rho_2}, \tag{A4.50}$$

since

$$\int\limits_0^\infty mJ_1(mr)dm = -\frac{\partial}{\partial r}\int\limits_0^\infty J_0(mr)dm = \frac{1}{r^2}.$$

In the same manner the leading terms of the electric field (Eq. (A4.22)) give

$$E_{1r}^* = 0 \text{ and } E_{1z}^* \approx -\frac{\rho_1 I dz}{4\pi r^3}(1 + K_{12}) = -\frac{\rho_1 I dz}{2\pi r^3}\frac{\rho_1}{\rho_1 + \rho_2} \tag{A4.51}$$

because

$$\int\limits_0^\infty m^2 J_0(mr)dm = \int\limits_0^\infty m^2 \exp(-mz)dm = \frac{\partial^2}{\partial z^2}\left(r^2 + z^2\right)^{-1/2}, \quad \text{if } z \to 0.$$

Thus in the near zone and plane $z = 0$, we have

$$E_{1z}^* \approx -\frac{\rho_1 I dz}{4\pi r^3}(1 + K_{12}) = -\frac{\rho_1 I dz}{2\pi r^3}\frac{\rho_1}{\rho_1 + \rho_2}, \quad B_\varphi^* \approx \frac{\mu_0 I dz}{2\pi r^2}\frac{\rho_1}{\rho_1 + \rho_2}. \tag{A4.52}$$

It is obvious, that the latter describes the field, provided that an influence of the upper boundary is neglected, while its contribution is taken into account by the following terms of the series describing functions C_1 and D_1. Note that the same result can be obtained using the residual theorem, since these functions have poles on the complex plane of variable integration m.

As concerns the wave zone when a separation exceeds the wavelength we may expect that the field arrives propagating mainly through a more resistive medium. Dependence of the field magnitude on the distance from the dipole is shown in Figure A4.4(a) and (b). At the same time the values of normalized field near the origin are well defined by Eqs (A4.50) and (A4.51). Of course, with an increase of the frequency an influence of the upper boundary becomes smaller and due to attenuation it can be neglected when $\lambda_1/h_1 < 10$.

Next consider the transient responses of the electric and magnetic fields when the dipole current is turned off, which have the same features as in the case of a uniform half-space. For instance at the early stage when time tends to zero the field coincides with that of the time invariant field and, therefore, near the dipole it is almost described by Eqs (A4.50) and (A4.51). With an increase of time there is a moment when the vertical component of the electric field changes sign and at the late stage it is described as

$$E_z \approx \frac{Idz}{20\pi(\pi)^{1/2}} h_1(h_1 + z) \frac{\gamma_2^{3/2} \mu_0^{5/2}}{t^{5/2}}. \tag{A4.53}$$

It is essential that their behavior at relatively large times is described by Eq. (A4.46), when ρ_1 is replaced by ρ_2, provided that inequality Eq. (A4.47) is held. In other words, due to diffusion an influence of resistivity of the upper medium is negligible at the late stage, and the component E_z coincides with that in a half-space with the resistivity of the lowest medium. In general, in the case of a layered medium we also have

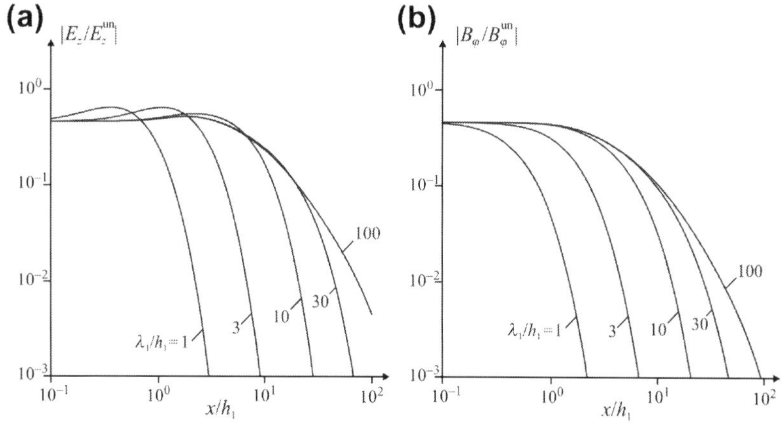

Figure A4.4 Normalized electric and magnetic fields as functions of spacing in two-layer structure, when $\rho_2/\rho_1 = 3.33$. Curve index is λ_1/h_1.

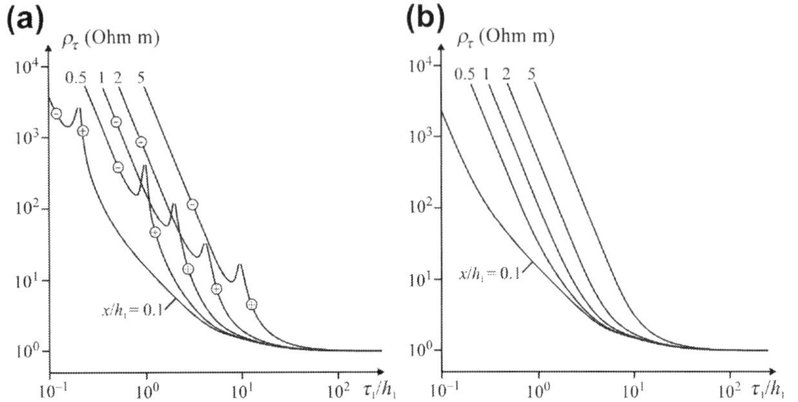

Figure A4.5 Apparent resistivity curves for the electric and magnetic fields in two-layer structure ($\rho_1 = 0.3$ Ohm m, $\rho_2 = 1$ Ohm m). Index of curves is x/h_1.

$$E_z \approx \frac{I\mathrm{d}z}{20\pi(\pi)^{1/2}} h_1(h_1 + z)\frac{\gamma_N^{3/2}\mu_0^{5/2}}{t^{5/2}}. \tag{A4.54}$$

As results of calculations show that at the late stage the magnetic field is

$$B_\varphi \approx \frac{\mu_0 I\mathrm{d}z}{40\pi(\pi)^{1/2}} h_1(h_1 + z)r\frac{\gamma_2^{5/2}\mu_0^{5/2}}{t^{5/2}}\frac{\gamma_1}{\gamma_2} \tag{A4.55}$$

and, correspondingly, an influence of the upper layer remains at all times. It turns out that in a general of the N-layered medium, we have

$$B_\varphi \approx \frac{\mu_0 I\mathrm{d}z}{40\pi(\pi)^{1/2}} h_1(h_1 + z)r\frac{\gamma_N^{5/2}\mu_0^{5/2}}{t^{5/2}}\frac{\gamma_1}{\gamma_N}. \tag{A4.56}$$

Thus, measuring each of these fields in principle it is possible to carry out soundings, regardless of separation between the dipole and observation point. Note that expressions for the electric and magnetic fields, Eqs (A4.54) and (A4.56), at points of the upper layer obey the second Maxwell equation: curl $\mathbf{B} = \mu_0\gamma_1\mathbf{E}$. To illustrate a behavior of the fields in the two-layered medium, we represent them in the form of the apparent resistivity, Figure A4.5, making use the expression for the field at the late stage for uniform half-space.

A4.4. INFLUENCE OF THE PLANE *T*

Now we study the influence of the thin resistive layer and for illustration consider a medium when $\rho_1 = \rho_2$. Substituting Eq. (A4.16) into boundary conditions, including Eq. (A4.14), we obtain the system of equations with respect to unknown coefficients:

$$C_0 \exp(-m_0h_1) = \frac{m}{m_1}\exp(-m_1h_1) + C_1 \exp(-m_1h_1) + D_1 \exp(m_1h_1),$$

$$\rho_0m_0C_0 \exp(-m_0h_1) = \rho_1[m \exp(-m_1h_1) + m_1C_1 \exp(-m_1h_1) - m_1D_1 \exp(m_1h_1)], \tag{A4.57}$$

$$\frac{m}{m_1}\exp(-m_1H) + C_1 \exp(m_1H) + D_1 \exp(-m_1H) = D_2 \exp(-m_1H),$$

$$m \exp(-m_1H) - m_1C_1 \exp(m_1H) + m_1D_1 \exp(-m_1H) = \left(m_1 + T\gamma_1m^2\right)D_2 \exp(-m_1H).$$

Eliminating from the first two equations the unknown C_0, we have

$$m_{01}C_1 \exp(-2m_1h_1) + D_1 = -\frac{m}{m_1}m_{01} \exp(-2m_1h_1). \tag{A4.58}$$

The last two equations of the set Eq. (A4.57) give

$$-T\gamma_1m^2\frac{m}{m_1}\exp(-2m_1H) = \left(2m_1 + T\gamma_1m^2\right)C_1 + T\gamma_1D_1m^2 \exp(-2m_1H) \tag{A4.59}$$

From Eqs (A4.58) and (A4.59) we find coefficients, characterizing the field between the sea surface and plate T

$$C_1 = T\gamma_1m^2\frac{m}{m_1}\exp(-2m_1H)\frac{-1 + m_{01} \exp(-2m_1h_1)}{2m_1 + T\gamma_1m^2 - T\gamma_1m^2m_{01} \exp(-2m_1H)\exp(-2m_1h_1)}$$

and

$$D_1 = -\frac{m}{m_1}m_{01} \exp(-2m_1h_1)\frac{2m_1 + T\gamma_1m^2[1 - \exp(-2m_1H)]}{2m_1 + T\gamma_1m^2 - T\gamma_1m^2m_{01} \exp(-2m_1H)\exp(-2m_1h_1)} \tag{A4.60}$$

A4.4.1. DC Soundings

Suppose that we perform DC geometrical soundings with an array consisting of vertical electric dipole and number of receivers measuring the components of the field E_z and B_φ. Inasmuch as signals are measured on the sea bottom we may expect that at small separations the field mainly depends on the resistivity ρ_1. Then an influence of the plane T becomes stronger and after it with further increase of the offset the field tends again to that in a uniform half-space. For illustration consider dependence of the ratio of the total electric and magnetic fields to their normal values (plane T is absent) on the array offset x, Figure A4.6. Index of curves is the parameter T/T_0, where $T_0 = \rho_1 H$ and the ratio x/h_1 is plotted along the horizontal axis. All curves are calculated for the same ratio $H/h_1 = 1$, and the apparent resistivity is introduced as

$$\frac{\rho_a}{\rho_1} = \frac{E_z}{E_z^N} \quad \text{or} \quad \frac{\rho_a}{\rho_1} = \frac{B_\varphi}{B_\varphi^N}$$

where E_z^N and B_φ^N are normal fields given by Eq. (A4.33).

As we see behavior of the apparent resistivity curves for the electric and magnetic field is almost similar and for illustration let us consider the case of the electric field.

As we know, the field in presence of T-plane is created by source dipole, its mirror reflection with respect to seawater/air boundary and the secondary dipoles is distributed over the T-plane. By definition the vertical component of the field generated by the single dipole is proportional to

$$E_N^1 = \frac{1}{4\pi}\left[\frac{3z^2}{(x^2+z^2)^{5/2}} - \frac{1}{(x^2+z^2)^{3/2}}\right],$$

and its behavior is shown in Figure A4.7.

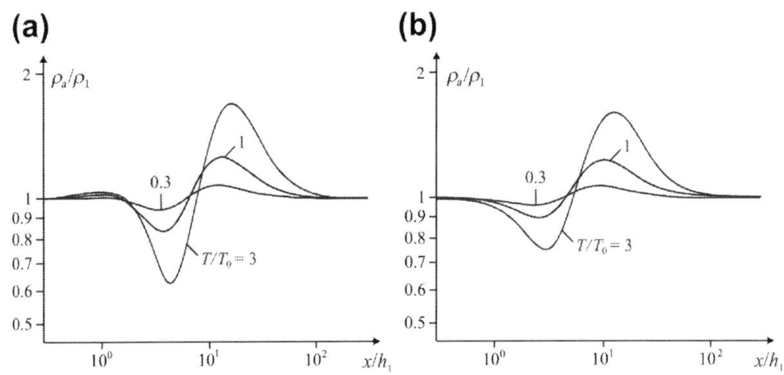

Figure A4.6 Apparent resistivity curves for the electric field E_z (a) and magnetic field B_φ (b).

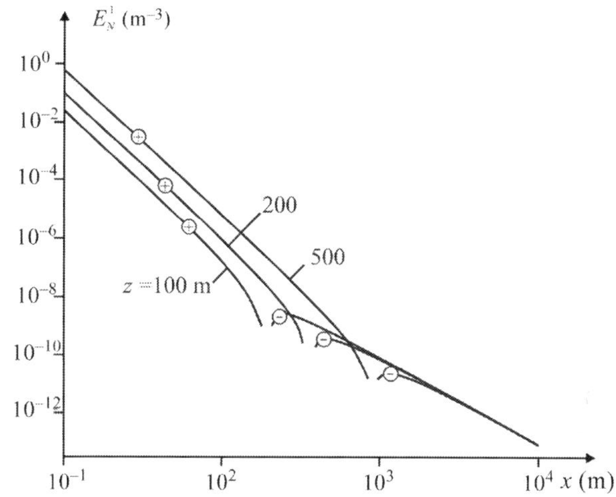

Figure A4.7 Illustration for explanation of DC sounding apparent resistivity behavior. E_N^1 as function of offset x. Curve index is depth z.

It is essential that at some point, where $x = 2^{1/2}z$ the component E_z changes its sign. Now, as the first approximation, assume that the dipoles at the plane T arise due to the dipole, located at the origin of coordinates and its mirror reflection with respect the upper boundary. Note that moments of these dipoles have opposite direction. At the beginning consider the secondary field due to the first dipole with the moment directed downwards. In the internal area, where $x < 2^{1/2} z$, it generates the secondary dipoles directed opposite to the z-axis, while in the external area of the plane T dipoles have opposite direction. It is almost obvious that, if the observation point at the plane $z = 0$ is located relatively close to the z-axis the dipoles of the internal area give the main contribution and the normal and secondary fields E_z have the same direction. Correspondingly, the ratio ρ_a/ρ_1 exceeds unity. In contrast, far away from the z-axis the influence of dipoles of the external area is dominant. This means that the direction of the normal and secondary fields is different, and as a result the ratio ρ_a/ρ_1 is less than unity. Such behavior of the apparent resistivity can be observed, if the upper boundary would be absent. In our case we have to take into account the effect of the vertical component, caused by the second dipole, which is a mirror reflection of the first one. Since this dipole is located at greater distance from the plane T and its moment has the opposite direction, it causes a different distribution of the secondary dipoles. For this reason, we observe two points where $\rho_a = \rho_1$, and correspondingly more complicated behavior of the curves of the apparent resistivity.

A4.4.2. Transient Soundings

Next, assume that transient soundings are performed at different distances from the dipole and for illustration consider the case when $T/T_0 = 3$ and $H/h_1 = 1$. Apparent resistivity curves are shown in Figure A4.8, and their behavior in the quasi-stationary approximation is almost obvious. At the first instance, when the dipole current is turned off, electrical charges arise around electrodes, as well as the current between them, and they preserve the same electric and magnetic field as before. Because of transformation of the electromagnetic field into heat, intensity of these charges and the field becomes smaller, and we begin to observe diffusion.

It is natural that at the early stage the left asymptote of the curves corresponds to the stationary field, and it contains some information about plane T. Its influence is very small, when observation point is close to the dipole and also it vanishes at relatively large distances. At the late stage, that corresponds to the right asymptote of the curves, the apparent resistivity approaches to that of the underlying medium, and the effect from the plane T becomes negligible. Because of diffusion there is a time interval when the magnitude of the dipole moments at this plane becomes maximal and, therefore, the secondary field becomes relatively large. It may be proper to notice that a change of a sign of the transient responses does not have any relation to the presence of the plate T. This is the feature of the field of the electric dipole, located beneath the upper boundary. As is seen from the curves, the influence of the thin resistive layer is more noticeable at relatively small distances from the primary dipole.

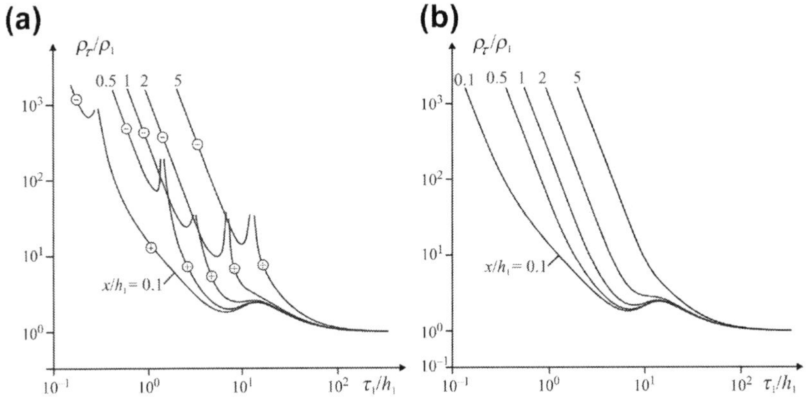

Figure A4.8 Apparent resistivity curves the electric field E_z (a), and the magnetic field B_φ (b) in presence of T-plane Index of curves is the ratio x/h_1.

REFERENCES AND FURTHER READING

[1] P.O. Barsukov, E.B. Fainberg, A mobile time domain sounding system for shallow water, First Break 31 (10) (2013) 53–63.
[2] A. Chave, C. Cox, Controlled electromagnetic sources for measuring electric conductivity beneath the oceans. Forward problem and model study, J. Geophys. Res. 87 (1982) 5327–5338.
[3] B.S. Svetov, Foundation of Geoelectrics, LKI, Moscow, 2008.

INDEX

Note: Page numbers with "*f*" denote figures; "*t*" tables.

C

D

E

F

G

H

R

S

T

U

V

W

Z

FSC
www.fsc.org
MIX
Papier | Fördert
gute Waldnutzung
FSC® C083411

Zeitfracht Medien GmbH
Ferdinand-Jühlke-Straße 7
99095 Erfurt, Deutschland
produktsicherheit@kolibri360.de